S. K. Upadhyay

Seismic Reflection Processing

Springer
Berlin
Heidelberg
New York
Hong Kong
London
Milan
Paris
Tokyo

S. K. Upadhyay

Seismic Reflection Processing

With Special Reference to Anisotropy

With 300 Figures and 15 Tables

Springer

S. K. Upadhyay
Indian Institute of Technology Roorkee
94 Vigyan Kunj
Roorkee – 247 667
Distt. – Haridwar (Uttranchal Pradesh)
India
e-mail: skues@iitr.ernet.in

and

S. K. Upadhyay
Department of Earth Sciences
Indian Institute of Technology Roorkee
Roorkee – 247 667
Distt. – Haridwar (Uttranchal Pradesh)
India
e-mail: s_upd@rediffmail.com

ISBN 3-540-40875-4 Springer-Verlag Berlin Heidelberg New York

Library of Congress Control Number: 2004103474
Bibliographic information published by Die Deutsche Bibliothek
Die Deutsche Bibliothek lists this publication in the Deutsche Nationalbibliografie;
detailed bibliographic data is available in the Internet at <http://dnb.ddb.de>.

Springer-Verlag is a part of Springer Science+Business Media
springeronline.com

Production: PRO EDIT GmbH, Heidelberg, Germany
Cover-Design: Erich Kirchner, Heidelberg, Germany
Typesetting: K + V Fotosatz, Beerfelden, Germany

Printed on acid-free paper 32/3141Di – 5 4 3 2 1 0

This book is dedicated to the founders and pioneering contributors, C. G. Knott, K. Zoeppritz, M. Muskat, M. W. Meres, C. H. Dix, J. G. Hagedoorn, O. Koefoed...., of Reflection Seismology, which is truly today's Energy Resource Seismology.

Preface

Routine reflection processing is based on the assumption of isotropy of the medium. It is now well realized that developments in isotropic reflection processing have reached a saturation stage, and something special needs to be done to improve accuracy. Anisotropic processing of reflections provides the answer to this. 'SEISMIC REFLECTION PROCESSING: With Special Reference to Anisotropy', aims at bridging the existing gap.

This book puts together the physical concepts, mathematical details and methodology for achieving the best results for reservoir modeling either under condition of isotropy or anisotropy. Further, the most common form of anisotropy – the transverse isotropy, is dealt with in detail. Besides, practical aspects in Reservoir Engineering – such as interval isotropic or anisotropic properties of layered media; identifying lithology, pore-fluid types and saturation; and determining crack/fracture-orientations and density, etc. form the core of discussions.

It has the following distinguishing features:

- **Emphasis on accurate reflection processing**
- **Integration and elucidation of concepts in isotropy with anisotropy**
- **Promoting easy understanding**
- **Bringing together the concept and theme based lead case studies on the benefits of exploring fracture-induced anisotropy**
- **Inspiring futuristic vision.**

This book is written with the aim to serve primarily the graduate/postgraduate students and research workers in geophysics desirous of pursuing a career in seismic investigations for Hydrocarbon Energy Resources. Members of the various Societies of Exploration/Petroleum Geophysics would benefit from this book. It should be particularly useful to professional geophysicists and engineers involved in interpreting seismic reflection data from geologically complex areas exhibiting fracturing conditions in the reservoir, in search of hydrocarbons, geothermal or ground water resources.

This book is concept oriented. It discusses the latest research breakthroughs in isotropic and anisotropic processing, focuses on requisite precision over isotropic reflection processing, and projects investigation of fracture-induced anisotropy as a new tool for modeling different types of reservoirs.

This book has been composed keeping in view the current research developments and appreciation by all categories of readers. This book should prove to be a very timely publication to teach a regular course in 'Exploration Seismology' to graduate/postgraduate students of geophysics at the present level of developments both in isotropic and anisotropic reflection processing and interpretation.

The substance of organization of this book is presented in Chapter 1 (Fig. 1.8). Both basic and practical aspects, for conditions of isotropy or anisotropy, are discussed with suitable emphasis.

This book is the outcome of my involvement in teaching courses and research guidance in the area of seismic prospecting for many years, at the University of Roorkee, Roorkee (now Indian Institute of Technology Roorkee, Roorkee). The realization of the phenomenal growth during the last decade in the field of anisotropic reflection processing has inspired me to write a book which is complete to the current level of developments. Although rapid advancements are taking place in this field, it has been my continued effort to incorporate the most recent developments, as far as possible.

My sincere thanks are due to the authors of research papers or books, publishers, and organizations for granting permissions to include their published materials in this book. Wherever permission could not be received due to changed address or some other reason, complete reference of the published material used, is given and the same is gratefully acknowledged here.

I have drawn from a number of sources from past to the most recent, and made deliberate efforts to present the ideas in their real forms. However, in case concise presentation has led to distortions of original concepts, suggestions from authors or readers in this regard would be gratefully acknowledged.

My special thanks are due to Dr. Wolfgang Engel, Executive Editor Geosciences, Springer Verlag, Heidelberg, Germany and his colleagues, Helen Rachner, Janet Sterritt-Brunner, Judith-Diemer, Constanze Sonntag and Susanna Pohl (formerly at Springer Verlag) and all others for their continued involvement in the copy editing process, timely publication and excellent quality production of this book.

Sarvesh Kumar Sharma, a member of the technical staff in the Department of Earth Sciences, Indian Institute of Technology Roorkee, Roorkee has provided constant continued support in typing in an excellent manner the difficult text material, including a variety of equations and symbols. Perhaps, his unique enduring patience has made my dream of completing this project come true. I have no words to express my appreciation for his cooperation and performance as a colleague.

N.K. Varshney of the National Institute of Hydrology, Roorkee provided much needed cooperation and all manner of support for the punctual production of excellent tracings. I would like to express my greatest appreciation to him. Santosh Kumar Mishra and Y.D. Gupta (retired) of the Central Building Research Institute Roorkee also traced some figures. I also express my appreciation to them.

I received special support in various forms from Professor R.P. Gupta, Professor, Department of Earth Sciences, Indian Institute of Technology Roorkee, Roorkee from the initial stage to the completion stage of this project. It is indeed a pleasure and fortune to have such an excellent colleague.

Dr. J.G. Negi (retired) Scientist and subsequently Emeritus Scientist at the National Geophysical Research Institute, Hyderabad, presently serving as Director General, Institute of Seismological Research, Gandhinagar, Gujrat, India, and my Guru (Ph. D. thesis supervisor) was always a source of inspiration for me. He also read Chapter 15 and made useful suggestions for which I would sincerely express my gratitude.

The completion of this book has been a long journey occupying me almost full time, for nearly five years with the most intensive efforts. My family members, Sharada (wife), Kaushal and Kundan (sons), and daughter Anita Mishra and son-in-law R.K. Mishra provided continued support for the completion of this project in its best form. In fact, their contributions are no less than mine. My (late) parents, M. Devi and B.D. Upadhyay, and Sharada's father Dr. S.N. Chaturvedi always encouraged me to write an excellent book as a contribution to the society. Their words have always inspired me.

This book should fulfill needs, and generate new ideas and interests in graduate students, academicians, research scientists, and professionals in Hydrocarbon Exploration Industries for better achievement of their goals. I sincerely hope that readers will value it as a timely and useful contribution to science.

Roorkee, June 2004 S.K. Upadhyay

Contents

CHAPTER 1

Introduction 1

1.1 Seismic Reflections

We know that two types of elastic waves (P and S) can travel through the body of an elastic medium. The particle motion and velocities of these waves are distinctly different. During propagation the particle motion of the longitudinal wave (or P-wave) is along the direction of the raypath, and of the shear wave (or S wave) is perpendicular to the raypath. Figure 1.1 a is a depiction of a simple reflection path geometry for near normal incidence. It assumes a horizontal reflector, and shows raypaths of a P-wave from an explosion and of S-waves from a horizontal vibrator. G is the detector, which may be sensitive to vertical or horizontal vibration. For a near normal raypath, the reflection coefficient for P- and S waves depend on compressional (P) and shear (S) wave velocities and densities on two sides of the reflector. It is the acoustic impedance contrast (i.e. ($\rho_2\alpha_2$–$\rho_1\alpha_1$) for a P-wave and ($\rho_2\beta_2$–$\rho_1\beta_1$) for the S-wave), which controls the absolute value of the normal incidence reflection coefficient.

The geometric reflection raypath as shown in Fig. 1.1 a obeys Snell's law. A more general situation is depicted in Fig. 1.1 b. It shows raypaths from a source to a number of points along a reflector and then back to the receiver. Both Fig. 1.1 a and b assume a homogeneous medium and consequently straight-line raypaths. A simple theory of seismic diffractions (Trorey 1970) is based on treating every segment along a reflector as a secondary source of scattered waves. The superposition of the arrivals, from linear reflecting segments along an interface, at any geophone location on the surface should generate a synthesized seismogram. This procedure for generating a synthetic seismogram is found more efficient where computation of the reflection angle is complicated, requiring iterative computations. For example, Alkhalifah (1995) made use of the *Kirchoff summation* procedure for generating *synthetic seismograms* for a transversely isotropic medium. It does not require explicit computations of reflection angles. This procedure is more rigorous compared to computations based on geometric raypaths. Nevertheless, geometric raypaths are widely used for seismic modeling, mainly because of ease in formulations.

Our interest lies in *primary reflections* arising from the principal reflectors. The primary reflection path is from seismic source to the reflector and back to the surface. The real reflection records, both in land and marine surveys, contain primary as well as multiple reflections. Multiple reflections may be categorized as

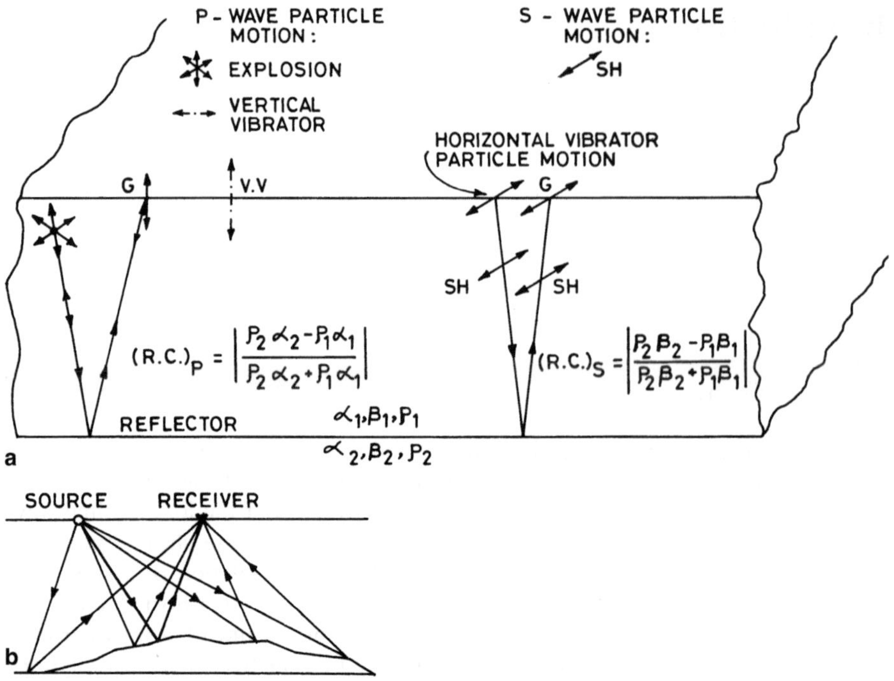

Fig. 1.1 a, b. a Schematic diagram showing longitudinal and shear wave reflections, and particle motions for a near normal incidence raypath. α, β, and ρ stand for P-wave velocity, S-wave velocity and density respectively. **b** Various raypaths from source to the receiver. The *thick line* is the geometric reflection raypath that obeys Snell's law (Source: after Alkhalifah 1995)

short- or long-path multiples. Short-path multiples are linked with thin beds, such as with the low velocity layer (or weathered layer) near the surface in land seismic surveys. Multiples are generated by reflections from the top and/or bottom of a thin layer, on the path of the primary reflection, and affect the waveform of the primary reflection events recorded on a seismogram. Short-path multiples may be categorized as *ghost*, *near surface multiples* or *peg leg multiples* depending on how they originate (Fig. 1.2 a–c).

The ghost type multiple reflection is of special interest to us as they strongly influence the primary reflection wavelet. They arise both in land as well as in marine seismic surveys. Figure 1.2 a shows the origin of ghost. With the shot below the weathered layer part of the seismic energy travels downwards, gets reflected from the principal reflector and reaches the ground surface as a primary reflection, whereas for the ghost the initial path of the energy is from seismic source to the base of the low velocity layer, followed by reflection from the bottom of the low velocity layer and finally after reflection from the principal reflector to the surface. The interference effect between the primary reflection event and the ghost depends upon the fraction of wavelength represented by the *effective path difference* between the two. The effective path difference between the

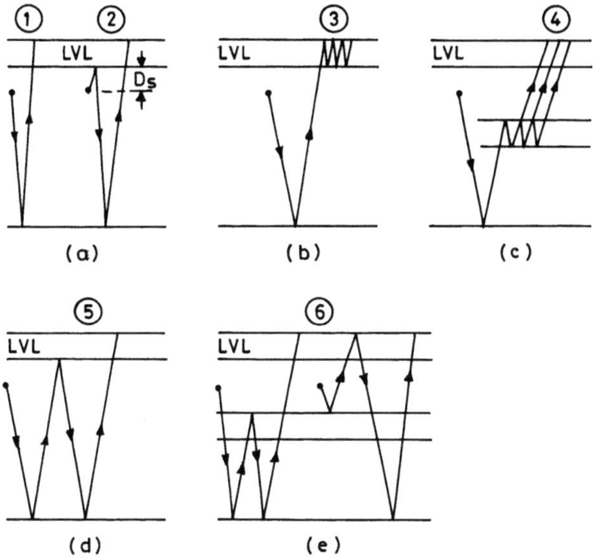

Fig. 1.2 a–e. a–c Short-path multiples; **d, e** long-path multiples. *(1)* Primary reflection; *(2)* ghost; *(3)* near surface multiples; *(4)* peg-leg multiples; *(5)* simple longpath multiple; *(6)* inter formational long-path multiples (Source: after Schoenberger and Levin 1978)

primary reflection and the ghost is ((1/2) λ + 2 D_S) where (1/2) λ arises due to 180° phase change (equivalent to half a wavelength) at the bottom of the low velocity layer, and 2 D_S represents the additional path of the ghost below the bottom of the low velocity layer to the shot level. It is to be noted that the seismic source wavelet contains a range of frequencies, and therefore the interference effect is expected to vary for different frequency components. The overall wavelet shape would vary as the depth D_S of the shot is varied. It should be mentioned here that a relatively small change in D_S might give rise to large variations in the reflection character. With this consideration the depth of the shot below the base of the weathered layer (in land seismic) and below the water surface (in marine seismic) is maintained as nearly constant as possible.

Figure 1.2 d, e shows the possible long-path multiple in case of land seismics. Strong long-path multiples are likely to be confused as the primary reflection event. These arise from strong reflectors at the surface, the sea floor, or in case of land seismics from the base of the low velocity layer (LVL), where the acoustic impedance contrast is very large. Long-path multiples involve at least two reflections at depth. If the reflection coefficients involved were large, the long-path multiples would be observed as distinctive events in the record. For example, the reflection coefficient at the base of the low velocity layer may be as high as 0.7 and a value of 0.2 may be assigned for very strong reflectors at depth. Thus, two reflections from depth and one from the low velocity layer would yield an absolute value of *effective reflection coefficient* for this type of multiple as $0.2\times0.7\times0.2=0.03$, which lies in the typical range of reflection coefficient values

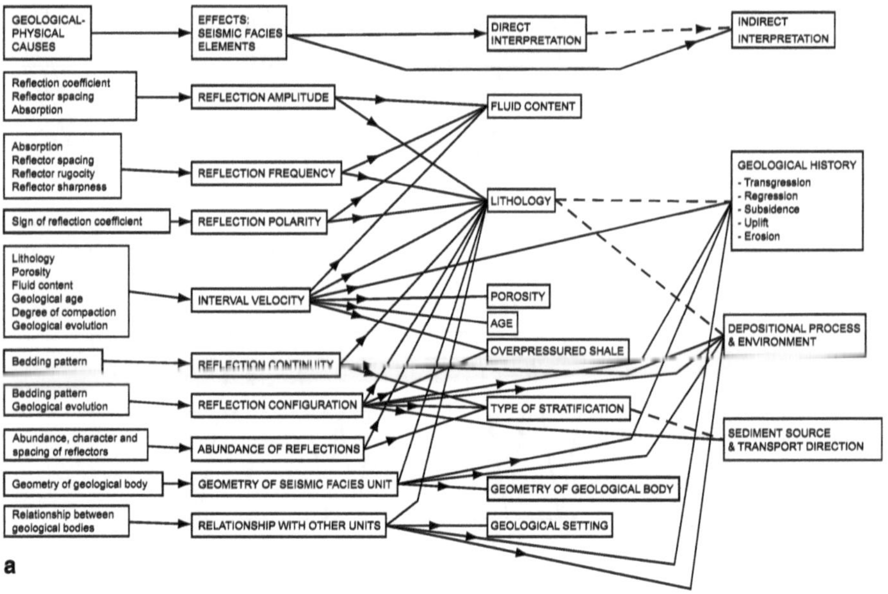

a

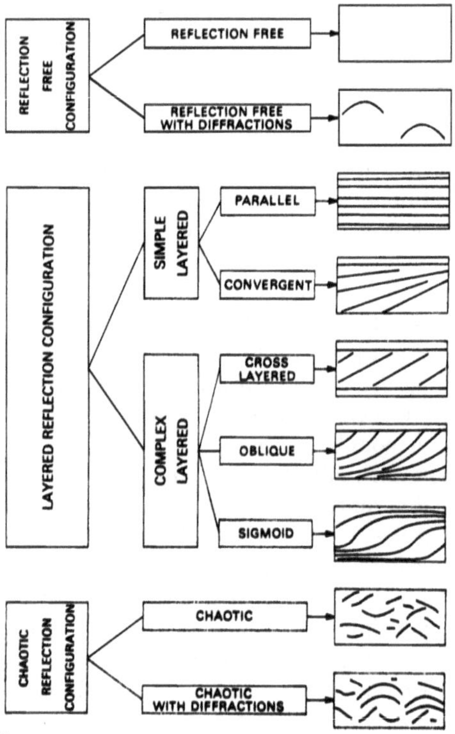

b

Fig. 1.3 a–c. a The relationship between geological-physical causes, seismic facies and interpretation. **b** Basic types of reflection configurations. **c** Theoretical model of a delta – (*upper*) geomorphology of a delta, (*middle*) theoretical model of the mechanism of formation of a delta, (*bottom*) expected reflection configurations of a deltaic sequence (Source: Roksandić 1978, from Geophysical Prospecting with permission)

Fig. 1.3 c

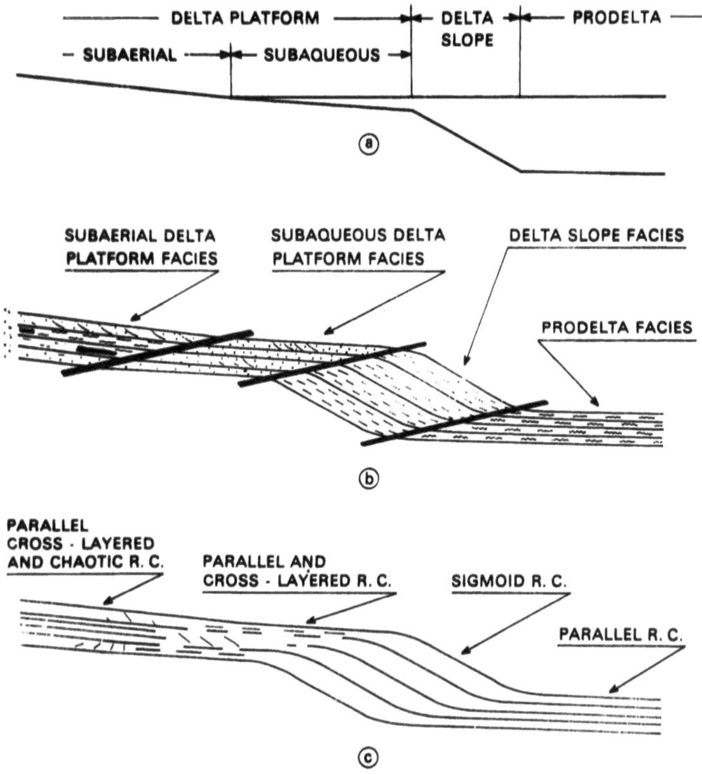

for the primary reflection. Therefore careful interpretation requires distinguishing such multiples to avoid the risk of interpreting them as the primary reflection from deeper horizons.

In marine seismic work, commonly occurring coherent noise, known by different names such as *ringing*, *singing*, or *water reverberations*, arise due to multiple reflections between the top and bottom of the water layer. Since the reflection coefficients at the top and bottom of the water layer are large, a considerable amount of seismic energy gets trapped in the water layer. The reverberating energy is sometimes reinforced by the reflections. Depending on the water depth there is a resonance effect for some frequencies resulting in enhancement of certain frequencies. As a result of this the appearance of the reflection seismogram becomes very sinusoidal. Under its influence, even preliminary analysis of the record, like picking up of reflection times and calculation of moveout corrections is inaccurate.

For details on multiple reflections, publications by Backus (1959) and Schoenberger and Levin (1978), or textbooks by Sheriff and Geldart (1982, pp 105–110) and Waters (1987, p 425) may be referred.

It is fundamental to develop a picture of how best reflections may be utilized to establish subsurface geology. Roksandić (1978) gave seismic facies analysis concepts that still form the fundamentals for most of the present day interpretations. These basic qualitative modeling concepts are helpful, and are therefore reproduced from Roksandić (1978) in Fig. 1.3 a–c. Figure 1.3 a gives the relationship between geological physical causes, seismic facies elements and interpretation. Figure 1.3 b gives the basic types of reflection configurations. Figure 1.3 c gives an example of a theoretical model of a *delta*, including expected reflection configurations of a deltaic sequence.

Seismic reflections contain a very wide spectrum of geological information. However, to make the best use of them requires processing them suitably to enable giving an interpretation as realistic as could be possible. We explain in Section 1.2 the necessity for processing reflections. Before this we give through Figs. 1.4 and 1.5 geological settings (structural and stratigraphic), which are considered classic for the occurrence of oil. These include dipping and curved geological boundaries, folds and faults, domal structures and uplifts, thin or thick beds and a variety of combinations of rock and rock fluid properties. Reflection

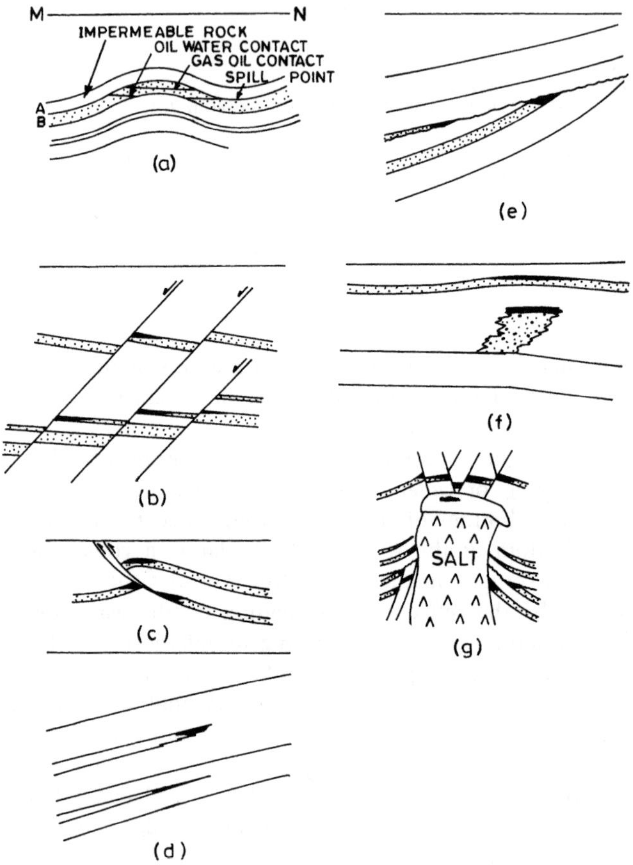

Fig. 1.4 a–g. Various sedimentary structures which produce hydrocarbon traps (Source: after Sheriff and Geldart 1983, Exploration Seismology vol 2: 84. © Cambridge University Press 1983)

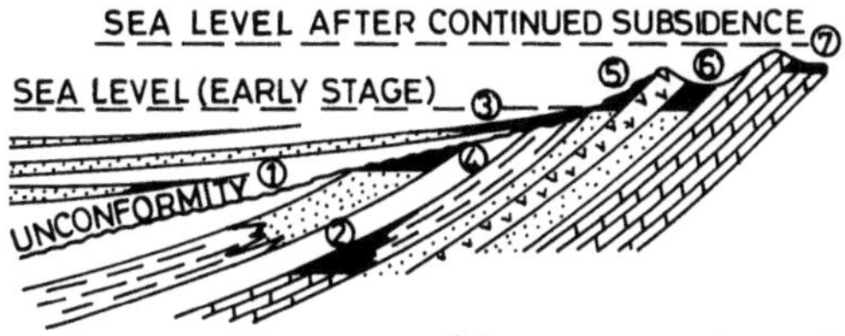

Fig. 1.5. True stratigraphic traps, unconformity traps, and paleogeomorphic traps (Source: Halbouty 1982, p 3 © 1982 The American Association of Petroleum Geologists. Reprinted by permission of the AAPG whose permission is required for further use)

seismology aims at delineating these features accurately by suitably planning field surveys, developing analytical formulations and processing algorithms to describe subsurface geology in its finest detail, with emphasis on rock and fluid properties of the reservoir rocks and their areal or volume extensions.

1.2 Necessity for Processing Reflections

Necessity for routine processing arises because of:

- simultaneous recording of noise along with reflected signals,
- modification of the reflection signal due to propagation effects (spherical spreading, anisotropic wave propagation, and/or viscoelastic dissipation),
- variation of reflection signal amplitude by seismic source/receiver directivity,
- interference (or tuning effect) between reflected signals or between the primary reflections and multiples,
- necessity to bring in focusing of the weak reflection events (stacking),
- artifacts in interpretation results, produced as a result of the method of data acquisition or processing or both,
- necessity to search for clues or patterns in the record (like diffractions, correlations or phase relationships),
- necessity to assign real locations of the origin of reflected signals (migration),
- necessity to appropriately account for geometry of reflectors, or reflector curvature, and
- need for preparing depth sections (velocity analysis), etc.

Special processing of reflection data is essential under the condition of anisotropy, inhomogeneity and/or viscoelasticity. As Chapter 3 and also part of Chapter 4 are devoted to providing a wider coverage of the different types of routine processing we limit our discussion here by giving two specific examples. The first illustrates the reason for the consideration of specialized processing for converted waves, and the second the failure of processing for typical anisotropic and structural conditions.

1.2.1
Specialized Processing of Converted Waves

Figure 1.6 shows a simple case of a isotropic, homogeneous single layer. The incident wave is a P-wave and the converted wave (or C- wave) is a S-wave. It may be verified that moveout of the C-wave is not hyperbolic. For this simple geometry, the exact traveltime (t_c (x)) and exact emergence offset x are given by:

$$t_c(x) = t_p(x) + t_s(x) = \frac{Z}{V_P \cos\alpha_P(x)} + \frac{Z}{V_S \cos\alpha_S(x)} \tag{1.1}$$

and

$$x = V_P t_p \sin\alpha_P + V_S t_S \sin\alpha_S \tag{1.2}$$

[Thomsen 1999, Eqs. (1) and (1)] where t_P is one way oblique traveltime through the layer for the P-wave; correspondingly t_s is one-way shear wave time. α_P and α_s in Fig. 1.6 are governed by Snell's law

$$\frac{\sin\alpha_P}{V_P} = \frac{\sin\alpha_S}{V_S} = p = \frac{\partial t_c}{\partial x} \tag{1.3}$$

[Thomsen 1999, Eq. (1)] where p is the ray parameter which is constant along the ray, and t_c is the time of arrival of converted wave (C-wave) at offset x.

Using (1.3), Eq. (1.2) for emergence offset x may also be written as

$$x = pV_p^2 t_p + pV_s^2 t_s \tag{1.4}$$

Equations (1.1)–(1.4) are exact expressions for the most simple case of a single, isotropic, homogeneous layer. Complicated cases of layering including isotropic or anisotropic layers necessitate using approximations in place of exact expressions. Using Taylor series approximation, Tsvankin and Thomsen (1994) gave the following expression for t_c^2 (x):

$$t_c^2(x) = t_{c0}^2 + \frac{x^2}{V_{c2}^2} + A_4 x^4 + \dots \tag{1.5}$$

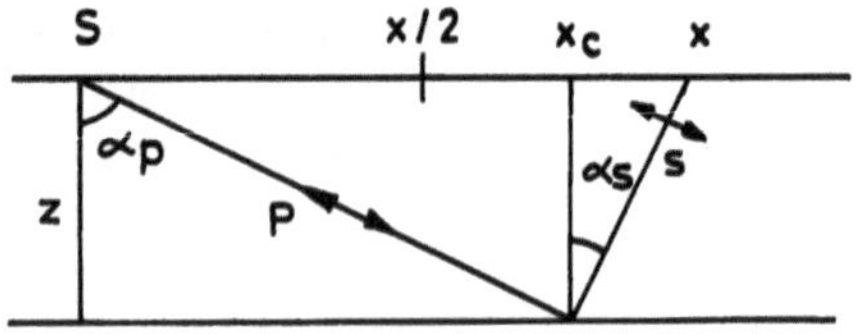

Fig. 1.6. Canonical converted wave schematic (Source: after Thomsen 1999, p 679)

In Eq. (1.5), t_{c0} is a two-way C-wave zero-offset time, given by

$$t_{co} = t_{po} + t_{so} = t_{po}\left(1 + t_{so}/t_{po}\right) = t_{po}(1 + \gamma) \tag{1.5 a}$$

[Thomsen 1999, Eq. (5)] with

$$\gamma \equiv \frac{V_P}{V_S} = \frac{z/t_{po}}{z/t_{so}} = \frac{t_{so}}{t_{po}} \tag{1.5 b}$$

where t_{po} and t_{so} are pure mode one-way times (taken to correspond to vertical times here); V_{c2} is the C-wave short spread moveout velocity.

A further improved version of Eq. (1.5) was given by Thomsen (1999) in the form:

$$t_c^2(x) = t_{c0}^2 + \frac{x^2}{V_{c2}^2} + \frac{A_4 x^4}{1 + A_5 x^2} \tag{1.6}$$

A_5 is not a free parameter and depends on A_4, V_{c2} and V_{P2}. For details on the quartic parameter A_4 reference may be made to Tsvankin and Thomsen (1994). Equation (1.6) possesses correct limiting behavior at both short and long offsets and varies smoothly in between (Thomsen 1999).

From Fig. 1.6, the conversion point offset x_c and total offset x are related in the asymptotic limit of vertical travel (i.e. of small values of offset divided by thickness x/z) for a homogeneous and isotropic layer:

$$\frac{x_{co}}{x} = \frac{\gamma}{1 + \gamma} \tag{1.7}$$

where

$$\gamma = V_P/V_S$$

Equation (1.7) was originally contributed in the classic publication by Tessmer and Behle (1988), which was extended by Thomsen (1999) to the more realistic case of many layers that may be isotropic or anisotropic.

The significant points to note for the processing of converted wave data are the following (Thomsen 1999):

- There is more critical dependence of converted wave processing on physical assumptions concerning rock velocities than the pure mode processing. This is for the reason (refer, for example, to Eq. 1.7) that even the offset (x_{co}) of the imaged point depends on the physical parameters of the medium. In case of pure mode processing, the image-point offset is determined geometrically rather than physically. This difference leads to fundamental differences in processing strategies and tactics for converted waves compared to pure mode waves.

- For the case of a layered anisotropic media an effective velocity ratio $\gamma_{eff}=\gamma_2^2/\gamma_0$ govern most of the behavior of the conversion point offset, where $\gamma_0 \equiv (V_P)_{aver}/(V_S)_{aver}$ is the ratio of average vertical velocities, and γ_2 is the corresponding ratio of short-spread moveout velocities (for details refer to Thomsen 1999). This aspect is directly related to appropriate data acquisition design. If the data acquisition for converted waves were based on γ_0 instead of γ_{eff}, it would result in sub-optimal data collection for converted waves.
- In the case where the subsurface possesses lateral variability and/or azimuthal anisotropy, converted wave data are not invariant under the exchange of source and receiver positions – split spread may have asymmetric moveout. This aspect is more stringent for 3-D seismic surveys, and an inadequate consideration of the converted wave velocity field may lead to imaging errors. As the physical properties of the medium plays a larger role for converted waves, the inhomogeneity and anisotropy of the medium are much more crucial for converted waves than for the pure P-mode.
- In pure mode propagation, physical characterization may follow imaging whereas in the case of using converted waves, one cannot form a proper image of the subsurface without careful consideration of the physical parameters. Thomsen (1999, p 688) recommends C-wave acquisition, processing and interpretation to proceed together for greatest effectiveness.

1.2.2 Failure of Processing for Typical Anisotropic and Structural Conditions

Figure 1.7 a and b relate to the situation of a tilted symmetry axis of a transversely isotropic medium towards a steeply dipping reflector (shown on the extreme right in these figures). This type of situation may arise, for example, under actual field conditions in which uptilted sediments, such as a shale formation, lie on the flanks of a salt dome (refer Fig. 1.4 g). This type of geological setting may give rise to a tilted symmetry axis (taken normal to the bedding plane) towards the steeply dipping flank of the salt body. Tsvankin (1997), considering this situation, showed that under this type of relationship between symmetry axis and steeply dipping reflector we may have serious problems in imaging the steeply dipping reflector, even if the processing algorithm is able to adequately handle transverse isotropy. The reason for this is explained in the following text (Tsvankin 1997).

Referring to Fig. 1.7 a there is a limiting value θ_{max} for the angle θ between the P-wavefront normal and the vertical corresponding to which the ray direction is horizontal. This happens because of an increase in the phase and group velocity from the symmetry axis towards the horizontal. In the angular range $\theta_{max}<\theta<90°$, there is no wavefront normal, and therefore, *no specular zero-offset-reflections* exist for dips larger than $\phi_{max}=\theta_{max}$.

Figure 1.7 b shows the results of anisotropic ray tracing (Rüger and Alkhalifah 1996) for a symmetry axis tilt of 20° from the vertical towards an interface dipping at 80°. Results in Fig. 1.7 a (Tsvankin 1997) are substantiated by Fig. 1.7 b.

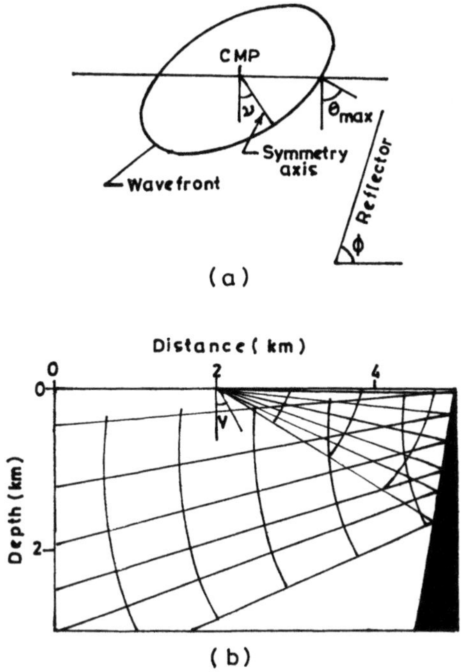

Fig. 1.7 a, b. a The P-wavefront for a transversely isotropic medium with a symmetry axis tilted towards the reflector. The increase in phase and group velocity away from the symmetry axis in this model reduces the angular range of the wavefront normal in the segment of the wavefront propagating towards the reflector. The maximum phase (wavefront) angle in the *right lower quadrant* is $\theta_{max} < 90°$ (Source: Tsvankin 1997, p 485. From Geophysical Prospecting with permission). **b** Rays and wavefronts of the P-wave excited by a point source at the surface and reflected from an interface dipping at 80°. The model parameters are $\nu = 20°$, $\varepsilon = 0.25$ and $\delta = 0.05$ (Source: Tsvankin 1997, p 485. Figure 1.7b after Rüger and Alkhalifah 1996. From Geophysical Prospecting with permission)

As the reflector dip exceeds ϕ_{max} all rays emanating from the source travel downwards after reflection and do not return to the surface. This shows that no specular reflection exists even for non-zero offsets, and illustrates that a CMP gather at the surface will not record any *specular energy*. This explains that even if we have an algorithm for processing reflections in the presence of anisotropy, it would not work in this type of special situation.

1.3 The Scope and Organization of the Book

Routine reflection processing is based on an assumption of isotropy of the medium. It is now well realized that developments in isotropic reflection processing have reached a saturation stage, and something special needs to be done to improve accuracy. Anisotropic processing of reflections provides the answer to this.

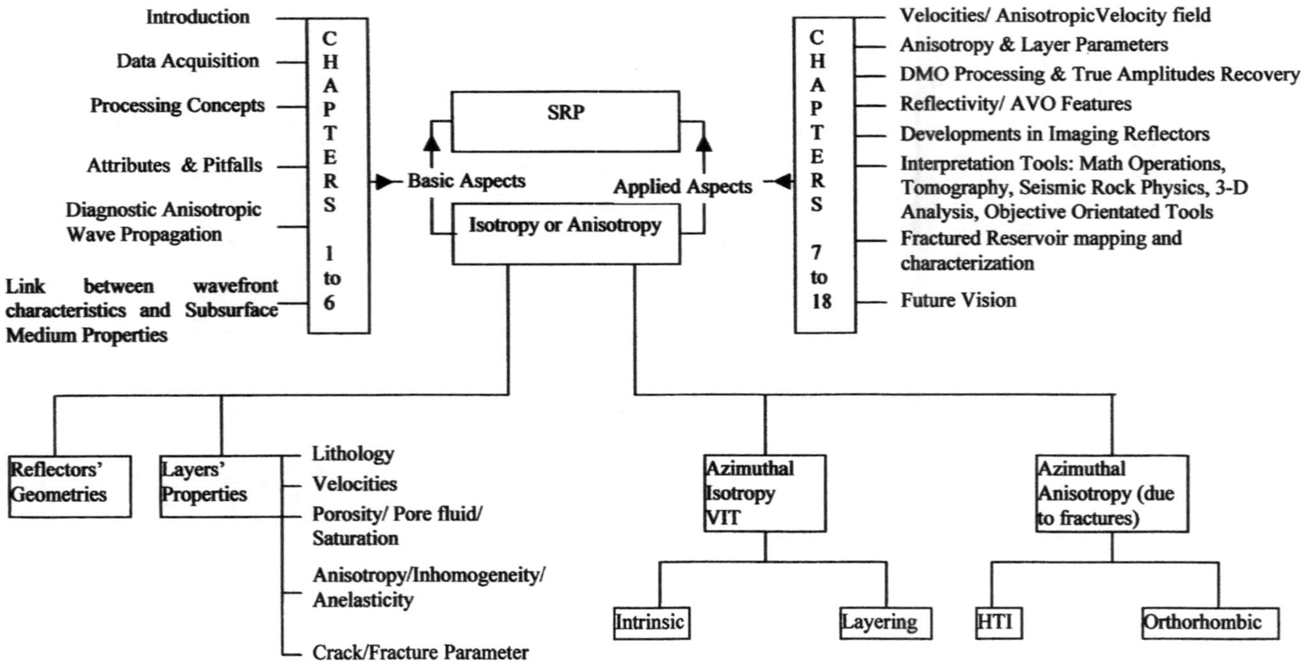

SRP : Seismic Reflection Processing
VTI : Vertically Transversely Isotropic
HTI : Horizontally Transversely Isotropic

Fig. 1.8. Substance of the organization of the book: *Seismic Reflection Processing*

The proposed book – 'SEISMIC REFLECTION PROCESSING: With Special Reference to Anisotropy' – aims at bridging the existing gap.

The proposed book puts together the physical concepts, mathematical details and methodology for achieving best results for reservoir modeling either under a condition of isotropy or anisotropy. Further, the most common form of anisotropy – the transverse isotropy is dealt with in detail. Besides, practical aspects in Reservoir Engineering – like interval isotropic or anisotropic properties of layered media; identifying lithology, pore-fluid types and saturation; and determining crack/fracture-orientations and density, etc. – form the core of discussions.

Figure 1.8 depicts significant features of the organization of the book. Two aspects, namely basic and applied, of seismic reflection processing (SRP) both for isotropy and anisotropy are identified as vital considerations for organizing the presentation of material of the book. The first six chapters deal with the basic aspects and subsequent chapters with the applied aspects.

Chapter 1 introduces seismic reflections and the necessity for processing these, and sets out a theme and layout for the book. Chapters 2 through 5 specifically discuss data acquisition; processing concepts; attributes and pitfalls of routine reflection processing; anisotropy models of sedimentary sections, quantification of anisotropy, and diagnostic features of anisotropic wave propagation; and most importantly Chapter 6 lays the foundation for geophysical applications by discussing the link between wavefront characteristics and subsurface medium properties.

Chapters 7, 8, and 9 deal with reflection time analysis. Hyperbolic, skewed hyperbolic reflection traveltime relations for isotropic and anisotropic media, and their application for the evaluation of stacking and interval velocities for multilayered media are discussed. The concept of *velocity estimation by beam stack* applicable to geologically complex areas is briefly covered as part of Chapter 7. Chapter 10 discusses the accurate estimation of anisotropy and layer parameters. Chapter 11 is devoted to describing conceptual developments of dip moveout processing and true amplitude imaging. Chapters 12 and 13 aim at discussing basic reflection theory for anisotropic models and AVO characteristic developments for application oriented interpretations. Chapter 14 discusses classical developments in reflection data migration both for isotropic and anisotropic media. Chapter 15 is exclusively devoted to imaging concepts by making use of *integral solutions* of the wave equations. Chapter 16 explains the role of various interpretation tools by including conventional mathematical methods, and modern developments in tomography, seismic rock physics and the use of 3-D data analysis. Chapter 17 presents methodologies and illustrative field system examples for characterization of fractured reservoirs. Chapter 18 concludes the objectives of this book by identifying perspective areas, which call for theme based approaches.

CHAPTER 2

Reflection Data Acquisition

2

2.1
Components of a Reflection Survey Unit

2.1.1
Seismic Sources

Seismic sources are used to generate and transmit vibration energy into the ground. Their choice depends on various factors, like survey area (whether land or sea), depth to be probed, and inhabitance, etc. Commonly used seismic sources in land areas are explosives, except in populated areas with a large density of dwellings. For seismic surveys in marine areas, a source like 'sparker' or 'air-gun' is used.

In seismic exploration work, explosives are loaded at the bottom of a hole drilled to penetrate below the weathered layer. For maximum downward transmission of energy released due to an explosion, the hole is filled with heavy drilling mud or water. An electric blasting cap is used to initiate an explosion. The blasting cap is placed inside one of the explosive charges. Explosion of the cap detonates the entire explosive charge. The blaster provides current to the blasting cap. There is a device in the blaster that generates an electric pulse at the instant the explosion begins. The *time break pulse* is used to fix the instant of explosion, taken as $t=0$. The time break pulse is recorded with the seismic data, for subsequent calculation of the reflection times, etc.

An explosion generates a wide range of frequencies. The detonation velocity of an explosion is of the order of 7000 m/s and is complete within a few hundreds of microseconds. The seismic pulses resulting due to the explosions have very steep fronts compared with pulses from other energy sources. This is a desirable feature for subsequent seismic wave analysis. Pressure in the detonation front is extremely high, measured in hundreds of thousands of atmospheres. After the explosion is complete, the pressure of hot gases although lower is still very high. The enormous pressure is able to break, fracture, and pulverize the rock material surrounding the charge. The result is formation of an actual cavity in the rock. Outward of this cavity, a shell of intense particle compression starts propagating. Even outside the actual cavity, the stress level is too high for well-behaved acoustic transmission. Under the influence of non-linear losses and geometrical spreading, the disturbance comes down to a level beyond which further propaga-

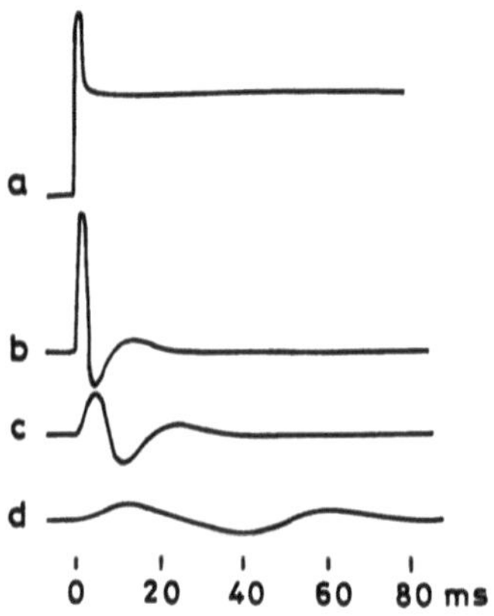

Fig. 2.1. Pressure waveform at various stages in the history of the pulse (Source: Anstey 1970, Seismic Prospecting Instruments vol 1: Signal characteristics and instrument specifications, p 51. © 1970 by Gebrüder Borntraeger, Berlin Stuttgart)

tion is substantially linear. A surface may be imagined beyond which the disturbance propagation is linear. This surface is called the '*equivalent cavity*'. Diameter of the actual cavity is much smaller compared to that of an equivalent cavity.

From the point of view of seismic analysis, it is important to understand the form of the disturbance that is initiated, which subsequently propagates outwards. Figure 2.1 a shows the pressure in the detonation front and the sustained pressure of hot gases close to the actual cavity. Surrounding the actual cavity, non-linear processes dominate, and the depiction of a pressure waveform is difficult. Just outside the equivalent cavity, the pressure waveform has a form as shown in Fig. 2.1 b. This is the input to the largely linear earth material. Beyond the equivalent cavity, two very important factors act to change the waveform: the *geometrical spreading* and the *frequency selective absorption.* As a result, pressure waveform changes continuously as the disturbance propagates. Due to geometrical spreading, the amplitude falls off, and as a result of *frequency selective absorption*, progressive loss of high frequencies takes place, which results in an increase of the duration of disturbance. The picture at a distance of about 100 m or so is depicted in Fig. 2.1 c. Figure 2.1 d depicts the waveform at a great distance.

We expect a pulse waveform rich in high frequencies from materials capable of supporting large stresses (with consequent small equivalent cavity). Propagation through a less competent rock mass would make the waveform rich in low frequencies.

2.1.1.1 Vibroseis

The Continental Oil Company introduced an entirely different type of source of seismic energy in 1960s. The principle behind the Vibroseis method was invented by Crawford, Doty and Lee (1960). In contrast to the impulsive pulse generated by a conventional explosive charge, the *Vibroseis* generates an oscillatory waveform of controlled frequencies for a fixed duration of time.

The Vibroseis system is truck mounted. The vibration plate is lowered to bring it in firm contact with the ground. The entire weight of the truck presses the plate against the ground. The vibration plate is actuated by a hydraulic servo cyl-

Fig. 2.2. Vibroseis oscillatory upsweep and downsweep signals

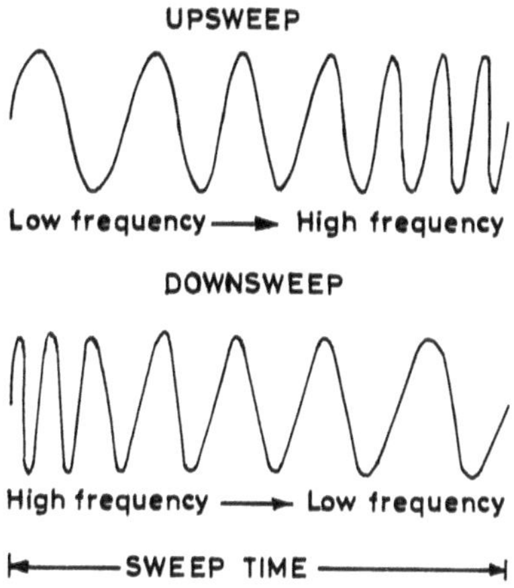

inder system. The controlled vibration of the plate injects into the Earth an oscillatory waveform of changing frequencies, for a fixed duration of time. This is known as *sweep*. A sweep, for example, may consist of progressively varying frequencies, say, from 7 to 70 Hz over a period of 10 seconds. A sweep may be an *upsweep* or *downsweep* depending on whether frequency increases or decreases with time (Fig. 2.2). The duration of vibration is referred to as *sweep time*.

Vibroseis has emerged as a highly successful alternative to the explosive sources. This is so for various reasons. For example, it is much easier to deploy a Vibroseis system than explosives, which require special arrangements for storage and transportation, and drilling of shot holes for their use. As Vibroseis is a prolonged low-energy source it is much safer than explosives, and for this reason it may be used along roads and even within cities. Further, sweep rates and frequencies may be selected in an optimal manner to suit the geological conditions of an area. Additionally, energy injected into the Earth may be increased as necessary by coupling a number of vibrators together and making them vibrate in phase from a master vibrator. Also, it is feasible to sum (vertically stack together) the results of several sweeps to increase the energy of seismograms. These features are the prime factors for success of the Vibroseis system over the explosive sources.

The variable pressure exerted by the vibrating plate on the ground is expressible as

$$P(t) = A(t)\sin 2\pi t\left(f_0 + \left(\frac{df}{dt}\right)t\right) \tag{2.1}$$

[Sheriff and Geldart 1982, vol 1, Eq. (5.3)]

df/dt may be either positive or negative and constant for the case of a *linear sweep*.

The duration of the ensuing sweep normally lasts for 7 to 35 s and the frequency swept may be from 12 to 60 Hz (or vice-versa).

2.1.1.2
Marine Seismic Sources

Development of sources of seismic energy for surveys in marine areas have resulted from the following considerations:

1. The source should not heavily disturb or injure marine life.
2. The bubble effect should be kept to a minimum.
3. The system should be such as to be repeatable sequentially at a short interval of time.
4. The system should be usable at a constant depth and should generate minimum drag on the ship carrying it.

Explosion was the earliest source of seismic energy used for marine seismic surveys. The main drawbacks of using it were that it inflicted much loss of marine life, and was associated with the bubble effect. New sources were invented which reduced to a very large extent both these effects. Amongst these sources are the air-, water-, or gas- guns for deeper seismic surveys, and Sparker or Boomer for shallower surveys.

From the point of view of seismic analysis, it is very significant to know about the form of the source pulses generated by various seismic sources used in marine reflection work.

The airgun, a widely used source of seismic energy, works on the principle that it discharges air into water under very high pressure. Air pressure of 14 Mpa

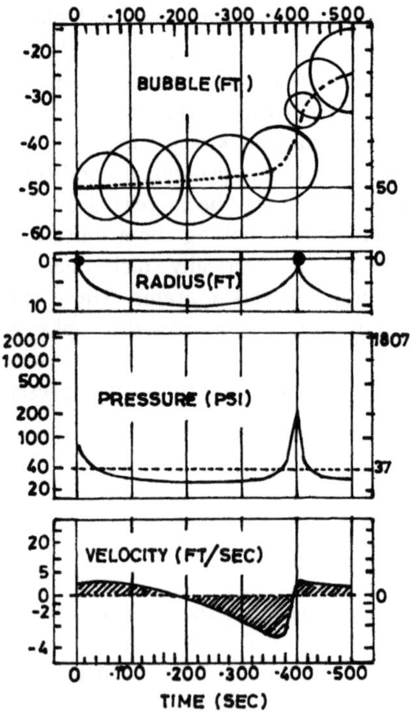

Fig. 2.3. Pressure and velocity signatures for the explosion of 22 kg (50 lb) of TNT taken at a distance of 10 m (33 ft). The depth below the water surface was 16 m (50 ft) (Source: Seismic-Energy Sources. United Bendix 1968. Reproduced from Waters 1987: Reflection Seismology, p 90. Third Edition. © 1987 John Wiley and Sons Inc, New York)

is most common, although pressures even up to 70 Mpa are used. Injection of high-pressure air into water creates sound and the formation of a bubble, which oscillates and simultaneously rises towards the water surface (Fig. 2.3). Seismic waves are generated with each oscillation of the bubble. Following the injection of air at very high pressure and formation of the bubble, the force works outward, which increases the size of the bubble. At the same time with increase in size of the bubble, the net outward force decreases and becomes zero and the gas pressure becomes equal to the hydrostatic pressure. However, as the water has achieved its maximum outward velocity it continues to move outward. While doing so it decelerates because the net force is now directed inward (hydrostatic pressure is larger than the gas pressure). This net inward force collapses the bubble with a consequential sharp increase in gas pressure. The process repeats itself and each oscillation generates seismic waves. Thus waves travel downwards, get reflected and generate additional seismic records every 0.2 to 0.4 seconds. The records produced by these reflections are superimposed on each other, and it is not possible to make out from which oscillation, a particular reflection event is associated. Obviously, the bubble oscillation is an undesirable effect, which needs to be minimized.

The bubble effect was also an undesirable aspect of the underwater explosion method. To minimize the bubble effect with explosion, the explosion was fired, say, within 2 m of the water surface. However, this practice proved unsuccessful, as it was insufficient to generate enough useful seismic energy.

The bubble oscillation effect may be reduced if a number of air guns having different chamber volumes are fired together, so that the first pulse for all of them align, but the later pulses due to bubble oscillations do not. This happens because the frequency of oscillation is controlled largely by the chamber volume, which is different for the different air guns used. A schematic picture depicting pulse shapes in the two cases of a single air-gun, and an array of air-guns of different sizes is given in Fig. 2.4 a, b. The attenuated bubble effect is seen in Fig. 2.4 b.

The bubble effect is also attenuated in the case of Sleeve Exploders (also called the Aquapulse and Deltapulse). This type of source makes use of a propane-oxygen mixture, which is exploded by a spark plug. The mixture is exploded inside a closed flexible chamber. A heavy rubber sleeve (the boot) is fastened around a steel frame, which contains the explosive mixture. On explosion, the products of explosion expand the rubber boot; and the valve opens to allow the gases to the surface on contraction of the boot. This has the effect of attenuating the bubble effect. A sleeve exploder waveform assumes the shape as shown in Fig. 2.4 c.

There is another category of seismic source, which makes use of *implosion.* The imploders work on the principle of generating a region of very low pressure; the collapse of water in the low-pressure region gives rise to a seismic shock wave. The seismic pulse generated is relatively free of the spurious bubble effect. The waveforms of two such imploders, the flexichoc and watergun, are shown in Fig. 2.4 d and e.

There is another category of seismic source, the vaporchoc or the steamgun. In this case, superheated steam is stored in a steam tank submerged in water. On the opening of a valve, the steam emerges in water where it forms a bubble. As

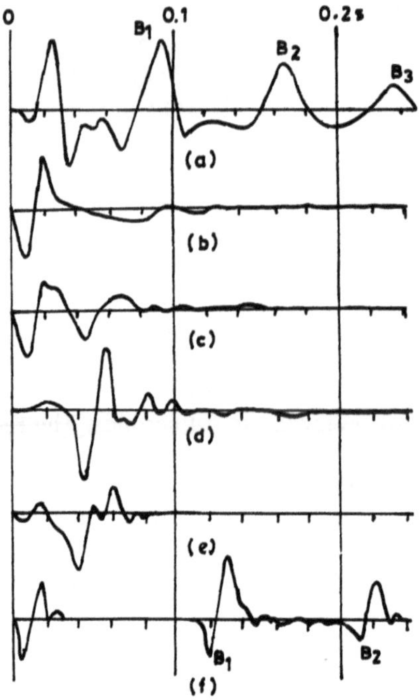

Fig. 2.4a–f. Far-field waveshapes generated by marine seismic sources: **a** single 120 in^3 airgun; **b** array of airguns of different sizes selected to attenuate bubble effects by destructive interference; **c** sleeve exploder; **d** Flexichoc; **e** watergun; **f** maxipulse. *Curves* are intended to show features of waveshape, not amplitude relationships. *B* indicates bubble effects; the interval between successive bubbles becomes smaller with time (Source: **a** and **e** are from McQuillin et al. 1979; **b, c** and **f** from Wood et al. 1978, **d** from manufacturers literature. In: Sheriff and Geldart 1982, p 180. Cambridge University Press 1982)

the steam condenses the bubble collapses and finally disappears. There is no bubble oscillation. The steam injection produces a forerunner pulse when the valve is opened. However, the main seismic pulse results from the collapse of the bubble.

In the Maxipulse method, bubble oscillations (Fig. 2.4f) are recorded, and this information is used to 'deconvolve' the bubble effect in seismic data processing. The pressure detector is used to record the initial pulse and subsequent bubble waveforms. It is observed that the first bubble collapse produces more energy than the initial explosion.

In Aquaseis, the use of a linear charge and low explosive per unit length reduces the bubble effect.

Sparkers are ideal sources for shallow seismic investigations. Sparkers emit acoustic energy by electric discharge in seawater. The energy released by sparkers is only a few percent of that released by air guns. However, the energy is sufficient to penetrate several hundred meters into the sea floor. Pulses without a bubble effect may be produced by using a number of electrodes and adjusting their spacing. For details on the mechanism of operation of different types of seismic sources reference may be made to Sheriff and Geldart (1982), Verma (1986), Waters (1987), and Pritchett (1990).

2.1.2
Detectors

Detectors are the devices designed to detect the vibratory motions produced by the reflected signals, either on the ground surface or in water. Their design and response are made compatible with the characteristics of the desired reflected signals.

For ground reflection surveys, detectors called *geophones* are used. For surveys in marine areas *hydrophones* are deployed. The lower part of the geophone (Fig. 2.5 a) is a pointed peg, and with this peg the geophone is firmly fixed into the ground. The aim is to pick up the ground vibrations accurately. It is possible to design geophones to respond to either vertical or horizontal ground motion. The usual types of geophones used are designed to respond to the (almost) vertical ground motion of P-waves generated by near vertical reflection paths.

The mechanical design of geophones is very simple. Figure 2.5 b shows a cross-sectional view of a geophone. It consists of a cylindrical magnet with a cylindrical hole in its outer part. Within the hole a moving coil is suspended by light flat springs. With the geophone firmly fixed into the ground, the magnet follows the ground motion whereas the coil tries to remain stationary because of its inertia. The relative motion between the coil and magnet is able to generate an electrical voltage. The voltage output is proportional to the relative velocity of the coil with respect to the magnet, the number of turns in the coil, the coil radius and intensity of the magnetic field in the gap where the coil is suspended. The unit for the geophone response to the ground motion is volts per cm per second. A typical response value may be 100 mv/cm/s. The geophone output responds to the velocity of ground particle motion.

The response of the geophone (Fig. 2.5 c) is measured as a function of frequency of the signal. An ideal response is one, which does not give preference to selected frequencies of the signal (unless specifically designed). Therefore, a flat response for the complete frequency range of the signal is the desired characteristic of the geophone response.

In fact, the geophone has its own natural frequency. If the signal frequency were close to the natural frequency of the geophone, the response would become very large as a result of resonance. The natural frequency of the geophone depends on the mass (m) of the suspended inertial member, and stiffness constant (k) of the spring and is given by

$$f = \frac{1}{2\pi}\sqrt{\frac{k}{m}} \tag{2.1}$$

The reflected signals reach geophones in quick succession, and it is desirable that the geophone responds in such a manner that after receiving a signal and giving an output in the form of an electrical voltage, it's oscillatory movement does not last for long and is able to respond to other signals as they affect the ground vibrations at the position of the geophone. This becomes possible by introducing a damping of the moving system of the geophone.

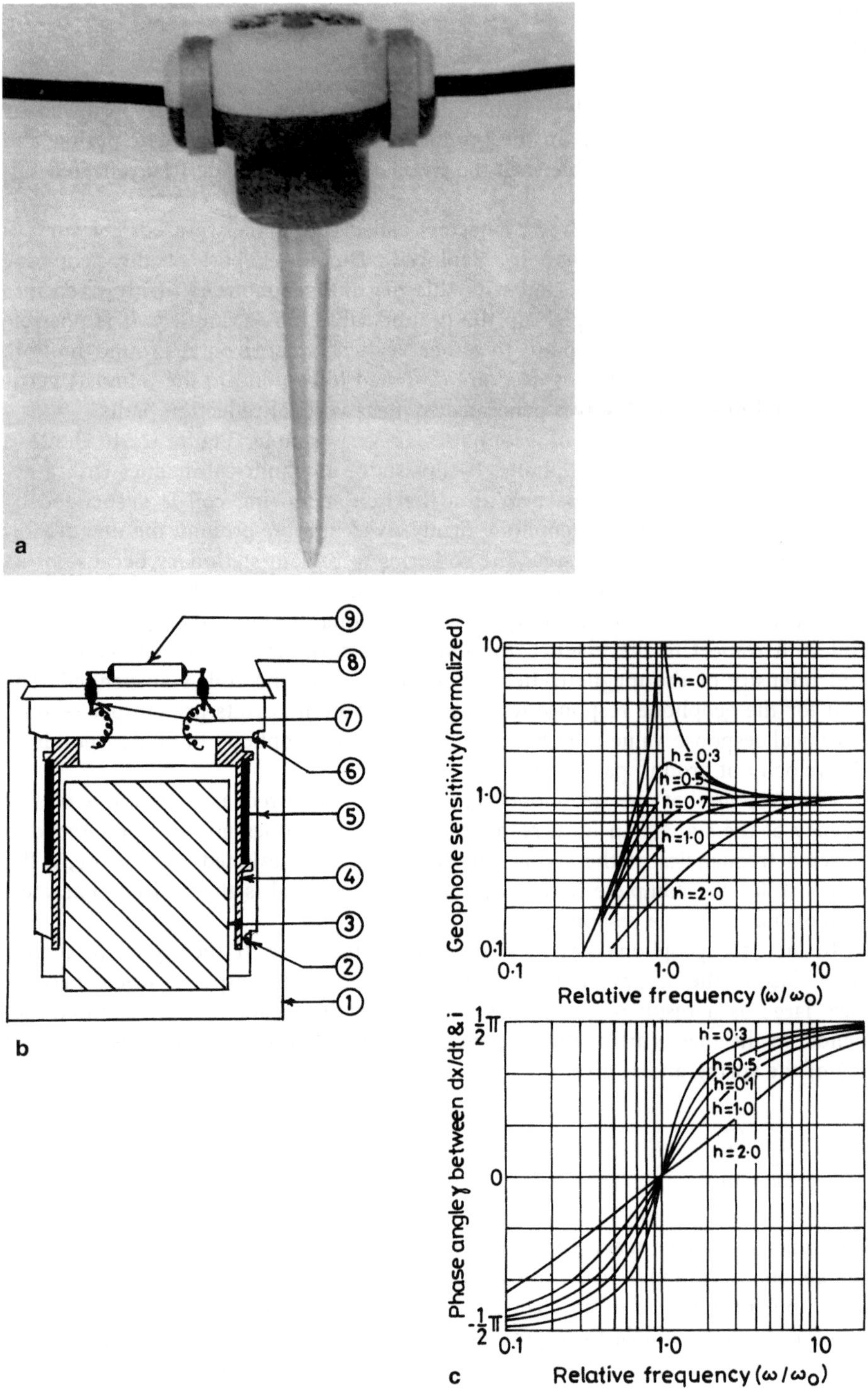
a
⑨
⑧
⑦
⑥
⑤
④
③
②
①
b
Geophone sensitivity (normalized)
10
1·0
0·1
h=0
h=0·3
h=0·5
h=0·7
h=1·0
h=2·0
0·1
1·0
10
Relative frequency (ω/ω₀)
Phase angle γ between dx/dt & i
½π
0
-½π
h=0·3
h=0·5
h=0·1
h=1·0
h=2·0
0·1
1·0
10
c
Relative frequency (ω/ω₀)

In the beginning, the damping was produced by filling the case with oil. Present day geophones have electrical damping produced by eddy current effects, and the degree of damping is governed by a resistor connected across terminals of the geophone. The damping is linearly dependent upon the velocity of the moving element.

Critical damping is defined as the maximum amount of damping that would just eliminate the oscillatory character of the response. At a value of 0.707 of the critical value, the peak in the geophone response almost disappears (Fig. 2.5c).

Depending on the type of seismic survey, a choice of geophones is made. Refraction survey geophones have frequencies below 10 Hz. For reflection work, geophones with a wide range of natural frequencies (say, from 4.5 to 100 Hz) are manufactured.

Ideally, we would expect high fidelity of the geophone. In the case of the geophone, the input is the ground motion and we would like the geophones own response not to affect the input waveform, i.e. the output of the geophone should closely match the input waveform. In the frequency domain, the corresponding criterion for *perfect fidelity* is that the amplitude-frequency response should be flat over the frequency range of the input, and that the phase-frequency response should be rectilinear (passing through zero or an integral multiple of 2π on the phase axis) over the same range (Anstey 1970, p 33).

Hydrophones (also referred to as pressure phones) are devices designed for use in appreciable depths of water. The sensitive device within the hydrophone is the piezoelectric crystal or comparable ceramic element that transforms the pressure variation due to a seismic signal into an electrical voltage. A conventional hydrophone is shown in Fig. 2.6a. It has a small cylindrical drum on which piezoelectric elements are mounted symmetrically on its two bases. The sensor elements are insensitive to water depth (static pressure) and to acceleration effects of the *streamer* (the pipe housing the hydrophones). As the ceramic elements have a very high impedance, it has to be lowered to a few hundred ohms to match with that of the connecting lines.

Hydrophones respond to the instantaneous water pressure associated with the seismic signals, which is proportional to the velocity of water particles set in motion by the signal. Typical response curves of a hydrophone are shown in Fig. 2.6b. The response is expressed as volts/bar. A flat response is exhibited for high frequencies (20 to 300 Hz), and a drop in response for low frequencies. The high frequency band is also the useful frequency range of signals in marine seismic surveys.

Fig. 2.5a–c. a Conventional geophone for reflection surveys (Geospace); **b** cross-section of a conventional moving coil, radial field geophone: (*1*) case, also serves as magnetic pole, (*2*) bottom suspension, (*3*) magnet, (*4*) coil support, (*5*) coil, (*6*) top suspension, (*7*) terminal seals, (*8*) sealing compound, (*9*) damping resistor; and **c** geophone response and its dependence on damping factor (Source for **a** and **b**: after Evenden et al. (1971), Seismic prospecting instruments, vol 2, Instrument Performance and Testing, © 1970 by Gebrüder Borntraeger, Berlin Stuttgart; source for **c**: after Dennison 1953. From Geophysical Prospecting with permission)

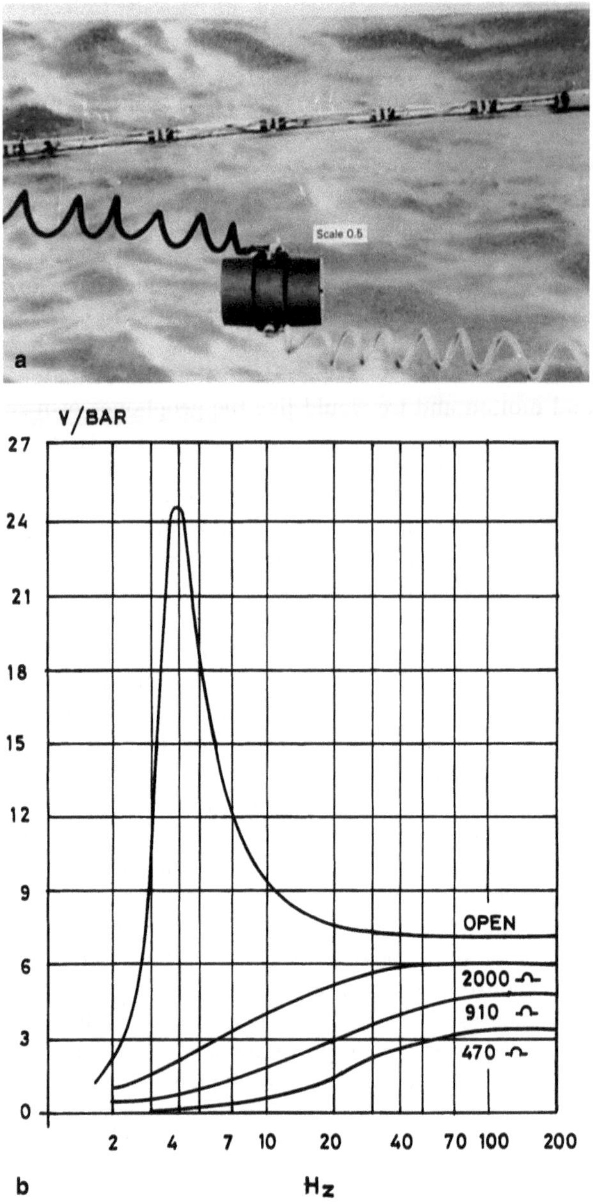

Fig. 2.6a, b. a Conventional hydrophone for reflection surveys (scale 0.5) and streamer (scale 0.05). (Technip Geoproduction document); **b** amplitude frequency response curves of a conventional hydrophone for reflection surveys (48 hydrophones in three parallel series) (IFP document). (Reproduced from Lavergne 1989, p 77. © 1989 Editions Technip, Paris by permission)

2.1.3 Amplifiers, Filters, Analog and Digital Recorders

In designing amplifiers, the primary task is to accommodate very strong to very weak signals of the ground motion. Ground motion in the vicinity of a shot point (immediately after the explosion) may be a million times in amplitude compared to amplitudes of the signals that have traveled to the deep reflectors and back, and thus arrive almost at the end of the recording. A million time range of amplitudes is much larger in comparison to the dynamic range (100 to 200 in terms of amplitude ratio) of the analog magnetic tapes used for recording. Therefore, for proper storage on the tape, the million range of seismic signals must be compressed in the amplifier. Signal compression is even more stringent for photographic paper recording for which the dynamic range is much lower than that for analog magnetic tape recording. Analog amplifiers include special circuitry in the form of an automatic gain control (AGC), which enables expansion or suppression of the gain level during recording. By making use of AGC, relative constancy (2:1) of the output voltage for a million range (10^6–1) of the input voltage may be obtained (Fig. 2.7).

Filtering of the seismic signals from geophones is often necessary before they are recorded. For example, low frequency, high amplitude noise, like surface waves, has to be pre-filtered before recording to avoid a saturation effect in magnetic tape recording. As a low frequency cut-off is needed to filter surface waves, a high frequency cut-off serves the purpose of preventing aliasing – a spurious effect associated with the sampling process (refer to Section 3.2.2).

Figure 2.8 shows the response curves of seismic filters. The purpose of the filters is to allow the passing of only a range of frequencies, called the *passband*. The passband is defined by lower and upper cut-off frequencies, which are the frequencies at which the gain drops by 3 dB (i.e. 30% of amplitude or 50% of power). In Fig. 2.8 various response curves are identified by their low and high cut-off frequencies. The curve marked 'out' is the amplifier response without a filter.

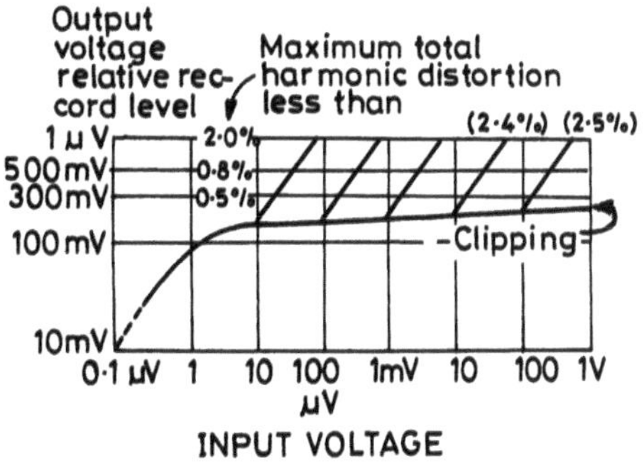

Fig. 2.7. Characteristic curves for automatic gain control illustrating relative constancy of output for 10^6 to 1 range of input voltages. *Diagonal lines* represent response to and distortion from simulated noise burstouts with a 35-Hz input signal (SIE, Inc) (In: Dobrin 1976, p 66. © 1976, 1960 by McGraw-Hill, Inc)

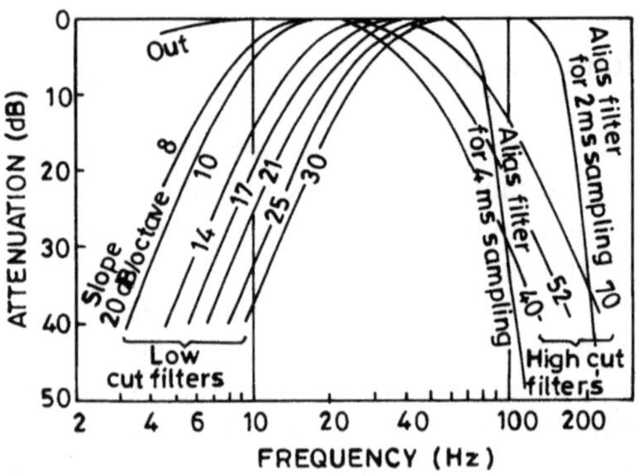

Fig. 2.8. Response of seismic filters (Source: Sheriff and Geldart 1982, vol 1, p 168. © Cambridge University Press 1982)

Filtering is effected both in field recording as well as in the processing of the recorded data. Analog filtering is also carried out in digital recording to avoid the recording of spurious signals.

In the early days of seismic recording, the output of the amplifier was sent for recording on photographic paper. During the early 1950s, magnetic tape recording came into existence – initially in the form of analog magnetic tape recording, and later in early 1960s fully developed into digital magnetic tape recording.

In analog magnetic tape recording, the tape moves past a band of *'recording heads'*. There is one head for recording one trace of a seismic channel. In addition, there is one head for timing and several more for auxiliary data channels.

In the early stages of analog recording, the usual practice was to send outputs from the amplifiers directly to the recording heads. In the subsequent stage of development, direct recording was replaced by frequency modulation and pulse width modulation. In the former case, the magnetic tape moves past *'heads'*, and gets magnetized according to the current in the recording heads, which in turn is governed by the signal strength. Therefore, the intensity of magnetization of the tape is proportional to the signal strength. With frequency and pulse width modulation recordings, a wider range of signal strengths could be handled in addition to lesser problems from noise. In these cases, as the amplifier output has been modulated, the geophone ground motion is obtained upon playback by demodulation.

In analog recording, the signal is represented by a voltage (or other quantity), which varies continuously with time. In digital recording, the signal is represented by a series of binary numbers, which are the outputs of the geophones sampled at regular intervals of time, normally 2 to 4 ms. The binary form description of digital recording is discussed in Section 3.3. The effectiveness of digital recording far exceeds the analog magnetic tape recording in terms of data acquisition. However, the early stage, i.e. the geophone output and the last stage, i.e. display of recording, continue to exist in analog form only.

The whole process of digital recording is controlled by an electronic 'clock', which is a crystal-controlled oscillator that operates in the megacycles range. It generates a continuous series of pulses whose spacings are accurately maintained. Measurement of time is based on the counting of these pulses. Further, the functioning of the multiplexer and formatter components of digital recording are dependent on the counting of the clock pulses. When the count reaches some predetermined value, electronic switches are made to operate.

We discuss briefly the functioning of analog-to-digital form presentation. To illustrate the procedure we assume that there is an unknown voltage and we seek to find its digital equivalent.

The digital form presentation is by use of the two digits 0 and 1. We may fix a total range of voltage to be measured and represent it as 1 unit. If we take 10 V as one unit, on the binary scale it is 2°. Similarly, 5 V and 2.5 V are 2^{-1} and 2^{-2}, respectively. We may think of the smallest voltage increments that need to be measured as 2^{-N}, and then any voltage value may be measured in terms of this increment. Thus, a voltage read, as 1001 in a digital presentation would correspond to 9 (decimal) increments.

Let us explore some ways, how a total number of increments of the smallest measurable (2^{-N}) voltage assumed may be found for an unknown voltage. There are two alternate ways in which this may be done. The first is the sequential counter method, and the second is the sequential adder method. The second method is much faster. The sequential counter method uses comparison of the unknown voltage with the requisite voltage rising in the form of a stair-step, i.e.

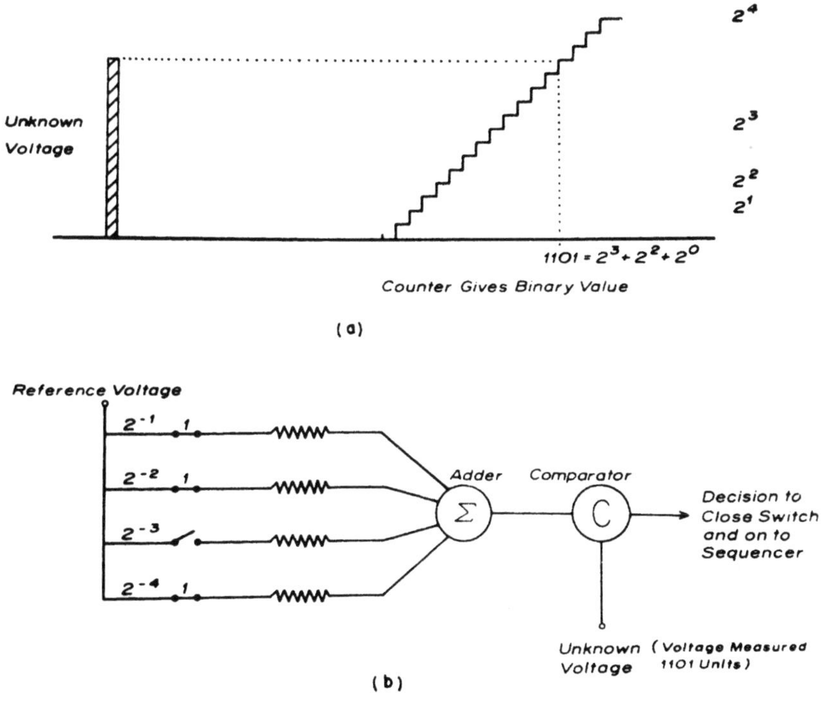

Fig. 2.9 a, b. Digitization methods: **a** sequential counter method; **b** the faster sequential adder method (Source: Waters 1987, p 206. © 1987 by John Wiley and Sons, Inc, New York. This material is used by permission of John Wiley and Sons, Inc)

the voltage increases in equal increments (Fig. 2.9 a) according to the stepping of a binary counter. When the comparator indicates equality (or change of sign) between the unknown voltage and the voltage found by comparison, the counter reading gives a measure of the unknown voltage. This method requires comparison from 0 to the unknown value of the voltage each time and is therefore more time consuming.

Figure 2.9 b illustrates the principle behind the sequential adder method. The method operates by dividing the total reference voltage into different ranges on a binary scale, and having a system in which the binary related voltages may be added together and compared with the unknown voltage. The comparison initiates by testing the most significant bit first. If it is less than the unknown voltage, a switch is set (as shown by a closed switch in the Figure) for keeping it permanently connected to the adder. Subsequently, the next significant bit is tested, and the process continues till all the bits have been tested and added. The final arrangement of switches set gives the measure of the unknown voltage.

Although the above method is still sequential, it involves a much lesser number of operations compared to the sequential counter method. Neither of these methods is suitable for achieving very high speeds of digitization. Parallel circuitry rather than sequential enables us to achieve digitization speeds of the order of 50 to 100 MHz.

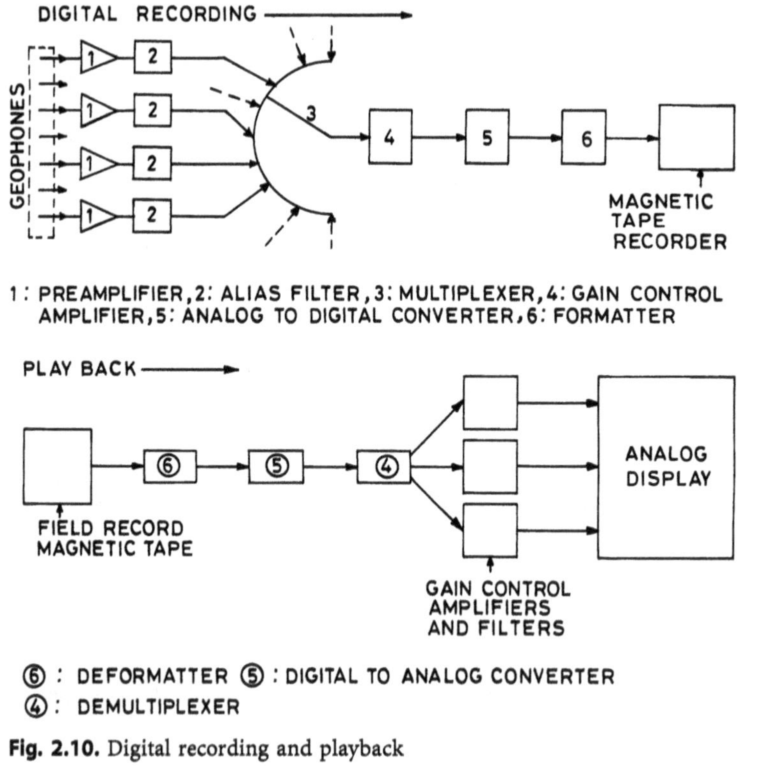

Fig. 2.10. Digital recording and playback

With a little diversion, it may be mentioned again that the role of analog filters in digital recording is to select the passband for the frequencies of seismic signals and to reject noise as much as possible. Figure 2.10 shows the components in digital recording and playback.

The multiplexer is an electronic switch to measure the amplitudes of different channels at regular intervals of time (called the sampling interval) and to feed them to a single output channel. For low-level multiplexing, the multiplexer is placed before the gain control amplifier. Multiplexed data is still in analogue form. The gain control amplifier is one of the most significant components in digital recording for achieving a high degree of accuracy. In binary gain control amplifiers, the gain changes by a factor of 2 (i.e. 6 dB). In instantaneous - floating - point recording the adjustment of gain is instantaneous to make voltage as close as possible to the recorder full-scale voltage. The output voltage and the gain are registered in coded form. Therefore, the input voltage of the seismic signal may be obtained by dividing the output voltage by the gain (gain normally exists in the form of 2^n). The analog-to-digital converter converts the signal into a binary-code number.

The formatter formats the digitized information before it is recorded on magnetic tape. One is required to have knowledge of the format in which formatting has been done to be able to access the data for reading from magnetic tape. A typical number for tape recording may be 1,600 to 6,250 bytes per inch at a tape speed of 40 to 120 in/s.

Playback works in the reverse order of *recording* and yields analog form presentation of the recorded digital data. It is an essential aspect for having visual inspection of data quality.

2.2 Surface Shot-Geophone Spreads

2.2.1 For 2-D Surveys

Spread is an arrangement of geophones (with respect to seismic source), which is used to record seismic waves in a reflection survey. Generally, the spread is along a straight line, as the interpretation of seismic records obtained in this case is easier. Basic types of straight-line spreads are shown in Fig. 2.11. Type 1 is an end-on spread, or an in-line offset spread. The seismic source is at one end, and receivers are only on one side of the source. For the *in-line offset spread*, the source is moved away from the nearest receiver because the seismic energy level may be high enough for useful recording by nearby geophones. Split spread or straddle spread (type 2) has been the most commonly used spread before the coming into operation of digital recording. In this case, the source is at the center and geophones are symmetrically placed on either side of it. Geophones may be offset with respect to the source when so desired. Records from a split spread reflection survey are used to detect the dipping reflectors and to find the amount

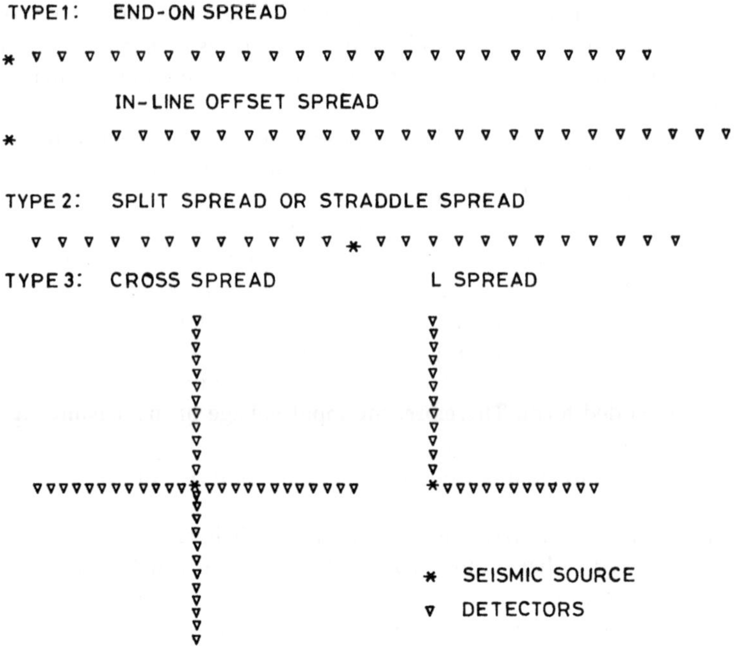

Fig. 2.11. Basic types of straight-line spreads

of their dip (refer to Subsection 3.5.8.2). However, the true dip of the reflector would only be obtained when the split-dip profile is perpendicular to the strike of the reflector. In other cases it is only the apparent dip that would be measured in the direction of the seismic profile line. If the strike direction is unknown, the value of true dip (both amount and direction) may be determined by measuring apparent dips in two perpendicular directions. This is possible by using *cross spread* (type 3). 'L spread' uses two lines of geophones in the form of an L with a seismic source at the intersection of two perpendicular lines of geophones. To save survey time, cross-spread may be replaced by L spread.

Let us analyze the suitability of commonly used linear spreads for the geological information we are searching for. It is true that linear spreads (as in conventional CDP surveys) give geological information of a cross-section in two dimensions only. This cross-section information is only a specific sample of the complete geological information that exists in three dimensions. The need for more information in the perpendicular direction of a cross-section is felt more keenly when it is realized that stratigraphic trap problems involve rapid changes in sedimentation, like the presence of channel sands, offshore sands, and reefs etc. Realization of this aspect has led to the development of techniques, which examine three-dimensional variations of geological information. We include two such methods below, which generate more than one line of CDP reflecting points.

Figure 2.12 a is an example of development over a linear CDP survey method. This shows two parallel CDP lines recorded simultaneously (Davis 1975). This

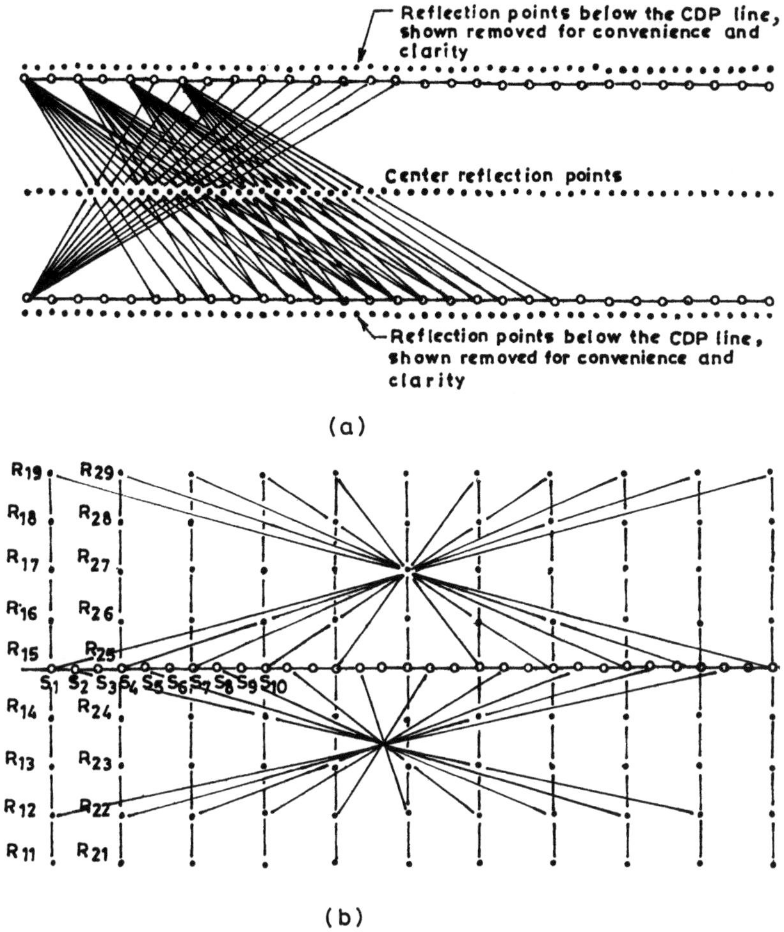

Fig. 2.12 a, b. a The method of formation of CDP reflecting points between two simultaneously recorded lines. The number of folds of stack is arbitrary, depending on source-receiver relation and the number of traces recorded; **b** the seismic swath arrangement. Sources stay along the central horizontal line. Receivers occupy lines at right angles. If 99 traces can be recorded simultaneously they will yield a 10-fold stack or more (Source: Waters 1987, pp 186–187. Reproduced by permission. © 1987 by John Wiley and Sons Inc, New York)

modification over the conventional single CDP line enables us to generate a row of depth points not only below CDP lines but also between these lines. The central line of depth points is covered by both sets of shot points, and therefore these points have double stack compared to points below the individual CDP lines. With this arrangement we not only have geological information along more than one cross-section but also increased information about the depth points along the central line.

Figure 2.12b illustrates a *seismic-swath arrangement*. An ideal basic consideration for revelation of geological information would be that points of information be equally spaced in two perpendicular directions. *Seismic swath* is a development in this direction, i.e. the recording points are on lines at right angles to the line of sources. Figure 2.12b shows how a swath of reflecting points may be generated by this arrangement. It also illustrates how, by suitable choice of records of specific sources and receivers, a ten-fold stack for a common depth point becomes possible.

We would find an even wider generalization of the seismic swath method in the following section on 3-D seismic surveys.

2.2.2
For 3-D Surveys

If we imagine the subsurface reflection paths, the most generalized situation would require that they might be oriented arbitrarily in three-dimensional space. The common assumption that reflection path lies in the vertical plane may be a large limitation of the real situation. The generality demands that the three-dimensional array of reflection travel times be measured precisely in an attempt to locate reflectors exactly in three-dimensional space.

Figure 2.13a shows the pattern for laying out sources and receivers for three-dimensional reflection surveys. In this pattern of square grids formed by two perpendicular sets of equally spaced profile lines, the receivers are placed at equal spacings along a square grid of profiles. The sources are placed in turn at each receiver point and all other receiver points obtain records. With this configuration, by appropriate choice of records from a number of source-receiver pairs, multifold coverage of a large number of regularly spaced mid-points (as shown in the figure) can be obtained.

Figure 2.13b shows grid points, corresponding both to source and receiver positions, for a three-dimensional seismic reflection survey. It may be seen that profiles, shown by straight lines, which connect different alignments of grid points, intersect at a common point. This common point is the common mid-point for suitable pairs of sources and receivers along any of these intersecting profiles. Therefore, the data from all these profiles may be included in three-dimensional CDP gather.

Although the voluminous increase in handling of 3-D data is obvious, the results obtained should be of far-reaching significance compared to the limited information that becomes available by the use of 2-D methods.

2.3
Practical Considerations for Acquisition of Reflection Data

In reflection surveys, arrays of geophones are used to detect reflections from various points on the reflector. Looking to the simple geometry as in Fig. 2.14 it is clear that the distance between the detector arrays is taken as double the spac-

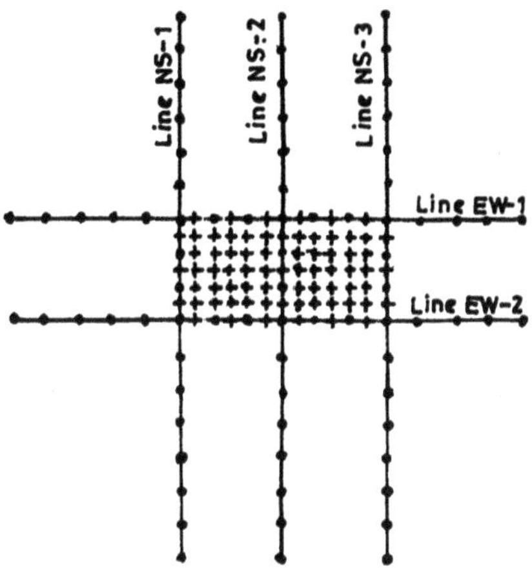

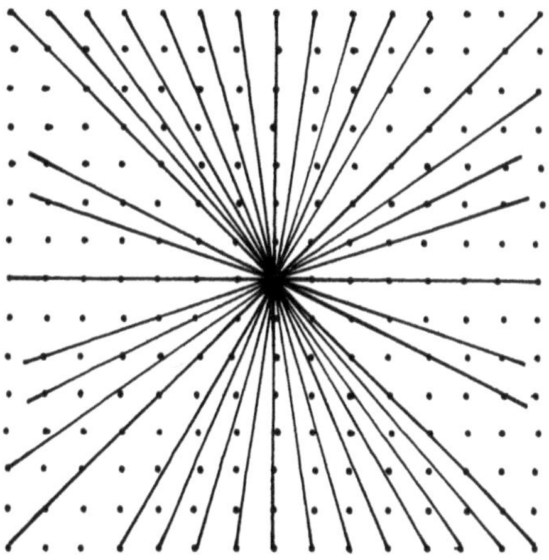

Fig. 2.13 a, b. a Diagram showing source and receiver locations (.) and common midpoints (+) for a three-dimensional seismic reflection survey; **b** grid showing source and receiver positions for a three-dimensional seismic reflection survey. Profiles connecting different alignments of grid points intersect at the common mid-point for all the source-receiver pairs that are included in a three-dimensional CDP gather (Source: Robinson and Coruh 1988, p 157, 216. © 1988, by John Wiley and Sons, Inc, New York. This material is used by permission of John Wiley and Sons, Inc)

ing between consecutive reflection points required on a horizontal reflector. Numerical value of spacing between detector arrays would depend on how closely we want to scan points on the reflector. Detector array spacing between 30 and 100 m is commonly used. The spread length, which is the distance between the seismic source and the farthest array, is commonly taken between 300 and 3,000 m.

The primary considerations for choosing the *spread length* are mode conversion, angle of incidence on reflectors and satisfactory estimation of slant path correction (*NMO correction*). In general, the spread lengths are short when the shallow reflectors are the target. For deep reflectors, the spread lengths are large.

For short spreads of 360 m, a split spread arrangement with 12 detector arrays on each side of the source at a spacing of 30 m may be deployed. For long spreads, 24 arrays on the same side of the shot at a spacing of 100 m between successive arrays may be taken as is done in an 'end-on spread'. The total length for a long spread may be 2,400 m for a 24-channel recording.

The reflection record would, in general, contain different types of waves, like primary reflections, multiple reflections, surface waves, air waves, mode-converted waves, direct or near surface compressional waves, refracted waves and diffractions (Fig. 2.15 a). The direct compressional waves and the surface waves have distinctly different velocities, and have straight-line paths from source to detector arrays. Air wave velocity is 330 m/s, velocity of surface waves is between 400 and

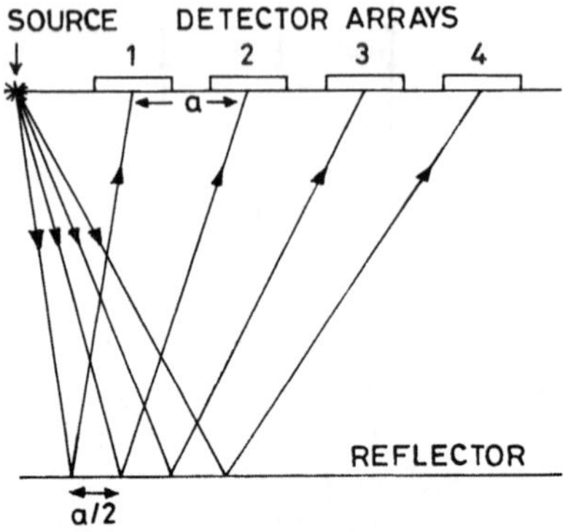

Fig. 2.14. The detector array spacing on the surface is double the resolution required on the reflector (Source: Anstey 1970, p 81. © 1970 Gebrüder Borntraeger, Berlin Stuttgart)

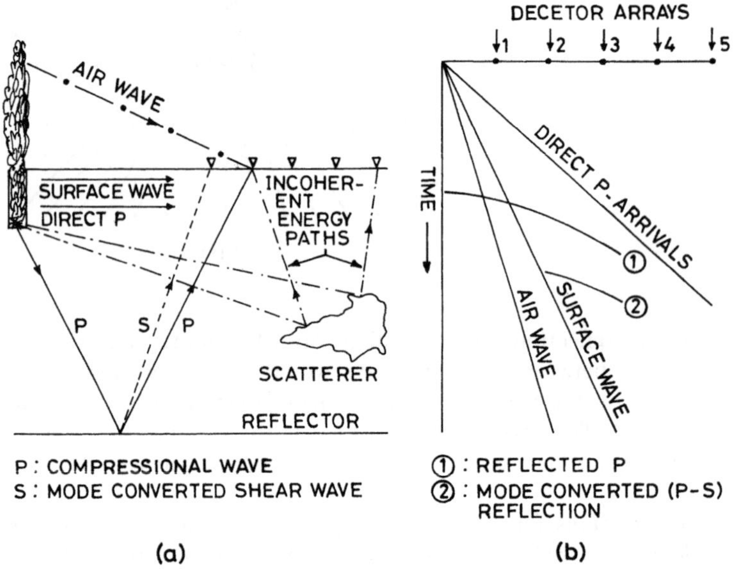

Fig. 2.15 a, b. a Schematic illustration of paths for various types of wave energy reaching the detectors (geophones); **b** distance versus arrival time graph for some representative wave types (Source for **b**; after Anstey 1970, p 80. © 1970 Gebrüder Borntraeger, Berlin Stuttgart)

800 m/s. The velocity of compressional direct arrival would depend upon rock type but would be higher than velocities of air waves and surface waves. Air waves, surface waves and direct compressional waves are coherent arrivals, have fixed velocities and may be easily followed in the reflection record along straight lines with different slopes (Fig. 2.15b). In addition to these coherent waves, there are incoherent arrivals also in the form of scattered energies from various directions. For our purpose, *primary reflections* constitute the useful signals, and all the other types of arrivals are noise.

Some types of noise may be weak and have very little influence on reflections while other types of noise are strong enough to substantially influence the recording of primary reflections.

Most of the research studies (field survey techniques, instrumentation, and processing) in the past aimed at achieving best records of reflected events, by suppressing or attenuating particular types of noise. This subject is too wide for a thorough coverage; and an attempt is made in the following subsections to analyze some characteristics of signals vis-à-vis noise, which are helpful in improving record quality.

2.3.1 Signal and Noise

We remarked in the previous paragraph that surface waves travel along a horizontal path on the ground surface with a fixed velocity – the true velocity. In contrast, the reflections travel along near vertical paths with a velocity depending on the formations' velocities (Fig. 2.16a). The upcoming reflections have curved wavefronts associated with them (Fig. 2.16b). As the wavefront is curved, the reflected disturbance is not expected to reach all detector arrays simultaneously. There would be a very small time lag at far detector arrays as compared to arrays near the seismic source. This means that the reflected disturbance has an *apparent velocity* along the line of detector arrays. The true velocity of reflections on their near-vertical upward paths and apparent velocity along the horizontal line are different from each other. It may be further noticed that apparent velocity of reflected waves along the common horizontal line (line of detector arrays) is different from the true velocity of surface waves. Therefore, this criterion is one which may be utilized to filter out surface waves from the records. Discussion of this aspect (velocity filtering) would be presented in Section 3.4.2.

Another important characteristic to distinguish reflected signals from noise (the surface waves) are their frequency range. In fact, frequency ranges of both the signal and noise depend on near surface geology, as well as on absorption characteristics of formations (for reflections only) through which the reflected wave is returned to the ground surface. For example, for a small seismic source in well-consolidated rock around the source, the seismic spectrum peak may be within 100 to 200 Hz. On the other hand, a large explosive source in sand may give a spectral peak of deep reflections within the frequency range of 10 to 20 Hz. In general, it may be taken that the frequency range of reflections may be

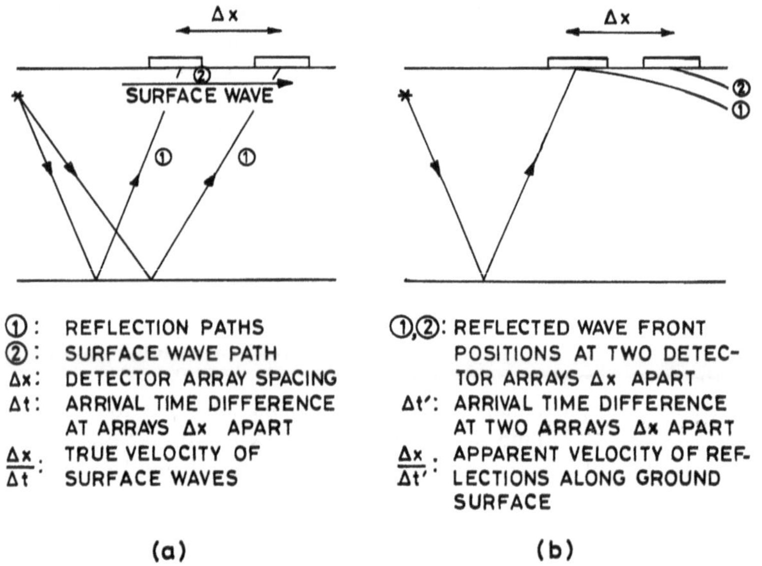

Fig. 2.16a, b. **a** Paths for primary reflection and surface wave; **b** wavefront positions, *1* and *2*, of primary reflection on ground surface. True velocity of surface wave and apparent velocity of primary reflection differ from each other

from a few hertz to few hundred hertz. In the case of surface waves, the dominant frequencies are below 20 Hz.

Information on frequency range (and hence the wavelength range) of surface waves in a particular area is utilized in designing detector arrays. This would be discussed in the following Subsection 2.3.2. However, it may be mentioned here that detector arrays attenuate the surface waves, only partly. Therefore, provision is also made in amplifiers of the seismic units to filter out low-frequency surface waves.

Besides apparent velocity and frequency range, it is significant to understand the amplitude behavior of reflected signals as a function of arrival time. Figure 2.17 shows the amplitude of envelope of the reflection signals above the average ambient noise level as 0 dB. *Ambient noise* is the microseismic noise, which originate from earthquakes, vehicular traffic, vibration of trees, wind action and waves in the oceans. The microseismic noise has spectral peaks well below 10 Hz. Thus, they are separated from the frequency range of useful reflection signals.

Returning to Fig. 2.17 it shows the observed amplitude – time relations, and thus includes in it effects of reflection and transmission coefficients, absorption and multiple reflections. The amplitude goes down the ambient level of the noise after a few seconds. The humps in the two curves are the strong reflections. The amplitude decay curve controls the design of instruments to accommodate in the record large amplitude ranges of the order of 100 dB as well as significant variations in the rate of decay of amplitudes.

With the above introductory characteristics of signals and representative noise, we discuss below field methods of reducing coherent and incoherent noise.

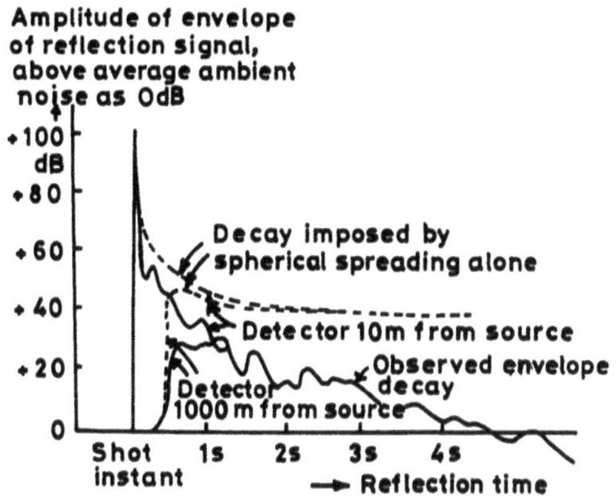

Fig. 2.17. The amplitude of reflection signal as a function of time, for close and distant detectors. Shown *dashed* are the decays in amplitudes to be expected from spherical spreading alone. The 0 dB is taken to be the average amplitude of the ambient noise (Source: Anstey 1970, p 68. © 1970 by Gebrüder Borntraeger, Berlin Stuttgart)

2.3.2 Reduction of Noise

Reduction of the influence of noise may be planned for in the different stages of data acquisition, instrument setting and processing of records. The later two, i.e. instrument setting (filtering action) and processing (deconvolution) would be discussed in Chapter 3. Here we outline briefly how a detector array is useful in the reduction of surface wave noise.

The general principle of the detector array design may be discussed with the help of Fig. 2.18 a, b for a linear and areal array of detectors, respectively. Figure 2.18 a shows the effectiveness of using a number of geophones together for a single seismic channel to attenuate a wave of a particular frequency (or wavelength). Three of the geophones sense the motion of the crest, and the other three of the trough. As the combined output of all six geophones is fed to one channel the net output motion is expected to be close to zero. Since the reflections arrive on near vertical paths, all the geophones in a group are affected by one type of motion. Therefore, it may be concluded that use of a group of detectors together should increase the *signal to noise ratio.*

The above description is only an idealized example, by considering only one wavelength of noise. In field conditions there is a range of wavelength of surface waves, and the number of geophones in an array and the spacing between the geophones has to be carefully chosen so that for given range of ratio of wavelength and detector spacing (covering the entire range of noise wavelengths), the response to noise is at a minimum.

The general principle for the linear array may be stated as – 'Detectors spacing should not be more than half a wavelength of the shortest wavelength present in the surface waves'.

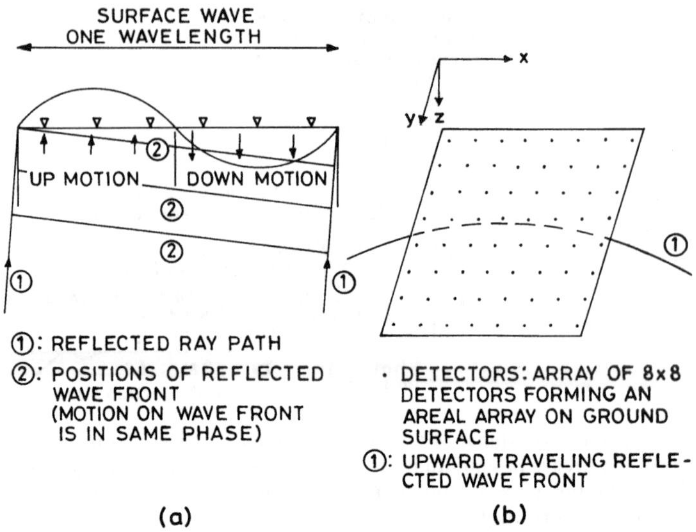

Fig. 2.18 a, b. a Schematic illustration showing effectiveness of using a number of geophones together to attenuate surface waves and increase signal to noise ratio; **b** areal array to attenuate uncorrelated noise

The ambient noise which is the uncorrelated noise may reach the site of the geophones from different directions. As ambient noise is uncorrelated we may use an areal array (see Fig. 2.18 b), instead of a linear array (as in the case of surface waves). The basis for the areal array is that noise is uncorrelated from one geophone to another, whereas the reflected signal is in phase at all geophones. Thus contributions from all geophones would cancel noise and add up the reflected signal, giving rise to an increase of the *signal to noise* ratio by $\sqrt{n}$ where n is number of geophones in the array.

2.4 VSP Geometries

Vertical Seismic Profiling is performed by having a source of seismic energy on the ground surface, and a detecting unit in the well. Recording is done every 50 to 100 ft as the detecting unit is pulled up the hole. The seismic source may be near the top of the well or displaced some distance from the well. In the later case, it would be referred to as small or large offset VSP. Vertical Seismic Profiling may also be planned by having seismic sources in various azimuths on the ground surface.

Data from different types of VSP configurations serve specific purposes. If the surface source is located near the well, VSP data may be utilized for estimating reflectivities of various reflectors, and to study deformation of the pulse during propagation. Small offset VSP data is an excellent tool to optimize processing in

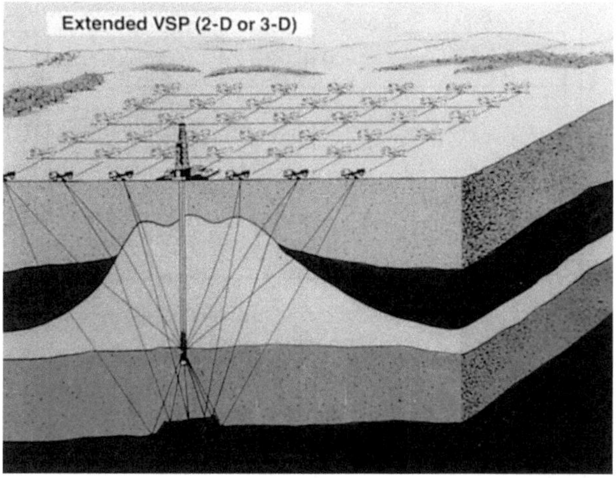

Fig. 2.19. Three-dimensional extended VSP to delineate a structure around the well (Prakla-Seismos document) (Reproduced from Lavergne 1989, p 137. Reproduced by permission. © 1989 Edition Technip, Paris)

reflection surveys. Large offset VSP data can give detailed information about structures like domes, reefs, faults and fracture orientations around the well. Figure 2.19 shows schematically the usefulness of extended VSP for 3-D structural information around the well.

When vibrational energy is generated on the ground surface, it may reach the geophone in the well directly, or after reflections from the subsurface layer boundaries interfaces. The reflected category of waves may be identified as descending or ascending waves. Figure 2.20 a shows selected possible paths for *direct* (top left), *descending* (bottom) and *ascending waves* (top middle and right). Figure 2.20 b is a VSP record showing in it the *descending P-wave*, the *descending multiples* and *ascending reflected P-waves*. In addition to these, VSP record may contain *fluid waves*. These waves originate in the mud of the well and propagate at a velocity of about 1,500 m/s. In general, they do not provide any useful information about the formation. *Fluid waves* may be removed from the record by *filtering*.

It is possible to separate out ascending and *descending* waves by suitable processing. The role of ascending waves is in identifying individual reflections. The descending waves help in studying signal deformations and for identifying secondary reflections.

Figure 2.20 c illustrates the separated picture of ascending and descending waves. The separation is effected by a downward shift of 't_0' for ascending waves and an upward shift of 't_0' for descending waves, where 't_0' stands for the time of first arrival.

The applications of VSP data are many. Some of these are:

- Calibration of surface seismic reflection data.
- Identification and removal of secondary reflections.
- Providing best information for synthetic seismograms both for timing relationships and waveform character analysis.

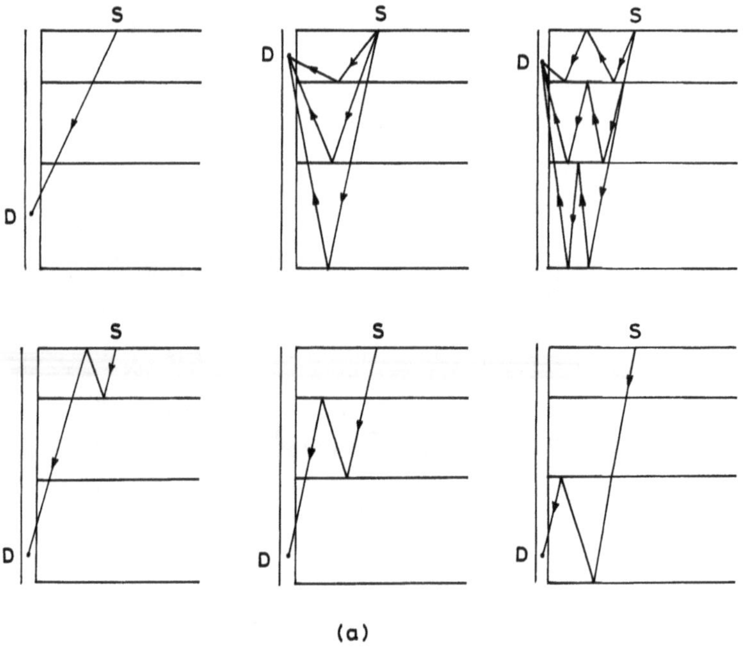

Fig. 2.20 a–c. a Schematic illustrations for: direct wave (*top left*), ascending waves (*top middle* and *right*) and descending waves (*bottom*); **b** raw VSP record showing a descending P-wave, descending multiples and ascending reflected P-waves (SSL Document); **c** separation of descending waves and ascending waves in the record (SSL Document) (Source for **b** and **c**: Lavergne 1989, p 139, 142. Reproduced by permission © 1989 Editions Technip, Paris)

- Delineation of structures around the well and assessment of petrophysical properties of reservoirs.

Application of VSP data for mapping fractures, delineating anisotropy of the subsurface medium would be specifically discussed in Chapters 16 and 17. Combined use of P and S wave VSP data has unique applications for reservoir characterization and modeling.

For detailed treatment of VSP data, reference may be made to Galperin (1973) and Hardage and Toksöz (1983).

2.5 Seismic Surveys Using Shear Waves

Conventional seismic surveys make use of P-waves. However, for some specific investigations, shear wave data alone may provide the desired result. For example, shear modulus – an important parameter required for designing the foundation of structures, may be obtained from the velocity ($V_S = \sqrt{(\mu/\rho)}$, where μ is shear

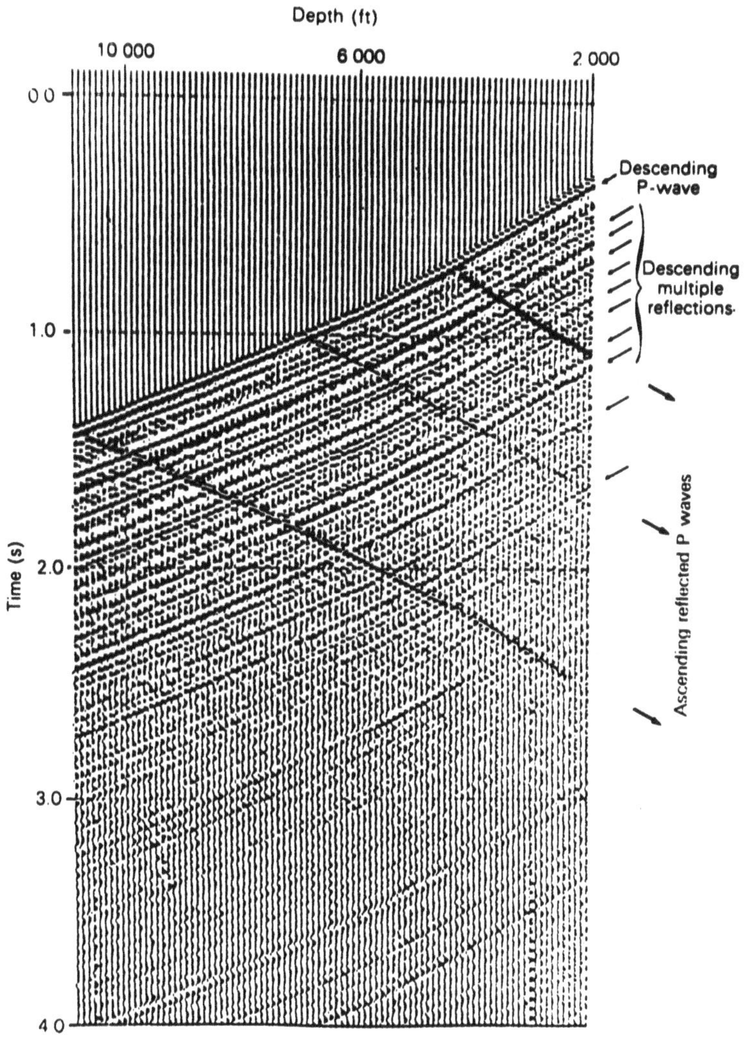

Fig. 2.20 b

modulus) of shear waves. The P waves velocity measurement would not serve the purpose, as it is dependent on two parameters, λ as well as μ.

Apart from the use of shear waves in engineering geophysics, shear waves also provide valuable information in reservoir geophysics when interpreted together with P wave data.

Shear wave seismic surveys differ in the choice of seismic sources and detectors. Although shear waves are formed in conventional P wave surveys also, and exist in record as mode-converted waves they may not be very useful for analysis purposes. Therefore, special shear wave sources are designed which can efficiently generate the shear waves. Shear wave sources are, for example, shear vibrators,

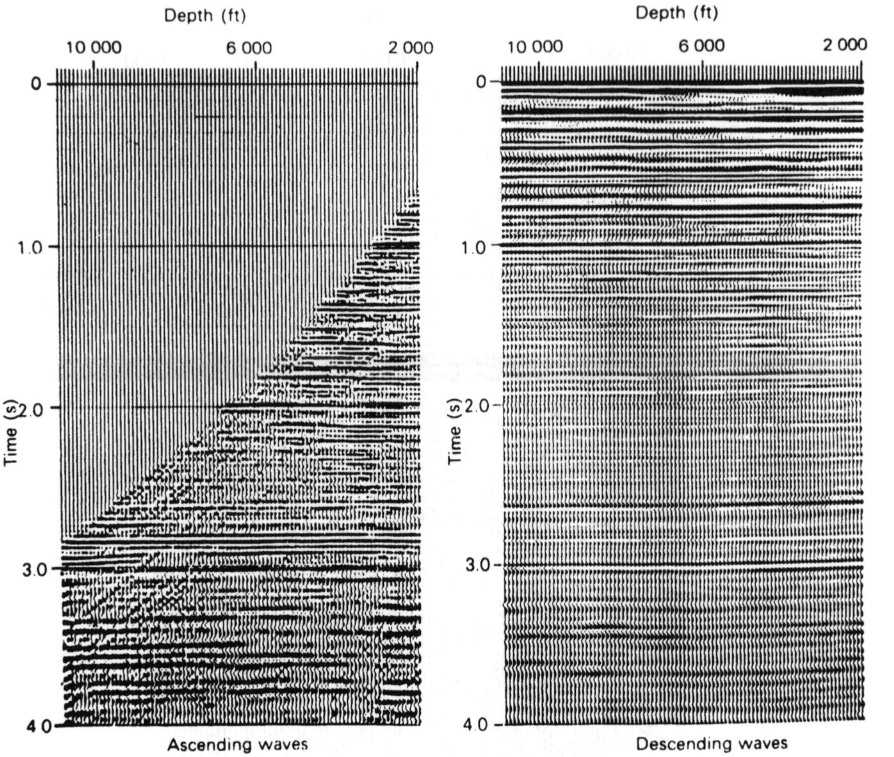

Fig. 2.20 c

inclined impact sources, horizontal hammers, and special arrays of P wave sources for maximizing shear wave energy generation. Shear wave sources make use of shearing action or asymmetry for generation of shear waves. A shear wave detector is of low frequency (4.5, 6, 8 Hz, etc.) which has its coil oriented horizontally so that shear waves polarized parallel to the coil direction are detected. The conventional cables, CDP shooting methods and recording system remain the same as for P wave surveys.

Display of the shear wave record is done in the same way as for the P wave. Figure 2.21 a, b gives a comparison of P and S wave records.

Shear waves may be of SH- or SV-type. For propagation along the ground surface, particle motion of SH waves are horizontal and are polarized perpendicular to the seismic line (taken as the direction of wave propagation); SV wave polarization is vertical and lies in the vertical plane along the seismic line.

In reservoir geophysics, combined interpretation of P wave and S wave data may find notable applications in direct detection of hydrocarbons and identification of lithology. An important problem in reservoir geophysics is to distinguish the lithological change indicator (LCI) from the hydrocarbon indicator (HCI). Ensley (1985) and Robertson and Pritchett (1985) used the P wave, and shear wave stacks to make a distinction between LCI and HCI. Lithologic change is in-

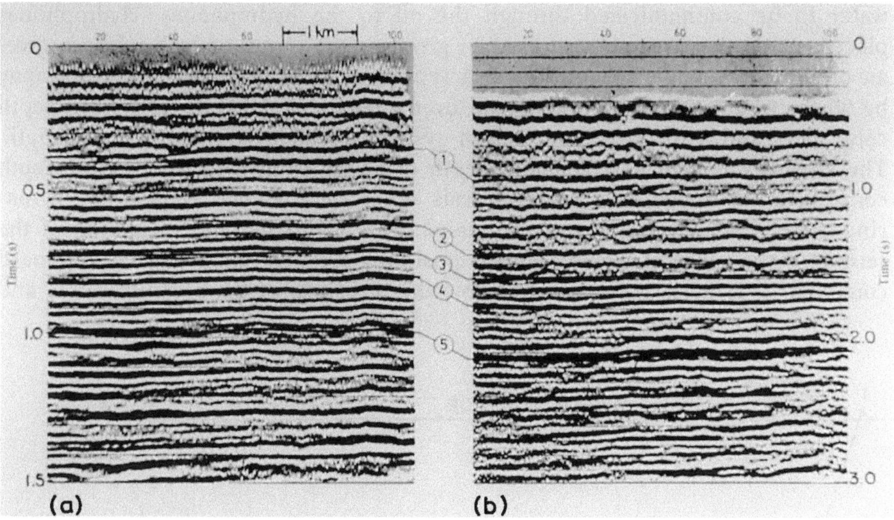

Fig. 2.21. Comparison of P- and S-wave records (CGG): **a** P-wave record; **b** S-wave record; the S-wave record is displayed at double the timing speed to make comparison of events easier (In: Sheriff and Geldart 1982, vol 1, p 204. © Cambridge University Press, 1982. Reproduced by permission)

dicated by a large amplitude change both on P wave and S wave stacks. On the other hand, in the case of hydrocarbons (change from water or oil to gas) a large amplitude (bright spot) is indicated for the P wave only. For the same situation no change in S wave amplitude would be found. This happens because the presence of gas in pore space affects the P wave reflection coefficient, whereas for S wave it does not.

The determination of lithology makes use of colocated P and shear wave stacks. Distinct Vp/Vs ratios of sand (between 1.5 and 1.9) and shales (2.1 to 2.4) helps to distinguish sands from shales (refer for example to McCormack et al. 1982).

The full potential of shear wave investigations remains largely an unexplored subject.

2.6 Marine Seismic Surveys

A simplified schematic diagram of a marine seismic survey system is shown by Fig. 2.22 a. Although every component of the system has its unique role, the most important components may be identified as the *streamer* (which contains the detecting units – the *hydrophones*), and the navigation system (which gives the exact location of the ship). The streamer is a long pipe (of the order of 2,400–3,000 m) 5 to 8 cm in diameter, and is filled with oil. The streamer is towed with the ship. The walls of the streamer are flexible to allow pressure variations in water to be communicated through the oil to the hydrophones. Hydrophones placed inside the streamer are used in groups. One group of hydrophones gives an output for a single seismic trace. A typical number in marine recording may be 96 traces at a 25 m-interval spread over a length of 2,400 m. There are depth controllers and a directional detection system spread over the streamer length. The streamer is kept at a constant depth of about 10 m with the help of depth controllers positioned at regular intervals along the streamer. Acquisition in marine surveys is continuous and automated. The navigation and triggering of the seismic disturbance at predetermined positions are controlled with a shipboard computer. It is most important to know the positions of the seismic source and

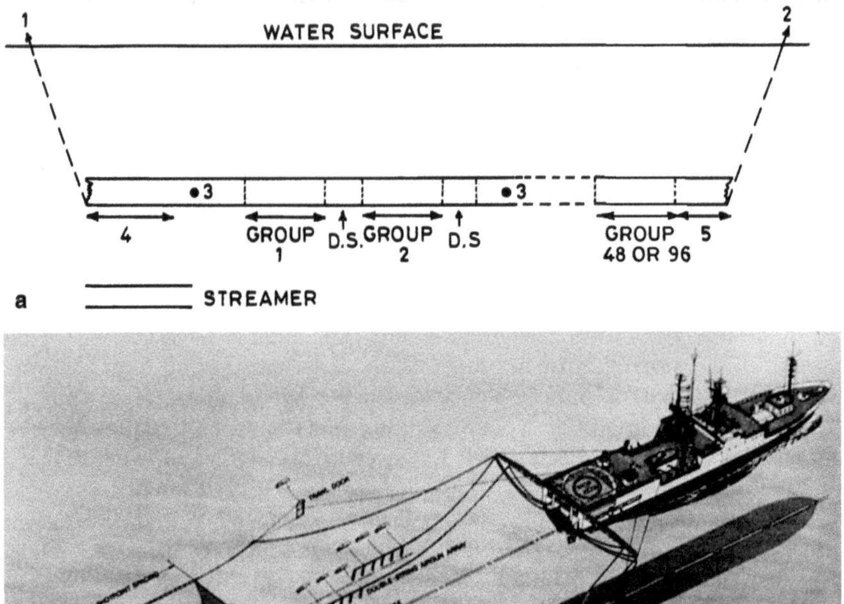

Fig. 2.22 a, b. a Schematic diagram for marine seismic survey system. *Group 1, 2...96* contains 20–100 hydrophones; two adjacent groups are separated by a dead section; *1* ships position; *2* tail buoy; *3* the depth controllers; *4, 5* the compliant sections to isolate the streamer from jerks. **b** A vessel equipped with the multiple source arrays and streamers used for three-dimensional reflection surveying at sea (Prakle-Seismos AG) (In: Robinson and Coruh 1988, p 158. Reproduced by permission. © 1988 by John Wiley and Sons Inc, New York)

detectors accurately to be able to locate exactly the positions of subsurface reflectors. The location of the ship is determined by radio positioning or by use of satellite positioning methods (refer, for example, to Dobrin and Savit 1988, pp 141–151). The radio positioning system is based on measuring the distance between the ship and two or more fixed locations. The range of the radio positioning system is up to 100 miles. Satellite positioning methods are useful when the ship is far from the coast. Integration of different types of data on speed of ship (the data from a Doppler Sonar), the heading of the ship (data from a gyrocompass), and position of the ship is done by a shipboard computer to compute the exact position of the streamer in water at all times, to be able to find the position of subsurface reflection points for each shot point.

Marine seismic surveys may be conducted either along a profile line already chosen, or along parallel lines for detailed three-dimensional studies. For three-dimensional surveys at sea, multiple air-gun source arrays and several parallel streamers may be used simultaneously (Fig. 2.22b). Three-dimensional surveys demand much increased cost, effort and time. Therefore, generally such a survey is recommended for relatively small areas, which are identified as prospective on the basis of previous two-dimensional survey.

CHAPTER 3

Reflection Seismogram and Steps in Processing 3

3.1 Creation of a Seismogram

The concept of creation of a seismogram is fundamental to the processing and interpretation of reflection seismogram.

Given the impulse response, it is possible to calculate the response for any other pulse form. The output of any linear system is called its *impulse response* when the input given to the system is a spike (Anstey 1970, p 35). For the digital evaluation of the response the primary requirement is that the pulse be sampled at the same rate as the impulse response.

Figure 3.1 shows how the process works. The impulse response is shown on the left and the pulse on the right, their values are known with the same sampling rate. The response is obtained by replacing each delta pulse in the impulse response by an appropriately scaled version of the pulse, and adding the results of overlapping delta impulses at each sample time. For each delta pulse in im-

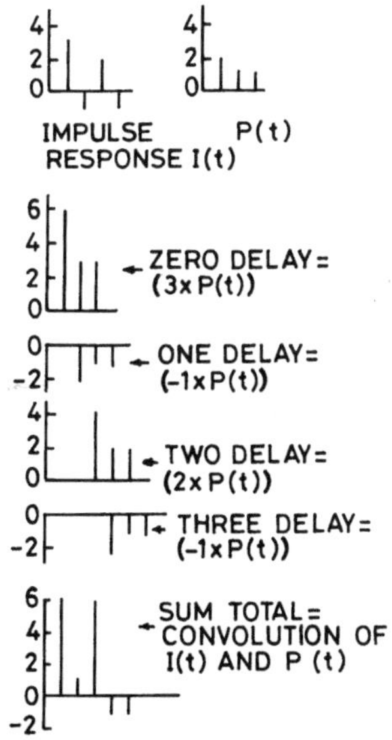

Fig. 3.1. Digitized impulse response I(t) and pulse P(t). Net output response as an outcome of superposition/ or convolution of I(t) and P(t)

pulse response, an appropriate delay for the pulse response has to be considered. Finally, the summation of overlapping delta-pulses at each sampling time would yield the net response for the pulse form considered.

The graphical depiction in Fig. 3.1 illustrates how the net response is an outcome of *superposition.*

The same result would be obtained, if we take the digitized pulse form in reverse order, slide it each time by one sampling interval against the impulse response, multiply and sum the values for each point and in this way complete the process till the pulse has covered all points of the impulse response. This process is known as *convolution.*

In fact, convolution and superposition describe one physical reality – the interaction of an input waveform with impulse response to produce an output waveform (Anstey 1970, p 38). *Superposition* or *convolution* is a powerful tool for *waveform manipulation.*

So far our discussion has been somewhat specific considering an illustrative example in digitized form. We will now describe the impulse response in a more general way and its response to any input waveform. We would assume *linearity* of the system. A linear operation of any system is described as one whose action is the same whatever the magnitude or polarity of the input waveform (Anstey 1970, p 17).

The concept of superposition (or convolution) may easily be described for any input waveform (not necessarily digital data with restriction on sampling rate as in the previous example). The superposition concept is stated as follows (Anstey 1970, p 36):

'For any linear system and for any input waveform, the output waveform is the superposition of all the impulse responses obtained by regarding the input waveform as a succession of spikes.'

The above point is well illustrated by Fig. 3.2 a, b. In Fig. 3.2 a the impulse response (second from top) is given, and superposed versions of the impulse re-

Fig. 3.2 a, b. a Analogue form impulse response (*second from top*). Output response (*bottom*) obtained by superposition of impulse responses for two different strength spikes; **b** output response from a succession of spikes of different strengths (Source: Anstey 1970, p 36. © 1970 by Gebrüder Borntraeger, Berlin Stuttgart)

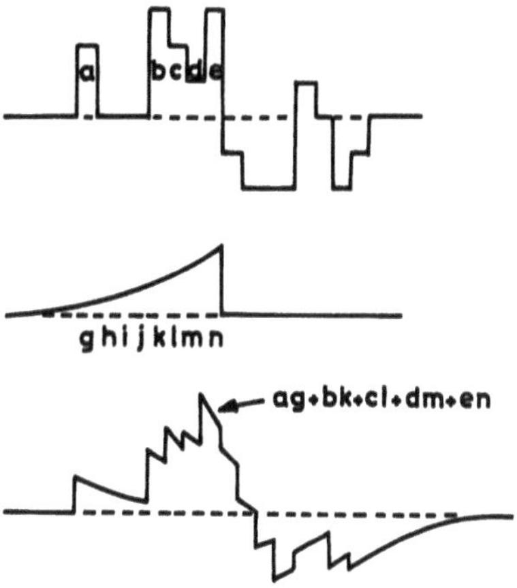

Fig. 3.3. Convolution applied to input of Fig. 3.2b produces the same output as obtained by superposition (*bottom* of Fig. 3.2b) (Source: Anstey 1970, p 38. © 1970 by Gebrüder Borntraeger, Berlin Stuttgart)

sponses for two spikes of different strengths are shown at the bottom. The responses scale down according to the strengths of the spikes. Since the responses for two spikes are superimposed on each other, the net result (shown by the continuous line) is obtained by summation of the individual response. The result in Fig. 3.2b is fully generalized. The input waveform is represented by a succession of contiguous spikes of different strengths and/or polarity. The lower part of the Fig. 3.2b shows the individual response of each spike. If the spike is negative, its scaled response is also negative. After summation of all responses (shown by dashed lines), the total output response is obtained. This is depicted by the continuous line.

Exactly the same output waveform would be obtained by a convolution process (Fig. 3.3). The impulse response is folded back (the middle part) and slid past the input waveform (shown as a succession of spikes). In one position of the impulse response, the result of convolution is shown by the arrow. The net output result (shown at the bottom) is exactly the same as obtained by considering superposition (compare with Fig. 3.2b).

Using this background information concerning convolution (or superposition) we will discuss in the following subsection the composition of an ideal seismogram with a spike as input.

3.1.1
An Ideal Seismogram – Impulse Response of a Spike

We call a seismogram an *ideal seismogram* when the input given to the Earth is a spike of unit strength. The Earth is assumed to be a stack of plane parallel layers with reflecting boundaries between them. Computation of the seismogram is straightforward in this case. We consider the effects of reflection, and two-way transmission coefficients for the boundaries between a reflector and the ground surface. We may also include the effects of geometrical spreading and attenuation loss on reflected pulse amplitudes. All types of multiple reflections should also be considered, but to simplify the presentation we omit it.

If the sequence of reflection coefficients of the reflecting boundaries is represented by

$$\{r_1, r_2, r_3, \ldots r_j\}$$

the following sequence of spikes should be observed on the ground surface, as a function of time

$$\left\{r_1, r_2\left(1-r_1^2\right), r_3\left(1-r_1^2\right)\left(1-r_2^2\right), \ldots, r_j \prod_{k=1}^{j-1}\left(1-r_k^2\right)\right\}$$

considering only the effects of reflection and two-way transmission coefficient of the layer boundaries.

This sequence of impulses would be of varying amplitudes and would be separated in time (two way time) depending on the thickness of the layers and their velocities. The resulting picture of an ideal seismogram should be similar to one shown in Fig. 3.4.

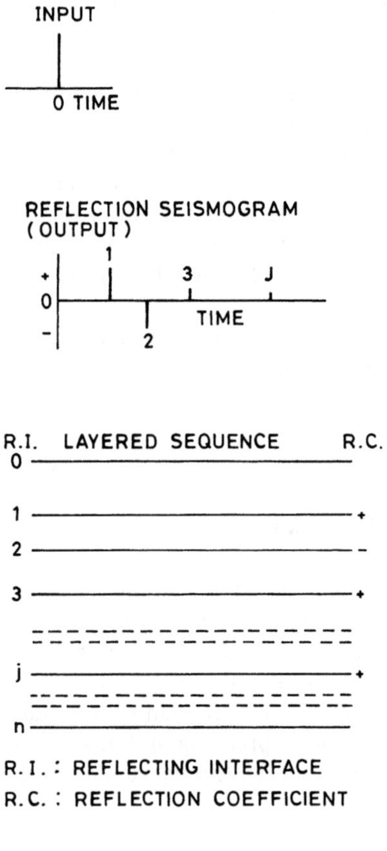

Fig. 3.4. Schematic diagram illustrating unit spike input (*top*) to a layered geological sequence (*bottom*). An ideal seismogram (*middle*) as an impulse response of a spike

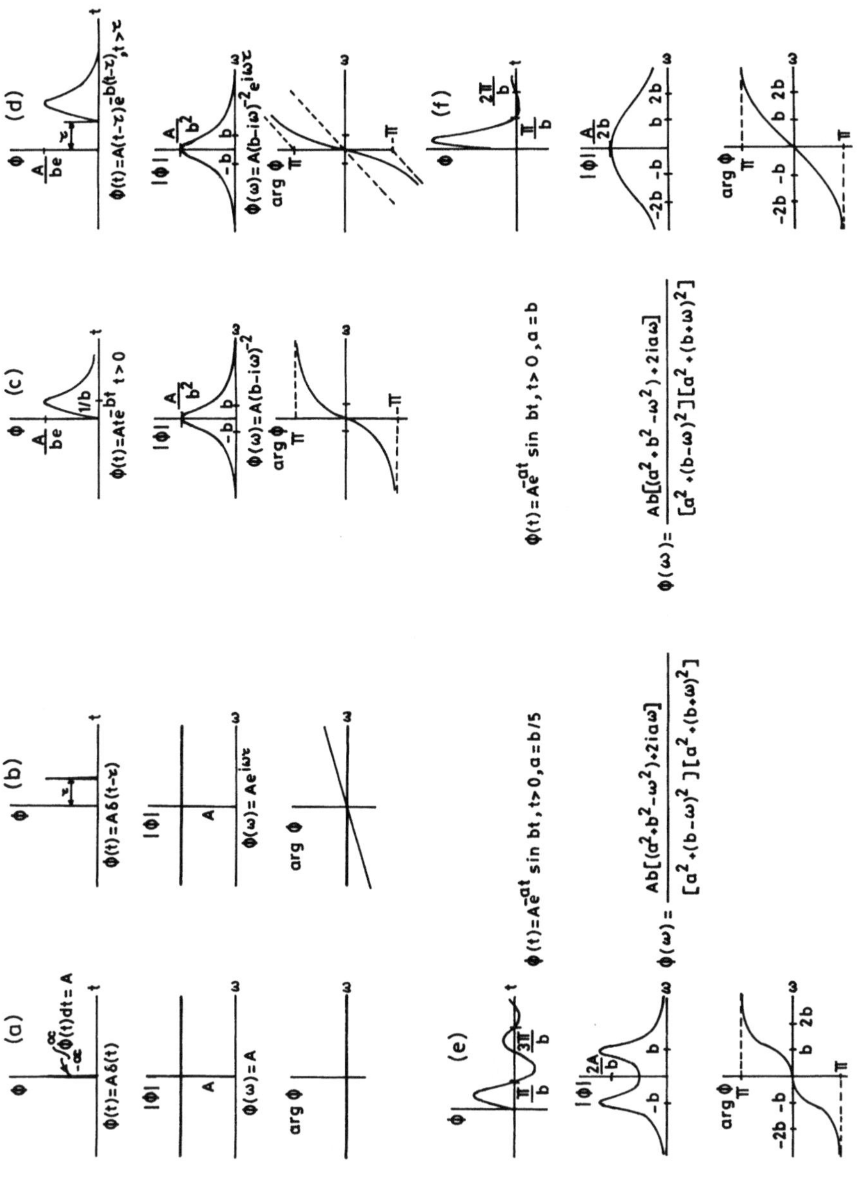

Fig. 3.5. Some simple transients and their Fourier spectra (Source: Grant and West 1965, p 32, 33. © 1965 McGraw-Hill, Inc)

If we also take into account the effects of geometrical spreading and attenuation loss, the impulse sequence of reflected pulses would have to be multiplied by the factors accounting for these effects.

3.1.2
Transient Seismic Pulse and Sweep Signal

In practice the seismic pulse is never a unit spike or a delta function. Transient seismic pulses occupy a very short but finite time duration, and occur only once. Figure 2.1a (Chapter 2) is an example of the *transient seismic pulse.* Various factors decide the forms of the transient pulse emitted by an explosion in the rock mass. Some of the primary factors are charge size, charge depth and material surrounding the charge. A discussion of pulse forms and their Fourier spectra, the result of varying charge size, charge depth and surrounding rock material, may be found in Anstey (1970, pp 55–56).

The unit spike or a delta function is an ideal tool for developing the modeling concepts for generating synthetic seismograms.

It is important to understand some of basic features of a delta pulse and also of other transient waveforms.

Delta function may be thought of as a limit of rectangular function, when the time interval is made to shrink to an infinitesimal value with the product of amplitude and time duration staying constant. Figure 3.5 presents the mathematical expressions and graphical depictions in the time and frequency (amplitude and phase spectra) domain for the *delta pulse* and representative types of *transient waveforms.* These may be helpful in the generation of synthetic seismograms, and for exercises on waveform manipulations.

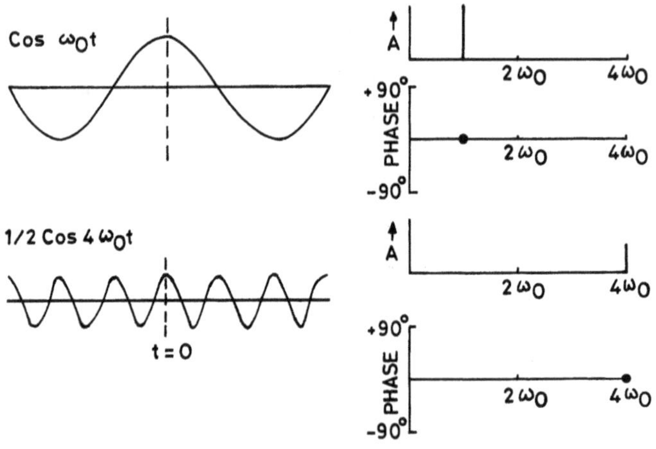

Fig. 3.6. Repetitive cosine waveforms of different frequencies and amplitudes (*left*), and their corresponding amplitude and phase spectra (*right*)

Transient waveforms have a continuous spectra as against line spectra for repetitive waveforms, like a sine or cosine waveform. Examples of spectra for repetitive waveforms are shown in Fig. 3.6.

3.1.2.1 Convolutional Model of a Seismogram

The convolutional model of seismograms describes the reflection seismogram as an outcome of convolution (superposition) of a reflectivity function with section multiple train (Robinson 1981):

$$\text{Seismogram} = \text{reflectivity function} * \text{section multiple train}$$

The inherent validity of the convolutional model is based on an assumption of small values of reflection coefficients.

The *section multiple train* is a consequence of all possible reverberations occurring within the whole of the crustal section. To simplify presentation, we think of the crustal section as consisting of layers (as many as we desire) such

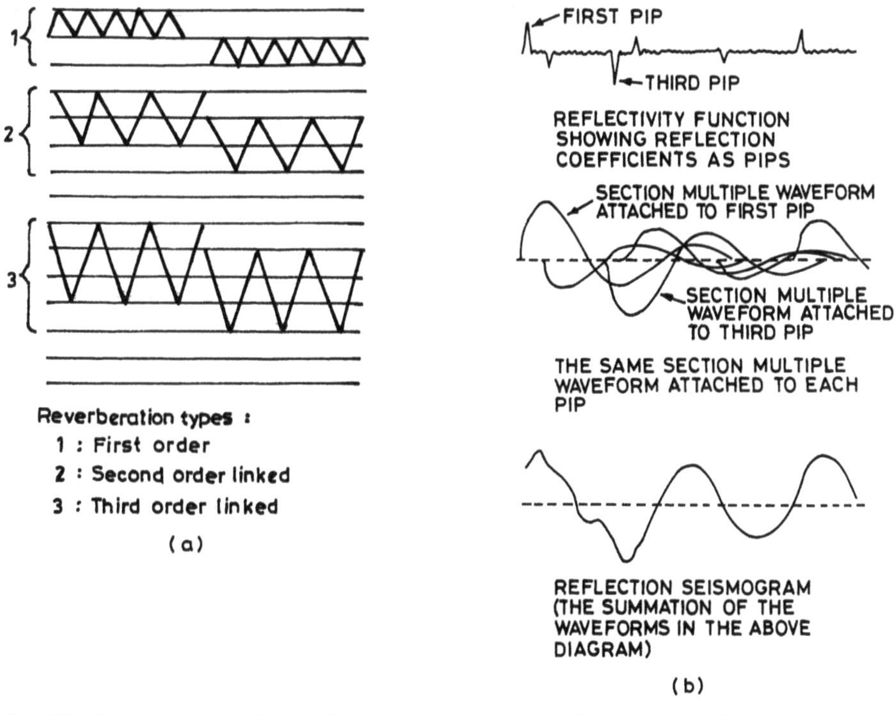

Fig. 3.7 a, b. a Reverberations within a geologic section of sedimentary layers; **b** convolutional model of reflection seismogram for a spike source (Source: Robinson 1981, p 84. In: Developments in Geophysical Exploration Methods – 2. Edited by A.A. Fitch. © Applied Science Publishers Ltd 1981)

that round-trip travel time in each layer is one time unit. The hypothetical layers are assumed to possess zero values of reflection coefficients.

First order, second order, third order reverberations, etc. are defined as those for which round trip travel times are one, two, and three time units, respectively. As shown in Fig. 3.7a the first order reverberation occurs within one layer, second order within two layers and so on. Therefore, for N layers there are (N–1) different second order linked reverberations, associated with following pairs of layers: (1, 2), (2, 3), (3, 4),...(N–1, N). Likewise (N–2) third order-linked reverberations are attached with layers: (1, 2, 3), (2, 3, 4), (3, 4, 5), etc.

Figure 3.7b shows diagrammatically the essential features of a convolutional model. The upper part of this figure shows the *reflectivity function* (the sequence of reflection coefficients) assuming a spike source. The middle part shows the common *section multiple train* (i.e. the response) associated with each reflection coefficient. The response for each pip is scaled. The bottom part of the figure shows the reflection seismogram, as a result of superposition of scaled versions of section multiple train shown in the middle part of the figure.

In this presentation of the convolutional model of the reflection seismogram, many multiple events associated with sedimentary interfaces are replaced by a *common section multiple train* for each interface, which is the multiple train associated with the entire layered medium. For quantitative description of the results, reference may be made to Robinson (1981, pp 96–106).

3.2 Analog and Digital Form Seismograms

An analog form seismogram is a continuous record of ground motion. A digital seismogram has in it a sequence of numbers corresponding to amplitudes of ground motion sampled at equal intervals of time. Figure 3.8 shows these seismo-

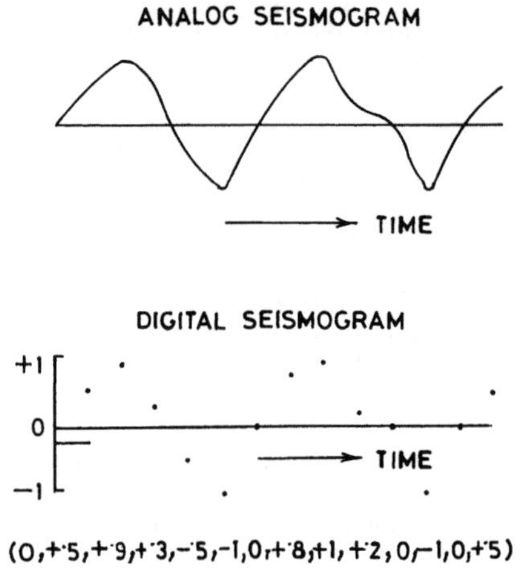

Fig. 3.8. Schematic illustration showing analog (*continuous*) and digital (*discrete*) seismograms. Digitized values are given in brackets

grams. If we plot the discrete digital values of amplitudes and join these, a seismogram like that of the analog seismogram would be obtained.

As mentioned in Section 2.1.3, analog recordings used to be initially done on photographic paper. Magnetic tape recordings, analog and digital forms, are the later developments.

The *dynamic range* of a seismic recording system is one of its most significant features. It is defined as the ratio of the strongest to the weakest signal that may be usefully recorded. The dynamic range of photographic recording was limited to about 24 dB. In analog magnetic tape recording, it increased to as much as 45 dB.

Amplitude ratios are expressed in terms of a decibel value as:

$$20 \log_{10} A_2/A_1 = \text{the decibel value}$$

where A_2/A_1 is the ratio of amplitudes.

24 dB corresponds to an amplitude ratio of about 16, and 45 dB to a ratio of about 178.

If we analyze the range of ground motions that reach the geophones, the amplitude ratio of the strongest to weakest ground motion may reach 100 dB – in terms of amplitude ratio it corresponds to 100,000. Therefore, photographic and analog magnetic tape recordings were inadequate due to their much lower dynamic range. The achievement of a large dynamic range (in excess of 100 dB) is the very special feature of digital recording units.

3.2.1 Frequency Bandwidth of a Reflected Wavelet

The reflection or transmission process does not involve any loss of frequencies. *Geometrical spreading* reduces the amplitude values but does not influence frequencies. *Frequency selective absorption*, however, causes loss of high frequencies. Therefore, this should affect the frequency bandwidth of the reflected wavelet.

In the context of bandwidth of the migrated *image wavelet* on a migrated section, Black et al. (1993) noted that the image wavelet should depend on all factors which influence the ray path, such as dip, velocity, depth and offset. Figure 3.9 shows an image of two wavelets with the same reflectivity: the inclined raypath has a more stretched (lower frequencies) image wavelet than the vertical raypath. These raypaths are associated with dipping and horizontal reflectors, both having the same reflectivity. The spectral depiction of these wavelets is given in the lower part. This concept has an important implication for *true amplitude imaging*. We would return to its discussion again in Chapter 11.

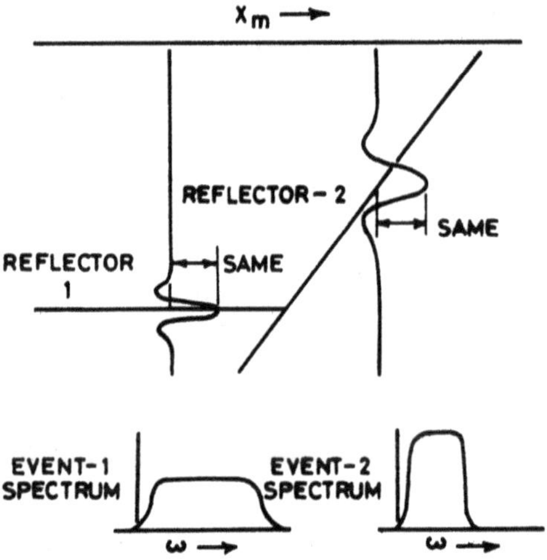

Fig. 3.9. Images of two events with the same reflectivity (Source: Black et al. 1993. From Geophysics with permission)

3.2.2 Use of Sampling Theorem

Sampling theorem discusses how an analogue form signal should be sampled so that, as a result of sampling, no information is lost about the signal. It is stated as: no information is lost by regular sampling provided that the sampling frequency is greater than twice the highest frequency component actually present in the waveform being sampled (refer, for example, to Telford et al. 1976, p 376). It amounts to saying that there must be more than two samples per cycle of the highest frequency being sampled. Two samples per highest frequency cycle is the absolute minimum. This is because two points may lie at the nodes, or at some constant value of the waveform. In fact, four samples per cycle should be used for all frequencies of interest and at least two for the highest frequency present.

What happens if we do not have more than two samples per highest frequency cycle? It may be shown that as a result of this, artificial low frequencies would get created. This effect of creation of artificial low frequencies as a result of inadequate sampling is known as *aliasing*. We explain *aliasing* with the help of an example.

Figure 3.10 shows a sinusoidal wave which is sampled at three different time intervals, i.e. less than half a period, greater than half a period, and greater than a full period of the sinusoidal waveform. It may be seen that for the cases in which the sampling rate is greater than half a period, lower frequency waveforms (shown by dashed lines) are artificially created. This is because adequate sampling requires that there must be more than two samples per cycle and this condition is obviously not fulfilled in these two later cases.

It is possible to calculate the frequencies of newly created waveforms (i.e. the alias frequencies) as a result of inadequate sampling. If Δt is the sampling interval, *sampling frequency* is $1/\Delta t$. Half of the sampling frequency is called the *Nyquist frequency*. We may denote it by f_N. Any frequency greater than f_N by an amount equal

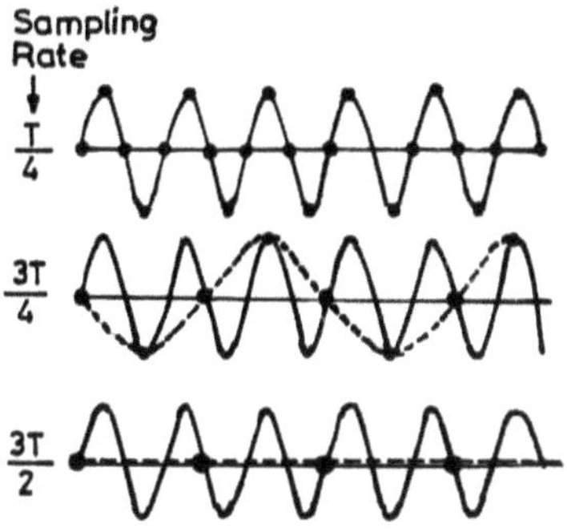

Fig. 3.10. Sampling of a sinusoidal waveform at less than a half period, greater than half and full periods, and the resulting low frequency aliased waveforms shown by the dashed lines

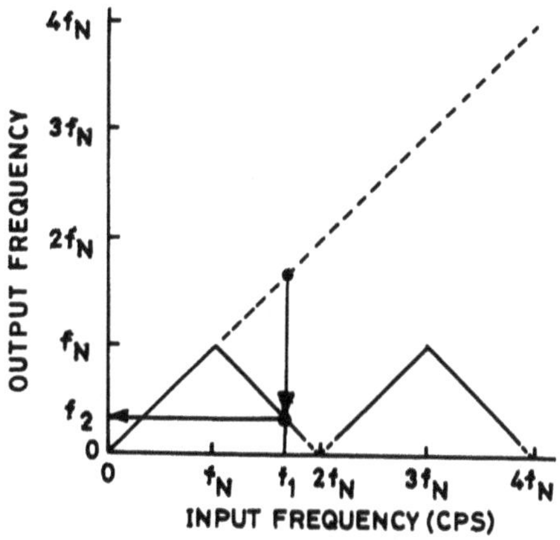

Fig. 3.11. Relationship between input and output frequencies (Source: after Peterson et al. 1966: A pictorial digital atlas. © United Geophysical Corporation, Pasadena, California)

to Δf in the signal would appear as a new lower frequency equal to (f_N–Δf). Suppose that sampling interval is 4 ms. This gives a sampling frequency of 250 Hz, and a Nyquist frequency equal to 125 Hz. If there is any frequency greater than 125 Hz in the signal, say it is 200 (=125+75) Hz, then this 200 Hz frequency component would appear as a lower frequency equal to 50 (=125–75) Hz.

The relationship between input and output frequencies is shown by Fig. 3.11. For frequencies less than the Nyquist frequency (f_N Hz in this case), the output frequency is the same as the input frequency. Above f_N Hz, the output frequency decreases linearly with input frequency and at 2 f_N Hz it is zero. Beyond 2 f_N Hz, the pattern repeats as that shown at zero Hz reaching to a maximum at 3 f_N Hz and zero at 4 f_N Hz.

Alias signals, which appear in the frequency band we are interested in, would appear to be legitimate signals. To circumvent this, alias filters are used in the amplifiers before sampling by the multiplexer for removal of frequency components higher than the Nyquist frequency. It is essential that filtering must be done before sampling for the reason that, after sampling, it is not possible to distinguish alias signals. *Alias filters* bear a finite slope (often 72 dB per octave); also the cut-off frequency is an octave below the Nyquist frequency to prevent aliasing. Supposing that samples are needed at 2 ms interval, this gives a sampling frequency of 500 Hz and a Nyquist frequency of 250 Hz. In this case, an analog alias filter is included in the circuit, which reject frequencies severely above 125 Hz.

Aliasing is an inherent property of all systems that sample. The sampling may be either in the time domain, space domain, or any other domain (like frequency).

Sometimes it may be required to resample the sampled data at a lower rate. This may be desired in view of processing fewer data samples. In this case also, *alias filtering* must be done before re-sampling the data.

Our discussion so far has been restricted to sampling of ground motion at a particular detector in the time domain. We may sample the wave motion spatially. In fact, spatial sampling is automatically being done when we use a number of detectors at spacing of Δx to sample a continuous wave motion in space. For time sampling we have $1/\Delta t$ samples per unit time; and in the space sampling we have $1/\Delta x$ samples per unit distance. As we have the Nyquist frequency ($1/2\Delta t$) for time sampling, we have the Nyquist wave number for spatial sampling:

$$\frac{K_N}{2\pi} = \frac{1}{\lambda_N} = \frac{1}{2\Delta x} \tag{3.1}$$

This gives the maximum number of waves per two unit lengths. Aliasing occurs when the Nyquist wave number is exceeded, i.e. when $K > K_N$ or $\lambda < \lambda_N$.

It may be mentioned that the Nyquist frequency gives a maximum number of waves per two units of time interval, i.e. $2\Delta t$. Aliasing is caused for components that have frequencies higher than f_N ($= 1/2\Delta t$).

The effectiveness of the sampling theorem may be illustrated by considering the corresponding time domain and frequency domain operations of sampling and reconstruction of waveform from the sampled values, as is illustrated by Fig. 3.12.

Sampling of a waveform may be done by multiplying the waveform by a *comb function*. A comb function consists of equally spaced unit impulses:

$$\mathrm{comb}(t) = \sum_{-\infty}^{+\infty} \delta(t - p\Delta) \tag{3.2}$$

Here p is an integer and Δ is fixed time interval of sampling. In the example being discussed it is 0.004 s. If the comb is having elements every 0.004 s, its transform has elements every 1/0.004 Hz (i.e. 250 Hz) as shown on the right side in Fig. 3.12.

When the comb function is multiplied with the continuous waveform, say g(t), a series of values of $g(p\Delta)$ are obtained:

$$g_t = \sum_{-\infty}^{+\infty} g(p\Delta)\delta(t - p\Delta) \tag{3.3}$$

Sampling of the function results in repetition of the original spectrum of the continuous waveform. Multiplication of the repeated spectrum with a *box car func-*

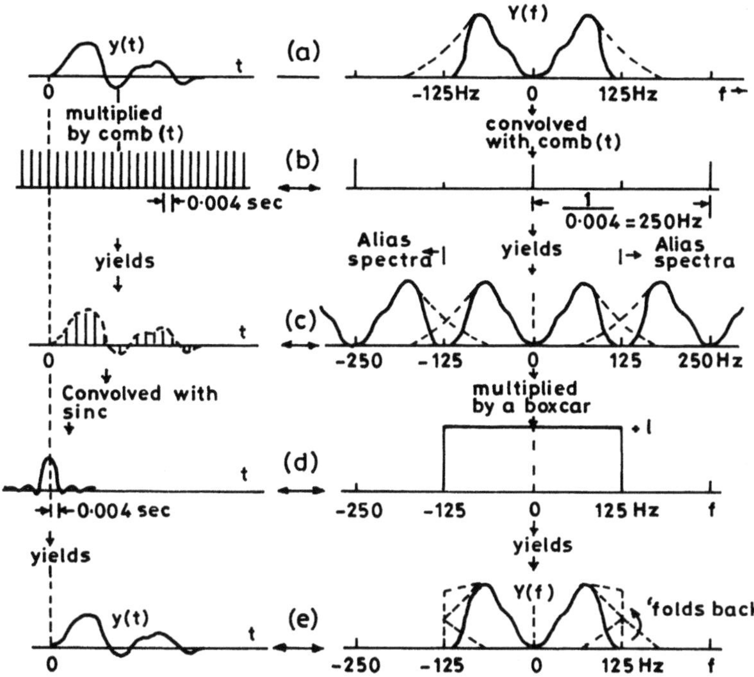

Fig. 3.12. Sampling and reconstruction of a waveform (Source: Sheriff and Geldart 1983, p 30. Reproduced by permission. © Cambridge University Press 1983)

tion would produce the spectrum of the continuous waveform as at the top of the figure. In the time domain the corresponding operation is convolution of sampled values by an appropriate *sinc function*, which yields the original waveform shown at the top. Recovery of the original waveform both in the time domain and in the frequency domain may be observed.

If the continuous waveform had frequencies higher than the Nyquist frequency (taken here as 125 Hz), the corresponding spectra would be as shown by the dashed lines. It is observed that overlap of spectra occurs in this case, which was not there earlier, when signal frequencies were below the Nyquist frequency. The spectra of frequencies higher than 125 Hz would fold back to appear as spectra of lower frequencies, shown in the figure as a dashed line at the bottom right. Because of these spurious effects in the spectrum, it would not be possible to reconstruct the original waveform from sampled data if the continuous waveform contains frequencies higher than the Nyquist frequency.

3.3 Digital Recording – Multiplexed Data

Digital recording uses binary form numbers to store the signal strength received by various geophones. Any number (either a whole number or fraction) may be represented in binary form by using only two digits, 1 and 0. Decimal form representation of any number requires digits from 0 to 9. Therefore, storage of binary form numbers on magnetic tape has it's distinct advantage because it requires a combination of only two digits 0 and 1. In fact, a combination of either magnetized (corresponding to 1) or non-magnetized (corresponding to zero) small areas on the magnetic tape is able to register any desired number on the magnetic tape. A binary digit is called a *bit*, and a collection of 'bits' is called a *byte*.

As an example, let us consider a decimal form number 261. In binary form, 261 would be written as 100000101, which is a short form for:

$$1 \times 2^8 + 0 \times 2^7 + 0 \times 2^6 + 0 \times 2^5 + 0 \times 2^4 + 0 \times 2^3 + 1 \times 2^2 + 0 \times 2^1 + 1 \times 2^0$$

The power of 2 may be continued for representing fractional numbers, as 1/2 may be represented as 2^{-1}, 1/4 by 2^{-2}, 1/8 by 2^{-3}, etc. This enables us to write a binary form number like 10111 for a decimal fractional number 0.90625.

Since the signals amplitude may be positive, or negative and may have been amplified, there also exists provision for recording sign, gain or scaling factor, and the digitizer output. In the logarithmic form, the digitizer output is mantissa and amplifier-scaling factor is exponent, and one 'bit' is allocated for sign. The general form of the binary number may look like

(1)	0101,	0111001
↓	↓	↓
sign	expon.	mantissa

By evaluating the exponent, the mantissa may be shifted to the left to express the complete number in a more compact form.

Normally, the binary numbers are stored in rows across a magnetic tape. Magnetic tapes with nine heads, generally record on 1/2-inch tape, with 21 heads on a 1-inch tape. Excluding two of nine heads, and four of twenty-one heads, which are used for other purposes, the rest of the heads are used to write the numbers.

Figure 3.13 shows the schematic pattern of a record for 7 heads recording on 1/2-inch tape. Each one of these heads is in contact with the tape. Each contact area is shown as a square. Each square is referred to as a 'bit', and each row of squares a *byte*. One *bit* accommodates one digit of a binary number. Depending on magnetization or non-magnetization of the area on the tape, digit 1 or 0 is inferred. The total number of bits assigned to record a binary number constitutes a *word*. In the illustration taken, a word is made of 14 bits, in two rows or bytes on the tape. A '*word*' as large as 32 bits is sometimes needed in exploration seismology.

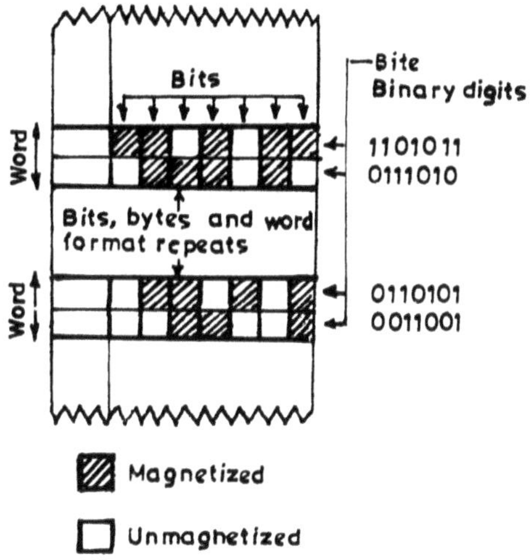

Fig. 3.13. Schematic illustration of digital magnetic tape recording

Multiplexed data is a result of sampling the multitrace seismic records. The sampling is done through a high-speed electronic switch, called the *multiplexer.* It may be done directly in the field or in the data processing center by replaying the analog form magnetic tape record to generate electric signals. The sampling is done for all traces within one digitization interval decided beforehand.

After the first samples of all traces have been read, the process is repeated for the second, third and subsequent samples till all the samples for all traces have been read.

To get an idea of how frequent multiplexed data are read we take an example of a seismic unit with 48 channels, and using a digitization interval of 4 ms. Within this 4 ms, amplitude values of each of the 48 channels should be read sequentially, then the procedure should be repeated for the second, third and all subsequent rounds. Therefore, reading and channel-to-channel switching time is of the order of 4/48 ms or 80 μs.

The digitization interval is decided by what is the maximum frequency that should be retained, keeping in view the type of geological problem that needs to be solved. For example, in coal exploration seismic work, using very high frequencies, the digitization interval may be as low as 0.005 s. And within this interval sample values of all channels have to be read to generate multiplexed data.

3.4 Processing Domains

Seismic surveys generate a very large volume of data. It contains in it useful information about reflected signals embedded in various types of contaminating coherent and incoherent noise. An important aspect of processing is to remove or suppress the contaminating effects to bring out signals in focus. A large volume of data mostly necessitates close scrutiny and sorting of the data, and many times statistical algorithms provide the appropriate approach.

The ultimate goal of processing is to accurately map the subsurface structures, stratification and to infer lithology and the possible presence of hydrocarbons. The raw data is in the form of traces – the record of scattered (reflected) energy from subsurface inhomogeneities as a function of time, one trace for each detector (or groups of detector) location. In order to convert reflection times to depths, an accurate knowledge of velocity function (variation of velocity with depth) is a must.

Reflected signals undergo various types of *filtering effects* produced by the Earth's material. To remove these effects, *inverse-filtering* or *deconvolution*' need be performed. An appropriate mathematical filter operator should be designed to accomplish this objective. Other type of processing is related with correct positioning of reflected events in the subsurface medium from where they originated in the seismic experiment. This aspect is dealt with in a processing step called '*migration*'.

In fact seismic processing may be viewed as an inverse scattering problem. The known information is the scattered wavefield data, and the ultimate goal is to map exact details of the scattering medium.

To discuss the processing domains, we may first consider the scalar wave equation. It is a second order differential equation in the space-time domain:

$$\frac{\partial^2 \mathrm{v}}{\partial \mathrm{x}^2} + \frac{\partial^2 \mathrm{v}}{\partial \mathrm{y}^2} + \frac{\partial^2 \mathrm{v}}{\partial \mathrm{z}^2} - \frac{1}{\mathrm{V}^2}\frac{\partial^2 \mathrm{v}}{\partial \mathrm{t}^2} = 0 \tag{3.4}$$

This equation is a description of the wave motion, which describes a natural way of transporting energy from one point to another. In Eq. (3.4) the wave motion is described with respect to coordinate (x, y, z) of any point of the medium, and time t; x, y, z and t are independent variables and the quantity (the disturbance), undergoing a change with respect to coordinates of the point and with time, is the dependent variable. V is the velocity of disturbance propagation in the medium.

In the form of Eq. (3.4), it is multi-dimensional problem to be solved. In its frequency domain representation, it appears as a considerably simplified equation. For example, the differential equation transforms to a simple algebraic equation in the frequency domain.

In the frequency domain we express our analysis or results in terms of angular frequency ω, and angular wave numbers k_x, k_y, k_z (also called angular spatial frequencies). In the space-time domain, t is expressed in seconds, and x, z in meters, corresponding units for ω would be radians per second, and for k_x, k_y, k_z radians per meter.

2-D forms of Eq. (3.4) in the space-time and frequency domains are

$$\frac{\partial^2 \mathrm{v}}{\partial \mathrm{x}^2} + \frac{\partial^2 \mathrm{v}}{\partial \mathrm{z}^2} - \frac{1}{\mathrm{V}^2}\frac{\partial^2 \mathrm{v}}{\partial \mathrm{t}^2} = 0 \tag{3.5}$$

and

$$\mathrm{k}_z^2 + \mathrm{k}_z^2 = \frac{\omega^2}{\mathrm{V}^2} \tag{3.6}$$

The transformed form [Eq. (3.6)] is easily obtained, if one remembers the differential operators transformation rules: $\partial/\partial x$ corresponds to $-jk_x$ and $\partial^2/\partial x^2$ to $(-jk_x)^2 = -k_x^2$. Similarly, corresponding transformations for $\partial/\partial z$ and $\partial/\partial t$ are $(-jk_z)$ and $(+j\omega)$, respectively. Transformation for second derivatives follow in the manner it is applied for $\partial^2/\partial x^2$.

Equation (3.6) is obtained by taking the Fourier transform of Eq. (3.5) and by replacing the space/time derivatives operators with corresponding frequency domain operators.

Equation (3.6) is called the *dispersion relation.* It expresses the relationship between k_x, k_z, ω and V. The interrelationship between these quantities makes it possible to extrapolate wavefield data over space or time.

The two domains of data processing, namely, space-time domain and frequency domains are equivalent and it is possible to analyze and interpret the results in either of these domains.

Take an example of a plane wave propagating with velocity V and an angle ϕ to the vertical z-axis (Fig. 3.14). The positions of plane wavefronts are shown at time t=0 and t=1. In the space-time domain the *apparent velocity* values are $V/\sin\phi$ along the horizontal x-axis and $V/\cos\phi$ along the vertical z-axis. Corresponding values in the frequency domain are ω/k_x and ω/k_z. Actual velocity V is given by ω/k and we have the following relations (refer, for example, to Robinson, Durrani and Peardon 1986, pp 236–237):

$$\frac{\sin\phi}{V} = \frac{k_x}{\omega} \quad \text{and} \quad \frac{\cos\phi}{V} = \frac{k_z}{\omega} \tag{3.7 a}$$

Solving for ϕ we get

$$\sin\phi = \frac{Vk_x}{\omega} \quad \text{and} \quad \cos\phi = \frac{Vk_z}{\omega} \tag{3.7 b}$$

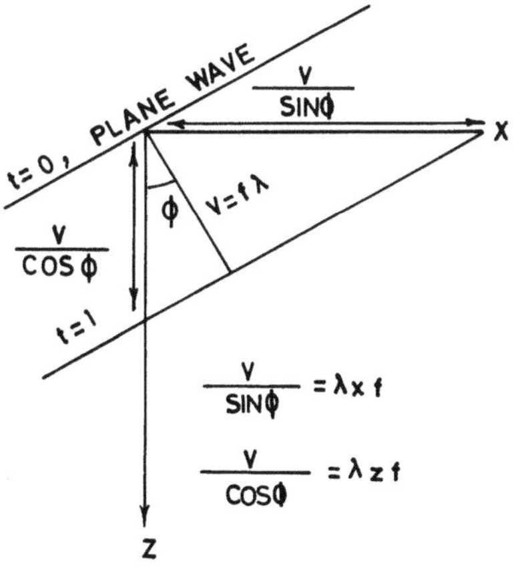

Fig. 3.14. Spatial depiction of plane wave and related quantities; λ_x and λ_z are wavelengths in the x and z directions, f is the temporal frequency

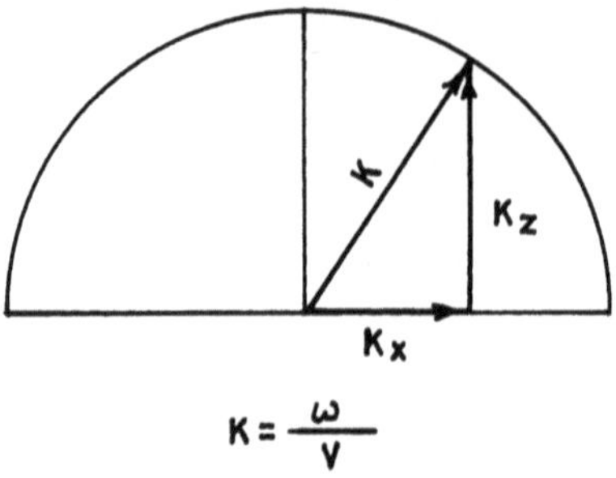

Fig. 3.15. Schematic depiction of dispersion relation

These equations relate the physical angle ϕ with the velocity and the frequency-domain quantities k_x, k_z and ω. From Eq. (3.7b) it is easy to write the dispersion Eq. (3.6). The graph of Eq. (3.6) is a circle in the k_x, k_z plane (Fig. 3.15). Any point on a constant ω curve defines a set of spatial angular frequencies k_x, k_z and a particular direction ϕ of wave propagation.

3.4.1 Space-Time Domain

An example of space-time domain processing is provided by the concept of calculating the wavefield at successively lower levels from the discretely sampled (at trace spacing) values of the wavefield along the surface. Referring to Fig. 3.16 the wavefield at Z_4 is obtained as a weighted superposition of values at nearby stations at a shallower depth, expressed as

$$\phi(Z_4) = \sum w_{4-i}\phi(X_{4-i}) \tag{3.8}$$

(Sheriff 1989, p 434) where ϕ represents the waveform function (at a particular time) and w is a weighting function which takes into consideration different phase shifts due to different distances involved, i.e. from Z_4 to X_4, Z_4 to X_3, Z_4 to X_2, and so on.

The operation represented by Eq. (3.8) is a process of superposition – a value at ($Z+\Delta Z$) is the weighted superposition of the values at shallower depth Z. The net result corresponds to a process of convolution. This particular operation may be carried out in other domains also. In the conventional space-time representation of a seismic section, the space axis is the location of seismic traces and the time-axis is the arrival time. We may do different types of transformations, like

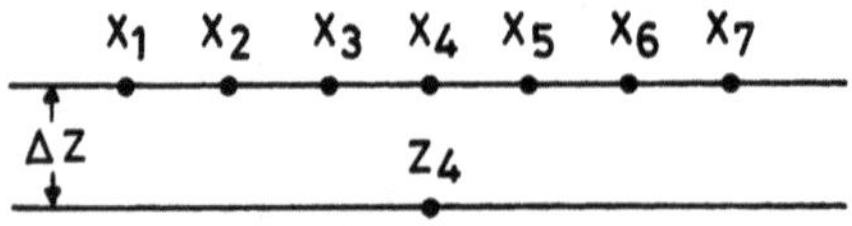

Fig. 3.16. Schematic showing that wavefield at a depth point is the weighted superposition of wavefield values at nearby stations at a shallower depth (after Sheriff 1989, p 434)

transforming both space and time axes or one of these axes. Accordingly, we would be dealing in the frequency-wave number domain, frequency-space domain, or the wave number-time domain. After doing the processing in these domains the results may be transformed back to the space-time domain. To take full advantage of a particular domain processing, sometimes data are transformed back and forth during the migration process to maximize the advantages in different domains and at the same time minimize the disadvantages.

The characteristic of the data being migrated may help in the choice of a suitable method. For example, for wave shape retention from steeply dipping reflectors, frequency-wave number domain migration is better suited, but to cope with velocity variation, space-time domain migration may be a better choice.

3.4.2 Frequency Domain

A seismic cross-section is composed of a series of seismic traces obtained at equidistant points along a line on the Earth's surface. The quantity sampled for each seismic trace is the amplitude value. An amplitude value of j-th seismic trace at time i Δt may be represented by A (i Δt, j Δx) where Δt is the sampling time and Δx is the spacing of detectors (or of traces). A typical matrix of amplitude values may have entries of the size of 1500×800.

A single seismic trace may be Fourier-analyzed, and may be represented in an alternative way by its complex spectrum. The outcome of Fourier-analysis is the amplitude and phase values of the constituent frequencies.

If we consider a two-dimensional data set, it would consist of sampled values of individual traces in columns. The rows would represent sampled values at equidistant points along a line on the ground surface–like the first sampled values of amplitudes of all traces in the first row, the second amplitude values of all traces in the second row, and so on.

It we denote f(t) and $F(\omega)$ as a Fourier-transform pair of a single trace data, f(x, t) and $F(k, \omega)$ would represent the corresponding transform pair for two-dimensional data.

A two-dimensional Fourier transform may be done as a combination of two one-dimensional Fourier transforms:

$$f(x,t) \Leftrightarrow F(x,\omega)$$

and

$$F(x,\omega) \Leftrightarrow G(k,\omega)$$

That is, in the first transform, each column is transformed from t to ω, and next rows of $F(x, \omega)$ are transformed from x to k_x.

The arrows directions reveal reversibility of each transformation.

In the case of 2-D cross-section data, the sampling considerations apply equally to adequate sampling in spatial (x) and temporal (t) directions. In this case therefore, two Nyquist frequencies would be involved. Because of the two dimensions involved it is possible to select values of ω and k, so that

$$\frac{\omega}{k} = \frac{2\pi f}{2\pi/\lambda} = f\lambda = V(\text{velocity}) \tag{3.9}$$

is constant. Therefore in the (k, ω) representation, a straight line passing through the coordinate origin would represent a particular velocity. This aspect makes it possible to divide the (k–ω) plot into regions in which the velocity is greater than or less than a particular value. Based on this property of the (k–ω) plot, it is possible to design a *filter* to serve a specific purpose.

In the (k–ω) plot the vertical axis would represent an infinite velocity. Velocities of the reflected signal lie in a wedge about the vertical ω-axis. Coherent noise has a generally lower velocity and on the k–ω plot they occupy a different zone (Fig. 3.17 a). In the k–ω plot removal of the sloping band eliminates all energy that propagates across the record at a particular velocity.

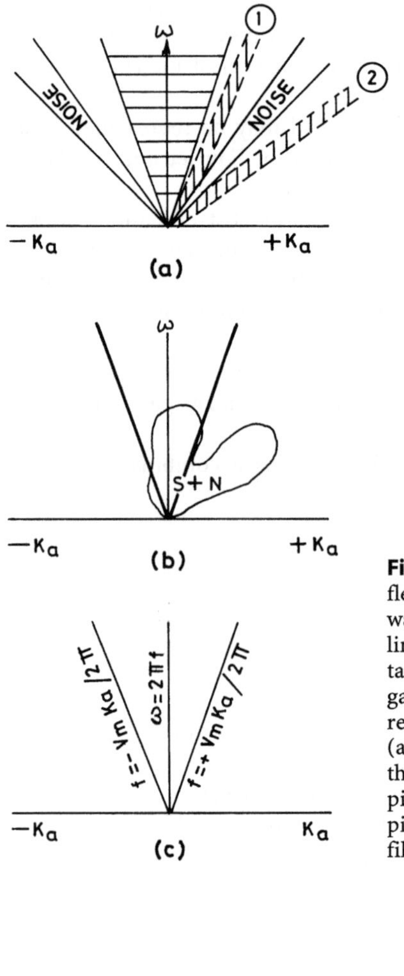

Fig. 3.17 a–c. Schematic illustration depicting. **a** Reflected signals and coherent noise in the frequency wave number domain. Within the dashed parallel lines energy propagates with one velocity. Elementary waves defined by (ω, k) sets in band 1 propagate with higher velocity than in band 2; **b** original reflection data contains signal and noise (S+N) both (after Waters 1987). The region enclosed within the thick lines represents the acceptance region of the pie-slice filtering process (Embree et al. 1963); **c** depiction of frequency-wave number apparent velocity filter (after Seiscom Delta)

Figure 3.17b shows the occupancy of original data in the k–ω plane, and the acceptance region of the *pie-slice process* or *filter*. In this type of filtering, the reflected events are assumed to have high apparent velocity, and on application of pie-slice filtering, the reflected events would be left behind if a small wedge of energy on the k–ω plane is conserved (about ω-axis), and the rest of the k–ω plane is given zero amplitude. Low-velocity energy is considerably cut by the pie-slice method. The pie-slice process may also be carried out in the time-domain.

The apparent velocity filter in ω–k domain may be defined as (Sheriff and Geldart 1983, p 56):

$$\begin{aligned} F(\omega, k) &= 1 \quad \text{for} \quad |k_a| < {}^{2\pi f}\!/\!_{V_m} \\ &= 0 \quad \text{for} \quad |k_a| > {}^{2\pi f}\!/\!_{V_m} \end{aligned} \tag{3.10}$$

where V_m is some minimum velocity for reflected signals. This filter would pass signal but reject noise (Fig. 3.17c).

As there is a problem of aliasing with temporal frequencies due to inadequate sampling in the time domain, there is also a problem of aliasing with spatial sampling in the wave number domain. Spatial sampling involves aliasing of data for wave number values exceeding the Nyquist wave number.

3.5 Various Kinds of Processing and Their Purposes

3.5.1 Frequency Filtering

In order to separate noise from useful reflected signal, criteria of their frequency range and velocity may be utilized. If the low frequency, high-energy noise, such as surface waves, has frequencies which are well separated from reflection signal frequencies, they may be filtered out during initial recording itself. However, it is found that noise spectrum often overlaps the signal spectrum, and for this reason the frequency filtering is of limited value only.

Figure 3.18 gives a comparison of the performance of digital and electric analog filters. Analog filters, in general, have a gentle slope. Digital filters, on the other hand, may be programmed to have almost vertical slopes, which would help better extraction of the reflection signal. This is evident from Fig. 3.18a–c. By digital filtering of frequencies, the signal spectrum is well discriminated against the noise spectrum.

The filter response may be expressed either in the time domain or in the frequency domain. If the response is known in one domain it is possible to express it in the other domain. For accomplishing the filter operation, both the seismic signal and filter response characteristics should be expressed in the common domain, either in the time domain or in the frequency domain.

The purpose of seismic amplifiers is to reproduce the input received by it with the minimum of distortion, and therefore the gain should be constant for the en-

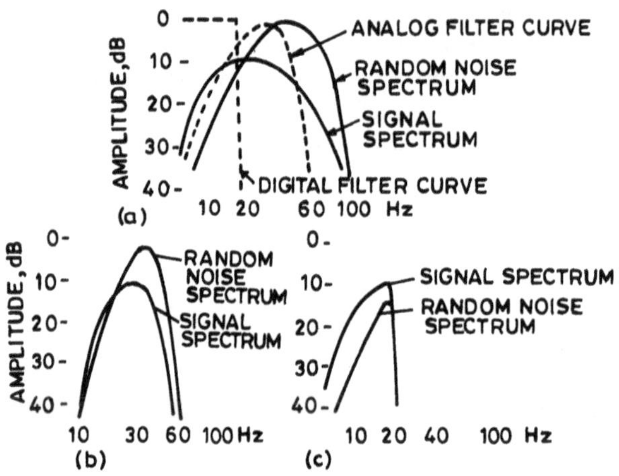

Fig. 3.18 a–c. Comparison of performance of digital and electric analog filters designed for extracting reflections from high frequency scattered noise: **a** before filtering; **b** after optimum analog filtering; **c** after optimum digital filtering (Source: Dobrin 1976: Introduction to Geophysical Prospecting 3rd edn, p 156. © 1976, 1960 by McGraw Hill Inc)

tire frequency range of interest. For reflection work, the frequency range of interest is 10–100 Hz and for refraction work 1–50 Hz. The response of amplifiers, in most cases, is flat from about 1 to 200 Hz or more. For high-resolution work for solving engineering or mining problems to map the upper 200 m or so, the high-resolution amplifiers have a uniform response up to 300 Hz and sometimes up to 500 Hz.

Because of preferred higher frequency attenuation, peg-leg multiples and other natural filtering processes, the frequency spectrum of the reflected signal tends to be lower with increasing arrival time of reflections. With this consideration, the pass band of the amplifier filter would be desired to be shifted towards lower frequencies for the later part of the record obtained at increased arrival times of reflections. It calls for the design of a time-varying filter. Design of a time-varying filter requires the operation of different filters or a combination of filters during different time ranges with weights of filters changing in time (refer, for example, to Waters 1987, p 238).

The purpose of the low-cut filter in seismic amplifiers is to supplement geophone filtering to remove very low frequency noise in areas of dominant ground roll (the surface waves) in land seismic surveys. The high frequency filtering is aimed at controlling *aliasing*. A typical slope is 72 dB per octave. *Filter effect* is also produced by timing errors *in stacking* (refer, for example, to Sheriff 1989, p 331).

Frequency filters are characterized as *low pass*, *high pass* or *band pass* according to their discrimination characteristics. When they discriminate against frequencies above, or below a certain limiting frequency, or outside of a given band of frequencies, they are respectively referred to as low pass, high pass or band pass filters. Mathematically, their frequency amplitude response is expressed as (refer, for example, to Sheriff and Geldart 1983, vol 2, p 187):

$$
\begin{aligned}
[F(\omega)]_{\text{low pass}} &= +1 \quad |\omega| < |\omega_0| \\
&= 0 \quad |\omega| > |\omega_0| \\
[F(\omega)]_{\text{high pass}} &= 0 \quad |\omega| < |\omega_0| \\
&= +1 \quad |\omega| > |\omega_0| \\
[F(\omega)]_{\text{band pass}} &= +1 \quad |\omega_1| < |\omega| < |\omega_2| \\
&= 0 \quad |\omega_1| > |\omega| \\
&\quad \text{or} \quad |\omega| > |\omega_2|
\end{aligned}
$$

Reflection signals generally dominate the noise only within a limited frequency band. To increase the signal-to-noise ratio, the filter should pass the frequencies within that range where the signals dominate (refer, for example, to Fig. 3.18 to observe the frequency range where signals dominate). On an experimental basis, record sections filtered by a succession of band pass filters, are compared with each other and with unfiltered record to find out the optimum band pass limits.

Ground roll in land surveys or streamer noise in marine surveys are reduced by choosing a low cut filter at around 8 Hz. A high cut filter-limiting frequency could be 62.5, 125 or 250 Hz. The choice of a high cut filter depends on the sampling frequency used. For 4 ms sampling (i.e. 250 Hz sampling frequency), the high cut-off frequency is taken as 62.5 Hz with a slope of 72 dB per octave. This frequency is adopted for conventional seismic surveys. For high-resolution surveys, the sampling is done at 2 ms or 1 ms and the corresponding cut-off frequencies take the values of 125 Hz or 250 Hz, respectively.

3.5.2 Demultiplexing

The multiplexed and digital (binary mode) data, consisting of a sequence of amplitude values, exists as a series of numbers on a single channel:

$$\{a_{11}, a_{21}, \ldots a_{N1}, a_{12}, a_{22}, \ldots a_{N2}, \ldots, a_{NM}\}$$

In this form, the first subscript stands for the channel number, and the second for sample number. Thus, the first sample of all N channels, second sample of all N channels ..., and Mth sample of all N channels occur sequentially.

However, for most digital processing, it is necessary to demultiplex the data existing on a single channel into different single channel sequences. The demultiplexed single channel sequences are ordered separate arrangement of samples of each individual channel. The demultiplexed arrangement of data appears as:

(Trace 1) $a_{11}, a_{12}, \ldots a_{1M}$

(Trace 2) $a_{21}, a_{22}, \ldots a_{2M}$

(Trace N) $a_{N1}, a_{N2}, \ldots a_{MNM}$

Thus, demultiplexing accomplishes separation and ordered arrangement of individual seismic channel data.

Demultiplexed data is often recorded on a new magnetic tape–called the *demultiplexed tape*. Data such as the profile number, shot number, trace number etc. are also entered on the demultiplexed tape. This information acts as identification 'headers' of the particular data set.

3.5.3
Gain Recovery and Attenuation Correction

Amplitude of reflected waves, for near normal incidence, depends on acoustic impedance contrast at the reflecting interface. However, there are various factors, which work to obscure the amplitude information of reflected waves. A diagrammatic presentation of such factors is given in Fig. 3.19.

The whole subject of amplitude vs. offset (AVO) interpretation is based on how accurately we can correct for factors other than the reflectivity (or the reflection coefficient) associated with acoustic impedance contrast. A very detailed discussion of this topic is given in Chapter 13.

Independent effects of *geometrical spreading* and *absorption* account for significant loss of amplitudes. The loss due to the absorption effect lies generally in the range of 0.1 to 0.5 dB per wavelength of the path traveled in the Earth. It depends on the frequency of the seismic signal. The absorption effect varies from one type of Earth material to another. A sharp crack would continue to remain

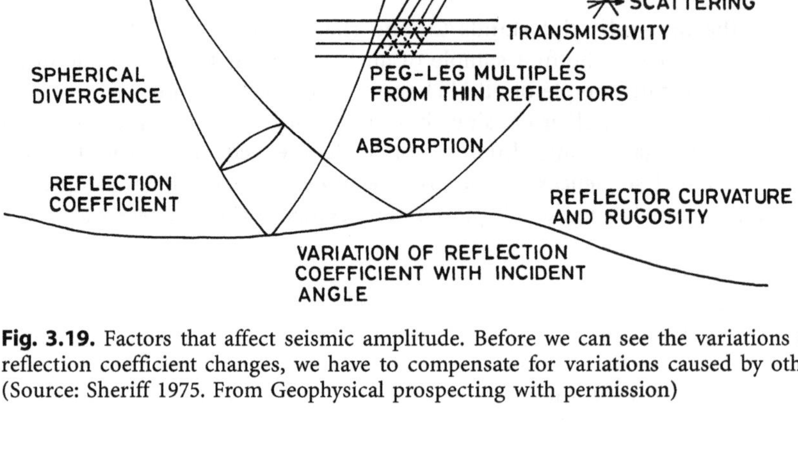

Fig. 3.19. Factors that affect seismic amplitude. Before we can see the variations caused by reflection coefficient changes, we have to compensate for variations caused by other factors (Source: Sheriff 1975. From Geophysical prospecting with permission)

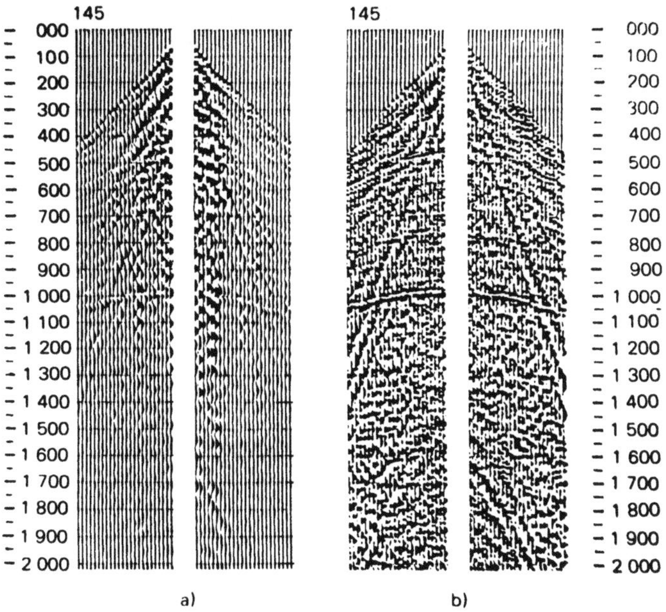

Fig. 3.20 a, b. Record before and after attenuation correction (IFP document). **a** Raw recording, source at the center; **b** after low-cut filtering and attenuation correction by automatic gain control (Source: Lavergne 1989, p 107. Reproduced by permission. © 1989 Editions Technip, Paris)

sharp in compact igneous rocks, like granite or basalt, but would degenerate to a thud in loose material, like sand or weathered rock. In the unconsolidated (weathered) layer, the attenuation may even exceed 10 dB per wavelength.

Figure 3.20 gives a feel of importance of this correction. The record before applying attenuation correction hardly shows any reflection event of interest, whereas the processed record improves quality of the reflected events significantly.

Attenuation correction for deeper reflections has to play a very significant role, as reflected signal strength weakens with increasing reflector depth.

Attenuation effects are responsible for the continual change of pressure pulse as the disturbance travels outward from a seismic source (refer to Fig. 2.1).

3.5.4 Deconvolution Before Stacking

Deconvolution before stacking aims at removing stretching of the seismic pulse. The stretching may be caused in various ways, including the bubble effect in marine surveys, the effect of ghost in ground surveys, or due to multiple reflections or reverberations in the subsurface structure.

Figure 3.21 shows a short pulse followed by two or three rebounds. This type of pulse may be observed in marine surveys. Figure 3.22 shows the stretched

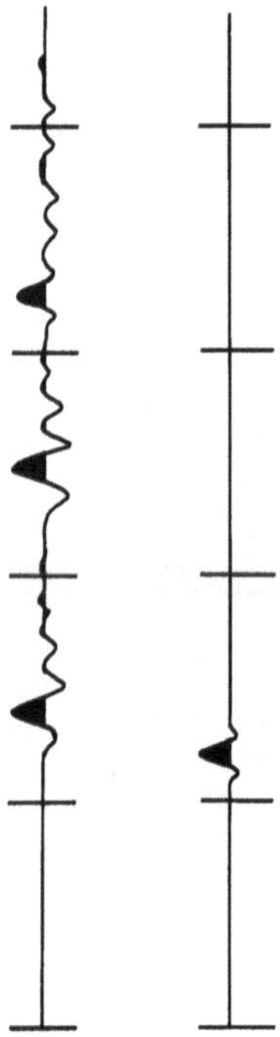

Fig. 3.21 a, b. Pulse deconvolution before stacking. Marine Survey (GSI document). **a** Initial pulse (*left*); **b** deconvolved pulse (*right*) (Source: Lavergne 1989, p 108. Reproduced by permission. © 1989 Editions Technip, Paris)

pulse formed due to immediate arrival of ghost, soon after the primary reflection, from the bottom of a weathered layer. The deconvolution operator is designed to spike these signals in Figs. 3.21 and 3.22 into short pulses. The required operator which destretches the recorded pulse may be calculated using the least square method. The only requirement to calculate this operator is that the signal delivered by the source be known. In marine seismic surveys, it may be recorded by placing hydrophones near the source of seismic energy. It may also be calculated by the autocorrelation of seismic traces obtained for the same source (refer also to Section 3.5.12).

Let the sample values of the initial signal be denoted by s_i $(i=1, 2, 3\ldots l)$; desired final pulse by d_i $(i=1, 2, 3\ldots m)$, and the desired deconvolution operator be f_i $(i=1, 2,\ldots n)$.

The *least square method* calculates the sum of the square of differences between the desired pulse d_i and the pulse obtained by convolving s_i with f_i (i.e. $s_i * f_i$), and minimizing it by equating the partial derivatives with respect to variables f_i to zero. The resulting equations are the following:

$$\sum = \sum_{i=1}^{m} [d_i - (s_i * f_i)]^2 \tag{3.11 a}$$

$$\frac{\partial}{\partial f_j} \sum_{i=1}^{m} [d_i - (s_i * f_i)]^2 = 0 \tag{3.11 b}$$

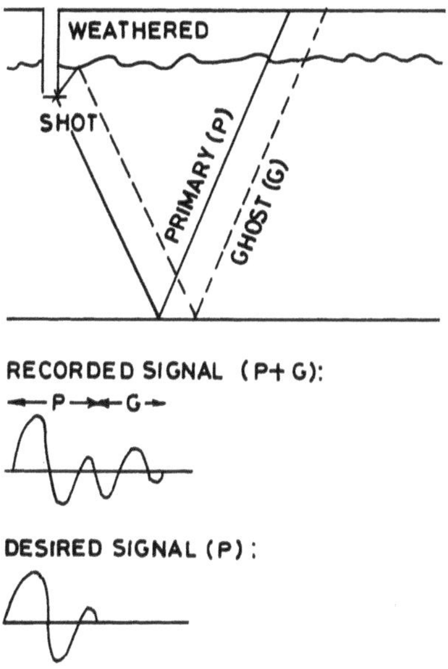

Fig. 3.22. Schematic illustration showing primary, ghost reflection, recorded signal (as their combination) and desired primary reflection signal (after Robinson 1967. © Hafner, New York)

Equation (3.11 b) consists of n equations which enables solving for n unknowns, $f_1, f_2, \ldots f_n$ of the *deconvolution operator.*

Knowing the filter function f_i, the seismic traces x_i may be convolved with f_i to obtain deconvolved traces Z_i:

$$Z_i = x_i * f_i \tag{3.12}$$

Deconvolved traces would exhibit in them the unstretched pulse form, free from the effects of rebounds (in marine surveys) and ghost (in land surveys) (refer to the bottom parts of Figs. 3.21 and 3.22).

Marine surveys require that the deconvolution operator f_i be computed at each shot to account for source pulse form variations due to source depth fluctuations.

3.5.5 Static Correction

Static correction, mainly required in land reflection surveys, aims at correcting traveltime anomalies which arise due to elevation differences of geophones, and variations of thickness of the near surface low velocity layer (LVL). By the application of static correction, the source of seismic energy and detectors are effectively brought to one common reference level, called the *datum plane* (Fig. 3.23). Accuracy in applying this correction is essential, as near surface travel time variations of reflection times due to the detector's elevation differences and effects of

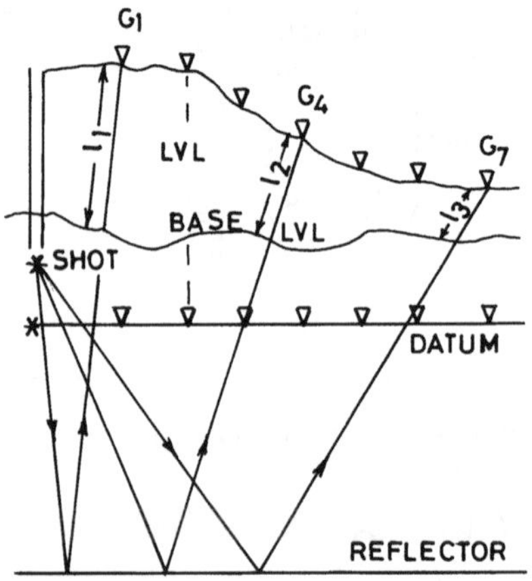

Fig. 3.23. Schematic illustration showing the necessity for static correction due to elevation differences of shot and geophones and different reflection path lengths (l_1, l_2, l_3) in the low velocity layer. Correction brings shot and geophone on a common level to the datum. Velocity below the datum level is the same

the low velocity weathered layer may be of the same order as that produced by subsurface reliefs of interest. Usually, velocity in the LVL may be in the range 300–800 m/s, compared to 1,500–2,500 m/s below it. The low velocity of this layer is attributed to the presence of air in pore spaces. The base of the LVL is many a time the water table. The thickness of the low velocity layer normally varies between 3–60 m. The thickness and velocity in the low velocity layer may change rapidly.

The low velocity layer is characterized by the following features (Sheriff 1989, pp 338–339):

- large impedance contrast with material below it,
- acts as a variable low pass filter. Therefore it would produce a change in upgoing reflected waveform,
- Raypaths in the low velocity layer are nearly vertical irrespective of their directions below it,
- thickness and velocity vary – as a result travel time of waves passing through it would vary.

It is obvious that shot effectiveness would be more when fired below the low velocity layer.

There are two steps in making accurate static correction: (i) *field statics* and (ii) *residual statics*. Field statics aims at adjusting the reflection times to what they would be if the shot and geophones were at one common reference level below the weathered zone. It may be easily calculated if the elevations of the shot and geophones and velocity and thickness of the weathered layer are known. For determination of velocity and thickness of the weathered layer, short refraction profiles are run. Also, accurate measurements of travel time in the weathered zone are done by firing a shot at the base of the weathered zone and recording waves with a detector on the ground surface. The use of both these approaches together enables better accuracy in calculating *field statics*. After the application of 'field statics', 'residual' travel time statics often continue to exist due to localized velocity variations below the receivers and the seismic sources. Although

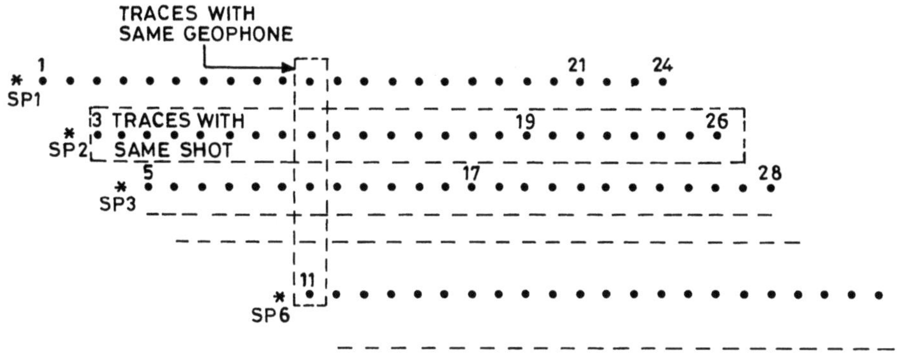

Fig. 3.24. Common-depth point profiles (CDP). The *symbols* (✶) and (●) represent source point and geophone groups respectively to achieve 6-fold coverage. Traces with the same shot point and the same geophone are shown by horizontal and vertical dashed line boxes. Traces from the following pairs of shot point (SP) and detector (D) come from the same subsurface depth point on the reflector: SP1D21, SP2D19, SP3D17, SP4D15, SP5D13, SP6D11 and thus provide 6-fold coverage (after Sheriff and Geldart 1982, p 138)

these anomalies may be of the order of a few milliseconds, these must be corrected for uniform, homogeneous recordings and accurate interpretation. These corrections are referred as '*residual statics*'.

For residual statics, seismic traces obtained by the same shot point and different detectors are compared after applying the normal moveout corrections. Time shifts observed in reflection arrival times arise essentially due to travel time anomalies in the neighborhood of each detector, as the condition near the shot is common for all detectors. Residual statics are evaluated at the geophones using these time shifts. Cross-correlation is the standard technique to achieve this objective. Residual statics at shot points are computed by using different shots and a common detector.

Figure 3.24 shows CDP sixfold geometry. The CDP survey is carried out along a straight profile although for clarity the shooting arrangement is shown on displaced lines. Time shifts common to the same shot lie along the row and the time shift common to the same geophone lie along columns.

3.5.6
Specific Selection of Seismic Traces – Different Types of Gathers Including CMP Gathers

Common Depth Point (CDP) survey progresses along a line by shifting positions of the shot points and connecting to an appropriate set of geophone groups for recording the subsurface reflections. Referring to Fig. 3.24 it shows a CDP survey arrangement for achieving sixfold coverage of subsurface points by shifting shot positions by two geophone group spacings. The fold number, specifying the number of times reflections are sampled from subsurface points is given by (Parasnis 1986, p 275):

$$\text{fold} = \frac{\frac{1}{2}(\text{Channel spacing} \times \text{number of channels})}{\text{Shot spacing}}$$

Since very large numbers of traces are involved, stacking charts (Morgan, 1970) are used to keep track of appropriate sets of traces for stacking or other types of processing. Besides the use for proper stacking, the stacking charts are also needed in making corrections.

Different types of data gathers may be described by considering Fig. 3.25a and b. Referring to Fig. 3.25a, we have the mid point coordinate (m) and half-offset coordinate (h) expressed in terms of source coordinate (s) and receiver coordinate (g):

$$m = \frac{g+s}{2}, \quad \text{and} \quad h = \frac{g-s}{2} \tag{3.13}$$

Figure 3.25b depicts in it the acquisition geometry, and method of data presentation with respect to source, and receiver coordinates, as well as the mid-point and half-offset coordinates.

The boat moves along a straight line and tows a source and a streamer cable of receivers. As the boat moves by a fixed spacing, a seismic source is activated and recording is done by the receivers.

Complete data from a single seismic line is a function of three coordinates s, g, and t (time). As data occupies a three-dimensional volume, it is usually dis-

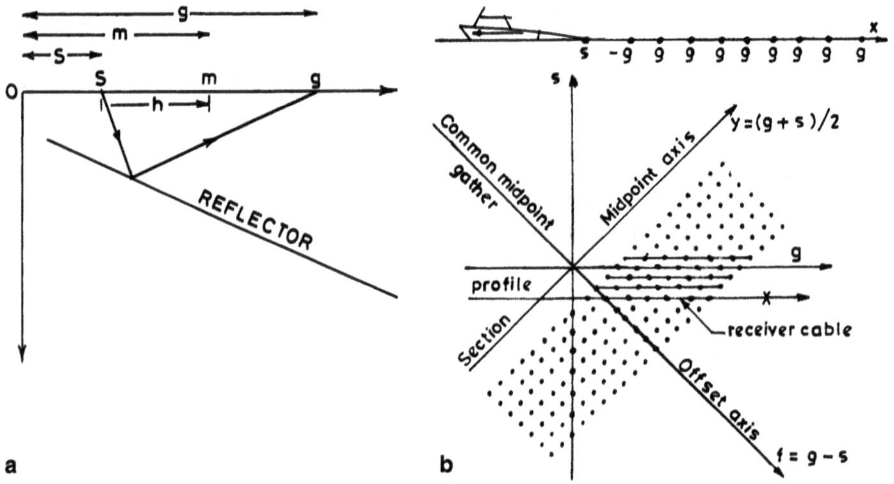

Fig. 3.25a,b. **a** Reflections may be referred with respect to source-receiver (S, g) position or mid-point (m) position and half offset (h); **b** the relationship among sound source coordinates S, geophone sound receiver coordinate g, offset coordinate f=g–s, and mid point coordinate y=(g+s)/2. Theoreticians generally use s and g as coordinates of wave pressure field, but interpreters generally use f and y (Source: Claerbout 1976, p 229. © 1976 by McGraw Hill Inc)

played through cross-sections. They are referred to as sections, gathers or profiles. Referring to Fig. 3.25b, we have the following classifications:

(s = constant, g, t): Common Source Point Gather (CSP gather)
(s, g = constant, t): Common Receiver Point Gather (CRP gather)
(m = constant, h, t): Common Midpoint Gather (CMP gather)
(m, h = constant, t): Common half-offset point gather (CHP gather).

3.5.7 Velocity Analysis

If the bedding planes are horizontal, a t^2–x^2 plot of reflection data may be used to determine the RMS velocities to various reflectors. In this case, the computation is general and does not require specific reference to a particular *trace array*, like a *common shot point, common detector point*, or a *common mid-point array*. However, for random stratification with beds having varying dips, only CMP trace arrays may be used for RMS velocities calculations to achieve a good degree of accuracy (see, for example, Levin 1971). For small dips of around 15°, the error in RMS velocity determination is not expected to be above 3%.

CMP stacking is discussed in Section 3.5.9. What is important here is the assumption that CMP traces are an outcome of data processing in which source

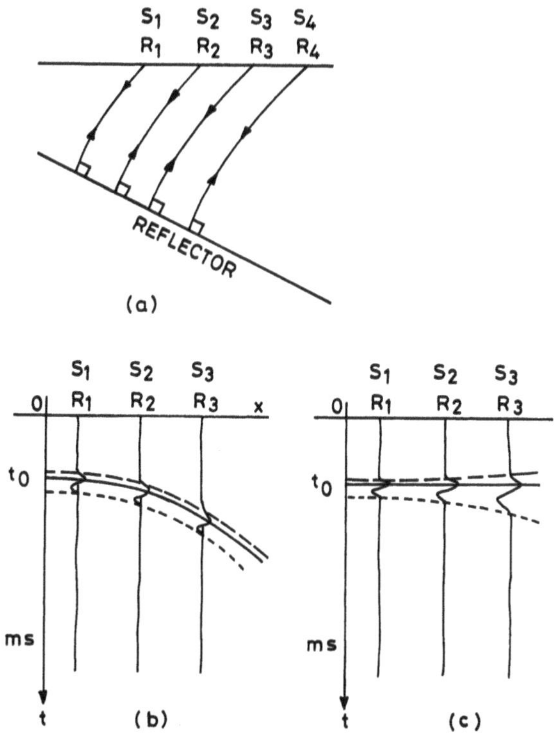

Fig. 3.26 a–c. a Raypath simulation of CMP stacked traces after correction; **b** reflected pulses before NMO correction; **c** reflected pulses after NMO correction (Source: Lavergne 1989, pp 117–118. Reproduced by permission. © 1989 Editions Technip, Paris)

and receiver are coincident and ray path to and return path from the reflector are coincident, with normal incidence at the reflector (Fig. 3.26 a). With reference to Fig. 3.26 b, the x–t plot gives a hyperbolic curve. The difference Δt between t and value on time-distance curve at origin (t_0) is the *normal moveout* (NMO). The normal moveout helps to determine the value of RMS velocity to a particular reflector ((Lavergne 1989, Eq. 4.31)]:

$$\Delta t = t - t_0 \cong \frac{x^2}{(VRMS)^2} \frac{1}{t + t_0} \tag{3.14}$$

After a correct application of normal moveout, the time-distance display of reflected events in the field record should look horizontal. It is clear that the value of normal moveout (Δt) depends on assumption of RMS velocity to a reflector. From Eq. (3.14) it may be inferred that large NMO corrections for distant traces and shallow reflector would produce a deformation of the reflected pulse, referred to as *NMO stretch* (Fig. 3.26 c). This is so because NMO correction is larger for the first arch of the signal than for the second. A 30 ms pulse, for example, may be stretched to 35 ms after NMO correction (compare Fig. 3.26 b and c).

Figure 3.27 is a display of the result of *velocity analysis*. A series of possible values of velocities are tried for data in CMP gather, and some criterion on how well the event is aligned, after application of NMO correction, for each trial value is calculated. For example, the criterion may be how large is the stacked energy? In the velocity analysis, the criterion value is displayed as a function of the *arrival time* and the *stacking velocity*. The values are often contoured. In Fig. 3.27 velocity analysis results display contours of *semblance*. *Semblance* is the ratio of the energy of the stacked trace and energy available for stacking. The contours join the values of velocity and time that emphasize stacked events. However, a large value of coherence should not always be interpreted as primary reflections. In fact, separate analysis of results of velocity analysis needs to be done to select values corresponding to primary reflections. Multiples are usually associated with

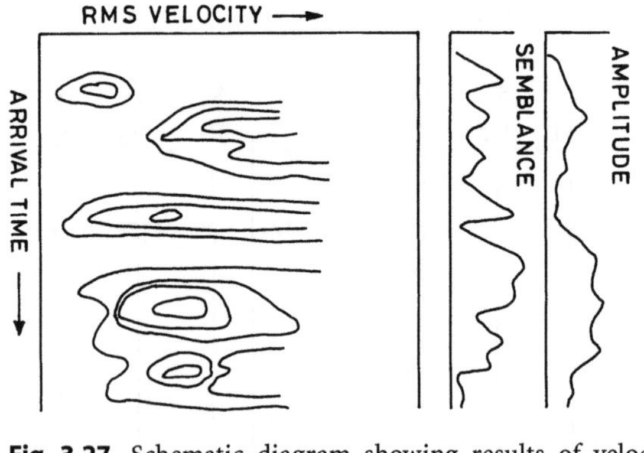

Fig. 3.27. Schematic diagram showing results of velocity analyses. Contours are of semblance, and graphs at right show maximum semblance and stacked amplitude at each arrival time (from Grant-Norpac. In: Sheriff 1989, p 396)

smaller stacking velocities for same arrival times as primary reflections. Therefore, larger values of *coherence* are given more weight in identification of primary reflection events. Velocity analysis results may be used for determining *interval velocities*. An educative discussion of this exercise may be found in Sheriff (1989, pp 395–401).

The main objective of *velocity analysis* is to determine the amount of normal moveout, which needs to be removed to maximize stacking of events that are considered to be primaries. An auxiliary objective is the identification of lithology.

In this section, use of NMO correction has been discussed in connection with velocity analysis. A more complete discussion on *normal moveout* (NMO) and *dip moveout* (DMO) would be found in Section 3.5.8 on moveout corrections.

3.5.8
Corrections

3.5.8.1
Normal Moveout

The reflection path for a single isotropic homogeneous layer with horizontal reflector is ABC (Fig. 3.28). Using Pythagoras theorem we have

$$BC^2 = CO^2 + OB^2,$$

$$\left(\frac{Vt}{2}\right)^2 = \left(\frac{x}{2}\right)^2 + h^2, \tag{3.15 a}$$

$$t^2 = \frac{4h^2}{V^2} + \frac{x^2}{V^2} = t_0^2 + \frac{x^2}{V^2}$$

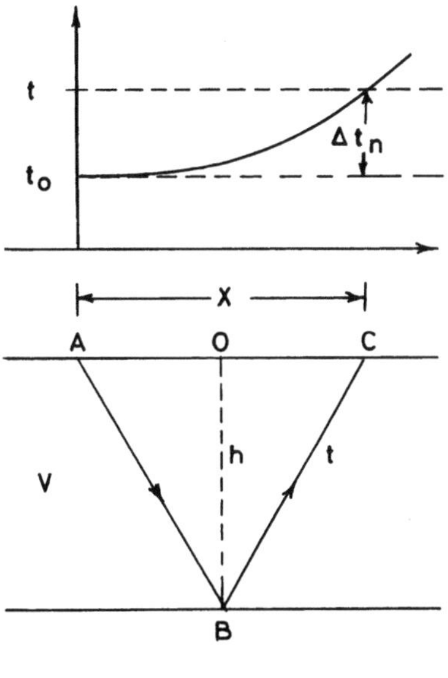

Fig. 3.28. Schematic illustration of hyperbolic time-distance (t–x) graph over a horizontal reflector, and a definition of normal moveout (Δt_n)

t is the two-way time for slant path ABC; and t_0 is the two way time for vertical path BO.

Normal moveout is the difference between slant path reflection time t and vertical path reflection time t_0:

$$\Delta t_n = \sqrt{\frac{x^2}{V^2} + t_0^2} - t_0 \tag{3.15 b}$$

When we consider a layered medium having different velocities, the ray paths from source to reflection points and reflection points to surface geophones are not straight lines. The ray paths change direction at each boundary according to the velocity relationship of adjoining layers, in accordance with Snell's law. The Fermat's least time incidence and reflection paths for a single layer are straight lines, allowing use of Pythagoras theorem. However, this does not hold for a multilayer medium. A relation like Eq. (3.15 a) is strictly not a relation for a layered medium.

However, in routine practice, Eqs. (3.15 a) and (3.15 b) are adopted in their new forms by replacing V by V_{RMS}, where V_{RMS} is the square root of time average of the velocity squared.

For a medium in which velocity is a function of depth only, V_{RMS} is given by

$$V_{RMS}^2 = \frac{1}{T}\int_0^T V_i^2(t)dt \tag{3.16}$$

Integral is along a round trip path of a reflected seismic ray received by a geophone close to the source.

Assumption of straight line paths from source to reflection point to the receiver, and use of V_{RMS} in place of V allows Eqs. (3.15a) and (3.15b) to take the following forms for a layered medium:

$$t^2 = t_0^2 + \frac{x^2}{V_{RMS}^2} \tag{3.17 a}$$

$$\Delta t_n = \sqrt{\left(t_0^2 + \frac{x^2}{VRMS^2}\right)} - t_0 \tag{3.17 b}$$

Equations (3.17 a) and (3.17 b) are only an approximation for a multilayer medium and curved interfaces.

Equations (3.15 a) or (3.17 a) describe a hyperbola in the (t–x) plane.

Bazelaire (1988) gave an improved version of the *hyperbolic equation*, which holds well for a plane parallel multilayered medium or for curved interfaces. He developed this equation using principles of geometrical optics. It takes the following form:

$$(t + t_r)^2 = (t_0 + t_r)^2 + \frac{x^2}{V_1^2} \tag{3.18}$$

[Bazelaire 1988, Eq. (2)] where t_r is regarded as a delay, and V_1 is the velocity of the input medium.

Equation (3.18) describes a hyperbola whose center lies at a point, $(-t_r)$, on the time axis. The hyperbola given by Eq. (3.17a) has its center at the origin. The condition for two families of hyperbolas to be oscilating is: $V_1 = V_{RMS}$. This suggests use of Eq. (3.18) for NMO correction provided V_1 (input medium velocity) is taken in place of RMS velocity.

Equation (3.18) may be further explained with the help of a two plane parallel layer model and a convex circular reflector model considered in Bazelaire (1988).

Referring to Fig. 3.29a ABCDE is the Fermat's least time path – the *actual ray path*, and ABFDE – the *equivalent ray path* for a two parallel plane layer model. The assumption on replacing the actual path taking by an equivalent straight-line path implies taking a single velocity V_1 instead of distinct velocities V_1 and V_2. F is the image of an object at C on the optical axis as would be observed by an observer on the ground surface. The observer finds one medium with velocity V_1 and an object at F.

The key concept on which Eq. (3.18) is derived is that actual moveout time and equivalent moveout time are identical, i.e.,

$$T_{CE} - T_{CO} = T_{FE} - T_{FO} \tag{3.19a}$$

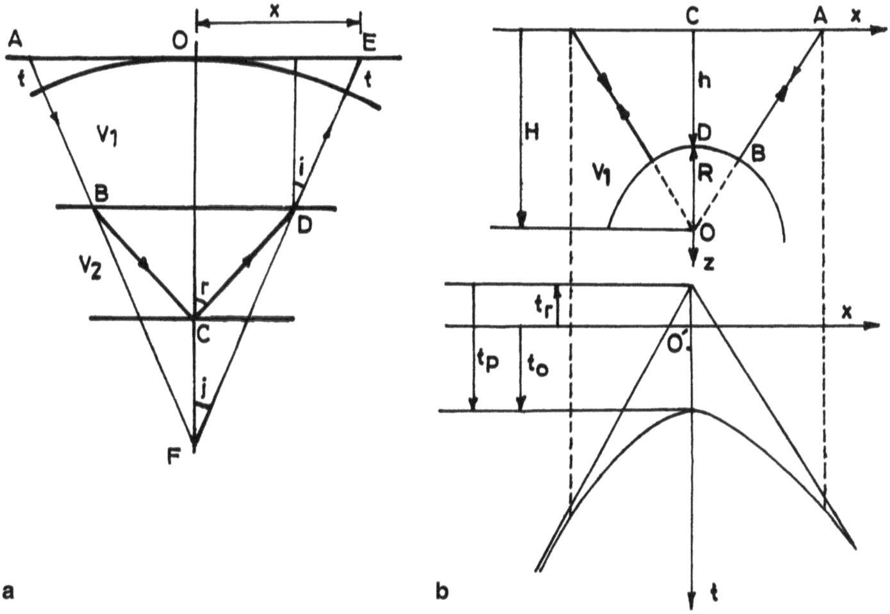

Fig. 3.29a, b. a Actual raypath ABCDE and equivalent raypath ABFDE for two media, plane and parallel layers with small aperture; **b** t_p (time of focusing depth) analysis generalized for pattern recognition after stack. Depth model (*top*) and its time equivalent (*bottom*). Raypath normal to a convex reflector of center O and radius R. h is the depth to the top of the structure. $2h = V_1 t_0$, $H = V_1 t_p$, $R = V_1 t_r$ (Source: Bazelaire 1988. From Geophysics with permission)

and

$$T_{FE} - T_{CE} = T_{FO} - T_{CO} \tag{3.19 b}$$

As equivalent ray path FE is a straight line and triangle FOE a right angle at O, we may write a relation like Eqs. (3.15) or (3.17 a) as,

$$T_{FE}^2 = T_{FO}^2 + \frac{x^2}{V_1^2} \tag{3.20}$$

which describe a hyperbola for the *equivalent ray path.*

Under the assumption of small aperture data and parallel plane layers Eq. (3.20) reduces to Eq. (3.18).

The assumption on replacing the actual path ABCDE by the equivalent straight-line path ABFDE implies a single velocity V_1 instead of two-distinct velocities V_1 and V_2.

Figure 3.29 b shows a model of a convex reflector, and a case of a coincident source-receiver position corresponding to a *stacked section.* We may write

$$CO^2 + CA^2 = OA^2$$

or

$$\left(t + \frac{2R}{V_1}\right)^2 = \left(t_0 + \frac{2R}{V_1}\right)^2 + \frac{x^2}{V_1^2} \tag{3.21 a}$$

where

$$t = \frac{2AB}{V_1}, \quad t_r = \frac{2R}{V_1}, \quad t_0 = \frac{2CD}{V_1}, \quad x = 2AC$$

Equation (3.21a) may be written as

$$(t + t_r)^2 = (t_0 + t_r)^2 + \frac{x^2}{V_1^2}$$

or

$$(t + t_p - t_0)^2 = t_p^2 + \frac{x^2}{V_1^2} \tag{3.21 b}$$

[Bazelaire 1988, Eq. (2)] where t_p is the time of focusing depth.

It is the same form hyperbola (with a delay) as for *before stack case.*

Bazelaire concludes that *normal moveout* based on *Fermat's principle* is more accurate than the one derived on the basis of Pythagoras theorem.

3.5.8.2
Dip Moveout

Referring to Fig. 3.30a, it may be proved that (for proof of Eq. (3.22) refer to Sheriff 1989, p 348):

$$\sin\alpha \approx \frac{1}{2}V\left(\frac{\Delta t_d}{\Delta x}\right) \tag{3.22}$$

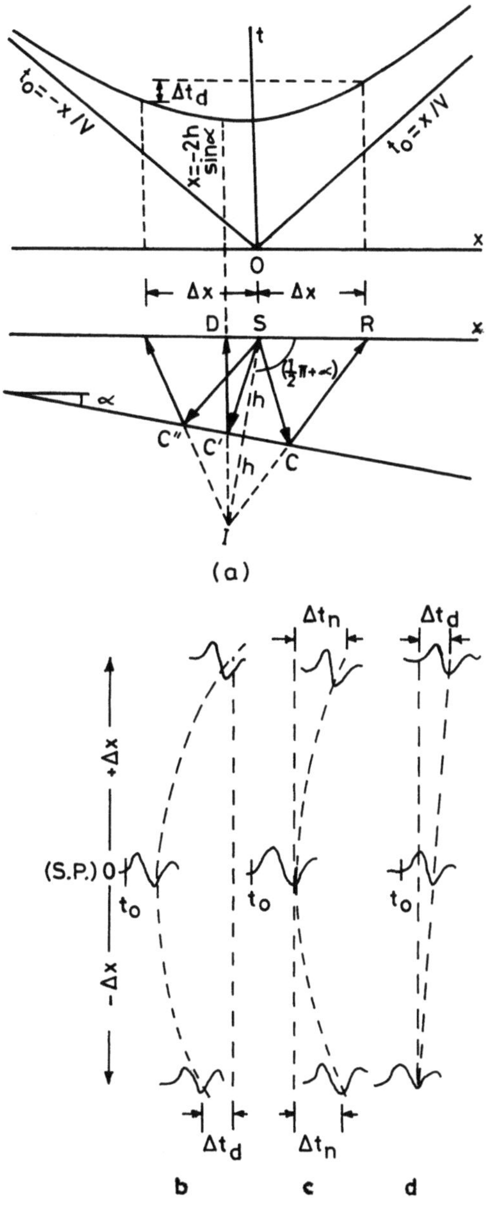

Fig. 3.30 a–d. Schematic diagram showing: **a** travel time curve over a dipping reflector and definition of dip moveout; **b** asymmetrical alignment of reflected events over a dipping reflector; **c** alignment of reflected events over a horizontal reflector; **d** alignment due to the effect of dip alone (after Sheriff, Geldart 1982, pp 81, 83. © Cambridge University Press 1982)

Here a is the angle of dip; Δt_d is the reflection time difference for two points at distance Δx on either side of the shot point.

The quantity $\Delta t_d/\Delta x$ is the *dip-moveout*. The dimensionality of *dip-moveout* is time/distance, and for *normal moveout* it is time.

Figure 3.30b–d demonstrates the effect of *dip moveout*. In Fig. 3.30b, the reflections from a dipping bed are shown. The alignment of reflected events is asymmetrical. Figure 3.30c shows what would be observed if the bed were horizontal. The alignment is curved and symmetrical due to *normal moveout* alone. Subtracting the normal moveout shown in Fig. 3.30c from the arrival times in Fig. 3.30b, the results shown in Fig. 3.30d would be obtained. Figure 3.30d shows the alignment due to the effect of dip alone. It is a straight line.

It may be mentioned that for the case of a horizontal reflector, if t_1, t_2, x_1 and x_2 are the travel times and offsets to the two geophones, we have

$$\Delta t = t_1 - t_2 \approx \frac{(x_2^2 - x_1^2)}{2V^2 t_0} \tag{3.23 a}$$

If one geophone is at the shot point, Δt is known as *normal moveout*, and is given by:

$$\Delta t_n \approx \frac{x^2}{2V^2 t_0} \tag{3.23 b}$$

NMO, (Δt_n) was depicted in Fig. 3.28 showing a reflection travel-time curve for a horizontal reflector.

In order to put the travel-time equations for the horizontal and dipping reflector in the same form (as Eq. 3.15a), consider the geometry shown in Fig. 3.31a. S and R are seismic source and receiver, respectively. P is the mid point between S and R. T is the time taken for the path SQR, Q is the point of reflection. X is the distance between source and receiver. S′ is the image of S in the reflecting surface. V is the velocity in the dipping layer. T_0 is the time for the return path PQ′ perpendicular to the dipping interface.

$$T = \frac{SQ + QR}{V} = \frac{S'R}{V} \tag{3.24 a}$$

$$T_0 = \frac{2PQ'}{V} \tag{3.24 b}$$

We define a time T_S from seismic source along a path normal to the reflecting surface as shown in Fig. 3.31a.

$$T_s = \frac{SS'}{V} = \frac{2SM}{V} \tag{3.24 c}$$

T_0 and T_S are related as

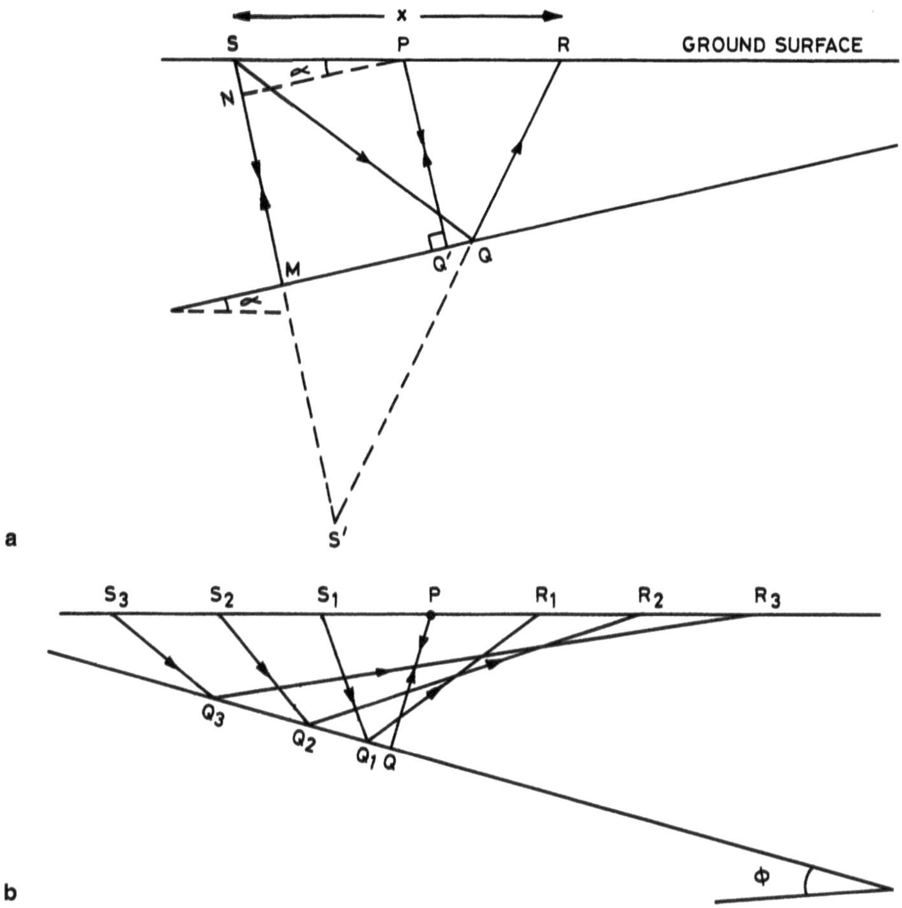

Fig. 3.31 a, b. Schematic diagram for: **a** time-distance relation over a dipping reflector in terms of normal incidence time at mid point P between source S and receiver R; **b** different sources (S_i)-receivers (R_i) reflection over a dipping reflector; the reflection points Q_1, Q_2, Q_3 are distributed on the reflector

$$T_0 = T_S - \frac{X\sin\alpha}{V} \tag{3.25 a}$$

for a receiver in an up-dip direction. For a receiver in the down-dip direction the corresponding relation is

$$T_0 = T_S + \frac{X\sin\alpha}{V} \tag{3.25 b}$$

Considering triangle SS'R,

$$S'R^2 = SS'^2 + SR^2 - 2SR \cdot SS'\sin\alpha \tag{3.26 a}$$

Dividing by V^2 throughout, Eq. (3.26 a) may be written as

$$T^2 = T_S^2 + \frac{X^2}{V^2} - \frac{2T_S \cdot X\sin\alpha}{V} \qquad (3.26\,b)$$

The corresponding result for the down-dip direction is

$$T^2 = T_S^2 + \frac{X^2}{V^2} + \frac{2T_S \cdot X\sin\alpha}{V} \qquad (3.26\,c)$$

T_s may be replaced by its value in terms of T_0 (Eq. 3.25 a). We have

$$T^2 = T_0^2 + \frac{X^2}{\left(\frac{V}{\cos\alpha}\right)^2} \qquad (3.27)$$

We note the following from Eq. (3.27):

If we plot T vs. X, it would be a hyperbola for multiple coverage for a set of sources and receivers placed symmetrically about the mid-point P.

An important point to note in this case is that for different sets (S_i, R_i) of source receiver positions symmetrically placed about mid point P, reflections do not come from one reflection point. They are distributed (Q_1, Q_2, Q_3) on the reflector. This is schematically illustrated in Fig. 3.31 b.

Symbolizing ΔT for $(T-T_0)$ we have for a dipping reflector with dip angle α:

$$(\Delta T)_{dip} = -T_0 + \left[T_0^2 + \frac{X^2}{\left(\frac{V}{\cos\alpha}\right)^2}\right]^{\frac{1}{2}} \qquad (3.28\,a)$$

For a horizontal reflector, it is

$$(\Delta T)_{horiz.} = -T_0 + \left[T_0^2 + \frac{X^2}{V^2}\right]^{\frac{1}{2}} \qquad (3.28\,b)$$

The meaning of T_0 in Eqs. (3.28 a) and (3.28 b) are different. In Eq. (3.28a), it is the two way time at a mid point along a path PQ′ perpendicular to the dipping reflector (Fig. 3.31 a). In Eq. (3.28 b) it is the vertical time below the mid-point for a horizontal reflector.

Therefore, when the *normal moveout* correction is applied for a dipping reflector assuming it to be a horizontal reflector, the reflected pulses from the dipping reflector would not be aligned and would not stack, as would be the case for a horizontal reflector. This is illustrated with the help of Fig. 3.32. Figure 3.32 a shows that seismic traces producing a common reflection point on a flat reflector would produce distinct reflection points for a dipping reflector for the same sets

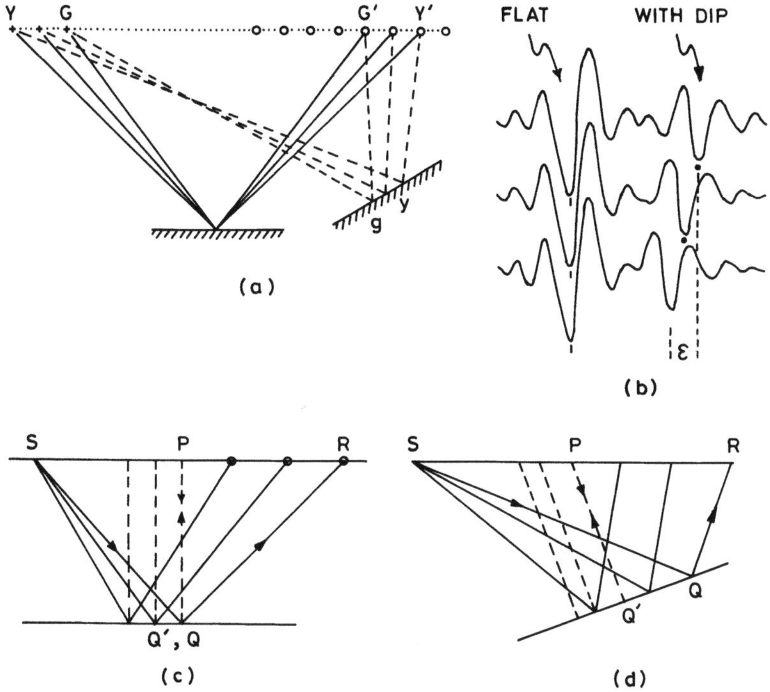

Fig. 3.32 a–d. Schematic illustrations of: **a** traces producing a common reflection point on a flat reflector, produce distinct reflection points on a dipping reflector; **b** seismic traces obtained from configuration in (**a**) and corrected for normal moveout and ready to be added to make a CDP stack; **c** and **d** NMO removal transforms the times over paths such as SQR to times for paths such as PQ′P even when the dip is present, to a very close approximation (Source: Cressman 1968. From Geophysics with permission)

of seismic sources and receivers. Figure 3.32b shows the seismic traces, as they would appear after applying a *normal moveout* correction assuming the reflector to be horizontal. The reflections for a horizontal reflector are properly corrected for *stacking*, but the reflection from a dipping reflector shows *remnant moveout* ε. This problem was investigated by Cressman (1968) who gave an expression for ε. The value of ε is given by

$$\varepsilon = \Delta T_{horiz} \sin^2 a \tag{3.29}$$

[Cressman 1968, Eq. (8)]

Thus, if the angle of dip 'a' is known, ε may be calculated. ε is the over correction introduced by assuming a horizontal bed. Thus by subtracting ε from NMO correction, the error due to dip may be largely removed. Thus, the *adjusted NMO* is

$$(\mathrm{NMO})_{adjusted} = (\Delta t - \varepsilon) = \Delta t \cos^2 a \tag{3.30}$$

Angle 'a' is the apparent dip in the direction of the seismic profile.

In conclusion, the NMO determined by ray tracing for horizontal layering must be multiplied by the square of the cosine of angle of dip. Any error in stacking may be compensated for by adjustment of NMO function.

The NMO corrected seismic traces may be treated as if they were obtained using shot and geophone side by side at the mid-point between actual positions of seismic sources and receivers. Referring to Fig. 3.32 c, d the NMO correction transforms the path such as SQR into the time for path PQ′P to a very close approximation even in the presence of dip. The effective reflection point (Q′) and actual reflection point (Q) are identical only for a horizontal reflecting boundary.

3.5.9 CMP Stacking

Stacking is a processing step to attenuate noise, and to increase signal-to-noise ratio. Particular types of stacking, like *vertical stacking*, *common-depth-point stacking*, or *up hole stacking*, may be implemented as is considered necessary. Vertical stacking is a procedure in which many records are combined for which both seismic source and detector locations remain the same. In this case, trace-to-trace corrections are normally not applied. Common-depth-point stacking involves stacking of records from common depth point and different source-receiver separations (offset). In this case, corrections have to be applied before stacking to effectively make the path of reflections common, i.e. of all reflections included for stacking.

In Section 3.5.6, CMP gather was defined as a collection of traces (for different sources and receivers) for which the mid point is common (m = constant). In Section 3.5.7, it was mentioned that, for CMP stacking, the CMP gather traces are corrected to make the positions of sources and receivers coincident, ray paths to and from the reflector common and normal incidence on the reflector. In the previous section, it was mentioned that NMO corrected traces may be treated as if they were obtained using shot- and geophones side by side at the mid points between the actual positions of the sources and receivers. Therefore, these traces may be stacked together to produce a stacked trace for one mid point. A collection of stacked traces for different mid points would be called a *stacked record section*.

The stacked record section, simulating the condition of coincident source-receiver position, is interpreted in terms of a hypothetical experiment – the *exploding reflector model.* The assumption of this model enables the use of the wave eqn. (Loewenthal et al. 1976) to interpret a stacked record section, which is really an outcome of several shots in the actual field experiment. A stacked record section is thought of as a result of a single experiment in which all points on the reflectors acted as the seismic source and got activated simultaneously at time t = 0 with seismic energy traveling perpendicular to reflectors in an upward direction and received by detectors on the ground surface. Thus, the several shot experimental data is interpreted in terms of a single hypothetical experiment – *the exploding reflector model* involving a single source and thereby enabling the use of the wave equation for forward and inverse modeling of surface recorded data.

We rewrite Eq. (3.27) in the commonly used form:

$$t^2 = t_0^2 + 4h^2 \frac{\cos^2\theta}{V^2}$$

where 'h' is the half-offset, 't' is the measured travel-time for a given offset, 't_0' is the time for presumed zero offset event at mid point; V is the constant velocity in the medium and θ is angle of dip (refer Fig. 3.31 a).

Equation (3.27) shows that the apparent velocity that best stacks the data from a dipping interface is given by medium velocity divided by cosine of angle of dip.

In general, to achieve the goal of imaging of the seismic data, the seismic data is first stacked and then imaged applying migration. It is assumed that the stacked section is equivalent to data that would be obtained using coincident source and receiver at the Earth's surface. When dip is present the NMO applied in common midpoint (CMP) stacking is not able to convert non-zero-offset data to true zero offset. Therefore, in this case, CMP stacking does not stack properly the dip-related events, and *post stack migration* is inaccurate.

One way to get out of this problem is the migration before stacking the data. This can circumvent fully the CMP stacking process, and would convert non-zero offset data directly to final image with no restriction on magnitude of the dips in the data, or offsets. *Migration before stack*, however, would be a costly process, obviously because it needs to migrate more data than in post stack migration.

To take care of the problem in stacking due to structural dip, a technique called *Dip Moveout (DMO) correction* has been devised which, in essence, converts non-zero-offset data to true zero offset after NMO has been applied using migration velocities. After application of DMO, dipping events stack coherently and can be imaged by use of post-stack migration algorithms. DMO is a smaller correction and therefore involves less computation than a full migration before stack.

Application of dip moveout corrects the prestack data in such a way that stacking velocity is no longer dependent on dip. This may be best illustrated by the results of a physical modeling experiment taken here from Forel and Gardner (1988).

Forel and Gardner (1988) considered a model as shown (Fig. 3.33 a). It consists of a horizontal interface and a ramp at an angle of 60°. Ten offset traces were obtained for each of 128 CMP gathers. The offset for the first trace was 304.8 m, and source-receiver pairs were set at increments of 304.8 m. The mid points were 12.1 m apart. Collections were made in the dip direction at a height of 1,219 m above the horizontal surface. The medium had a propagation velocity of 2,954 m/s.

The results of this model showed that, after correcting for *dip-moveout*, both horizontal and inclined reflectors may be stacked with the same velocity, i.e. after applying DMO, *stacking velocity* is independent of dip. The results are presented here in Fig. 3.33 b. This figure shows the effect of DMO on stacking velocity. Before application of DMO, the stacking velocity required for a horizontal reflector H is 2,950 m/s, and for ramp R it is 5,800 m/s. After DMO, both have a stacking velocity of 3,000 m/s.

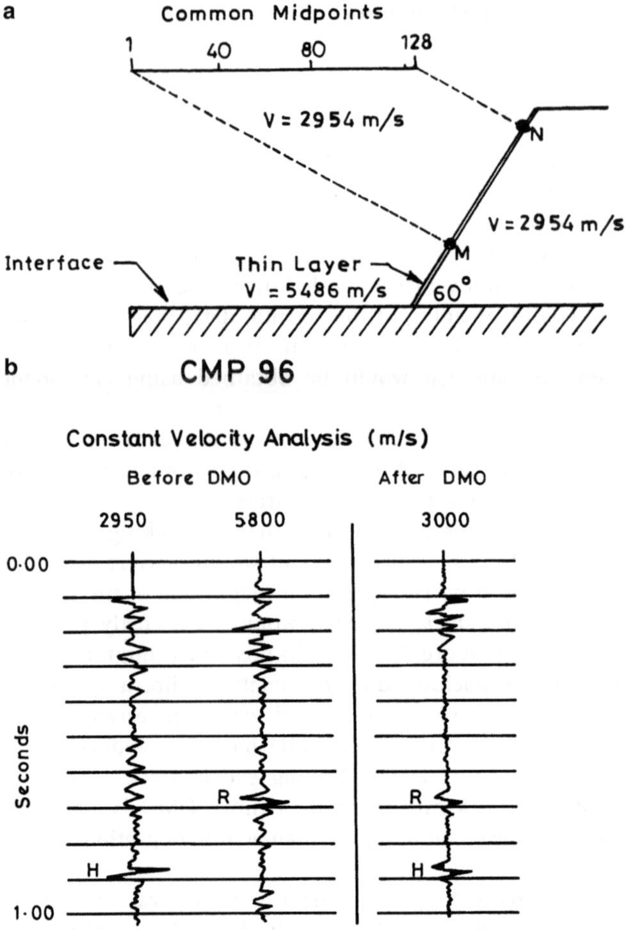

Fig. 3.33 a, b. **a** M and N are zero-offset reflection points on the ramp from midpoints 1 and 128, respectively; **b** the medium has a propagation velocity of 2,954 m/s. At CMP96 before DMO, the stacking velocity for the horizontal interface H is 2,950 m/s and ramp R has a stacking velocity of 5,800 m/s. After DMO both reflectors have a stacking velocity of 3,000 m/s (Source: Forel and Gardner 1988. From Geophysics with permission)

3.5.10 Deconvolution After Stacking

CMP stacking is not able to attenuate sufficiently well the effects of secondary reflections, like reverberations of the water layer. In order to reduce or eliminate the effects of reverberations, processing known as *inverse filtering* or *deconvolution* needs to be performed. By inverse filtering, the undesired filtering effect produced by the water layer may be cancelled.

Reverberations involve multiple reflections of waves, both at the source and receiver ends of the reflection path (Fig. 3.34). The multiple reflections (the

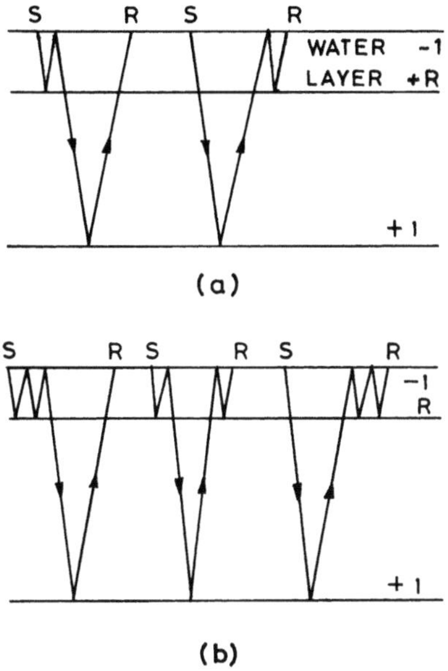

Fig. 3.34. Reverberations in the water layer: **a** Single bounce at source end (*left*), single bounce at receiver end (*right*); **b** possible paths for two bounces – two bounces at source end (*left*), two bounces at receiver end (*right*); and one bounce at source end and one bounce at receiver end (*middle*)

bounces) occur between the top and bottom of the water layer. Reverberations are mainly associated with marine shooting work. They are a type of noise. The reverberation effect in the record is also referred to as *ringing* or *singing*. The effect of reverberations is so strong on the record that no meaningful reflection analysis may be undertaken without effectively reducing or eliminating the effects of reverberations. On a *variable area record section*, reverberation effects appear as uniformly spaced stripes. After removal or suppression of the reverberation effect, the marine data quality is in general better than land seismic data. This is for the reason of uniformity of the medium surrounding the sources and receivers. Also, absence of an irregular and absorptive weathering layer, which is common in land seismic work, contributes to the better quality of marine seismic data.

Figure 3.35 a shows the frequency spectrum for reverberations of a pressure pulse in a water layer (Backus 1959). In this example, the fundamental frequency is 12 1/2 Hz, and odd harmonics are at 37 1/2, 62 1/2 and 87 1/2 Hz ... etc. The fundamental frequency is given by one fourth of the reciprocal of one-way travel time through the water layer, and higher harmonics frequencies are odd multiples of the fundamental frequency:

$$f_n = \frac{(2n-1)V_w}{4h_w} \tag{3.31}$$

Each peak in spectrum (corresponding to harmonics) may be related to repetitive event cycles in the reverberating signal. However, they cannot be resolved on the record because individual cycles are superimposed on each other.

Figure 3.35 b shows how the reflection spectrum would be modified due to reverberations in the water layer of different thickness. To get the output spectrum,

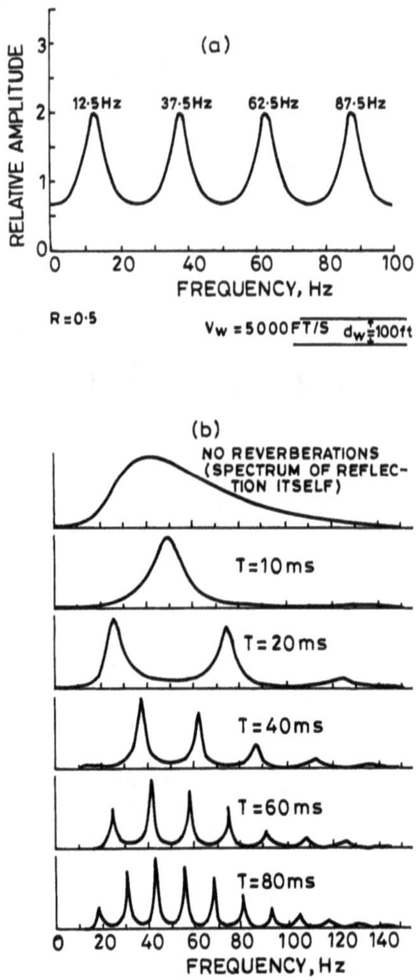

Fig. 3.35 a, b. a Frequency spectrum for reverberation of pressure pulse in a water layer 100 ft deep (after Backus 1959); **b** spectra of reverberating pulses in a water layer. T is the two-way thickness (in time) of the layer (from McDonal, Sengbush, Proc 7th World Petrol Cong Mexico, vol 2:591, 1967. In: Dobrin, Savit (1988, p 140–207), Introduction to Geophysical prospecting. © McGraw Hill Inc, New York)

the spectrum of the reflection reaching from below the water layer has to be multiplied with the spectrum of the reverberation system. The output spectrum for two way time T equal to 40 ms (two-way time-thickness of the water layer) shown in Fig. 3.35b is obtained by multiplying the spectrum of reflection itself (shown at the top) with the spectrum of the reverberations shown in Fig. 3.35a. It is obvious that reverberation associated with different water layer thicknesses would modify the output spectrum differently. This would have an effect on recorded *reflection pulse* form on the water surface.

In order to fix our ideas we assume that amplitude of a wave returning directly after reflection, without any bounce, to a hydrophone is unity. Let it's travel time be t. Reflection coefficients at the top and bottom of the water layer are assumed to be −1 (for upgoing paths of reverberations) and +R (for downgoing paths of reverberations). The round trip travel time (top to bottom and back) is taken to be $p\Delta$, where p is an integer. Picture of the reflected wave with one and two bounces and reflected by the same reflectors were shown in Fig. 3.34 a, b respectively. Referring to Fig. 3.34 a and b respectively, it is clear that two sets of path are possible for one bounce, either before or after the downward travel, and each arrive at the hydrophone at time $t+p\Delta$ with amplitude $-R$. Therefore, the total amplitude is $-2R$. Similarly, three reflections (refer to Fig. 3.34 b) involving two bounces from the top and bottom of the water layer arrive simultaneously at time $t+2p\Delta$, with each having an amplitude of R^2. Thus the total amplitude together is $3\ R^2$. Hence, the hydrophone at the surface would detect a series of successive signals in the sequence $1, -2R, 3R^2, -4R^3\ldots$ each at the interval of $p\Delta$. The *impulse response* of the water layer of various depths z $(=(1/2)\ V_w p\Delta)$ with water layer velocity V_w is (Sheriff and Geldart 1983, vol 2, p 33):

$$\begin{aligned} d_t &= \{1, -2R, 3R^2, -4R^3, 5R^4\ldots\} && (p=1) \\ &= \{1, 0, -2R, 0, 3R^2, 0, -4R^3, 0, 5R^4\ldots\} && (p=2) \\ &= \{1, 0, 0, -2R, 0, 0, 3R^2, 0, 0, -4R^3, 0, 0, 5R^4\ldots\} && (p=3) \end{aligned} \tag{3.32}$$

An *inverse filter*, i_t is one, which when convolved with the reverberatory output d_t would produce a spike δ_t i.e. suppress the reverberation effects and retain only the effect of primary reflection:

$$d_t * i_t = \delta_t \tag{3.33 a}$$

It may be shown that the inverse of the water layer filter is given by

$$\begin{aligned} i_t &= \{1, 2R, R^2\} && (p=1) \\ &= \{1, 0, 2R, 0, R^2\} && (p=2) \\ &= \{1, 0, 0, 2R, 0, 0, R^2\} && (p=3) \end{aligned} \tag{3.33 b}$$

There are only three non-zero terms in this filter, called the *Backus filter.*

3.5.11 Migration

The first stage in studying a reflection record is to identify reflections and read their arrival times on each seismic trace. Subsequent to this, a *cross-section* for all traces together is prepared. This is accomplished by marking shot locations on a horizontal line and plotting reflection events according to their arrival times vertically below corresponding shot points. *Cross-sections* are called *time-sections* if the vertical scale is linear with time. They are called *depth sections* if the vertical scale is linear with depth.

Further, a cross section is *unmigrated* when the reflections are plotted vertically below the shot points. For a *migrated section* it is assumed that the seismic line is normal to the strike, and dip moveout is representative of the true dip of the formation. Also, in a *migrated section*, the recorded reflection events are plotted in their actual location from where they originated and not vertically below the shot point as in an unmigrated section. An *unmigrated section* produces a distorted picture of the subsurface, and distortion increases with increase in the dip angle. For this reason, the unmigrated sections are used only in areas of gentle dip, or when an accurate structural picture is not a requirement. When the dip angle exceeds 5° or so, distortions in unmigrated sections are large enough for correct determination of the subsurface geometry.

A *record section* is a form of time section obtained by displaying successive records side by side. They represent a large amount of data in a compact form. Record sections are usually corrected for elevation, weathering and normal moveout (NMO).

In a *seismic section*, all the stacked traces of a CDP reflection profile are plotted side by side. A seismic section bears some resemblance to a *geological cross-section*. However, for a seismic section, time (instead of depth) is the vertical dimension. Even with this representation, positions of faults or folds are not difficult to interpret. However, a seismic section and its true geologic section may deviate substantially from each other under the influence of velocity relationships, diffraction arrivals, and multiple reflections. Out of these, spurious image produced by diffraction arrivals may be removed by a processing procedure called '*migration*'.

Common depth point (CDP) stacking of reflections does not take into account reflector dip. Each *CDP stacked trace* of a seismic section displays reflection in a position that corresponds to a vertical travel path. With this type of plotting, the reflectors are observable as aligned inclined events on seismic sections. To analyze the error or distortions that result because of treating the inclined reflectors as if they were horizontal, we take the case of a dipping reflector.

In Fig. 3.36 a, AB is a dipping reflector. C_1, C_2, C_3,... are coincident source-receiver positions on the ground surface. Ray paths normally incident on the reflector retrace the same paths after reflections. If the distances C_1R_1, C_2R_2,... etc. corresponding to travel times along these paths are plotted vertically below coincident source-receiver positions, a displaced image, A′B′, of reflector, AB, results.

In a similar way, if we plot the inclined path reflection times on a vertical line, the reflector would be imaged in a different position.

The distortions in the image of the reflector is a result of the standard way of preparing a seismic section, namely, plotting reflection times below shot point on a vertical line.

The dip distortion may be corrected by moving the reflection point from their positions on a vertical line, to the inclined line corresponding to actual travel path (Fig. 3.36 b).

The process of shifting reflections shown on vertical lines to the points that correctly image the reflector points is called '*migration*'.

As the reflection time is correct, the actual reflection point must be on a circular arc at the same distance from coincident source-receiver position and at an angle from vertical equal to the angle of dip. Therefore, having determined the dip angle, the true reflection point (s) may easily be determined. For theoretical details, reference may be made to Robinson and Coruh (1988, p 187–190).

Up to this point, the specific case of a dipping reflector and distortion in it's image has been considered. In nature, irregular and curved geological structures

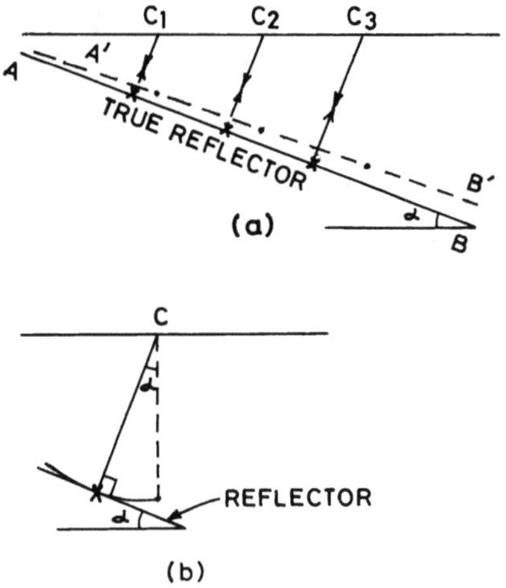

Fig. 3.36 a, b. a True reflector (AB) and its displaced image (A′B′). Cross (x) represents actual points R_1, R_2, R_3 on the reflector and the dot (.) their displaced image; **b** correcting dip distortion from the point (.) of the displaced image of the reflector to the true point on the reflector (after Robinson, Coruh 1988, pp 187–188)

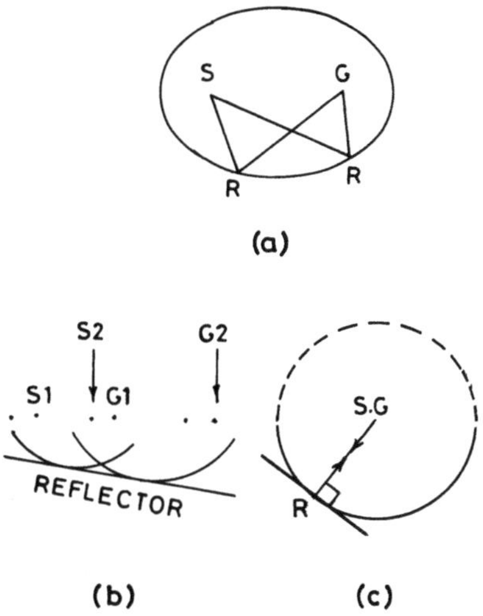

Fig. 3.37 a–c. Schematic diagram showing: **a** locus (*ellipse*) of reflecting surface for shot at S and geophone at G with travel time T_{SG}; **b** specification of reflector using two reflection times T_{S1G1} and T_{S2G2} to draw ellipses. Common tangent to ellipses approximates the reflector; **c** locus (*circle*) of reflections for coincident source (S) and receiver (G) (Fig. 3.37 a, b after Dobrin 1976, p 239. 3rd edn. © McGraw Hill Inc, New York)

are not uncommon. In the following, a more generalized concept of migration will be highlighted.

Consider a seismic source located at S and a geophone located at a point G (Fig. 3.37 a). Seismic waves originate at S, get reflected from a point on a surface below and arrive at G. Time taken for the reflected path is, say, T_{SG}. The question is to locate the position of the point, for example 'R', in space from which reflection took place and for which the travel time is T_{SG}. If the velocity of the wave in the medium is assumed to be V, we may write the following equation:

$$\frac{SR}{V} + \frac{RG}{V} = T_{SG} \tag{3.34 a}$$

where T_{SG} is the total time from source to geophone through the reflection point R.

From Eq. (3.34 a) we have

$$SR + RG = VT_{SG} \tag{3.34 b}$$

Given a fixed time T_{SG}, the locus of point R would describe an ellipse. This follows from the definition of the ellipse; namely, the sum of the distance from two fixed points is constant.

From Eq. (3.34b), it is clear that, from only one observation of the reflection time, it is not possible to locate the reflector. It would be possible to locate the reflector, from at least two sets of independent observations on reflection times. The reflector is tangent to the ellipses, which are loci of points corresponding to two reflection times (Fig. 3.37b).

Given that the source and receiver are coincident, the incident path and the reflected path, i.e. SR and RG in Fig. 3.37c, would be identical. In this case the locus of reflection points would lie on a circle.

If a reflected pulse occurs at a certain reflection time, theoretically it can originate from any point on a curve having a constant reflection time equal to the given reflection time. This concept is made use of in the *wavefront method* (Schneider 1971) of migration.

We have mentioned above that, for a coincident source-receiver position, a curve of constant reflection time would be a circle. For separated source-receiver positions, the curve of constant reflection times may be derived by drawing two sets of wavefronts from source and receiver positions, and joining such intersection points of wavefronts, at which times corresponding to two wavefronts add up to a constant value equal to the reflection time. The curve obtained by joining these points defines a curve, or a surface of equal reflection time. These follow from the fact that, irrespective of the dip of the reflector, reflection time would be the same whether A is the source and B is the receiver or A is the receiver and B is a source (refer to Hagedoorn 1954).

For coincident source-receiver positions, the ray path is normally incident to the reflecting surface in the case of *isotropy.* In this case, there is a definite relationship between two way reflection time and the reflecting surface. The reflecting surface is tangential to the equal time surface (wavefront) at the point corresponding to half of the two-way reflection time. Generalization of this idea re-

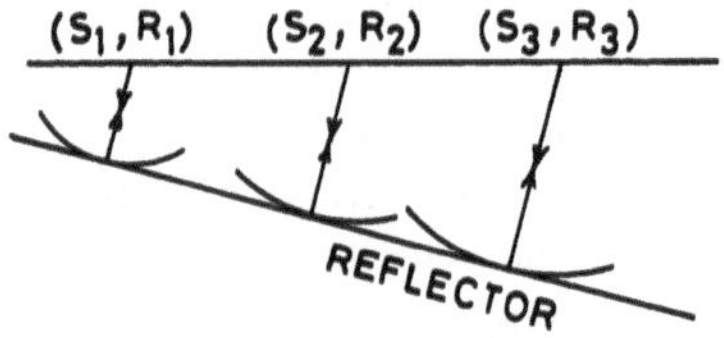

Fig. 3.38. Schematic showing (S_1, R_1), (S_2, R_2), (S_3, R_3) as coincident source-receiver positions. The reflector surface is tangential to wavefronts corresponding to half of the two-way times

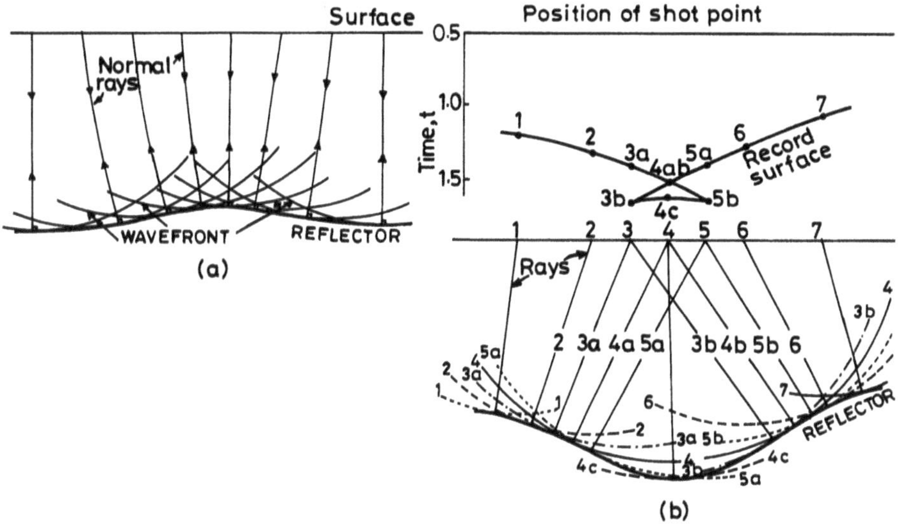

Fig. 3.39 a, b. a The reflector surface as the envelope of wavefronts drawn from shot points; **b** double reflections and cusps observed over a reflector more concave than the incident wave front (from Grant and West 1965, p 157. © 1965 by McGraw Hill Inc)

veals that the reflector would be the envelope of wavefronts corresponding to half of the two-way reflection times for different coincident source-receiver positions (Fig. 3.38).

For drawing wavefronts, we need to know two-way reflection times and the velocity function of the medium. In this case, *wavefronts* are the generating surface of the reflector. The position of the point of reflection is uniquely defined if the curvature of the reflector is smaller than the curvature of the wavefronts (Fig. 3.39a). If the curvature of the reflector is more, reflections may arise from more than one point that would be received by one and the same coincident source-receiver (Fig. 3.39b). In this case, an ambiguity arises about the actual point of reflection. Further, if the reflections were plotted vertically below the coincident source-receiver position it would result in two conflicting dip segments, and the presence of a cusp (upper part of Fig. 3.39b).

3.5.12 Wavelet Processing

Wavelet processing includes determining, assuming or operating on *effective wavelet* shape with different kinds of objectives (Sheriff and Geldart 1983, vol 2, p 44):

(i) making wavelet shape the same everywhere,
(ii) replacing the effective wavelet with a more desirable wavelet, or
(iii) separating reflectivity of the Earth from the seismic wavelet shape effect.

The aim in wavelet processing is to achieve better resolution of the *embedded wavelet.*

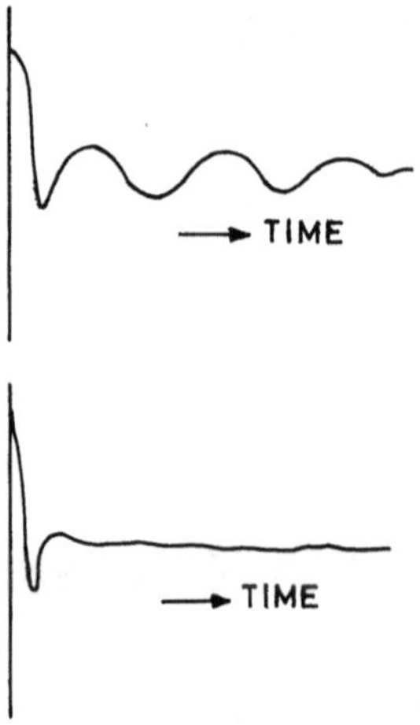

Fig. 3.40 a, b. Schematic diagram showing source wavelet forms before (*top*) and after (*bottom*) wavelet processing

Figure 3.40 shows effective source wavelet forms before and after *wavelet processing*. It shows the benefits of wavelet processing in the form of wavelet stabilization and increase in the frequency content (*wavelet compression*). *Wavelet processing* of the field records play a significant role in removing effects of variations in the *source waveshape*. Thus, it helps to emphasize wavelet shape variations produced by stratigraphic features. The effect of wavelet processing may be seen in terms of a broadening of *frequency contents*, sharpening of events and better resolution of *stratigraphic features*.

Type (i) *wavelet processing* is done on *pre-stack* records. The objective in this case is that all individual traces, which are to be stacked, bear the same effective wavelet shape. If this processing step is not implemented there is a likelihood that low-frequency components would be stacked better than high frequency components, and stacking would act as a filter attenuating higher frequencies. Retention of high frequency or short wavelength components of the reflected pulse would help smaller magnitude features to be focused better in the stacked record.

Wavelet processing begins by determining the *effective source wavelet shape*. It is considered to be a critical and difficult aspect of wavelet processing. Sometimes, the record of source wavelet is actually obtained near every seismic energy source in the marine seismic surveys, and is used for *deterministic wavelet processing*. More commonly, however, a statistical method is adopted. In this method, the *source wavelet* is determined from the auto-correlation function obtained by summing the auto-correlation of all traces obtained from the same source. The assumption is that the only common element for all traces taken is the source wavelet and therefore the auto-correlation sum represents the auto-correlation of the *source wavelet*.

Type 2 *wavelet processing* aims at correcting for filtering actions (particularly by phase-shifts) associated with different instrument response characteristics. The end product of this type of processing is to present data recorded with different types of detectors or recording instruments in a similar fashion or with better match to each other.

The process of changing *effective wave shape* to some arbitrary desired wave shape is accomplished by *Wiener filtering*. Usually the desired wave shape is taken to be *zero-phase*. It approximates a symmetric wavelet with energy concentrated in a center lobe. The most common *zero-phase wavelet* is the Ricker wavelet (Ricker 1940, 1953, 1977) (refer to Fig. 16.11c – top). In Wiener filtering, suitable filter coefficients are determined which when operated on an *effective wave shape* produces an output very similar to desired wave shape.

Type 3 *wavelet processing* aims at separating *reflectivity* of the Earth from seismic wavelet shape effects. The reflection seismogram output is considered a product of successive convolutions with different parts of the Earth through which the waves are propagated:

Reflection seismogram ≡ Pulse generated by a seismic source * impulse response of the zone near the shot (downward path) * impulse response of the reflecting sequence * impulse response of the zone near geophone (upward path), where * denotes convolved with.

We may consider an *'equivalent wavelet'* as one which combines all convolution effects together except the impulse response of the reflecting sequence. In short notation, we may write

$$d_t = w_t * e_t \tag{3.35}$$

where d_t is the reflection seismogram; e_t is the impulse response of reflecting sequence, and w_t is the *equivalent wavelet*. Type 3 wavelet processing aims at separating e_t (the impulse response of the reflecting sequence) from w_t (the equivalent wavelet obtained by combining together various convolution effects).

3.5.13 Displays of Seismic Sections

Magnetic tape data must be presented in a visual form for inspection, monitoring and interpretation. A *trace* in seismic recording is a graph of the motion sensed by a geophone (or a group of geophones) during the time of recording. A collection of traces makes a *seismic section*. Simple recording of amplitude of the motion against arrival time is called *wiggly trace mode recording* (Fig. 3.41a). *Variable area* is a presentation of the wiggly-trace where part of the area under a wiggly trace curve is blackened (Fig. 3.41b); sometimes the half of the wiggly cycle is blackened and sometimes the portion between the trace and some reference value called *bias* is blackened. Figure 3.41c is an example of variable area traces after filtering for attenuation of surface waves. Sometimes *variable density mode* is used in which light intensity is varied across the record in accordance with the signal variations (for example intensity may be high for signal peaks and low for signal troughs). This would make peaks darker and troughs lighter in the photographic record of the seismic data.

Display of three-dimensional data – amplitudes as a function of north-south and east-west coordinates and arrival times – involves special display problems because such a huge amount of data is involved. The amplitude display in this

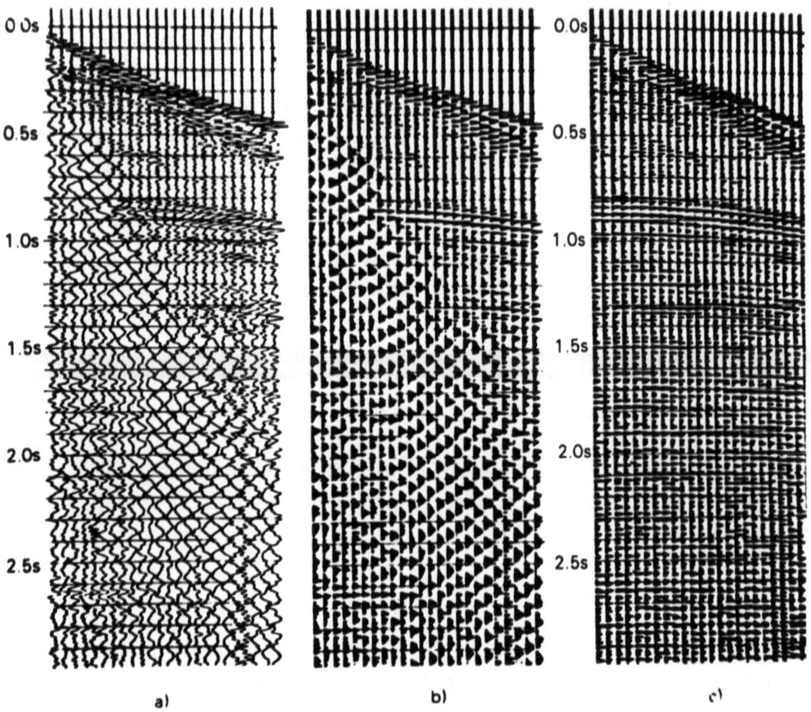

Fig. 3.41 a–c. Two-dimensional (x, t) seismic records: **a** Wiggle traces; **b** variable-area traces; **c** variable-area traces after filtering for attenuation of surface waves (Source: Anstey 1970: Signal characteristics and instrument specifications vol 1, p 86, Gebrüder Borntraeger, Berlin. In: Lavergne 1989: Seismic methods. Reproduced by permission. © 1989 Editions Technip, Paris)

case occupies volume (Fig. 3.42 a). In many cases, the data display is done on a series of seismic sections for simpler analysis. Figure 3.42 a shows the data on NS and EW sections. Sections in arbitrary directions may also be displayed. Data for a whole area is displayed as a *time slice map* (Fig. 3.42 b). Time is the vertical scale in this display, and an amplitude display on *time slices* at various times, at an interval of a few milliseconds, is depicted (Fig. 3.42 c). From the time slice map, *time contour maps* may be prepared for analysis and interpretation. The use and application of three-dimensional seismic data is currently the most fascinating field of seismic interpretations.

3.6 Information Content of Seismic Sections

A seismic section is the product of a seismic reflection survey. It has the appearance as shown in Fig. 3.43. The horizontal axis corresponds to the profile line and the vertical axis is the two-way vertical reflection time. An individual seismo-

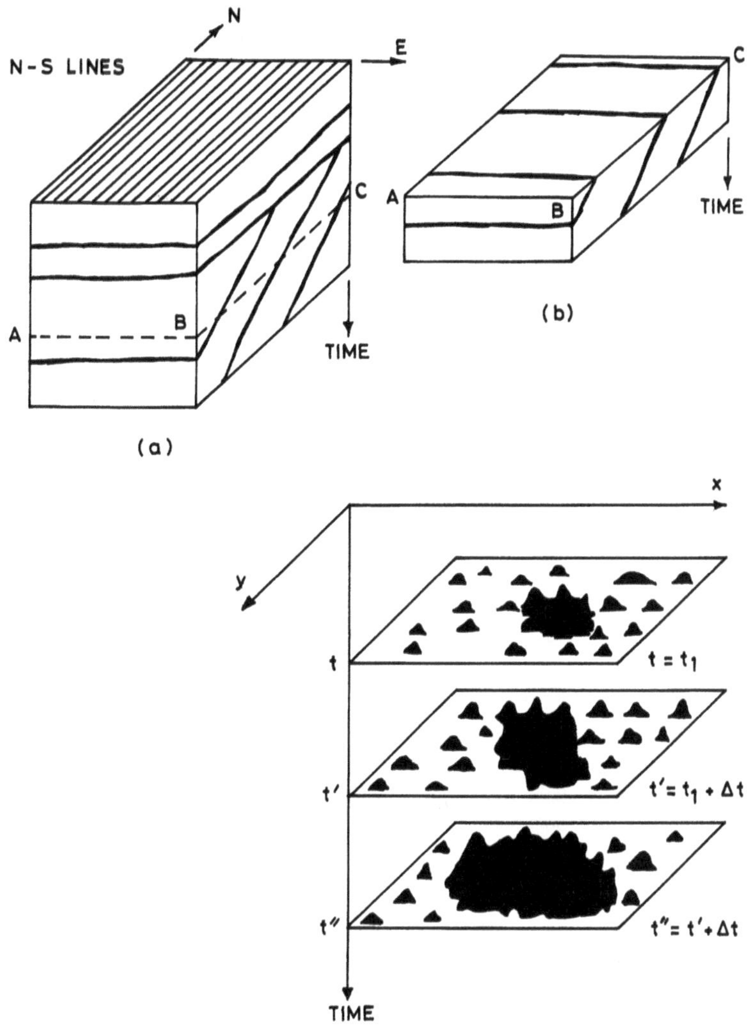

Fig. 3.42 a–c. Schematic diagram to illustrate: **a** three-dimensional data obtained from a set of closely spaced N-S seismic lines. Sectional displays of data along NS and EW sections are also shown; **b** data set as would be observed with the upper part removed. The display on the top of the remaining portion is called the time slice (seiscrop) map; **c** schematic diagram showing times slices at times t_1, $t_1+\Delta t$, and $t_1+2\Delta t$

gram trace corresponds to an appropriate selection of source-receiver position. Although Fig. 3.43 gives an appearance of a geological cross-section bearing in it thrust faults and folds, it is truly not a geological cross-section. To convert this seismic section to a correct geological cross-section, interval velocities and layer thickness have to be determined from arrival times of reflected waves.

Seismic sections displaying variations of reflection arrival times along a profile are referred to as *time sections*. It is possible to prepare seismic sections that dis-

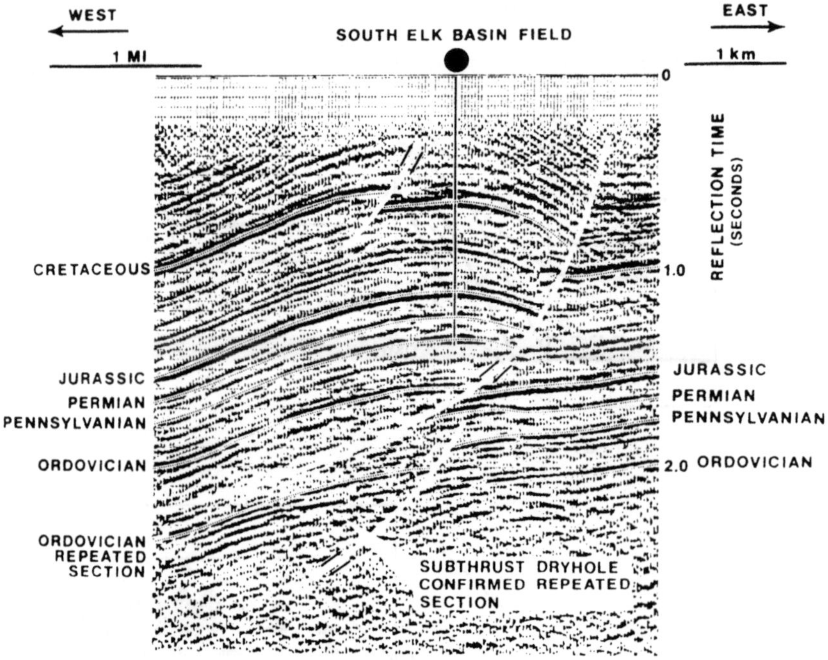

Fig. 3.43. A seismic section that indicates folded and thrust-faulted structures. This is the conventional representation of a seismic reflection data after lengthy processing steps. The horizontal axis at the top corresponds to the profile, and the vertical axis is the two-way vertical reflection time. The data are obtained by the powerful method called stacking. This type of data may be viewed as a subsurface cross section in reflection time (Conoco Inc) (in Robinson and Coruh 1988, p 83. © 1988, by John Wiley and Sons Inc, New York)

play reflector depths. At each mid-point of a CDP reflection profile, RMS velocities are determined and used in the Dix equation [refer to Eq. (4.24)] to determine interval velocities. These values are used for the calculation of layer thickness and reflector depths below the common mid-points. In the subsequent step, reflected wavelets are shifted to appropriate reflector depths to generate depth traces. A computer controlled plotting machine is programmed to generate depth traces. A seismic section prepared from depth traces is called a *depth section.* In the depth section the wavelets image the reflector positions. In this type of section, the vertical scale is depth. It may be observed that the position of reflections on time sections depends both on velocity differences as well as the thickness of the layers. However, in a depth section wavelet positions are governed by layer thickness only. Precise velocity estimation is the key factor for correctness of the *depth sections.*

Most interpretations begin by a study of *time sections.* However, if good estimates of velocity are available, the interpretation of depth sections provides definite advantages for determination of structural and stratigraphic details.

3.6.1
Geological Structural Details

Sedimentary structures that are suitable as *hydrocarbon traps* were mentioned in Section 1.1 (refer to Fig. 1.4). A seismic section over these structures may sometimes be clear enough to interpret unambiguously the presence of a suitable structure underneath. In general, discontinuities in reflections represent faults, and undulating reflections represent folded beds. However, in certain cases an unambiguous interpretation becomes difficult. In such cases, one interpretation may be preferred which gets support from other known geological or geophysical information in the area. Preparation of *synthetic seismic sections* may prove helpful in interpreting complicated seismic sections.

It is possible to prepare structural contour maps, if the data from several profiles crossing an area are available. Through such a presentation, valuable three-dimensional fault-patterns may be observed.

Sometimes it becomes necessary to use higher frequencies for *increased resolutions.* A number of educative examples with explanation of records, obtained by using high frequencies, may be found in Waters (1987, pp 509–516).

3.6.2
Geological Stratigraphic Details: Concept of Seismic Stratigraphy

Mention of stratigraphic traps was made in Section 1.1 (refer to Fig. 1.5). Not only structural situations create hydrocarbon-trapping conditions but stratigraphic changes may also give rise to such favorable situations. Location of stratigraphic traps by seismic method is, however, more difficult. *Seismic stratigraphy* aims at determining the nature of rocks and their fluid contents – the additional aspects of sedimentary rocks beyond their structural aspects. Further, *seismic stratigraphy* is based on the premise that reflections are able to depict unconformities and/or the attitude of *depositional time surfaces. Unconformities* reveal gaps in the sequence of rock deposits and are often among the best of reflectors. They also show some angularity between the bedding below and above. Stratigraphic traps often associated with unconformities, with unconformities acting as a seal, involve pinchout of sediments, and stratigraphic variations of sediments laid down on unconformity. *Seismic sequences* may be defined by their bounding unconformities. Different seismic sequences are separated on the basis of angularities among reflections. Specific patterns of reflections, like a continuity of reflections, are particularly helpful in *stratigraphic correlation* over a large area.

Appendix 3A:
Problems of Detection of Thin Bed, Horizontal and Vertical Resolution

Detection deals with the possibility of obtaining reflections from thin beds. The minimum thickness of a layer to produce a visible reflection in the reflection record is called the *detection limit.* It depends on the signal-to-noise ratio, wavelength and velocity contrast. Under the most favorable conditions, the detection limit may be in the range of $\lambda/30$ where λ is wavelength in the layer. In practice, however, the detection limit is around $\lambda/4$.

Vertical resolution deals with the possibility of separating two adjacent reflecting surfaces at depth from the reflection record. How the vertical distance between two surfaces compares with the wavelength goes to control the vertical resolution. The *resolvable limit* is the minimum thickness of the bed that permits the effects of the top and bottom of the bed to be seen separately on the reflection record. It is about 1/4 of the wavelength. It is expected that the use of small wavelengths or high frequencies should increase the resolving limit. However, normally we are not able to record high frequencies. Reason for this is that combining outputs of several geophones or of several sources produces a filtering effect on high frequencies. Figure 3A.1 explains this point. It reveals that the combination of small time-shifts has little effect on low frequencies. However, the combination results in out-of-phase addition producing a filtering effect on high frequencies. This result is expected as these minor time shifts are an appreciable fraction of high frequency cycles, which results in cancellation. The use of a single source and single geophone improves the *high frequency response* and resolving limit, but at the cost of less energy, lower sensitivity and a decreased ability to discriminate against noise.

It is observed that high frequency components are much better recorded in the well log record than on a seismic trace.

The other reason for loss of high frequencies in the seismic records is *attenuation loss* produced by *selective absorption* of high frequencies by the Earth material on the downward and upward path of the reflected wave (refer, for example, to Anstey 1970, p 51). Reflected pulses from the bottom and top of thin layers which may be separated from each other at the level of the reflector may

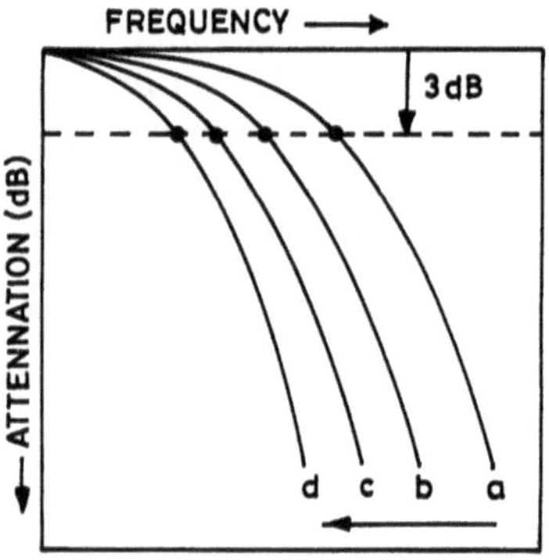

Fig. 3A.1. Schematic diagram to illustrate the filtering effect as a result of timing errors in stacking. Curves a, b, c, d denote increasing rms averages (of the order of a few milliseconds) of the time differences among the traces being stacked, assuming random time shifts. At the 3-dB level, a lower value of rms average of time differences produces a higher frequency filter. For example, 1 ms is equivalent to a 125 Hz filter, 2 ms to a 62 Hz filter and so on (after Sheriff 1989, p 331)

look merged on the ground surface because of the loss of high frequencies and consequent *pulse broadening* of the two pulses.

An example of the spectral response effect of attenuation of Earth material may be found in Anstey (1970, p 57).

It is obvious that if the wavelength is small compared to the thickness of the layer, and the seismic pulse has a small number of oscillations, the reflection record should permit easy separation of reflections from the top and bottom of a thin layer. If these conditions are not fulfilled it would be problematic to separate reflections from the two surfaces of thin layers.

The other aspect of resolving power is to understand how large a structure must be in the horizontal extent to be observable on a seismic record. This aspect is explained by the concept of the *Fresnel zone.*

The Fresnel zone concept is based on principles of physical optics in which image formation is considered by illumination of the entire reflecting surface by

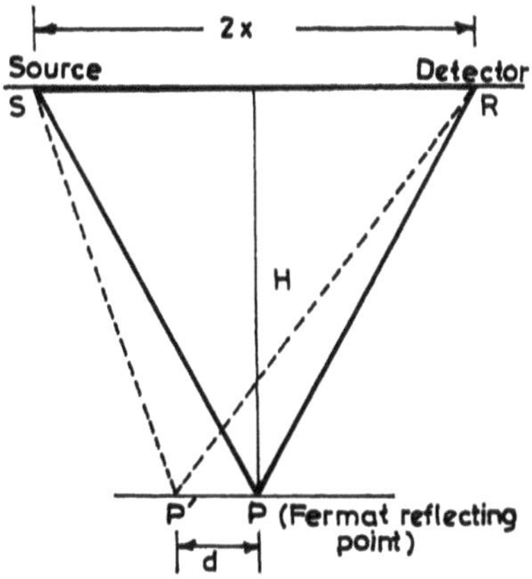

Fig. 3A.2. Schematic diagram for determination of the Fresnel zone for a reflection from a plane boundary (Source: after Waters 1987: Reflection Seismology, 3rd edn, p 18. © 1987 John Wiley and Sons Inc)

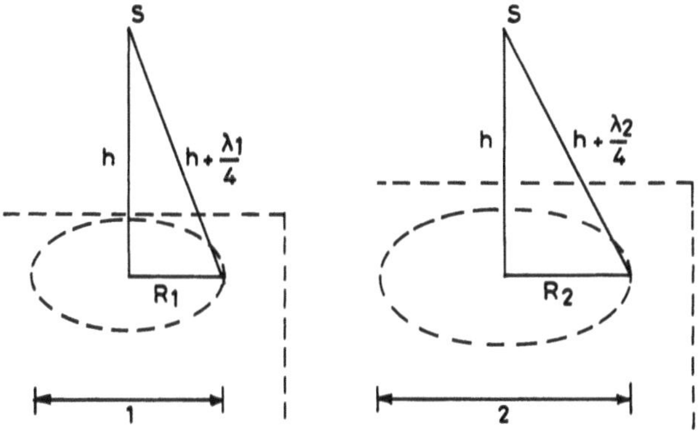

Fig. 3A.3. Schematic diagram showing Fresnel zones on the reflector for high and low frequencies. 1 is the diameter of the circular area on the reflector for a high frequency; 2 is for a lower frequency (after Sheriff 1978, p 124)

the source, and by summation of the scattered waves (refer to Fig. 1.1b) produced by all parts of the reflecting surface. Only those portions of the reflecting surface that produces waves substantially in-phase at the receiver make a substantial contribution to the image. The out-of-phase waves cancel each other.

Now, consider a situation for reflection from a plane boundary (Fig. 3A.2). The source-receiver separation is 2x. P is the reflection point for the Fermat path. Consider that P′ is the intersection of another ray with the horizontal reflecting surface. P′R is the scattered ray path from new point P′ on illuminated surface to the detector.

The path difference between SPR and SP′R may be shown to be equal to:

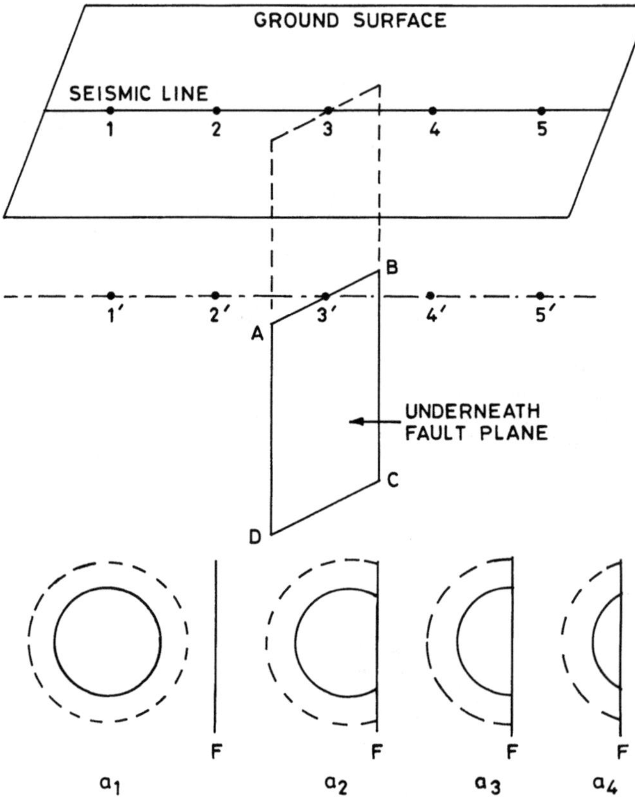

Fig. 3A.4a$_1$–a$_4$. (*Upper*) Geometry of fault and seismic line on the ground surface. 1, 2, 3, 4 depict source-detector locations on a seismic line. 1′, 2′, 3′, 4′ are points vertically below 1, 2, 3, 4; 3′ is over the fault plane ABCD. (*Lower*) Fresnel zone conditions for high and low frequency as reflection point approaches the fault edge F. **a$_1$** is when the reflection point is away, **a$_2$** near, **a$_3$** over, and **a$_4$** away from the fault edge F. This change of Fresnel zone contributions results in changes of waveshape as the reflection point approaches a fault plane, or moves away from it (Source: after Sheriff 1978, p 127)

$$\delta l = \left[H^2 + (x+d)^2\right]^{\frac{1}{2}} + \left[H^2 + (x-d)^2\right]^{\frac{1}{2}} - 2\left[H^2 + x^2\right]^{\frac{1}{2}}$$

Or letting $x = aH$, and $d = bH$, we have (3A.1)

$$\frac{\delta l}{H} = \left[1 + (a+b)^2\right]^{\frac{1}{2}} + \left[1 + (a-b)^2\right]^{\frac{1}{2}} - 2\left[1 + a^2\right]^{\frac{1}{2}}$$

If d is small such that the path difference δl does not exceed one quarter of the wavelength of any individual single frequency of the incidence pulse from S, the waves would interface constructively. Choice of d has to satisfy this criterion for a positive contribution to wavefield amplitude at R. In fact, all points on the reflecting surface from both sides of Fermat's reflection point that satisfy this criteria would contribute to the reflected pulse amplitude at R. It may easily be verified that lower frequency waves would come from a large area, called the *Fresnel zone* on the reflector. This is depicted by considering Fig. 3A.3. The concept of the Fresnel zone disposes of the idea that seismic reflections come from a point or even a small area of subsurface reflector. The above simplified description is given for two dimensions. Rigorous treatment would require considering contributions to scattered waves arriving at R from all points in three-dimensional space.

The concept of the contribution from the Fresnel zone leads to an understanding of the change of wave shape as the reflection point approaches a fault (Fig. 3A.4a_1–a_4). As the reflection point approaches the fault, the Fresnel zone for high and low frequencies change leading to a variable contribution received by the geophone and thereby a change in the wave shape is produced.

CHAPTER 4

Attributes of Routine Reflection Processing and Pitfalls

4.1 Introduction

Improved reflection data quality and processing procedures have enabled significant new developments in geological interpretations. In this respect, not only precise structural interpretations are possible but valuable conclusions may be deduced about lithology, fluid content, porosity, stratification, depositional processes and environments, sediment transport direction and geological evolution, etc.

Table 4.1 (Sangree and Widmier 1979) gives a summary of seismic facies parameters and their geologic interpretations.

The general principle of parallelism of bedding and reflections holds. However, there are certain exceptions to this principle, for example, from non-stratigraphic, coherent seismic events, such as diffraction patterns, or sideswipe reflections originating from outside the seismic profile line (Tucker and Yorston 1973). Exceptions also arise from fluid-contacts as, for example, a gas-water contact in a thick sand. Characteristic of source signal may also produce an exception. Figure 4.1 a

Table 4.1. Geologic interpretation of seismic parameters (Source: Sangree and Widmier 1979, p 135)

Seismic facies parameters	Geologic interpretation
Reflection configuration	Stratification patterns Depositional processes Erosion and paleotopography
Reflection continuity	Bedding continuity Depositional processes
Reflection amplitude	Velocity-density contrast Bed spacing Fluid content
Reflection frequency	Bed spacing Fluid content
External form and areal association	Gross depositional environment Sediment source Geologic setting

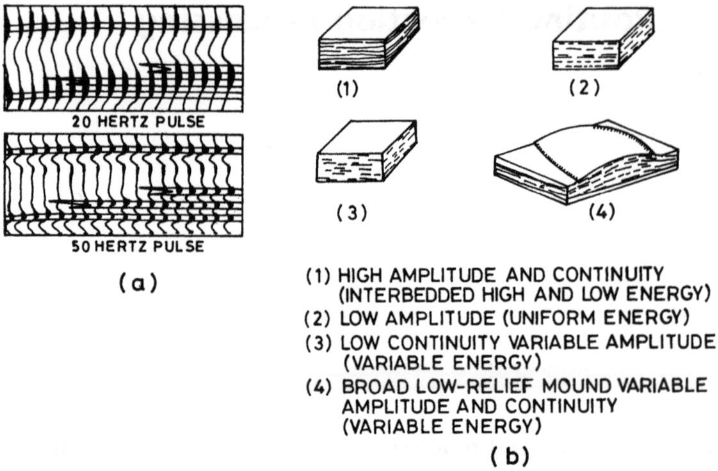

Fig. 4.1 a, b. a Parallelism of beds and reflections. Four sand beds are resolved by a 50 Hz pulse, but not by a 20 Hz pulse. On the synthetic seismic section with a 20 Hz pulse, the lower high amplitude reflection moves up and across the lower sand to the second and third sands; **b** shelf seismic facies types (Source: Sangree and Widmier 1979)

shows a computer model with several sand bodies interbedded in massive shales (Sangree and Widmier 1979). The synthetic seismic section is superimposed for a pulse of 20 and 50 Hz. It is observed that with a 50 Hz high-resolution pulse, it is possible to track continuity of the pulse with upper and lower sands. Also the lateral facies change of the two middle sand beds is predictable by a loss of amplitude in the shale. For the same model, with a low-resolution 20 Hz pulse, the reflected pulses from the lower three sands interfere to produce a reflection waveform which drifts across the depositional units, and violates the principle of parallelism. In general, *bed resolution* is a function of *reflection frequency.*

After a description of sedimentary facies becomes possible, an interpretation of sedimentary processes and environmental facies enables us to predict the sand potential of the facies. The facies would be called *sand prone* if the environmental facies have sufficiently high energy to transport and deposit significant quantities of sand. If the depositional energy is low and insufficient to form significant sand accumulations, then the facies are referred to as *shale prone.* However, it is not true that the sand prone facies would always have sand deposits; in fact, it may consist entirely of shale and silts as no source of sand was available for transport to the depositional area. Thus, the sand prone criterion is insufficient for sand prediction, and other techniques like calibration of interval velocity have to be used to verify the actual presence of bedded sand. Examples of reflection characteristics of high energy, low energy, uniform energy and variable energy type deposits are depicted in Fig. 4.1 b.

4.2 Seismic Attributes of Reflection Data

By *seismic attributes* of reflection data we mean the various characteristic qualities of the data: like amplitude, phase, frequency, polarity, and velocity etc. These attributes help locate hydrocarbon-bearing beds and are useful as hydrocarbon indicators. We discuss below for interpretation purposes the physical and geological controls of these attributes.

Reflection co-efficient (R.C.) for normal incidence (or for angles of incidence on an interface close to zero) is given by

$$\begin{aligned} \text{R.C.} &= \frac{\text{amplitude of the reflected wave}}{\text{amplitude of the incident wave}} \\ &= \frac{\rho_2 V_2 - \rho_1 V_1}{\rho_2 V_2 + \rho_1 V_1} \end{aligned} \tag{4.1 a}$$

where ρ_1, V_1 are the density and velocity respectively of the medium of the incident ray, and ρ_2, V_2 are the density and velocity of the other medium in contact with the incident ray medium. The value of the reflection coefficient would, therefore, change its sign depending upon whether the incidence is considered from above or from below the reflecting boundary.

Product of density and velocity is termed as *acoustic impedance*. Thus, there is an acoustic impedance ρV_p for compressional waves, and acoustic impedance ρV_s for shear waves. It is possible to prove that acoustic impedance in these forms is equal to the ratio of stress to particle velocity for plane waves in a homogeneous medium. In the case of compressional waves and shear waves, the stress would be dilatational stress and distortional stress, respectively.

Equation (4.1 a) shows how characteristics of the rock materials above and below an interface influence the amplitude of the reflected wave for normal incidence, which is usually assumed in reflection processing. It also shows that an interface associated with large acoustic impedance contrast would reflect strongly. The contrast may be positive or negative. If the contrast is positive, an incident *compressional disturbance* would be reflected as compression. If the contrast is negative, an incident compressional disturbance is reflected as rarefaction.

Geologically, it is very rare that rock materials with very large contrasting properties are in contact with each other. This sets an overall limit on the reflection coefficient of between zero and ± 0.1 for real geological interfaces as actually encountered in the Earth. For interfaces at large depths, where the velocities tend to be very high thus making the denominator in Eq. (4.17 a) large, very strong reflections are unlikely. Common field observations confirm the view that *strong reflectors* tend to occur at relatively shallower depths. It may also be observed from Eq. (4.1 a) that the range of rock velocities encountered in the Earth is much greater than the range of densities. On this basis, sometimes an approximation is taken by saying that reflection coefficient could be defined as the difference of velocities divided by the sum of velocities around a geological interface, by ignoring the too small variation of density. However, it is to be remembered that materials like water, coal and salt have very different densities compared to other Earth materials, and for interfaces involving these materials density cannot be ignored.

The reflection process does not involve any energy loss. Therefore, the transmitted energy should equal incident energy minus the reflected energy, and the transmission coefficient at normal incidence may be found from the expression for the reflection coefficient. As our concern is to measure reflected wave amplitude on the ground surface, we consider the two-way transmission coefficient of each interface between the reflector and the ground surface. The two-way transmission coefficient corresponds to incident downward and reflected upward paths through an interface for any reflection. If we assume that the incidence wave amplitude is one, the amplitude recorded on the ground surface after reflection is the multiplication of the reflection coefficient with two-way transmission coefficients of each interface between the surface and the reflector. The *two-way transmission coefficient* of an interface is given by

$$(\mathrm{T.C.})_{2\,\mathrm{way}} = \frac{4\rho_1\rho_2 V_1 V_2}{(\rho_2 V_2 + \rho_1 V_1)^2} \tag{4.1 b}$$

As for reflection coefficient, it is possible to assign values for the two-way transmission coefficient (T.C.) for real geologic interfaces. The two-way transmission coefficient lies between 1 and 0.99. However, it is possible to come across interfaces with greater contrast–with reflection coefficients as high as ±0.3 and a two-way transmission coefficient as low as +0.9.

We now discuss two of the seismic attributes, reflection coefficient and velocity, which are affected by *water saturation* in the porous sand bed. The model considered is an oil or gas sand formation which has an interface with overlying shale.

The effect on reflectivity and velocity of water saturation in an oil or gas sand reservoir was investigated by Domenico (1974). The model considered for study is an oil or gas sand bed which has an interface with overlying shale at depths of 2,000, 6,000 and 10,000 feet respectively. Sand reservoir characteristics taken for

Table 4.2. Sand reservoir characteristics (Source: Domenico 1974, p 761)

	Depth (ft)		
	2,000	6,000	10,000
Porosity (ϕ)	0.39	0.33	0.26
Temperature (°F)	95	135	175
Differential pressure (psi)	1,060	3,180	5,300
Densities (gm cm^{-3})			
Sand grain (ρ_s)	2.650	2.650	2.650
Water (ρ_ω)	1.097	1.089	1.083
Oil (ρ_o)	0.755	0.749	0.742
Gas (ρ_g)	0.023	0.103	0.156

Table 4.3. Sand reservoir Poisson's ratio and constituent compressibilities (Source: Domenico 1974, p 763)

	Depth (ft)		
	2,000	6,000	10,000
Poisson's ratio (σ)	0.39	0.24	0.24
Compressibilities (ft^2 poundal^{-1})*			
Matrix material (c_s)	3.72×10^{-11}	3.72×10^{-11}	3.72×10^{-11}
Water (c_w)	6.82×10^{-10}	6.25×10^{-10}	6.18×10^{-10}
Oil (c_o)	1.51×10^{-9}	2.22×10^{-9}	3.09×10^{-9}
Gas (c_g)	2.37×10^{-7}	7.14×10^{-8}	3.89×10^{-8}
Pore-volume (c_p)	1.64×10^{-9}	0.82×10^{-9}	0.58×10^{-9}
Bulk (c_b)	6.82×10^{-10}	3.07×10^{-10}	1.90×10^{-10}

* Compressibility dimensions are the inverse of pressure dimensions. One ft^2 poundal^{-1} = 6.72×10^{-2} cm^2 dyne^{-1} or to 463×10^{-4} psi^{-1}

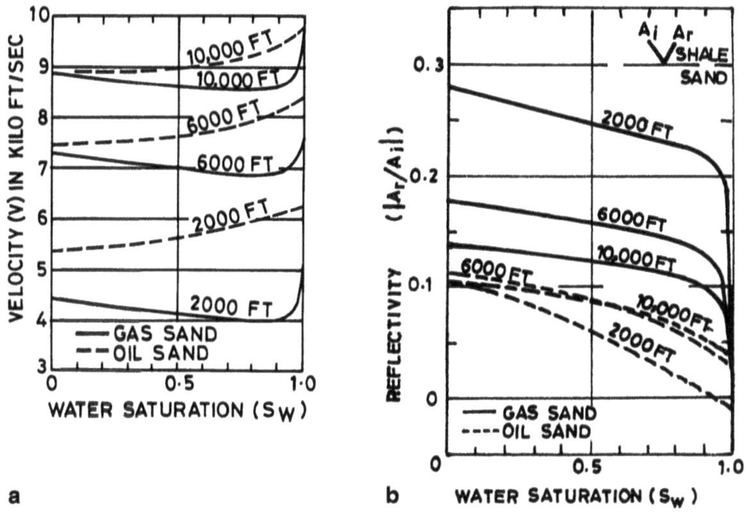

Fig. 4.2 a, b. a Longitudinal velocity V as a function of water saturation S_w for gas and oil sands at depths of 2,000, 6,000 and 10,000 ft; **b** absolute value of the ratio of the reflected wave amplitude A_r to incident wave amplitude A_i at shale/gas sand and shale/ oil-sand interfaces as a function of water saturation S_w for depths of 2,000, 6,000 and 10,000 ft (Source: Domenico 1974)

study in Domenico's investigations are given in Tables 4.2 and 4.3. Variations of longitudinal velocity and reflectivity are shown in Fig. 4.2 a, b.

The reflectivity varies with water saturation. Basic assumptions on a sand reservoir are: reservoir matrix consists of quartz grains and that pore sizes and fluid constituents (oil and water, and gas and water) have even distribution. The following characteristic features of reflectivity evolved from Domenico's analysis:

(1) the reflectivity of the shale/oil-sand interface decreases moderately with increase of water saturation (Fig. 4.2 b),
(2) the reflectivity of the shale/gas sand interface also decreases moderately from a completely gas-saturated reservoir to a water saturation of approximately 0.95; after which the decrease is abrupt to the reflectivity of a completely water saturated sand reservoir (Fig. 4.2 b). It shows that a small quantity of gas (about 5%) would produce an appreciable increase in the reflection amplitude, and also additional gas should not change the reflection amplitude significantly.

In Domenico's investigation the reflectivity R is taken as the absolute value of the reflection coefficient defined as

$$R = \left|\frac{A_r}{A_i}\right| = \left|\frac{\rho_b^{'}V^{'} - \rho_b V}{\rho_b^{'}V^{'} + \rho_b V}\right| \tag{4.1c}$$

[Domenico 1974, Eq. (1)]

Here, ρ_b is the bulk density of the sand reservoir, and ρ_b' is the bulk density of the overlying shale, V is the longitudinal velocity in the sand reservoir, and V' is the longitudinal velocity in overlying shale. Equation (4.1 c) is valid for normal incidence, corresponding to a plane wave front parallel to a plane interface.

The *bulk density* of the sand reservoir, ρ_b, is taken as weighted-by-volume average of the constituent densities, and is expressed as,

$$\rho_b = \phi S_\omega \rho_\omega + \phi(1 - S_\omega)\rho + (1 - \phi)\rho_s \tag{4.2}$$

[Domenico 1974, Eq. (2)] where

ϕ fractional porosity of sand
S_ω fractional volume of pores filled with water
ρ_ω water density
ρ_s sand grain density
ρ gas density ρ_g or oil density ρ_o.

For evaluation of reflectivity, the bulk density and velocity for shale were taken from well log data in Domenico's investigations.

For the velocity in porous media, the most widely used expression is the *time-average equation* proposed by Wyllie et al. (1956):

$$V = \frac{\phi}{V_f} + \frac{1 - \phi}{V_s} \tag{4.3}$$

where V_f is the velocity of the pore fluid and V_s is the velocity of the matrix solid material.

The longitudinal velocity V at zero frequency in a fluid-filled porous solid is given in Geertsma (1961) as:

$$V^2 = \left[\left(\frac{\beta}{C_s} + \frac{4}{3}G_b\right) + \frac{(1-\beta)^2}{(1-\phi-\beta)C_s + \phi C_f}\right]\frac{1}{\rho_b} \tag{4.4}$$

where C_s is the *compressibility* of the reservoir matrix material, C_b is the *compressibility* of the empty reservoir bulk material, β is the ratio C_s/C_b; G_b is the *shear modulus* of the reservoir bulk material, and C_f is the compressibility of the fluid.

In Eq. (4.4), values of the shear modulus G_b, bulk compressibility C_b, and fluid compressibility C_f are given by

$$G_b = \frac{3(1-2\sigma)}{2C_b(1+\sigma)} \tag{4.4 a}$$

$$C_b = \phi C_p + C_s, \tag{4.4 b}$$

and

$$C_f = S_\omega C_\omega + (1 - S_\omega)C \tag{4.4 c}$$

where σ is the Poisson's ratio, ϕ is porosity, C_s is the reservoir matrix material compressibility (taken as that of quartz), C_p is the pore volume compressibility, C_ω is the water compressibility, and C is the oil compressibility (C_o) for oil sand, or the gas compressibility (C_g) for gas sand.

Equations (4.2) and (4.4) were used by Domenico (1974) to compute bulk density ρ_b and the velocity V of gas– or oil bearing unconsolidated sand reservoir as a function of water saturation, S_ω.

Amplitude variations are affected by the velocity density contrast at the interfaces. However, the spacing of reflecting surfaces controls strongly the amplitude and frequency of the seismic response. Frequency signifies the *characteristic frequency* of a seismic event, and is inverse of the cycle breadth as measured on a seismic section. To explain the effect of reflecting surface spacing on amplitude and frequency of seismic response, we take an illustrative example considered in Sangree and Widmier (1979).

Figure 4.3 a shows a model of five sand beds, having a velocity higher than that of the enclosing shale. Density values are kept constant to simplify the discussions. In the last but one column, individual reflected wavelets for each interface are shown. Low contrast or weak reflection coefficient interfaces are associated with small-amplitude reflections; and a high degree of velocity-density contrast results in much higher amplitude. The upper two thick sand beds show the effects of a velocity-density contrast in a clear manner. The lower three beds are thin and are suitable for showing the effect of the spacing of the top and bottom of the reflecting bed, i.e. the effect of bed thickness on the *composite waveform*. For the thinnest bed, both the amplitude and peak-to-peak cycle breadth of the composite waveform are found reduced below those of individual wavelets. The fourth bed from the top satisfies the condition of *constructive interference* for reflected wavelets from the top and bottom of the bed. The reflection pulse cycle breadth for this bed is the same as that of the wavelet but the composite

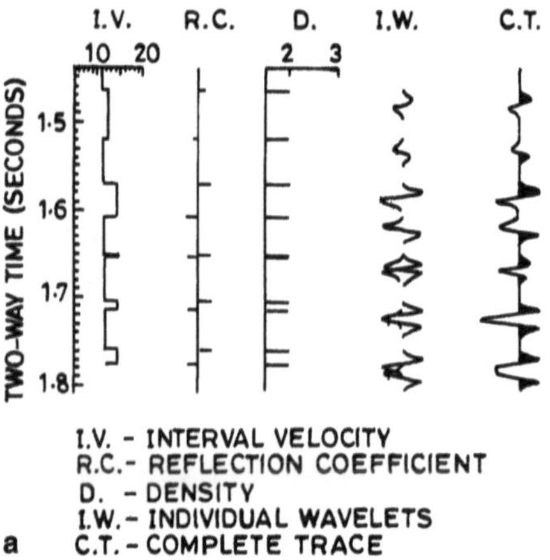

a

Fig. 4.3 a, b. Composition of reflections, illustrating the effect on a composite synthetic seismic trace of varying bed thickness and velocity contrasts (Source: Sangree and Widmier 1979). **b** Variations in amplitude response due to changes in the pore fluid of the reservoir (Source: Meckel and Nath 1977, p 423. © AAPG 1977. Reprinted by permission of AAPG whose permission is required for further use)

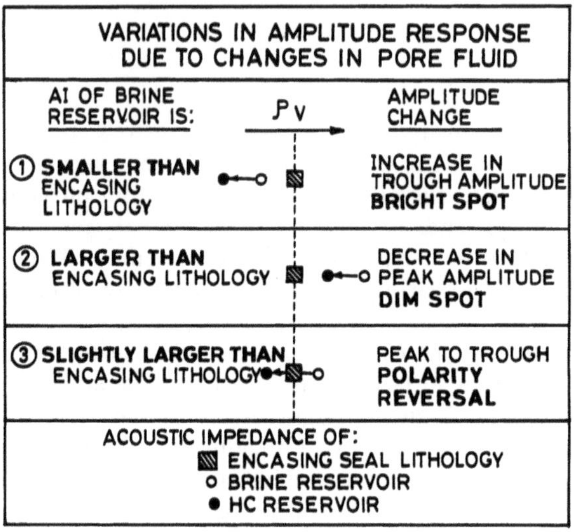

b

waveform amplitude is enhanced in comparison to that of individual reflected wavelets from either the top or from the bottom. This effect is important as it shows that high amplitude on seismic sections may not be the result of large velocity-density contrasts between individual beds. On the other hand, they represent beds with relatively moderate reflection coefficients, which have optimum thickness for strong constructive addition of top and bottom reflected waveforms. *Optimum bed thickness* corresponds to 1/4 of a reflection cycle breadth on a seismic section. The lowest bed has a thickness larger than those of the upper two beds. It is found that in the composite waveform cycle breadth is increased and amplitude has an intermediate value between those for the next two upper beds. *Wide cycle breadth reflections* on seismic sections interspersed with reflections of less cycle breadth tells that the bedding is thick enough and exceeds values of optimum thickness for constructive addition.

Figure 4.3b shows the variations in amplitude response due to changes in the pore fluid of the reservoir. Depending on whether the acoustic impedance (ρV) of a brine filled reservoir rock is smaller than, larger than, or only slightly larger than that of the incasing rock, it produces the effect of a *bright spot, dim spot* or a *reversal of polarity* which are considered as potential *Hydrocarbon indicators* (HCI). A bright spot is used to signify local strong reflections where gas (or oil) is locally concentrated. This may occur, for example, when gas replaces brine in sand encased in shale. A dim spot situation may arise in a limestone reservoir capped with shale. In this case, a strong reflection is expected where there are no hydrocarbon concentrations due to the considerably larger acoustic impedance of the brine-saturated reservoir than that of the encasing rock. However, slight weakening of the reflections is there where oil accumulates, and considerable weakening takes place due to accumulations of gas resulting in a dim spot. In another case where the acoustic impedance of reservoir rock is only slightly greater than that of the encasing rock, the lowering of acoustic impedance because of the replacement of brine by hydrocarbons may reverse the reflection polarity from positive to negative, giving rise to polarity reversal as another hydrocarbon indicator.

4.3
Complex Trace Analysis to Determine Values of Attributes

In *complex trace analysis*, a conventional seismic trace is viewed as the real component of a complex trace, which can be uniquely calculated under usual conditions. A complex trace has the advantage that it allows unique separation of envelope amplitude and phase information, and the calculation of instantaneous frequencies becomes possible.

The conceptual basis and application of complex trace analysis was originally contributed by Taner, Koehler and Sheriff (1979). The presentation in this section is based on their work. Complex trace analysis brings out a natural separation of amplitude and phase information. In complex trace analysis amplitude and phase (called *attributes*) are measured. The amplitude attribute is called *reflection strength*. Phase information is itself an attribute, and also a basis for *instantaneous frequency* measurement. Amplitude and phase information are combined to have additional attributes, like *weighted average frequency* and *apparent polarity*. Color displays are particularly effective in complex trace analysis. Further, in a seismic section a color display is very helpful in interpretation. For details on the application of seismic attributes and standard practices of their color displays, reference may be made to Taner and Sheriff (1977), Landraum et al. (1994), Roden and Sepulvede (1999).

Complex trace analysis assumes a seismic trace f (t) as a real part of an analytical signal or complex trace, $F(t) = f(t) + jf^* (t)$. Imaginary component $f^*(t)$ is uniquely determinable from f(t), if f(t) satisfies certain conditions.

A real seismic trace f(t) may be expressed as the time-dependent amplitude A(t), and a time-dependent phase, θ(t).

Taking f(t) = A(t) cos θ(t), f*(t) is A(t) sin θ(t).
Thus, we have F (t)=A (t) $e^{j\theta (t)}$
If f (t) and f* (t) are known, one may solve for A (t) and θ (t):

$$A(t) = \left[f^2(t) + f*^2(t)\right]^{\frac{1}{2}} = |F(t)| \quad (4.5\,a)$$

and

$$\theta(t) = \tan^{-1}[f*(t)/f(t)] \quad (4.5\,b)$$

[Taner et al. 1979, Eqs. (4) and (5)]
A(t) is called the *reflection strength* and θ(t) the *instantaneous phase.*
The rate of change of time-dependent phase gives a time-dependent frequency.

$$\frac{d\theta(t)}{dt} = \omega(t) \quad (4.5\,c)$$

Apparent polarity is defined by the sign of f(t) when A(t) has a local maximum.
The theory for calculating the quadrature component f*(t) from f(t) may be referred to in Taner et al. (1979).

Reflection strength may have it's maximum at phase points other than peaks or troughs of the real trace. *High reflection strength* is often associated with major lithologic changes (for example, across unconformities, changes in sea level or depositional environments) or with gas accumulations.

The instantaneous phase brings out the continuity of events. Instantaneous phase value is associated with a point in time, and is different from *phase* as a function of frequency. As phase is independent of reflection strength, it makes weak coherent events clearer. Phase displays emphasize discontinuities, faults, pinchouts, angularities, and events with dip attitudes interfering with each other.

Like instantaneous phase, instantaneous frequency is a value associated with a point in time. It works as a useful correlation tool, or for location of pinchout and edges of the hydrocarbon water interfaces. *Low-frequency shadows* are often found associated with fracture zone in brittle rocks, or from reflectors immediately below the petroliferous zones.

Apparent polarity measurements are particularly sensitive to data quality. Apparent polarity analysis assumes a single reflector, a zero-phase wavelet and no ambiguity due to *phase inversion.* Gas accumulations in clastic sediments show negative polarity from reservoir top reflections and positive polarity for reflections from gas-oil or gas-water interfaces. Polarity is a useful tool to distinguish between various kinds of *bright spots.*

Weighted average frequency is useful to emphasize frequency of the stronger reflection events. It also smoothes irregularities caused due to noise.

An example of real and quadrature traces, phase, and instantaneous frequency is given in Fig. 4.4. It shows an actual trace (a), quadrature trace (b), phase (c) and instantaneous frequency (d). The complex trace F(t) may be thought of as the trace in complex space of a vector continually changing its length and rotating, and thus generating an irregular helix as in Fig. 4.4e. We then think of A(t)

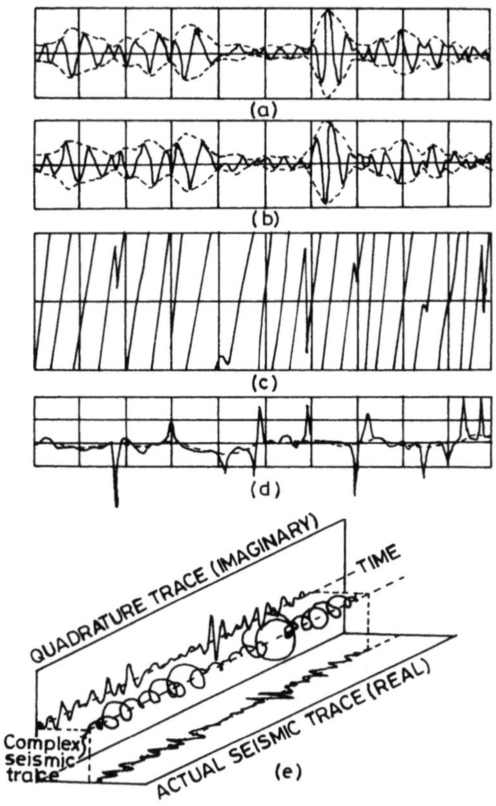

Fig. 4.4 a–d. Real **a** and quadrature **b** traces for a portion of an actual seismic trace. Also shown is the envelope (*dotted curve* in (**a**, **b**)); phase **c**; instantaneous frequency **d** (and weighted average frequency (*dotted curve* in **d**. Isometric diagram of a portion of an actual seismic trace **e** (Source: Taner et al. 1979)

as the time varying modulus and $\theta(t)$ as the time varying argument of this vector. For excellent details on applications and examples, the readers are advised to refer to the original contribution by Taner, Koehler, and Sheriff (1979).

4.4 Pitfalls in Routine Reflection Processing due to Conditions of Anisotropy: Effects on

4.4.1 Waveform and Arrival Times

Reflection surveys use a localized source of seismic energy. It is therefore important to have an idea of how anisotropy of the medium would affect radiation from a localized source. For illustration purposes, an idealized source – a point force may be considered. This problem was investigated in detail in White (1982). Here, some results of White's investigations are included. Our purpose is to initiate an introductory explanation diagrammatically, on the importance of considering anisotropy in waveform interpretations.

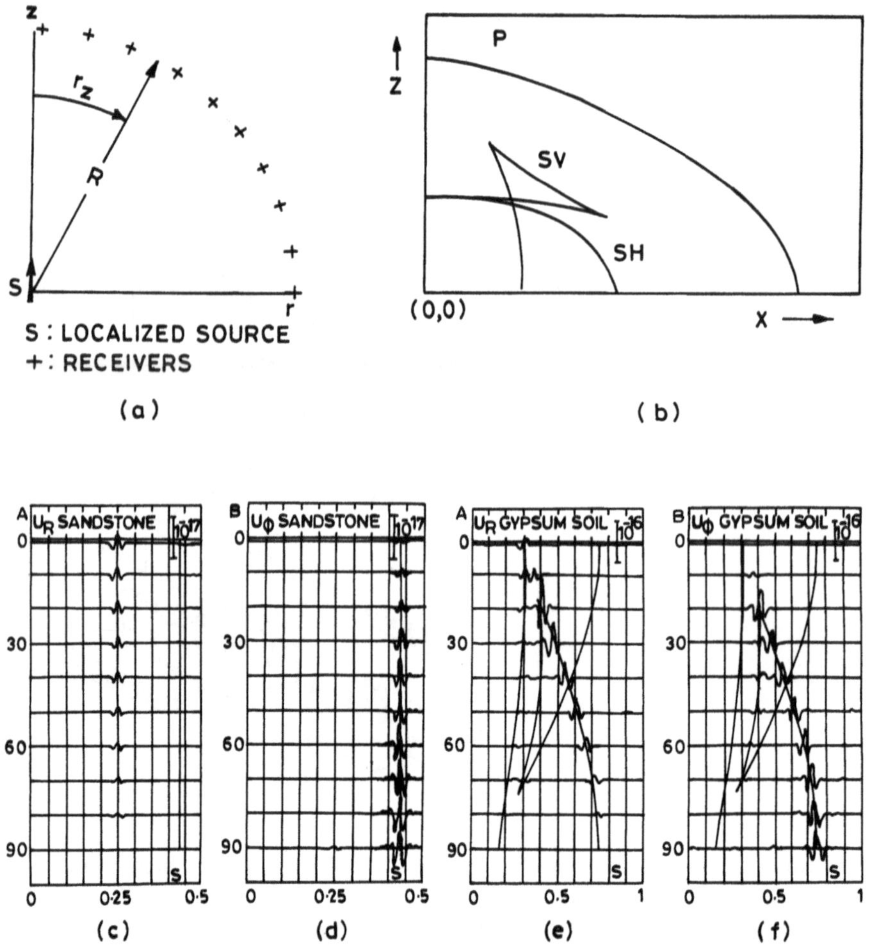

Fig. 4.5 a–f. a Source and receiver geometry; **b** a quadrant depiction of wave surfaces for the Plexiglas-aluminum model; **c** radial displacement due to a point force in (isotropic) sandstone, R = 1.0 km; **d** tangential displacement due to a point force in sandstone, R = 1.0 km; **e** radial displacement due to a point force in anisotropic gypsum soil medium, R = 0.6 km; **f** tangential displacement due to a point force in gypsum soil medium, R = 0.6 km (Source: White 1982)

Assume a point force seismic source (Fig. 4.5 a) that executes a simple symmetrical transient in an isotropic (sandstone) or anisotropic medium. The later incorporates a highly anisotropic gypsum soil medium. The purpose is a comparison of waveform characteristics of radial and tangential displacements in these media.

The form of a wave surface in a transversely isotropic Plexiglas-aluminum model is shown in Fig. 4.5 b. The wave surface is a plot of group velocity against angle (α) between the group velocity vector and the z-axis. It is observed that P and SH wave surfaces are not circular as in the isotropic case, and the SV wave surface has cusps associated with it.

Figure 4.5 c–f depicts the radial and tangential displacements in the contrasting media. For an isotropic sandstone, the arrival times for pure longitudinal and transverse waves lie on straight lines. Their amplitudes variations have cos α and sin α dependence. For an anisotropic Plexiglas-aluminum model, the arrivals are on curved lines, the SV wave has cusps, and quasi-longitudinal and quasi-transverse waves do not show cos α and sin α dependence. The waveform, in general, duplicates the force transient; and is also independent of distance (not depicted in these diagrams).

The above example shows that anisotropy of the medium should play a role to modify the velocity of the medium, arrival times and the picture of amplitude variations on incidence and reflection paths of a seismic pulse. Anisotropy may also cause multiple arrivals of a pulse, as, for example, in the case of SV-waves. The presence of cusps is a special feature that has a bearing on the characteristics of amplitude decay with distance in the direction of cusps and within the two cusps. All these features point towards the necessity for the consideration of anisotropy for an accurate seismic interpretation.

4.4.2 Evaluation of Moveout Corrections

Moveout corrections were discussed in Section 3.5.8. Evaluation of these corrections is based on travel time equations for CMP geometry. These equations assume the following forms for a horizontal (Eq. 3.15 a) and a dipping (dip angle α) reflector (Eq. 3.27), respectively:

$$T^2 = T_0^2 + \frac{X^2}{V^2}$$

and

$$T^2 = T_0^2 + \frac{X^2}{(V/\cos\alpha)^2}$$

In these equations velocity has a constant value in all directions.

Anisotropic media are characterized by directional variations of velocity. It is, therefore, immediately significant to know the equivalent form of the above equations for an anisotropic medium. Grechka and Tsvankin (1998) developed a normal moveout equation for an anisotropic, inhomogeneous medium and described a 3-D picture of normal moveout velocity. The presentation below is based on their contribution.

It is clear that an anisotropic medium with angular velocity variations should influence the normal moveout (NMO), and therefore the NMO value should have in it information about the anisotropic velocity field.

Figure 4.6 is a 3-D display of a zero-offset ray, CMP lines, the coordinates on the ground surface and the angle ψ between the CMP line and x_1-axis. The as-

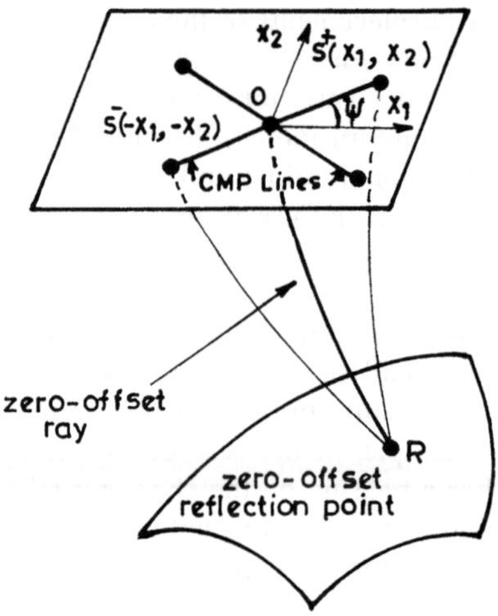

Fig. 4.6. In the derivation of NMO velocity, the CMP reflection raypath from S^+ to S^- (not shown in the plot) can be replaced with a non-specular raypath going through the zero-offset reflection point (Source: after Grechka and Tsvankin 1998. From Geophysics with permission)

sumption is made that reflection point dispersal (i.e. change of position of reflection point with offset) has no influence on NMO velocity (Hubral and Krey 1980). Accordingly, the difference between the true (specular) reflection point and zero-offset reflection point R is ignored.

Taking t (x_1, x_2) as one way travel-time between zero-offset reflection point R and a point (x_1, x_2) on the ground surface, we may express it as a double-Taylor series in the vicinity of CMP:

$$\begin{aligned} t\pm &\equiv t(\pm x_1, \pm x_2) \\ &\equiv t_0 \pm \frac{\partial t}{\partial x_1} x_1 \pm \frac{\partial t}{\partial x_2} x_2 + \frac{\partial^2 t}{\partial x_1^2}\frac{x_1^2}{2} + \frac{\partial^2 t}{\partial x_1 \partial x_2} x_1 x_2 + \frac{\partial^2 t}{\partial x_2^2}\frac{x_2^2}{2} + \ldots \end{aligned} \tag{4.6}$$

[Grechka and Tsvankin 1998, Eq. (A-1)] where $t+=t(+x_1, +x_2)$, $t-=t(-x_1, -x_2)$; t_0 is one way travel time; $(\pm x_1, \pm x_2)$ are coordinates of source S^+ and receiver S^-, and derivatives are evaluated at common mid point location O (Fig. 4.6).

For the 2-D case, the form of Eq. (4.6) was first suggested in Hale et al. (1992).

Squared two-way CMP travel time t^2 (x_1, x_2) may be written from Eq. (4.6) as

$$\begin{aligned} t^2(x_1, x_2) &\equiv (t_+ + t_-)^2 \\ &= \left(t_0'\right)^2 + 4\left(M_{11}x_1^2 + 2M_{12}x_1x_2 + M_{22}x_2^2\right) \end{aligned} \tag{4.7}$$

where $t_0' = 2\,t_0$ is the two way zero-offset travel time.

M_{ij} is a symmetric matrix given by

$$M_{ij} = t_0\left(\partial^2 t / \partial x_i \partial x_j\right)$$

A simple transformation of x_1, x_2 in terms of the source-receiver *half-offset* h, the square of the two-way CMP travel time is expressible in conventional equations like (3.15a) and (3.27) in the form:

$$t^2(h,\psi) = \left(t_0'\right)^2 + \frac{4h^2}{V_{nmo}^2(\psi)} + \dots \tag{4.8}$$

[Grechka and Tsvankin 1998, Eq. (A-5)] where

$$V_{nmo}^2(\psi) = \frac{2}{t_0'}\left[\frac{\partial p_1}{\partial x_1}\cos^2\psi + \frac{\partial p_2}{\partial x_2}\sin^2\psi + \left(\frac{\partial p_1}{\partial x_2} + \frac{\partial p_2}{\partial x_1}\right)\sin\psi\cos\psi\right]^{-1} \tag{4.8 a}$$

$\partial p_1/\partial x_1$, $\partial p_1/\partial x_2$, $\partial p_2/\partial x_1$, $\partial p_2/\partial x_2$ are horizontal components of the *slowness vector* **p** of rays emanating from the zero-offset reflection point; p_1 and p_2 are measured at the ground surface.

A very detailed discussion of NMO velocity (V_{nmo} (ψ)) with illustrative examples for VTI and orthorhombic media is given in Grechka and Tsvankin (1998). All details are obviously difficult to be covered. However, results in the Grechka and Tsvankin investigation should provide useful guidelines for orienting interpretation tools to account for anisotropy of the medium.

4.4.3 Stacking Velocity

By Eq. (3.27), it is clear that the stacking velocity under isotropic conditions for a dipping reflector is given by $V/\cos\theta$ where θ is the angle of dip.

Under a condition of anisotropy, velocity is not constant. It varies with ray-direction. Therefore, entirely different sets of travel-time equations result when considering anisotropy. In the following, results on travel time for *transverse isotropy* are given from derivations in Levin (1990).

A transversely isotropic medium has an *axis of symmetry*, and in the plane perpendicular to this axis there is no variation of velocity with direction. Ray velocity or phase velocity varies with the angle the ray direction or phase velocity direction makes with the symmetry axis. On geological considerations the symmetry axis may be assumed either perpendicular to the ground surface or to the reflector (Levin 1990). These cases are illustrated by Fig. 4.7 a and b.

Under the situation when the symmetry axis is perpendicular to the reflector, the angles of incidence and reflection are equal, both angles being measured from the direction of the symmetry axis. When the symmetry axis is not perpendicular to the reflecting surface, Snells law takes a complicated form that includes the plane wave angle of incidence and reflection, the dip of the reflecting surface and velocities in incident and reflected plane wave directions.

Let us say that the reflecting plane is dipping at the angle γ from the horizontal, and suppose that the symmetry axis is perpendicular to the ground surface as in Fig. 4.7 a. In this case, Snell's law gives

$$\frac{\sin(\theta_I + \lambda)}{V(\theta_I)} = \frac{\sin(\theta_R - \lambda)}{V(\theta_R)} \tag{4.9}$$

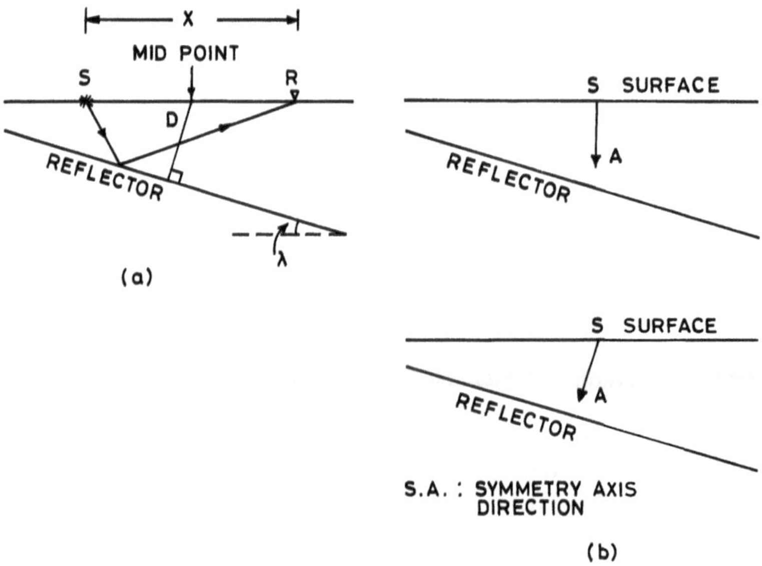

Fig. 4.7 a, b. a Schematic diagram showing a reflection raypath in a transversely isotropic dipping layer; **b** symmetry axis perpendicular to the surface (*upper part*) and to the reflector (*lower part*) (after Levin 1990, from Geophysics with permission)

θ_I and θ_R are angles between the propagation direction of incident and reflected waves and the symmetry axis, respectively, and correspond to the angles of incidence and reflection. $V(\theta_I)$ and $V(\theta_R)$ are the phase velocities in these directions. θ_I and θ_R are in general not equal.

For a given angle of incidence θ_I, it is possible to calculate $V(\theta_I)$ from the known relation between θ_I and V [Levin 1990, Eq. (A-1)]:

$$2\rho\,[V(\theta)]^2 = 2L + (A-L)\sin^2\theta + (C-L)\cos^2\theta + \left\{ \begin{array}{l} [(A-L)\sin^2\theta + (C-L)\cos^2\theta]^2 \\ +[(F+L)^2 - (A-L)(C-L)\sin^2 2\theta] \end{array} \right\}^{\frac{1}{2}} \tag{4.10}$$

In Eq. (4.10), A, C, F, and L are elastic stiffness constants of a transversely isotropic medium.

Denoting the ratio sin $(\theta_I + \gamma)/V(\theta_I)$ as γ, we may find θ_R by finding the root of the equation:

$$V(\theta_R) - \frac{[\sin(\theta_R - \lambda)]}{\gamma} = 0 \tag{4.11}$$

Knowing θ_I and θ_R, the corresponding incident and reflected ray angles ϕ_I and ϕ_R may be calculated from the relation

$$\tan\phi = \frac{\left[V\sin\theta + \left(\frac{dV}{d\theta}\right)\cos\theta\right]}{\left[V\cos\theta - \left(\frac{dV}{d\theta}\right)\sin\theta\right]} \tag{4.12}$$

Angles ϕ_I and ϕ_R are also measured from the symmetry axis.

The ray velocity $V(\phi)$ may be calculated from the well known relation:

$$V^2(\phi) = V^2(\theta) + \left(\frac{dV}{d\theta}\right)^2 \tag{4.13}$$

In the following, we consider the travel-time for CMP geometry (see Fig. 4.7 a, b) for two different orientations of symmetry axis.

Consider first that the axis of symmetry is perpendicular to the reflector (Fig. 4.7 b). In this case the angles of incidence and reflection are equal, and we have

$$\tan\phi = \left(\frac{X\cos\lambda}{2D}\right) \tag{4.14}$$

The equation of travel time in this case is

$$t^2 = \left[\frac{2D}{V^2(\phi)}\right] + \left[\frac{X\cos\lambda}{V(\phi)}\right]^2 \tag{4.15}$$

[Levin 1990, Eq. [A-9)]

To get the value of $V(\phi)$, we start from $V(\theta)$. For an assumed θ, we determine ϕ from Eq. (4.12). We vary θ and find a ϕ value which satisfies Eq. (4.14). From ϕ value we calculate $V(\phi)$. With this knowledge of $V(\phi)$ Eq. (4.15) may be utilized to calculate time 't' for a given spread length X centered around the mid point (compare the form of Eq. (4.15) with the earlier derived Eq. (3.27) for the isotropic case). We may find V_{stack} and t_0 by fitting a least square straight line to a t^2vs x^2 plot.

We next consider the case when the symmetry axis is vertical and perpendicular to the ground surface.

In this case, we note that given the values for D, ϕ_I and ϕ_R, and the spread length X centered about the mid point, it is possible to calculate the point on the ground surface where the reflected ray reaches the surface. In general, that point where the ray reaches the surface would not be at the same distance from the mid point as the source is on the other side of the mid-point. We change ϕ_I and hence ϕ_R until the ray emergence point coincides within some preset accuracy with the receiver position. Mathematically it is equivalent to finding the root of the equation

$$[\cos\phi_R\cos(\phi_I + \lambda)]^{-1}\begin{bmatrix} D\sin(\phi_I + \phi_R) - x\cos\phi_I\phi_R\cos\lambda \\ -\frac{x}{2}\sin(\phi_R - \phi_I)\sin\lambda \end{bmatrix} = 0 \tag{4.16}$$

[Levin 1990, Eq. (A-6)]

Having determined the correct values of all angles, the travel-time may be calculated from

$$t = \left\{ \frac{\left[D - \frac{x}{2}\sin\lambda\right]}{\cos(\phi_I + \lambda)} \right\} \left\{ [V(\phi_I)]^{-1} + \frac{\cos\phi_I}{[V(\phi_R)\cos\phi_R]} \right\} \tag{4.17}$$

[Levin 1990, Eq. (A-7)]

Equation (4.17) may be utilized to generate the (x, t) pair for a CMP gather by incrementing 'X' in equal steps.

Levin (1990) analyzed Eqs. (4.15) and (4.17) to find the dependence of the *stacking velocity* and 't_0' on the angle of dip for two cases of different orientations of *symmetry axis*. We present these results in Fig. 4.8 a, b. In their calculation D was assumed to be 3,000, X varied from 100 to 3,000 ft. For the properties of anisotropic sedimentary rocks, reference may be made to Levin (1990, Table 1).

The conclusions of Levin's (1990) study as revealed from Fig. 4.8 a b are:

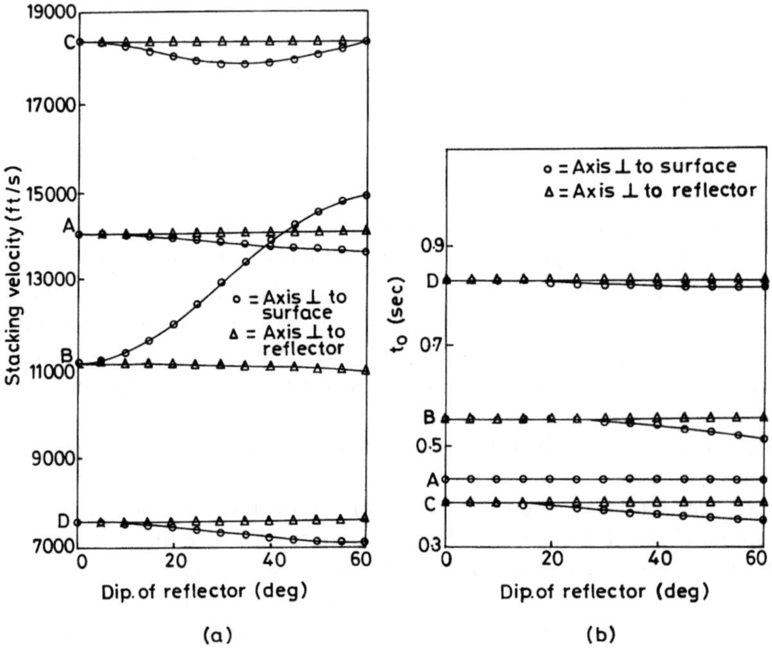

Fig. 4.8 a, b. a 2-D CMP stacking velocities corrected with the cosine of the dip angle as a function of dip of the reflecting plane for a transversely isotropic solid. Values indicated by *circles* are those when the symmetry axis is vertical. Values indicated by *triangles* are those when the symmetry axis is perpendicular to the reflector. A, B, C, D represent different types of sedimentary rocks – shale, sandstone, shale-limestone; **b** t_0 values corresponding to the stacking velocities of Fig. 4.8 a (Source: Levin 1990. From Geophysics with permission)

(i) In the case where the symmetry axis is perpendicular to the reflector, the CMP stacking velocity for P-wave reflections from a dipping plane, after correction by the cosine of the dip angle, is almost independent of the dip angle. This is also expected by observing similar forms of Eq. (3.27) for an isotropic medium and Eq. (4.15) for an anisotropic medium with a symmetry axis perpendicular to the reflector. If the symmetry axis is perpendicular to the surface, the stacking velocity may increase, decrease, or go through an extremum as dip increases – the actual behavior depending on elastic constants of the solids.
(ii) The change in t_0 is small compared to *stacking velocity.*

Singh (1995) and Kumar (2000) developed computer programs to study the influence of anisotropy (transverse isotropy), on stacking velocity.

4.4.4 NMO Velocity

Normal Moveout (NMO) Velocity is defined by the slope of t^2 vs x^2 for small offset x:

$$\frac{1}{(\mathrm{VNMO})^2} = \left(\frac{dt^2}{dx^2}\right)_{x\to 0} \tag{4.18}$$

For a single isotropic layer the t^2 vs x^2 graph for a horizontal reflector would be a straight line, and the Normal Moveout Velocity would be equal to the layer velocity.

To illustrate how the NMO velocity would be affected by anisotropy, Thomson (1986) considered a simple one layer seismic reflection experiment (Fig. 4.9 a).

With reference to Fig. 4.9 a, travel time 't' at offset x may be written as

$$\left[u(\phi)\frac{t(\phi)}{2}\right]^2 = \left[u(0)\frac{\tau}{2}\right]^2 + \left[\frac{x}{2}\right]^2 \tag{4.19 a}$$

't' is the two way travel-time for path SPR; and 'τ' is the two way vertical travel time.

Equation (4.19a) gives

$$t^2(\phi) = \left[\frac{u(0)}{u(\phi)}\right]^2 \left[\tau^2 + \frac{x^2}{u^2(0)}\right]$$

Because of the ϕ dependence of travel time, the plot of t^2 vs x^2 would give a curved line. The slope of this line is given by

$$\begin{aligned}\frac{dt^2}{dx^2} &= \frac{1}{u^2(\phi)} - \frac{t^2}{u^2}\frac{du^2}{dx^2} \\ &= \frac{1}{u^2(\phi)}\left[1 - \frac{2\cos^2\phi}{u(\phi)}\frac{du(\phi)}{d\sin^2\phi}\right]\end{aligned} \tag{4.19 b}$$

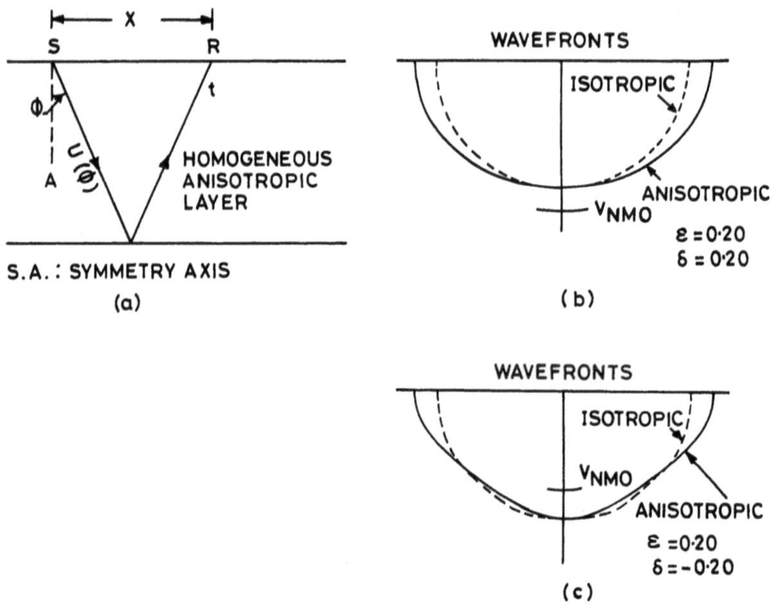

Fig. 4.9 a–c. a Reflection in a single anisotropic layer. Ray velocity is $U(\phi)$; ϕ is the angle the ray makes with the symmetry axis taken as vertical (along the z-axis); **b** elliptical wavefront for the case $\delta = \varepsilon$. The wavefront marked V_{NMO} is a segment of the wavefront that would be inferred from isotropic moveout analysis of reflected energy. $V_{NMO} > V_{vert}$; **c** plausible anisotropic wavefront for $\delta = -\varepsilon$. The curve marked V_{NMO} is a segment of the wavefront that would be inferred from isotropic moveout analysis of reflected energy. $V_{NMO} < V_{vert}$. since $\delta < 0$ (Source: Thomsen 1986. Fig. 4.9b, **c** reproduced from Geophysics by permission)

Accordingly, Normal Moveout Velocity (VNMO) would be

$$\begin{aligned}\frac{1}{(\mathrm{VNMO})^2} &= \lim_{x\to 0}\left(\frac{dt^2}{dx^2}\right)\\ &= \frac{1}{u^2(0)}\left[1 - \frac{2}{u(0)}\frac{du(\phi)}{d\sin^2\phi}\right]_0\end{aligned} \tag{4.20}$$

[Thomsen 1986, Eq. (25)]

Equation (4.20) is a short-spread NMO velocity for an anisotropic layer. The second term within the bracket is usually not zero, and therefore the NMO velocity is not the vertical velocity u(0). Thomsen (1986) gave the concise form of the NMO velocity for P-waves for transverse isotropy (T.I.) in terms of anisotropy parameters δ:

$$\begin{aligned}&(\mathrm{VNMO})_{\text{P-wave}}\\ &= \alpha_0\sqrt{1+2\delta}\end{aligned} \tag{4.20 a}$$

For weak anisotropy, Eq. (4.20 a) reduces to

$$(\text{VNMO})_{\text{P-wave}} = \alpha_0(1+\delta) \tag{4.20 b}$$

Thomsen (1986) also derived the NMO velocities for SV and SH waves for weak anisotropy in the form

$$(\text{VNMO})_{\text{SV-wave}} = \beta_0\left[1+\frac{\alpha_0^2}{\beta_0^2}(\varepsilon-\delta)\right] \tag{4.21 a}$$

$$(\text{VNMO})_{\text{SH-wave}} = \beta_0(1+\gamma) \tag{4.21 b}$$

Anisotropy parameters 'δ', ε and γ were defined in Thomsen (1986) by the ratio of elastic constants of a transversely isotropic medium as

$$\delta = \frac{(C_{13}+C_{44})^2-(C_{33}-C_{44})^2}{2C_{33}(C_{33}-C_{44})} \tag{4.22 a}$$

$$\varepsilon = \frac{C_{11}-C_{33}}{2C_{33}} \tag{4.22 b}$$

$$\gamma = \frac{C_{66}-C_{44}}{2C_{44}} \tag{4.22 c}$$

δ and ε are considered the dominant parameters for vertical and horizontal propagations, respectively.

Supposing that a reflection experiment has been carried out as in Fig. 4.9 a, and isotropic analysis of the surface reflection experiment is done, the curve marked VNMO in Fig. 4.9 b and c would be inferred as a segment of the wavefront. Actual positions of anisotropic and isotropic wavefronts are also shown for comparison with the position of the wavefront inferred from VNMO.

It may be mentioned that, for weak anisotropy, the Normal Moveout Velocity (VNMO) for P-waves does not equal either the vertical velocity α_0 $(=c_{33}/\rho)\ \frac{1}{2}$, or the horizontal velocity $\alpha_0(1+\varepsilon)$. Neither it is necessarily intermediate between these values. This is because δ and ε may have opposite signs under field conditions.

Here the effect on NMO velocity has been considered for the most simple case of one layer with a horizontal boundary, only to illustrate significance of considering anisotropy. A generalized discussion of NMO velocity for layered media and dipping reflectors is given in Chapter 6.

4.4.5
Interval Velocity

In order to find out a relationship between the interval velocity and routinely measured quantities VRMS (or VNMO) and zero-offset time (t_0), we first take the case of isotropy, and a plane, parallel homogeneous multilayered structure (Fig. 4.10a). In this case, for near normal incidence (short seismic source-geophone offset) we have

$$(\mathrm{VNMO})^2 = \frac{1}{t_0}\sum_{j=1}^{n} v_j^2 \Delta t_j \equiv (\mathrm{VRMS})^2 \tag{4.23}$$

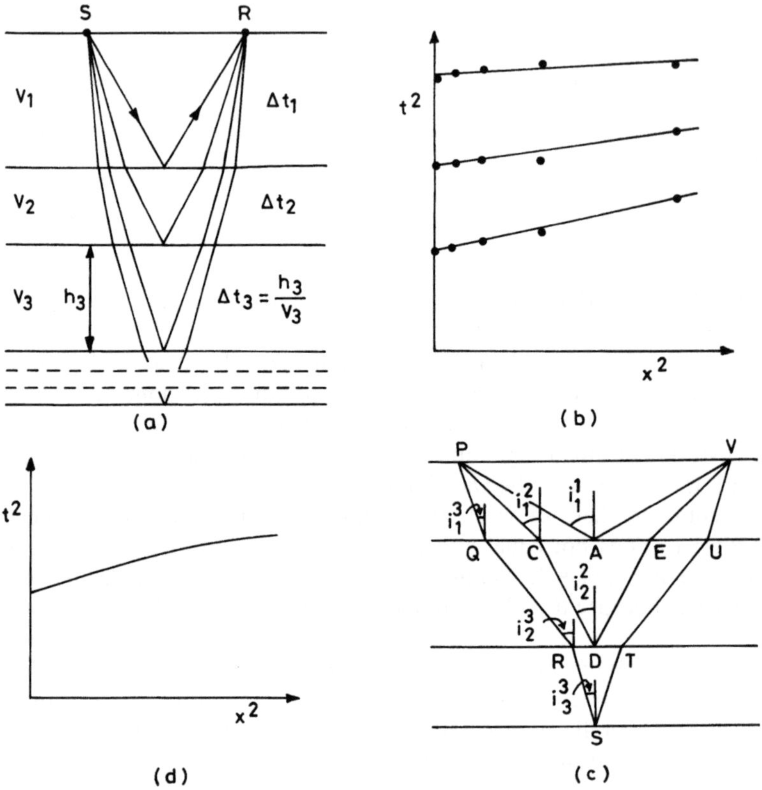

Fig. 4.10a–d. **a** Near vertical raypath in plane, parallel, isotropic, homogeneous layers; **b** (t^2–x^2) fit to find RMS velocities to various reflectors for case **a**; **c** ray geometry for paths much offset from the vertical; **d** upward convex (t^2–x^2) graph for a transversely isotropic-k medium (refer to Chapter 9)

In Eq. (4.23), v_j represents velocity (interval) of jth layer; Δt_j the vertical time in jth layer, and t_0 the total vertical time to the nth reflector.

Equation (4.23) is useful for finding the *interval velocity* from knowledge of normal moveout (VNMO), or root mean square velocity (VRMS).

In CDP surveys, two quantities are routinely measured. These are the *stacking velocity* and reflection time 't_0' corresponding to zero-distance between a shot and geophone. If the width of the common mid-point gather is not too large stacking velocity is a good approximation to normal moveout velocity.

For a horizontal layered system, a reflection time-distance (x, t) and x^2–t^2 graph for reflection from various horizons may be prepared. From the x^2–t^2 graph, RMS velocities to various horizons may be obtained (Fig. 4.10b). The interval velocity for n-th layer is given by

$$v_n^2 = \left(\frac{(VRMS)_n^2 t_{0,n} - (VRMS)_{n-1}^2 t_{0,n-1}}{t_{0,n} - t_{0,n-1}} \right) \tag{4.24}$$

where $(VRMS)_n$ is the RMS velocity and $t_{0,n}$ is the vertical two-way travel time to the nth reflector. A similar meaning is assigned to $(VRMS)_{n-1}$ and $t_{0,n-1}$.

Using Eq. (4.24) $V_2, V_3, \ldots V_{n-1}, V_n$ may be obtained.

The thickness h_n is given by

$$h_n = v_n \times (t_{0,n} - t_{0,n-1}) \tag{4.25 a}$$

The total depth of the n-th reflector is given by

$$H_n = \sum_{i=1}^{n} h_i \tag{4.25 b}$$

Average velocity to any reflector (say n) is defined for a zero-offset (vertical) path. If v_i is the velocity of the i-th layer and Δt_i is the vertical (zero) offset time, the average velocity to the n-th reflector is given by

$$(VAVG)_n = \frac{\sum_{i=1}^{n} v_i \Delta t_i}{\sum_{i=1}^{n} \Delta t_i} \tag{4.26}$$

The depth to the reflector 'n', say – 'H_n', is given by

$$H_n = \frac{(VAVG)_n \times t_{0,n}}{2} \tag{4.26 a}$$

'$t_{0,n}$' is the zero offset vertical reflection time for boundary 'n'.

The average velocity expressed by Eq. (4.26) is not suitable for wave paths like PAV, PCDEV, PQRSTUV (Fig. 4.10c), which are much offset from the vertical. The ray path length in any layer is not the layer thickness as it is for the zero-offset vertical path. The ray length varies according to the angle at which the ray refracts into the layer and with the layer thickness. According to Snell's law, the angle of refraction is more for higher velocity. This means that, for layers of equal thickness, the ray path length is longer in a higher velocity medium.

To analyze travel time along offset paths, a *weighted average velocity* is to be found which accounts for changes in path directions as well as layer thicknesses. The *weighted average velocity* is called the root mean square velocity (VRMS). This is defined for the path PAV as

$$(v_1)_{rms} = \left(\frac{v_1^2 \delta t_1}{\delta t_1}\right)^{\frac{1}{2}} \tag{4.27}$$

where δt_1 is travel time along PA or AV.

$(v_1)_{rms} = v_1$, as the path is entirely in the first layer.

$$\begin{aligned} \text{As } \cos i_1^1 &= \frac{h_1}{PA} = \frac{\Delta t_1}{\delta t_1} \\ \text{or} \quad \delta t_1 &= \frac{\Delta t_1}{\cos i_1^1} \end{aligned} \tag{4.27 a}$$

we have

$$(v_1)_{rms} = \left(\frac{\dfrac{v_1^2 \Delta t_1}{\cos i_1^1}}{\dfrac{\Delta t_1}{\cos i_1^1}}\right)^{\frac{1}{2}} \tag{4.27 b}$$

Weighted average velocity expressed in this way takes account of path direction by the cosine term and of layer thickness by 'Δt_1' term.

For the offset reflection path to the second reflector, we have

$$\begin{aligned} \cos i_1^2 &= \frac{h_1}{PC} = \frac{\Delta t_1}{\delta t_1} \quad \text{or} \quad \delta t_1 = \frac{\Delta t_1}{\cos i_1^2} \\ \cos i_2^2 &= \frac{h_2}{CD} = \frac{\Delta t_2}{\delta t_2} \quad \text{or} \quad \delta t_2 = \frac{\Delta t_2}{\cos i_2^2} \end{aligned} \tag{4.28}$$

As these forms of δt_1 in Eqs. (4.27 a) and (4.28) are the same, rms velocity for the 2nd reflector may be defined as

$$(v_2)_{rms} = \left(\frac{\dfrac{v_1^2 \Delta t_1}{\cos i_1^2} + \dfrac{v_2^2 \Delta t_2}{\cos i_2^2}}{\dfrac{\Delta t_1}{\cos i_1^2} + \dfrac{\Delta t_2}{\cos i_2^2}}\right)^{\frac{1}{2}} \tag{4.29}$$

Or, for n-layers, we have

$$(v_n)_{rms} = \left(\frac{\sum_{k=1}^{n} \frac{v_k^2 \Delta t_k}{\cos i_k^n}}{\sum_{k=1}^{n} \frac{\Delta t_k}{\cos i_k^n}} \right)^{\frac{1}{2}} \tag{4.30}$$

Equation (4.30) expresses the RMS velocity to nth reflector as the weighted average velocity along the path of the reflected wave. It shows how RMS velocity is controlled by thickness, and travel path directions through the layers above the reflector.

For a single layer the expression for reflection is (refer Fig. 3.28)

$$t_x^2 = t_0^2 + \frac{x^2}{v_1^2} \tag{4.31 a}$$

For the layered medium 't_0' is replaced by '$t_{0,n}$' and v_1 by VRMS

$$t_{x,n}^2 = t_{0,n}^2 + \frac{x^2}{(\mathrm{VRMS})^2} \tag{4.31 b}$$

For a plane horizontal layered medium, Eq. (4.31 b) gives a hyperbola for $t_{x,n}$ vs. x. We may plot $t_{x,n}^2$ vs x^2 and find the RMS velocity from the slope of this line. In this way, the RMS velocity may be determined to each reflector (refer to Fig. 4.10 b).

Except for the upper most reflector, t^2 vs x^2 points would not fall on a straight line. However, a straight line may be fitted which gives a good approximation to the actual VRMS value.

Therefore, two quantities in Eq. (4.24) which are needed to determine the interval velocities for a plane layered medium are t_0 and the root mean square velocity (VRMS) to each reflector, which may be determined from the plot of $t_{x,n}^2$ vs. x^2.

It is immediately relevant to mention here the nature of the t^2–x^2 plot and its questionable use for the case of an anisotropic medium.

Hake et al. (1984) showed that for transverse isotropy (k-medium) the t^2–x^2 curves are convex upwards (Fig. 4.10 d). Many sedimentary rocks approximate to a transversely isotropic k-medium. Therefore, the t^2–x^2 plot in this case cannot be used to determine interval velocities using Eq. (4.24) for isotropic horizontally layered mediums (Dix 1955). Also, interval velocities calculated from t^2–x^2 curves in this case do not always approximate vertical velocities and therefore cannot be used to calculate depths to the reflectors.

Detailed discussion of the Dix type Eq. (4.24) for anisotropic media for determining interval velocities and interval anisotropic properties would be taken up in Chapters 7 to 10 and 15.

4.4.6
Reflectivity

Anisotropy influences the reflectivity (meant here as the overall reflection process) in a number of ways. For example:

(i) the angle of reflection equals the angle of incidence if the symmetry axis is perpendicular to the dipping reflector. However, for a vertical symmetry axis, appropriate incidence and reflection angles have to be redefined suitably (Levin 1990),

(ii) for a common midpoint (CMP) geometry, the normal incidence (or zero offset) ray may not be perpendicular to the reflector (Uren et al. 1990),

(iii) the reflection coefficient, both from the top and bottom of an anisotropic layer would be different from that for an isotropic layer (Banik 1987; Tsvankin 1995),

(iv) for a reflection from the bottom of an anisotropic layer, anisotropy would distort the angular amplitude distribution of the incident wavefront (Tsvankin 1995),

(v) for transverse isotropy with a horizontal axis of symmetry and P-P, SV-SV and SH-SH reflections from a horizontal plane, the reflected ray lies in the plane of incidence, but the plane wave direction would generally differ from the ray direction.

Further, for P-SV reflection, the reflection points are in the plane of incidence only for profiles parallel to or perpendicular to the symmetry direction (Levin 1994).

Regarding item (i) reference may be made to Fig. 4.7b and Eq. (4.9) for Snell's law in the case of a symmetry axis perpendicular to the ground surface.

Item (ii) may be illustrated with the help of Fig. 4.11a–c.

First, we consider the change in ray path produced by anisotropy. Figure 4.11a shows incident and reflected wavefronts and ray paths for an isotropic medium. Figure 4.11b shows the corresponding case of elliptical anisotropy. It may be noted that the elliptical wavefronts changed position of image point and direction of the reflected rays. Point P is the point where the incident elliptical wavefront touches the ground first. The image would be formed at equal distance at Q. The deviations of reflected ray paths are obviously an effect of anisotropy.

Next, we consider the zero-offset ray for an elliptically anisotropic medium and a dipping reflector (Fig. 4.11c). The image point is found by considering elliptical wavefronts. The wavefront touches the dipping reflector first at F. The distances SF and FI are equal. S and R are shot and receivers. The reflected ray path is on IR, and the complete ray path is SPR. The zero-offset ray is MTM; MT, SF and the construction line RG are all parallel. The zero-offset ray is not normal to the reflector as should be the case for an isotropic layer. It meets the reflector at an angle different from 90°.

Similar results hold for a transversely isotropic solid. If θ is the direction of phase velocity with respect to z-axis, the direction (ψ) of the group velocity vector is expressed as

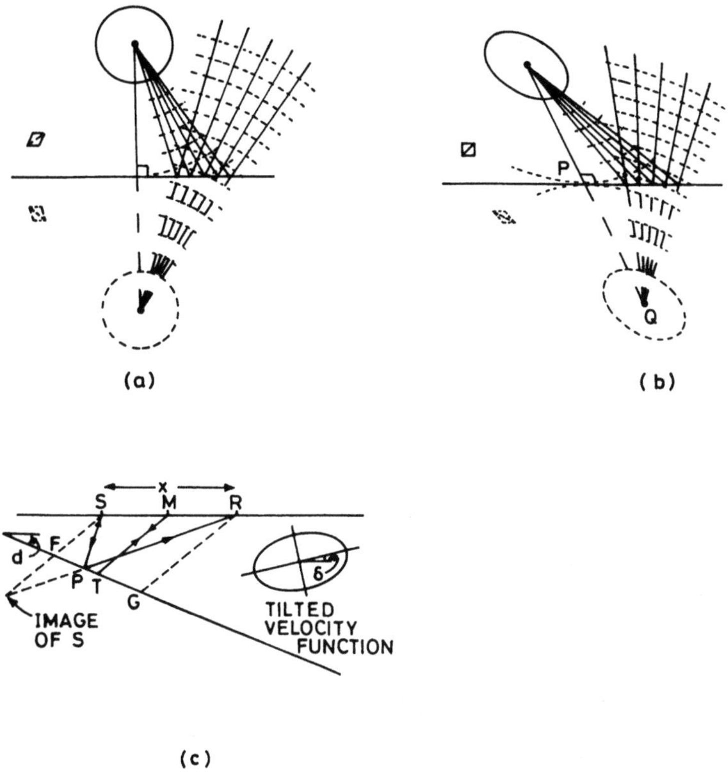

Fig. 4.11 a–c. Reflection ray path for: **a** isotropy; **b** elliptical anisotropy; **c** over a dipping reflector for separated source and receiver positions (S, R) and coincident source-receiver position (M). Note that the zero-offset (normal incidence) ray is not perpendicular to the reflector (Source: Dellinger and Muir 1988 for **a** and **b**; Uren et al. 1990 for **c**)

$$\tan\psi = \frac{\tan\theta + \frac{1}{V}\frac{dV}{d\theta}}{1 - \frac{\tan\theta}{V}\frac{dV}{d\theta}} \tag{4.32}$$

[Tsvankin 1995, Eq. (6)] where V is the phase velocity.

Phase velocity and ray velocity directions, in general, differ as a result of anisotropy, and the normal incidence (zero offset) ray is not necessarily perpendicular to the reflector.

Item (iii) mentions a change of reflection coefficient. Reflection coefficient $R(\theta)$ under the condition of weak transverse isotropy may be thought of as consisting of isotropic and anisotropic parts:

$$R(\theta) = R_{iso}(\theta) + R_{anis}(\theta) \tag{4.33 a}$$

(Banik 1997) where $R_{iso}(\theta)$ is the reflection coefficient in the absence of anisotropy (anisotropy parameter, ε and δ, are each equal to zero). $R_{anis}(\theta)$ is given by:

$$
\begin{aligned}
R_{aniso}(\theta) &= \frac{1}{2}(\delta_2 - \delta_1)\sin^2\theta \\
&+ \frac{1}{2}(\varepsilon_2 - \varepsilon_1)\sin^2\theta\tan^2\theta
\end{aligned}
\tag{4.33 b}
$$

[Tsvankin 1995, Eq. (12)]

In the expression for $R_{anis}(\theta)$, subscripts 1 and 2 refer to media above and below the reflector.

It is clear that anisotropy would produce a change in the values of the reflection coefficient depending on the contrast in anisotropy parameters; and the total value of the reflection coefficient $R(\theta)$ would differ from $R_{iso}(\theta)$.

In respect of item (iv), Tsvankin and Chesnokov (1990) showed that the distribution of energy along wavefronts in anisotropic media is substantially distorted by focusing and *defocusing* of energy. Increase of energy (focusing) occurs in the parts of wavefronts with a high concentration of group velocity vectors of elementary plane waves, comprising point-source radiation. A low concentration of group velocity vectors results in *defocusing*. In many instances, focusing occurs near velocity maxima, and defocusing is associated with velocity minima.

Figure 4.12 shows a construction depicting in it phase velocity vectors, group velocity vectors, the *wave surface* and related *cusps* for a transversely isotropic medium. Points denoted by 1, 2, 3 etc. are end points of phase velocity vectors (V), and 1′, 2′, 3′ – etc. are end points of group velocity vector(s) obtained by the relation

$$
S^2 = V^2 + \left(\frac{dV}{d\phi}\right)^2 \tag{4.34}
$$

[Postma 1955, Eq. (19)]

Figure 4.12 shows concentration of group velocity vectors (ends represented by 4′ to 9′ and 11′ to 15′) in the direction of cusps.

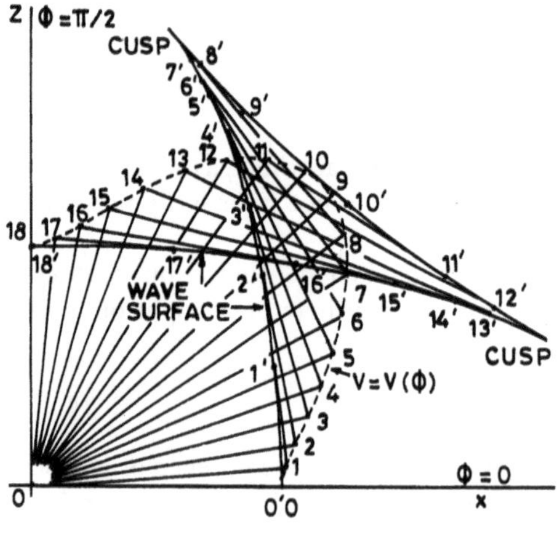

Fig. 4.12. Relation between a wave surface with cusps and polar representation of the normal velocity. Corresponding points are represented by the same numerals (Source: Postma 1955)

Results on distortion of point source radiation by velocity anisotropy (Tsvankin 1995) would be discussed in more detail in Chapter 13. The important conclusion of Tsvankin's (1995) analysis is that, in the interpretation of AVO anomalies under conditions of anisotropy, an integrated approach should be followed that takes into account not only the effect of the reflection coefficient but also the anisotropic wave propagation above the reflector.

Regarding reflection point behavior (item v), it may be mentioned that, for a horizontal reflector and isotropic medium, the reflection point is a point on the reflector vertically below the mid point between shot and receiver on the ground surface. This does not always hold if the medium is anisotropic. Reflection point behavior in a transversely isotropic medium with a horizontal axis of symmetry was studied in Levin (1994). The model considered is a horizontal reflector at a depth of 3,000 (arbitrary units). Anisotropic media considered are representative of weakly and strongly anisotropic solids. Source receiver separations occupy symmetric positions about the origin varying from 200 to 6,000 (arbitrary units). Profile directions were taken in steps of 5° from 0 to 90°; the 0° profile is in the direction of symmetry axis.

Computational results in Levin bring out the following significant points:

(i) For P-P, SV-SV or SH-SH reflections, it is the rays that lie in the plane of incidence, i.e. the plane defined by the profile direction and the normal to the reflector.

Except for a profile in the symmetry direction or perpendicular to the symmetry direction, the *plane wave vector* does not lie in the plane of incidence.

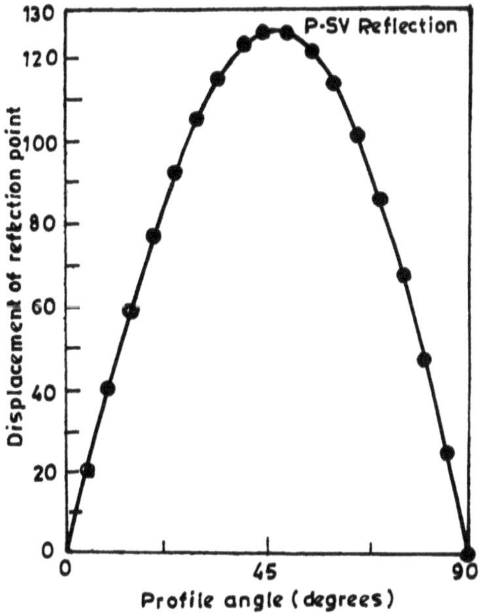

Fig. 4.13. Displacement of the reflection point from the profile line as a function of profile direction. Depths to the reflector 3,000. Source-to-receiver separation 3,000 (Source: Levin 1994. Reproduced from Geophysics with permission)

(ii) For P-SV reflections, the reflection point lies in the plane of incidence only for the plane parallel or perpendicular to symmetry direction.

Figure 4.13 gives a plot of perpendicular offset of the reflection points for source-receiver separation of 3,000. The maximum offset of around 125 is close to 4 % of source-receiver separation. Displacements of P-SV reflection points from the plane of incidence may go unnoticed. This is an effect produced by anisotropy of the medium.

4.4.7 Imaging

To illustrate the effect of anisotropy on imaging we first take an example of the point diffractor model, and subsequently generalize it to an exploding dipping reflector model.

Consider a *point diffractor* in an elliptically anisotropic medium with coincident sources-receivers on the surface (Fig. 4.14 a (bottom)). The ray velocity (V_ϕ) variation along the paths joining the point diffractor and coincident source-receiver positions is shown at the top of the figure. The ray velocity $V\phi$ is a function of direction from source to receiver point. Its value is given by

$$V_\phi = \frac{V_1 V_2}{\sqrt{(V_2^2 \cos^2\phi + V_1^2 \sin^2\phi)}} \tag{4.35}$$

[Levin 1978; Uren et al. 1990, Eq. (1)]

where V_ϕ is the ray velocity at angle ϕ to the minor axis of the elliptical velocity function and V_1, V_2 are velocities along minor and major axes of the elliptical velocity function. The ray velocity for the isotropic model is shown by the dashed line (Fig. 4.14 a (top)), taken equal to V_2, the horizontal component of the elliptical velocity function. The time distance (t–x) graph for isotropic and anisotropic model conditions is shown in the middle part of Fig. 4.14 a. In the case of an isotropic medium, the minimum of time occurs right over the point diffractor. For an anisotropic condition, this would, in general, depend on the orientation of the elliptic velocity function.

The travel time between the source point (x_A, y_A) and receiver point (x, y) for an arbitrary orientation of elliptic function is given by

$$\begin{aligned} t^2 = {} & (x - x_A)^2 \alpha + (y - y_A)^2 \gamma + \\ & 2(x - x_A)(y - y_A)\beta \end{aligned} \tag{4.36}$$

[Uren et al. 1990, Eq. (3)] where

$$\alpha = \frac{\cos^2\delta}{V_2^2} + \frac{\sin^2\delta}{V_1^2} \tag{4.36 a}$$

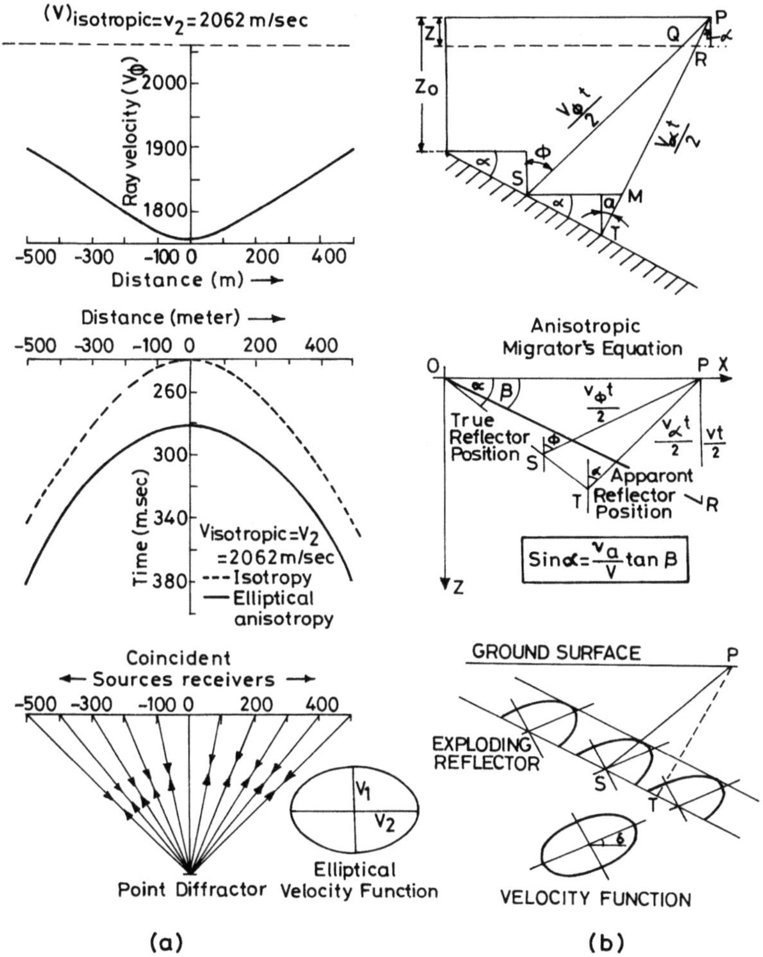

Fig. 4.14 a, b. a Point diffractor model in an elliptically anisotropic medium (*bottom*); time-section for coincident source-receivers (*middle*), and variation of ray velocity (*top*). **b** (*bottom*) Huygen's principle may be used to propagate waves from an exploding reflector in an anisotropic medium; (*middle*) the apparent and true positions of a reflection event on seismic depth sections before and after migration are related by the migrator's equation. In an anisotropic medium, it is assumed that the depth section before migration is depth converted with an isotropic velocity V; (*top*) actual ray path SP, assumed raypath TP for the isotropic case, and shift ST in the position of the reflection point by assumption of isotropy (Source: for Fig. 4.14b (*bottom* and *middle*): Uren et al. 1990. Source for Fig. 4.14b (*top*): Kumar 2000)

$$\gamma = \frac{\sin^2\delta}{V_2^2} + \frac{\cos^2\delta}{V_1^2} \tag{4.36 b}$$

$$\beta = \left(\frac{1}{V_2^2} - \frac{1}{V_1^2}\right)\cos\delta\sin\delta \tag{4.36 c}$$

and V_1, V_2 are the velocities along the minor and major axes of the elliptical velocity function which is inclined at an angle δ to the surface of the Earth.

If the fixed seismic source is at $(0, y_A)$ and receivers are at (x, 0), Eq. (4.37) may be written as

$$t^2 = \alpha x^2 - 2\beta y_A x + \gamma y_A^2 \tag{4.37}$$

At the point x, where t is minimum, we have

$$x|_{t=\min} = \frac{\beta y_A}{\alpha} \tag{4.38}$$

It may be noted that the point x where the time is minimum depends on the source depth and the function β and α, but does not depend on γ. Since α and β are functions of V_1, V_2 and δ, it shows that the distance x from the point above the source where the time is minimum is dependent only on the characteristic of the elliptical velocity function and the depth of the source.

The time-distance (t–x) graph for a point diffractor in an anisotropic medium would be different from that in an isotropic medium. With this simple example, it is demonstrated that the modeling results (both forward and inverse) would be affected if anisotropy is not taken into consideration.

Figure 4.14b illustrates the application of Huygen's principle for constructing a wavefront for the *exploding reflector model* for anisotropic medium (Uren et al. 1990). Due to anisotropy, there are two velocities associated with a particular wave-front: the phase velocity V_α normal to the wavefront, and the ray velocity V_ϕ. S is the source of elliptical wavefront on the exploding reflector, P is the receiver on surface, PT is the perpendicular to the reflector. The phase and ray velocities, V_α and $V\phi$, are given by

$$V_\alpha = \frac{TP}{t} \tag{4.39 a}$$

$$V\phi = \frac{SP}{t} \tag{4.39 b}$$

In Eq. (4.39), t is the travel time from seismic source S to receiver point P.

It is to be noted that for an isotropic medium both the phase and ray directions coincide (i.e. along TP perpendicular to the reflector, the phase and ray velocities are equal). This is not so for an anisotropic medium. This would significantly affect the position of the imaged point for an anisotropic medium by assumption of isotropy. For an isotropic medium our target image point is T; for an anisotropic medium it is S from which the elliptical wave front starts and spreads out to reach the point P. ST is the error in imaged position of the reflector point using stacked data and assuming isotropy. This is on the assumption that phase velocity V_α is numerically equal to V, the isotropic velocity used for getting the *depth section* in the migration process. Byun (1982) refers to it as the *side-slip* effect. An expression for the error was derived in Kumar (2000) and the numerical results are illustrated with an example. We would discuss only its basis here. However, first the migrator's equation for an anisotropic medium is given to substantiate the concepts involved.

Figure 4.14b (middle) shows the apparent position of reflector (OR) before migration under assumption of isotropic velocity (V) for the time to depth conversion. The true position of the reflector is OT. ϕ is the ray direction, and a is the phase direction with respect to the vertical. The relationship between a and β, i.e. angles that the true reflector and apparent reflectors make with the horizontal, may be found by considering the right angle triangles OTP and OPR:

$$\sin a = \frac{V_a}{V} \tan \beta \tag{4.40}$$

[Uren et al. 1990, Eq. (7)]

Equation (4.40) is the *migrator's equation* in an anisotropic medium.

In the isotropic case, this relation is

$$\sin a = \tan \beta$$

Figure 4.14b (top) shows the path TP for the isotropic condition for a wave started from the reflector point T to the receiver point P on the surface. TP is the perpendicular to reflector, corresponding to a stacked data case for coincident source receiver. If the medium is anisotropic, TP is not the actual path of energy travel but it is SP because the wave front is elliptical and not circular as in case of isotropy.

We have from Figure 4.14b (top):

$$ST = \frac{V_a t}{2} \tan(\phi - a) \tag{4.41 a}$$

Also

$$\begin{aligned} ST &= SP \sin(\phi - a) \\ &= V_\phi t/2 \cdot \sin(\phi - a) \end{aligned} \tag{4.41 b}$$

Shift ST is expressed in terms of phase and ray velocities by Eqs. (4.41 a) and (4.41 b), respectively.

Taking the similar triangle PQR and PSM we have

$$\frac{QR}{SM} = \frac{PR}{PM} \tag{4.42}$$

Further,

$$\begin{aligned} SM &= \frac{ST}{\cos a} \\ &= V_\phi \frac{t}{2} \frac{\sin(\phi - a)}{\cos a} \end{aligned} \tag{4.43 a}$$

$$\begin{aligned} PM &= PT - MT \\ &= V_a \frac{t}{2} - V_\phi \frac{t}{2} \frac{\sin(\phi - a) \sin a}{\cos a} \end{aligned} \tag{4.43 b}$$

and

$$PR = \frac{Z}{\cos a} \tag{4.43c}$$

Substitution of Eq. (4.43 a) to (4.43 c) in (4.42) gives

$$QR = \frac{V_\phi Z \sin(\phi - a)}{V_a \cos^2 a - V_\phi \sin a \cos a \sin(\phi - a)} \tag{4.44 a}$$

Using Eq. (4.41 b) for ST, QR may also be expressed as

$$QR = \frac{Z \tan(\phi - a)}{\cos^2 a[1 - \tan(\phi - a) \tan a]} \tag{4.44 b}$$

QR is the error in the correct position of the imaged point at depth Z below the observation plane in the downward continued field, corresponding to error ST on the reflector [see Fig. 4.14 b (top)]. This error results because energy traveled along ST due to anisotropy whereas its path is taken as TP considering isotropy.

CHAPTER 5

Anisotropy Models of Sedimentary Sections and Characteristics of Wave Propagation

5

5.1
Introduction

A material is said to be anisotropic if its properties, when measured at the same point, change with direction. *Anisotropy* is exhibited in its pure form in crystals. A crystal's properties are governed by the periodic structure of its atoms. The study of anisotropic behavior of crystals has greatly helped understanding seismic anisotropy of sedimentary rocks and fractured zones. With this perspective, some salient points of anisotropy in crystals are introduced first.

A collection of points in a periodic arrangement is referred to as a *lattice*. A lattice could be created by translation of a point. By repeated translation of a point in one direction, two directions and three directions, one-dimensional, two-dimensional and three-dimensional lattices could be produced (Fig. 5.1 a–c).

Periodic repetition of points may also be created by rotation about a line and reflection in a plane.

Plane and space lattices could be produced by the operation of translation, rotation, reflection and their combinations.

Crystal faces are referred to in three or four lines or axes, which pass through a common point, the *center of the crystal*. These lines are called the *crystallographic*

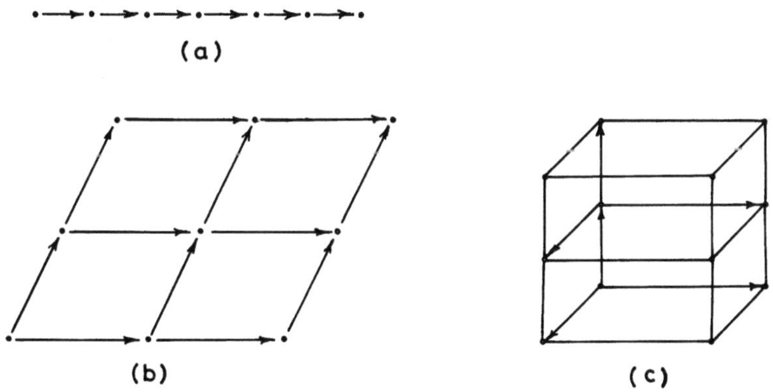

Fig. 5.1 a–c. Generation of: **a** one-dimensional; **b** two-dimensional; and **c** three-dimensional lattice, by translation of a point

axes. The direction of these lines is generally parallel to the edges of a unit cell, which is the same as parallel to the edges of possible faces of the crystal.

It may be remarked that the physical properties of an isotropic material can be represented by scalar quantities. In a crystalline solid, the properties of the material depend on the direction. Supposing a force or a field is applied to a crystal. The response of the crystal would depend not only on the magnitude of the force or field, but also on the direction in which it is applied relative to reference directions in the crystal. Since crystals are built on a periodic structure of atoms, they have the same symmetry as this structure. Because of symmetry, sets of all directions break up into subsets. The physical properties represented by scalar quantities in isotropic material are therefore replaced by a finite set of quantities in an anisotropic solid, which are direction-dependent.

It is important to note that variation of physical properties with direction in a crystal depends on the symmetry of the crystal. This may be illustrated by considering a *unit cell* in the form of a parallelogram (Fig. 5.2a) and square (Fig. 5.2b) in a plane lattice.

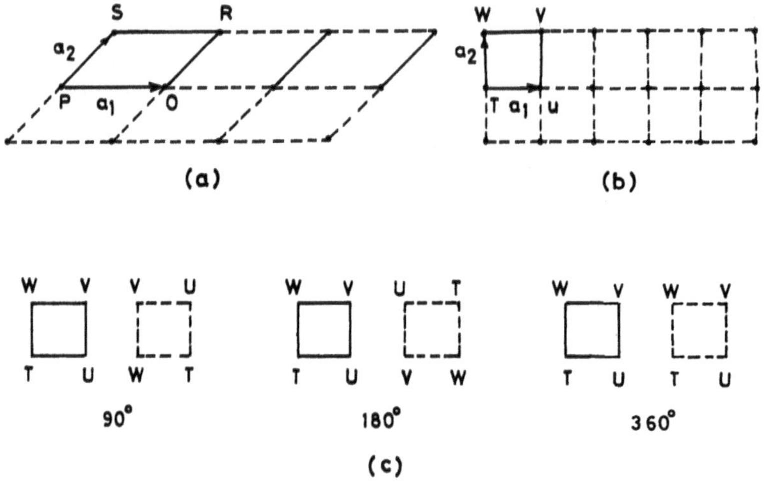

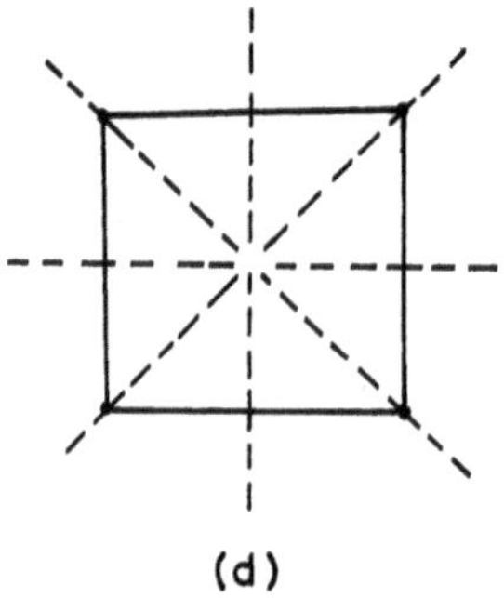

Fig. 5.2 a–d. a Unit cell in the form of a parallelogram; **b** unit cell in the form of a square; **c** rotational symmetry of 90, 180 and 360° about the center of a unit cell in the form of a square; and **d** symmetry planes shown by *dashed lines* for a unit cell in the form of a square

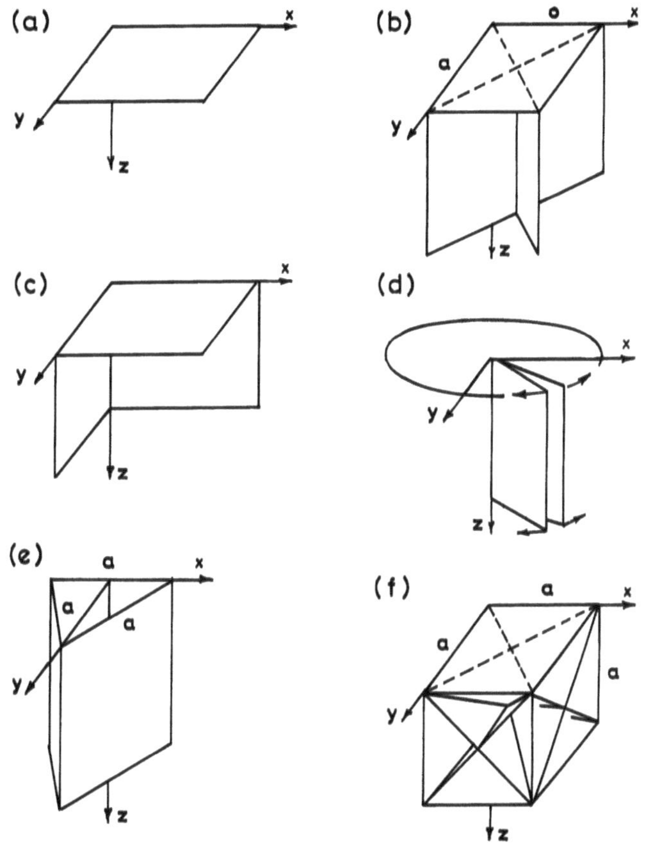

Fig. 5.3 a–f. Symmetry planes of the more common symmetry systems: **a** monoclinic; **b** tetragonal; **c** orthorhombic; **d** hexagonal; **e** trigonal; and **f** cubic (Crampin 1984, p 22)

Let us consider a symmetry operation on a *unit cell* in the form of a square in Fig. 5.2 b. This is shown in Fig. 5.2 c.

The unit cell in Fig. 5.2 b possess a symmetry of $\pm 90°$ rotation, $\pm 180°$ rotation, and $\pm 360°$ rotation. This means that the square obtained from rotations by these angles is identical to the original square and may be superposed exactly on the original square. Other than these symmetry properties, it contains reflection plains shown by the dashed lines in Fig. 5.2 d which are the planes of symmetry, i.e. a reflection of all points of the square on one side of the reflecting plane would produce a new square which superposes on the original. The parallelogram unit cell in Fig. 5.2 a does not possess all these symmetries.

Consider an imaginary two-dimensional crystal made up of metallic atoms located on lattice points in Fig. 5.2 a, b. If we measure the tensile stiffness along $\mathbf{a}_1$ and $\mathbf{a}_2$, they would in general not be equal for a lattice in the form of a parallelogram. It would show anisotropic behavior, i.e. directional dependence. In the case of a square lattice, if we consider metallic atoms located on a square lattice in the crystal, then because of rotational symmetry of 90° and 180°, tensile stiffness along $\mathbf{a}_1$ would be equal to tensile stiffness along $\mathbf{a}_2$ (Fig. 5.2b).

A general form of *compliance and stiffness* matrices may be obtained by the consideration of the microscopic property, the medium. If the two directions of a medium were symmetrically equivalent, identical stresses applied along these directions would produce the same amount of strain, which means that certain compliance constants are equal. This implies that if the medium itself is symmetric with respect to a particular transformation of coordinates the *stiffness* and *compliance matrices* would themselves be unchanged by the same transformation.

Solids having lattices like those in crystals may be constructed in only 32 distinct ways. These 32 ways are called 32 symmetry classes, which may be arranged in seven *symmetry systems* using group theory (see, for example, Auld 1973). These seven *symmetry systems* are triclinic, monoclinic, orthorhombic, tetragonal, trigonal, hexagonal and cubic.

Representatives of 32 crystal classes differ in symmetry properties. Symmetry properties define the way by which a material may be transformed either physically or conceptually to make it appear the same after transformation as before.

Apart from the rotation and reflection symmetries mentioned above, there are other symmetry operations that may be generated by simple combinations of rotation and reflection. Symmetry properties such as axis of rotation, plane of reflection or center of inversion go to define a symmetry class or system. For example, considering a monoclinic solid, it may have at most (1) a single two-fold axis of rotation, (2) a single mirror plane, or (3) a single two-fold axis ⊥r to a single mirror plane plus a center of inversion. These three sets of symmetry properties are the three *symmetry classes* belonging to monoclinic system.

Crystals have 32 symmetry classes. However, in principle, there are more than 32 symmetry classes for rocks. An example is a transversely isotropic (T.I.) rock mass. A transversely isotropic solid has an infinite fold axis–a symmetry possessed by no crystal. A transversely isotropic solid may be created, for example, by a single set of plane parallel cracks in an otherwise isotropic rock (refer to Section 5.3.2 on *azimuthal anisotropy*). The *axis* of infinite fold is perpendicular to the aligned crack planes. In fact, hexagonal symmetry in a crystal with a single six-fold axis is a symmetry class of the T.I. system.

Further, by introducing sets of oriented cracks into a homogeneous isotropic rock, it is possible to create rocks belonging to all 32 crystal symmetry classes, and some more. The introduction of two sets of homogeneous, plane parallel cracks into an isotropic rock can create a member of the *monoclinic, orthorhombic* or *tetragonal symmetry system*. Three sets of parallel planar cracks can produce a *triclinic, trigonal, hexagonal* or *cubic system*. For details on this subject, reference may be made to an excellent contribution by Winterstein (1990).

It may be mentioned here that, among the symmetry systems, transverse isotropy (T.I.) is of large significance to geophysicists. This is because dominant anisotropy of horizontally layered sedimentary rocks may be modeled as transversely isotropic with a *vertical axis of symmetry,* referred to as a VTI medium (Section 5.3.1). Also, the presence of a single set of vertical cracks in a non-layered rock system would result in *azimuthal anisotropy,* referred to as a HTI medium with a *horizontal axis of symmetry* (refer to Section 5.3.2 for details).

Some symmetry systems may be more common in sedimentary rocks than others. For example, transverse isotropy is one such likely system. The presence of a multiple set of aligned cracks suggest that the *monoclinic* and *triclinic system* may also be common (c.f. Nelson 1985). The *orthorhombic system*, which can be generated by a single set of vertical cracks in a transversely isotropic (T.I.) background medium, may also be found in sedimentary rocks. Some basic details of the orthorhombic system are given in Section 5.5.

All rocks possess sets of cracks or layering or both, and many rocks with such features are effectively homogeneous over distances of the order of seismic wavelengths or larger. Causes of *anisotropy* in sedimentary rocks may be layering, aligned cracks, in-situ stresses, and orientation of microscopic mineral particles such as clay. Winterstein and Paulson (1990) show that orientation of mineral particles produce the dominant anisotropic effect in shale. Shale constitutes almost seventy five percent of sedimentary rocks, and play an important role in fluid flow and seismic wave propagation in sedimentary basins. Studies of *velocity anisotropy* in shale have been made in laboratories (see, for example, Kaarsberg 1959, Podio et al. 1968, Lo et al. 1986, Yin 1993, Hornby et al. 1994, Johnston and Christensen 1995). In-situ studies of shale may be found in Winterstein and Paulsson (1990). Vernik and Liu (1997) discuss in detail the physical properties of shale and cause-effect of *velocity anisotropy* in shale.

5.2 Description of Elastic Behavior

In contrast to crystallographers who study the symmetry property of solid crystals, a seismologists' concern is on the symmetries of elastic properties. The symmetries of elastic properties may be explored by studying *elastic stiffness tensors* for different symmetry systems.

In some specific seismic investigations, the symmetry of elastic property is explored. As an example, in an isotropic medium with vertical plane parallel fractures, elastic properties in the vertical plane containing the fracture would differ from properties in any other vertical plane. The aim in seismic investigations would, therefore, be to detect this difference, and interpret it appropriately on the basis of elasticity theory.

Crampin (1984) concluded that anisotropic symmetry systems can be very readily classified on the basis of the relative arrangement of (mirror) symmetry planes. In fact, in seismic investigations, the arrangement of symmetry planes with respect to free surface and with respect to the direction of wave propagation have major effects on seismic phenomenon observed. Crampin (1984) proved the following theorem relating orientation of symmetry plane with fourth order tensor of elastic constants (which describe the elastic properties of the medium):

'The plane $x_p = 0$ is a plane of (mirror) symmetry if and only if $c_{jkmn} = 0$ whenever one or three of j, k, m, n is equal to p.'

Here, c_{jkmn} is a fourth-order tensor of elastic stiffness constants. This theorem implies that a plane in an anisotropic solid will possess mirror symmetry if the elastic constants are unchanged by reflection in the plane.

To illustrate the use of the above theorem, we consider Hooke's law dealing with elastic properties of the medium:

$$\begin{aligned} &\sigma_{ij} = c_{ijkl} S_{kl} \\ &i, j, k, l = 1, 2, 3 \end{aligned} \tag{5.1}$$

In this equation, c_{ijkl} is the stiffness tensor, which specifies the elastic properties of the solid. Eq. (5.1) shows that there are 81 (=3×3×3×3) elastic stiffness constants relating nine components of strain tensor to nine components of stress tensor. However, all of the 81 elastic constants are not independent. The following condition on stiffness constants

$$c_{jkmn} = c_{kjmn} = c_{jknm} = c_{kjnm}$$

reduces the number of independent constants to 36. Also the condition

$$c_{jkmn} = c_{mnjk}$$

further reduces the number of constants to 21. Twenty-one is the maximum number of elastic stiffness constants for any medium. For details on the reduction of the number of elastic constants to 21, reference may be made to standard textbooks, like Auld (1973, vol. 1) or Love (1944). Usually, the number is much less than 21 due to additional restrictions on the microscopic nature of the medium.

We may express the elastic stiffness tensor in the form of a 6×6 symmetric matrix using abbreviated notations (refer Section 5.2.1). Thus, we have

$$\begin{bmatrix} c_{11\,11} & c_{11\,22} & \cdots & c_{11\,31} & c_{11\,12} \\ c_{22\,11} & c_{22\,22} & \cdots & c_{22\,31} & c_{22\,12} \\ c_{33\,11} & c_{33\,22} & \cdots & c_{33\,31} & c_{33\,12} \\ c_{23\,11} & c_{23\,12} & \cdots & c_{23\,31} & c_{23\,12} \\ c_{31\,11} & c_{31\,22} & \cdots & c_{31\,31} & c_{31\,12} \\ c_{12\,11} & c_{12\,22} & \cdots & c_{12\,31} & c_{12\,12} \end{bmatrix} = \begin{bmatrix} c_{11} & c_{12} & c_{13} & c_{14} & c_{15} & c_{16} \\ c_{12} & c_{22} & c_{23} & c_{24} & c_{25} & c_{26} \\ c_{13} & c_{23} & c_{33} & c_{34} & c_{35} & c_{36} \\ c_{14} & c_{24} & c_{34} & c_{44} & c_{45} & c_{46} \\ c_{15} & c_{25} & c_{35} & c_{45} & c_{55} & c_{56} \\ c_{16} & c_{26} & c_{36} & c_{46} & c_{56} & c_{66} \end{bmatrix}$$

Love (1944) used capital letters in place of $c_{\alpha\beta}$ (α, β=1, 2, 3, 4, 5, 6). In Love's notation the above symmetric matrix is given by

$$\begin{bmatrix} A & H & G & P & Q & R \\ H & B & F & S & T & U \\ G & F & C & V & W & X \\ P & S & V & L & K & J \\ Q & T & W & K & M & I \\ R & U & X & J & I & N \end{bmatrix}$$

The symmetry of the matrix evolves from the consideration of *strain energy*, and holds for solids of all symmetries.

We next consider forms of stiffness tensor for materials of different symmetry systems. Comparing stiffness tensors of different symmetry systems is to compare their corresponding 6×6 stiffness matrices. Form of the stiffness tensor of a particular symmetry system means the arrangement of non-zero constants in the 6×6 matrix and any constraints that may exist among constants in the 6×6 matrix when it is in the frame of maximum symmetry.

The orientation of symmetry planes of common symmetry systems is given in Fig. 5.3.

According to Crampin (1984, p 22) 'wave motion in any particular symmetry plane is very similar to motion in any other symmetry plane, and wave motion in any off-symmetry plane has generalized behavior which is not dependent on the particular symmetry system.'

To illustrate use of the theorem given by Crampin (1984), quoted earlier in this section, for writing a stiffness matrix for any symmetry system, we take case of the orthorhombic system illustrated in Fig. 5.3. In this case $x_1=0$, $x_2=0$, $x_3=0$ are the symmetry planes. Using a theorem relating orientation of symmetry planes with stiffness constants, the form of the stiffness matrix for the *orthorhombic system* is:

$$\begin{bmatrix} c_{11} & c_{12} & c_{13} & 0 & 0 & 0 \\ c_{21} & c_{22} & c_{23} & 0 & 0 & 0 \\ c_{31} & c_{32} & c_{33} & 0 & 0 & 0 \\ 0 & 0 & 0 & c_{44} & 0 & 0 \\ 0 & 0 & 0 & 0 & c_{55} & 0 \\ 0 & 0 & 0 & 0 & 0 & c_{66} \end{bmatrix}$$

The orthorhombic system has nine independent elastic constants.

The stiffness matrix for other symmetry systems may be obtained in a similar way using the theorem proposed by Crampin (1984). However, the tensor of elastic constants must be rotated to get the symmetry plane normal to one of the coordinate axes for the application of Crampin's theorem.

The form of the stiffness tensor for a transversely isotropic (T.I.) solid is:

$$\begin{bmatrix} c_{11} & c_{12} & c_{13} & 0 & 0 & 0 \\ & c_{22} & c_{23} & 0 & 0 & 0 \\ & & c_{33} & 0 & 0 & 0 \\ & \text{(SYM)} & & c_{44} & 0 & 0 \\ & & & & c_{55} & 0 \\ & & & & & c_{66} \end{bmatrix}$$

with constraints $c_{11}=c_{22}$, $c_{13}=c_{23}$, $c_{44}=c_{55}$, and $c_{12}=c_{11}-2c_{66}$. The constraints come from the requirement that the tensor be the same after a symmetry operation, as it was before.

Isotropic tensors have the same non-zero locations as T.I. tensors, and there are more symmetry constraints than for T.I. tensors, such as: $c_{11}=c_{22}=c_{33}$, $c_{44}=c_{55}=c_{66}$, and $c_{12}=c_{13}=c_{23}=c_{11}-2c_{66}$. In terms of Lame's constants for isotropic solids

$$\gamma = c_{12}, \mu = c_{44}$$

It may be observed that orthorhombic tensors occupy the same non-zero positions as T.I. tensors with no constraints. Hexagonal tensors have the same form as T.I. tensors, with only constraint

$$c_{66} = \frac{c_{11} - c_{12}}{2}$$

In acoustic vibrations, the deformation of the medium is described by the strain field, **S** (**r**, t), and elastic restoring forces are defined by stress field $\sigma(\mathbf{r}, t)$, where **r** and t are space and time variables, respectively.

Elastic restoring forces that try to bring back the particle to its equilibrium position through vibrational motion are like the systems of masses and springs.

It is pertinent to find a relationship between elastic deformation and the corresponding restoring forces. In other words, we should define microscopic "spring constants" of the medium.

Various stages in the deformation of the body may be identified. For example, for small deformations strain is linearly proportional to applied stress (Hooke's law). With increasing deformation, the relationship between stress and strain becomes non-linear, but the body returns to its initial position when stress is removed. This region of linear and non-linear deformation is included in elastic deformation. Increasing the strain beyond a certain limit, called elastic limit, leads to permanent deformation and ultimately to fracturing (Fig. 5.4).

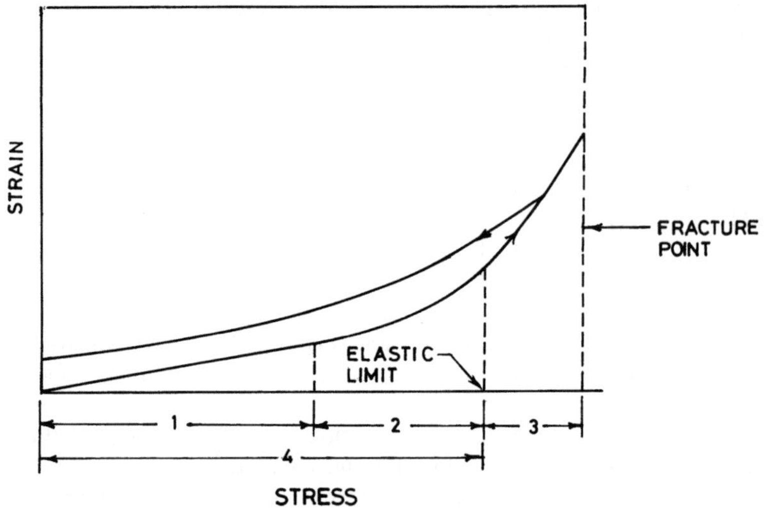

Fig. 5.4. Typical stress strain relation for a solid material: *1* is the stress range of linear deformation; *2* is the range for non-linear deformation; *3* is for plastic deformation; and *4* is the range of elastic deformation. (Source: Auld 1973, p 58)

5.2.1
Elastic Constitutive Relations: Stiffness and Compliance Constants

Hooke's law states that strain is linearly proportional to stress, or conversely, the stress is linearly proportional to strain. We first take the second form. It states that each component of stress is a general function of all strain components. For example:

$$\begin{aligned}\sigma_{xx} = {} & c_{xxxx}S_{xx} + c_{xxxy}S_{xy} + c_{xxxz}S_{xz} \\ & + c_{xxyx}S_{yx} + c_{xxyy}S_{yy} + c_{xxyz}S_{yz} \\ & + c_{xxzx}S_{zx} + c_{xxzy}S_{zy} + c_{xxzz}S_{zz}\end{aligned} \tag{5.1 a}$$

Also, strains may be expressed as general linear functions of all components of stress:

$$\begin{aligned}& S_{ij} = s_{ijkl}\sigma_{kl} \\ & i, j, k, l = x, y, z\end{aligned} \tag{5.2}$$

s_{ijkl} are called compliance constants.

Elastic stiffness constants have small values for easily deformed materials and large values for very rigid materials. Compliance constants are a measure of the deformability of the medium, and consequently they have large values for easily deformed material, and small values for rigid materials.

Relations (5.1) and (5.2) are called *elastic constitutive relations*.

Analysis of Eq. (5.1) shows that stiffness has the dimension of stress (newtons/m^2) because strain is dimensionless. Compliance has a dimension of m^2/newton. The range of values of stiffness is 0.1×10^{10} newtons/m^2 (for rubber-like material) to 10×10^{10} newtons/m^2 (for metal-like materials). Corresponding compliances range from $1{,}000\times10^{-12}$ to 10×10^{-12}. The stress at elastic limit (strain $\approx 10^{-4}$ to 10^{-3}) is in range of 10^7 to 10^8 newtons/m^2 for rigid materials (Auld 1973, p 60).

Because of the stress and strain symmetry condition, the following relations hold in stiffness and compliance constants:

$$c_{ijkl} = c_{jikl} \tag{5.3 a}$$

$$c_{ijkl} = c_{ijlk} \tag{5.3 b}$$

and

$$s_{ijkl} = s_{jikl} \tag{5.4 a}$$

$$s_{ijkl} = s_{ijlk} \tag{5.4 b}$$

Under these constraints, the four subscripts may be reduced to two, by the use of abbreviated subscript notation.

The relationship between stiffness in full notation and abbreviated notation is

$$c_{ijkl} = c_{IJ} \tag{5.5}$$

Expression for Hooke's law in abbreviated notation is

$$\begin{aligned} &\sigma_I = c_{IJ} S_J \\ &I, J = 1, 2, 3, 4, 5, 6 \end{aligned} \tag{5.6}$$

Also we have

$$\begin{aligned} &S_I = s_{IJ} \sigma_J \\ &I, J = 1, 2, 3, 4, 5, 6 \end{aligned} \tag{5.7}$$

where

$$s_{IJ} = s_{ijkl} \times \left\{ \begin{array}{lll} 1 \text{ for I and} & J = 1, 2, 3 \\ 2 \text{ for I or} & J = 4, 5, 6 \\ 4 \text{ for I and} & J = 4, 5, 6 \end{array} \right\}$$

[Auld 1973, Eq. (3.19)]

From Eqs. (5.6) and (5.7) we have

$$[s] = [c]^{-1} \tag{5.8}$$

The compliance matrix [s] is the inverse of stiffness matrix [c], or

$$[s][c] = [c][s] = [I] \tag{5.9}$$

[I] is the identity matrix. Equation (5.9) is useful for converting stiffness constants to compliance constants and viceversa.

In acoustic vibration and wave propagation problems, it is required to transform certain quantities into a new coordinate system suitable to the geometry of a particular problem. The transformation required may be from a rectangular coordinate to cylindrical coordinate or to a spherical coordinate; or, it may be a rotation of coordinate axes. We consider here only the transformation with respect to rotation of coordinate axes for first, second and fourth rank tensor quantities.

Let us assume that x, y and x′, y′ represent the old and new coordinate axes. The relative orientation of the two axes is defined by the direction cosine a_{ij} as depicted in Fig. 5.5. It is cosine of the angle between the new axis i and old axis j. By convention the first subscript i represent the new axis and second subscript j the old axis. The matrix form of a_{ij} is

$$[a] = \begin{bmatrix} a_{xx} & a_{xy} & a_{xz} \\ a_{yx} & a_{yy} & a_{yz} \\ a_{zx} & a_{zy} & a_{zz} \end{bmatrix} \tag{5.10}$$

Consider first the transformation of a first rank tensor (vector). Its components in the old coordinate system are V_x and V_y, and in the new coordinate system V'_x

and V'_y. The components along the system of new axes may be obtained by taking projections of original components on relevant axes (Fig. 5.5). For example:

$$V_{x'} = V_x \cos \phi_{x'x} + V_y \cos \phi_{x'y} \tag{5.11}$$

In terms of elements of the transformation matrix it may be written as

$$V_{x'} = a_{xx} V_x + a_{xy} V_y \tag{5.12}$$

The general law for transforming a vector 'V' is

$$\begin{aligned} V_i &= a_{ij} V_j \\ i, j &= x, y, z \end{aligned} \tag{5.13 a}$$

The transformation expressed by Eq. (5.13 a) gives a new description of the initial vector V. The magnitude of V is invariant under the transformation.

The transformation law in the form of Eq. (5.13 a) is general and applies to any vector quantity.

The corresponding transformation of second rank tensor quantities, like stress or strain, follows from Eq. (5.13 a). For example, for stress it may be written as

$$\begin{aligned} \sigma'_{ij} &= a_{ik} a_{jl} \sigma_{kl} \\ i, j, k, l &= x, y, z \end{aligned} \tag{5.13 b}$$

where σ_{kl} are the components of stress in the old coordinates and σ'_{ij} in the new coordinates.

Similarly, th

c_{mnop} and s_{mnop} in Eqs. (5.13 c) and (5.13 d) are the transformed values in the new coordinate system of the fourth rank tensors c_{ijkl} and s_{ijkl}, respectively.

Equations (5.13 c) and (5.13 d) may be used to write stiffness and compliance constants in the new rectangular coordinate system (x′, y′, z′) from their given values in old coordinate system (x, y, z). The relationship is through four transformation matrices of a_{mi}, a_{nj}, a_{ok} and a_{pl}. The general transformation procedure should be clear by looking to forms of transformation for the first rank tensor [Eq. (5.13 a)], second rank tensor [Eq. (5.13 b)] and fourth rank tensors [Eqs. (5.13 c and d)].

5.2.2
Bond Transformation Matrix

It is a technique of transforming stress and strain, or stiffness and compliance without using tensor notations. Transformation is implemented through matrix multiplication. The advantage of a bond transformation matrix for stiffness and compliance is that it is applied directly on stiffness or compliance constants given in *abbreviated subscript notation.* The algebra involved is much less, as it uses a suitable matrix multiplication. For example, if σ' and σ are transformed stress and initial stress, they are related by a 6×6 *Bond stress transformation matrix* [M].

$$[\sigma'] = [M][\sigma] \tag{5.14 a}$$

Similarly, for strain

$$[S'] = [N][S] \tag{5.14 b}$$

[M] and [N] in Eqs. (5.14 a) and (5.14 b) are Bond-Transformation Matrices for stress and strain, respectively.

Further, if [c] is *original stiffness matrix* and [c′] is *transformed stiffness matrix*, then these are related as

$$[c'] = [M][c]\left[\tilde{M}\right] \tag{5.15 a}$$

Similarly, compliance transformation is given by

$$[s'] = [N][s]\left[\tilde{N}\right] \tag{5.15 b}$$

Algorithm 5.1

Method of Writing Elements of Bond Transformation Matrix

The form of a transformation matrix may be easily derived by expressing transformed stress components and original components in abbreviated notation. For example, the transformed component of stress in abbreviated notations is expressed as

$$\begin{aligned}\sigma_1' =& a_{xx}^2\sigma_1 + a_{xy}^2\sigma_2 + a_{xz}^2\sigma_3 + 2a_{xy}a_{xz}\sigma_4 \\ & + 2a_{xx}a_{xz}\sigma_5 + 2a_{xx}a_{xy}\sigma_6\end{aligned} \tag{5.16}$$

In writing Eq. (5.16), $\sigma_{ij}=\sigma_{ji}$ is assumed.

In a similar way, all six equations may be written to express transformed stress components in terms of original stress components through transformation matrix elements a_{xx}, a_{xy} etc. These six equations may be written in compact form as

$$[\sigma' H] = [M_{HI}][\sigma_I] \tag{5.17}$$

[Auld 1973, Eq. (3.31)]

$$H, I = 1, 2, 3, 4, 5, 6$$

The form of a 6×6 transformation matrix $[M_{HI}]$ is

$$[M_{HI}] = \begin{bmatrix} A & B \\ C & D \end{bmatrix} \tag{5.18}$$

where A, B, C and D have an array of 3×3 elements in the following forms:

$$\begin{aligned}
A \equiv \; & \begin{matrix} a_{xx}^2 & a_{xy}^2 & a_{xz}^2 \\ a_{yx}^2 & a_{yy}^2 & a_{yz}^2 \\ a_{zx}^2 & a_{zy}^2 & a_{zz}^2 \end{matrix} \\
B \equiv \; & \begin{matrix} 2a_{xy}a_{xz} & 2a_{xz}a_{xx} & 2a_{xx}a_{xy} \\ 2a_{yy}a_{yz} & 2a_{yz}a_{yx} & 2a_{yx}a_{yy} \\ 2a_{zy}a_{zz} & 2a_{zz}a_{zx} & 2a_{zx}a_{zy} \end{matrix} \\
C \equiv \; & \begin{matrix} a_{yx}a_{zx} & a_{yy}a_{zy} & a_{yz}a_{zz} \\ a_{zx}a_{xx} & a_{zy}a_{xy} & a_{zz}a_{xz} \\ a_{xx}a_{yx} & a_{xy}a_{yy} & a_{xz}a_{yz} \end{matrix} \\
D \equiv \; & \begin{matrix} a_{yy}a_{zz} + a_{yz}a_{zy} & a_{yx}a_{zz} + a_{yz}a_{zx} & a_{yy}a_{zx} + a_{yx}a_{zy} \\ a_{xy}a_{zz} + a_{xz}a_{zy} & a_{xz}a_{zx} + a_{xx}a_{zz} & a_{xx}a_{zy} + a_{xy}a_{zx} \\ a_{xy}a_{yz} + a_{xz}a_{yy} & a_{xz}a_{yx} + a_{xx}a_{yz} & a_{xx}a_{yy} + a_{xy}a_{yx} \end{matrix}
\end{aligned} \tag{5.19}$$

(Auld 1973)

If the form of [a] is remembered, it is possible to write the elements of A, B, C and D. For example, A is the square of the corresponding term in [a] in Eq. (5.10). For writing B, C, D from [a] the procedure as given in Table 5.1 may be followed:

Similarly the form of strain transformation matrix may be derived. In this case transformation matrix [N] is the same as [M] except that factor 2 multiplied in elements of B is now multiplied in the elements of C.

Table 5.1. Method of writing elements B, C and D from [a] in Eq. (5.19) (Source: after Auld 1973, p 74)

To get elements of	Method
C	Strike out the corresponding element in [a] and take the product of remaining two terms in the same column
B	Use row in [a] and follow the same procedure as for C, and multiply by 2
D	Strike out both row and column in [a] corresponding to the same term and take the sum of the cross product in the remaining 2×2 array of elements of [a]

Exercise 5.1

Elastic Stiffness Constants in a Rotated Coordinate

Before including an exercise on the use of a bond transformation matrix for writing elastic stiffness constants in a rotated coordinate, we first obtain a suitable matrix equation governing this transformation.

Hooke's law may be written as

$$[\sigma] = [c][S]$$

Using the Bond stress transformation matrix we may write

$$[\sigma'] = [M][\sigma] = [M][c][S] \tag{5.20}$$

Replacing [S] by transformed strain matrix [S′] we have

$$[\sigma'] = [M][c][N]^{-1}[S'] \tag{5.21}$$

Equation (5.21) is a relation between transformed stress and strain. From this equation it follows that the transformed elastic stiffness constant matrix [c′] is given by

$$[c'] = [M][c][N]^{-1} \tag{5.22}$$

Similarly, the *transformed compliance matrix* [s′] is given by

$$[s'] = [N][s][M]^{-1} \tag{5.23}$$

Since

$$[N]^{-1} = [\tilde{M}]$$

we have

$$[c'] = [M][c][\tilde{M}] \quad (5.24)$$

$$[s'] = [N][s][\tilde{N}] \quad (5.25)$$

Equations (5.24) and (5.25) may be used to write *transformed values of stiffness and compliance constants* using matrix multiplication.

Now consider, as an example, a medium with the following non-zero stiffness constants (Auld 1973, p 60):

$$\begin{aligned}
&c_{xxxx} = c_{yyyy} = c_{zzzz} \\
&c_{xxyy} = c_{yyxx} = c_{xxzz} = c_{zzxx} = c_{yyzz} = c_{zzyy} \\
&c_{yxyx} = c_{yxxy} = c_{zyyz} = c_{zyzy} \\
&\quad = c_{xzxz} = c_{xzzx} = c_{zxxz} = c_{zxzx} \\
&\quad = c_{xyxy} = c_{xyyx} = c_{yxxy} = c_{yxyx}
\end{aligned} \quad (5.26)$$

For writing a value of the stiffnesses in a rotated coordinate system with clockwise rotation by an angle ξ about the z-axis, we make use of Eq. (5.13 c)

$$c_{mnop} = a_{mi} a_{nj} a_{ok} a_{pl} c_{ijkl}$$

and

$$[a] = \begin{bmatrix} \cos\xi & \sin\xi & 0 \\ -\sin\xi & \cos\xi & 0 \\ 0 & 0 & 1 \end{bmatrix} \quad (5.27)$$

The result is

$$\begin{aligned}
c'_{xxxx} &= a_{xx}^4 c_{xxxx} + a_{xy}^4 c_{yyyy} + a_{xy}^4 c_{zzzz} \\
&\quad + a_{xx}^2 a_{xy}^2 c_{xxyy} + a_{xy}^2 a_{xx}^2 c_{yyxx} + \dots \\
&= (\cos^4\xi) c_{xxxx} + (\sin^4\xi) c_{xxxx} + (2\cos^2\xi \sin^2\xi) c_{xxyy} + \dots
\end{aligned} \quad (5.28)$$

Similarly, other stiffness constants may be written in rotated coordinates.

Now let us consider a material with the following stiffness matrix (Auld 1973, p 71)

$$c = \begin{bmatrix} c_{11} & c_{12} & c_{12} & 0 & 0 & 0 \\ c_{12} & c_{11} & c_{12} & 0 & 0 & 0 \\ c_{12} & c_{12} & c_{11} & 0 & 0 & 0 \\ 0 & 0 & 0 & c_{44} & 0 & 0 \\ 0 & 0 & 0 & 0 & c_{44} & 0 \\ 0 & 0 & 0 & 0 & 0 & c_{44} \end{bmatrix} \quad (5.29)$$

Suppose these stiffness constants are to be expressed in a rotated coordinate system with rotation about the z-axis. In this case [a] is

$$[a] = \begin{bmatrix} \cos\xi & \sin\xi & 0 \\ -\sin\xi & \cos\xi & 0 \\ 0 & 0 & 1 \end{bmatrix}$$

and the bond-transformation matrix from Eq. (5.19) is:

$$[M] = \begin{bmatrix} \cos^2\xi & \sin^2\xi & 0 & 0 & 0 & \sin^2\xi \\ \sin^2\xi & \cos^2\xi & c_{12} & 0 & 0 & -\sin^2\xi \\ 0 & 0 & 1 & 0 & 0 & 0 \\ 0 & 0 & 0 & \cos\xi & -\sin\xi & 0 \\ 0 & 0 & 0 & \sin\xi & \cos\xi & 0 \\ -\dfrac{\sin^2\xi}{2} & \dfrac{\sin^2\xi}{2} & 0 & 0 & 0 & \cos^2\xi \end{bmatrix} \tag{5.30}$$

To obtain stiffness constants in the new coordinate system we make use of Eq. (5.24), i.e. the stiffness matrix is multiplied on the left by [M] and the right by $[M]^T$ giving (Auld 1973, p 77)

$$[c'] = \begin{bmatrix} c'_{11} & c'_{12} & c'_{13} & 0 & 0 & c'_{16} \\ c'_{12} & c'_{11} & c'_{13} & 0 & 0 & -c'_{16} \\ c'_{13} & c'_{13} & c'_{33} & 0 & 0 & 0 \\ 0 & 0 & 0 & c'_{44} & 0 & 0 \\ 0 & 0 & 0 & 0 & c'_{44} & 0 \\ c'_{16} & -c'_{16} & 0 & 0 & 0 & c'_{66} \end{bmatrix} \tag{5.31}$$

where,

$$\begin{aligned} c'_{11} &= c_{11} - \left(\frac{c_{11} - c_{12}}{2} - c_{44}\right)\sin^2\xi \\ c'_{12} &= c_{12} + \left(\frac{c_{11} - c_{12}}{2} - c_{44}\right)\sin^2\xi \\ c'_{13} &= c_{12} \\ c'_{16} &= -\left(\frac{c_{11} - c_{12}}{2} - c_{44}\right)\sin^2\xi\cos^2\xi \\ c'_{33} &= c_{11} \\ c'_{44} &= c_{44} \\ c'_{66} &= c_{44} + \left(\frac{c_{11} - c_{12}}{2} - c_{44}\right)\sin^2\xi \end{aligned} \tag{5.31 a}$$

There is a notable difference between elements of [c′] and [c]. Original elements of the stiffness matrix that were equal are now unequal for the transformed elements. Also some zero elements in the original stiffness matrix are nonzero in the transformed stiffness matrix.

In this section explanation of concepts and methodologies is derived from detailed discussions in Auld (1973).

5.3 Occurrence of Anisotropy and Models for Sedimentary Sections

5.3.1 Layering Induced Anisotropy: VTI Media

The process of formation of sedimentary rocks generates layering, and *cyclic sedimentation* produces *periodic layering.* Figure 5.6a shows an elementary volume of a stratified periodic medium consisting of alternating rock types 1 and 2.

On a proper scale, the elastic behavior of a periodic layered structure would be like a homogeneous transversely isotropic (T.I.) medium–having a symmetry axis perpendicular to the plane of stratification. The isotropic constituents of the layered structure are each characterized by two *elastic stiffness constants* called *Lames constants,* λ_1 and μ_1, and λ_2 and μ_2, respectively. The equivalent homogeneous transversely isotropic medium, however, possesses five independent elastic constants: c_{11}, c_{12}, c_{13}, c_{33}, c_{44} (including c_{66}, where $c_{66}=c_{11}-c_{12}/2$).

Values of elastic constants c_{11}, c_{12}... etc. may be explicitly calculated knowing the thicknesses and elastic constants of isotropic constituents of the medium. These are given by the following equations:

$$c_{11} = \frac{1}{L}\left\{ \frac{(h_1+h_2)^2(\lambda_1+2\mu_1)(\lambda_2+2\mu_2)+4h_1h_2(\mu_1-\mu_2)}{[(\lambda_1+\mu_1)-(\lambda_2+\mu_2)]} \right\} \tag{5.32 a}$$

$$c_{12} = \frac{1}{L}\{(h_1+h_2)^2\lambda_1\lambda_2 + 2(\lambda_1 h_1+\lambda_2 h_2)(\mu_2 h_1+\mu_1 h_2)\} \tag{5.32 b}$$

$$c_{13} = \frac{1}{L}\{(h_1+h_2)[\lambda_1 h_1(\lambda_2+2\mu_2)+\lambda_2 h_2(\lambda_1+2\mu_1)]\} \tag{5.32 c}$$

$$c_{33} = \frac{1}{L}\{(h_1+h_2)^2(\lambda_1+2\mu_1)(\lambda_2+2\mu_2)\} \tag{5.32 d}$$

$$c_{44} = \frac{(h_1+h_2)\mu_1\mu_2}{(h_1\mu_2+h_2\mu_1)} \tag{5.32 e}$$

$$c_{66} = \frac{(\mu_1 h_1+\mu_2 h_2)}{(h_1+h_2)} \tag{5.32 f}$$

[Postma 1955, Eq. (7)] where L is given by

$$L = (h_1+h_2)[h_1(\lambda_2+2\mu_2)+h_2(\lambda_1+2\mu_1)]$$

Derivation of Eq. (5.32) is based on the establishment of stress-strain relations for an elementary volume of the periodic layered structure, as in Fig. 5.6a. The coefficients connecting the six stress components with six strain components are expressible in terms of thicknesses and elastic constants (Lame's constants) of isotropic layers.

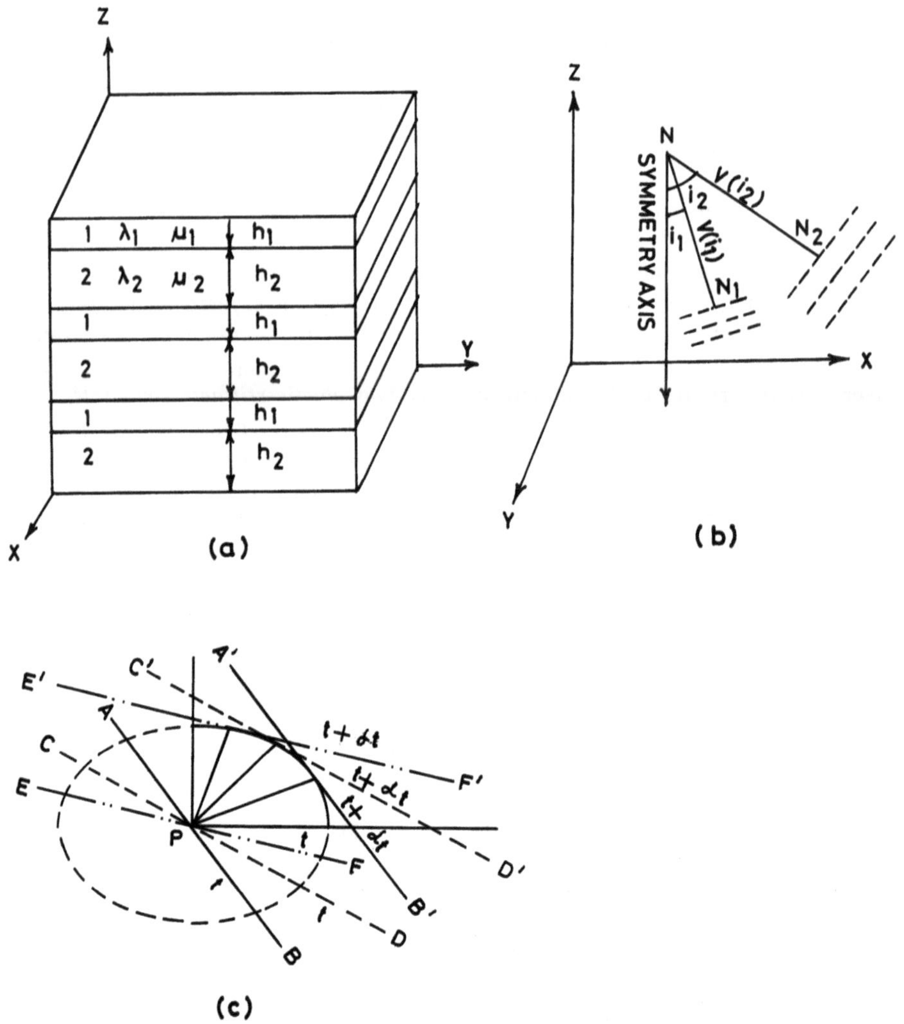

Fig. 5.6a–c. a Elementary volume of a stratified periodic medium; **b** plane wave velocity (V) is a function of the direction of the normal (NN_1 or NN_2 etc.) to the wave front (*dashed lines*); **c** wave surface generated by taking plane wave fronts in all directions

Making use of stress-strain relations and equations of motion for a homogeneous T.I. medium, equations describing propagation of plane waves may be derived. The plane wave velocity would be found to be a function of the direction of normal to the wavefront (Fig. 5.6b).

This result of velocity dependence on direction is in complete contrast with behavior in isotropic constituents in which velocity is independent of direction.

The directional dependence of velocity in a periodic layered structure as in Fig. 5.6a may be further analyzed in the light of Huygen's principle. Consider the position of a plane wavefront at time 't' (Fig. 5.6c). The secondary wavefront emanating from point P on a plane wavefront at time t is tangent to the plane wavefront at time t+dt. Now consider another plane wavefront in a different direction, passing through point P at time t. The secondary wavefront must again be tangent to the plane wavefront at time (t+dt). Therefore, the secondary wavefront is the envelope of plane waves having all possible directions at t+dt. For a homogeneous medium, the envelope of all plane wavefronts at time (t+1) is referred to as the *wave surface* (Love 1944). The wave surface for SH type motion would be an ellipsoid obtained by rotation of the ellipse

$$\frac{x^2}{c_{66}/\rho} + \frac{z^2}{c_{44}/\rho} = 1$$

about the z-axis.

The question of anisotropic behavior of a *non-periodic layered structure* was systematically examined in Backus (1962). It also dealt with the situation when individual layers may be intrinsically anisotropic as against isotropic layers considered in Postma (1955). Backus showed that a horizontally stratified medium whose Lame's parameters and density have position-independent averages over a vertical distance 'l' behaves like a homogeneous transversely isotropic medium with a vertical symmetry axis for wavelengths greater than 'l'. Mathematical steps described below follows from Backus (1962). Suppose (x, y, z) is the position coordinate in a Cartesions coordinate system. Let w(z) be a continuous weighting function which averages over a length 'l'. Further, w(z) has the following properties:

$$\begin{aligned} & w(z) \geq 0 \qquad w(\pm\infty) = 0 \\ & \int_{-\infty}^{+\infty} w(z)dz = 1 \int_{-\infty}^{+\infty} zw(z)dz = 0 \\ & \int_{-\infty}^{+\infty} z^2 w(z)dz = l'^2 \end{aligned} \tag{5.33}$$

With the above definition of the weighting function, the average of a function f(z) over a distance l' around position z is given by

$$\langle f\rangle(z) - \int_{-\infty}^{+\infty} w(\zeta - z)f(\zeta)d\zeta \tag{5.34}$$

It is a moving average and depends on position z.

Also if 'f' is a function of x, y, z, three equations of the following type hold:

$$\left\langle \frac{\partial f}{\partial x} \right\rangle = \frac{\partial}{\partial x}\langle f\rangle \tag{5.35}$$

In addition to averaging definitions of f(z) and its derivatives with respect to coordinate axes x, y, z, it is assumed that if f(z) is nearly constant when z changes by no more than l′ while another function g(z) may vary by a large fraction over this distance, then approximately

$$\langle fg \rangle = f \langle g \rangle \tag{5.36}$$

The above averaging Eqs. (5.34), (5.35) and (5.36) are applied to a stress-strain relation in the case of a T.I. medium to yield an equivalent set of averaged stress-strain relation from averaged displacement values.

Taking that for each z the medium is T.I. with a vertical axis of symmetry, the stress strain relations may be written as (Stoneley 1949):

$$\sigma_{11} = c_{11}\frac{\partial u_1}{\partial x} + c_{12}\frac{\partial u_2}{\partial y} + c_{13}\frac{\partial u_3}{\partial z} \tag{5.37 a}$$

$$\sigma_{22} = c_{12}\frac{\partial u_1}{\partial x} + c_{11}\frac{\partial u_2}{\partial y} + c_{13}\frac{\partial u_3}{\partial z} \tag{5.37 b}$$

$$\sigma_{33} = c_{13}\frac{\partial u_1}{\partial x} + c_{13}\frac{\partial u_2}{\partial y} + c_{33}\frac{\partial u_3}{\mathrm{ialz}} \tag{5.37c}$$

$$\sigma_{23} = c_{44}\left(\frac{\partial u_2}{\partial z} + \frac{\partial u_3}{\partial y}\right) \tag{5.37 d}$$

$$\sigma_{13} = c_{44}\left(\frac{\partial u_1}{\partial z} + \frac{\partial u_3}{\partial x}\right) \tag{5.37 e}$$

$$\sigma_{12} = c_{66}\left(\frac{\partial u_1}{\partial y} + \frac{\partial u_2}{\partial x}\right) \tag{5.37 f}$$

Here σ_{11}, σ_{12}... are stress components; c_{11}, c_{12}... are elastic stiffness constants; u_1, u_2, u_3 are components of displacements in x, y and z directions. The relation holds

$$c_{66} = \frac{c_{11} - c_{12}}{2}$$

Now consider an infinite horizontal slab of vertical thickness $L \gg l'$, consisting of discrete horizontal layers so thin that, when averaged over l′, all properties of the slab are nearly independent of z. Also, all properties are independent of x and y. Stress-strain equations for a homogeneous T.I. medium as in Eqs. (5.37 a–f) are written in a form that does not contain the product of rapidly varying field variables, and rapidly varying elastic parameters. When this set of equations are averaged over l′ by means of weighting function (Eq. 5.33), and using the definitions Eqs. (5.35) and (5.36), it is possible to write stress-strain relations in terms of averaged displacements and *effective elastic constants*. This yields the following set of *averaged stress–strain relation* (again ⟨ ⟩ represents averaged values):

$$\langle\sigma_{11}\rangle = \overline{c_{11}}\frac{\partial}{\partial x}\langle u_1\rangle + \overline{c_{12}}\frac{\partial}{\partial y}\langle u_2\rangle + \overline{c_{13}}\frac{\partial}{\partial z}\langle u_3\rangle \tag{5.38 a}$$

$$\langle\sigma_{22}\rangle = \overline{c_{12}}\frac{\partial}{\partial x}\langle u_1\rangle + \overline{c_{11}}\frac{\partial}{\partial y}\langle u_2\rangle + \overline{c_{13}}\frac{\partial}{\partial z}\langle u_3\rangle \tag{5.38 b}$$

$$\langle\sigma_{33}\rangle = \overline{c_{13}}\frac{\partial}{\partial x}\langle u_1\rangle + \overline{c_{13}}\frac{\partial}{\partial y}\langle u_2\rangle + \overline{c_{33}}\frac{\partial}{\partial y}\langle u_3\rangle \tag{5.38 c}$$

$$\langle\sigma_{23}\rangle = \overline{c_{44}}\left(\frac{\partial}{\partial z}\langle u_2\rangle + \frac{\partial}{\partial y}\langle u_3\rangle\right) \tag{5.38 d}$$

$$\langle\sigma_{13}\rangle = \overline{c_{44}}\left(\frac{\partial}{\partial z}\langle u_1\rangle + \frac{\partial}{\partial x}\langle u_3\rangle\right) \tag{5.38 e}$$

$$\langle\sigma_{12}\rangle = \overline{c_{66}}\left(\frac{\partial}{\partial y}\langle u_1\rangle + \frac{\partial}{\partial x}\langle u_2\rangle\right) \tag{5.38 f}$$

where $\overline{c_{11}}, \overline{c_{12}}, \ldots \overline{c_{66}}$ represent the effective elastic constants [Backus 1962, Eq. (8)].

The above derivations for averaged values of stress and displacements was done in Backus (1962). Backus derivation is a smoothing operation, which gives averaged values of stress-strain coefficients in the following form:

$$\overline{c_{11}} = \left\langle c_{11} - c_{13}^2 c_{33}^{-1}\right\rangle + \left\langle c_{33}^{-1}\right\rangle^{-1}\left\langle c_{13}c_{33}^{-1}\right\rangle^2 \tag{5.39 a}$$

$$\overline{c_{12}} = \left\langle c_{12} - c_{13}^2 c_{33}^{-1}\right\rangle + \left\langle c_{33}^{-1}\right\rangle^{-1}\left\langle c_{13}c_{33}^{-1}\right\rangle^2 \tag{5.39 b}$$

$$\overline{c_{33}} = \left\langle c_{33}^{-1}\right\rangle^{-1} \tag{5.39 c}$$

$$\overline{c_{13}} = \left\langle c_{33}^{-1}\right\rangle^{-1}\left\langle c_{13}c_{33}^{-1}\right\rangle \tag{5.39 d}$$

$$\overline{c_{44}} = \left\langle c_{44}^{-1}\right\rangle^{-1} \tag{5.39 e}$$

$$\overline{c_{66}} = \langle m\rangle \tag{5.39 f}$$

[Backus 1962, Eq. (9)]

This solid with effective elastic constants $\overline{c_{11}}, c_{12} \ldots \overline{c_{66}}$ is the long wave equivalent to the original more strongly layered solid.

Also, it may be noted that *effective elastic constants* are not simply averages of corresponding elastic constants in the more inhomogeneous solid. It is appropriate to discuss a velocity picture with *effective elastic constants* and compare it with the elastic constant for the original T.I. layered medium. For the T.I. medium, the following velocity relations hold:

Vertical direction, compressional wave

$$V_P^2 = \frac{c_{33}}{\rho} \tag{5.40 a}$$

Vertical direction, shear wave

$$V_S^2 = \frac{c_{44}}{\rho} \tag{5.40 b}$$

Horizonal propagation, compressional wave:

$$V_P^2 = \frac{c_{11}}{\rho}$$

Horizonal propagation, SV wave:

$$V_{SV}^2 = \frac{c_{44}}{\rho}$$

Horizonal propagation, SH wave:

$$V_{SH}^2 = \frac{c_{66}}{\rho}$$

In the layered transversely isotropic medium corresponding values of velocities for waves with wavelength much longer than l′ would be:

Vertical propagation, compresional wave:

$$\left(\overline{V}_P^2\right) = \frac{\overline{c_{33}}}{\langle\rho\rangle} \tag{5.41 a}$$

Vertical propagation, shear wave:

$$\left(\overline{V}_S^2\right) = \frac{\overline{c_{44}}}{\langle\rho\rangle} \tag{5.41 b}$$

Horizontal propagation, compressional wave:

$$\left(\overline{V}_P^2\right) = \frac{\overline{c_{11}}}{\langle\rho\rangle} \tag{5.42 a}$$

Horizontal propagation, SH wave:

$$\left(\overline{V}_S^2\right) = \frac{\overline{c_{66}}}{\langle\rho\rangle} \tag{5.42 b}$$

Horizontal propagation, SV wave:

$$\left(\overline{V}_S^2\right) = \frac{\overline{c_{44}}}{\langle\rho\rangle} \tag{5.42 c}$$

$\langle \rho \rangle$ represents averaged value of densities over length l′.

If the layering is of isotropic layers, we have

$$c_{11} = c_{33} = (\lambda + 2\mu)$$

$$c_{12} = c_{13} = 1$$

$$c_{44} = c_{66} = \mu$$

in Eqs. (5.39 a–f).

Carcione et al. (1991) discussed the effect of layering on wave propagation. It has been observed in this investigation that layering induces scattering with a wave dispersion or anisotropic effect depending on wavelength to layer thickness ratio. Numerical modeling results show smoothed T.I. behavior at long wavelengths, and a dispersive character of wavefield at short wavelengths. This investigation quantifies the minimum ratio of seismic wavelength to layer thickness for which the *long wavelength approximation* is valid.

Anisotropy and dispersion in a periodically layered medium was investigated in detail in Helbig (1984). Helbig's conclusions confirm that a periodically layered medium exhibits transverse isotropy provided the wavelength is long compared to spatial period of layer sequence. The approach of proving it is, however, different from the methods described earlier in this section on the basis of stress-strain relations. It is based on evaluating the dispersion relation for a periodically layered medium, and evaluating it for a low frequency (long wavelengths) limit. Both approaches yield the same replacement medium. More details on how the wave propagation exhibit characters of anisotropy or dispersion would be found in Section 5.7.2.

As elastic anisotropy in a layered medium influences velocity and depth interpretations, likewise resistivity, anisotropy distorts interpretations of apparent resistivity measurements. Sri Niwas (1973) analyzed theoretical basis of such interpretations.

Problem 5.1

Assume a periodic layer model of a isotropic wet sand and clay layer with velocities and densities as given in the following Table 5.2:

Further assume that thickness of wet sand (h_1) is 1, and of clay formation (h_2) is 3.

Table 5.2. Velocities and densities of constituent formations in the periodic layer model

Formation types	Compressional velocity V_P (m/s)	Shear velocity V_s (m/s)	Density ρ (gm/cm^3)
Wet sand	2,800	800	2.0
Clay (Lavergne 1989, p 13)	1,800	500	2.2

(i) Determine the values of the elastic stiffness constants c_{11}, c_{12}, c_{13}, c_{33}, c_{44}, and c_{66} ($=(c_{11}-c_{12})/2$) of the equivalent transversely isotropic model.
(ii) Given that the stiffness constants account for velocities in different directions as given below,
c_{33}: vertical P-wave velocity,
c_{44}: vertical SH and SV wave velocities,
c_{11}: horizontal P-wave velocity,
c_{44}: horizontal SV-wave velocity,
c_{66}: horizontal SH-wave velocity,

determine the vertical and horizontal P, SH and SV-wave velocities.
Hint:
- From the values of shear wave velocity and density, determine shear-modulii μ.
- From compressional velocities, and μ values determine values of λ (λ and μ are Lame's constants).
- With known values of λ, μ, h_1 and h_2, use the Postma relation (Eq. 5.32) to determine the values of elastic stiffness constants.
- Knowing the elastic stiffness constants, velocity values are given by the square root of the stiffness constant divided by density.

The average density value may be taken as

$$\langle\rho\rangle = \frac{\rho_1 h_1 + \rho_2 h_2}{(h_1 + h_2)}$$

5.3.2
Azimuthal Anisotropy: HTI Media

By *azimuthal anisotropy* is meant the variation of one or more properties of a homogeneous material with azimuth (Winterstein 1990). Crampin (1985) concluded that most upper crustal media and some hydrocarbon reservoir rocks show evidence of the presence of azimuthal anisotropy. Azimuthal anisotropy occurs as a result of the natural fracturing of rocks, often aligned in accordance with the dominant stress direction. An example of a very simple type of azimuthally anisotropic medium is a transversely isotropic model with a *horizontal axis of rotational symmetry* (HTI). This model is used to describe the vertically aligned penny shaped cracks in an isotropic matrix (Thomsen 1995). Figure 5.7a shows a *transversely isotropic model* with a vertical and horizontal axis of rotational symmetry, referred to as VTI and HTI media, respectively.

A HTI model is shown in Fig. 5.7b. The horizontal symmetry axis is taken in the x_1-direction. The HTI medium possess two vertical symmetry planes. The vertical plane containing the symmetry axis is referred to as the *symmetry axis plane*. A plane perpendicular to the symmetry axis is referred to as the *isotropy plane*. For the case of HTI media resulting from vertically aligned cracks, the *isotropy plane* is coincident with the *fracture plane*. Therefore, in Fig. 5.7b, the

Fig. 5.7 a, b. a Symmetry plane (x_1, x_3) of a transversely isotropic medium with the symmetry axis pointing either in the x_3 (VTI) or in the x_1 (HTI) directions. The slowness surface in the symmetry axis plane remains the same as the HTI medium is replaced with the equivalent VTI model. (Source: Tsvankin 1997, pp 615–617. Reproduced from Geophysics with permission). **b** Sketch of a HTI model. Shear waves polarized parallel and normal to the isotropy plane have different vertical velocities (Source: Tsvankin 1997. From Geophysics with permission)

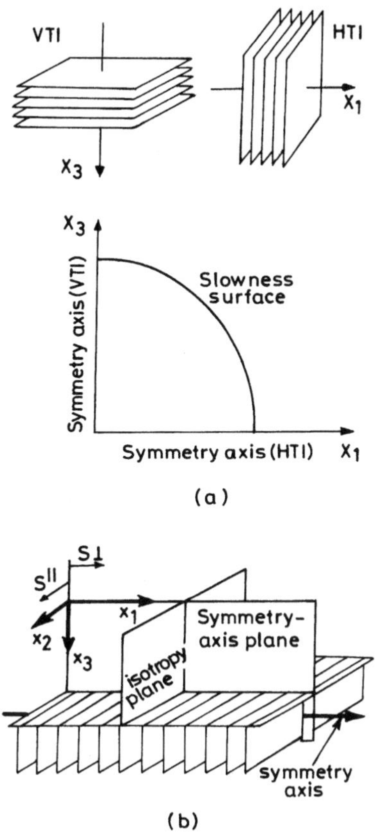

[x_1, x_3] plane containing the symmetry axis is the *symmetry axis plane*, and the [x_1, x_2] plane perpendicular to the symmetry axis is the *isotropy plane*. In the isotropy plane, the kinematic signatures and polarizations of all three types of waves are describable by isotropic equations. The velocities and polarizations in the symmetry axis plane may be found by analogy with a VTI medium.

A shear wave propagating in the isotropy plane may travel with two different velocities depending on whether its polarization is confined in the isotropy plane or is perpendicular to it. The *split shear waves* in the HTI medium are denoted by S^{11} and $S^{\perp}$. The polarization in s^{11} is parallel to the crack plane. $S^{\perp}$ at vertical incidence is polarized normal to crack plane. S^{11} is called the *fast shear wave* as it propagates faster than $S^{\perp}$ at vertical incidence (Crampin 1985).

5.3.3 Limited Equivalence between VTI and HTI Media

There is a limited equivalence (analogy) between a VTI medium and a HTI medium. The *Christoffel equation* (Section 5.7), which describes the wave propagation, generally differs for VTI and HTI media. However, this equation is *identical* for wave propagation in the symmetry axis plane of the HTI medium, and the vertical symmetry planes of VTI models. As a consequence of this, all equations describing kinematic properties and polarizations for transversely isotropic media with a vertical symmetry axis (VTI) are identical for the symmetry axis plane of HTI media (Rüger 1997). This important observation has led to the determination of a new set of dimensionless anisotropy parameters similar to Thomsen's (1986) anisotropy parameters (refer to Section 5.6 on weak anisotropy). In brief, a HTI medium is described conveniently by vertical P-wave velocity 'α', the vertical velocity of a shear wave polarized parallel to the isotropy plane β ($=\beta^{11}$), and three anisotropy parameters $\delta^{(V)}$, $\varepsilon^{(V)}$, and $\gamma^{(V)}$. $\delta^{(V)}$, $\varepsilon^{(V)}$, and $\gamma^{(V)}$ are Thomsen parameters defined with respect to the vertical in the same way as in VTI media. The derivation of these parameters follows from the analysis of Christoffel equations. These parameters represent the *effective parameters* for describing seismic signatures in HTI media. For example, 'α' is the *normal moveout velocity* (VNMO) obtained from seismic surveys aligned with isotropy plane; $\delta^{(V)}$ determines the difference between the NMO velocity in the isotropy plane and symmetry axis plane (Tsvankin 1997). In addition to these, use is also made of $\beta^{\perp}$ (the slow vertical shear wave velocity) and symmetry axis velocities V_{PO} and V_{iso}. Rüger (1998) discussed anisotropy parameters used to study HTI media and their relation to generic Thomsen parameters. The generic Thomsen parameters are defined in a *rotated coordinate system* in which the symmetry direction coincides with the x_3-direction.

Referring to Fig. 5.7b, for wave propagation in the symmetry axis plane (x_1, x_3 plane), all equations describing velocities, travel time, polarization and stresses are identical for media with a symmetry axis pointing either in the x_1 (HTI medium) or the x_3 (VTI medium) directions. Figure 5.7a shows a slowness surface in the symmetry axis plane. The slowness surface in the symmetry axis plane remains the same if the HTI model is replaced by an "equivalent" VTI model. As the VTI model has been the subject of extensive study in exploration seismology, the results valid for HTI models in a symmetry axis plane may be written directly from known results for the VTI model.

5.3.3.1
Matrix Representation of Elastic Stiffness Constants

Consider VTI and HTI models of Fig. 5.7 a. Both for VTI and HTI media, there exist five independent *elastic stiffness constants.* With the x_3-axis pointing downwards, the *elastic stiffness constants matrix* (6×6) for *VTI and HTI media* are given by (e.g. Musgrave 1970, Rüger 1997)

$$[c]_{VTI} = \begin{bmatrix} c_{11} & (c_{11} - 2c_{66}) & c_{13} & & & \\ (c_{11} - 2c_{66}) & c_{11} & c_{13} & & & \\ c_{13} & c_{13} & c_{33} & & & \\ & & & c_{55} & & \\ & & & & c_{55} & \\ & & & & & c_{66} \end{bmatrix} \quad (5.43\,a)$$

$$[c]_{HTI} = \begin{bmatrix} c_{11} & (c_{11} - 2c_{66}) & c_{13} & & & \\ c_{13} & c_{33} & (c_{33} - 2c_{44}) & & & \\ c_{13} & (c_{33} - 2c_{44}) & c_{33} & & & \\ & & & c_{44} & & \\ & & & & c_{55} & \\ & & & & & c_{55} \end{bmatrix} \quad (5.43\,b)$$

[Tsvankin 1997, Eq. (3)]

It may be observed that even though the VTI and HTI models belong to the same symmetry class (a transversely isotropic medium with a vertical symmetry axis and a horizontal symmetry axis), respectively the stiffness matrices differ from each other since the directions of the symmetry axes are different. For example, c_{55} is identical to c_{44} in the VTI model; and c_{55} equals c_{66} in the HTI model.

The method for determining the equivalent VTI medium for a HTI medium is simple. It consists of inserting the matrix for VTI and HTI media in the Christoffel equation and comparing the resulting forms.

The Christoffel equation is written in the form (e.g. Musgrave 1970, Tsvankin 1997)

$$\left[G_{ik} - \rho V^2 \delta_{lk}\right] U_k = 0 \quad (5.44)$$

where ρ is density, G_{ik} is the *Christoffel matrix* given by

$$G_{ik} = c_{ijkl} n_j n_l \quad (5.44\,a)$$

n is the unit vector in the *slowness direction.*

Values of G_{ik} for wave propagation in the symmetry plane (x_1, x_3) of a transversely isotropic medium ($n_2 = 0$) with the symmetry axis pointing in the x_1 direction are given by

$$\begin{aligned} G_{11} &= c_{11}n_1^2 + c_{55}n_3^2 \\ G_{33} &= c_{55}n_1^2 + c_{33}n_3^2 \\ G_{13} &= (c_{13} + c_{55})n_1n_3 \\ G_{22} &= c_{66}n_1^2 + c_{44}n_3^2 \end{aligned} \tag{5.44 b}$$

(Tsvankin 1997)

Matrix (G_{ik}–$\rho V^2 \delta_{ik}$) in the symmetry plane (x_1, x_3) is given by

$$G_{ik} - \rho V^2\delta_{ik} = \begin{bmatrix} c_{11}n_1^2 + c_{55}n_3^2 - \rho V^2 & 0 & (c_{13} + c_{55})n_1n_3 \\ 0 & c_{66}n_1^2 + c_{44}n_3^2 - \rho V^2 & 0 \\ (c_{13} + c_{55})n_1n_3 & 0 & c_{55}n_1^2 + c_{33}n_3^2 - \rho V^2 \end{bmatrix} \tag{5.45}$$

[Tsvankin 1997, Eq. (8)]

Equation (5.45) with Eq. (5.44) as a whole is identical to corresponding equations in the (x_1, x_3) plane of the VTI media. As a consequence of this, the well-known phase velocity equations for P-SV waves in VTI media may be used to describe phase velocities of P-$S^{\perp}$ waves in the symmetry axis plane of HTI media (refer to Fig. 5.7b). Since P-$S^{\perp}$ and S^{11} waves in the symmetry axis plane are decoupled from each other, two equivalent VTI models may be used to describe P-$S^{\perp}$ and S^{11} propagation in the symmetry plane. The *VTI equivalent model* for P-$S^{\perp}$ waves would be characterized by c_{11}, c_{33}, c_{55}, c_{13}; and other model for S^{11} would include c_{44} and c_{66}.

5.4 Thomsen Parameters for HTI Media (or Equivalent VTI Media)

Tsvankin (1996, 1997) introduced the *Thomsen parameters* for *equivalent VTI* medium. These parameters are

$$\varepsilon^{(V)} \equiv \frac{c_{11} - c_{33}}{2c_{33}} \tag{5.46 a}$$

$$\delta^{(V)} \equiv \frac{(c_{13} + c_{55})^2 - (c_{33} + c_{55})^2}{2c_{33}(c_{33} - c_{55})} \tag{5.46 b}$$

$$\gamma^{(V)} \equiv \frac{c_{66} + c_{44}}{2c_{44}} \tag{5.46 c}$$

[Tsvankin 1997, Eqs. (9)–(11)]

Superscript 'V' stands for *equivalent VTI medium.*

The other two "isotropic" quantities for wave propagation in the symmetry axis plane of HTI media are the vertical phase (and group) velocities of P-wave and of $S^{\perp}$ wave.

$$V_{P_{vert}} = \sqrt{\frac{c_{33}}{\rho}} \tag{5.46 d}$$

$$V_{S^{\perp}_{vert}} = \sqrt{\frac{c_{55}}{\rho}} \tag{5.46 e}$$

The vertical velocity of shear wave (S^{11}) is given by

$$V^{11}_{Svert} = \sqrt{\frac{c_{44}}{\rho}} = \frac{V_{S^{\perp}_{vert}}}{\sqrt{1+2\gamma^{(V)}}} \tag{5.46 f}$$

For obtaining the NMO velocities, polarizations and plane wave reflection coefficients in the symmetry axis plane of HTI media, it is only required to substitute the above-defined Thomsen parameters into the known VTI equations. In terms of Thomsen parameters, P and $S^{\perp}$ velocities and polarizations are described by $V_{P_{vert}}, V_{S^{\perp}_{vert}}$, $\varepsilon^{(V)}$, and $\delta^{(V)}$. Velocity of the S^{11} wave depends on $V_{S^{\perp}_{vert}}$ and $\gamma^{(V)}$ (Tsvankin 1997). The anisotropy coefficient of the "equivalent" VTI medium is found responsible for azimuthal variation of *normal moveout velocity* (VNMO) as well (Tsvankin 1997).

5.5 Thomsen Parameters for Orthorhombic Media

The HTI model is useful for studying first-order influence of *azimuthal anisotropy*. More realistic models may be described by the *orthorhombic symmetry system*. An orthorhombic medium resulting from the combination of a VTI background medium with a system of aligned vertical cracks is shown in Fig. 5.8. This orthorhombic model has two vertical symmetry planes: (x_1, x_3) and (x_2, x_3). As in the case of the HTI medium, study of symmetry plane Christoffel equations in the (x_1, x_3) plane and the (x_2, x_3) plane of the orthorhombic media helps to relate wave propagation with VTI media. The symmetry plane propagation in orthorhombic media may be fully described by known VTI equations. All conclusions on limited equivalence of VTI and HTI media are also valid for symmetry-plane propagation in orthorhombic media. This allows new dimensionless aniso-

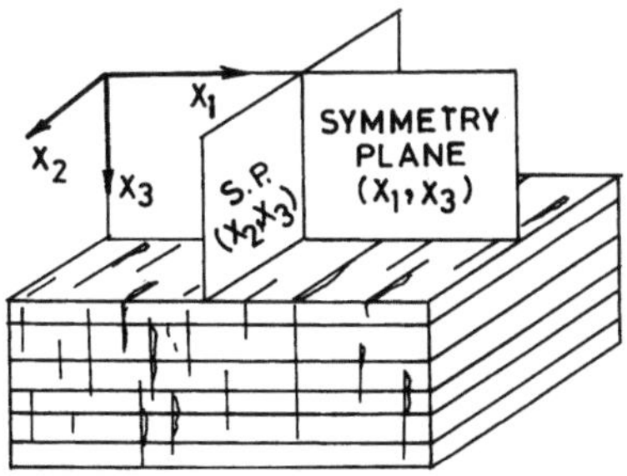

Fig. 5.8. Sketch of an orthorhombic model created by combining horizontal layering with a system of parallel vertical cracks. Two vertical symmetry planes (S.P.) are determined by crack orientations (Source: Rüger 1998, p 942. Reproduced from Geophysics with permission)

tropy parameters (Tsvankin 1996), to be defined similarly to well known Thomsen (1986) coefficients ε, δ, and γ:

1. $\varepsilon^{(2)}$ – the VTI parameter ε in the symmetry plane (x_1, x_3)

$$\varepsilon^{(2)} \equiv \frac{c_{11} - c_{33}}{2c_{33}} \tag{5.47 a}$$

2. $\delta^{(2)}$ – the VTI parameter δ in (x_1–x_3) plane

$$\delta^{(2)} \equiv \frac{(c_{13} + c_{55})^2 - (c_{33} - c_{55})^2}{2c_{33}(c_{33} - c_{55})} \tag{5.47 b}$$

3. $\gamma^{(2)}$ – the VTI parameter γ in (x_1–x_3) plane.

$$\gamma^{(2)} \equiv \frac{c_{66} - c_{44}}{2c_{44}} \tag{5.47 c}$$

4. $\varepsilon^{(1)}$ – the VTI parameter ε in (x_2–x_3) plane

$$\delta^{(1)} \equiv \frac{c_{22} - c_{33}}{2c_{33}} \tag{5.47 d}$$

5. $\delta^{(1)}$ – the VTI parameter δ in (x_2–x_3) plane

$$\varepsilon^{(1)} \equiv \frac{(c_{23} + c_{44})^2 - (c_{33} - c_{44})^2}{2c_{33}(c_{33} - c_{44})} \tag{5.47 e}$$

6. $\gamma^{(1)}$ – the VTI parameter γ in (x_2–x_3) plane

$$\gamma^{(1)} \equiv \frac{c_{66} - c_{55}}{2c_{55}} \tag{5.47 f}$$

7. $\delta^{(3)}$ – the VTI parameter in the (x_1–x_2) plane (x_1 taken as the symmetry axis):

$$\delta^{(3)} \equiv \frac{(c_{12} + c_{66})^2 - (c_{11} - c_{66})^2}{2c_{11}(c_{11} - c_{66})} \tag{5.47 g}$$

Seven dimensionless parameters, as defined above for the equivalent VTI model with vertical velocities α and β as defined below, can replace the nine independent stiffness constants of the orthorhombic model.

8. α – the vertical velocities of a P-wave

$$\alpha \equiv \sqrt{\frac{c_{33}}{\rho}} \tag{5.47 h}$$

9. β – the vertical velocity of an S-wave polarized in the x_2-direction

$$\beta \equiv \sqrt{\frac{c_{44}}{\rho}} \tag{5.47 i}$$

Both vertical and horizontal transverse isotropy are special cases of *orthorhombic media*. An orthorhombic medium reduces to a VTI medium if properties in all vertical planes are identical, and the velocity of each mode in the (x_1–x_2) plane is constant. Another special case is T.I. with a horizontal axis of symmetry. If the symmetry axis is along the x_1-direction, parameters $\varepsilon^{(2)}$, $\delta^{(2)}$, $\gamma^{(2)}$ coincide with coefficients $\varepsilon^{(V)}$, $\delta^{(V)}$, $\gamma^{(V)}$ defined in Eqs. 5.46 (a to c). These parameters help to write expressions for orthorhombic media from known expressions in VTI media.

5.6 Weak Anisotropy

In exploration for hydrocarbons, an assumption of isotropy is most commonly made. However, experimental results show that most crustal rocks possess anisotropy. Krey and Helbig (1956) observed that, in near vertical seismic profiling, the most common form of anisotropy is the transverse isotropy with angular dependence of velocity. The angular dependence of velocity in real occurrence manifests itself as uncertainty in depth estimates to each reflector interpreted assuming isotropy.

Thomsen (1986) observed that in most cases of interest to geophysicists, the anisotropy is weak (10–20%). Starting from an exact equation for transversely isotropic media he developed equations for *weak anisotropy*.

It was shown in Thomsen (1986) that the common measure of anisotropy, i.e. contrasting vertical and horizontal velocities, is not a useful measure for near vertical P-wave propagation, which is of interest to exploration geophysicists. Alternatively, it was shown that a critical measure of anisotropy (δ) for near vertical P-wave propagation does not depend on horizontal velocity at all. Since the assumption of weak anisotropy simplifies the equations to a great extent for their application in problems related to petroleum exploration, these results are presented below.

5.6.1 Phase- and Group-Velocities in VTI Media and Their Linearized Forms

The elastic stiffness matrix for isotropic and transversely isotropic (equation) media are

$$[c]_{\text{Isotropic}} = \begin{bmatrix} c_{33} & (c_{33}-2c_{44}) & (c_{33}-2c_{44}) & & & \\ & c_{33} & (c_{33}-2c_{44}) & & & \\ & & c_{33} & & & \\ & & & c_{44} & & \\ & & & & c_{44} & \\ & & & & & c_{44} \end{bmatrix} \tag{5.48 a}$$

[Thomsen 1986, Eq. (3)]

$$[c]_{TI} = \begin{bmatrix} c_{11} & (c_{11} - 2c_{66}) & c_{13} & & & \\ & c_{11} & c_{13} & & & \\ & & c_{33} & & & \\ & & & c_{44} & & \\ & & & & c_{44} & \\ & & & & & c_{66} \end{bmatrix} \tag{5.48 b}$$

[Thomsen 1986, Eq. (5)]

It may be noted that isotropy is a degenerate case of transverse isotropy with replacements as

$$c_{11} \rightarrow c_{33}$$
$$c_{66} \rightarrow c_{44}$$
$$c_{13} \rightarrow (c_{33} - 2c_{44})$$

c_{33} and c_{44} are related to the commonly referred Lame's parameters 'λ' and 'μ' and bulk modulus 'k' of an isotropic medium as

$$c_{33} = \lambda + 2\mu = k + \frac{4}{3}\mu$$

and

$$c_{44} = \mu$$

Use of the elements of stiffness matrices for T.I. medium in equation of motion yields the wave equation. For each direction of propagation, three independent solutions to wave equation is found (e.g. Daley and Hron 1977: one quasi-longitudinal, one transverse, and one quasi-transverse). The exactly transverse wave is denoted by SH. The directional dependence of three *phase velocities* for transversely isotropic (T.I.) medium is given by

$$\rho V_P^2(\theta) = \frac{1}{2}\left[c_{33} + c_{44} + (c_{11} - c_{33})\sin^2\theta + D(\theta)\right] \tag{5.49 a}$$

$$\rho V_{SV}^2(\theta) = \frac{1}{2}\left[c_{33} + c_{44} + (c_{11} - c_{33})\sin^2\theta - D(\theta)\right] \tag{5.49 b}$$

and

$$\rho V_{SH}^2(\theta) = c_{66}\sin^2\theta + c_{44}\cos^2\theta \tag{5.49 c}$$

[Thomsen 1986, Eq. (7)]

In Eqs. (5.49 a–c), ρ is the density and θ is the angle between wavefront normal and the symmetry (vertical) axis. Also, D (θ) stands for quadratic combination:

$$D(\theta) \equiv \left\{ \begin{array}{l} (c_{33} - c_{44})^2 + 2\left[2(c_{13} + c_{44})^2 - (c_{33} - c_{44})(c_{11} + c_{33} - 2c_{44})\right]\sin^2\theta \\ + \left[(c_{11} + c_{33} - 2c_{44})^2 - 4(c_{13} + c_{44})^2\right]\sin^4\theta \end{array} \right\}^{\frac{1}{2}} \tag{5.49 d}$$

The expression for $D(\theta)$ is algebraically complex and produces difficulty in analyzing exploration data for anisotropic models. Thomsen (1986) devised a method under the assumption of weak anisotropy to write Eqs. (5.49 a–c) in a suitable form for applications in hydrocarbon exploration.

Thomsen (1986) recast Eqs. (5.49 a–c) for T. I. media, which has five *elastic moduli* into those having only two elastic moduli (or, equivalently, vertical P- and S-velocities) and three *measures of anisotropy.* The three *measures of anisotropy* (δ, ε, and γ) were chosen to be suitable combinations of elastic moduli satisfying the criteria:

- simplify phase velocity Eqs. (5.49 a–c),
- are non-dimensional (to enable expressing anisotropy as a percentage),
- reduce to zero in a degenerate case of isotropy,

so that materials having small values ($\ll 1$) of anisotropy may be denoted as *weakly anisotropic.*

Satisfying the above criteria, Thomsen (1986) suggested the following alternatives to the five stiffness constants for T. I. media:

$$\varepsilon \equiv \frac{c_{11} - c_{33}}{2c_{33}} \tag{5.50 a}$$

$$\gamma \equiv \frac{c_{66} - c_{44}}{2c_{44}} \tag{5.50 b}$$

$$\delta^* \equiv \frac{1}{2c_{33}^2}\left[\begin{array}{l} 2(c_{13} + c_{44})^2 \\ -(c_{33} - c_{44})(c_{11} + c_{33} - 2c_{44}) \end{array}\right] \tag{5.50 c}$$

$$\alpha_0 = \sqrt{c_{33}/\rho} \tag{5.50 d}$$

and

$$\beta_0 = \sqrt{c_{44}/\rho} \tag{5.50 e}$$

Thomsen (1986) gave an exhaustive list of anisotropy parameters for a number of sedimentary rocks. The listed values of anisotropy parameters show that, for most rocks, the anisotropy is in the weak-to-moderate range (less than 20%).

Using notations of Eqs. (5.50 a–e), expressions for $V(\theta)$ take the following form:

$$V_P^2(\theta) = \alpha_0^2\left[1 + \varepsilon \sin^2\theta + D^*(\theta)\right] \tag{5.51 a}$$

$$V_{SV}^2(\theta) = \beta_0^2\left[1 + \frac{\alpha_0^2}{\beta_0^2}\varepsilon\sin^2\theta - \frac{\alpha_0^2}{\beta_0^2}D^*(\theta)\right] \tag{5.51 b}$$

$$V_{SH}^2(\theta) = \beta_0^2\left[1 + 2\gamma\sin^2\theta\right] \tag{5.51 c}$$

with

$$D^*(\theta) \equiv \frac{1}{2}\left(1 - \frac{\beta_0^2}{\alpha_0^2}\right) \times \left\{\left[1 + \frac{4\delta^*}{\left(1 - \frac{\beta_0^2}{\alpha_0^2}\right)^2}\sin^2\theta\cos^2\theta + \frac{4\left(1 - \frac{\beta_0^2}{\alpha_0^2} + \varepsilon\right)\varepsilon}{\left(1 - \frac{\beta_0^2}{\alpha_0^2}\right)^2}\sin^4\theta\right]^{\frac{1}{2}} - 1\right\} \tag{5.51 d}$$

On expanding D* in the Taylor series and retaining terms linear in ε, δ^* and γ, D* may be written as

$$D^* \approx \frac{\delta^*}{\left(1 - \frac{\beta_0^2}{\alpha_0^2}\right)}\sin^2\theta\cos^2\theta + \varepsilon\sin^4\theta \tag{5.51 e}$$

Substituting (5.51e) in (5.51 a and b) and further linearizing, yields

$$V_P(\theta) = \alpha_0\left[1 + \delta\sin^2\theta\cos^2\theta + \varepsilon\sin^4\theta\right] \tag{5.52 a}$$

$$V_{SV}(\theta) = \beta_0\left[1 + \frac{\alpha_0^2}{\beta_0^2}(\varepsilon - \delta)\sin^2\theta\cos^2\theta\right] \tag{5.52 b}$$

and

$$V_{SH}(\theta) = \beta_0\left[1 + \gamma\sin^2\theta\right] \tag{5.52 c}$$

[Thomsen 1986, Eqs. (16 a–c)]

Equations (5.52 a–c) for the *quasi-longitudinal*, *quasi-transverse*, and *pure shear motions* respectively are obtained as a result of expanding exact expressions (Eqs. 5.51 a–c) for phase velocity in a Taylor series in the small parameters ε, δ and γ at a fixed θ, and by retaining only linear terms in these parameters.

In Eqs. (5.52 a) and (5.52 b), δ^* is replaced by δ defined as

$$\delta \equiv \frac{1}{2}\left[\varepsilon + \frac{\delta^*}{\left(1 - \frac{\beta_0^2}{\alpha_0^2}\right)}\right] \tag{5.52 d}$$
$$= \frac{(c_{13} + c_{44})^2 - (c_{33} - c_{44})^2}{2c_{33}(c_{33} - c_{44})}$$

Equations (5.52 a–c) are the simplified equations for *phase velocities* under assumption of *weak anisotropy.*

At a small angle of θ (as in reflection prospecting), the δ term in Eq. (5.52 a) dominates the most anisotropic effects. In this context it may be noted that the measured anisotropy on sedimentary rocks shows that the three parameters ε, δ, and γ are usually of the same order of magnitude (Thomsen 1986). For this reason in Eq. (5.52 a) the second term is not as small as the last one, because for small angles of θ, $\sin^2\theta \cos^2\theta$ is not nearly as small as $\sin^4\theta$. As near vertical propagation is dominated by δ, the near horizontal propagation is dominated by ε (as $\cos \pi/2=0$). For horizontal direction

$$V_P\left(\frac{\pi}{2}\right) = \alpha_0(1 + \varepsilon) \tag{5.53 a}$$

α_0 is the vertical P wave velocity. We have

$$\varepsilon = \frac{V_P\left(\frac{\pi}{2}\right) - \alpha_0}{\alpha_0} \tag{5.53 b}$$

ε is the fractional difference between vertical and horizontal P velocities. Prior to Thomsen's (1986) results, ε had been popularly referred to as anisotropy of rock.

The parameter δ in Eq. (5.52 d) does not have stiffness constant c_{11}, and is thus totally independent of horizontal velocity. δ may be positive or negative. Normally, P wave horizontal velocity is greater than vertical P velocity, as ε is positive. An elliptically anisotropic medium is characterized by the conditions $\varepsilon=\delta$ (c.f. Daley and Hron 1979).

For *anisotropy* arising due to fine layering of isotropic materials, Berryman (1979) showed that δ is strictly smaller than ε.

The magnitude of group velocity is given by (Postma 1955)

$$U^2(\phi(\theta)) = V^2(\theta) + \left(\frac{dV}{d\theta}\right)^2 \tag{5.54 a}$$

where ϕ is the angle from the source point to the point on the wavefront (Fig. 5.9). Both angles ϕ and θ are shown in this figure.

Under the *weak anisotropy or linear approximation*, the relationship between the phase angle θ and ray angle ϕ is given by (Thomsen 1986):

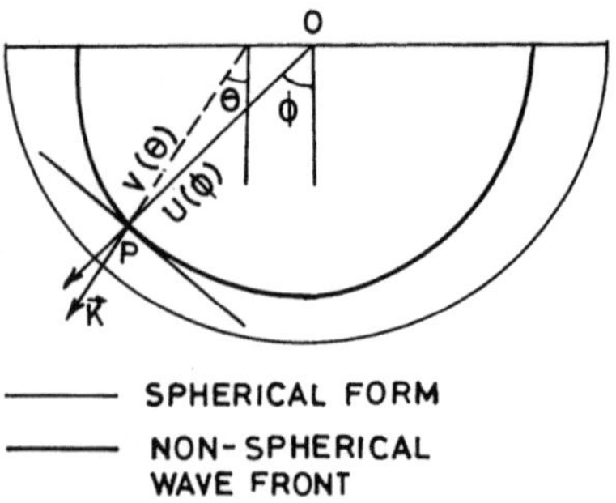

Fig. 5.9. Schematic illustration showing part of non-spherical wavefront from source point O. Joining O to a point on the wavefront defines group or ray direction ϕ. The wave front is locally perpendicular to the wave vector **K**. The phase velocity (V(θ)) direction is along **K**. The phase angle is θ. Ray angle ϕ is a function of phase angle θ. The magnitude of ray velocity (U) is related to the phase velocity (V) and its derivative by Eq. (5.54 a) (Source: after Thomsen 1986)

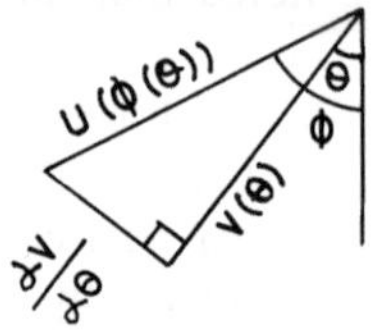

$$\tan\phi = \tan\theta\left[1 + \frac{1}{\sin\theta\cos\theta}\cdot\frac{1}{V(\theta)}\frac{dV}{d\theta}\right] \tag{5.54 b}$$

This equation explicitly written for quasi-P waves, quasi-SV waves and SH waves, respectively, are

$$\tan\phi_P = \tan\theta_P\left[1 + 2\delta + 4(\varepsilon - \delta)\sin^2\theta_P\right] \tag{5.55 a}$$

$$\tan\phi_{SV} = \tan\theta_{SV}\left[1 + 2\left(\frac{\alpha_0}{\beta_0}\right)^2(\varepsilon - \delta)\left(1 - 2\sin^2\theta_{SV}\right)\right] \tag{5.55 b}$$

and

$$\tan\phi_{SH} = \tan\theta_{SH}(1 + 2\gamma) \tag{5.55 c}$$

Use of Eqs. (5.52 a–c) with Eqs. (5.54 a) yields the group velocity at any ray angle, for quasi-P, quasi SV and SH-motion. This gives the following general expression for the group velocity by retaining only linear terms in the small parameters ε, δ, and γ (Sena 1991)

$$U^{-2}(\phi) = a_0 + a_1\sin^2\phi + a_2\sin^4\phi \tag{5.56}$$

where a_0, a_1 and a_2 are given by
(i) for a quasi-P wave:

$$a_0 = \alpha_0^{-2},\ a_1 = -2\delta\alpha_0^{-2},\ a_2 = 2(\delta - \varepsilon)\alpha_0^{-2} \tag{5.56 a}$$

(ii) for a quasi-SV wave:

$$\begin{aligned} a_0 &= \beta_0^{-2}, a_1 = 2\beta_0^{-2}\left(\frac{\alpha_0}{\beta_0}\right)^2(\delta-\varepsilon) \\ a_2 &= -2\beta_0^{-2}\left(\frac{\alpha_0}{\beta_0}\right)^2(\delta-\varepsilon) \end{aligned} \tag{5.56 b}$$

(iii) for a SH wave:

$$a_0 = \beta_0^{-2}, a_1 = -2\gamma\beta_0^{-2}, a_2 = 0 \tag{5.56 c}$$

Equations (5.56 a–c) are a general linearized expression for group velocity, that hold for weak anisotropy and a condition of transverse isotropy corresponding to the case of azimuthal isotropy. From Eqs. (5.56 a–c) it is possible to get a complete description of an *azimuthal isotropy* medium with five parameters – three of which are related to the quasi-P ray surface, and two to the S-ray surface. Thus, it is possible to obtain an estimate of five elastic constants of azimuthal isotropy from five measurement parameters (Sena 1991).

In the following subsection equations for azimuthal anisotropy are presented.

5.6.2 Linearized Form of Group (Ray) Velocity for HTI Media

Azimuthal isotropy refers to a vertical axis of symmetry and a horizontal isotropy plane. By *azimuthal anisotropy* we mean here the horizontal axis of symmetry and the vertical plane of symmetry. Alford (1986) and Wills et al. (1986) advanced evidence of azimuthal anisotropy in some areas and showed that *azimuthal anisotropy* may have a significant impact on the quality of *shear-wave* data. Expressions for group velocity for *azimuthal anisotropy* have been derived by Sena (1991). Sena assumed that for the case of *azimuthal anisotropy*, expression (5.56) is given in the *natural coordinate frame* F′ where the axis of symmetry is horizontal with respect to survey frame F and it makes an angle ϕ' with the x-axis (Fig. 5.10).

The natural coordinate frame and survey frame are related by two sequential rotations: a clockwise rotation of ϕ' about the x′ axis, followed by a clockwise rotation of $\pi/2$ about the new Y-axis. The *rotation matrix* (refer to Sections 5.2.1 and 5.2.2) is

$$\underset{\approx}{R} = \begin{pmatrix} 0 & -\sin\phi' & \cos\phi' \\ 0 & \cos\phi' & \sin\phi' \\ -1 & 0 & 0 \end{pmatrix} \tag{5.57 a}$$

[Sena 1991, Eq. (18)]

Thus, a vector $\mathbf{v}'$ given in a *natural coordinate frame* may be transformed to a *survey frame* via the operation [refer to Eq. (5.13 a)]

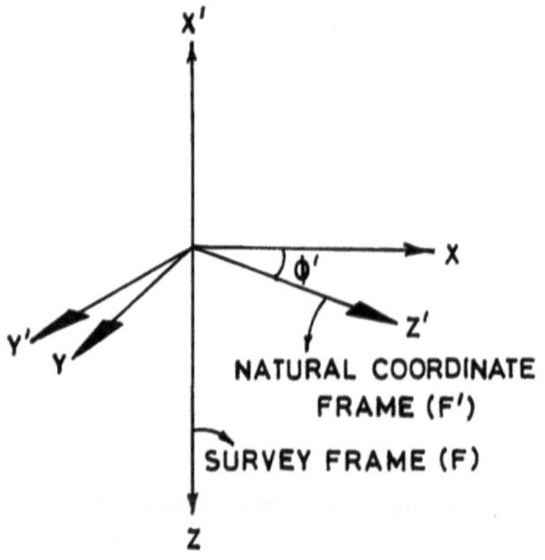

Fig. 5.10. Relationship between natural coordinate frame F′ and survey frame F (Source: Sena 1991, p 2092)

$$v = \underset{\approx}{\mathbf{R}} v' \tag{5.57 b}$$

This operation may be performed on the *ray vector in a natural coordinate frame* and keeping the linear terms in anisotropy parameters, the magnitude of *group velocity* in the *survey frame* (Eq. 5.56) may be obtained with coefficients a_0, a_1, a_2 now assuming the values:

(a) quasi P-wave

$$a_0 = \alpha_0^{-2}(1 - 2\varepsilon) \tag{5.58 a}$$

$$a_1 = 2\alpha_0^{-2}(2\varepsilon - \delta)\cos^2(\phi - \phi') \tag{5.58 b}$$

$$a_2 = 2\alpha_0^{-2}(\delta - \varepsilon)\cos^2(\phi - \phi') \tag{5.58 c}$$

(b) quasi SV-wave

$$a_0 = \beta_0^{-2} \tag{5.50 a}$$

$$a_1 = 2\beta_0^{-2}\left(\frac{\alpha_0}{\beta_0}\right)^2(\delta - \varepsilon)\cos^2(\phi - \phi') \tag{5.59 b}$$

$$a_2 = -2\beta_0^{-2}\left(\frac{\alpha_0}{\beta_0}\right)^2(\delta - \varepsilon)\cos^4(\phi - \phi') \tag{5.59 c}$$

(c) SH wave

$$a_0 = \beta_0^{-2}(1 - 2\gamma) \tag{5.60 a}$$

$$a_1 = 2\beta_0^{-2}\gamma\cos^2(\phi - \phi') \tag{5.60 b}$$

$$a_2 = 0 \tag{5.60 c}$$

'ϕ' and 'ϕ'' are azimuths of the observation point and horizontal axis of symmetry with respect to the x-axis in the survey frame.

As explained in Sena (1991), extension of equation sets (5.58), (5.59) and (5.60) to the case of arbitrary orientation of symmetry axis (not necessarily horizontal), may be obtained by writing Eq. (5.57 a) for an additional rotation. Then the resulting matrix would have an additional angle, corresponding to the *co-latitudinal angle* of the axis of symmetry.

Compared to the case of *azimuthal isotropy*, in the case of *azimuthal anisotropy* an additional parameter is involved in the group velocity equation, which is the azimuth (ϕ') of the horizontal axis of symmetry. Therefore, for an *azimuthally anisotropic medium*, more than five measurement parameters would be required for a complete description of the medium. The treatment in this subsection is based on Sena's (1991) results. The method of investigating anisotropy parameters, using Sena's (1991) theory, briefly presented above is discussed in Chapter 8.

5.7 Wave Equation in Elastic Media: The Christoffel Equation

The Christoffel equation is the wave equation in elastic media, and is commonly used for calculating velocity and slowness surfaces.

5.7.1 Plane Wave Equation in an Isotropic Solid

We consider propagation of a uniform plane wave along a direction l, in a source-free region

$$\mathbf{l} = \hat{\mathbf{x}}\mathbf{l}_x + \hat{\mathbf{y}}\mathbf{l}_y + \hat{\mathbf{z}}\mathbf{l}_z \tag{5.61}$$

A plane wave propagating in direction l has field proportional to $e^{i(\omega t - \mathbf{kl.r})}$. In this case the acoustic wave equation takes the form:

$$k^2\left(\underline{\mathrm{lcl}}^{\mathrm{T}}\right)\mathbf{v} = \rho\omega^2\mathbf{v} \tag{5.62 a}$$

In this equation l is the direction cosine matrix given by

$$\begin{bmatrix} l_x & 0 & 0 & 0 & l_z & l_y \\ 0 & l_y & 0 & l_z & 0 & l_x \\ 0 & 0 & l_z & l_y & l_x & 0 \end{bmatrix} \tag{5.62 b}$$

and l^T is the transpose of l; k is the wave number; ρ is the density; and ω is the angular frequency.

Equation (5.62 a) is referred to as the *Christoffel equation* and the quantity (lcl^T) is called the *Christoffel matrix*.

Using G_{ij} for the Christoffel matrix the wave equation becomes

$$k^2 \Gamma_{ij} v_j = \rho \omega^2 v_i \tag{5.62 c}$$

The elements of the Christoffel matrix are functions of plane wave propagation direction (**l**) and stiffness constants (**c**) of the medium.

Equation (5.62 c) applies for propagation of plane waves in both isotropic and anisotropic media. Discussion of this for an isotropic solid is given first.

For an isotropic medium three coordinate axes and three coordinate planes are equivalent. This results in the same response of the medium to a compressive stress applied along any coordinate axis. Also, shear strain in a coordinate plane due to shear stress in that plane is the same for all coordinate planes. Interpretation of these equivalence properties for all coordinate axes and planes lead to a stiffness constant matrix of the following form for an isotropic medium [Auld 1973, Eq. (6.12)]

$$[c] = \begin{bmatrix} c_{11} & c_{12} & c_{12} & 0 & 0 & 0 \\ c_{12} & c_{11} & c_{12} & 0 & 0 & 0 \\ c_{12} & c_{12} & c_{11} & 0 & 0 & 0 \\ 0 & 0 & 0 & c_{44} & 0 & 0 \\ 0 & 0 & 0 & 0 & c_{44} & 0 \\ 0 & 0 & 0 & 0 & 0 & c_{44} \end{bmatrix} \tag{5.63}$$

This is also the form for a cubic crystal stiffness matrix. To ensure isotropy of the medium the form of [c] should remain unchanged by an arbitrary rotation of coordinate axes. The general elastic isotropy condition is found to be

$$c_{12} = c_{11} - 2c_{44} \tag{5.63 a}$$

Equations (5.63) and (5.63 a) together show that there are only two independent elastic constants for an isotropic medium. These are referred to as Lame's constants λ and μ,

$$\lambda = c_{12}$$
$$\mu = c_{44}$$

This gives

$$c_{11} = \lambda + 2\mu = c_{12} + 2c_{44}$$

'c_{11}' and 'c_{44}' are directly related to the velocity of compressional and shear waves in an isotropic medium which becomes evident from the analysis of the *Christoffel equation (wave equation)* for an isotropic medium.

From Eqs. (5.62 a) and (5.62 b) the *Christoffel matrix* for an isotropic medium for a plane wave propagating in direction **l** is given by

$$[\Gamma_{ij}] = \begin{bmatrix} c_{11}l_x^2 + c_{44}(1 - l_x^2) & (c_{12} + c_{44})l_x l_y & (c_{12} + c_{44})l_x l_z \\ (c_{12} + c_{44})l_y l_x & c_{11}l_y^2 + c_{44}\left(1 - l_y^2\right) & (c_{12} + c_{44})l_y l_z \\ (c_{12} + c_{44})l_z l_x & (c_{12} + c_{44})l_z l_y & c_{11}l_z^2 + c_{44}(1 - l_z^2) \end{bmatrix} \quad (5.64)$$

(Auld 1973, p 383) where $c_{12}=c_{11}-2c_{44}$.

The form of the Christoffel matrix for a cubic system is the same as in Eq. (5.64). However, constraint $c_{12}=c_{11}-2c_{44}$ holds for an isotropic solid only. For an isotropic medium the wave solutions are the same for all directions of propagation. The choice naturally falls on choosing a direction which simplifies the analysis. Considering propagation along the z axis makes the Christoffel matrix (5.64) purely diagonal, and the Christoffel equation from (5.62 c) is

$$k^2 \begin{bmatrix} c_{44} & 0 & 0 \\ 0 & c_{44} & 0 \\ 0 & 0 & c_{11} \end{bmatrix} \begin{bmatrix} v_x \\ v_y \\ v_z \end{bmatrix} = \rho\omega^2 \begin{bmatrix} v_x \\ v_y \\ v_z \end{bmatrix} \quad (5.65)$$

Consider a z-propagating shear wave polarized in the x- or y-directions, represented as

$$(v)_{\text{shear wave}} = \hat{x}ve^{i(\omega t - kz)} \quad (5.66\,a)$$

$$(v)_{\text{shear wave}} = \hat{y}ve^{i(\omega t - kz)} \quad (5.66\,b)$$

From the Christoffel Eq. (5.65), both the waves satisfy the dispersion relation

$$k^2 c_{44} = k^2\mu = \rho\omega^2 \quad (5.67)$$

These shear waves are *pure mode* and *degenerate.* A shear wave propagating along the same direction with the same velocity is called *degenerate.* Their property of being degenerate gives the possibility that they may be combined to produce *circular* or *elliptical polarization* (refer to Appendix 5 A).

For propagation in a general direction **l**, the shear wave polarization would be in two perpendicular planes. One of these is defined by the plane defined by the unit vector of particle polarization motion and **l**, and the other plane is perpendicular to it.

For a compressional wave propagating along the z-direction we have only the v_z component ($(v)_{\text{longitudinal}}$):

$$(v)_{\text{long.}} = \hat{z}ve^{i(\omega t - kz)} \quad (5.68\,a)$$

and the *dispersion relation*, is satisfied.

$$k^2 c_{11} = \lambda + 2\mu = \rho\omega^2 \quad (5.68\,b)$$

Equation (5.68 b) is satisfied for all propagation directions in an isotropic medium.

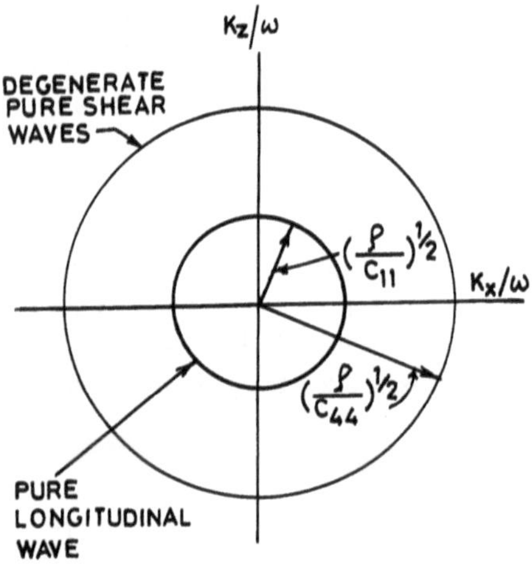

Fig. 5.11. Slowness (inverse velocity) surfaces in an arbitrary plane in an isotropic medium (Source: Auld 1973, p 386)

Therefore, in an isotropic medium, there are only two velocities of plane wave propagation along any direction. These are the longitudinal and shear waves with velocities:

$$V_{\text{long}} = \sqrt{\frac{c_{11}}{\rho}} = \sqrt{\frac{\lambda + 2\mu}{\rho}} \tag{5.69 a}$$

$$V_{\text{shear}} = \sqrt{\frac{c_{44}}{\rho}} = \sqrt{\frac{\mu}{\rho}} \tag{5.69 b}$$

The form of slowness (inverse velocity) surfaces are two concentric spheres (Fig. 5.11).

5.7.2
Plane Wave Equation in an Anisotropic Solid

The most general type of anisotropic solid falls under the triclinic system. Such a solid has no effective symmetry, and has all possible 21 elastic stiffness constants. The *stiffness matrix* in this case is given by

$$[c]_{\text{tricl}} = \begin{bmatrix} c_{11} & c_{12} & c_{13} & c_{14} & c_{15} & c_{16} \\ c_{12} & c_{22} & c_{23} & c_{24} & c_{25} & c_{26} \\ c_{13} & c_{23} & c_{33} & c_{34} & c_{35} & c_{36} \\ c_{14} & c_{24} & c_{34} & c_{44} & c_{45} & c_{46} \\ c_{15} & c_{25} & c_{35} & c_{45} & c_{55} & c_{56} \\ c_{16} & c_{26} & c_{36} & c_{46} & c_{56} & c_{66} \end{bmatrix} \tag{5.70}$$

The matrix elements are referred to coordinate axes x, y, z which coincide with crystal axes X, Y, Z.

The Christoffel matrix, G_{ij} ($=lcl^T$) [refer to Eq. (5.62 c)] for plane wave propagation in arbitrary direction l is given by

$$\Gamma_{ij} = \begin{bmatrix} p_1 & p_2 & p_3 \\ p_2 & p_4 & p_5 \\ p_3 & p_5 & p_6 \end{bmatrix} \tag{5.71}$$

where

$$\begin{aligned}
p_1 &= c_{11}l_x^2 + c_{66}l_y^2 + c_{55}l_z^2 + 2c_{56}l_yl_z + 2c_{15}l_zl_x + 2c_{16}l_xl_y \\
p_2 &= c_{16}l_x^2 + c_{26}l_y^2 + c_{45}l_z^2 + (c_{46} + c_{25})l_yl_z + (c_{14} + c_{56})l_zl_x \\
&\quad + (c_{12} + c_{66})l_xl_y \\
p_3 &= c_{15}l_x^2 + c_{46}l_y^2 + c_{35}l_z^2 + (c_{45} + c_{36})l_yl_z + (c_{13} + c_{55})l_zl_x \\
&\quad + (c_{14} + c_{56})l_xl_y \\
p_4 &= c_{66}l_x^2 + c_{22}l_y^2 + c_{44}l_z^2 + 2c_{24}l_yl_z + 2c_{46}l_zl_x + 2c_{26}l_xl_y \\
p_5 &= c_{56}l_x^2 + c_{24}l_y^2 + c_{34}l_z^2 + (c_{44} + c_{23})l_yl_z + (c_{36} + c_{45})l_zl_x \\
&\quad + (c_{25} + c_{46})l_xl_y \\
p_6 &= c_{55}l_x^2 + c_{44}l_y^2 + c_{33}l_z^2 + 2c_{34}l_yl_z + 2c_{35}l_zl_x + 2c_{45}l_xl_y
\end{aligned} \tag{5.71 a}$$

[Auld 1973, Eq. (7.12)]

Plane wave propagation vector **k** has components k_x, k_y, k_z along the x, y, z direction, and points in direction l, having direction cosines l_x, l_y, l_z, given by

$$l_x = \frac{k_x}{k}, l_y = \frac{k_y}{k}, l_z = \frac{k_z}{k} \tag{5.72}$$

A general form of the Christoffel matrix (5.71) may be utilized to study wave propagation in any other crystal system by correctly choosing stiffness constants for a particular crystal system under study.

Two very commonly used crystal systems to model anisotropy in the Earth are transverse isotropy (T.I.) generated as a result of fine layering of sedimentary rocks, and the orthorhombic system resulting from aligned vertical plane parallel cracks in a transversely isotropic rock. The specific form of Christoffel equation (wave equation) for these systems result from Eq. (5.71).

The form of stiffness matrix for the orthorhombic system having nine stiffness constants is

$$[c]_{ortho} = \begin{bmatrix} c_{11} & c_{12} & c_{13} & 0 & 0 & 0 \\ c_{12} & c_{22} & c_{23} & 0 & 0 & 0 \\ c_{13} & c_{23} & c_{33} & 0 & 0 & 0 \\ 0 & 0 & 0 & c_{44} & 0 & 0 \\ 0 & 0 & 0 & 0 & c_{55} & 0 \\ 0 & 0 & 0 & 0 & 0 & c_{66} \end{bmatrix} \tag{5.73}$$

A transversely isotropic (T. I.) solid has stiffness constants in the same position as an orthorhombic solid with the following constraints due to its symmetry

$$c_{11} = c_{22}, c_{13} = c_{23}, c_{44} = c_{55}, c_{12} = c_{11} - 2c_{66}$$

This leads to five independent stiffness constants for a *T. I. solid* as compared to nine for an *orthorhombic solid.* The form of stiffness constant for the T. I. solid, is given by

$$[c]_{T.I.} = \begin{bmatrix} c_{11} & c_{12} & c_{13} & 0 & 0 & 0 \\ c_{12} & c_{11} & c_{13} & 0 & 0 & 0 \\ c_{13} & c_{13} & c_{33} & 0 & 0 & 0 \\ 0 & 0 & 0 & c_{55} & 0 & 0 \\ 0 & 0 & 0 & 0 & c_{55} & 0 \\ 0 & 0 & 0 & 0 & 0 & c_{66} \end{bmatrix} \tag{5.74}$$

Using the above form of stiffness constant matrix $[c]_{T.I.}$, in the Christoffel matrix (5.71), the Christoffel equation for plane wave propagation in an arbitrary direction **l** may be written as

$$k^2 \begin{bmatrix} c_{11}l_x^2 + c_{66}l_y^2 + c_{55}l_z^2 & (c_{12} + c_{66})l_x l_y & (c_{13} + c_{55})l_x l_z \\ (c_{12} + c_{66})l_x l_y & c_{66}l_x^2 + c_{11}l_y^2 + c_{55}l_z^2 & (c_{55} + c_{13})l_y l_z \\ (c_{13} + c_{55})l_z l_x & (c_{44} + c_{13})l_y l_z & c_{55}l_x^2 + c_{55}l_y^2 + c_{33}l_z^2 \end{bmatrix} \begin{bmatrix} v_x \\ v_y \\ v_z \end{bmatrix} = \rho\omega^2 \begin{bmatrix} v_x \\ v_y \\ v_z \end{bmatrix} \tag{5.75}$$

Proceeding exactly in the same way, the Christoffel equation for the orthorhombic system may be obtained as

$$k^2 \begin{bmatrix} c_{11}l_x^2 + c_{66}l_y^2 + c_{55}l_z^2 & (c_{12} + c_{66})l_x l_y & (c_{13} + c_{55})l_x l_z \\ (c_{12} + c_{66})l_x l_y & c_{66}l_x^2 + c_{22}l_y^2 + c_{44}l_z^2 & (c_{44} + c_{23})l_y l_z \\ (c_{13} + c_{55})l_z l_x & (c_{44} + c_{23})l_y l_z & c_{55}l_x^2 + c_{44}l_y^2 + c_{33}l_z^2 \end{bmatrix} \begin{bmatrix} v_x \\ v_y \\ v_z \end{bmatrix} = \rho\omega^2 \begin{bmatrix} v_x \\ v_y \\ v_z \end{bmatrix} \tag{5.76}$$

(Auld 1973, p 385)

The procedure for obtaining the Christoffel equation for other crystal systems is the same as in the above examples and may be obtained in a straightforward way. Christoffel equations (5.75) and (5.76) written for an arbitrary direction of wave propagation may be specialized for specific directions by appropriately choosing direction cosines values.

The Christoffel equation has a general form [see for example, Eqs. (5.62 c) and (5.71)]:

$$k^2 \begin{bmatrix} p_1 & p_2 & p_3 \\ p_2 & p_4 & p_5 \\ p_3 & p_5 & p_6 \end{bmatrix} \begin{bmatrix} v_x \\ v_y \\ v_z \end{bmatrix} = \rho\omega^2 \begin{bmatrix} v_x \\ v_y \\ v_z \end{bmatrix} \tag{5.77}$$

The dispersion relation is obtained by equating the characteristic determinant of Eq. (5.77) equal to zero. It is a relation in angular frequency ω and wave vector **k**.

To study the propagation characteristics in an anisotropic solid, Eq. (5.62 c) may be rewritten as

$$\left[k^2\Gamma_{ij} - \rho\omega^2\delta_{ij}\right]\left[v_j\right] = 0 \tag{5.77 a}$$

δ_{ij} is the *Kronecker-delta* having the property: $\delta_{ij}=0$ for i*j and $\delta_{ij}=1$ for i=j.

The dispersion equation is:

$$\Omega\left(\omega, k_x, k_y, k_z\right) = \left|k^2\Gamma_{ij}\left(l_x, l_y, l_z\right) - \rho\omega^2\delta_{ij}\right| = 0 \tag{5.78}$$

[Auld 1973, Eq. (7.14)]

The above dispersion equation may be solved for the *wave vector surface* and *slowness* (or *inverse velocity*) *surface.* If ω is fixed, it defines a surface in k-space that yields 1/k as a function of direction **l**. This surface is the *wave vector surface.* Alternatively, we may express in terms of variable k/ω (= 1/V, the inverse of the phase velocity), and get the *slowness (or inverse velocity) surface* as a function of propagation direction. From Eq. (5.78), it is clear that **k** is proportional to ω. The *wave vector surface* scales with ω, whereas the inverse velocity surface is independent of ω. For shapes of inverse velocity surfaces in various crystal systems reference may be made to Musgrave (1970) or Auld (1973).

Equation (5.77 a) has information on particle velocity polarization also. We would discuss it in the next Section 5.8. At this stage it is important to analyze the information contained in the dispersion relation (5.78).

To initiate the discussion, we begin with the dispersion Eq. (5.78) for a plane acoustic wave:

$$\Omega\left(\omega, k_x, k_y, k_z\right) = 0$$

This connects frequency (ω) with wave vector (**k**), from which various velocity surfaces may be derived.

For a given ω, it describes a *wave vector surface* in k-space. For isotropic propagation, the *wave number* is normally a scalar quantity. However, in an anisotropic medium, it is vector **k**.

Geometric representation of Eq. (5.78) may be given in a plane. This is shown in Fig. 5.12 a (Helbig 1984).

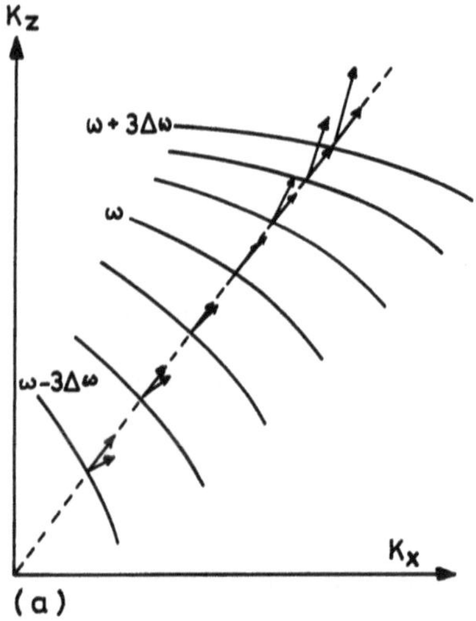

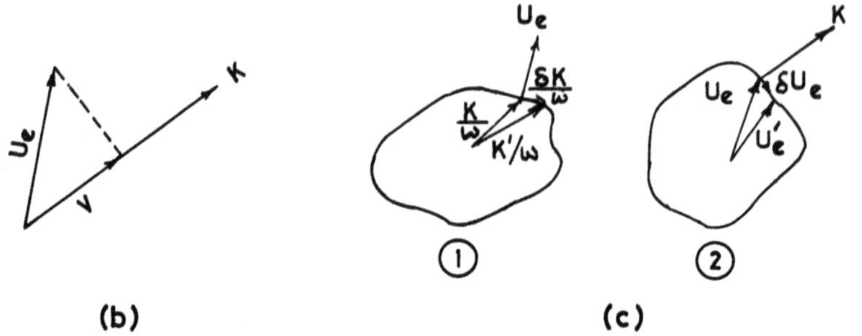

Fig. 5.12 a–c. a Geometrical representation of dispersion and anisotropy: Lines of constant ω in the K_x, K_z plane. The phase velocity **V** is represented by an *arrow* – in the direction of vector **K** with length ω/k, the group velocity **u** by an *arrow* perpendicular to the lines of constant ω with magnitudes dω/dk in this direction; in other words $\mathbf{u} = \text{grad}^{(k)}\ \omega$. In this hypothetical case different elementary waves with the same wave normal but different frequency travel with different (phase) velocity. For an impulse generated by a confined source, different spectral components travel not only with different group velocity, but also in different directions (Source: Helbig 1984, p 365. Reproduced from Geophysics with permission). **b** Relationship between phase velocity V and energy velocity **U**. **c** Relationship between slowness surface and ray surfaces (Source for Fig. 5.12b and c: Auld 1973, p 224)

The *phase velocity* is defined as the propagation velocity of a point of constant phase (ωt–kx = constant):

$$V = \omega/k.$$

The phase velocity may also be written in the form

$$V = \frac{\omega \mathbf{k}}{(\mathbf{k} \cdot \mathbf{k})} \tag{5.79}$$

which shows that the phase velocity has the direction of propagation vector **k**.

The group velocity **U** has a direction of outer normal to surfaces of constant ω and the magnitude is $d\omega/dk$ taken in this direction (Fig. 5.12a).

$$\mathbf{U} = \mathrm{grad}^{(k)}\omega \tag{5.80}$$

It is worthwhile to mention that determination of surfaces of constant 'ω' may be simple in the cases where the dispersion relation is expressed in an explicit form as

$$\omega = f\left(k_x, k_y, k_z\right) \tag{5.81}$$

In this case, the group velocity is given by

$$\mathbf{U} = \hat{\mathbf{x}}\frac{\partial\omega}{\partial k_x} + \hat{\mathbf{y}}\frac{\partial\omega}{\partial k_y} + \hat{\mathbf{z}}\frac{\partial\omega}{\partial k_z} \tag{5.82}$$

In certain cases of dispersion expressed by Eq. (5.78), ω cannot be explicitly expressed in terms of k_x, k_y, k_z. In these cases *implicit differentiation* is required to obtain the group velocity, **U**.

Thus, from Eq. (5.78),

$$\Omega\left(\omega, k_x, k_y, k_z\right) = 0, \tag{5.83}$$

and

$$\left(\frac{\partial\Omega}{\partial\omega}\delta\omega + \frac{\partial\Omega}{\partial k_x}\delta k_x\right)_{k_y,k_z} = 0 \tag{5.83 a}$$

or

$$\left(\frac{\partial\omega}{\partial k_x}\right)_{k_y,k_z} = -\frac{\dfrac{\partial\Omega}{\partial k_x}}{\dfrac{\partial\Omega}{\partial\omega}} \tag{5.83 b}$$

Also, we would have similar expressions for $(\partial\omega/\partial k_y)_{kx,\ kz}$ and $(\partial\omega/\partial k_z)_{kx,\ ky}$.

Thus, the group velocity **U** in Eq. (5.82) may be evaluated by,

$$\mathbf{U} = -\frac{\nabla_k\Omega}{\dfrac{\partial\Omega}{\partial\omega}} \tag{5.84}$$

[Auld 1973, Eq. (7.78)]

The gradient of Ω or $(\nabla_k\Omega)$ is taken with respect to variables k_x, k_y, k_z. To reproduce a figure like Fig. 5.12a in this case, one needs to calculate ω for given values of k_x, k_y, and k_z and to contour data to obtain surfaces of constant 'ω'.

The relationship between wave vector and slowness vector is given by

$$\mathbf{K} = \omega \mathbf{n} \tag{5.85}$$

The surfaces of equal ω are scaled versions of slowness surfaces. The angular wave vector $\mathbf{k}'$ and circular wave vector $\mathbf{k}$ are related as

$$\mathrm{K}' = \mathrm{f}\mathbf{n} = \frac{\mathbf{k}}{2\pi} \tag{5.86}$$

The energy velocity (V_e) and group velocity (**U**) are identical for acoustic waves.

The relationship between wave vector (**K**), phase velocity (**V**) and energy velocity U_e is given by Fig. 5.12 b:

$$\mathbf{U_e} \cdot \mathbf{K} = \mathbf{V} \tag{5.87}$$

Group velocity is the *velocity of modulation* on a wave. To understand it, we may construct a *modulated wave* by assuming two waves with slightly differing values of ω and k, and propagating in the x-direction (Auld 1973, p 227):

$$\begin{aligned}&\cos(\omega t - kx) + \cos[(\omega + \delta\omega)t - (k + \delta k)x] \\ &= 2\cos\left\{\left(\omega + \frac{\delta\omega}{2}\right)t - \left(k + \frac{\delta k}{2}\right)x\right\}\cos\left(\frac{\delta\omega t}{2} - \frac{\delta kx}{2}\right)\end{aligned} \tag{5.88}$$

The first part in the flower bracket in Eq. (5.88) represents a carrier wave and the second part a modulation envelope.

The propagation velocity of the carrier is the phase velocity

$$V = \frac{\omega + \frac{\delta\omega}{2}}{k + \frac{\delta k}{2}} \to \frac{\omega}{k} \tag{5.88 a}$$

and velocity of the modulation envelope is the group velocity

$$U = \frac{\delta\omega}{\delta k} \to \frac{\partial\omega}{\partial k} \tag{5.88 b}$$

We may generalize Eq. (5.88 b) by considering a band of frequencies and a general direction of the ray. The treatment below for this case follows from Helbig (1984).

Consider a wavefield represented by

$$u(x, y, z, t) \equiv u(\mathbf{x}) = \int_{k_x=-\infty}^{+\infty} \int_{k_y=-\infty}^{+\infty} \int_{k_z=-\infty}^{+\infty} u(\mathbf{k}) e^{j(\mathbf{k}\cdot\mathbf{x} - \omega t)} dk_x dk_y dk_z \tag{5.89}$$

with

$$\mathbf{k} = \left(k_x, k_y, k_z\right)^T$$

In Eq. (5.89), u (x, y, z, t) is the amplitude of the wavefield, $e^{j\phi}$ with $\phi = (\mathbf{k} \cdot \mathbf{x} - \omega t)$ represent elementary waves with wave vector **k**, and circular frequency ω; u (k_x, k_y, k_z) is the wave number spectrum of the amplitude of elementary waves. Equation (5.89) represents a wavefield in a small volume, where significant amplitudes occur that move along a ray.

To obtain an equation for the velocity with which the envelope of the wavefield moves, Eq. (5.89) may be written as

$$u(\mathbf{x}) = e^{j(\mathbf{k}_0 \cdot \mathbf{x} - \omega_0 t)} \int\int\int \left(u(\mathbf{k}) e^{j[(\mathbf{k} - \mathbf{k}_0) \cdot \mathbf{x} - (\omega - \omega_0)t]} dk_x dk_y dk_z \right) \tag{5.90 a}$$

or

$$u(\mathbf{x}) = e^{j(\mathbf{k}_0 \cdot \mathbf{x} - \omega_0 t)} \int\int\int \left(u(\mathbf{k}) e^{j(\Delta \mathbf{k} \cdot \mathbf{x} - \Delta \omega t)} dk_x dk_y dk_z \right) \tag{5.90 b}$$

where $\mathbf{k}_0$ is the centroid of the spectrum, and ω_0 is the circular frequency related to k_0 by the dispersion equation.

$$\Omega(\omega, k_x, k_y, k_z) = 0 \tag{5.ah}$$

Equation (5.90 b) may be interpreted as an elementary wave with wave normal $\mathbf{k}_0/k_0$ and phase velocity $\mathbf{v} = (\omega/k_0)\, \mathbf{k}/k_0$ (the carrier wave) modulated by time-dependent envelope, given by the triple integral.

Taylor-series expansion of $\Delta\omega$ may be taken in the form

$$\begin{aligned} \Delta\omega = {} & \frac{\partial \omega}{\partial k_x} \Delta k_x + \frac{\partial \omega}{\partial k_y} \Delta k_y + \frac{\partial \omega}{\partial k_z} \Delta k_z \\ & + \frac{1}{2} \left(\frac{\partial^2 \omega}{\partial k_x^2} \Delta k_x^2 + \frac{\partial^2 \omega}{\partial k_y^2} \Delta k_y^2 + \frac{\partial^2 \omega}{\partial k_z^2} \Delta k_z^2 \right) + \ldots \end{aligned} \tag{5.91}$$

The integrand in Eq. (5.90 a) oscillates strongly except in the vicinity of the point where the exponent vanishes, i.e. $\mathbf{k} = \mathbf{k}_0$ and $\omega \mp \omega_0$. Therefore, the significant contribution to the integral comes only from the first term of the series in Eq. (5.91). We thus obtain *group velocity* (or *signal velocity*) from the equation

$$\begin{aligned} & \Delta \mathbf{k} \cdot \mathbf{x} - \text{grad}^{(k)} \omega \Delta \mathbf{k} t = 0 \\ & \text{or} \\ & U = \text{grad}^{(k)} \omega \end{aligned} \tag{5.92}$$

We have discussed phase velocity (V), energy velocity (V_e) and their relationship through wave vector (**k**). A plot of phase velocity versus wave vector direction is called the *normal surface* or the *phase velocity surface.* A plot of energy velocity versus energy flow direction would give the *ray velocity surface* or the *wave surface* (Fig. 5.12 c).

Separate *phase* and *ray velocity* surfaces exist for each wave P, S_1, and S_2. In isotropic media the two S-wave ray-surfaces coincide. The group velocity surface and phase velocity surfaces are the same.

The phase-velocity S-wave surfaces (S_1 and S_2) may intersect for a T.I. solid. The *phase velocity surface* is considered more fundamental than the *group velocity surface* because the latter may be derived from the former. The *phase velocity surface* is smoother than the *group velocity surface* and never has cusps.

The inverse of the phase velocity surface obtained by taking reciprocals of all points on the phase velocity surface is called the *slowness surface.* It may also be calculated directly from the Christoffel equation. Slowness vectors are parallel to phase velocity vectors.

Wave surface and *phase velocity surfaces* are constructed in velocity space and have dimension of $[LT^{-1}]$ whereas the slowness surface is constructed in slowness space and has a dimension of $[L^{-1}T]$.

The dependence of phase velocity on wave vector (k) may show up as anisotropy or dispersion or as both. Wave vector (**k**) magnitude is proportional to the product of frequency and slowness, and its direction is perpendicular to the wavefront. If the slowness (or velocity) depends only on the magnitude of **k**, we call it *dispersion*; if it depends only on the direction of wave vector (**k**), we call it *anisotropy.* In the case of dispersion, the *phase velocity* and *velocity of energy transport* have the same direction but different magnitudes. The magnitude and direction of phase velocity and the velocity of energy transport are both different in the case of *anisotropy* without *dispersion.* In this case, phase velocity is the projection of velocity of energy transport in the direction of wave normal (refer to Fig. 5.12 b). Further, different combinations of anisotropy and dispersion may exist together, i.e. anisotropy may be frequency dependent, and dispersion may be direction dependent.

Anisotropy and dispersion may exist together in natural geological occurrences. Considering spatially periodic structures like crystals and periodically layered media of the Earth, if the wavelength is much longer as compared to spatial period, this results in anisotropy. If the wavelength is of the order of the spatial period, the material is characterized by strong dispersion. Spatial period in crystals is of the order of nanometer, which is much less than the wavelength of elastic waves. Thus crystals behave as an anisotropic, non-dispersive medium. The sequence of formations of Schist and Shale, having layers on a scale below 1 or 2 m, may be treated as homogeneous, non-dispersive, anisotropic media. For a spatial period just below the resolution of seismic observation, the layer sequence possesses anisotropy and dispersion.

Dispersion is observed in a frequency-dependent gradient (i.e. different distance between adjacent isolines for different ω). Anisotropy results in deviation of iso-ω lines from circles (Helbig 1984).

Effects of anisotropy on dispersion of Love waves were studied by Negi and Upadhyay (1968), Negi and Upadhyay (1968-II), Negi and Upadhyay (1971), Upadhyay (1972), Upadhyay and Gupta (1972-III).

Problems 5.2

Write down the Christoffel equation (wave equation) for the cubic crystal symmetry system. Assume that crystal axes coincide with coordinate axes.

Further, assuming that the plane wave propagates along one of the crystal axis, determine the velocities of propagation of P and S waves from the characteristic equation.

Generalize the result for any direction of wave propagation in a plane perpendicular to one of the axes.

Hint: Make use of the elastic stiffness constants matrix to write the Christoffel equation for the cubic system. Write down the characteristic equation for wave propagation along the crystal axis. Factorize this equation to obtain its different roots. The roots yield the values of velocities.

To generalize the result for any propagation in a plane, choose one of the direction cosines equal to zero. This would define a plane in which the propagation of waves may be analyzed. Repeat the procedure of writing the characteristic equation, factorizing it and finally getting its roots giving the wave velocities in the chosen direction in the plane.

5.8 Particle Motion Polarization

Returning to Eq. (5.78) for discussion on particle motion polarization, it may be mentioned that after having solved Eq. (5.78) and obtained the value of k^2/ω^2 (either through numerical computation or analytically), *particle velocity polarization* may be obtained from Eq. (5.77). It may be illustrated for the general case of a *triclinic solid.*

From Eqs. (5.77) and (5.71 a), the *characteristic equation* for a triclinic system is a third degree equation in k^2/ω^2 given by

$$\begin{aligned}&\left(p_1\frac{k^2}{\omega^2}-\rho\right)\left(p_4\frac{k^2}{\omega^2}-\rho\right)\left(p_6\frac{k^2}{\omega^2}-\rho\right)\\&=\frac{k^4}{\omega^4}\left\{\left(p_1p_5^2+p_4p_3^2+p_6p_2^2-2p_2p_3p_5\right)\frac{k^2}{\omega^2}-\rho\left(p_2^2+p_3^2+p_5^2\right)\right\}\end{aligned} \tag{5.93 a}$$

(Auld 1973, p 212)

This is obtained by expansion of the determinant $|K^2\,\Gamma ij-\omega^2\delta ij|=0$ for the triclinic system. In its initial form this determinant is

$$\begin{vmatrix} p_1k^2-\rho\omega^2 & p_2k^2 & p_3k^2\\ p_2k^2 & p_4k^2-\rho\omega^2 & p_5k^2\\ p_3k^2 & p_5k^2 & p_6k^2-\rho\omega^2\end{vmatrix}=0 \tag{5.94}$$

Assuming that the inverse of phase velocity, k/γ, has been found, we may proceed to analyze the particle polarization. The first two lines from the Christoffel equation (5.77) are

$$\left(p_1 - \frac{\rho\omega^2}{k^2}\right)v_x + p_2 v_y = -p_3 v_z \tag{5.95}$$

$$p_2 v_x + \left(p_4 - \frac{\rho\omega^2}{k^2}\right)v_y = -p_5 v_z \tag{5.96}$$

From Eqs. (5.95) and (5.96), by elimination of v_z, we get (Auld 1973)

$$\frac{v_y}{v_x} = \frac{p_5\left(p_1 - \frac{\rho\omega^2}{k^2}\right) - p_2 p_3}{p_3\left(p_4 - \frac{\rho\omega^2}{k^2}\right) - p_2 p_5} \tag{5.97 a}$$

Also

$$\frac{v_z}{v_x} = \frac{\left(p_1 - \frac{\rho\omega^2}{k^2}\right)\left(p_4 - \frac{\rho\omega^2}{k^2}\right) - p_2^2}{p_3\left(p_4 - \frac{\rho\omega^2}{k^2}\right) - p_5 p_2} \tag{5.97 b}$$

Equations (5.97 a) and (5.97 b) enable determination of *particle polarization ratios* in terms of the *phase velocity* (ω/k), *propagation direction* (l_x, l_y, l_z) and the *stiffness constants* of the medium. They would represent *quasi-longitudinal and quasi shear waves.* Quasi means similar but not exactly.

For a certain direction of propagation, the determinant factorizes into three factors in k^2/ω^2. In such cases, the wave solutions are *pure longitudinal* and *pure shear.* The meaning of *quasi-longitudinal and quasi-shear* waves is further explained below.

In an isotropic medium, the *wave polarization* is either perpendicular or parallel to the direction of wave propagation. However, this is true for some special directions only in the case of an anisotropic medium. The wave polarization deviates from the propagation direction in the sense that they are strictly neither parallel nor perpendicular to the propagation direction. These waves are denoted by qP, qSV, qSH, or qP, qS_1, qS_2. On the other hand, waves having their polarization strictly parallel or perpendicular to the phase velocity direction are called pure modes.

Pure modes exist for the direction of propagation along the *symmetry axis.* One pure, and two quasi waves exist for a direction of propagation in the symmetry plane. All quasi waves exist for an arbitrary direction of wave propagation in an anisotropic medium (refer to, for example, Winterstein 1990).

It may be further remarked that for a given direction of propagation **l**, Eq. (5.77 a) poses an eigen value problem which has three real solutions (eigen values) for squares of phase velocity ω^2/k^2 or slowness k^2/ω^2. Insertion of a non-degenerate eigen value back into the Christoffel equation gives the ratio of components of particle velocity **v**, but not **v** itself. From these ratios, the polarization direction may be calculated and this gives one of three *eigen vectors* (refer to, for example, Daley and Hron 1997). These polarization directions are, in general, neither parallel nor perpendicular to the direction of propagation **l** (i.e. the solution represents quasi-compressional or quasi-shear). If **k** denotes the wave propagation vector, and **u** is the particle displacement, this implies that $\mathbf{k}\cdot\mathbf{u}\neq 0$, and $\mathbf{k}\times\mathbf{u}\neq 0$. For a pure P wave, we require that $\mathbf{k}\times\mathbf{u}=0$ and for a pure S-wave $\mathbf{k}\cdot\mathbf{u}=0$.

In an isotropic medium the compressional and shear waves exist in their pure form, i.e. the particle motion polarization is strictly either in the direction of wave propagation (for a compressional wave) or perpendicular to it (for a shear wave).

Study of body wave and surface wave polarization is significant both from the point of view of understanding propagation in anisotropic media, and providing, in *polarization anomalies*, a sensitive diagnostic phenomenon for recognizing the presence of anisotropy and mapping its characteristics.

5.8.1 P-Wave Polarization

It may be proved considering the symmetry of the Christoffel matrix

$$[\Gamma] = \begin{bmatrix} p_1 & p_2 & p_3 \\ p_2 & p_4 & p_5 \\ p_3 & p_5 & p_6 \end{bmatrix}$$

that, for a particular direction of propagation, the three waves have *particle velocity vectors* mutually orthogonal to each other. In an isotropic medium, one vector is strictly along the direction of propagation, and the other two are strictly transverse. These are *pure longitudinal waves* and *pure shear waves*. However, in anisotropic media the *particle motion polarization* direction, in general, makes an angle with the *propagation vector*, and *ray or group-velocity vector*. The deviation may be a diagnostic feature for identifying and mapping anisotropy.

5.8.2 Shear-Wave Delay and Polarization-Anomalies

Figure 5.13 shows a schematic representation of *shear wave splitting*. Shear wave splitting is also called *shear-wave birefringence* and *shear-wave double refraction*. It shows that, on entering an anisotropic region, the shear wave splits into two phases with polarization orthogonal with respect to the propagation direction. The polarizations are fixed for a particular direction through anisotropy. The two shear waves with fixed polarizations travel at different velocities and at the other end of the anisotropic region arrive at different times. The *delay* is proportional to the degree of differential shear-wave anisotropy in that particular direction and to the length of path through the anisotropic region. On exit from the aniso-

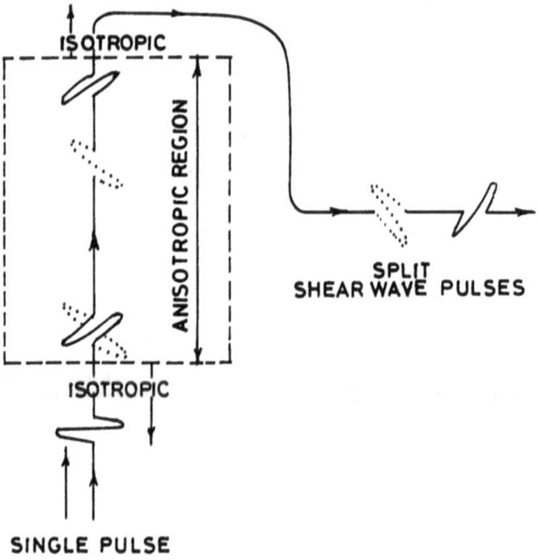

Fig. 5.13. Schematic diagram showing shear wave splitting (after Crampin 1985)

tropic region into the isotropic region, the time-delay between the two differently polarized shear-wave trains will be *preserved unchanged* for isotropic propagation outside. This is so because the velocity of shear waves in isotropic solids is independent of the polarization. The recognition of *split arrivals* in a shear wave train is strong evidence of anisotropy somewhere along the propagation path.

Shear-wave splitting is routinely observed in three component Vertical Seismic Profiling (VSP). Crampin and McGonigle (1981) recognized shear-wave-splitting as the most diagnostic effect of wave propagation in effectively anisotropic rocks, which could also be used to determine parameters of anisotropy.

There are two classes of problems with measurement and interpretation of split-shear waves. One class is related to evaluating shear-wave-splitting associated with earthquake generated shear-waves. The other class is interpreting split shear waves in controlled seismic experiments like VSP, cross-hole seismic surveys etc. Our emphasis in the presentation here is on split shear waves in a VSP experiment.

Shear-wave splitting, associated with earthquake-generated shear waves, has been studied mainly through the use of *delay* and *polarization stereograms*, and by transformation of a recorded seismogram. For studying shear-wave splitting observed in controlled VSP experiments the use of *propagator matrices* with VSP measurements has been suggested to evaluate local birefringence properties.

Delays and polarizations are presented in the form of stereograms called the *delay stereogram* and *polarization stereogram*. Crampin (1981) observes that such stereograms are a powerful technique for estimating anisotropy from observations. For methods of determining delay and polarization stereograms, reference may be made to Crampin and McGonigle (1981).

For excellent details on *polarization singularities*, reference should be made to Crampin (1981) and Winterstein (1990).

A seismic wavefield recorded on the ground is a picture of superposed wave groups, like direct, reflected, refracted, converted and scattered body and surface waves. The record of seismic stations equipped with narrow-band seismograph response is a smooth seismogram. In contrast to this the broadband digital seismograph records, recorded at high sampling rate, show pronounced *multipulse seismograms* suitable for the analysis of various wave groups. The separation and identification of various wave groups can be best accomplished by transforming the recorded seismogram from the coordinate system of the recording seismometers to the apparent *local wave coordinate* system of the different wave groups.

In heterogeneous, anisotropic media, the seismic wave particle motion is three dimensionally polarized, and the orientation of each wave group in a composite seismogram is required for investigating heterogeneous and anisotropic mediums. In the following text, the basis of transformation of the recorded seismogram is given (refer to Aki, Rechards 1980, pp 103–104, Plesinger et al. 1986).

5.8.2.1
Transformation of the Seismogram

Conventional seismic records are for a right-handed 3-component coordinate system, viz vertical, east and north. Let $S_z(t)$, $S_E(t)$ and $S_N(t)$ represent the -3 component recorded seismogram. This is transformed to a local wave coordinate system. The local wave coordinate system is related to the recorded coordinate system by the angles θ and i, where θ ($0° \leq \theta \leq 360°$) is the back-azimuth to the epicentre of the projection of $\mathbf{e}_L$ into the horizontal plane against north. $\mathbf{e}_L$ is the *unit vector* of a local wave group (i.e. for the P-direction), and i ($0° \leq i \leq 90°$) is the incidence angle between e_L and e_z where e_z is the *unit vector* in the vertical z-direction.

The transformed seismogram u_L, u_q, u_T, in a local ray system coordinate, is related to the recorded seismogram S_Z, S_E, S_N, through angles θ and i by the matrix relation:

$$\begin{pmatrix} u_L \\ u_Q \\ u_T \end{pmatrix} = \begin{pmatrix} \cos i & -\sin i \sin\theta & -\sin i \cos\theta \\ \sin i & \cos i \sin\theta & \cos i \cos\theta \\ 0 & -\cos\theta & \sin\theta \end{pmatrix} \begin{pmatrix} S_Z \\ S_E \\ S_N \end{pmatrix} \tag{5.98}$$

(Plešinger et al. 1986)

Elements in the first, second and third lines are components of unit vectors (e_L, e_Q, e_T) in the *local wave group* coordinate in the recording system (e_Z, e_E, e_N).

Transverse unit vector $\mathbf{e}_T$ (SH-direction) results from the vector product of $\mathbf{e}_L$ and $\mathbf{e}_Z$; and other transverse vector $\mathbf{e}_Q$ (SV-direction) from $\mathbf{e}_T \times \mathbf{e}_L$. Thus ($\mathbf{e}_L$, $\mathbf{e}_Q$, $\mathbf{e}_T$) form a right-handed system:

$$\mathbf{e}_T = \frac{\mathbf{e}_L \times \mathbf{e}_Z}{|\mathbf{e}_L \times \mathbf{e}_Z|},$$

$$\mathbf{e}_Q = \frac{\mathbf{e}_L \times \mathbf{e}_T}{|\mathbf{e}_L \times \mathbf{e}_T|}$$

Plešinger et al. (1986) illustrate the wide use of calculating particle motion diagrams in a rotated local wave group coordinate system from the records of Broad

Band Stations (Gräfenberg, West Germany), KHC (Kăsperske, Hony, Czechoslovakia) and KSP (Ksiaz, Poland). The particle motion diagrams are used for the discrimination of wave types, for the determination of onset time, and for pulse duration. They arrive at the conclusion that *rotated seismogram*, *particle motion* diagrams, and the *component product* help in understanding the resolution of *wave splitting*, and for understanding fine structure generation of quasi-P and quasi-S waves as well as 'inclined', 'tilted' and sloping Rayleigh waves.

Rotation of the recorded seismogram has been found to be a simple way to recognize *shear-wave splitting* (Savage et al. 1990). Shih, Meyer and Schneider (1989) developed an *automated technique* for estimating polarization of *leading split shear-wave arrival.*

In the following text, the basis for the interpretation of VSP measurement in terms of anisotropy (or shear wave splitting) is discussed.

5.8.2.2
Use of Propagator Matrices

Examples of two shear-waves with horizontal and mutually perpendicular polarizations traveling in a vertical direction with different velocities (called *shear wave splitting* or *shear wave birefringence*) has been described in the publications of Crampin (1985), Alford (1986), Naville (1986), Lefeuvre et al. (1992), and Winterstein and Meadows (1991).

Lefeuvre et al. (1992) formulated a method, based on the use of the *propagator matrix*, to measure the local *shear wave birefringence* properties over any depth interval. In this method no information is required about layering above the zone of interest, in particular about the *weathered zone.* The method is based on acquisition of two shear-wave vertical seismic profiles (VSPs), each with different *initial polarizations* of shear wave. The formulation in Lefeuvre et al. (1992) deals with the estimation of a 2×2 *propagator matrix.*

The *propagator matrix* represents a *linear operator* between two states of polarization. For vertical wave propagation it is the *transfer function* between states of polarization at depths Z_1 and Z_2 (Fig. 5.14a). In this figure the *propagator matrix* describes the propagation of downgoing shear waves between depths z_1 and z_2. Further, *transfer function* between *polarization of upgoing* and *downgoing* waves at a particular depth, Z_I, describes the *reflection* of the shear wave on a given interface at depth Z_r ($Z_i \leq Z_r$).

The transfer function can therefore be interpreted either in terms of the *birefringence property* for a given depth interval (called *transmission analysis*), or for finding the *reflection matrix* in VSP measurement (called *reflection analysis*). The reflection matrix may be interpreted in terms of anisotropy below Z_i (including depths, below well bottom) or for anisotropy of reflector itself.

In their interesting publication, Lefeuvre et al. (1992) give the mathematical basis for expressing *transfer functions* in time and frequency domains. A brief discussion of the theory follows.

Lefeuvre et al. (1992) consider propagation of a downgoing shear wave between levels Z_1 and Z_2. The polarizations at Z_1 and Z_2 in the horizontal plane of the direct wave are assumed in the form

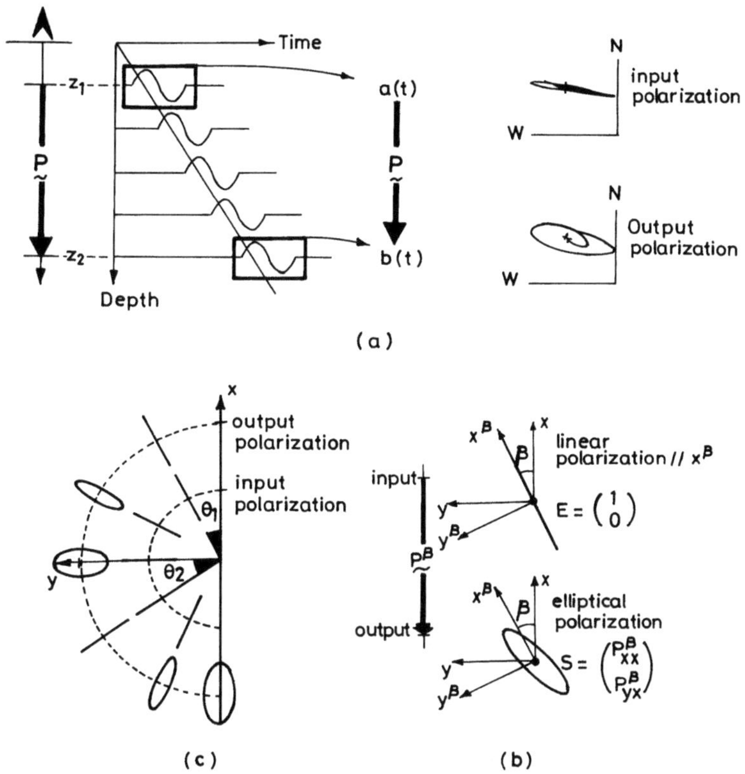

Fig. 5.14 a–c. a Polarization pattern and definition of the shear-wave transfer matrix (propagator matrix, **P**) in transmission. **P** describes the changes in polarization between two given depths; **b** physical meaning of the vector (P_{xx}, P_{yx}) which composes the propagator matrix **P** for a particular frequency; **c** the two azimuthal eigen directions θ_1 and θ_2, varying respectively between 0 and 90°, and 90 and 180°. In a simple birefringent medium $\theta_1 = \theta_2$ (Source: Lefeuvre 1992. Reproduced from Geophysics with permission)

$$a(t) = \begin{pmatrix} x_1(t) \\ y_1(t) \end{pmatrix} \tag{5.99 a}$$

$$b(t) = \begin{pmatrix} x_2(t) \\ y_2(t) \end{pmatrix} \tag{5.99 b}$$

where x(t) and y(t) are two horizontal coordinate functions of polarization.

Time domain definition of the *propagator matrix* P(t) is given as (Fig. 5.13 a)

$$b(t) = P(t) \times a(t) \tag{5.100 a}$$

The corresponding frequency domain representation is

$$\begin{pmatrix} X_2(f) \\ Y_2(f) \end{pmatrix} = \begin{pmatrix} P_{xx}(f) & P_{xy}(f) \\ P_{yx}(f) & P_{yy}(f) \end{pmatrix} \cdot \begin{pmatrix} X_1(f) \\ Y_1(f) \end{pmatrix} \tag{5.100 b}$$

For a direct downgoing wave at two different depths, the propagator matrix may be interpreted in terms of natural directions of *polarization*, *delays* (and *attenuation*) and thereby about the *birefringence property* in any depth interval. Depending on the properties of the *propagator matrix*, initial *linear input polarization* may result in *elliptical output polarization* (Fig. 5.14b).

With reference to Fig. 5.14b we have the output polarization **B** as

$$B = \begin{pmatrix} P^{\beta}_{xx}(f) \\ P^{\beta}_{yx}(f) \end{pmatrix} = \begin{pmatrix} P^{\beta}_{xx}(f) & P^{\beta}_{xy}(f) \\ P^{\beta}_{yx}(f) & P^{\beta}_{yy}(f) \end{pmatrix} \cdot \begin{pmatrix} 1 \\ 0 \end{pmatrix} \tag{5.100 c}$$

[Lefeuvre et al. 1992, Eq. (24)]

In Eq. (5.100c), ($P^{\beta}_{xx}(f)$, $P^{\beta}_{yx}(f)$), and ($P^{\beta}_{xy}(f)$, $P^{\beta}_{yy}(f)$) are the *output polarization* of two input waves *linearly polarized* respectively for a frequency f parallel to directions, Ox^{β} and Oy^{β}, propagating through a medium characterized by the *propagator matrix* $P^{\beta}(f)$ over the interval (Z_2–Z_1).

Lefeuvre et al. (1992) estimated the *propagator matrix* in the frequency domain, and in an initial coordinate system (O, Ox, Oy) assumed known. Designating $P^{\circ}(f)$ as the initial estimate at frequency f, the propagator matrix $P^{\beta}(f)$ in a coordinate system (O, Ox^{β}, Oy^{β}) making an angle β with the initial system is

$$\mathbf{P}^{\beta}(f) = \mathbf{Rot}(\beta) \cdot \mathbf{P}^{\circ}(f) \cdot \mathbf{Rot}^{-1}(\beta) \tag{5.101 a}$$

Rot (β) denotes the *rotation matrix*,

$$\mathbf{Rot}(\beta) = \begin{pmatrix} \cos\beta & \sin\beta \\ -\sin\beta & \cos\beta \end{pmatrix} \tag{5.101 b}$$

Rotation is performed for all angles from 0° to 90°, and those directions in which a *linearly polarized* shear wave propagates keeping its linear polarization (Fig. 5.14c). For a simple, birefringence medium $\theta_1 = \theta_2$. In this direction, the form of the *propagator matrix* in the frequency domain is

$$\mathbf{P}^{\theta}(f) = \begin{pmatrix} P^{\theta}_{xx}(f) & 0 \\ 0 & P^{\theta}_{yy}(f) \end{pmatrix} \tag{5.102}$$

In Eq. (5.102), $P^{\theta}_{xx}(f)$ and $P^{\theta}_{yy}(f)$ are scalar transfer functions in eigen directions characterized by θ and $\theta + 90°$.

Assuming plane wave propagation, the two *transfer functions* are

$$P^{\theta}_{xx}(f) = A_1(f)\exp(-i2\pi f t_1) \tag{5.102 a}$$

$$P^{\theta}_{yy}(f) = A_2(f)\exp(-i2\pi f t_2) \tag{5.102 b}$$

[Lefeuvre et al. 1992, Eqs. (26) and (27)] where $A_1(f)$, and $A_2(f)$ are amplitude factors, and t_1, t_2 are corresponding travel times of shear waves over depth interval (Z_2–Z_1). The sign of (t_1–t_2) gives the *fast shear wave*.

By shifting the depth interval Δz from the top of recorded depths to the bottom, it is possible to find continuous measurements as a function of depth. For a detailed discussion of the method with illustrated synthetic and real data, reference may be made to the original contribution of Lefeuvre et al. (1992), which is followed in our presentation above.

Problem 5.3

Assume that three components (a_i, $i=1, 2, 3$) corresponding to vertical, N-S and E-W components of particle ground motion, at various sampling times, are given on the ground surface for a ray path from underneath arriving in azimuth $\theta=\theta_1$ and at an angle $i=i_1$ (incidence angle: angle with vertical).

Write a computer program to generate an array of numbers for particle motions in a local ray coordinate system at various sampling times of ground motion from the given 3-component input values in vertical, N-S and E-W directions.

Hint: Make use of Eq. (5.98). For the computer program, read the values for azimuth and incidence angle of ray path for a fixed location of seismic source and recording station. Treat a_i ($i=1, 2, 3$) as the variable input of ground motion; compute the product of matrices [Eq. (5.98)] to yield output values of ground motion in local ray coordinates. The values may be displayed in graphical form to produce a picture of the *rotated seismogram.*

Appendix 5A: Existence of Degenerate and Non-degenerate Shear Waves – Shear Wave Splitting

Consider a x-polarized y propagating shear wave. Its stress and strain fields are (Auld 1973, p 68)

$$\sigma_6 = \sigma_{xy} = \sin(\omega t - ky) \tag{5A.1}$$

$$S_6 = 2S_{xy} = \frac{k^2}{\rho\omega^2}\sin(\omega t - ky) \tag{5A.2}$$

The fields of stress and strain are related by *elastic constitutive equations* through the elastic stiffness matrix. Accordingly, considering the elastic stiffness matrix for the cubic system we obtain

$$\sigma_6 = c_{44}S_6 \tag{5A.3}$$

In Eq. (5A.3), c_{66} is taken equal to c_{44} as applicable to the cubic crystal symmetry system.

It gives for a x-polarized, y-propagating shear wave (along the cubic crystal axis y):

$$c_{44}k^2 = \rho\omega^2$$

Or, phase Velocity (V_S)=

$$\frac{\omega}{k} = (c_{44}/\rho)^{\frac{1}{2}} \qquad (5A.4)$$

We may alternatively consider a z-polarized shear wave propagating along the y-direction. This would also possess the same phase velocity:

$$(V_s)^{\text{Z-polarization}}_{\text{Y-propagating}} = (V_s)^{\text{X-polarization}}_{\text{Y-propagating}} = (c_{44}/\rho)^{\frac{1}{2}} \qquad (5A.5)$$

Shear waves propagating along the same direction with the same velocity but having different polarizations are called degenerate. Degenerate waves may be combined in an arbitrary way to produce a wave motion with a variety of polarizations. This aspect is of fundamental significance.

Consider, for example, x-polarized and z-polarized shear waves propagating in the direction of the y-axis of the crystal. Let us combine these waves with arbitrary amplitudes but the same *time phase.*

The particle displacement fields are

$$\mathbf{u}_x = \hat{\mathbf{x}}A\cos(\omega t - ky) \qquad (5A.6)$$

$$\mathbf{u}_z = \hat{\mathbf{z}}B\cos(\omega t - ky) \qquad (5A.7)$$

for individual waves; and for the combined wave

$$\mathbf{u} = (\hat{\mathbf{x}}A + \hat{\mathbf{z}}B)\cos(\omega t - ky) \qquad (5A.8)$$

Equation (5A.8) is a wave propagating along the y-direction with the same velocity $(c_{44}/\rho)^{\frac{1}{2}}$, but having a linearly polarized particle displacement along the direction (**x**A + **z**B) (Fig. 5A.1 a (middle)). Further if x- and z-polarized waves are combined with equal amplitudes (A = B = 1) and 90 ° phase shift, i.e.

$$\mathbf{u} = \hat{\mathbf{x}}\cos(\omega t - ky) + \hat{\mathbf{z}}\sin(\omega t - ky) \qquad (5A.9)$$

the *particle displacement* is *circularly polarized* (Fig. 5A.1 a (bottom left)). With arbitrary combinations of amplitude and phase, elliptical polarization of particle displacement of various orientations would be produced (Fig. 5A.1 a (bottom right)).

In the above example, shear waves have identical phase velocity, i.e. $(c_{44}/\rho)^{\frac{1}{2}}$, but their combined polarizations are different from individual polarizations. It may be linear, circular or elliptical depending on the amplitude and phase relationship of *degenerate* shear waves.

Let us now consider a situation in which shear waves are propagating along the same direction with different polarizations and different velocities. We take the example discussed in Auld (1973, pp 84–85) for a cubic crystal.

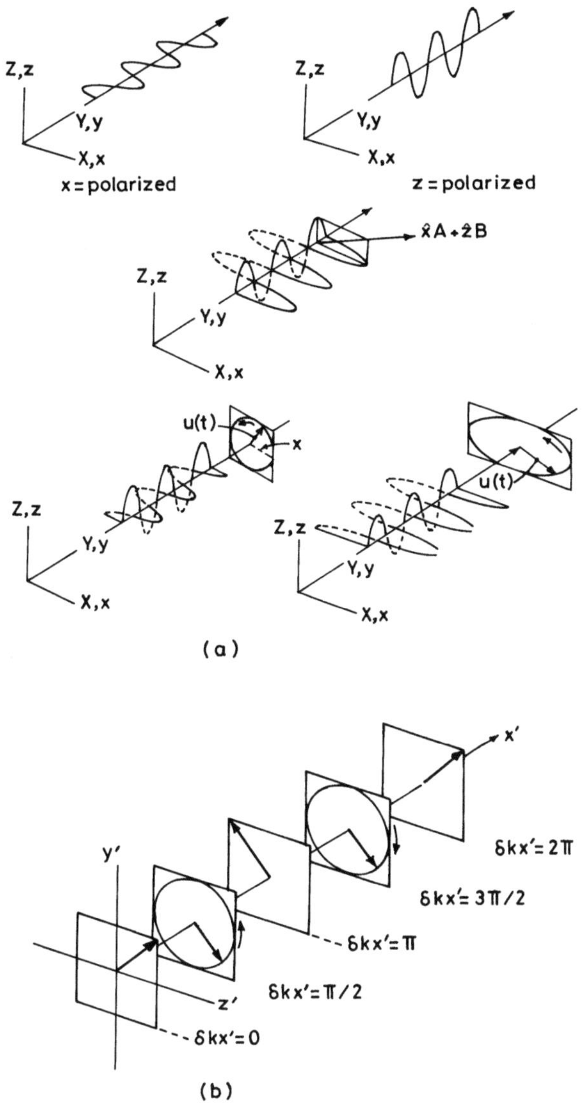

Fig. 5A.1 a, b. a Combination of degenerate shear waves to form arbitrary particle displacement polarization; **b** polarization transformations produced by combining non-degenerate shear waves in a birefringent medium (Source: Auld 1973, pp 69–85)

We consider two polarizations of shear waves:

(i) polarized along edge (Z-direction, [001]) of cubic crystal,

(ii) polarized along face diagonal (Y′, [$\bar{1}$10]) of cubic crystal,

both waves propagating along the face diagonal (x′, [110]) of the crystal with different velocities.

For a shear wave polarized along the Z-direction (cube edge) and propagating along the direction of the face diagonal (x′), the velocity is given by:

$$(V_s) = \left(\frac{c_{44}}{\rho}\right)^{\frac{1}{2}} \tag{5A.10}$$

[Auld 1973, Eq. (3.54)]

For shear wave polarized along the face diagonal (y′) the velocity is

$$(V_s) = \left(\frac{c_{11} - c_{12}}{2\rho}\right)^{\frac{1}{2}} \tag{5A.11}$$

Equation (5A.10) gives the shear wave velocity for propagation along a face diagonal with polarization of the wave along the cube edge. Equation (5A.11) gives the velocity of a shear wave propagating along the face diagonal (x′) and having a polarization along a face diagonal (y′).

These results show that, for propagation along a face diagonal, there are two shear wave velocities, one for cube edge (z-axis) polarization and the other for face diagonal (y′) polarization. Shear waves having this property are called *non-degenerate* and the medium is called *birefringent. Thus, a birefringent medium may be defined as a homogeneous solid, which causes two waves of different polarizations to travel with different speeds in the same direction.*

The effect of *birefringence* for shear waves propagating along the face diagonal is that these waves cannot be combined to form an unchanging polarization pattern, as was the case discussed above for *degenerate waves.*

In the case of a *birefringent* medium, because of the velocity difference, the phase relationship of two waves is not constant. As the waves propagate along the x′ direction, the two waves shift phase, and there is a change of polarization along the direction of propagation.

To illustrate the point, consider shear waves propagating along the crystal face diagonal with equal amplitude and phase in the beginning. The combined particle displacement is

$$\begin{aligned} \mathbf{u} = \hat{y}' \cos\left(\omega t - \frac{2\pi}{\lambda_1} x'\right) + \\ \hat{z}' \cos\left(\omega t - \frac{2\pi}{\lambda_2} x'\right) \end{aligned} \tag{5A.12}$$

Here, λ_1 is the wavelength of the face diagonal polarized wave, and λ_2 is the wavelength of the face edge polarized wave. At $x' = 0$, the y′ and z′ components are in time-phase and particle displacement is linearly polarized.

As the two waves are propagating along x′ with different velocities, at a certain point on x′ there would be phase shifts of $\pi/2$, π, $3\pi/2$, 2π. At those points we have

$$\delta k x' = 2\pi\left(\frac{1}{\lambda_1} - \frac{1}{\lambda_2}\right) x' = \frac{\pi}{2}, \pi, \frac{3\pi}{2}, 2\pi \tag{5A.13}$$

respectively.

At points where $\delta k\ x' = \pi/2$, π, $3\pi/2$ and 2π are satisfied, the particle displacement would be circularly polarized in a counter clockwise direction, linearly polarized at a right angle to initial polarization, circularly polarized in a clockwise sense about x′, and the original linear polarization as in the beginning, respectively. This is illustrated in Fig. 5A.1 b. These polarization changes would be repeated along x′ with a change of observation point along x′.

Cumulative S wave splitting is an effect of birefringence in each of several media, each medium being birefringent in a different way (differences in S-wave polarization) and by different amounts.

For geophysicists, the amount of birefringence is the fractional difference in wave speeds for travel in an observed direction. It is usually expressed as a percentage. When S wave birefringence is measured in vertical or horizontal directions, it is referred to as *vertical refringence* and *horizontal refringence*, respectively.

CHAPTER 6

Raypath, Wavefront Curvature, Normal Moveout Velocity (VNMO) and Subsurface Medium Properties 6

When a seismic disturbance is created near the Earth's surface it propagates downward and laterally. On meeting an elastic discontinuity, the incident wave energy is partitioned; a part of the energy returns back as a reflected wave and another part is transmitted as a refracted wave. Our interest in seismic (reflection) prospecting centers on the reflected wave. The reflected and transmitted waves propagate on wave fronts or wave surfaces. A direction perpendicular to the wave front is taken as the ray direction in an isotropic medium. The ray is only a mathematical simplification. The issue of physical significance is the wave front. The wave front or the wave surface has associated with it useful information about the subsurface medium. The ray directions (ray paths) of reflected and refracted waves are dependent on medium properties.

The subsurface medium velocities control the geometrical properties of the wave front or wave surface. However, direct study of the wave surface is not possible. Their properties are inferred indirectly from surface recorded reflection seismograms. Wave fronts represent equal travel time surfaces. The time by which a reflected wave front is identified represents the total time from seismic source to the reflector to the surface geophone. A simple relationship exists between the radius of curvature of an emerging wave front on the ground surface with the second derivative of travel time of the reflected wave emerging on ground surface. Further, the radius of curvature of the emerging wave front is related to the *Normal Moveout Velocity* (VNMO): a quantity routinely measured in *Common Depth Point* (CDP) surveys. The significance of evaluation of VNMO is that it is expressible in terms of *interval velocities or velocity function* of the subsurface. Therefore analysis of VNMO should enable determination of *interval velocities* or subsurface medium properties.

In this chapter, a simple theory of law governing the directions of rays corresponding to reflected and refracted waves; and also the relationship governing certain properties of wave surface and the medium properties are given. Finally, interpretation of VNMO in terms of subsurface medium properties under conditions of isotropy and anisotropy is discussed.

6.1 Conventional and Generalized Forms of Snell's Law

(i) Referring to Fig. 6.1, if the displacement in the incident wave front is oblique to the interface, it would produce shearing as well as compressive stresses due to discontinuity in elastic properties. Thus, the transmitted and reflected stress fields would contain both P and S waves. Snell's law governs the directions in which the reflected and transmitted waves propagate. According to this law, the apparent wave number in a direction parallel with the boundary separating two media is the same for incident P or S waves, reflected P and S waves, and transmitted P and S waves.

Referring to Fig. 6.1, we have

$$\begin{aligned}(K_{p1}\sin i_1)_{\text{incident P}} &= (K_{p1}\sin i_2)_{\text{reflected P}}\\ &= (K_{p2}\sin i_3)_{\text{transmitted P}}\\ &= (K_{s1}\sin i_4)_{\text{reflected S}}\\ &= (K_{s2}\sin i_5)_{\text{transmitted S}}\end{aligned} \tag{6.1}$$

where K_{p1} is the wave number of both incident P and reflected P-wave in the medium 1. A similar meaning is assigned to K_{p2}, K_{s1} and K_{s2}. Angles i_1, i_2, i_3, i_4 and i_5 are the angles the incident P, reflected P, transmitted P, reflected S and transmitted S wavefronts make with the boundary, respectively. These are also the angles that corresponding ray directions make with the normal to the boundary. Rays are taken perpendicular to the wavefronts. The wave number K is given by

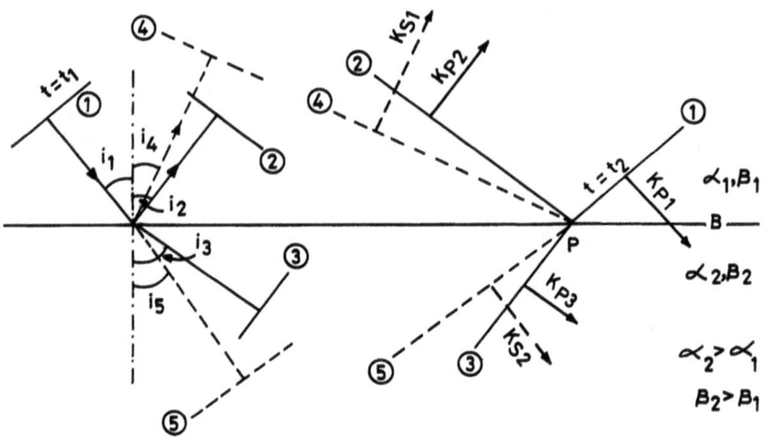

Fig. 6.1. B is an elastic discontinuity boundary. *1* is an incident plane wavefront at $t=t_1$. *2*, *3*, *4*, and *5* show relative orientations of reflected and transmitted wavefronts with respect to incident wavefront 1. *2* and *3* are reflected and transmitted P-wavefronts; *4* and *5* are reflected and transmitted S-wavefronts. Propagation vectors are shown normal to wavefronts

$$K_p = \frac{\omega}{\alpha} \tag{6.1 a}$$

(for a P wave) and

$$K_s = \frac{\omega}{\beta} \tag{6.1 b}$$

(for a S wave)

Accordingly, we get

$$\frac{\omega}{\alpha_1}\sin i_1 = \frac{\omega}{\alpha_1}\sin i_2 = \frac{\omega}{\alpha_2}\sin i_3 = \frac{\omega}{\beta_1}\sin i_4 = \frac{\omega}{\beta_2}\sin i_5$$

or

$$\frac{\sin i_1}{\alpha_1} = \frac{\sin i_2}{\alpha_1} \Rightarrow i_1 = i_2 \text{(angle of incidence = angle of reflection)} \tag{6.1 c}$$

and

$$\frac{\sin i_1}{\alpha_1} = \frac{\sin i_3}{\alpha_2} = \frac{\sin i_4}{\beta_1} = \frac{\sin i_5}{\beta_2}$$

Equation (6.1 c) is the simplest form of Snell's law for an isotropic medium. In this type of medium the velocities of P and S waves are independent of the direction of wave propagation. Also, the ray direction and the phase velocity direction (direction of wave propagation vector, **K**) are the same (Fig. 6.2 a). This, in general,

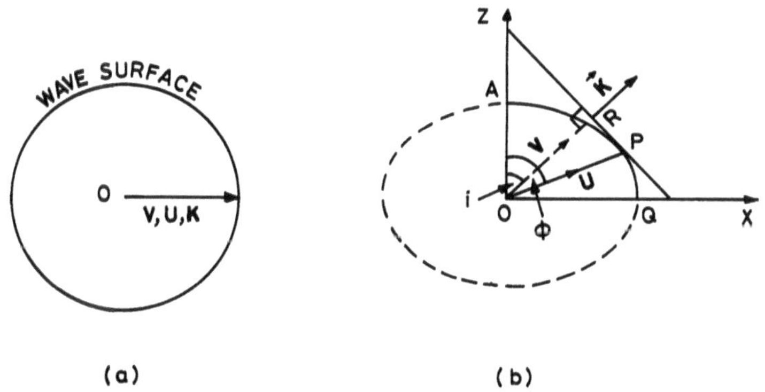

Fig. 6.2 a, b. a A spherical wave surface from a point disturbance in an isotropic, homogeneous medium. Phase velocity (V), and ray (or group) velocity (U) are scalars. Directions of both phase- and ray-velocities is the same as the direction of propagation vector **K**. **b** APQ is part of an elliptical wave surface; the tangent at P is the wavefront; OP is the direction of ray (or group) velocity (U); OR is the direction of phase velocity. The phase velocity (V) direction coincides with the direction of propagation vector **K**

does not hold for an anisotropic medium. For example, consider the case of SH-wave propagation in an elliptically anisotropic medium. The wave surface has an elliptical form APQ (APQ is only a part of the wave surface, Fig. 6.2b). Joining O with a point P on the wave surface gives the direction of ray velocity or group velocity (**U**). A tangent is drawn at P to APQ and perpendicular OR is dropped on the tangent line. Then, OR gives the direction of wave vector **K** and the magnitude and direction of the phase velocity **V**. Throughout in this chapter the ray directions are shown by a continuous line and phase direction by a dashed line. In general, the angle i (angle between the symmetry axis OA and the wave propagation vector '**K**') and the ray angle 'ϕ' (angle between the symmetry axis 'OA' and direction of propagation of energy 'OP') are not equal for an anisotropic medium.

With the above simple distinction between the wave propagation in isotropic and anisotropic mediums it is clear that *Snell's law* for an anisotropic medium should be expressed separately considering *phase velocity* '**V**' and the direction 'i' of wave vector '**K**', and ray (or group) velocity 'U' and the direction 'ϕ' of the *energy flux vector.*

Although plane wave fronts are idealization to the actual shape of the wave fronts or wave surface, they make the mathematical treatment simpler, and tractable, and is able to explain the essential details of the propagation process. In what follows below, a plane wave assumption is made.

With reference to Fig. 6.3a the Snell's law in terms of phase velocities, V_1 and V_2, and angle of incidence 'i_1' and refraction 'i_2' is expressed as

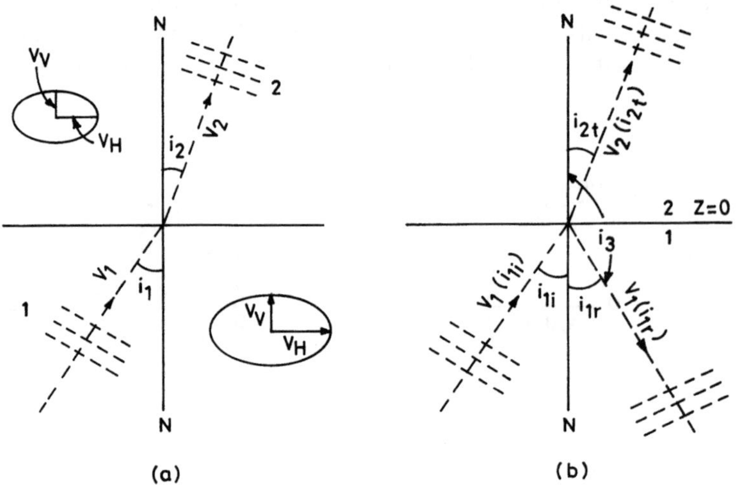

Fig. 6.3a, b. a Refraction between elliptically anisotropic media. NN is normal to the interface; V_H and V_V are horizontal and vertical components of a elliptical velocity function for two media in contact; *1* and *2* are positions of incident and refracted plane wavefronts. **b** SH wave reflection and refraction between two elliptically anisotropic media. Subscripts 1, 2, i, r, and t to angle i correspond to different media (1, 2) and incident, reflected and transmitted phase velocity directions (i, r, t), respectively

$$\frac{\sin i_1}{V_1} = \frac{\sin i_2}{V_2} \tag{6.2}$$

This form of Snell's law is in terms of angles of the normals to the plane incident and refracted wavefronts with normal to the boundary and the phase velocities in these directions.

Since phase velocity V (refer Fig. 6.2 b) is given by

$$(V(i))^2 = V_V^2 \cos^2 i + V_H^2 \sin^2 i$$

Equation (6.2) is also expressible as

$$\left\{ \frac{\sin i_1}{(V_V^2 \cos^2 i_1 + V_H^2 \sin^2 i_1)^{\frac{1}{2}}} \right\}_{\text{medium 1}} = \left\{ \frac{\sin i_2}{(V_V^2 \cos^2 i_2 + V_H^2 \sin^2 i_2)^{\frac{1}{2}}} \right\}_{\text{medium 2}} \tag{6.2 a}$$

[Sato and Lapwood 1968, Eq. (5.3)]
where V_V and V_H take respective values for medium 1 and 2.

Levin (1978) arrives at the Snell's law governing reflected and refracted plane SH waves in media with ellipsoidal velocity dependence and separated by interface z=0 (Fig. 6.3b):

$$\frac{\sin i_{1i}}{V_1(i_{1i})} = \frac{\sin i_{1r}}{V_1(i_{1r})} = \frac{\sin i_{2t}}{V_2(i_{2t})} \tag{6.3}$$

[Levin 1978, Eq. (A-17)]
where i_{1i}, i_{1r}, and i_{2t} are angles of incidence, reflection and refraction; $V_1(i_{1i})$, $V_1(i_{1r})$ and $V_2(i_{2t})$ are phase velocities in these directions; suffixes 1 and 2 denote the relevant medium.

Derivations of Snell's law in the form (6.2 a) or (6.3) is simple. The incident wave is taken as a plane SH-wave. Expressions for reflected and refracted waves are written. Continuity conditions of displacement due to incident and reflected waves in medium 1 and transmitted wave in medium 2, and continuity of normal stress (p_{xz}) at z = 0 is utilized to arrive at the conditions [refer to Fig. 6.3 and Eq. (6.2)]:

$$i_3 = \pi - i_{1i} \quad \text{and} \quad \left(\frac{\sin i_1}{V_1}\right)_{\text{medium 1}} = \left(\frac{\sin i_2}{V_2}\right)_{\text{medium 2}} \tag{6.4}$$

[Sato and Lapwood 1968, Eq. (5.2)]

Thus, the quantity sin i/V is a constant for incident, reflected and transmitted waves. It is to be noted that V is dependent on angles i_1, i_2, and is not constant as in an isotropic medium.

It may be mentioned here that for a transversely isotropic (T.I) medium with vertical symmetry axis (z axis), values of the phase velocity in the vertical and horizontal direction for SH waves are V_V $(= \sqrt{(C_{44}/\rho)}$ and V_H $(= \sqrt{(c_{11} - c_{12})/2\rho)}$ respectively, where c_{44}, c_{11} and c_{12} are elastic stiffness constants of the T. I. medium.

In order to express Snell's law in terms of ray or group velocity 'U', and ray angle 'ϕ' with the symmetry axis OA (refer to Fig. 6.2b), an expression for ray velocity is to be found. *Ray velocity* is the velocity with which energy propagates in the ray direction 'ϕ'.

The wave surface APB is the envelope of plane wave fronts for various angles 'i'. The equation of the *wave surface* is found by eliminating 'i' between the equation of the wave front (assumed to have passed origin O at time t=0), i.e.

$$f(i) \equiv x \sin i + z \cos i - Vt = 0 \tag{6.5a}$$

and

$$\frac{\partial f}{\partial i} = 0 \tag{6.5b}$$

Any point 'P' on the wave surface has it's parametric representation as

$$x_P = \frac{V_H^2 \sin i}{V} t \tag{6.6a}$$

$$z_P = \frac{V_V^2 \cos i}{V} t \tag{6.6b}$$

The radius vector OP increases at the rate

$$U = \frac{r}{t} = \frac{(V_V^4 \cos^2 i + V_H^4 \sin^2 i)^{\frac{1}{2}}}{V} \tag{6.7}$$

[Sato and Lapwood 1968, Eq. (3.5)] giving the ray or group velocity in direction 'ϕ' in terms of the horizontal (V_H) and vertical component (V_V), of phase velocity 'V' and angle 'i'.

To express *Snell's law* in terms of ray angles and ray velocity, use is made of the relation (refer to Fig. 6.2b)

$$VU \sin \phi = V_H^2 \sin i \tag{6.8a}$$

or

$$\frac{\sin i}{V} = \frac{U \sin \phi}{V_H^2} \tag{6.8b}$$

Therefore, from Eq. (6.2), the modified form of *Snell's law* for refracted SH waves in terms of ray angles becomes

$$\frac{U_1 \sin \phi_2}{V_{H_1}^2} = \frac{U_2 \sin \phi_2}{V_{H_2}^2} \tag{6.9a}$$

[Sato and Lapwood 1968, Eq. (5.6)]

In Eq. (6.9 a) subscript 1 refers to quantities for an incidence wave, and 2 for a refracted or transmitted wave.

A more general situation is to consider the incidence of a P or SV plane wave from above or below the boundary of two ellipsoidally anisotropic media. In this case, the boundary conditions require continuity of vertical and horizontal components of displacement and normal and shear stresses. Snell's law and the treatment of reflection and transmission coefficients for this case were developed by Daley and Hron (1977, 1979). At the interface the following law holds:

$$\frac{\sin i_i}{V_i} = \frac{\sin i_v}{V_v}$$

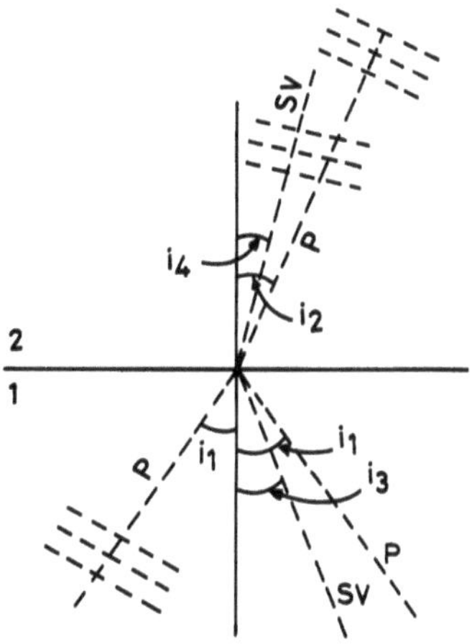

Fig. 6.4. For an incident P-wavefront in medium 1, $i_i = i_1$; for a resultant P-wavefront in medium 1, $i_v = i_1$. For a resultant SV wavefront in medium 1, $i_v = i_3$; and for the resultant SV wavefront in medium 2, $i_v = i_4$

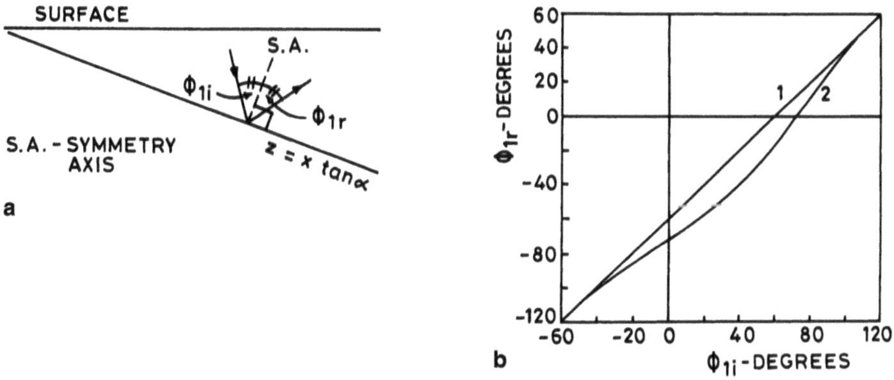

Fig. 6.5 a, b. a The geometry of an anisotropic layer with a dipping reflector, and a symmetry axis perpendicular to the reflector. **b** Variation of angle of reflection ϕ_{1r} vs angle of incidence ϕ_{1i} from an isotropic or anisotropic layer with the dipping reflector at 30°. The *straight line 1* at 45° relates to isotropy; and *curved variation 2* to a transversely isotropic medium, the ratio of horizontal to vertical velocity is 1.2 (Levin 1978)

In this equation, subscripts ‘i’ and ‘v’ stand for incident and resultant wavefront. For the incident and resultant P wavefront in medium 1 and 2, i, v=1 and i,v=2 respectively. Also i, v=3 and 4 are assigned for the incident and resultant SV wavefront in medium 1 and 2 respectively. Figure 6.4 shows the use of this convention for incident P and resultant P and SV waves in medium 1 and 2.

The modified Snell's law for ellipsoidal anisotropy for the general case of P or SV wave incidence, and resultant P or SV reflected or refracted waves is given by

$$\frac{U_i \sin \phi_i}{V_{H_i}^2} = \frac{U_v \sin \phi_v}{V_{H_v}^2} \tag{6.9 b}$$

[Daley, Hron 1979, Eq. (19)].

(ii) In the above part, incident, reflected and refracted plane wave fronts with respect to the horizontal interface and vertical *symmetry axis* normal to the interface were considered. A dipping interface is assumed in this part (Fig. 6.5 a).

If the axis of symmetry is perpendicular to the dipping reflector, the angle of incidence (ϕ_{1i}) equals the angle of reflection (ϕ_{1r}) (Fig. 6.5 a). This, in general, need be redefined when the axis of symmetry is perpendicular to the surface. For treating the problem of a dipping interface, Levin (1978) assumed a single layer above a reflecting plane: $z = x \tan a$.

A plane wave is assumed incident on the dipping plane at an angle ‘ϕ_i’ with respect to the vertical. Taking ‘ϕ_r’ as the corresponding angle of reflection, Levin (1978) showed that

$$\tan \phi_r = \frac{\left[V_H^2 \tan a(2 + \tan \phi_i \tan a) - V_v^2 \tan \phi_i\right]}{\left[V_H^2 \tan^2 a - V_v^2(1 + 2\tan a \tan \phi_i)\right]} \tag{6.10 a}$$

[Levin 1978, Eq. (4)]

A depiction of variation of ‘ϕ_r’ with ‘ϕ_i’ is given in the Fig. 6.5 b.

For an isotropic medium, ϕ_r with ϕ_i are related as

$$\phi_r = \phi_i - 2a \tag{6.10 b}$$

For the elliptical velocity function, ϕ_r and ϕ_i are related by Eq. (6.10 a). This is proved in Levin (1978) by making use of equal phase on a dipping plane for plane incident and reflected waves. The incident and reflected SH waves are taken of the form

$$u_i = A \exp\left[2\pi\, \mathrm{i} f\left(\left(x \sin i_i / V(i_i)\right) - \left(z \cos i_i / V(i_i)\right) - t\right)\right] \tag{6.11 a}$$

$$u_r = B \exp\left[2\pi\, \mathrm{i} f\left(\left(x \sin i_r / V(i_r)\right) + \left(z \cos i_r / V(i_r)\right) - t\right)\right] \tag{6.11 b}$$

Here u_i and u_r represent displacement field of incident and reflected plane SH waves in the same medium. ‘i_i’ and ‘i_r’ are angles of phase velocity directions with respect to the symmetry axis taken as vertical; $V(i_i)$ and $V(i_r)$ are phase velocities in the direction i_i and i_r.

Equality of phase of incident and reflected waves on the dipping surface $z = x \tan a$ gives

$$\frac{\sin i_i - \cos i_i \tan a}{V(i_i)} = \frac{\sin i_r + \cos i_r \tan a}{V(i_r)} \tag{6.12}$$

For an isotropic medium $V(i_i) = V(i_r) = V$. In this case Eq. (6.12) reduces to

$$\tan a = \tan \frac{i_i - i_r}{2}$$

or (6.13)

$$i_r = i_i - 2a$$

Using the relation $\tan \phi = (V_H/V_v)^2 \tan i$ (Jolly 1956) for ellipsoidal anisotropy and taking $V_H/V_v = 1$ for the isotropic case, we get Eq. (6.10 b):

$$\phi_r = \phi_i - 2a$$

To get Eq. (6.10 a), values of $V(i_i)$ and $V(i_r)$ are to be substituted, giving

$$\frac{(\sin i_i - \cos i_i \tan a)^2}{(V_H^2 \sin^2 i_i + V_v^2 \cos^2 i_i)} = \frac{(\sin i_r + \cos i_r \tan a)^2}{(V_H^2 \sin^2 i_r + V_v^2 \cos^2 i_r)} \tag{6.14}$$

This equation is a quadratic in $\tan i_r$. The appropriate sign of the quadratic root is determined from the condition that for $V_H = V_v = V$ for an isotropic medium, it should yield $i_r = i_i - 2a$. Further, using relation

$$\tan \phi = \left(\frac{V_H}{V_v}\right)^2 \tan i$$

for ellipsoidal anisotropy, i_i and i_r are changed to ϕ_i and ϕ_r. This yields Eq. (6.10 a) – a relation between reflected ray angle (ϕ_r) and incident ray angle (ϕ_i) for a dipping reflector and vertical symmetry axis in the case of ellipsoidal anisotropy.

The equation for transmitted waves, corresponding to Eq. (6.10) for reflected waves is given by

$$\cot^2 \phi_{2t} = V_{V2}^2 \left[V_{H1}^4 \cot^2 \phi_{1i} + V_{V1}^2 (V_{H1}^2 - V_{H2}^2)\right] / \left[V_{H2}^4 V_{V1}^2\right] \tag{6.15}$$

[Levin 1978, Eq. (8)]

Here 1 stands for the incident wave in medium 1, and 2 for the transmitted wave in medium 2. 'ϕ' correspondingly stands for the ray angle; 'ϕ_{1i}' for the incident ray angle and 'ϕ_{2t}' for the transmitted ray angle.

(iii) In Sections 6.1 (i) and 6.1 (ii), Snell's law (Eqs. (6.2, 6.2 a, 6.9 a and 6.9 b)] is expressed in terms of angles and phase or group (or ray) velocities. These equations hold for ideal conditions of isotropy and ellipsoidal anisotropy. Nevertheless, these are useful relations for understanding differences in wave propagation characteristics under conditions of isotropy and anisotropy. They are also useful relations for ray tracing in layered media.

An alternative approach is to express *Snell's law* in terms of wave characteristics (*frequency* 'ω' and *wave vector* '**K**') of incident, reflected and refracted waves. Snell's law for reflection in this case requires that the incoming and reflected rays have the same frequency, and the component of the wave vector '**K**' parallel to the interface for reflected wave is the same as that of incoming wave.

If 'ω' and 'ω'' are frequencies and '**K**' and '**K**'' are wave vectors for incident and reflected waves, these conditions may be written as (Fig. 6.6)

$$\omega(\mathbf{K}') = \omega(\mathbf{K}), \tag{6.16 a}$$

and

$$\mathbf{K}' - \mathbf{n}(\mathbf{n}\cdot\mathbf{K}') = \mathbf{K} - \mathbf{n}(\mathbf{n}\cdot\mathbf{K}) \tag{6.16 b}$$

[Blair and Korringa 1987, Eqs. (4) and (5)]

Eq. (6.16 a) may also be written as

$$\mathbf{U}'\cdot\mathbf{K}' = \mathbf{U}\cdot\mathbf{K} \tag{6.17 a}$$

$\mathbf{U}$ and $\mathbf{U}'$ are ray (group) velocities for incident and reflected waves, respectively.

Also, defining $\delta\mathbf{k} = \mathbf{k}' - \mathbf{k}$, Eq. (6.16 b) may be written as

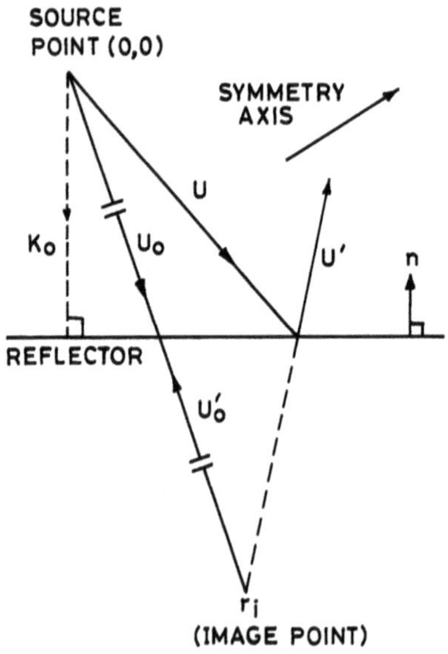

Fig. 6.6. U_0 is the group velocity vector through the source point (0, 0) for which the propagation vector $\mathbf{K}_0$ is perpendicular to the reflector. The image of the source point is formed at image point $\mathbf{r}_i$, which is as far behind the interface as the source point (0, 0) is in front of it. The group velocity $\mathbf{U}'$ for any other ray vector $\mathbf{U}$ appears to come from the image point (Source: Blair and Corringa 1987)

$$\frac{\delta\mathbf{K}}{(\mathbf{n}\cdot\delta\mathbf{K})} = \mathbf{n} \tag{6.17b}$$

Equation (6.17 b) states that vector δK possess the same direction (i.e. that of **n**) for all rays.

Equations (6.17 a) and (6.17 b) together constitute the Snell's law for reflection for a completely general case, including converted waves.

Blair and Korringa (1987) proved a theorem for SH waves in transversely isotropic medium. The theorem states (Blair and Korringa 1987, p 1563): 'SH waves emanating from a point source in a homogeneous transversely isotropic medium are reflected from a planar interface between the transversely isotropic medium and another homogeneous medium in such a way that they define a reflective image that is free of aberrations, regardless of the relative orientation of the *elastic symmetry axis* and the interface. It is an image for rays in the direction of *group velocity vectors* not the *slowness vectors.* The image is located on a line through the source point in the direction of the group velocity of a wave for which the slowness vector is perpendicular to the interface. The distance measured along this line, of the image behind the interface is equal to that of the source point in front. This is illustrated with the help of Fig. 6.6. The direction and position of image point (r_i) is shown. A wave with group velocity '**U**' is reflected with group velocity '**U**′', which appears to come from image point '$\mathbf{r}_i$' situated on the ray with group velocity '**U**' whose *slowness vector* ($\mathbf{K}_0$) is perpendicular to the reflector. The arbitrary direction of the symmetry axis of a transversely isotropic medium with respect to the interface is also shown.

6.2 The Wave Surface, Ray Parameter and Ray Tracing

In an anisotropic medium, the direction of propagation of the disturbance (or the ray velocity direction) generally differs from the direction of movement of the wavefront (or the phase velocity direction). Ray velocity and phase velocity directions define two surfaces as the disturbance propagates in an anisotropic medium. Consequently, a *wave surface* is defined as the loci of points at a given time reached by the disturbance produced by an impulsive point source at the origin at zero time. A *phase velocity surface* is defined as the loci of radial vectors, which are perpendicular to the plane wavefronts of the wave surface.

The relationship between the wave surface and the phase velocity surface is shown in Fig. 6.7 a for a medium with elliptical velocity dependence. An impulsive disturbance produced at O at time t=0 in such a medium generates a wave surface AP_1P_2B at time t=1 s of the elliptical form; *phase velocity surface* is AR_1R_2B.

From the Fig. 6.7 it is clear that the phase velocity and ray velocity are related as

$$V(i) = U(\phi)\cos(\phi - i) \tag{6.18}$$

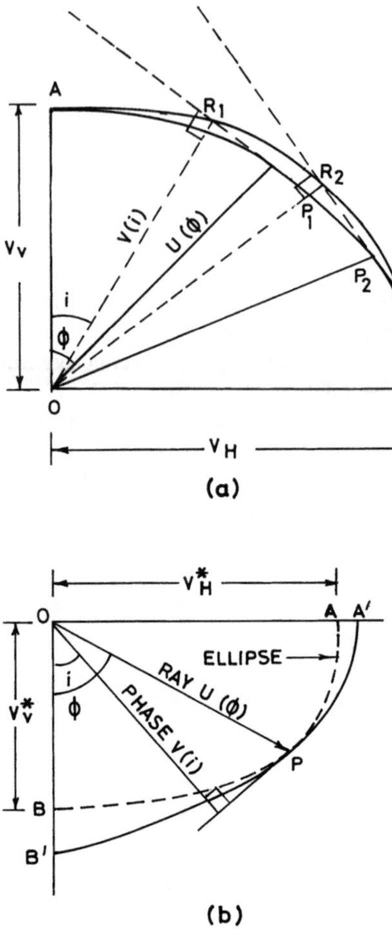

Fig. 6.7 a, b. a The relationship between the wave surface (AP_1P_2B) and phase velocity surface (AR_1R_2B). The phase velocity surface is obtained by construction from the wave surface. Tangents at points P_1 and P_2 on the wave surface represent plane wavefronts. Perpendicular from O on tangents cut at R_1 and R_2. R_1 and R_2 are points on the phase velocity surface AR_1R_2B (after Byun 1982, p 622); **b** Elliptical equivalence of wave surface (after Byun 1984, p 1910)

For the elliptical velocity function, angles 'i' and 'ϕ' are related as

$$\tan \mathrm{i} = \left(\frac{V_V}{V_H}\right)^2 \tan \phi$$

Eliminating ϕ from the above two equations, we can express the phase velocity V (i) as

$$V^2(\mathrm{i}) = V_V^2 \cos^2 \mathrm{i} + V_H^2 \sin^2 \mathrm{i} \tag{6.19}$$

[Byun 1982, Eq. (4)]

The wave surface generated by an impulsive point source at time zero at 'O' would be an ellipse at unit time given by

$$U^{-2}(\phi) = V_V^{-2} \cos^2 \phi + V_H^{-2} \sin^2 \phi \tag{6.20}$$

The forms of Eqs. (6.19) and (6.20) are strikingly similar.

It may be mentioned here that in field conditions the wave surface for a P wave would, in general, not be an ellipse. Nevertheless, the local behavior of the wave surface could be described by an elliptical surface and corresponding to this elliptical surface, ray velocity and phase velocity values could be assigned. The non-elliptical wave surface, the tangential ellipse fitted to the wave surface corresponding to ray direction ϕ at point P of the wave surface are shown in Fig. 6.7b. For the elliptical fit to the wave surface at P, Snell's law in the form of Eqs. (6.9a) or (6.9b) holds.

The relationship between the ray velocity $U(\phi)$ for the actual wave surface A'PB' to parameters of the elliptical fit surface APB are

$$\left(V_H^*\right)^2 = U^2(\phi)\cos(\phi - i)\cos\phi/\sin i \tag{6.21a}$$

$$\left(V_V^*\right)^2 = U^2(\phi)\cos(\phi - i)\sin\phi/\cos i \tag{6.21b}$$

[Byun 1984, Eqs. (9a,b)]

Knowing V_H^* and V_V^* for the fitting elliptical velocity function for a ray OP of actual wave surface A'PB', the ray parameter 'p' for ray OP may now be determined in terms of 'V_H^*' and 'V_V^*'. Therefore, in the discussion that follows below, the equivalent quantities 'V_H^* 'and 'V_V^*' would be used.

Figure 6.8 shows the geometry. AB is an interface between two media. NN is normal to the interface. The ray angle ϕ and phase angle i measured from the V_V^* axis is shown. It is to be noted that the ray incidence angle 'α' and phase incidence angle 'β' are different for anisotropic media.

The *ray parameter* 'p' of a given ray (Fig. 6.8) along an interface is defined as

$$p = \frac{\sin(i + \gamma)}{V^*(i)} = \text{constant} \tag{6.22}$$

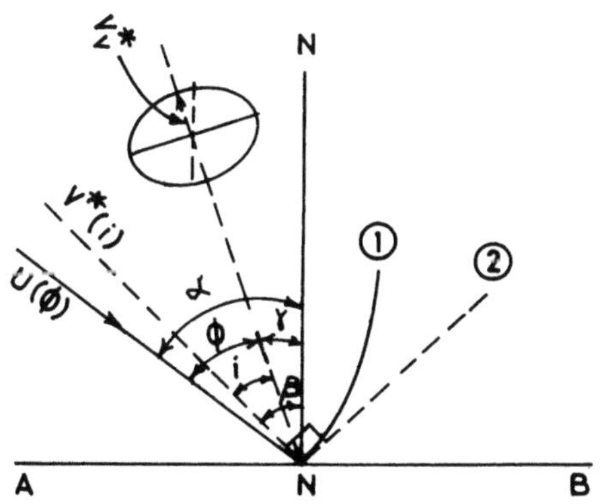

Fig. 6.8. Geometry of ray and corresponding phase incident on interface. Ray angle ϕ and phase angle i are measured from the V_V^* axis of the elliptical velocity function. The ray incidence angle α and phase angle β are measured from normal NN to the interface AB. At N (a point on the interface), the wave surface and plane wavefront are shown by curve 1 and *straight line 2*, respectively (after Byun 1982, p 1622)

[Byun 1982, Eq. (6)]

Using Eq. (6.19), we have

$$p = \frac{\sin\gamma + \cos\gamma \tan i}{\sqrt{\left(V_V^*\right)^2 + \left(V_H^*\right)^2 \tan^2 i}} \tag{6.23}$$

Changing i to ϕ using

$$\tan i = \left(\frac{V_V^*}{V_H^*}\right)^2 \tan\phi \text{ gives}$$

$$p = \frac{\left(V_V^*\right)^{-2} \sin\gamma + \left(V_H^*\right)^{-2} \cos\gamma \tan\phi}{\sqrt{\left(V_V^*\right)^{-2} + \left(V_H^*\right)^{-2} \tan^2\phi}} \tag{6.24}$$

As $U^{-2}(\phi) = V_V^{-2}\cos^2\phi + V_H^{-2}\sin^2\phi$, we have

$$p = U(\phi)\left(\left(V_V^*\right)^{-2} \sin\gamma \cos\phi + \left(V_H^*\right)^{-2} \cos\gamma \sin\phi\right) \tag{6.25}$$

[Byun 1982, Eq. (9)]

This is a relation expressing *ray parameter* ‘p’ in terms of ray velocity ‘$U(\phi)$’, ray angle ‘ϕ’ and parameters (V_H^* and V_V^*) of the elliptically-fitted wave surface, for the ray OP of part (PB′) of actual wave surface (refer to Fig. 6.7b).

Squaring Eq. (6.24) and solving for tan ϕ, we get

$$\tan\phi = \left(\frac{V_H^*}{V_V^*}\right)^2 \frac{\sin\gamma\cos\gamma \pm p\sqrt{\left(V_V^*\right)^2\cos^2\gamma + \left(V_H^*\right)^2\sin^2\gamma - \left(pV_V^*V_H^*\right)^2}}{p^2\left(V_H^*\right)^2 - \cos^2\gamma} \tag{6.26}$$

[Byun 1982, Eq. (10)]

Taking $p=0$ in Eq. (6.26) yields

$$\tan\phi = -\left(\frac{V_H^*}{V_V^*}\right)^2 \tan\gamma$$

This shows that the ray incidence angle $a=\phi+\gamma$ would be zero only under the conditions for which $V_V^*=V_H^*$, i.e. the medium is isotropic, or when $\gamma=0$, i.e. a transversely isotropic medium. Thus, the normal ray, i.e. the ray for which the corresponding phase is normal to the interface, is not in general normal to the interface. This would result in “side-slip” of zero offset reflection events laterally on the surface seismic section (Byun 1982, p 1623).

Algorithm 6.1: A Method of Ray Tracing

Given the elliptical velocity function parameters γ, V_H, and V_v for each of two layers bounded by an interface, tracing of the ray incident on the interface at angle α_1 can be accomplished by the following steps outlined in Byun (1982, p 1623):

(i) ray angle 'ϕ_1' in medium 1 may be determined by using the relation $\alpha_1 = \phi_1 + \gamma_1$.
(ii) ray parameter 'p' may be calculated using Eq. (6.24).
(iii) Ray angle 'ϕ_2' for the lower medium may be determined using the relation (6.26), and using the condition that the ray parameter 'p' is constant across the interface.
(iv) Refraction angle 'α_2' of the ray in the lower medium may be determined using the relation $\alpha_2 = \phi_2 + \gamma_2$.

6.3 Surface Emerging Wavefront and Normal Moveout Velocity (VNMO)

Snell's law governs the relationship between the angles of incident, reflected and refracted waves including converted waves for a given velocity model. The incident, reflected and refracted waves have different independent wavefronts. In seismic surveys, the wavefront generated by a seismic source is refracted through the interfaces in the Earth and is reflected back to the ground surface. The final shape of the wavefront on the ground surface is a consequence of refraction and reflection processes on the initial wave front generated at the seismic source. This emphasizes the fact that the properties of the emerging wave front on the ground surface are dependent on the past history of refraction and reflection underneath. Therefore, *the emerging wavefront properties must possess an intimate relationship to the parameters of the real Earth model.* We consider separately isotropic and anisotropic conditions in the discussions that follow:

6.3.1 Arbitrarily Dipping Layered Media with Isotropic or Ellipsoidal Velocity Dependencies

Like Snell's law for a ray path discussed in Section 6.1 there is law governing the radius of curvature of the incident and refracted wavefront. For *isotropic media* this is expressed in the following form (Fig. 6.9):

$$R_t = \left(\frac{\cos^2 i_t}{\cos^2 i_i}\right)\left(\frac{V_i}{V_t}\right) R_i \tag{6.27}$$

[Shah 1973, Eq. (16)]

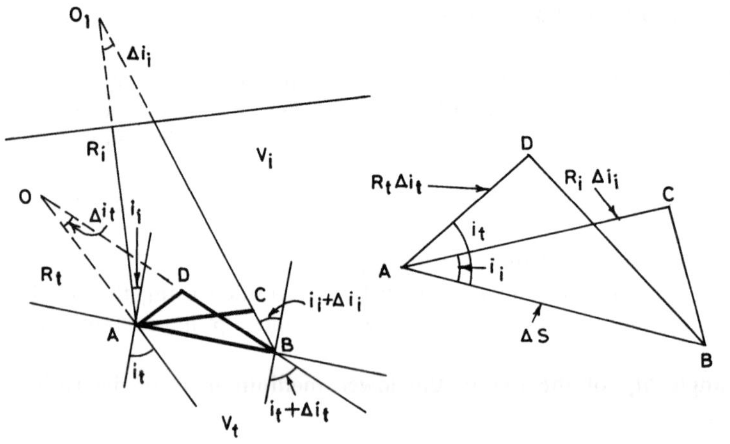

Fig. 6.9. Geometry of incident and the transmitted wavefronts at an interface. AB is a part of the interface between media with velocities V_i and V_t. The angle of incidence at A is i_i and at B it is $(i_i + \Delta i_i)$. O_1 is the center of curvature obtained by extending two rays at A and B. O_1A is the principal radius of curvature equal to R_i. AC is the tangent to the incident wavefront at A. Similarly, O is the center of curvature of the refracted rays and the principal radius of curvature is $OA = R_t$. AD is the tangent to the transmitted wavefront (after Shah 1973, p 818)

where R_i and R_t are the radii of curvature of the incident and refracted (transmitted) wave fronts; 'i_i' and 'i_t' are angles of incident and refracted rays (also these are the angles which incident and refracted wavefronts make with the refractor); 'V_i' and 'V_t' are velocities in the medium of the incident and refracted wave, respectively.

Equation (6.27) is a general form applicable to both incident and transmitted compressional and shear waves with appropriate choice of velocities 'V_i' and 'V_t'.

When a P-wave is reflected as P-wave $V_i = V_r$ and $i_i = i_r$ we have from Eq. (6.27)

$$R_r = R_i \tag{6.27 a}$$

Subscript 'r' stands for the reflected wave.

Equations (6.27) and (6.27 a) are the key equations. For a given seismic source and geophone position repeated use of Eqs. (6.27) and (6.27 a) would allow us to calculate the final radius of curvature 'R' of the wave front emerging on the ground surface for an arbitrarily dipping layered model. Shah (1973) derived an expression for the radius of curvature 'R' of the reflected P-wave front emerging on the ground surface, which has been normally incident on the n-th subsurface interface. Figure 6.10 shows it for the third interface. The normally incident ray retraces its path backwards and reaches the position of the seismic source, i.e. at the coincident source-receiver position. The radius of curvature (R) for a normal incidence reflection from the nth interface is given by

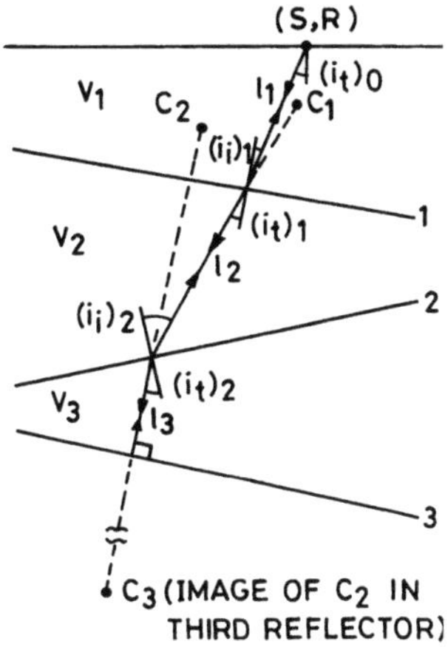

Fig. 6.10. Normal incidence ray in a three-layer model. C_1 and C_2 are centers of curvature of the wavefronts transmitted in the second and third layer; C_3 is the center of curvature of the wavefront reflected from the third subsurface interface (after Shah 1973, p 819)

$$R = \frac{1}{V_1}\left[\begin{array}{l} V_1^2\Delta t_1 + \left(\dfrac{\cos^2 i_i}{\cos^2 i_t}\right)_1 V_2^2\Delta t_2 + \ldots \\ + \ldots \\ + \left\{\left(\dfrac{\cos^2 i_i}{\cos^2 i_t}\right)_1 \left(\dfrac{\cos^2 i_i}{\cos^2 i_t}\right)_2 \cdots \left(\dfrac{\cos^2 i_i}{\cos^2 i_t}\right)_{j--1}\right\} V_j^2\Delta t_j \\ + \ldots \\ + \left\{\left(\dfrac{\cos^2 i_i}{\cos^2 i_t}\right)_1 \left(\dfrac{\cos^2 i_i}{\cos^2 i_t}\right)_2 \cdots \left(\dfrac{\cos^2 i_i}{\cos^2 i_t}\right)_{n-1}\right\} V_n^2\Delta t_n \end{array}\right] \tag{6.28}$$

[Shah 1973, Eq. (19)]

In Eq. (6.28), 'Δt_j' is the two way travel time in the j-th layer for the normally incident ray path with primary reflection from the nth interface, and 'V_j' is the velocity (Fig. 6.10). If l_j is the length of the ray in the j-th layer

$$2l_j = V_j\Delta t_j (j = 1, 2, 3 \ldots n) \tag{6.29 a}$$

The total travel time for a normally incident ray path is

$$t_0 = \sum_{i=1}^{n} \Delta t_j \tag{6.29 b}$$

Also i_i and i_t refer to the same *normal incidence ray* from the nth interface. Subscripts 1, 2, 3... n–1 outside the brackets in Eq. (6.28) refer to the interface number for which the ratio ($\cos^2 i_i/\cos^2 i_t$) is to be evaluated.

It is to be noted that the *angle of emergence* at the ground surface '$(i_t)_0$' may only be determined from the surface reflection data. The other angles of incidence and refraction at various interfaces and time(s) in each layer have to be determined by ray tracing. The method of finding the emergence angle on the ground surface is explained below.

In surface CDP reflection surveys, it is the reflection travel time, which is measured. Shah (1973) showed that the radius of curvature of the emerging wavefront on the ground surface is related to the second derivative of the travel time 't' with respect to geophone position 'g' by the relation

$$t_{gg}(g = s) = \frac{\cos^2(i_t)_0}{RV_1} \tag{6.30}$$

[Shah 1973, Eq. (8)]

Here 'g' stands for the coordinate of the geophone and 's' for the shot. '$t_{gg}(g=s)$' stands for the second derivative of travel time for the *coincident source-receiver position.* Referring to Fig. 6.11 '$(i_t)_0$' in Eq. (6.30) may be determined from

$$\sin(i_t)_0 = V_1 \frac{\Delta t}{\Delta g} = V_1 \frac{\partial t}{\partial g} = V_1 t_g(g = s) \tag{6.31}$$

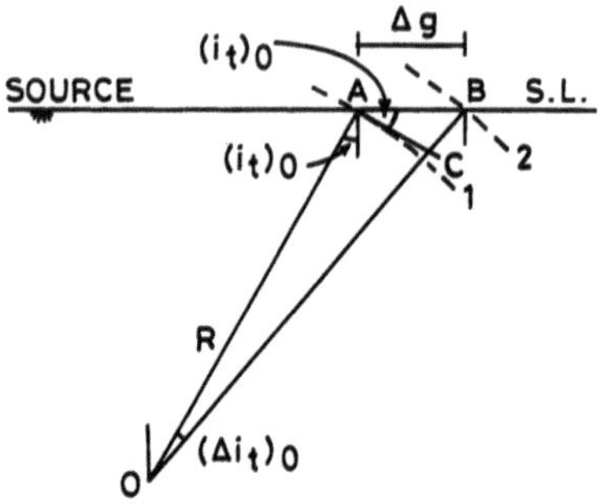

Fig. 6.11. Geometry of wavefront approaching receivers. S. L. is the seismic line. *1* and *2* are portions of wavefront at times t and $(t+\Delta t)$ (after Shah 1973, p 816)

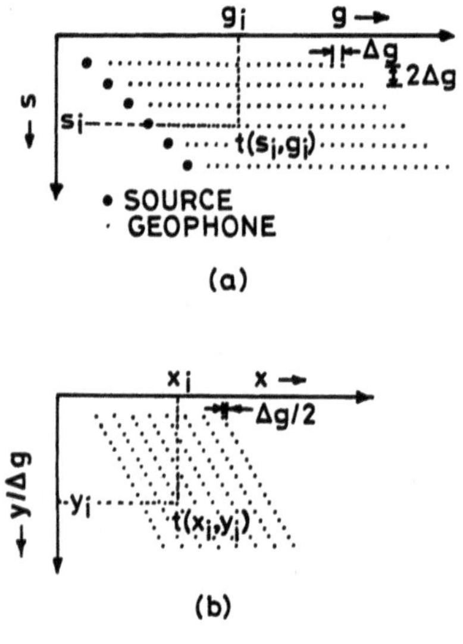

Fig. 6.12 a, b. a Grid of points over which time (t) is measured in the (s, g) coordinate system. **b** Geometry of points over which (t) is measured in the mid-point-offset (x, y) coordinate (after Shah 1973, p 815)

With the shot and geophone positions denoted by 's' and 'g', mid-point coordinate (x) and seismic source-geophone offset (y) are defined as (Fig. 6.12 a, b)

$$x = \frac{s+g}{2} \quad \text{and} \quad y = g - s$$

Reflection times may be expressed in terms of (x, y). Shah (1973) showed that a simple relationship exists between time derivatives with respect to (g, s) and (x, y). Denoting t_x for $\partial t/\partial x$, tg for $\partial t/\partial g$, tgg for $\partial^2 t/\partial g^2$ Shah (1973) gave the following relationships:

$$t_g = (t_x/2) + t_y \tag{6.32 a}$$

$$t_{gg}(g = s) = \frac{t_{xx}(x, 0)}{4} + t_{yy}(x, 0) \tag{6.32 b}$$

[Shah 1973, Eq. (3 b)]

Since $t_y = 0$ at $y = 0$ or $g = s$, we have

$$t_g = t_x/2$$

from (6.32 c)

$$t_g = t_x/2 + t_y$$

Also as t_{xx} (x, 0) is equal to zero, we have

$$t_{gg}(g = s) = t_{yy}(x, 0) \tag{6.32 d}$$

t_{yy} (x, 0) is related to NMO velocity V_{NMO} by the relation

$$t_{yy}(x, 0) = \frac{1}{(V^2_{NMO} t_0)} \tag{6.33}$$

[Shah 1973, Eq. (5)]

As $t_g = t_x/2$, we have from Eq. (6.31)

$$\sin(i_t)_0 = \frac{V_1 t_x}{2} \tag{6.34}$$

This gives the *emergence angle* $(i_t)_0$ on the ground surface from the knowledge of t_x (*time dip*) from *stacked data*.

Now writing values of t_{gg} (g = s) and t_{yy} (x, 0) from Eqs. (6.30) and (6.33) we have from Eq. (6.32 d)

$$\frac{\cos^2(i_t)_0}{RV_1} = \frac{1}{(VNMO)^2 t_0}$$

or

$$R = \frac{\cos^2(i_t)_0}{V_1}(\mathrm{VNMO})^2 t_0 \tag{6.35}$$

When the normal ray path is vertical (for a horizontal layered system)

$$R = \frac{(\mathrm{VNMO})^2 t_0}{V_1}$$

Substitution of Eq. (6.28) in Eq. (6.35) gives

$$(\mathrm{VNMO})^2 t_0 = \frac{1}{\cos^2(i_t)_0} \sum_{j=1}^{n} \prod_{k=0}^{j-1} \left(\frac{\cos^2 i_i}{\cos^2 i_t}\right)_k V_j^2 \Delta t_j \tag{6.36}$$

$$\text{taking } \left(\frac{\cos^2 i_i}{\cos^2 i_t}\right)_0 = 1.$$

[Shah 1973, Eq. (20b)]

It is important to note that angles $(i_i)_k$, $(i_t)_k$ and Δt_j refer to the *normal ray path* of the *primary reflection* from the n-th interface.

Equation (6.36) is a relation giving normal moveout velocity (VNMO) in terms of velocity V_j of the isotropic layers for the normal incidence reflection from the nth reflector. Equation (6.36) has also been derived in Hubral (1976).

For a layered media with arbitrarily oriented interfaces and elliptical velocity dependencies within the layers, Byun (1982) derived *normal moveout velocity* (VNMO), similar to Eq. (6.36) for isotropic media [Byun 1982, Eq. (26)]. The formula is given by

$$(\mathrm{VNMO})^2 = \frac{1}{t_0 \cos^2(i_t)_0} \sum_{j=1}^{n} \left(\frac{V_H V_V}{V}\right)_j^2 \Delta t_j \prod_{k=1}^{j-1} \left(\frac{\cos^2 i_i}{\cos^2 i_t}\right)_k \tag{6.36a}$$

For derivations of Eq. (6.36a), the reader is advised to study Byun's (1982) publication.

Equations (6.36) and (6.36a) for *normal moveout velocity* (VNMO) are a Dix type formula, which is written in the form

$$(\mathrm{VNMO})^2 t_0 = \sum_{j=1}^{n} V_j^2 \Delta t_j \tag{6.36b}$$

for a horizontal layered isotropic medium with the *normal incidence ray* vertical and Δt_j representing the vertical time in the jth layer.

In contrast to the Dix Eq. (6.36b) for a horizontal layered medium, in Eqs. (6.36) and (6.36a), the normal moveout velocities (VNMO) (for layered arbitrarily dipping layers with isotropic or ellipsoidal velocity dependence) are a *weighted root mean square of velocities* in layers along the path of the *normal incidence ray.*

It is important to analyze the effects of *anisotropy* on *normal moveout velocity.* We take an example from Buyn (1984). Without giving details of numerical values of the Earth model considered, it suffices our purpose to give qualitatively the dependence of the emergence angle and NMO velocity on anisotropy. The two models (Fig. 6.13a) considered in Byun (1984) are sandstone-limestone and a gypsum- weathered layer. These models are characterized by five elastic constants as for the T.I. medium. The variation of ray emergence angle and diffractor velocity (NMO velocity corrected for the surface emergence angle) are shown in Fig. 6.13b,c,d. Figure 6.13b shows the change in *emergence angle* of the *normal ray* as a function of reflector dip. In the isotropic case, the emergence angle is equal to the dip angle of the reflector. However, in the anisotropic case, the emergence angle is significantly different from the dip angle of the reflector. In this

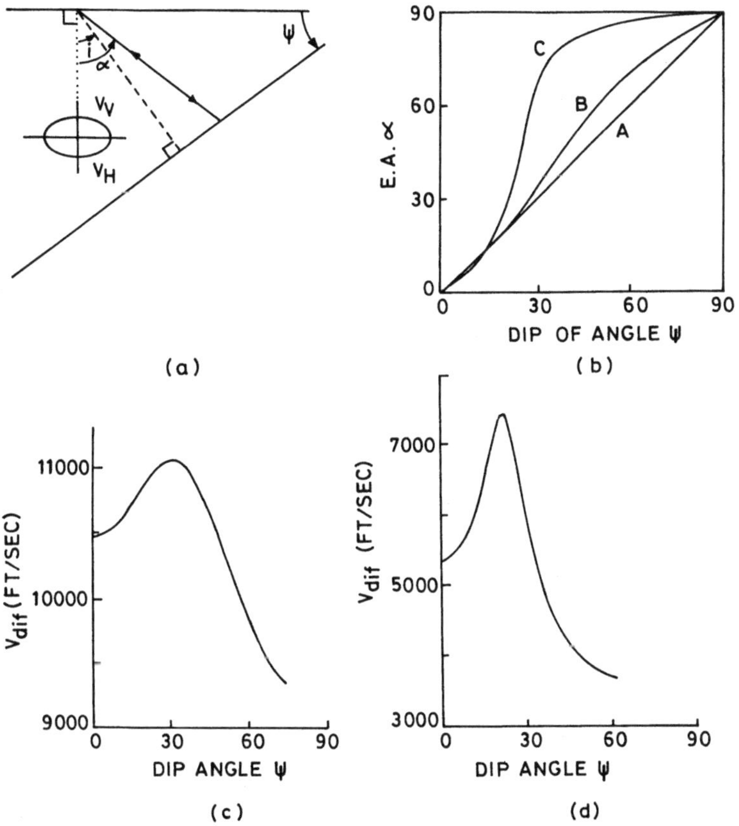

Fig. 6.13a–d. a Anisotropic dipping single layer model with dip angle ψ, normal incidence ray, phase angle i and emergence angle α; **b** surface emergence angle (E.A.) of the normal ray versus dip angle; **c** and **d** show variation of diffractor velocity versus dip angle. A is for isotropy; B and C are for limestone-sandstone and gypsum-weathered anisotropic materials (after Byun 1984)

figure, A shows the variation for isotropy; B for anisotropy for the limestone-shale model and C is for the gypsum-weathered anisotropic material.

The variation of diffractor velocity vs. dip angle is shown in Fig. 6.13c and d. Figure 6.13c is for the sandstone-limestone medium, and Fig. 6.13d is for the gypsum-weathered material. The ratio between V_{lower}/V_{upper} is almost 2; a large variation of VNMO with the dip angle of the reflector for an anisotropic model of the gypsum-weathered material.

Exercise 6.1

Take a scale of 1 cm = 200 m and draw the dipping reflectors on a vertical line from point A, at depths of 500, 1,000 and 2,000 m, respectively; dips of the first, second and third reflectors are 20, 20 and 25° in opposite directions. The velocity values in the layers are $V_1 = 3{,}000$ m/s, $V_2 = 4{,}000$ m/s, and $V_3 = 6{,}000$ m/s.

Assume A to be the position of the coincident seismic source and geophone position. Draw from point A the path of the *normal incidence ray* (ABCD) to the third reflector. Determine the radius of curvature of the surface emerging reflected wavefront at A; also determine normal moveout velocity (VNMO) to the third reflector.

Solution

Refer to Fig. 6.14 (not to scale). Draw first the dipping reflectors model according to the scale given. Next determine the angles β_2, α_2, β_1, α_1, and β_0, by using the values of angles of dips of the reflectors, with the condition that the ray is normally incident on the third reflector and by making use of Snell's law at the layer boundaries. Angle β_2 is determined first; then using Snell's law α_2 is found. For determining β_1, use may be made of the theorem that the sum of all angles of the quadrilateral HBRC is 360°. For finding α_1, Snell's law is again used. For determining β_0, the angle of emergence on the ground surface, the quadrilateral JAKB may be considered. The path lengths in the layers ($AB = l_1$, $BC = l_2$ and $CD = l_3$) may be found by measurement. Two-way time in each layer along the normal incidence ray path is found by dividing the path lengths (2-way) by the velocity value in the layer.

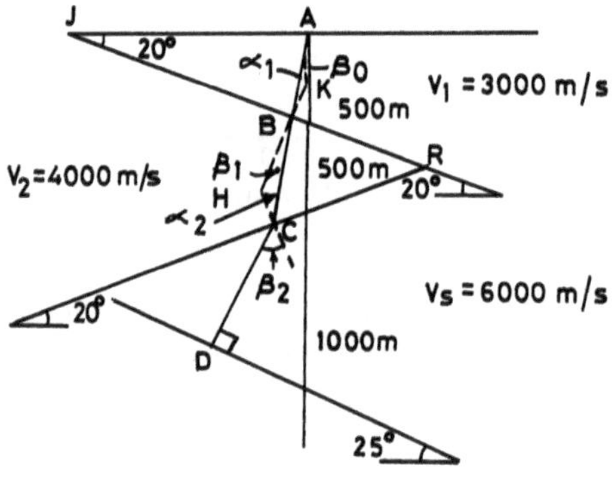

Fig. 6.14. Geometry of reflectors and normal incidence ray path

Table 6.1. Angles, pathlengths, two-way times in the layers for a normal incidence ray on the third reflector

Angles	Path length (m) in layers	Two way time (s) in layers
$\beta_0 = 11.2°$	$AB = l_1 = 480.76$	$\Delta t_1 = 2\,l_1/3000 = 0.32055$
$\alpha_1 = 8.88°$	$BC = l_2 = 615.38$	$\Delta t_2 = 2\,l_2/4000 = 0.30765$
$\beta_1 = 11.88°$	$CD = l_3 = 764.70$	$\Delta t_3 = 2\,l_3/6000 = 0.25495$
$\alpha_2 = 28.12°$		
$\beta_2 = 45°$		

With reference to Fig. 6.14, the respective values are tabulated in Table 6.1: Therefore, the total *normal incidence travel-time* ($\Delta t_1 + \Delta t_2 + \Delta t_3$) is 0.88315 s.

The radius of curvature of the wavefront emerging at the geophone is given by [refer to Eq. (6.28)]:

$$R = \frac{1}{V_1}\left[V_1^2 \Delta t_1 + \left(\frac{\cos^2\alpha_1}{\cos^2\beta_1}\right) V_2^2 \Delta t_2 + \left(\frac{\cos^2\alpha_1 \cos^2\alpha_2}{\cos^2\beta_1 \cos^2\beta_2}\right) V_3^2 \Delta t_3\right]$$

Substituting the numerical values, the radius of curvature (R) is found to be 7,484.9 m.

For determining the normal moveout velocity to the third reflector we use [refer to Eq. (6.36)]

$$(VNMO)^2 t_0 = \frac{1}{\cos^2\beta_0}\left[\begin{array}{l} V_1^2 \Delta t_1 + \left(\dfrac{\cos^2\alpha_1}{\cos^2\beta_1}\right) V_2^2 \Delta t_2 \\ + \left(\dfrac{\cos^2\alpha_1 \cos^2\alpha_2}{\cos^2\beta_1 \cos^2\beta_2}\right) V_3^2 \Delta t_3 \end{array}\right]$$

It gives the value of VNMO, for normal incidence ray to the third reflector, equal to 5,140.72 m/s.

Further Discussion of the Exercise

The above example pertains to a calculation of the various parameters for a zero source receiver separation and for a normal incidence ray on some subsurface reflector. Ursin (1978) used Eq. (6.27) for calculating the radius of curvature of transmitted and reflected wavefronts for a *finite source-receiver separation* (Fig. 6.15). Repeated application of Eq. (6.27) enables us to write the *radius of curvature of the reflected wavefront* at the receiver on the ground surface.

Starting with Eq. (6.27), we have

$$\frac{\cos^2 i_i}{R_i V_i} = \frac{\cos^2 i_t}{R_t V_t}$$

Letting R_j denote the radius of curvature of the wavefront leaving the jth layer, the following expressions may be written for the radius of curvature of transmitted wavefronts (Fig. 6.15),

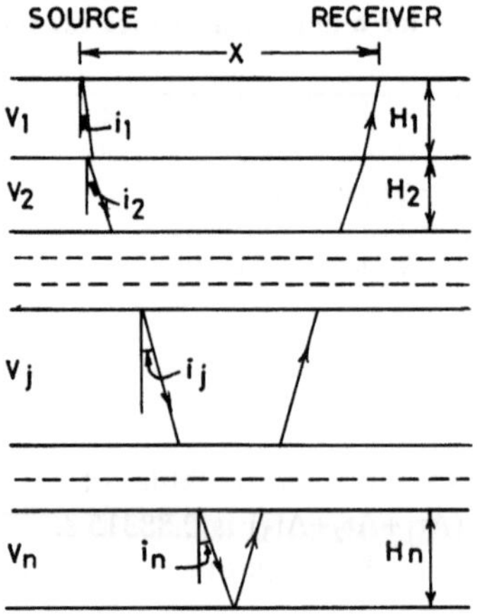

Fig. 6.15. Layered model and reflection ray path for a separated seismic source and receiver

$$R_1 = \frac{h_1}{\cos i_1}$$

$$R_2 = \frac{h_2}{\cos i_2} + h_1 \frac{V_1}{V_2} \frac{\cos^2 i_2}{\cos^3 i_1}$$

$$\vdots$$

$$R_n = \frac{h_n}{\cos i_n} + h_{n-1} \frac{V_{n-1}}{V_n} \frac{\cos^2 i_n}{\cos^3 i_{n-1}} + \cdots + h_1 \frac{V_1}{V_n} \frac{\cos^2 i_n}{\cos^3 i_1}$$

The radius of curvature of the reflected upgoing wave, leaving the j-th layer may be written as R_{2n-j+1}

$$R_{n+1} = \frac{h_n}{\cos i_n} + R_n \quad \text{(upgoing wave leaving nth layer)}$$

$$R_{n+2} = 2\frac{h_{n-1}}{\cos i_{n-1}} + 2h_n \frac{V_n}{V_{n-1}} \frac{\cos^2 i_{n-1}}{\cos^3 i_n} + \ldots\ldots\ldots\ldots + h_1 \frac{V_1}{V_{n-1}} \frac{\cos^2 i_{n-1}}{\cos^3 i_1} \quad \text{(upgoing wave leaving (n}-1\text{)th layer)}$$

$$\begin{aligned} R_{2n} &= 2\frac{h_1}{\cos i_1} + 2h_2 \frac{V_2}{V_1} \frac{\cos^2 i_1}{\cos^3 i_2} + \ldots \\ &\quad + 2h_n \frac{V_n}{V_1} \frac{\cos^2 i_1}{\cos^3 i_n} \\ &= 2\frac{\cos^2 i_1}{V_1} \sum_{j=1}^{n} \frac{h_j V_j}{\cos^3 i_j} \end{aligned} \tag{6.37}$$

[Ursin 1978, Eq. (4)]

R_{2n} is the radius of curvature of a wavefront received on the ground surface.

In a further development, Ursin (1978) expressed the radius of curvature and source-receiver separation in terms of a ray-parameter. By eliminating the ray-parameter, he gave an expression to compensate for the *geometrical spreading effect.* Hurbral (1978) pointed out in a discussion the limitations of Ursin's (1978) result. Ursin's (1978) result is valid for a line source (emanating cylindrical waves) perpendicular to the line of measurement. On the other hand Hubral's (1978) result holds for the point source normally used in practice.

The concept of the *image ray* is a useful parallel concept to that of the *normal incidence ray.* Referring to Fig. 6.16a depicting both the normal incidence (AB) and image ray (QP), we observe that the image ray starts from a *point scatterer* (Q) on the reflector, travels through the layer boundaries and reaches the surface at normal incidence. For any other path other than that of the image ray, the path time is more than that for the image ray (Fig. 6.16b). The path of the image ray

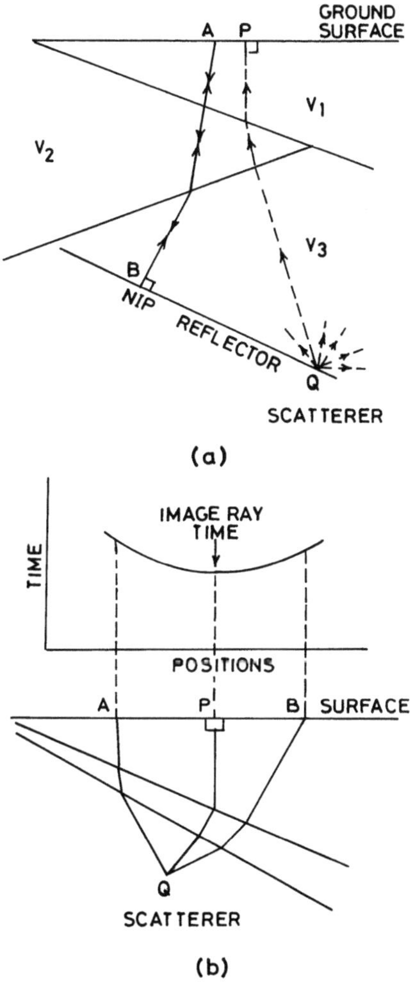

Fig. 6.16a, b. a Normal incidence ray path AB. Q is the point scatterer. QP is the image ray arriving at point P on the ground surface at 90°; **b** schematic diagram showing the image ray normally incident on the ground surface with minimum time

or any other ray originating from the scatterer on the reflector and received on the surface may be easily traced using Snell's law provided velocities and layer boundaries are known.

Hubral made an important observation that the apex of the *diffraction hyperbola* (refer to Chapter 14 for a definition) from the diffracting point lies on a ray that reaches the surface vertically. It is in fact the *image ray*. Reversing the argument, we say that the diffraction point would lie on a ray that departs vertically from the minimum time point on the surface, and subsequently refracts according to Snell's law at each boundary to finally reach to the point diffractor.

Making use of the concept of an *image ray*, migration velocity may be defined for depth migration (see, for example, Larner et al. 1981).

Problem 6.1

For the model assumed in Exercise 6.1, determine the Normal Moveout Velocity (VNMO) to the second reflector.

Hint: For determining the Normal Moveout Velocity (VNMO) to the second reflector, first draw the ray path normal to the second reflector. This path would be distinctly different from the normal incidence ray path to the third reflector. Determine the corresponding angles β_1, a_1, and β_0 (as were defined in Fig. 6.14). Determine the lengths of the normal incidence ray path in the first and second layer; also using the velocity values in the layers determine Δ_1 and Δ_2. Make use of the relation

$$R = \frac{1}{V_1}\left[V_1^2\Delta t_1 + \left(\frac{\cos^2 a_1}{\cos^2\beta_1}\right)V_2^2\Delta t_2\right]$$

to calculate the value of radius of curvature of the wavefront emerging at the geophone at A for the normal incidence ray to the second reflector. Calculate the Normal Moveout Velocity (VNMO) using the relation

$$(\mathrm{VNMO})^2 t_0 = \frac{1}{\cos^2\beta_0}\left[V_1^2\Delta t_1 + \left(\frac{\cos^2 a_1}{\cos^2\beta_1}\right)V_2^2\Delta t_2\right]$$

6.4 Normal Moveout Velocity for a Transversely Isotropic (T.I.) Medium

(i) Single Dipping T.I. Layer

A general form of NMO velocity (VNMO) for a short spread as a function of dip angle 'a' of the reflector in a homogeneous transversely isotropic (T.I.) medium is given by

Fig. 6.17. Phase velocity vector **V**, group velocity vector **U**, phase angle i, and group angle ϕ, and dipping reflector at an angle a (after Tsvankin 1995, p 269)

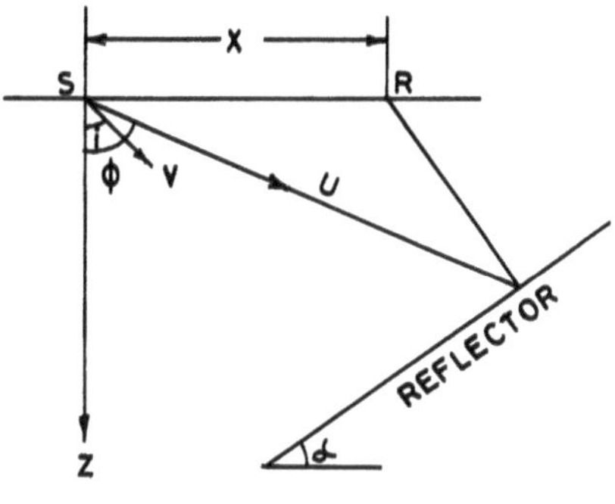

$$\mathrm{VNMO}(a) = \frac{V(a)}{\cos a} \frac{\sqrt{1 + \frac{1}{V(a)} \frac{d^2V}{di^2}}}{1 - \frac{\tan a}{V(a)} \frac{dV}{di}} \tag{6.38}$$

[Tsvankin 1995, Eq. (9)]

In Eq. (6.38), 'V' is *phase velocity* expressed as a function of phase angle 'i' measured from the vertical, 'a' is the dip angle of the reflector; derivatives are evaluated at dip angle a (Fig. 6.17). Further, Eq. (6.38) holds in the symmetry plane of any anisotropic medium, and is not restricted to any particular wave type. The incidence plane is taken as the dip plane of the reflector.

Specific forms of Eq. (6.38), when VNMO is expressed in terms of ray parameter 'p' (i.e. VNMO (p)) are:

(i) a single T.I. layer with a horizontal reflector [Eq. (4.20 a)]

$$\mathrm{VNMO}(0) = V_{PO}\sqrt{1 + 2\delta} \tag{6.38 a}$$

[Alkhalifah and Tsvankin 1995, Eq. (5)]

(ii) a single layer with elliptical anisotropy and a dipping reflector

$$\mathrm{VNMO}(p) = \frac{V_{nmo}(O)}{\sqrt{1 - p^2V^2_{nmo}(O)}} \tag{6.38 b}$$

[Alkhalifah and Tsvankin 1995, Eq. (6)]

(iii) a single layer with weak anisotropy i.e. $|\varepsilon| \ll 1$ and $|\delta| \ll 1$ and a dipping reflector. Weak anisotropy approximation is to infer behavior of non-elliptic velocity models

$$\mathrm{VNMO}(p) = \frac{V_{nmo}(O)}{\sqrt{1 - p^2V^2_{nmo}(O)}} \left[1 + (\varepsilon - \delta) \frac{p^2V^2_{nmo}(O)\{4p^4V^4_{nmo}(O) - 9p^2V^2_{nmo}(O) + 6\}}{1 - p^2V^2_{nmo}(O)}\right] \tag{6.38 c}$$

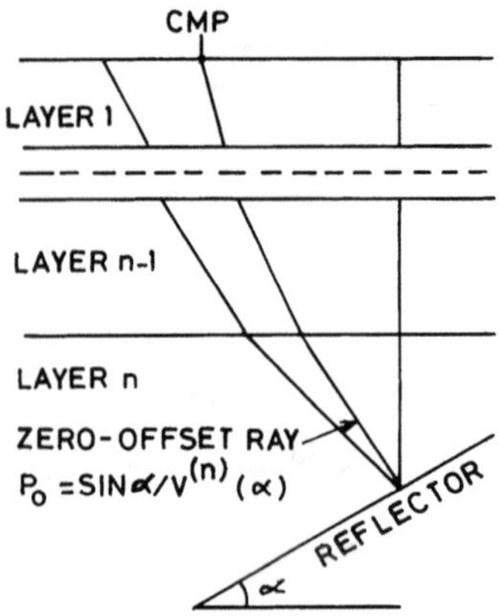

Fig. 6.18. A stratified anisotropic model including a dipping reflector beneath a stack of horizontal homogeneous layers. t_0^i is the two way travel time along a zero-offset ray in layer i (after Alkhalifah and Tsvankin 1995, p 1566)

[Alkhalifah and Tsvankin 1995, Eq. (7)]

(iv) a stack of plane homogeneous layers above a dipping reflector (Fig. 6.18)

$$V_{nmo}^2(p_0) \equiv \frac{1}{t_0}\sum_{i=1}^{n} t_0^{(i)}\left[V_{nmo}^{(i)}(p_0)\right]^2 \tag{6.38 d}$$

[Alkhalifah and Tsvankin 1995, Eq. (19)]

In Eq. (6.38 a) to (6.38 d) VNMO(p) is the NMO velocity as a function of the ray parameter 'p' in place of the dip angle 'a' in Eq. (6.38); VNMO(0) is NMO velocity for a zero dip (or horizontal) reflector [as in Eq. (6.38 a)]; 'i' stands for the quantities in the ith layer; 'p_0' corresponds to the ray-parameter for a zero-offset ray; ε and δ are anisotropy parameters.

Also, in Eqs. (6.38 a) to (6.38 d), ray parameter 'p' is the *horizontal slowness.* The ray parameter 'p' corresponding to a zero-offset reflection (say p_0) is given by

$$p(a) = \frac{1}{2}\frac{dt_0}{dx_0} = \frac{\sin a}{V(a)} \tag{6.39}$$

[Alkhalifah and Tsvankin 1995, Eq. (2)]

In Eq. (6.39) 't_0' is the two way travel time on the zero-offset (or stacked) section, and 'x_0' is the mid point position.

P-wave phase velocity 'V' in Eq. (6.38) depends on four Thomsen (1986) parameters: P- and S-wave vertical velocities 'V_{PO}' and 'V_{SO}', and dimensionless parameters 'ε' and 'δ'. Shear wave vertical velocity 'V_{SO}' has only negligible influence on P-wave velocity. Therefore, the remaining parameters 'V_{PO}', 'ε' and 'δ' need only be evaluated from the analysis of P-wave NMO velocity.

The advantage of expressing NMO velocity as a function of ray parameter 'p' may be found by observing Eqs. (6.38 b) and (6.38 c). VNMO (p) is related to VNMO(0) only or to VNMO(0) and the difference of ε and δ rather than three individual values of 'V_{PO}', 'ε' and 'δ'.

Equation (6.38 a) is useful to find 'δ' if VNMO(0), the NMO velocity for horizontal reflector and vertical velocity 'V_{PO}' are known. 'V_{PO}', for example, may be found from check shots or well logs. Equation (6.38 a) was derived in Thomsen (1986) using definition of VNMO [refer to Eq. (4.20 a)].

(ii) Layered T.I. Subsurface Model

The results on normal moveout velocity as a function of dip angle a (VNMO(a)) for a homogeneous transversely isotropic medium above a reflector was extended in Alkhalifah and Tsvankin (1995) to realistic *layered subsurface models.* The layered model considered is a stack of homogeneous layers above a dipping reflector (Fig. 6.18). The normal moveout velocity in this case is given by Eq. (6.38 d):

$$V_{nmo}^2(p_0) \equiv \frac{1}{t_0}\sum_{i=1}^{n} t_0^{(i)}\left[V_{nmo}^{(i)}(p_0)\right]^2$$

[Alkhalifah and Tsvankin 1995, Eq. (19)]

This shows that NMO velocity is the root mean square velocity of NMO velocities in each layer given by Eq. (6.38 d) and evaluated at parameter p_0:

$$p_0 = \frac{\sin a}{V^{(n)}(a)} \tag{6.39 a}$$

Also, travel time t_0 need be found along the *zero offset ray.*

The requirement of Eq. (6.38 d) is that the incidence plane should coincide with the dip plane of the reflector and with the plane of symmetry in all layers. This equation enables determination of NMO velocity for any layer 'j' in the form

$$\left[VNMO^{(j)}\right]^2 = \frac{t_0(j)(VNMO(j))^2 - t_0(j-1)(VNMO(j-1))^2}{t_0(j) - t_0(j-1)} \tag{6.39 b}$$

[Alkhalifah and Tsvankin 1995, Eq. (20)]

In Eq. (6.39 b) '$t_0(j-1)$' and '$t_0(j)$' are two way reflection times to the top and bottom of layer j, evaluated along the zero offset ray with $p = p_0$. These values of '$t_0(j-1)$' and '$t_0(j)$' are calculated along the zero-offset ray. The value of p_0 is determined by the dipping reflector and is given by

$$p_0 = \frac{\sin a}{V^{(n)}(a)}$$

where 'a' is the dip of the reflector; $V^{(n)}(a)$ is the phase velocity immediately above the dipping reflector. NMO velocities in Eq. (6.39 b) correspond to ray parameter value p_0.

It may be mentioned that unless the medium is T.I. with a vertical symmetry axis, the zero-offset ray may deviate from a vertical direction, and t_0 may be different from vertical time.

Equation (6.38 d) yields NMO velocity expressions given in Hake et al. (1984) and Tsvankin and Thomsen (1994) for a stack of horizontal T.I. layers. The difference between the Dix (1955) formula [Eq. (6.36 b)] and Eq. (6.38 d) is that in Eq. (6.38 d) 't_0' and velocities are to be evaluated at ray parameter value 'p_0' corresponding to dip angle (α) of the reflector ($p_0 = \sin\alpha / V^n(\alpha)$), whereas in Eq. (6.36 c) these quantities are taken for a vertical ray path.

The conventional NMO velocity analysis for horizontal, layered interfaces provides us with the NMO velocity and travel times corresponding to ray parameter value $p = 0$ (zero dip case). Use of Eq. (6.38 b) would then enable us to determine NMO velocity for any other ray parameter 'p' from zero dip NMO velocity in the case of isotropy or elliptical anisotropy. Application of Eq. (6.39 b) requires an evaluation of two-way travel time t_0 (p) from t_0 (o) for a horizontal reflector. 't_0(p)' is related to 't_0(o)' for isotropy or elliptical anisotropy as

$$t_0(p) = t_0(o)\sqrt{1 + p^2(\text{VNMO}(p))^2} \tag{6.40}$$

[Alkhalifah and Tsvankin 1995, Eq. (21)]

If the medium is T.I., one requires 'η' in addition to zero-dip NMO velocity, VNMO(0), to find VNMO (p). VNMO(0) and η suffice to find t_0(p) for a T.I. medium as in this case, we have,

$$t_0(p) = t_0(0) f[\eta, \text{VNMO}(0)] \tag{6.41}$$

[Alkhalifah and Tsvankin 1995, Eq. (22)]

The procedure to strip the influence of overburden on NMO velocity may be referred to in Alkhalifah and Tsvankin (1995) with reference to layered structure as in Fig. 6.18.

6.5 From Normal Moveout Velocity to Interval Velocities and Positioning of Reflectors: Isotropic Layers

The case of arbitrary dips with a two-dimensional case of n-reflectors was dealt with in Shah (1973) [Eq. (6.36)]:

$$(\text{VNMO})^2 t_0 = \frac{1}{\cos^2(i_t)_0} \sum_{j=1}^{n} \prod_{k=0}^{j-1} \left(\frac{\cos^2 i_i}{\cos^2 i_t}\right)_k V_j^2 \Delta t_j$$

In this equation t_0, i_i, i_t and t_k relate to the particular ray that is normally incident on nth interface (refer to Fig. 6.10).

It is possible to estimate VNMO from maximum coherency stacking (MCS) velocity for a short spread. For a short spread, $V_{NMO} \cong V_{MCS}$ is generally true.

Analyzing Eq. (6.36) interval velocities for various layers in an arbitrarily dipping layered system and the positions of reflectors may be determined. These values may be found by *stripping* successive layers. Assuming that positions of (n−1) interfaces and velocities between interfaces have been determined, the emergence angle $(i_t)_0$ for the ray, which has been reflected back from the nth interface, is given by [Eq. (6.34)]:

$$\sin(i_t)_0 = 0.5\,V_1(TD)_n$$

Here $(TD)_n$ is the time-dip (ms/km) of the n-th reflector on a *stacked seismic section*. The emergent ray may then be traced to the (n−1)-th reflector. Having found t_j, $(i_i)_j$ and $(i_t)_j$ ($j = 1, 2 \ldots n-1$), t_n (time in nth layer) may be found by subtracting all t_j from t_0. 'v_n' is found from Eq. (6.36). Knowing v_n and t_n, the position of the nth reflector normal to the ray in the nth layer may be easily determined.

Ray tracing would not be required in the case of a single reflector or n parallel reflectors. In the case of n parallel reflectors (Fig. 6.19 a)

$$(VNMO)^2 t_0 = \frac{1}{\cos^2(i_t)_0}\sum_{j=1}^{n} V_j^2 \Delta t_j$$

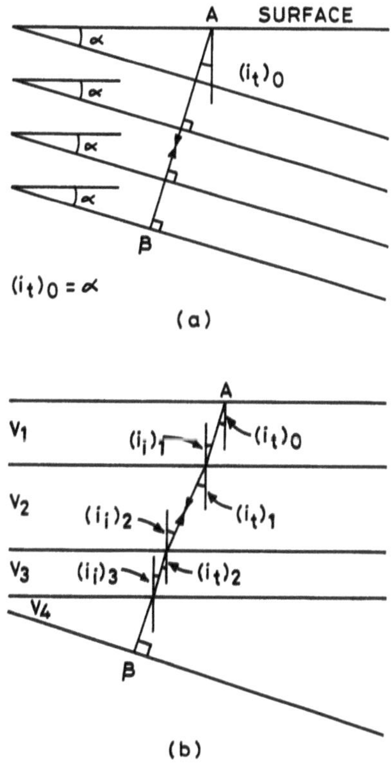

Fig. 6.19 a, b. a Parallel reflectors with a common normal incidence ray path; **b** dipping reflector beneath horizontal reflectors. The ray is normally incident on the dipping reflector

or (6.42)

$$\mathrm{VNMO} = \frac{\mathrm{VRMS}}{\cos(i_t)_0}$$

(Shah 1973)

$(i_t)_0$ is the emergence angle and is equal to the reflector dip. The emergence angle $(i_t)_0$ may be determined from a stacked section as in Eq. (6.34).

Consider a dipping reflector underlying a number of horizontal layers (Fig. 6.19b). In this case we have $(i_t)_0 = (i_i)_1$, $(i_t)_1 = (i_i)_2$ etc. and

$$\prod_{k=0}^{j-1} \left(\frac{\cos^2 i_i}{\cos^2 i_t} \right)_k = \frac{\cos^2(i_i)_1}{\cos^2(i_t)_{j-1}} = \frac{\cos^2(i_t)_0}{\cos^2(i_t)_{j-1}} \tag{6.43}$$

This gives

$$(\mathrm{VNMO})^2 t_0 = \sum_{j=1}^{n} \frac{V_j^2 \Delta t_j}{\cos^2(i_t)_{j--1}} \tag{6.44}$$

[Shah 1973, Eq. (21d)]

Using Snell's law with Eq. (6.34), we have

$$\sin(i_t)_{j-1} = \frac{V_j}{V_1} \sin(i_t)_0 = \frac{V_j t_x}{2}$$

Angles $(i_i)_j$ and $(i_t)_j$ may be computed without ray tracing for all values of j from the knowledge of time dip (TD) (ms/km) on the *stacked section.*

6.6
Normal Moveout Velocity (VNMO) and Reconstruction of the Velocity Field in Anisotropic Media

Section 6.5 dealt with the use of normal moveout velocity (VNMO) for determining interval velocities and positions of the reflectors in the case of isotropy, i.e. each layer is assumed to have a unique directionally independent velocity – *the interval velocity.*

Unlike isotropic velocity, in the case of anisotropy the *velocity field* has directional dependence.

The primary difficulty in seismic data processing under the conditions of anisotropy is the recovery of the velocity field from surface reflection data. Tsvankin (1995) and Alkhalifah and Tsvankin (1995) developed a method to determine the velocity field from *normal moveout velocity* (VNMO) measured as a function of the ray parameter for two different reflector dips. They also analyzed numerical implications involved in the *inversion process* for transversely isotropic (T.I.) media using P-wave NMO velocities. The discussion in this section is based on their findings.

VNMO (p) (normal moveout velocity for a dipping reflector) is controlled by the difference (ε–δ) as revealed by Eq. (6.38 c) for weak anisotropy ($\varepsilon \ll 1$, $\delta \ll 1$). Assuming that VNMO(o) and 'δ' have been determined, VNMO(p) for a single dipping reflector should in principle yield the value of 'ε', and it becomes possible to construct the velocity field VNMO(p) using Eq. (6.38 c).

Considering Eq. (6.38 b) – the case of elliptical anisotropy, the NMO velocity for ray parameter 'p' depends only on zero dip NMO velocity (VNMO(0)) with no dependence on V_{PO}, ε or δ. This gives results because of expressing NMO velocity as a function of the ray parameter instead of the dip angle of the reflector 'α' (see, for example, Tsvankin 1995). Equation (6.38 b) may be used to reconstruct the velocity field VNMO(p) from knowledge of VNMO(0). Alkhalifah and Tsvankin (1995) mention the important implication of this VNMO(p) dependence in dip moveout processing. The form of VNMO(p) for isotropic media coincides with Eq. (6.38 b). Alkhalifah and Tsvankin (1995) show that NMO velocity for a dipping reflector for general transverse isotropy (T. I.) is characterized by two parameters VNMO(0) and 'η', where

$$\eta \equiv 0.5\left(\frac{V_h^2}{(\mathrm{VNMO}(0))^2} - 1\right) = \frac{\varepsilon - \delta}{1 + 2\delta} \tag{6.45 a}$$

or

$$V_h = \mathrm{VNMO}(0)\sqrt{1 + 2\eta} \tag{6.45 b}$$

where V_h is the horizontal velocity.

Equation (6.45 b) shows that if an accurate value of horizontal velocity V_h is known (say, from head waves traveling along a horizontal reflector or from cross well tomography), it would give the value of 'η' knowing VNMO(0). From knowledge of VNMO(0) and parameter 'η' it should be possible to build a P-wave NMO velocity field as a function of ray parameter 'p' for a general T. I. medium. 'η' is also a suitable parameter to characterize *non-hyperbolic moveout*. For example, *P-wave moveout* for *long-spread* in T. I. media is described by the relation

$$t^2(X) - t_{PO}^2 + \frac{x^2}{(\mathrm{VNMO}(0))^2} \quad \frac{2\eta x^4}{(\mathrm{VNMO}(0))^2\left[t_{PO}^2(\mathrm{VNMO}(0))^2 + (1+2\eta)x^2\right]} \tag{6.46}$$

[Alkhalifah and Tsvankin 1995, Eq. (18)]

Therefore, long-spread P-wave moveout depends on the vertical travel-time 't_{po}', and two effective parameters, VNMO(0) and η. 'η' exhibits the deviation from hyperbolic moveout. For $\eta = 0$ (elliptical anisotropy), *moveout* is purely hyperbolic.

The normal moveout velocity corresponding to ray parameter 'p_0' should be combined with the normal moveout velocity for some other dip (it may be zero-dip NMO velocity) for the same layer to perform the single layer inversion for anisotropic parameters.

Alkhalifah and Tsvankin (1995) studied the sensitivity of NMO velocity to evaluation of anisotropic parameters. The main conclusion is that a wide range of Thomsen (1986) anisotropic parameters (V_{PO}, ε and δ) have almost the same NMO velocity. If one of these parameters is known, values of other two could be reliably recovered from NMO velocities for two different dips. In their exercise on numerical inversion, they considered models with $\varepsilon - \delta \geq 0$. These models are considered typical of subsurface formations. For a conventional spread length, limited by the distance between the common mid point (CMP) and the reflector, NMO velocity is well approximated by the stacking velocity routinely used in seismic processing. If only the surface reflection data is available, the exact vertical velocity may not be known, and in this case inversion of NMO velocities yield a family of *equivalent solutions (ES)*. This family of equivalent solutions has the same dip-dependent NMO velocity and satisfies the following relationships:

$$\mathrm{VNMO}(0) = V_{PO}\sqrt{1+2\delta} = \text{constant}$$

and

$$V_h = V_{PO}\sqrt{1+2\varepsilon} = \text{constant}$$

The inversion procedure can yield VNMO(0) or V_h and η but not individual values of vertical velocity and anisotropy parameters V_{pO}, ε and δ. 'η' is given by Eqs. (6.45 a) or (6.45 b).

The inversion of VNMO data from two distinct dips are sufficient to recover values of VNMO(0) and η. For weak anisotropy P-wave NMO velocity Eq. (6.38 c) is characterized by VNMO(0) and the value of $(\varepsilon - \delta)$. For transverse isotropy it is characterized by VNMO(0) and η.

Alkhalifah and Tsvankin (1995) also discuss the properties of *equivalent solutions* with illustrative examples. However details are too extensive to be presented here and the reader is advised to refer to the original publication.

Results on normal moveout velocity (VNMO) for the two-dimensional case have been developed for a three-dimensional case by Krey (1976) and Hubral (1976 a, b). In these cases also, the treatment is based on a normal raypath trajectory. It requires estimation of VNMO and the time-dip of each reflector along the profile (from the seismic section) and in some other direction, preferably at right angles to the profile. These approaches are similar to Shah (1973). *Interval velocity* V_n is related to VNMO through a number of matrix functions. *Dix type inversion* of non-hyperbolic moveout of the travel-time in layered media has been developed by Grechka and Tsvankin (1998), and provides the values of effective anisotropic parameters that control the non-hyperbolic moveout. Analytical developments of the steps may be found in the publication of Grechka and Tsvankin (1998).

CHAPTER 7

Reflection Time Analysis and Velocity for Isotropic Layered Media

7

In seismic interpretations, velocity plays a dominant role in the calculation of *dynamic correction*, and the conversion of time sections to depth sections. The problem of realistic velocity estimation owes its complexity to heterogeneity of the subsurface and the lack of knowledge of seismic ray paths. For a heterogeneous ground, the ray path deviates from a straight line. In fact, the ray path is the least time path (*Fermat's principle*). According to Fermat's principle: *a seismic signal follows the least time path of propagation from one point to another in a medium.*

The travel-time of seismic waves may be calculated by determining ray paths using Snell's law (Section 6.1). An alternative method makes use of Fermat's principle of least time. Use of this principle is illustrated below for a reflected wave in a multilayered medium, each layer being homogeneous and isotropic.

7.1 Reflected Wave Travel Time for Multi-Layer Isotropic Media: Use of Fermat's Principle of Least Time

Consider a layered medium and coordinate axes as in Fig. 7.1a. Assume that the layer thicknesses are $h_1, h_2, \ldots h_j \ldots h_n$ and corresponding velocities are $V_1, V_2, \ldots V_j \ldots V_n$. Let l_j be the ray path length in the j-th layer, and let its horizontal component be x_j and angle with vertical be ϕ_j (Fig. 7.1 b). We have,

$$l_j^2 = h_j^2 + x_j^2 \tag{7.1}$$

The total travel-time of a reflected wave from the bottom of the n-th layer is

$$T_x = 2\sum_{j=1}^{n} \frac{l_j}{V_j} \tag{7.2}$$

According to Fermat's principle, the total travel-time T_x should be minimum subject to the constraint

$$2\left(x_1 + x_2 + \ldots x_j + \ldots x_n\right) = x \tag{7.3}$$

The minimum requirement of T_x under the constraint (7.3) yields

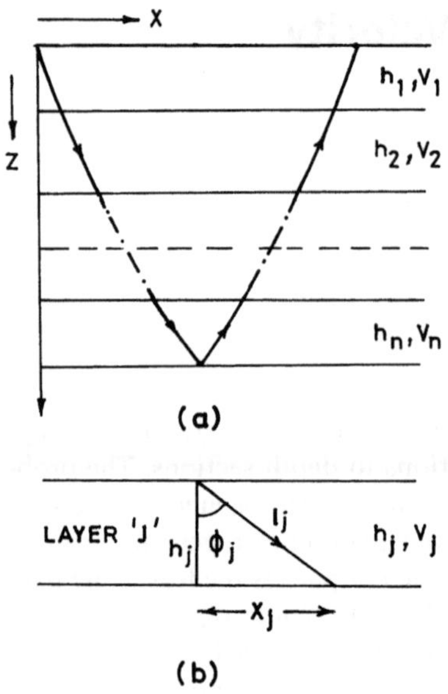

Fig. 7.1 a, b. a Geometry of a layered medium and reflection raypath; **b** raypath in the j-th layer

$$\frac{\delta T_x}{\delta x_j} = \frac{2x_j}{l_j V_j} = 2\lambda, \quad j = 1, 2, \ldots .n \tag{7.4}$$

where λ is the *Lagrange multiplier.*

Equation (7.4) with (7.1) gives

$$x_j^2 = \frac{\lambda^2 V_j^2 h_j^2}{1 - \lambda^2 V_j^2} \tag{7.5}$$

Substitution of Eq. (7.5) in Eqs. (7.2) and (7.3) yields

$$T_x = 2\sum_{j=1}^{n} \frac{h_j/V_j}{\sqrt{1 - \lambda^2 V_j^2}} \tag{7.6}$$

$$x = 2\lambda \sum_{j=1}^{n} \frac{V_j h_j}{\sqrt{1 - \lambda^2 V_j^2}} \tag{7.7}$$

(Taner and Koehler 1969)

Equations (7.6) and (7.7) are parametric forms of the time-distance relationship with λ as the parameter. Referring to Eq. (7.4) with Fig. 7.1 b gives the following, which is Snell's law:

$$\lambda = \frac{\sin\phi_j}{V_j} \tag{7.8}$$

$j = 1, 2, 3 \ldots n$

If the velocities of all layers are equal ($V_1 = V_2 = V_3 = \ldots V_n$), Eqs. (7.6) and (7.7) reduce to

$$T_x = \frac{2h}{\sqrt{1 - \lambda^2 V^2}} \tag{7.9}$$

and

$$x = \frac{2\lambda V h}{\sqrt{1 - \lambda^2 V^2}} \tag{7.10}$$

where $h = h_1 + h_2 + \ldots + h_n$.

Elimination of 'λ' gives a familiar form for a single layer:

$$T_x^2 = \frac{4h^2}{V^2} + \frac{x^2}{V^2} \tag{7.11}$$

From a general function-theoretical consideration (Copson 1935, pp 121–125) the generalization of Eq. (7.11) gives

$$T_x^2 = C_1 + C_2 x^2 + C_3 x^4 + C_4 x^6 + \ldots \tag{7.12}$$

[Taner and Koehler 1969, Eq. (4)]

The coefficients C_1, $C_2 \ldots C_n$ in Eq. (7.12) depend on the thickness of the layers and velocities in them.

To compare a 2-term travel-time relation for a single layer case with that of an n-layered case, the values of the first two coefficients C_1 and C_2 are

$$C_1 = \left(2 \sum_{j=1}^{n} \frac{h_j}{V_j} \right)^2 \tag{7.13}$$

and

$$C_2 = \sum_{j=1}^{n} \frac{h_j}{V_j} \bigg/ \sum_{j=1}^{n} V_j h_j \tag{7.14}$$

C_1 is the square of the two-way travel-time for the vertical ray. Further

$$\frac{1}{C_2} = (\mathrm{VRMS})^2 = \frac{\sum_{j=1}^{n} V_j h_j}{\sum_{j=1}^{n} \frac{h_j}{V_j}} = \frac{\sum_{j=1}^{n} \Delta t_j V_j^2}{\sum_{j=1}^{n} \Delta t_j} \tag{7.15}$$

$1/C_2$ is thus the *weighted average* of the squares of the *interval velocities*, and weights are equal to *vertical travel times*.

Comparing Eqs. (7.11) with (7.12) it is inferred that $1/V^2$ in Eq. (7.11) for a single layer is replaced in the general formula for the n-layered case by the reciprocal of time-weighted mean square velocity. This is a reiteration of the Dix (1955) formula (Section 7.2).

Derivation of equations in the Section 7.12 was originally contributed by Taner and Koehler (1969). Some of these results are included here.

7.2 Dix Equation for Interval Velocity

In Section 4.5, commonly used terms on velocity like *average velocity,* weighted *average velocity (RMS Velocity), stacking velocity, normal moveout velocity* (VNMO), and *interval velocity* were introduced. Here, the meaning of these terms is further explained particularly in context to the heterogeneous conditions of the subsurface, as a prelude to the Dix equation for interval velocity.

Assuming an ideal horizontally stratified ground (refer to Fig. 7.1 a), and taking 'V_k' to denote the velocity in the k-th layer, and 't_k' to denote the two-way travel time in it, the two way *normal incidence time* is defined by:

$$t_0 = \sum_{k=1}^{n} t_k = 2\sum_{k=1}^{n} \frac{h_k}{V_k} \tag{7.16}$$

where h_k is the thickness of the k-th layer. The *normal incidence time* refers to vertical ray path.

The average velocity and RMS velocity to the n-th interface are defined by the relations

$$(\mathrm{VAVG})_n = \frac{1}{t_0}\sum_{k=1}^{n} V_k t_k,$$

$$(\mathrm{VRMS}) = \left(\frac{1}{t_0}\sum_{k=1}^{n} V_k^2 t_k\right)^{\frac{1}{2}}$$

These velocities are related as

$$(\mathrm{VRMS})^2 - (\mathrm{VAVG})^2 = \frac{(\mathrm{VAVG})^2}{D^2}\sum_{k=1}^{n-1} h_k \sum_{j=k+1}^{n} h_j \frac{(V_k - V_j)^2}{V_k V_j} \tag{7.17}$$

(Al-Chalabi 1974) where D is the depth to the n-th interface.

For rapid velocity changes of the subsurface as is commonly observed in sonic logs, the average velocity and RMS velocities are defined in terms of *instantaneous velocity* V_i.

The *average velocity* (VAVG) is the time average of the *instantaneous velocity,* $V_i(t)$,

$$\text{VAVG} = \frac{1}{T}\int_0^T V_i(t)dt \tag{7.18}$$

The RMS velocity is the square root of the time average of the square of the *instantaneous velocity:*

$$(\text{VRMS})^2 = \frac{1}{T}\int_0^T V_i^2(t)dt \tag{7.19}$$

A measure of *velocity heterogeneity* 'g' is obtained from Eq. (7.17) in the form

$$\frac{(\text{VRMS})}{(\text{VAVG})} = (1+g)^{\frac{1}{2}} \tag{7.20}$$

where

$$g = \frac{1}{D^2}\sum_{k=1}^{n-1} h_k \sum_{j=k+1}^{n} h_j \frac{(V_k - V_j)^2}{V_k V_j} \tag{7.21}$$

[Al-Chalabi 1974, Eq. (4)]

'g' gives a measure of *velocity heterogeneity* in the ground, and is called the *heterogeneity factor.* It does not depend on the ordering of layers. For a layered medium it is positive and for homogeneous ground it's value is zero.

From Eq. (7.20), we find that the RMS velocity equals the average velocity for a homogeneous ground and exceeds the average velocity for a heterogeneous ground.

The *interval velocity* for the n-th layer defined for a short shot-geophone spread is given by [refer to Eq. (4.24)]

$$(V_n)^2_{\text{interval}} = \left\{ \frac{(\text{VRMS})_n^2 t_{0,n} - (\text{VRMS})_{n-1}^2 t_{0,n-1}}{t_{0,n} - t_{0,n-1}} \right\}$$

where $(\text{VRMS})_n$ is RMS velocity and $t_{0,n}$ is vertical two-way travel-time to n-th reflector. A similar meaning is assigned to $(\text{VRMS})_{n-1}$ and to $t_{0,n-1}$.

The method of calculating the '*interval velocity*' of the subsurface depth interval from surface reflection measurements was first given by Dix (1955). The Dix (1955) equation, reproduced as Eq. (7.28b), assumes that the distance between shot and geophone is small compared to the depth of investigation and the deviation of the ray at the interfaces between layers is neglected. Thus, when dealing with layer interfaces close to the surface with a heterogeneous conditions, care should be exercised when using the Dix (1955) formula.

Realizing the wide use of the Dix (1955) formula in seismic interpretations, derivation of this equation and its further generalization is given in the discussions that follow.

7.2.1 Vertical Ray Path

Dix (1955) contributed a method for calculating interval velocity from the RMS velocity determined from the x^2–t^2 graph.

Consider first a single isotropic homogeneous layer (Fig. 7.2 a). We have the following time-distance relation in this case

$$t^2 = t_0^2 + \frac{x^2}{V^2} \tag{7.22}$$

where t denotes the reflection time at a distance 'x', and V is the velocity.

On plotting t^2 vs. x^2, we get a straight line having a slope of $1/V^2$ and an intercept of t_0^2. The thickness of the layer is given by

$$\Delta z_1 = \frac{V t_0}{2} \tag{7.22 a}$$

Equations (7.22) may be written to apply to the second layer only, to enable determination of velocity and thickness of the second layer.

If a graph of t^2 vs. x^2 is plotted for two layers (Fig. 7.2 b and c), it would not be a straight line, and would produce a curve slightly concave towards small values of x^2.

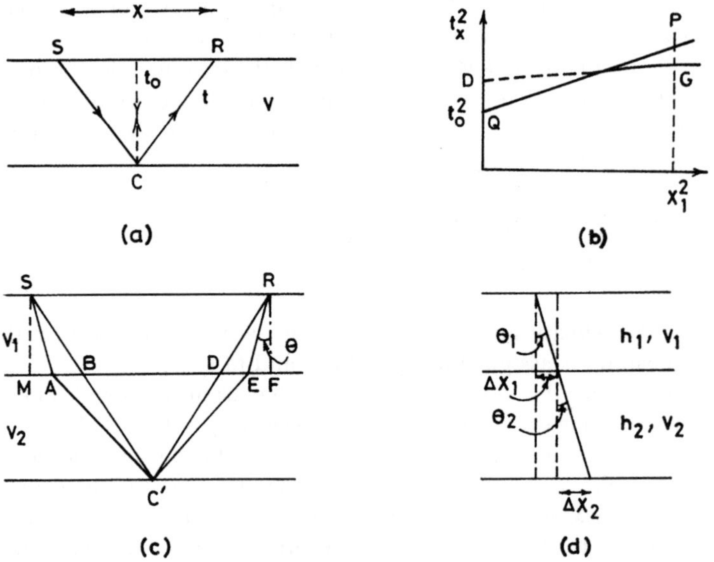

Fig. 7.2 a–d. **a** Reflection path in an isotropic layer; **b** (t^2–x^2) graph for two layers; **c** straight-line and minimum time reflection paths in a 2-layer medium; **d** geometry of a down-traveling ray for the case of an arbitrarily small x and no dip (Source for Fig. 7.2 b–d: after Dix 1955)

To get the t^2 vs. x^2 graph corresponding to the ray path in the second layer only, Dix (1955) describes the following procedure.

If we calculate the times over straight paths only, the t^2 vs. x^2 graph would always be a straight line, for example, for the paths SCR in a single layer model, and SBC′DR in a 2-layer model. However, the least time path for a geophone at R is SA-C′ER. The time for SAC′ER is smaller than the time for straight path SBC′DR. The points on the t^2 vs. x^2 plot corresponding to these paths would be at G and P respectively. The smaller the shot-receiver separation 'x', the smaller is this effect.

The problem of determining the velocity and thickness of the second layer may be solved by removing the effect of the first layer. The process is called *layer stripping*. For this we need to calculate the path lengths SA and ER. These may be calculated knowing angle θ. For angle 'θ', the following relation is satisfied:

$$\frac{\sin\theta}{V_1} = \frac{dt}{dx} \tag{7.23}$$

Knowing V_1 and dt/dx, θ may be calculated.

Derivative dt/dx in Eq. (7.23) may be approximated by $\Delta t/\Delta x$.

Dix (1955) showed that the replacement of dt/dx by $\Delta t/\Delta x$ leads to a large error in 'θ', resulting in a subsequent large error in the velocity V_2 of the second layer. As an alternative to this, it was suggested that dt/dx might be found for the straight line tangent GD at G (say at x_1, in Fig. 7.2b). For this tangent line, the following travel-time relation holds:

$$t^2 = t_0'^2 + \frac{x^2}{(VRMS)_2^2} \tag{7.24}$$

It's differentiation with respect to x at $x = x_1$ gives

$$t\frac{dt}{dx} = \frac{x}{(VRMS)_2^2} \tag{7.24a}$$

From Eqs. (7.23) and (7.24a), we have

$$\sin\theta = \frac{xV_1}{(VRMS)_2^2 t} \tag{7.25}$$

Equation (7.25) makes a possible determination of 'θ' with a higher precision than with the use of Eq. (7.23).

Knowing θ, the times for ray lengths, SA and ER, in the first layer may be found. Similarly, horizontal distances MA and EF may be found. The times for the paths in the first layer may be subtracted from the recorded time t for the path SAC′ER. Similarly, the distance AE may be found by subtracting MA and EF from x. These reduced values of time and distance correspond to path AC′E, fully in the second layer. A travel-time graph for the *reduced travel-time and distance* may be drawn, and the velocity and thickness of the second layer may be found by treating it as a single layer problem, as done for the top first layer.

The above explains the basis of the technique in a very clear manner. A simpler calculation may be done as explained below.

Consider Fig. 7.2 b and c. The x^2–t^2 graph for reflection from the second layer is shown in Fig. 7.2 b. The x^2–t^2 graph is curved, and the equation of the tangent at a given offset 'x' may be written as [Eq. (7.24)]:

$$t^2 = \frac{x^2}{(\mathrm{VRMS})_2^2} + t_0'^2$$

where $(\mathrm{VRMS})_2$ is the RMS velocity to the second reflector. This equation differs from the corresponding Eq. (7.22) for a single layer in having RMS velocity (VRMS) in place of average velocity V. This is the result of abrupt bending of the ray at the boundary between the first and second layer.

Assuming that Δt_1 and Δt_2 are one way vertical travel time through the first and second layer, and assuming that x is small we may write (Fig. 7.2 d)

$$\begin{aligned} \frac{1}{2}x &= \Delta x_1 + \Delta x_2 = h_1 \tan\theta_1 + h_2 \tan\theta_2 \\ &\approx V_1 \Delta t_1 \sin\theta_1 + V_2 \Delta t_2 \sin\theta_2 \\ &\approx \left(V_1^2 \Delta t_1 + V_2^2 \Delta t_2\right) \frac{\sin\theta_1}{V_1} \\ &\approx \left(V_1^2 \Delta t_1 + V_2^2 \Delta t_2\right) \frac{x}{(\mathrm{VRMS})_2^2 t} \end{aligned} \tag{7.26}$$

[from Eq. (7.25)]

Since $t \approx 2\,(\Delta t_1 + \Delta t_2)$, we have

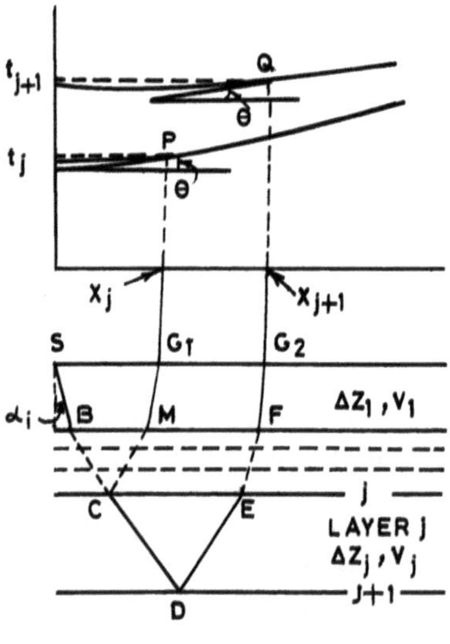

Fig. 7.3. Travel time and raypath diagrams depicting reflection from the top and bottom of layer j for the common ray parameter $p_i = \sin\alpha_i / V_1$ (Source: Nowroozi 1989, from Geophysics with permission)

$$(\mathrm{VRMS})_2^2 \approx \sum_{i=1}^{2} V_i^2 \Delta t_i \Big/ \sum_{i=1}^{2} \Delta t_i \tag{7.26 a}$$

Generalization to n horizontal beds gives the travel-time equation as

$$t^2 \approx \frac{x^2}{(\mathrm{VRMS})_n^2} + \left(\bar{\bar{t}}_0\right)^2 \tag{7.27}$$

with

$$(\mathrm{VRMS})_n^2 \approx \sum_{i=1}^{n} V_i^2 \Delta t_i \Big/ \sum_{i=1}^{n} \Delta t_i \tag{7.27 a}$$

Considering reflections from the top and bottom of the n-th layer, Eq. (7.27 a) may be written as

$$(\mathrm{VRMS})_{n-1}^2 \sum_{i=1}^{n-1} \Delta t_i = \sum_{i=1}^{n-1} V_i^2 \Delta t_i \tag{7.28 a}$$

and

$$(\mathrm{VRMS})_n^2 \sum_{i=1}^{n} \Delta t_i = \sum_{i=1}^{n} V_i^2 \Delta t_i \tag{7.28 b}$$

Subtracting Eq. (7.28 a) from (7.28 b) we get the interval velocity of the n-th layer:

$$(V_n)_{\mathrm{interval}}^2 = \left\{ \frac{(\mathrm{VRMS})_n^2 t_{0,n} - (\mathrm{VRMS})_{n-1}^2 t_{0,n-1}}{\left(t_{0,n} - t_{0,n-1}\right)} \right\} \tag{7.28 c}$$

where $t_{0,n} = S_{i=1}^{n} \Delta t_i$ and $t_{0,n-1} = S_{i=1}^{n-1} \Delta t_i$ and $(\mathrm{VRMS})_{n-1}$ and $(\mathrm{VRMS})_n$ are the RMS velocities to reflectors (n−1) and n, respectively.

Equation (7.28 c) given above is of particular significance because errors made in earlier or later velocity determinations have no effect on interval velocity of the n-th layer.

The value of interval velocity would depend upon the method by which it is calculated. Supposing that interval velocity is calculated using RMS velocity derived from the stacking velocity, and it is also calculated from average velocity from measurements in wells nearby. The interval velocity calculated from RMS velocity would be larger than that calculated using average velocity. The difference between the two interval velocities so determined would depend upon the heterogeneity factor defined by Eq. (7.21).

Cordier (1985) analyzed the errors in the determination of interval velocity due to the inaccuracy in the evaluation of RMS velocities. Cordier's analysis shows that even small errors of opposite signs in measurement of RMS velocities from the top and bottom of the layer may lead to a large error in the interval velocity value of the layer.

7.2.2
Non-Vertical Ray Path

The Dix (1955) Eq. (7.28b) is strictly valid for a vertical ray path. This equation enables determination of the interval velocity from routinely recorded seismic reflection data. The generalized form of the Dix equation was contributed to by Gonzalez-Serrano and Claerbout (1984), and Claerbout (1985).

Referring to Fig. 7.3, we have a seismic source at the top of a layered structure. Two sets of ray paths, $SBCMG_1$ and $SBCDEFG_2$ reflected from the top and bottom of layer 'j' are shown. It is to be noted that both these paths share a common path SBC. The reflection from point C from the top of layer 'j' reaches G_1, and the reflection from the bottom point D of layer 'j' reaches G_2. Coordinates of the points G_1 and G_2 are denoted by x_j and x_{j+1}, and the corresponding reflection times by t_j and t_{j+1}. It may be noted that both rays $SBCMG_1$ and $SBCDEFG_2$ have the common ray parameter $p_i = \sin a_i / V_1$.

Gonzalez-Serrano and Claerbout (1984) showed that the interval velocity for the j-th layer is given by

$$\left(V_j\right)^2_{\text{interval}} = \frac{x_{j+1} - x_j}{t_{j+1} - t_j} \cdot \frac{dx}{dt} \tag{7.29}$$

dx/dt is the inverse ray parameter p_i. (x_j, t_j) and (x_{j+1}, t_{j+1}) are coordinates of the two points P and Q, with the same slope dx/dt on reflections associated with interfaces j and j+1.

Nowroozi (1989) showed that two points on reflection travel time curves with the same ray parameter p_i share a common ray path that is reflected from the top and bottom of the layer. Also, that Eq. (7.29) is related to the equation for travel time in layer j. Further, the layer thickness Δz_j is given by

$$\Delta z_j^2 = \frac{\left[V_j^2 \left(t_{j+1} - t_j\right)^2 - \left(x_{j+1} - x_j\right)^2\right]}{4} \tag{7.30}$$

[Nowroozi 1989, Eq. (3)]

Equation (7.29) is *a generalized form of the Dix (1955) equation* as it is valid for any ray path. Further, Eq. (7.29) reduces to (7.28c) for a vertical ray path.

7.3 Stacking Velocity for a Layered Structure

Taner and Koehler (1969) gave the following time-distance relationship [Eq. (7.12)] for a layered medium with horizontal interfaces between the layers

$$T_x^2 = C_1 + C_2x^2 + C_3x^4 + \ldots C_jx^{2j-2} + \ldots$$

In this equation T_x is the travel time from the shot point to the n-th interface and after reflection back to the receiver at a distance x from the shot point; C_j are functions of the velocities and thicknesses of the layers

$$C_1 = \left(2\sum_{k=1}^{n}\frac{h_k}{V_k}\right)^2 = t_0^2$$

$$C_2 = \frac{t_0}{2\sum_{k=1}^{n} h_k V_k} = \frac{1}{(VRMS)^2}$$

The first two coefficients include t_0, which is the two way vertical travel time, and VRMS which is the RMS velocity.

In practice, only two terms of Eq. (7.12) are used assuming that x is small. In a truncated form, Eq. (7.12) is written as

$$T^2 = C_1 + C_2x^2 \tag{7.31}$$

Figure 7.4 shows the travel-time curves for the n-th reflector. Curve (1) in Fig. 7.4 is the true travel-time curve corresponding to Eq. (7.12). When we fit the best-fit hyperbola to curve 1 for the true travel time curve, the velocity that provides the best-fit hyperbola (curve 2) is effectively the *stacking velocity*, V_s.

Most methods of determining stacking velocity are based on two-term truncation of Eq. (7.12), i.e. on Eq. (7.31). The result of this truncation is that the *stacking velocity* calculated does not equal the true *root mean square (RMS) velocity* except in the case where ground is homogeneous.

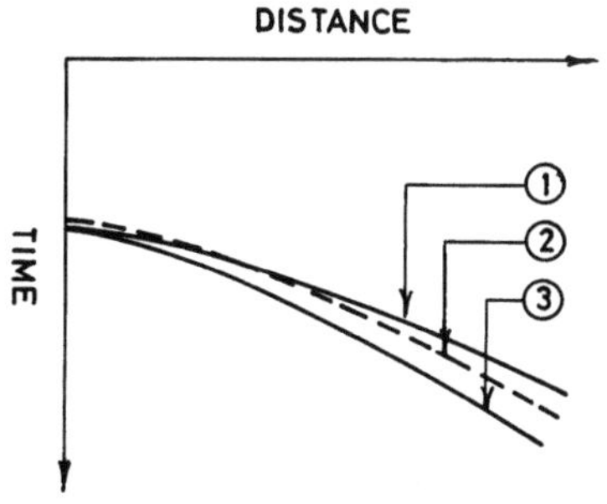

Fig. 7.4. Travel time curves for the n-th reflector: 1 is the true travel time curve; 2 is a best-fit hyperbola (corresponding to stacking velocity V_s); 3 is a hyperbola corresponding to true V_{rms} (Source: Al-Chalabi 1973, from Geophysical prospecting with permission)

The best-fit criterion may vary according to a particular method of determining the stacking velocity. Assuming that the best fit criterion is based on minimizing the function

$$f(C_1,C_2) = \sum_{i=1}^{m} (C_1 + C_2 x_i^2 - t_i^2)^2 \tag{7.32}$$

we have

$$V_s^2 = \frac{1}{C_2} = \frac{\left[m\sum_{i=1}^{m} x_i^4 - \left(\sum_{i=1}^{m} x_i^2\right)^2\right]}{\left[m\sum_{i=1}^{m} t_i^2 x_i^2 - \sum_{i=1}^{m} t_i^2 \sum_{i=1}^{m} x_i^2\right]} \tag{7.32 a}$$

(Al-Chalabi 1973) where 'm' is number of *stacked traces*; x_i and t_i are the spread length and the travel time corresponding to the i-th trace.

It is to be noted that the difference between stacking and RMS velocities is due to the difference between true travel time and the time computed by two-term truncation (refer to Fig. 7.4).

The accuracy of stacking velocity may be estimated from travel time data only, without knowledge of the actual velocity distribution. For details reference may be made to Al-Chalabi (1973).

RMS velocity is usually estimated from the *stacking velocity.* The difference (V_s–V_{RMS}) represents the *bias* in this estimate, in the absence of errors.

The bias is a non-negative quantity and increases as $\Sigma_{k=1}^{\infty} F_k(V,h)p^{2k}$ increases. F_k is a function of velocity and thickness. It increases as, heterogeneity factor g (Eq. (7.21) increases. 'p' is the ray parameter ($=\sin\theta_i/V_i$) where θ_i is the angle of incidence at the i-th layer. The ray parameter increases with increasing offset and decreases with increasing depth over the same ground.

We have the following relationship between stacking velocity (V_s), root mean square velocity (VRMS) and average velocity (VAVG):

$$(V_s) > (VRMS) > (VAVG)$$

7.4 Hyperbolic Velocity Analysis

7.4.1 Basis of Analysis

Different methods of velocity analysis have a common basis of identifying times of primary reflection on a set of seismic traces, arising from one *common depth point.* The reflected signals, r(t), from the common depth point are assumed identical (or slightly distorted) from one trace to another. Also, these signals are separated from each other by a certain time-interval and the length of this interval is a function of velocity characteristics of the subsurface medium. This is expressible as

$$f_i(t) = r(t - \tau_i), \quad i = 1, 2, \ldots M \tag{7.33}$$

where f_i (t) is the i-th trace of the set; τ_i is the time elapsed between emission of the signal and it's reception at the i-th trace; M is the number of traces in the set.

In addition to the reflected signal r(t), the seismic record also contains noise. The problem is to determine the arrival times of reflected coherent signals on different traces in the presence of noise. This problem is solved using a method involving *maximization of selected functions.* The numerical process of *maximization* is simplified by assuming that the transit time (t_x) of seismic energy from the seismic source to reflector to receiver point at a distance 'x' from the seismic source is given by a two term equation [Eq. (7.31)] in the form

$$t_x^2 = t_0^2 + \frac{x^2}{V_s^2} \tag{7.34}$$

where 't_0' is the travel time for a vertical path for a horizontally layered medium; and V_s is the stacking velocity. Equation (7.34) is an approximation, which holds good for small values of source-receiver distance 'x'.

The two-term approximation for travel time given by Eq. (7.34) greatly simplifies the maximization problem as the time interval (lag time, τ) of the reflected signals depends only on two parameters, t_0 and V_s. However, in the case of sloping interfaces, the lag time is a function of more than two parameters (Taner and Koehler 1969). With sloping interfaces the square of lag time (τ^2) would have both odd and even powers of x [compare with Eq. (7.12)].

Assumption of the validity of Eq. (7.34) for reflected signals from a common depth point is equivalent to seeking the reflected signals to be along a hyperbola defined by this equation. It is simple to calculate the value of the function given by this equation and to find its maximum value as a function of the stacking velocity, V_s.

The maximization of selected functions from an assemblage of seismic traces involve 'summation', 'cross-correlation', 'semblance' etc. to retrieve an improved estimate of desired quantity from records containing both signal and noise. Various investigators (for example, Trorey 1961, Simpson 1967, Schneider and Backus 1968, Taner and Koehler 1969, Neidel and Taner 1971) have made notable contributions in this regard.

The reflection data is in the time-distance space. Our aim is to transform this space into a 'time-stacking velocity' space, and measure the energy of reflections in this new space. A procedure called '*semblance*' gives effect to this transformation. Semblance acts as a filter. The following description based on the theory contributed by Taner and Koehler (1969) would define the basis of '*semblance*'. *Semblance* is one of several *coherency measures* (see, for example, Neidell and Taner 1971 for details).

7.4.2 The Semblance Coefficient

Referring to Eq. (7.34) as basis for calculation of the '*semblance coefficient*', its value is computed for each value of 't_0'. A curved data window, as in Fig. 7.5, is centered around the trial t (x, v, t_0) curve.

Index i and j stand for 'M' geophone positions, and time samples, respectively. For example A (j, i) is the j-th sample of the i-th trace. Considering odd number (n+1) of time points centered around the trial curve at the considered value of 't_0' (j=k), the *semblance coefficient* is given by

$$S_c = \frac{\sum_{j=k-\frac{n}{2}}^{k+\frac{n}{2}} \left| \sum_{i=1}^{M} A(j,i) \right|^2}{M \sum_{j=k-\frac{n}{2}}^{k+\frac{n}{2}} \sum_{i=1}^{M} |A(j,i)|^2} \tag{7.35}$$

[Kirlin 1992, Eq. (2)]

The *semblance coefficient* measures, over the set of traces concerned, the ratio of the signal energy to the total energy under the assumption that the sum of the noise on the different traces is zero at all times.

A hyperbolic search for semblance, as shown in Fig. 7.5 from among the appropriately gathered arrays of traces, is the basis for estimation of velocities. Measured values of *semblance* are displayed as *velocity spectra*. Interpretation of *velocity spectra* gives velocities of primary reflections as well as multiple events. Taner and Koehler (1969) discuss the fundamental principle for calculating *velocity spectra*. A brief discussion of *velocity spectra* follows.

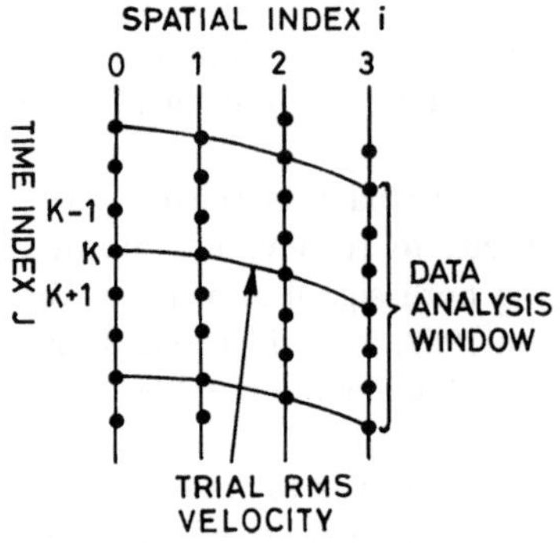

Fig. 7.5. Schematic illustration of data sampling scheme, centered on the trial wavefront curvature (trial rms velocity) intersecting time index j=k and offset index i=0 (Source: Kirlin 1992, p 1028, from Geophysics with permission)

7.4.3 Velocity Spectra

It is to be noted from Eq. (7.34) that each pair of (t_0, V_s) represents a hyperbola. Reflected pulses for a set of seismic traces obtained for a common depth point would fall on a hyperbola (curve 2 in Fig. 7.6a). This hyperbola corresponds to the pair of (t_0, V_s) for which the hyperbola passes through the reflected pulses. Keeping 't_0' fixed, if a lower or higher value of V_s is assumed, the hyperbola shown by dashed curves (1) and (3) would result. These curves do not pass through the actually reflected pulses (signals). As a result, assumption of a realistic value of V_s would produce the maximum summed up amplitude for curve 2 rather than for curve 1 or 3. This is illustrated in Fig. 7.6b.

The steps in computation of *velocity spectra* are the following (Cordier 1985):

- a reference time 't_0' is fixed;
- for this fixed value of 't_0' the number of values of V_s are associated usually at an interval of 100′/s. The limits within which V_s values are taken is determined by the local conditions. Each new pair of (t_0, V_s) defines one hyperbola
- the traces are passed through the semblance filter (Fig. 7.6c), which computes the variation of energy with V_s for the fixed value of 't_0'
- the above operation is repeated for a new value of t_0 (say t_0^1). The new value 't_0^1' equals to the preceding value 't_0' increased by a half width of the window used for the analysis
- the operation is terminated at the desired record time.

The result is presented in the form of spectral curves with two way time on one axis (say the x-axis) and stacking velocity on the downward pointing z-axis (Fig. 7.6d).

From the displays in the form of Fig. 7.6d, it is possible to identify velocities corresponding to primary as well multiple events. Cook and Taner (1969) and Taner and Koehler (1969) presented examples of such displays giving depth and details of geological information.

The basis of the type of *velocity analysis* described above is the determination of stacking velocities by measurement of the *coherency* of reflections along *hyperbolic moveout trajectories* in the offset. This conventional method becomes inaccurate in geologically complex areas because conversion of the stacking velocity to interval velocity assumes a horizontally stratified medium, and mild lateral velocity variations. Biondi (1992) proposed a velocity estimation by *beam stack*. The method may be applied where there are dipping reflectors and strong *lateral velocity variations*. The *beam stack* method measures local coherencies of reflections along *hyperbolic trajectories*. It is claimed in Biondi (1992) that the beam stack can provide information on *non-hyperbolic moveout* in the data. The concept of *velocity estimation* using the *beam stack* is presented in Appendix 7A. In a similar development, Kirlin (1992) presents a high-resolution wavefront or *spectral eigenstructure method*. It is claimed in Kirlin (1992) that the resolving power of *eigenstructure velocity estimators* is better than *conventional semblance*. The details of the method, however, are too extensive to be included here.

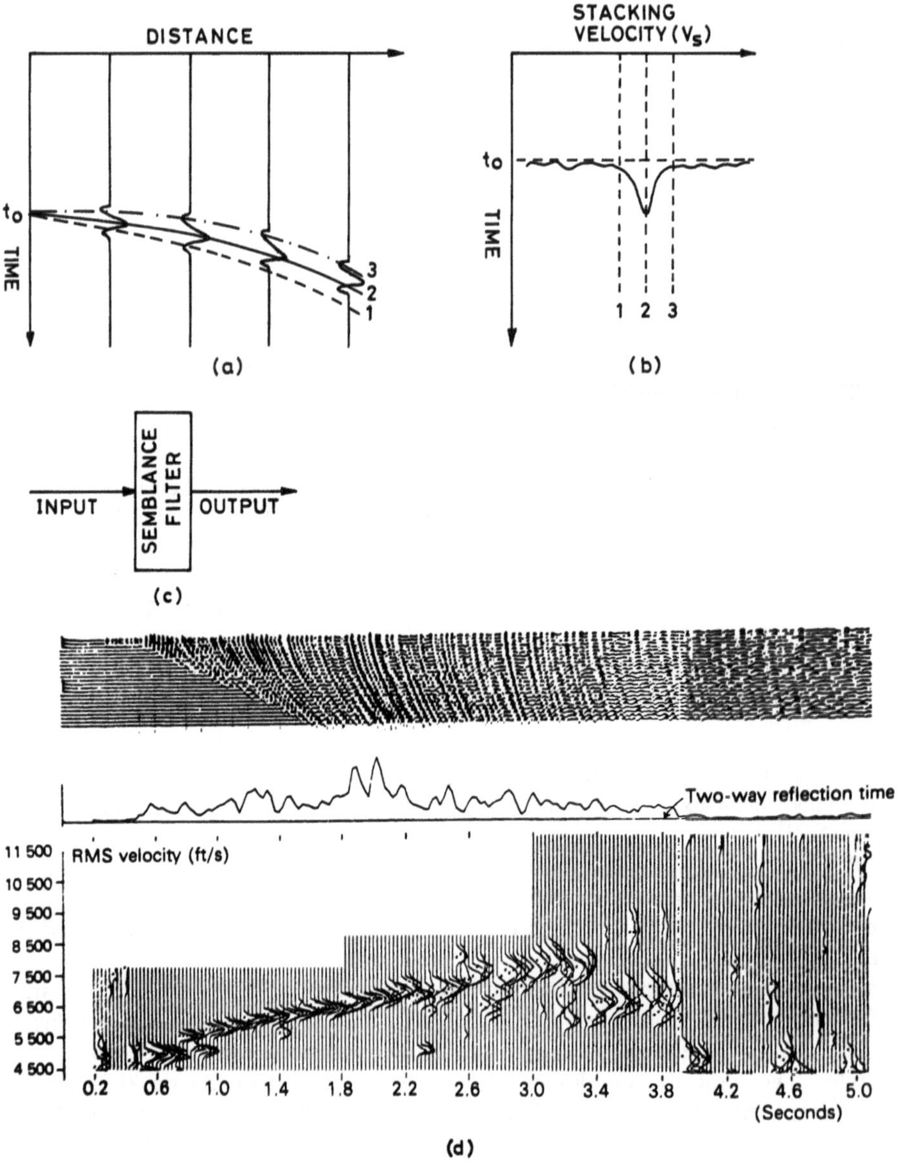

Fig. 7.6a–d. Steps in velocity analysis: **a** curve 2 is for appropriate values of t_0 and V_s for which a hyperbola passes through reflected pulses. Curve 1 and 3 are for lower and higher velocity values; **b** appropriate stacking velocity would produce maximum summed up amplitude; **c** semblance filter. The output is the variation of energy with V_s for a fixed value of t_0; **d** top – CMP gather; middle – maximum amplitude in stacked traces; bottom – stacked trace amplitudes as a function of stacking velocity in 100 ms window (Source: for Fig. 7.6a–c – SEISCOM. For Fig. 7.6d from Telford et al. 1976, Applied Geophysics, Cambridge University Press, Cambridge, UK. Reproduced from Lavergne 1989:Seismic methods. With permission © 1989 Editions Technip, Paris)

Semblance analysis has been successfully used for estimation of vertical and skew moveout velocities for *non-hyperbolic* reflection moveout curves in the case of an anisotropic medium. This aspect is covered in detail in Chapters 8 and 10.

Appendix 7A
Concept of Velocity Estimation using Beam Stack: Applicable to Geologically Complex Areas

The hyperbolic stacking trajectories are defined by the equation

$$t^2 = t_0^2 + \frac{4h^2}{V^2} \tag{7A.1}$$

where t_0 is the time at apex of the hyperbola, h is the half offset, and V is the stacking velocity.

In contrast to Eq. (7A.1), the *beam stack* is computed by measuring the coherency of the data along hyperbolic trajectories of the form

$$T_{ph}(h) = \sqrt{(t^*)^2 - h^* t^* p_h + \frac{h^2}{h^*} p_h t^*} \tag{7A.2}$$

[Biondi 1992, Eq. (5)]

The stacking trajectory given by Eq. (7A.2) is defined in terms of three parameters t^*, h^* and p_h, whereas in Eq. (7A.1) it is defined in terms of only two free parameters, t_0 and V.

In fact Eq. (7A.2) is an outcome of Eq. (7A.1). This results from the fact that a hyperbola from the family of hyperbolas defined by Eq. (7A.1) is fully determined when slope p_h of the tangent at the travel time t^* and half-offset h^* is fixed. Therefore, these three parameters (t^*, h^* and p_h) completely determine the *stacking trajectory*. To arrive at the form of Eq. (7A.2), let the ray parameter p_h be defined from Eq. (7A.1) as (Biondi 1992)

$$\left.\frac{dt}{dh}\right|_{h=h^*} = p_h = \frac{4h^*}{V^2 t^*} \tag{7A.a}$$

From this we get

$$V^2 = \frac{4h^*}{p_h t^*} \tag{7A.b}$$

and from Eq. (7A.1) the squared apex time t_0^2 is given by

$$t_0^2 = (t^*)^2 - h^* t^* p_h \tag{7A.c}$$

It may be noted that the velocity V and t_0 in Eq. (7A.1) are now expressed in terms of the three parameters t^*, h^* and p_h.

Substitution of t_0^2 and V^2 in Eq. (7A.1) would yield the form of Eq. (7A.2).

As the beam stack is a *local stacking operator*, defined in terms of t, h and p_h, its inherent usefulness lies in providing information on the presence of *non-hyperbolic moveout* in the reflection data. This aspect would be explained with an illustrative example from Biondi (1992).

When the travel time t is a hyperbolic function of half-offset h, Eq. (7A.1) is stratified, and two parameters, t_0 and V, are sufficient to describe the moveout of the reflection data. However, these parameters do not suffice to describe the non-hyperbolic moveout. In such cases, data may be decomposed with *local stacking operators* defined by three parameters: the time t, half-offset h of the central point of the trajectories and the time dip (dt/dh) at the central point (t, h). p_h (=dt/dh) is described as the *ray parameter* of the recorded reflection.

It may be mentioned that beam stacks are computed by measuring local coherency of the data along trajectories defined by Eq. (7A.2). The stacks are along *curved trajectories* as against stacks on *straight-line* trajectories in a slant stack. The advantage of the *curved trajectory stack* over the straight-line trajectory stack was illustrated in Biondi (1992: Fig. 1).

Figure 7A.1 (Biondi 1992, Fig. 2) shows the application of *beam stack* to *non-hyperbolic moveout data*. The left part of this figure is a window of marine CMP gather after hyperbolic normal moveout with a velocity of 2.79 km/s has been ap-

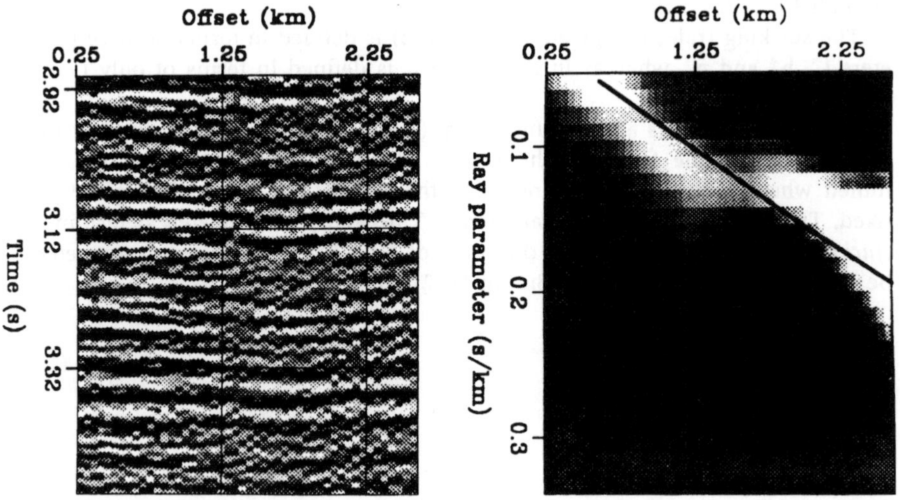

Fig. 7A.1. (Left) A window of CMP stack gathers after hyperbolic moveout with a velocity of 2.79 km/s has been applied. A low-velocity anomaly caused non-hyperbolic moveouts in the data. Residual time delays, of about a quarter of a cycle of the seismic wavelet, are uncorrected by the hyperbolic normal moveout. (Right) The beam stack semblance for the reflection at 3.12 s. The black line superimposed onto the semblance plot shows what the offsets would be as a function of the ray parameter if the moveout were perfectly hyperbolic (Source: Biondi 1992, p 1038)

plied. In this case, non-hyperbolic moveout was caused by a *low-velocity anomaly*, which could not be corrected for by the application of hyperbolic normal moveout. The right part of this figure shows *beam stack semblance* (a measure of coherency) as a function of offset and the ray parameter for a 3.12-second reflection. The black line on the semblance shows the offsets as a function of the ray parameter for hyperbolic moveout. The non-hyperbolic moveout is more evident in beam stacked data than in normally displayed offset-time domain data shown in the left part of the Fig. 7A.1.

Velocity estimation by beam stack is a *reflection tomography method* and may be applied in the case of dipping reflectors and strong lateral velocity variations. Further, it does not require the picking up of the travel time of events as needed in *travel time tomography*. It involves choosing velocity models, accomplishing ray tracing, and maximizing the beam stack energy at travel time and the surface location predicted by ray tracing. Mathematical details and application of the method with other useful features for *velocity estimation* may be found in Biondi (1992).

CHAPTER 8

Reflection Time Analysis and Velocities for Transversely Isotropic (T.I.) Media

In Chapter 7, various velocity terms relevant to the analysis of a layered isotropic medium were explained. It was shown that surface seismic reflection data could be utilized to find an interval property (velocity) of the subsurface medium. The method of velocity analysis assumed that the time-distance (t–x) relation is hyperbolic of the form [Eq. (7.27)]

$$t^2 = \tilde{\tilde{t}}_0^2 + \frac{x^2}{(VRMS)_n^2}$$

Anisotropic velocity conditions in the subsurface demand that the *hyperbolic velocity analysis* method be modified to account for the directional variation of velocity. In this chapter, the concept of *skewed hyperbolic analysis* is given which deals with time-distance (t–x) curves obtained for reflection surveys over a transversely isotropic medium. Results are given both for surface seismic reflection surveys and offset Vertical Seismic Profiling (VSP) geometry. The expressions for an offset VSP geometry are adoptable to surface seismic reflection surveys. Further, application of the skewed hyperbolic velocity analysis method for identification of lithology is discussed. The theoretical background for generalization of the method to investigate azimuthal isotropy or anisotropy in a multilayered medium is also presented.

Sena (1991) developed analytical expressions for multilayered azimuthally isotropic and anisotropic media. The derivation of travel time relations in this case is very similar to the case of multilayered isotropic media and makes use of Fermat's principle of least time (refer to Section 7.1). The presentation in this section is based on the results in Sena (1991).

8.1
Skewed Hyperbolic Travel Time Relations

8.1.1
For a Single Azimuthally Isotropic or Anisotropic Layer

Sena (1991) initiated discussions by first considering reflection survey geometry over a single azimuthally isotropic or anisotropic layer (Fig. 8.1).

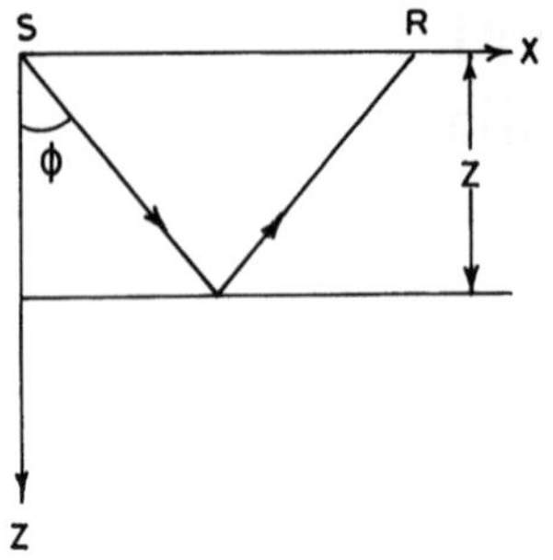

Fig. 8.1. Single layer reflection survey ray geometry. The layer may be azimuthally isotropic or anisotropic

The time along the ray path shown in Fig. 8.1 is given by

$$t = \frac{2z}{U(\phi)\cos\phi} \tag{8.1}$$

where group velocity (or ray velocity) is given by

$$U^2(\phi) = a_0 + a_1 \sin^2\phi + a^2 \sin^4\phi \tag{8.2}$$

[Sena 1991, Eq. (14)]

The expression for group velocity U (ϕ) is valid both for *azimuthally isotropic* and *anisotropic media*. However, values of a_0, a_1 and a_2 are different.

Substitution of Eq. (8.2) in Eq. (8.1) gives the travel-time equation for surface reflection survey geometry (Fig. 8.1) over a single anisotropic layer

$$t^2 = \left(\frac{2z}{V_v}\right)^2 + \left[\left(\frac{z}{V_\gamma}\right)^2 + \left(\frac{x}{2V_h}\right)^2\right] \frac{x^2}{z^2 + \left(\frac{x}{2}\right)^2} \tag{8.3}$$

[Sena 1991, Eq. (24)]

In Eq. (8.3) V_v is the *average vertical velocity* and V_h, V_γ are the *horizontal* and *skew moveout velocities*. Equation (8.3) conforms to the form in Buyn et al. [1989, Eq. (4)] for an offset VSP geometry – referred to in the text as Eq. (10.1) in Section 10.1.

In terms of coefficients a_0, a_1 and a_2, these velocities are

$$V_v^2 = \frac{1}{a_0} \tag{8.4 a}$$

$$V_\gamma^2 = \frac{1}{a_0 + a_1} \tag{8.4 b}$$

$$V_h^2 = \frac{1}{a_0 + a_1 + a_2} \tag{8.4 c}$$

Sena (1991) gave values of V_v, V_h, and V_γ in terms of elastic stiffness constants of a single azimuthally isotropic or anisotropic layer. For example, for a quasi-P wave, for azimuthal isotropy, these are

$$V_v^2 = \frac{c_{33}}{\rho}, V_h^2 = \frac{c_{11}}{\rho} \quad \text{and} \quad V_\gamma^2 = \frac{c_{33}}{\rho}(1 + 2\delta) \tag{8.5}$$

[Sena 1991, Eqs. (A-1) to (A-3)]
where

$$\delta = \frac{(c_{13} + c_{44})^2 - (c_{33} - c_{44})^2}{2c_{33}(c_{33} - c_{44})}$$

For an azimuthal anisotropy case, V_v is given by $\sqrt{c_{11}/r}$; V_h and V_g are a function of elastic constants and the orientation of the horizontal symmetry axis with respect to the seismic survey line. Corresponding expressions for quasi-SV and SH motions have also been derived in Sena (1991).

These expressions for V_v, V_h and V_γ are utilized in the *inversion procedure* to enable estimation of five elastic constants and orientations of the symmetry axis. It would be taken up for discussion in Section 10.2.2.

8.1.2
Generalization to a Multi-Layer Medium and the Relationship between Bulk Velocities and Interval Velocities

Sena (1991) derived a time-offset relationship for multilayered azimuthally isotropic and anisotropic media. In this case, each transversely isotropic layer is characterized by three velocity parameters V_{vj}, V_{hj} and $V_{\gamma j}$ for each wave type. The travel time for multilayered half-space is expressible as

$$t^2 = c_1' + c_2' x^2 + c_3' x^4 + c_4' x^6 + \ldots \tag{8.6}$$

Equation (8.6) has the same form as for an isotropic multilayered medium [compare with Eq. (7.12)]. However, the coefficients c_1', c_2', c_3' in the infinite series are redefined as

$$c_1' = \left(2\sum_{j=1}^{n} t_j\right)^2 \equiv T_0^2, \tag{8.6 a}$$

$$c_2' = \frac{\sum_{j=1}^{n} t_j}{\sum_{j=1}^{n} t_j V_{\gamma j}^2} \equiv \frac{1}{V_{\gamma RMS}}, \tag{8.6 b}$$

$$c_3' = \frac{1}{T_0^3 V_{\gamma RMS}^8} \left\{ \frac{1}{2} \sum_{j=1}^{n} t_j V_{\gamma j}^4 \left[4 \frac{V_{\gamma j}^4}{V_{vj}^2} \left(\frac{1}{V_{hj}^2} - \frac{1}{V_{\gamma j}^2} \right) - 1 \right] + \frac{1}{4} T_0 V_{\gamma RMS}^4 \right\} \tag{8.6 c}$$

In the above equations

$$t_j = \frac{h_j}{V_{vj}} \tag{8.6 d}$$

is the one way travel time in the j-th layer, and T_0 is the two-way vertical travel time through a stack of n layers, i.e.

$$T_0 = 2 \sum_{j=1}^{n} t_j \tag{8.6 e}$$

'$V_{\gamma RMS}$' was introduced by Sena (1991) as the *root mean square skew velocity.*

Sena's (1991) main result is the relationship between the bulk velocities V_v, V_h and V_γ for each reflector of a multilayered anisotropic medium with interval properties V_{vj}, V_{hj} and V_{ij} for every 'j' above the reflector. These are given by

$$V_v = \frac{2}{T_0} \sum_{j=1}^{n} t_j V_{vj} \tag{8.7 a}$$

$$V_\gamma^2 = V_{\gamma RMS}^2 = \frac{2}{T_0} \sum_{j=1}^{n} t_j V_{\gamma j}^2 \tag{8.7 b}$$

$$\frac{1}{V_h^2} = \frac{V_v^2}{T_0 V_\gamma^8} \left\{ \frac{1}{2} \sum_{j=1}^{n} t_j V_{\gamma j}^4 \left[4 \frac{V_{\gamma j}^4}{V_{vj}^2} \left(\frac{1}{V_{hj}^2} - \frac{1}{V_{\gamma j}^2} \right) - 1 \right] + \frac{1}{4} T_0 V_\gamma^4 \right\} + \frac{1}{V_\gamma^2} \tag{8.7 c}$$

Equations (8.7 a) to (8.7 c) enable computation of interval properties from the bulk velocities measured from travel time curves. The methodology for this inversion along with the determination of the orientation of the *symmetry axes* in various layers is discussed in chapter 10.

8.2 Three Velocity Parameters in Skewed Hyperbolic Analysis

To initiate our discussion, we explain first the phase velocity and ray velocity and their relationship to elastic stiffness parameters of a transversely isotropic medium for different types of waves. Formulae on ray velocity are subsequently utilized to derive the time-distance (t–x) relation for VSP geometry. This relation is found to be skewed hyperbolic which is expressible in terms of three velocity parameters.

8.2.1 Phase- and Ray-Velocities of SH and P Waves

A transversely isotropic medium is characterized by five independent elastic stiffness constants: A, C, F, L and N. The elastic stiffness matrix for a transversely isotropic (T.I.) medium has the following form:

$$[C]_{T.I.} = \begin{bmatrix} A & B & F & & & \\ B & A & F & & & \\ F & F & C & & & \\ & & & L & & \\ & & & & L & \\ & & & & & N \end{bmatrix}$$

where $A = c_{11}$, $B = c_{12}$, $C = c_{33}$, $F = c_{13}$, $L = c_{44}$ and $N = c_{66}$ in abbreviated notation of elastic stiffness constants (Section 5.2.1). Equality of $A = B + 2N$ results from isotropy in a horizontal plane with a vertical symmetry axis of a transversely isotropic (T.I.) medium.

In a T.I. medium, the phase velocity V_{SH} of the SH wave depends on only two elastic stiffness constants L and N and is given by

$$V_{SH}^2(i) = \left(\frac{L}{\rho}\right)\cos^2 i + \left(\frac{N}{\rho}\right)\sin^2 i \tag{8.8}$$

[White 1965, Eq. (2.38)]
where 'i' is the angle between the symmetry axis (vertical) and the phase velocity direction (Fig. 8.2 a).

Further, in a T.I. medium the SH-wave ray velocity U (ϕ) in a vertical plane is given by (Levin 1979, Byun 1982)

$$U^{-2}(\phi) = \frac{\cos^2\phi}{U_v^2} + \frac{\sin^2\phi}{U_h^2} \tag{8.9}$$

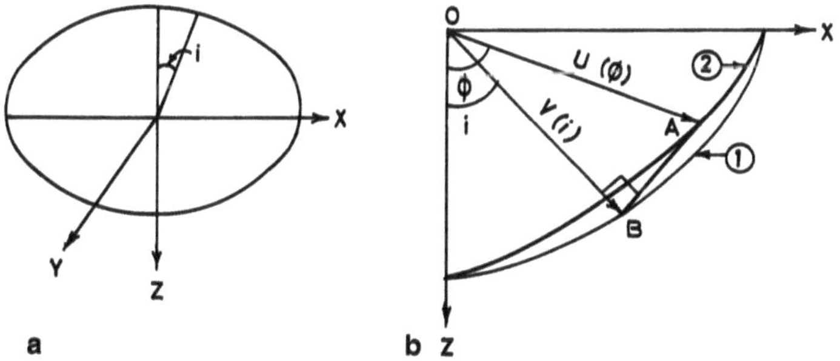

Fig. 8.2 a, b. **a** Phase velocity curve for SH-waves; **b** relationship between wave surface and phase velocity surface (Source for Fig. 8.2b: Byun 1982, p 622, from Geophysics with permission)

where ϕ is the angle between the ray and symmetry axis taken as vertical; $U(\phi)$ is the ray velocity in direction ϕ and U_v and U_h are vertical and horizontal velocities given by (Fig. 8.2b)

$$U_v = \sqrt{\frac{L}{\rho}} \tag{8.9a}$$

and

$$U_h = \sqrt{\frac{N}{\rho}} \tag{8.9b}$$

The geometrical relation between $U(\phi)$ and $V(i)$ is given by

$$V(i) = U(\phi)\cos(\phi - i)$$

In contrast to the simple expression for the phase velocity of SH waves, the phase velocity of a P-wave, $V_p(i)$, is a complicated function of four elastic stiffness coefficients A, C, F, L and is given by

$$\begin{aligned}2\rho V_p^2(i) = {} & (A+L)\sin^2 i + (C+L)\cos^2 i + \\ & + \sqrt{\left[(A-L)\sin^2 i - (C-L)\cos^2 i\right]^2 + 4(F+L)^2 \sin^2 i \cos^2 i}\end{aligned} \tag{8.10}$$

(Byun and Coorigan 1990)

The ray velocity for P waves for a T.I. medium cannot be expressed in a closed form as Eq. (8.9). Thomsen (1986) gave the expression for the ray velocity (refer to Section 5.6.1) by linearization of phase velocity in the case of *weak anisotropy.* A parametric representation of ray velocity was given by Byun (1984, p 1910). Further, Byun et al. (1989) proposed an approximation to ray velocity $U(\phi)$ by a truncated Fourier cosine transform

$$U_P^{-2}(\phi) = a_0^2 + a_1^2\cos^2\phi - a_2^2\cos^4\phi \tag{8.11}$$

[Byun et al. 1989, Eq. (3)]
where a_0, a_1 and a_2 are functions of four elastic stiffness constants of the medium; ϕ is the angle the ray makes with the symmetry axis taken as vertical (Fig. 8.3). Expressing ray velocity in the form of Eq. (8.11) has the advantage of it's compactness, as only three parameters, a_0, a_1 and a_2, are involved instead of four elastic stiffness constants for phase velocity. This equation is therefore more manageable for interpretation purposes. Byun et al. (1989) showed that even for highly anisotropic media, ray velocity expression (8.11) gives an excellent fit to P-wave ray velocity for a wide range of ray angles.

Taking a_1 and a_2 equal to zero in Eq. (8.11) corresponds to a purely isotropic medium; taking a_2 equal to zero corresponds to an *elliptically anisotropic medium.*

Fig. 8.3. Ray-geometry in source-receiver offset VSP geometry (Source: Byun et al. 1989)

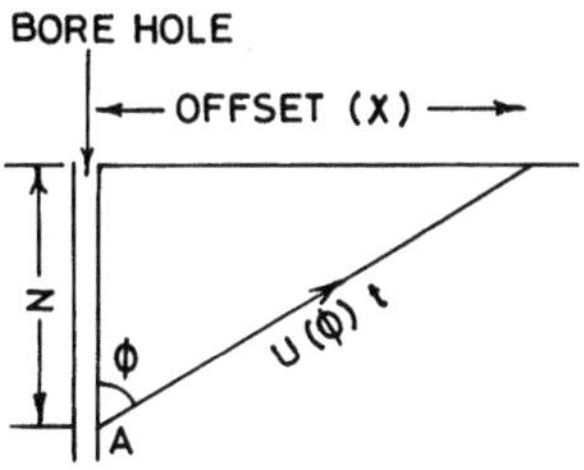

For elliptical anisotropy, the vertical and horizontal phase (and ray) velocities for SH-waves are

$$U_v = \left(a_0^2 + a_1^2\right)^{-\frac{1}{2}} \quad \text{and} \quad U_h = \frac{1}{a_0}$$

The ray velocity surface for the SH wave is elliptical. For P waves, the *ray velocity surface* is generally not perfectly elliptical but possess a *skewed elliptical shape.* By use of Eq. (8.11) for ray velocity, it is possible to find the skewness by deviation of horizontal moveout velocity V_h from skew moveout velocity V_γ. This would be explained below.

8.2.2 Skewed Hyperbolic Formula for Travel Time for VSP Geometry

Consider an offset VSP geometry as shown in Fig. 8.3. For a surface seismic survey, point A represents a reflection point, and 'x' would be half the seismic source receiver offset.

With reference to Fig. 8.3 we have

$$\tan\phi = \frac{x}{z} \tag{8.12}$$

The time from the seismic source at A to the receiver at C is

$$t^2 = \frac{z^2}{U^2(\phi)\cos^2\phi} \tag{8.13}$$

Using Eq. (8.12) with Eq. (8.11) in Eq. (8.13), we have

$$t^2 = \frac{a_0^2x^4 + \left(2a_0^2 + a_1^2\right)z^2x^2 + \left(a_0^2 + a_1^2 - a_2^2\right)z^4}{\left(x^2 + z^2\right)} \tag{8.14}$$

[Byun et al. 1989, Eq. (A-7)]

Introducing three observable parameters: *the average vertical velocity,* and *horizontal* and *skew moveout velocities* defined as (Byun et al. 1989)

$$V_v = \left(a_0^2 + a_1^2 - a_2^2\right)^{-\frac{1}{2}} \tag{8.15a}$$

$$V_h = {}^1\!/_{a_0} \tag{8.15b}$$

$$V_\gamma = \left(a_0^2 + a_2^2\right)^{-\frac{1}{2}} \tag{8.15c}$$

Equation (8.14) may be written as

$$t^2 = \frac{z^2}{V_v^2} + \left[\frac{z^2}{V_\gamma^2} + \frac{x^2}{V_h^2}\right] \frac{x^2}{(x^2 + z^2)} \tag{8.16}$$

[Byun et al. 1989, Eq. (4)]

Equation (8.16) is the *skewed hyperbolic formula* for the traveltime (t)-offset (x) relation. It involves three travel time measurement parameters: V_v, V_γ and V_h.

The skewness in Eq. (8.16) is governed by deviation of V_γ from V_h. When $V_\gamma = V_h$, Eq. (8.16) reduces to familiar form

$$t^2 = t_0^2 + \frac{x^2}{V_{st}^2} \tag{8.16a}$$

where $t_0 = z/V_v$ is the vertical travel-time and V_{st} is the stacking velocity.

Using $t_0 = z/V_v$, Eq. (8.16) may be written as

$$t^2 = t_0^2 + \left[t_0^2\left(\frac{V_v}{V_\gamma}\right)^2 + \frac{x^2}{V_h^2}\right] \frac{x^2}{(x^2 + t_0^2 V_v^2)} \tag{8.17}$$

[Byun et al. 1989, Eq. (6)]

This form is particularly suitable for performing a *semblance scan* over three velocity parameters V_v, V_h and V_γ, which reduces to two (V_h and V_γ) when the average vertical velocity is known as in case of VSP or check-shot surveys. Let us now discuss the three *velocity parameters* V_v, V_h and V_γ for a horizontal stack of different transversely isotropic layers.

8.2.3 Three Velocity Parameters in Skewed Hyperbolic Analysis

Let us first consider the case of SH-waves and geometry as shown in Fig. 8.3 corresponding to a offset VSP survey. For a single transversely isotropic layer, the time-distance curve for SH-waves observed on the surface is a hyperbola. Also, for a horizontal stack of transversely isotropic layers, the SH-wave arrival from the bottom of the stack may still be represented by a Dix-type hyperbola:

$$\tau^2 = \frac{z^2}{(U_v)^2_{\text{average}}} + \frac{x^2}{(U_h)^2_{\text{average}}} \tag{8.18}$$

[Byun and Corrigan 1990, Eq. (5)]

Here $(U_v)_{average}$ and $(U_h)_{average}$ are the average vertical and horizontal moveout velocities defined as

$$(U_v)_{average} = \left(\sum \tau_n U_{v,n}\right) \Big/ \tau_0 \tag{8.19 a}$$

and

$$(U_h)^2_{average} = \left(\sum \tau_n U^2_{h,n}\right) \Big/ \tau_0 \tag{8.19 b}$$

[Byun and Corrigan 1990, Eqs. (6 a, b)]

In Eqs. (8.19 a) and (8.19 b), Σ denotes summation over individual transversely isotropic layers (denoted by n). Also

$$\tau_0 = \sum \tau_n = \frac{z}{U_v} \tag{8.19 c}$$

is sum of SH-wave vertical travel-time t_n through the layers. 'z' is the depth to the borehole point from the surface.

In contrast to the travel-time relation [Eq. (8.18)] for SH waves, the P-wave time-distance relation [Eq. (8.16)] is written in the form

$$t^2 = \frac{z^2}{V_v^2} + \left[\frac{z^2}{V_\gamma^2} + \frac{x^2}{V_h^2}\right] \frac{x^2}{(x^2 + z^2)}$$

Comparing Eq. (8.18) with Eq. (8.16), we immediately find the complexity involved in the case of P waves where we have to deal with three velocity parameters, V_v, V_h, and V_γ. It may be mentioned here that for a single transversely isotropic layer V_v is the vertical ray (or phase) velocity given by $\sqrt{C/r}$. However V_h and V_γ are functions of the horizontal ray (or phase) velocity $\sqrt{A/r}$ and skew velocity $\sqrt{F/r}$ and depend on acquisition geometry.

For a multilayered transversely isotropic medium, the average vertical velocity is the total vertical depth divided by the total vertical travel-time

$$(V_v)_{average} = \frac{z}{t_0} = \left(\sum t_n V_{v,n}\right) \Big/ t_0 \tag{8.20}$$

where

$$V_{v,n} = \sqrt{C_n/\rho_n} \tag{8.20 a}$$

and t_0 is the sum of P-wave vertical travel-times t_n. V_h and V_γ are functions of the coefficients a_0, a_1 and a_2 and these coefficients are a function of four stiffness constants A, C, F and L of layers through which the ray passes.

Thus Eq. (8.16) is a generalization of the Dix hyperbolic formula [refer to Eq. (3.15 a)]:

$$t^2 = t_0^2 + \frac{x^2}{V^2}$$

and reduces to a hyperbola when $V_h = V_\gamma$. This equation describes a *skewed-hyperbola* in the (t–x) domain, and *skewness* is governed by deviation of the *skew moveout velocity* V_γ from the *horizontal moveout velocity* V_h. *V_γ gives a measure of skewness of the ray velocity surface from an ellipse.*

Three velocity parameters V_v, V_h and V_γ may be determined through a *coherency analysis* along a *skewed hyperbolic time-distance (t–x) curve*, as against hyperbolic coherency analysis (Section 7.4) for the isotropic case. If one of the velocities (say V_v) is known down to depth z, a two-dimensional semblance scan (refer to Fig. 8.6) may be performed over a range of the remaining two *velocity parameters* V_h and V_γ.

8.3
Apparent Anisotropy and Identification of Lithology

The skewed hyperbolic formula [Eq. (8.16)] is an extension of the Dix-type hyperbolic moveout formula [Eq. (8.16a)] and provides a new method for velocity analysis. In comparison to stacking velocity found by simple hyperbolic analysis, the three velocity parameters in Eq. (8.16) contain more physically meaningful geological information on anisotropy and/or velocity heterogeneity of the subsurface (Byun et al. 1989).

Further, the apparent anisotropy defined as the ratio between *horizontal moveout velocity* and *average vertical velocity* correlates well with *lithology.*

The skewed hyperbolic formula gives an excellent fit to the time-distance curve over a wide range of ray angles, better than that provided by the hyperbolic fit. This permits improved coherence of signal over large source-receiver offsets for more accurate moveout correction. These points were illustrated by Byun et al. (1989) by considering synthetic as well as field experiment data.

Byun et al. (1989) considered a transversely isotropic model consisting of five layers of equal thickness. The depth wise sequence of model layers is: anisotropic shale, limestone-shale, sand-anisotropic shale, sand-shale and gypsum soil. The top layer has *intrinsic anisotropy*, and the other layers are taken as long-wave

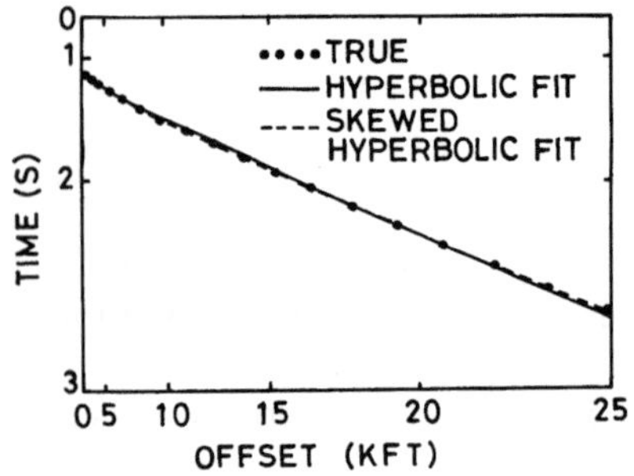

Fig. 8.4. Travel time curves from the bottom of anisotropic layer 5 to the surface (Source: Byun et al. 1989. From Geophysics with permission)

equivalent transversely isotropic media (refer to Section 5.3) consisting of thin alternating layers of two-different components. Layers 1, 3, 4 were taken to be weakly anisotropic, layer 2 moderately anisotropic, and layer 5 strongly anisotropic. Seismic sources were taken at depths of each interface. Using a P-wave ray-tracing program (Byun et al. 1984), travel-times were calculated at various distances of x (Fig. 8.4). A least square fit to the t^2–x^2 pair was made with Eq. (8.16a) (Dix-type hyperbolic fit) and Eq. (8.17) (skewed hyperbolic fit). The skewed hyperbolic fit was found to be better than the hyperbolic fit to the synthetic (t–x) data. The skewed hyperbolic Eq. (8.17) provides an excellent fit to the (t–x) data to maximum source-receiver offset – 2.5 times the source depth z.

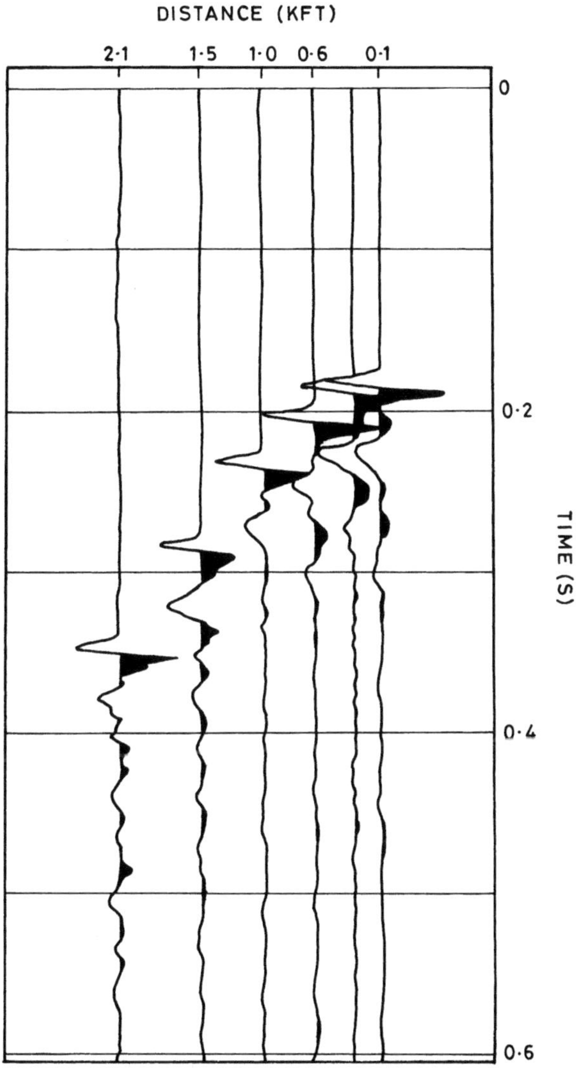

Fig. 8.5. Offset VSP gather at geophone depth of 333.5 m (Source: Byun et al. 1989. From Geophysics with permission)

Buyn et al. (1989) further analyzed actual VSP data from Sulfur Springs, Texas to probe the dependence or correlation of the three-velocity parameters V_v, V_γ and V_h and also V_{st} (for hyperbolic fit), and their ratios V_H/V_v, V_{st}/V_v and V_γ/V_v with *lithology*. The data used for this purpose was a multiple offset VSP, con-

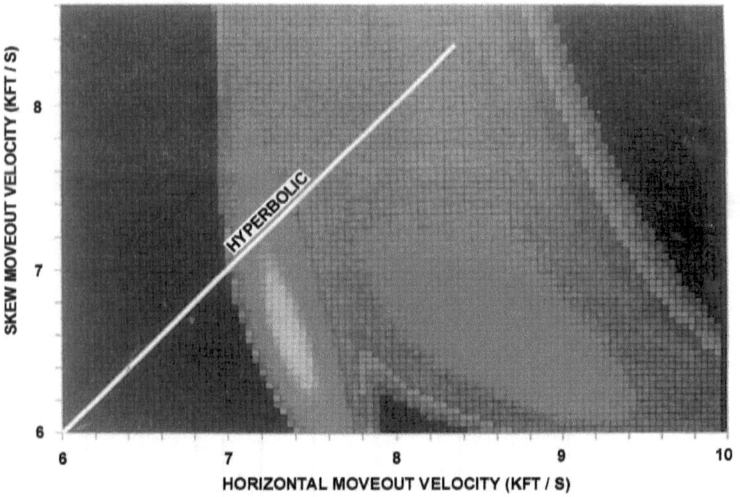

Fig. 8.6. Result of semblance analysis over V_h and V_γ for the VSP gather shown in Fig. 8.5. Straight line at 45° corresponds to hyperbolic moveout with $V_\gamma = V_h$. Least dark portion is the peak value of semblance and dark portion is the background noise level (Source: Byun et al. 1989. From Geophysics with permission)

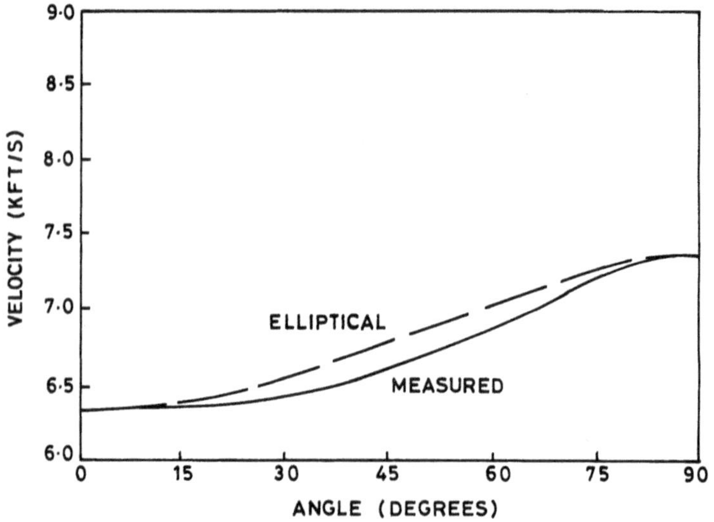

Fig. 8.7. Velocity versus ray angle estimated from the VSP gather shown in Fig. 8.5 (Source: Byun et al. 1989. From Geophysics with permission)

sisting of 76 geophone levels from 91.5 m and spaced at an interval of 7.6 m up to a depth of 663.4 m. The offsets of surface sources were at 30.5, 91.5, 183, 305, 457.5 and 640.5 m.

Figure 8.5 shows the receiver gathers from the receiver at a 335.5 m depth.

Figure 8.6 depicts the *semblance analysis* as a function of the horizontal velocity V_h and the skew velocity V_γ. Least dark part shows high semblance. The

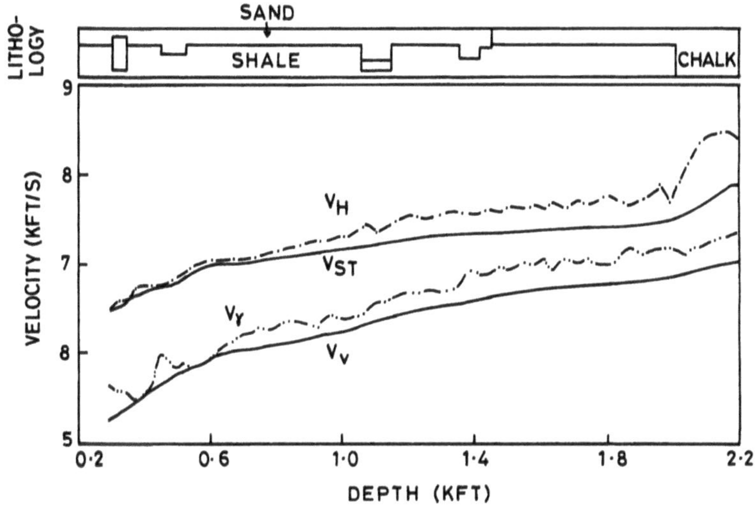

Fig. 8.8. Velocity versus depth estimated for the test site well. Lithology log is shown at the top (Source: Byun et al. 1989. From Geophysics with permission)

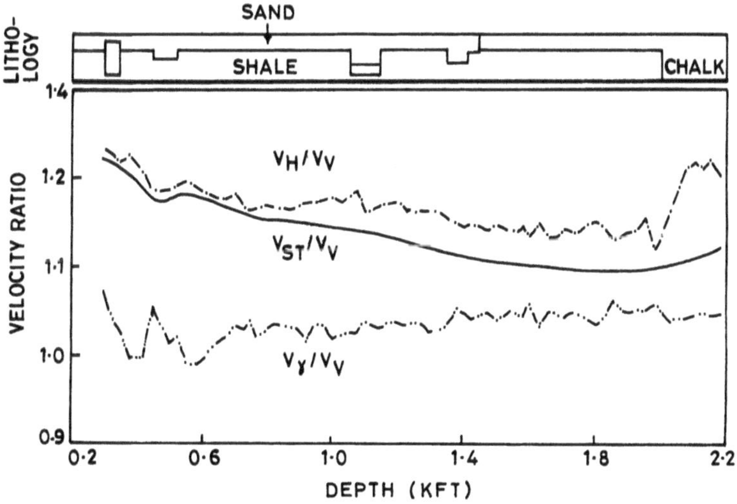

Fig. 8.9. Velocity ratio and lithology versus depth for the test site well (Source: Byun et al. 1989. From Geophysics with permission)

straight line at a 45° angle in the figure shows the *hyperbolic moveout,* i.e. $V_\gamma = V_h$. The semblance peak does not fall on the line for a hyperbolic fit. It is indicative of the fact that *skewed hyperbolic analysis* should improve *normal move-out correction* for *stacking.*

Figure 8.7 shows the variation of measured velocity with ray angle. The measured velocity is the *estimated ray velocity* from measurements of V_h and V_γ from the semblance analysis shown in Fig. 8.6. Figure 8.7 also includes a velocity variation with ray angle for the elliptical ray velocity surface fit (or hyperbolic t–x fit). The elliptical velocity fit was determined from the true vertical velocity V_v and the measured horizontal velocity V_h. The measured velocity is lower in the mid-angle range than for the elliptical fit.

Figure 8.8 shows velocities (V_v, V_h, V_γ, V_{st}) vs. depth. Stacking velocity (V_{st}) is a result of smoothing. The horizontal velocity, V_h, is found to be a better indicator of lithology changes.

Figure 8.9 shows the variation of velocity ratios with depth and lithology. Byun's et al. (1989) conclusion is that *apparent anisotropy,* as measured from the horizontal to vertical velocity ratio, correlates well with *lithology.*

CHAPTER 9

Analysis of t^2–x^2 Relations

In reflection prospecting, the travel time-offset (t–x) curves are used on a routine basis to calculate velocity and depth. Levin (1979) suggested that a more informative way is to examine (t–x) arrivals in the (t^2–x^2) plane. The hyperbola in the (t–x) plane gets transformed into a straight line in the (t^2–x^2) plane. The slope of the 't^2–x^2' graph is used to find the stacking velocity. By equating stacking velocity to RMS velocity (VRMS), interval velocities are calculated. This type of analysis holds in general for an isotropic medium but is not valid when one of the layers is anisotropic. It is pertinent to analyze the relationship between the true velocities and the velocities calculated from the t^2–x^2 curves for anisotropic models.

Analysis of t^2–x^2 curves for determination of velocity and elastic parameters of a transversely isotropic (T.I.) medium may be found in the investigations of Levin (1978, 1979, 1980, 1990), Crampin and Radovich (1982), Radovich and Levin (1982), and Hake et al. (1984). The derivations, discussion and results presented in this chapter are based on their investigations.

The reason that (straight-line) t^2–x^2 analysis works in isotropic media but not when the subsurface is anisotropic is the following. In an anisotropic medium, the velocity of plane waves generally differs from the velocity of waves from a point seismic source. The plane-wave surfaces and point-source surfaces, therefore, do not coincide. The point-source surface is the ray-surface or wave surface. It may be mentioned here that the arrival-time (t–x curve) observed on the ground surface is directly related to the wave surface of waves propagating in a medium above the reflecting interface. In an anisotropic medium the ray determines the time and point of arrival of energy at an interface, but the direction of the reflected or refracted phase is determined from Snell's law applied to phase velocity and direction (Crampin and Radovich 1982). The reason for this is *both incident and reflected waves must have the same apparent phase velocity along the interface*. This is illustrated in Fig. 9.1.

The wave surface is a sphere, the (t–x) curve a hyperbola and the (t^2–x^2) curve a straight line for an isotropic medium. Considering a planar isotropic layer of velocity V, thickness h and two way reflection time t from the bottom of a homogeneous, isotropic layer, we have the following equation relating travel time t with x, the distance between the seismic source and the geophone (Fig. 9.2 a):

$$V^2t^2 = 4h^2 + x^2 \tag{9.1}$$

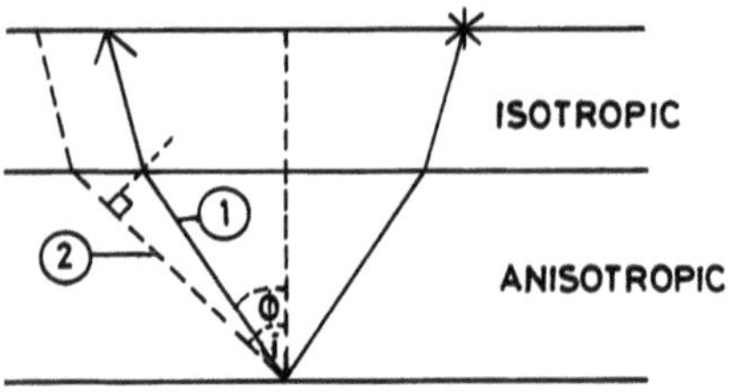

Fig. 9.1. The behavior of a ray at an anisotropic/isotropic interface: energy propagated along a ray (*solid line*) at the group velocity, but the behavior at the interface is controlled by the phase velocity and direction. *1* is the ray or group velocity direction and *2* is the phase velocity direction (Source: After Crampin, Radovich 1982, from Geophysics with permission)

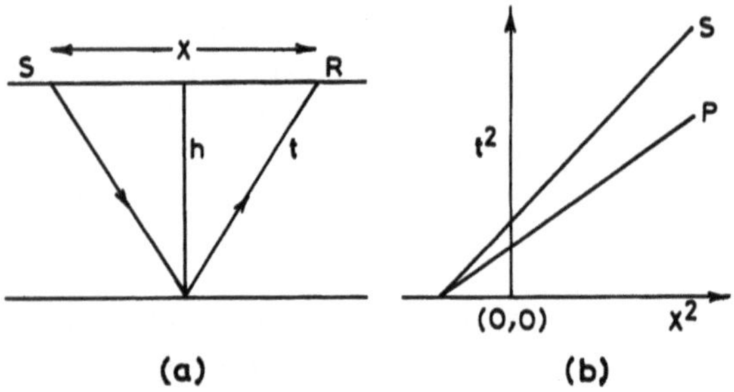

Fig. 9.2 a, b. a Isotropic medium reflection path geometry; **b** t^2–x^2 graph for CDP arrivals of P and S wave reflection from the bottom of a plane isotropic layer (Source: Crampin, Radovich 1982, from Geophysics with permission)

This equation holds both for P- and S-waves with appropriate constant values of V. The *common depth point* (CDP) gather both for P and S waves are hyperbola. The corresponding representation in the t^2–x^2 plane are straight lines which intersect at $t=0$ at the point $(0, -4\,h^2)$. The slope of the t^2–x^2 curve in this case is the squared slowness; the intercept with the t^2-axis is the square of the two way vertical travel time; and the intercept with the x^2-axis is the square of twice the reflector depth multiplied by minus one (Fig. 9.2 b).

Equation (9.1) is valid for an anisotropic medium also, but V is not constant and varies with the angle of incidence and azimuth if the layer possesses general anisotropy. Because of the direct relationship of the *wave surface* and the (t^2–x^2) curve, the effect of transverse isotropy (T.I.) on the (t^2–x^2) curve may be deduced from the wave surface. This is discussed below.

9.1 Single Anisotropic Layer Model

9.1.1 Wave Surfaces and t^2–x^2 Curves

It may be mentioned that for SH waves the wave surface is always elliptical and the (t^2–x^2) curve a straight line. For P- and SV waves the *wave surfaces* are generally not elliptical and the (t^2–x^2) curves are generally different from a straight line.

For an elliptical wave surface (Fig. 9.3 a) at time t, we have (Hake et al. 1984)

$$t^2 = \left(\frac{z}{V_z}\right)^2 + \left(\frac{x}{V_x}\right)^2 \tag{9.2}$$

where V_x is the wave propagation velocity in a horizontal direction and V_z is the velocity in the vertical direction.

From Eq. (9.2) it follows that the (t^2–x^2) graph is a straight line (Fig. 9.3 b). The slope of the (t^2–x^2) graph gives a square of the *horizontal slowness*. The *intercept* with the t^2-axis is again the square of the two-way vertical travel time, but the intercept with the x^2-axis is not equal to the square of twice the reflector depth.

When the wave surface differs from an ellipse the (t^2–x^2) curve is not a straight line. For an arbitrary wave surface, the (t^2–x^2) curve can be constructed by fitting an ellipse to every point on the wave surface, the axes of the ellipse being coincident with the horizontal and vertical axes of the wave surface (Hake et al. 1984). The slope of the tangent to the (t^2–x^2) curve at a point is given by the horizontal axis of this best fit ellipse. The intercept with the t^2-axis of the tangent is given by the vertical axis of the ellipse.

The above paragraph explains a general procedure for constructing the (t^2–x^2) curve by fitting ellipses to the wave surface.

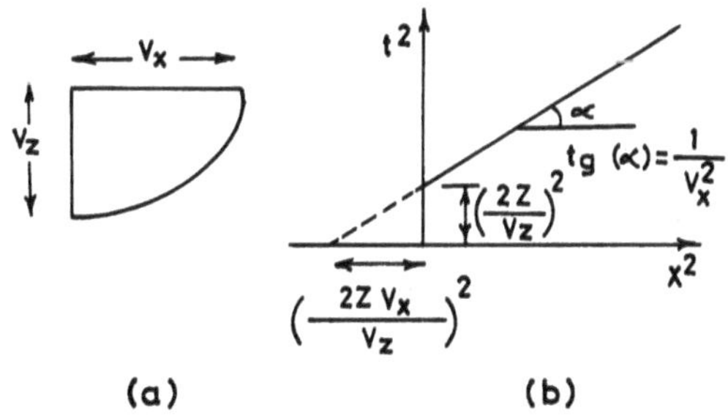

Fig. 9.3 a, b. **a** Part of an elliptical wave surface at unit time; V_x is the velocity in a horizontal direction, and V_z in a vertical direction; **b** (t^2–x^2) curve corresponding to the wave surface in **a** (Source: after Hake et al. 1984, from Geophysical Prospecting with permission)

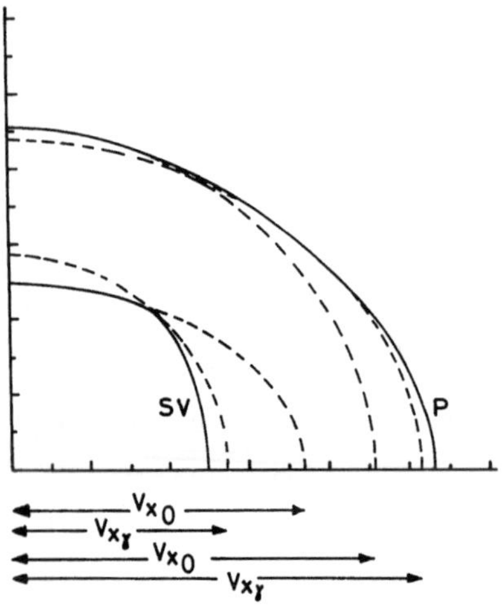

Fig. 9.4. Wave surfaces of P- and SV-waves from a homogeneous transversely isotropic medium (with $\lambda = 0.7$, $\tau = 0.3$, $h = 0$ and $k = 0$ – for definition of λ, τ, h and k refer to Section 9.3). *Dashed lines* represent best-fit ellipses at the wave surface for a vertical direction and for an arbitrary direction γ. V_{xo} is the horizontal axis of the ellipse for a vertical direction, $V_{x\gamma}$ for direction γ (Source: Hake et al. 1984, from Geophysical Prospecting with permission)

Hake et al. (1984) determined the form of the (t^2–x^2) curves for transversely isotropic (T. I.) media from wave surfaces. They considered transverse isotropy due to periodic layering and fitted the ellipse to the P and SV wave surfaces for two directions: one vertical and the other making an angle with the vertical (Fig. 9.4). For the P-wave and for vertical directions the horizontal axis of the best-fit ellipse is smaller than the horizontal axis for the other direction. This shows that the velocity calculated from the slope of the (t^2–x^2) curve increases with the offset, i.e. the t^2–x^2 curve is convex upward. For an SV-wave, it is the opposite. Hake et al. (1984) showed that this is true not only for the particular case of periodic layering considered, but is also true in general for transverse isotropy due to layering.

9.1.2
Curved t^2–x^2 Graph, Instantaneous Velocity and Intercept Time

It was explained above why the (t^2–x^2) curve is not a straight line for P and SV wave reflections from a horizontal interface in a transversely isotropic (T. I.) solid. To find the velocity from the curved (t^2–x^2) graph in these cases, one way-out is to fit a least-square straight line to the (t^2–x^2) data, and find the velocity from the slope of this line, and intercept time, t_0, from the intercept of this line with $x = 0$. Radovich and Levin (1982) suggested we consider the tangent to the curved (t^2–x^2) plot for an assigned 't' and 'x' and to find the slope and intercept of the tangent line and define an *instantaneous velocity* and *instantaneous intercept time* 't_0'. The *instantaneous velocity* and *instantaneous intercept time* provide insight into how the velocities change in transversely isotropic media, and what magnitude of change is expected.

The model considered by Radovich and Levin (1982) is a single transversely isotropic layer with a horizontal interface (Fig. 9.5). They obtained expressions for x, t, *instantaneous velocity* $(dx^2/dt^2)^{1/2}$ and *instantaneous intercept time* (t_0) in

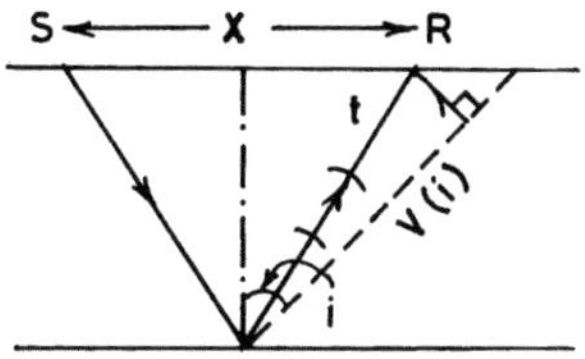

Fig. 9.5. Reflection in a single transversely isotropic layer. *Dot-dash line* is the axis of symmetry. Energy propagates along a ray (*continuous straight line*). Corresponding phase velocity (V_i) direction makes an angle i with the symmetry axis

terms of *plane wave velocity* V and *plane wave angle* i. Without giving derivations, the final result on squared *instantaneous velocity* is

$$\frac{dx^2}{dt^2} = V^2 + V\left(\frac{dV}{di}\right)\cot i = V^2 + \left(\frac{1}{2}\right)\left[\frac{d(V^2)}{di}\right]\cot i \tag{9.3}$$

[Radovich, Levin 1982, Eq. (A-10 a)]

Equation (9.3) gives at $i=0$ (by application of l' Hospital's rule) and $i=\pi/2$ the values of squared instantaneous velocities as

$$\frac{dx^2}{dt^2} = V^2 + \left(\frac{1}{2}\right)\left[d^2V^2/d\,i^2\right](i=0) \tag{9.4 a}$$

$$= V^2\left(i = \frac{\pi}{2}\right) \tag{9.4 b}$$

Knowing the slope of the (t^2–x^2) curve (for a given i), the intercept time 't_0' is given by

$$t_0^2 = t^2 - \left[d(t^2)/dx^2\right]_{t,x} \cdot x^2 \tag{9.5}$$

[Radovich, Levin 1982, Eq. (A-11)]

For a given plane wave angle 'i', the x^2 and t^2 are expressible as

$$x^2 = 4h^2 \frac{\left[V\sin i + \left(\frac{dV}{di}\right)\cos i\right]^2}{\left[V\cos i - \left(\frac{dV}{di}\right)\sin i\right]^2} \tag{9.6}$$

and

$$t^2 = \frac{4h^2}{\left[V\cos i - \left(\frac{dV}{di}\right)\sin i\right]^2} \tag{9.7}$$

[Radovich, Levin 1982, Eqs. (A-9) and (A-8)]

h is the depth to the horizontal plane reflector at the base of a single bed of a transversely isotropic solid.

It may be mentioned that travel time and distance (t, x) expressions are in terms of phase velocity V and phase angle i. Phase velocity V and phase angle i are not a measurable quantity. They are merely the parameters to vary seismic source and receiver offset (x) and corresponding travel time (t). By varying i, new values of 'x' and 't' may be determined and consequently the squared instantaneous velocity d^2x/dt^2 may be found corresponding to new values of 'x' and 't'.

It is desirable to find the deviation (δ^2) of $d(x^2)/dt^2$ at i (and at the corresponding x) from the value at $x=0$ and $x=\infty$.

Taking $f=d(x^2)/dt^2$, δ_0^2 and δ_∞^2 are expressed as

$$\delta_0^2 = \left[f^2(i) - f^2(0)\right]/f^2(0) \tag{9.8a}$$

and

$$\delta_\infty^2 = \left[f^2(i) - f^2(\infty)\right]/f^2(\infty) \tag{9.8b}$$

Radovich and Levin (1982) showed that

$$\delta_0^2 = (i^2/12)\{4[d^2(V^2)/di^2] + [d^4(V^2)/di^4]\} \times_{\text{at } i=0} \left\{V^2 + \left(\frac{1}{2}\right) d^2(V^2)\Big/di^2\right\}_{i=0}^{-1} \tag{9.9a}$$

and

$$\delta_\infty^2 = [(\pi/2) - i]^4 (1/6) \left\{ \begin{array}{l} [d^2(V^2)/di^2] + \\ \left(\frac{1}{4}\right)[d^4(V^2)/di^4] \end{array} \right\}_{i=\frac{\pi}{2}} \{V^2\}_{i=\frac{\pi}{2}}^{-1} \tag{9.9b}$$

[Radovich, Levin 1982, Eqs. (3) and (4)]

Equations (9.9a) and (9.9b) may be used to compute i. The inputs required are elastic constants of a transversely isotropic (T.I.) medium and V^2 (i). Knowledge of 'i' would enable us to find the corresponding 'x'.

As x increases, $d(x^2)/dt^2$ approaches the square of the velocity of travel in a horizontal direction. For P-waves, this information is not of much significance, as P-wave horizontal velocity has no exploration value. However, the SV-wave horizontal velocity is the same as the SH-wave vertical velocity. This information may be used to convert the SH-wave surface reflection times into depths.

We have restricted our discussions to the main formulae. However, these results are widely illustrated in Radovich and Levin (1982) with numerical analysis and graphical variations of ray velocity, phase velocity, instantaneous velocity with offset square for P and SV waves.

9.1.3 Interpreting t^2–x^2 Curves for Anisotropy

The theory for interpreting common depth point (CDP) gathers over a single anisotropic layer using (t^2–x^2) curves was contributed by Crampin and Radovich (1982). The interpretation gives the values of five elastic constants of a transversely isotropic layer. This is achieved by drawing tangents to the t^2–x^2 curves for P and SH waves and determining their slopes and intercepts. The significant steps of the Crampin and Radovich (1982) theory are given below.

There is a well-known relation between the phase velocity surface and wave surface cut by the symmetry plane (Postma 1955, Krey and Helbig 1956, Levin 1979). Denoting the phase velocity by V, phase angle by i, and the corresponding quantities for the ray as U and ϕ we have (Fig. 9.6)

$$U \sin\phi = V \sin i + \left(\frac{dV}{di}\right) \cos i \tag{9.10 a}$$

and

$$U \cos\phi = V \cos i - \left(\frac{dV}{di}\right) \sin i \tag{9.10 b}$$

From Eqs. (9.10 a) and (9.10 b) we have

$$U^2(\phi) = V^2(i) + \left(\frac{dV(i)}{di}\right)^2 \tag{9.11 a}$$

and

$$\tan\phi = \frac{\left[V \sin i + \frac{dV}{di} \cos i\right]}{\left[V \cos i - \frac{dV}{di} \sin i\right]} \tag{9.11 b}$$

Postma (1955) proved that Eqs. (9.11 a) and (9.11 b) hold for propagation in the symmetry plane of a T.I. medium. However, these equations are also valid in any other anisotropic symmetry plane (Crampin and Radovich 1982, p 325).

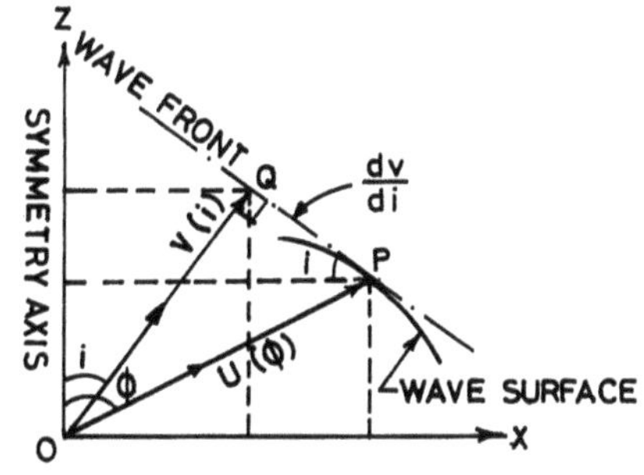

Fig. 9.6. Relationship between phase angle (i), phase velocity V(i), ray angle (ϕ) and ray (group) velocity (U(ϕ)). Wavefront is tangent at P to wave surface. OQ is ⊥r to the plane wavefront. OQ is the phase velocity direction for ray velocity direction OP

These equations show that the velocity surface and wave surfaces are very close to each other when dV/di is small. Crampin and Radovich (1982 a) showed that dV/di is small when overall anisotropy is weak, and also near the directions in which the two symmetry planes intersect.

Variation of phase velocity in a particular symmetry system is expressible in exact form (Auld 1973). The phase velocity function for a transversely isotropic media is known [White 1965; refer to Eq. (8.10)]. This function involves irrational quantities and is therefore difficult for algebraic manipulations. Crampin (1977) gave *reduced equations*, which are valid in symmetry planes and are *linear* in elastic constants.

Consider an anisotropic medium with three orthogonal symmetry planes $x_i = 0$, $i = 1, 2, 3$, with the x_3-axis taken as vertical. Crampin (1977) showed that *phase velocity* variation in one of the vertical symmetry planes (say, $x_2 = 0$) satisfy approximately the following *reduced equations*:

$$\rho V_P^2 = A + B\cos 2i + C\cos 4i \tag{9.12 a}$$

$$\rho V_{SV}^2 = D + E\cos 4i \tag{9.12 b}$$

$$\rho V_{SH}^2 = F + G\cos 2i \tag{9.12 c}$$

where

$$A = [3(c_{3333} + c_{1111}) + 2(c_{1133} + 2c_{1313})]/8 \tag{9.13 a}$$

$$B = (c_{3333} - c_{1111})/2 \tag{9.13 b}$$

$$C = [c_{3333} + c_{1111} - 2(c_{1133} + 2c_{1313})]/8 \tag{9.13 c}$$

$$D = [c_{3333} + c_{1111} - 2(c_{1133} - 2c_{1313})]/8 \tag{9.13 d}$$

$$E = -C \tag{9.13 e}$$

$$F = (c_{2323} + c_{1212})/2 \tag{9.13 f}$$

$$G = (c_{2323} - c_{1212})/2 \tag{9.13 g}$$

In the above Eqs. (9.12) and (9.13) ρ is density, c_{ijkl} are elastic constants, 'i' is the incidence angle – the angle of propagation of phase velocity measured from the vertical direction. The quasi nature of waves due to anisotropy is understood.

It may be mentioned that forms of phase velocity equations for P, SV and SH waves, expressed by Eqs. (9.12) and (9.13), are linear and simple, compared to the value of the exact phase velocity (Eq. 8.10) involving an irrational function.

The presence of the cos 4i term in Eq. (9.12) signifies that unless coefficients of cos 4i become zero, the P-arrivals would not fall on a hyperbola in a CDP gather. The coefficient C of the cos 4i term is often very small, and for this reason the hyperbola may be a good approximation for the common depth point (CDP) arrival curve for a transversely isotropic (T. I.) structure.

Crampin and Kirkwood (1981) showed that Eqs. (9.12 a–c) are a good approximation in symmetry planes in all symmetry systems except the trigonal symmetry system. These equations are also good approximations to the wave velocity (U), also for directions near the symmetry axes. On the consideration that anisotropy in the crust is likely to have a vertical symmetry axis, Eqs. (9.12 a–c) provide a reasonable basis for interpreting seismic reflection results (Crampin and Radovich 1982, p 325).

We now discuss the Eq. (9.1)

$$V^2t^2 = x^2 + 4h^2$$

for reflection from a single anisotropic bed with a vertical symmetry axis.

As mentioned, this equation is valid both for an isotropic or anisotropic layer. However, in the case of an anisotropic layer, the velocity V is not constant and varies with the angle of incidence.

In Crampin and Radovich (1982) Eqs. (9.12 a–c) were taken to hold both for phase velocity and wave velocity for P and SH waves for *weak anisotropy*, particularly near symmetry directions. For the SV wave, the separation of phase and wave velocity is more as compared to P and SH waves, which make them unsuitable for using in common depth point (CDP) gather studies. Values of V for the appropriate wave type may be substituted in Eq. (9.1) and the value of cos i may be expressed in terms of depth of the interface h and shot geophone separation x as $\cos i = 2\,h/(4h^2+x^2)^{1/2}$. This gives

(i) for P waves:

$$\begin{bmatrix} A + B(4h^2 - x^2)/(4h^2 + x^2) + \\ C(16h^4 - 24h^2x^2 + x^4)/(4h^2 + x^2)^2 \end{bmatrix} t^2 = (4h^2 + x^2) \tag{9.14 a}$$

(ii) for SH waves:

$$[F + G(4h^2 - x^2)/(4h^2 + x^2)]t^2 = (4h^2 + x^2) \tag{9.14 b}$$

(iii) for SV waves:

$$\left[D - C(16h^4 - 24h^2x^2 + x^4)/(4h^2 + x^2)^2\right]t^2 = (4h^2 + x^2) \tag{9.14 c}$$

[Crampin, Radovich 1982, Eq. (A-2)]

The density ρ is taken absorbed in coefficients A, B...G of Eq. (9.13).

Expanding Eq. (9.14), and neglecting powers of x greater than x^2 and squares of ratios B/A, C/A, C/D and G/F – which are reasonable assumptions, gives the t^2–x^2 relation in a more conventional straight line form:

$$t^2 = mx^2 + c$$

We have

(i) for P waves:

$$t^2 = \left[4(A - B - C)h^2 + (A + B + 7C)x^2\right]/A^2 \tag{9.15a}$$

(ii) for SH waves:

$$t^2 = \left[4(F - G)h^2 + (F + G)x^2\right]/F^2 \tag{9.15b}$$

(iii) for the SV wave:

$$t^2 = \left[4(D + C)h^2 + (D - 7C)x^2\right]/D^2 \tag{9.15c}$$

Equations (9.15a–c) represent equations of tangents to the transformed gather arrivals at $x^2 = 0$ in the (t^2, x^2) diagram. Equations of these tangent lines may be written in the general form

$$t^2 = a_i x^2 + b_i \quad \text{for} \quad i = P, SH \quad \text{and} \quad SV \tag{9.16}$$

where a_i and b_i are slopes and intercepts.

Figure 9.7 shows these tangents (I) and (II) for a P and SH wave in the t^2–x^2 diagram.

On equating Eq. (9.16) with Eq. (9.15) and neglecting squares of B/A, C/A, C/D and G/F, it is possible to express A, B, C, F and G [i.e. a combination of elastic coefficients – refer to Eqs. (9.13a–f)] in terms of slopes and intercepts, a_i and b_i, of the tangent lines of (t^2–x^2) curves. We have

(i) for the P-wave:

$$A - B - 7C = \frac{1}{a_P}; \tag{9.17a}$$

$$A + B + C = \frac{4h^2}{b_P} \tag{9.17b}$$

(ii) for the SH-wave:

$$F - G = \frac{1}{a_{SH}} \tag{9.17c}$$

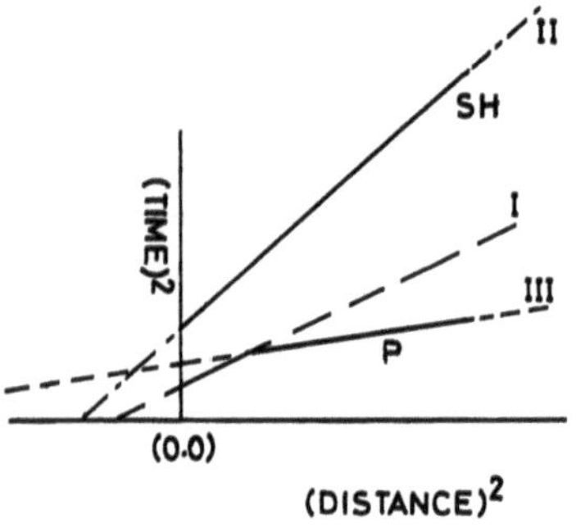

Fig. 9.7. Schematic illustration showing the system of tangents (I and II) and asymptote (III) in the (t^2–x^2) diagram that may be used to estimate elastic constants (Source: after Crampin and Radovich 1982, from Geophysics with permission)

$$F + G = \frac{4h^2}{b_{SH}} \tag{9.17d}$$

(iii) for the SV wave:

$$D + 7C = \frac{1}{a_{SV}} \tag{9.17e}$$

$$D - C = \frac{4h^2}{b_{SV}} \tag{9.17f}$$

[Crampin, Radovich 1982, Eq. (A-5)]

Application of Eqs. (9.17 a–f) requires knowledge of 'h', the thickness of the anisotropic layer. If it could be estimated separately (say, from well logs) Eqs. (9.17 a–f) could be solved for the elastic constants. It may be noted that (A+B+C) and (F+G) are squares of vertical velocities of P and SH waves. (F–G) is the square of the velocity of the SH-wave in the horizontal direction.

Crampin and Radovich (1982) gave an interpretation of the slopes and intercepts of tangents to arrivals in the (t^2–x^2) diagram. For details reference may be made to this publication. The final outcome of the analysis is that from measurements of slope and intercepts of tangents of P and SH waves at $x^2 = 0$ in the t^2–x^2 diagram, and by using asymptote of the P wave for long offset (marked III in Fig. 9.7), it is possible to solve for A, B, C, F and G, and substitution in Eq. (9.12) would yield values of elastic constants as

$$\begin{aligned} c_{1111} &= A - B + C \\ c_{3333} &= A + B + C \\ c_{1212} &= F - G \\ c_{2323} &= F + G \end{aligned}$$

and

$$c_{1133} + 2c_{1313} = A - 3C \tag{9.18}$$

[Crampin, Levin 1982, Eq. (B-1)]

Further a transversely isotropic (T.I.) medium possesses five independent elastic constants:

$$\begin{aligned} c_{1111} &= c_{2222}, c_{3333}, c_{1122} \\ c_{2233} &= c_{1133}, c_{2323} = c_{1313} \end{aligned}$$

and

$$c_{1212} = (c_{1111} - c_{1122})/2 \tag{9.18a}$$

The use of Eq. (9.18) with Eq. (9.18 a) enables determination of all five elastic constants of a transversely isotropic (T.I.) medium.

The assumptions in this method are: the symmetry axis is vertical, the anisotropy is weak (for validity of linear equations), and the measurements are made in vertical symmetry planes for seismic reflection lines. The method makes use of (t^2–x^2) data for both P and SH waves. Throughout the derivations, a single T.I. layer has been assumed.

Treatment of the (t^2–x^2) curve for multilayered media is considered in the next sections. First we consider elliptical anisotropy. In the subsequent section a multi-layered transversely isotropic (T.I.) medium is analyzed. Before discussing the case of multi-layered media, it is important to describe the main conclusions of the Crampin and Radovich (1982) study on anisotropy and CDP gathers.

Crampin and Radovich (1982) analyzed the effects of anisotropy on synthetic common depth point (CDP) travel time gathers, and concluded that the presence of anisotropy may affect significantly the interpretation. In their study, synthetic gather curves were generated for a number of models with an anisotropic layer underlying an isotropic medium. The gather curves show departure from hyperbola.

9.1.4 Velocity and Depth Errors under Assumption of Isotropy

Crampin and Radovich (1982) analyzed in detail the velocity estimates from tangents drawn to t^2–x^2 curves at short spread and long spread, and compared these estimates with actual vertical and horizontal velocities in the layer. They arrived at the following conclusions (Crampin and Radovich 1982, p 330):

(i) the velocity determined from the near-offset tangent may lie outside the range of true velocity variation, and may be much larger or smaller than the true vertical velocity.
The velocity estimate from a far-offset tangent gives a good estimate of horizontal velocity. However, in an anisotropic layer, this velocity may be substantially different from the vertical velocity;

(ii) the value of t_0 (intercept) is a good approximation to true vertical travel time, even for large offset gathers. If the anisotropy is not taken into consideration, this accurate vertical travel time from the far-offset tangent would be associated with the horizontal velocity, and the horizontal velocity may be significantly different from the vertical velocity. This would result in wrong estimates of depths.
These numerically arrived results by Crampin and Radovich (1982) suggest caution concerning considerable errors, which may result if anisotropic common depth point (CDP) gathers are interpreted assuming isotropy.

9.2
What Interval and Stacking Velocities Correspond to Under Conditions of Anisotropy

In an anisotropic medium, the velocity depends on the direction of propagation. The question arises as to which value the interval velocity corresponds to when computed by the usual Dix (1955) formula for an anisotropic medium. Denoyer de Segonzac and Laherrere (1959), Van der Stoep (1966) and Levin (1978) have examined this question. Using the Dix (1955) formula for computing interval velocity, one gets the horizontal velocity in an elliptically anisotropic medium, and stacking velocity is the vertical time weighted sum of the horizontal interval velocities (Levin 1978). We give below the main results from Levin (1978).

Consider a medium consisting of beds separated by horizontal interfaces. Let h_j be the thickness of the j-th bed. Let V_h and V_v be the horizontal and vertical velocities respectively in the j-th elliptically anisotropic layer (Fig. 9.8). The t^2–x^2 relation for this geometry is given by

$$t^2(x^2) = 4\left(\sum_{j=1}^{n} h_j \Big/ (V_v)_j\right)^2 + \left[\left(\sum_{j=1}^{n} h_j \Big/ (V_v)_j\right) \Big/ \sum_{j=1}^{n} h_j (V_h^2)_j \Big/ (V_v)_j\right] x^2 \tag{9.19}$$

[Levin 1978, Eq. (A-37)]

Treating Eq. (9.19), in the form

$$t^2 = t_0^2 + \frac{x^2}{(VRMS)^2}$$

the *root mean square velocity* for reflection from the n-th interface for a multi-layered elliptically anisotropic medium is given by

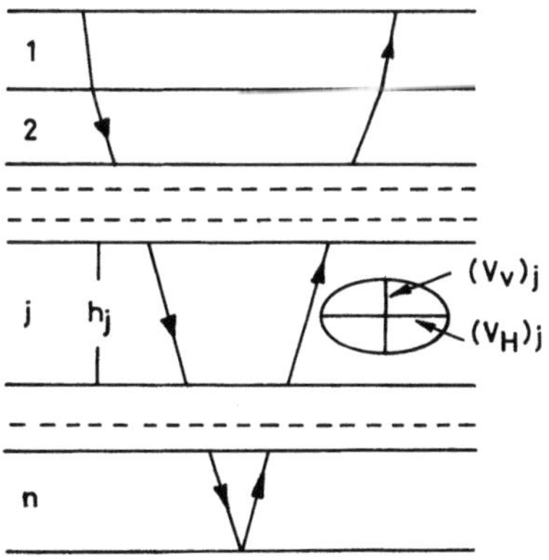

Fig. 9.8. Reflection in a layered medium having horizontal layers with elliptical velocity functions

$$
\begin{aligned}
(\mathrm{VRMS})_n^2 &= \left[\sum_{j=1}^{n}\left(h_j (V_h)_j^2 \Big/ (V_v)_j\right)\right] \Big/ \left[\sum_{j=1}^{n} h_j \Big/ (V_v)_j\right] \\
&= \left[\sum_{j=1}^{n} (\Delta t_v)_j (V_h)_j^2\right] \Big/ \left[\sum_{j=1}^{n} (\Delta t_v)_j\right]
\end{aligned}
\tag{9.20}
$$

[Levin 1978, Eq. (1)]

Equation (9.20) gives interval velocity [from Dix Eq. (7.28 c) for interval velocity:

$$
(V_h)_n^2 = \left[(T_v)_n (\mathrm{VRMS})_n^2 - (T_v)_{n-1} (\mathrm{VRMS})_{n-1}^2\right] \Big/ \left[(T_v)_n - (T_v)_{n-1}\right] \tag{9.21}
$$

[Levin 1978, Eq. (2)]
where

$$
(T_V)_n = \sum_{j=1}^{n} (\Delta T_V)_j
$$

Equation (9.21) tells us that the interval velocity calculated by the Dix method are the horizontal components only, and stacking velocity is the vertical time weighted sum of the horizontal interval velocities. In Eqs. (9.19) to (9.21), as defined in the beginning, h_j is the thickness of the j-th layer, $(V_v)_j$ and $(V_h)_j$ are the vertical and horizontal velocities of the elliptic velocity function in the j-th layer; $(\Delta t_v)_j$ is the travel time in the j-th layer for a vertical ray path.

9.3 Three-Term Taylor Series Approximation of t^2–x^2 Curve

Hake et al. (1984) analyzed the t^2–x^2 curve by approximating it by a Taylor series. The representation in the Taylor series is restricted to only three terms. These terms are expressed as a function of elastic parameters of the medium. A short description of the elastic parameters used follows.

A transversely isotropic medium is characterized by five independent elastic stiffness constants: c_{11}, c_{13}, c_{33}, c_{44} and c_{66}. These parameters along with density are used to discuss wave propagation in a transversely isotropic medium. Helbig (1981) introduced four normalized, dimensionless parameters: λ, τ, h, and k. These were defined in the following form:

$$
\begin{aligned}
\lambda &= \frac{c_{44}}{c_{66}};\ \tau = \frac{1}{2}\left(1 - \frac{c_{13}}{c_{33}}\right) \\
h &= \frac{1}{2}(2c_{44} - c_{33} + c_{13}) \Big/ c_{33}, \quad \text{and} \\
k &= \frac{1}{4}\left\{ \begin{matrix} c_{13}(c_{13} + 2c_{66}) + \\ c_{33}(2c_{66} - c_{11}) \end{matrix} \right\} \Big/ (c_{66}c_{33})
\end{aligned}
\tag{9.22}
$$

[Helbig 1981; Hake et al. 1984, Eq. (1)]

For transverse isotropy due to *periodic layering* the relationship between anisotropy parameters and layer parameters is describable in terms of the squared ratio of shear and compressional velocity (denoted by θ) and the shear modulus of constituent layers (denoted by μ).

These relations are

$$\begin{aligned} \lambda &= \frac{1}{\langle\mu\rangle \cdot \langle\frac{1}{\mu}\rangle} \\ \tau &= \langle\theta\rangle \\ h &= \frac{\langle\frac{\theta}{\mu}\rangle - \langle\theta\rangle \cdot \langle\frac{1}{\mu}\rangle}{<\frac{1}{\mu}>} \\ k &= \frac{\langle\theta \cdot \mu\rangle - \langle\theta\rangle \cdot \langle\mu\rangle}{\langle\mu\rangle} \end{aligned} \qquad (9.23)$$

[Hake et al. 1984, Eq. (2)]
where $\langle\rangle$ denotes the *weighted average*.

Parameter τ is the average of the squared ratio of the shear and compressional velocity of the component layers. Taking μ as constant, the averaging Eq. (9.23) gives $\lambda=1$ and $h=k=0$. This is the case of an isotropic medium. If θ is taken as constant, and μ variable, we have $\lambda<1$ and $h=k=0$. A medium possessing this property is called a *k-medium*. For sedimentary rocks, the *ratio of the shear to compressional wave velocity* varies very little and a *k-medium* is a good approximation in this case.

In the following, the main results of the theoretical development in Hake et al. (1984) are given.

9.3.1 Term-wise Interpretation for Isotropy, Anisotropy and Inhomogeneity

Consider Fig. 9.9. Each layer is isotropic or transversely isotropic (T.I.). Let the ray path be as shown and let the *group velocity* (or *wave velocity*) *vector* be **U**, and ϕ be the angle between the wave velocity vector **U** and the vertical. Let $\Delta t/2$ be the time along the ray path (AB) in a particular layer. Ray lengths along the horizontal and the vertical in a layer are denoted by $\Delta x/2$ and Δz, respectively.

Also, let **V** be the *phase velocity vector*, and **n** be the *slowness vector*. The *phase velocity* (or *normal velocity*) **V** is the velocity with which planes of equal phase travel perpendicular to the wavefront.

The direction of slowness vector **n** is the same as *phase velocity vector* **V** and the magnitude is the reciprocal of the magnitude of **V**. Denoting n_1 (= dt/dx) as the horizontal component of slowness **n** we have

$$n_1 = \frac{V_1}{|\mathbf{V}|^2}$$

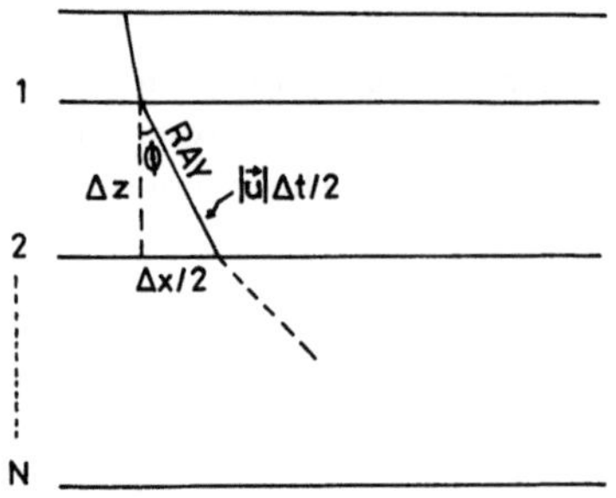

Fig. 9.9. Geometry of a ray segment (Source: Hake et al. 1984, from Geophysical Prospecting with permission)

The relationship between the arrival time t and offset x is taken as

$$t^2 = \lim_{x \to x_0} \sum_j A_j \left(x^2 - x_0^2\right)^j \tag{9.24 a}$$

[Hake et al. 1984, Eq. (A1)]
where A_j is defined by

$$A_j = \frac{1}{(F)j} \frac{d^j(t^2)}{(dx^2)^j} \tag{9.24 b}$$

(F)j stands for factorial j.

The Taylor series approximation at zero offset takes the form

$$t^2 = \lim_{x \to 0} \sum_j A_j x^{2j} \tag{9.24 c}$$

[Hake et al. 1984, Eq. (3)]
with A_0, A_1 and A_2 given by,

$$\lim_{x \to 0} A_0 = \left(\sum_i \Delta t_i\right)^2 \tag{9.25 a}$$

$$\lim_{x \to 0} A_1 = \frac{\sum_i \Delta t_i}{\sum_i R_i^2 \Delta t_i} \tag{9.25 b}$$

and

$$\lim_{x \to 0} A_2 = \frac{\sum_i \left(R_i^2 \Delta t_i\right)^2 - \sum_i \Delta t_i \sum_i \left(R_i^4 + M_i\right) \Delta t_i}{4\left(\sum_i R_i^2 \Delta t_i\right)^4} \tag{9.25 c}$$

In Eqs. (9.25 b) and (9.25 c), R_i^2 ($=u_{1i}/n_{1i}$) is the ratio of the horizontal components U_1 and n_1 of wave velocity **U** and slowness **n**, respectively. n_1 is equal to ray parameter p. Δt_i in Eqs. (9.25 a–c) is the two way vertical travel time in the layer i (i = 1, 2...N), and N is total number of layers above the reflector. Also, M_i in Eq. (9.25 c) is given by

$$M_i = \left(\frac{1}{n_1}\frac{d(R^2)}{dn_1}\right)_i$$

The *wave velocity* and *phase velocity* are liable to be calculated from known expressions for a particular direction of ray angle ϕ for P-, SV- and SH waves. Further, these velocities are expressible in terms of elastic parameters of the medium. Therefore R and M in Eqs. (9.25 b) and (9.25 c) may be expressed in terms of the parameter sets c_{11}, c_{13}, c_{33}, c_{44} and c_{66}, or h, k, λ and τ and density. These values for P-, SV- and SH waves in terms of parameters c_{11}, c_{13}, c_{33}, c_{44} and c_{66} are

(i) P-waves:

$$\lim_{x\to 0} R^2 = \frac{1}{\rho}\left(c_{11} - \frac{E^2}{C}\right) \tag{9.26 a}$$

$$\lim_{x\to 0} M = \frac{1}{\rho^2}\frac{4E^2A^2c_{33}}{C^3} \tag{9.26 b}$$

(ii) SV-waves:

$$\lim_{x\to 0} R^2 = \frac{1}{\rho}\left(c_{44} + \frac{E^2}{C}\right) \tag{9.26 c}$$

$$\lim_{x\to 0} M = -\frac{1}{\rho^2}\frac{4E^2A^2c_{44}}{C^3} \tag{9.26 d}$$

(iii) SH-waves:

$$R^2 = \frac{c_{66}}{\rho} - \left(\frac{c_{44}}{\rho\lambda}\right) \tag{9.26 e}$$

$$M = 0 \tag{9.26 f}$$

[Hake et al. 1984, Eqs. (7 c–f)]
where

$$A = c_{13} + c_{44},\ B = c_{11} - c_{44},$$
$$C = c_{33} - c_{44} \quad \text{and} \quad E^2 = BC - A^2$$

It is to be noted that the values of the three terms A_0, A_1 and A_2 in Eqs. (9.25a–c) are finally expressible in terms of layer parameters of transversely isotropic (T.I.) layers and the travel time of ray paths in the layers. A_0 in Eq. (9.25a), the intercept with the t^2-axis, is the square of the sum of vertical travel times. This is the same as for isotropic media. The interpretation of A_1 for isotropic media is that it is the reciprocal of the time average of the squared velocity given by

$$(A_1)_{\text{isotropy}} = \frac{1}{(\text{VRMS})^2} = \frac{\sum \Delta t_i}{\sum v_i^2 \Delta t_i}$$

For transversely isotropic (T.I.) media A_1 in Eq. (9.25b) is the reciprocal of the time average of R^2, the ratio of the horizontal component of wave velocity and slowness. Interpretation of R^2 is as follows. R^2 for SH waves is the horizontal velocity of SH-waves, i.e. $(c_{66}/\rho)^{1/2}$. For a k-medium for a P-wave

$$R = (c_{33}/\rho)^{\frac{1}{2}} = (c_{44}/\rho\tau)^{\frac{1}{2}}$$

and is the vertical velocity.

A_2 in Eq. (9.25c) gives the deviation of the t^2–x^2 graph from a straight line. A_2 is seen as a combination of two parts – one due to transverse isotropy, $A_2(\text{T.I.})$, and the other due to inhomogeneity, $A_2(\text{I})$. We have

$$\begin{aligned} A_2 &= A_2(\text{I}) + A_2(\text{T.I.}) \\ &= \frac{\left(\sum R_i^2 \Delta t_i\right)^2 - \sum \Delta t_i \sum R_i^4 \Delta t_i}{4\left(\sum R_i^2 \Delta t_i\right)^4} - \frac{\sum_i \Delta t_i \sum_i M_i \Delta t_i}{4\left(\sum R_i^2 \Delta t_i\right)^4} \end{aligned} \tag{9.27}$$

[Hake et al. 1984, Eqs. (11) and (12)]

Both *inhomogeneity* and *transverse isotropy* make the t^2–x^2 graph curved.

The sign of A_2 (T.I.) depends on the sign of M. Considering a T.I. medium for which $E^2 > 0$ [see Eqs. (9.26b) and (9.26d)]. M is positive for a P-wave and negative for a SV wave. E^2 is greater than zero for a transversely isotropic medium due to periodic layering (Berryman 1979). The effect of transverse isotropy on the t^2–x^2 graph for a P-wave is the same as that for inhomogeneity, i.e. both make the t^2–x^2 graph convex upward for a P-wave. In the case of a SV wave, the sign of the effect of transverse isotropy is opposite to that of the effect of inhomogeneity. Therefore, whether the t^2–x^2 graph is convex upward or downward would depend on the relative magnitudes of the two effects.

The above presentation is restricted to the discussion of the Taylor series approximation of the t^2–x^2 graph at zero offset.

For an arbitrary offset x_0, we take Eq. (9.24a)

$$t^2 = \lim_{x \to x_0} \sum_j A_j \left(x^2 - x_0^2\right)^j$$

where

$$\lim_{x \to x_0} A_0 = \left(\sum_j \Delta t_i \right)^2 ,$$

$$\lim_{x \to x_0} A_1 = \frac{\sum_j \Delta t_i}{\sum_j R_i^2 \Delta t_i}$$

with

$$R_i^2 = \frac{U_{1i}}{n_{1i}}$$

R_i^2 is the ratio of the horizontal component of *wave velocity* and *slowness* for a ray in the i-th layer arriving at the surface at offset x_0. R_i depends on the angle of incidence. Both inhomogeneity and R_i determine the slope of the t^2–x^2 graph at a certain offset x_0.

Apart from the three terms (A_0, A_1 and A_2) of the t^2–x^2 graph, another characteristic parameter is the *intercept* with the t^2-axis (t_0^2) of the tangent to the t^2–x^2 graph at offset x_0. The intercept with the t^2-axis (t_0^2) is given by

$$t_0^2 = \frac{4z^2}{U_3/n_3} \tag{9.28}$$

(Hake et al. 1984)
where z is the depth of the reflector. For $x = 0$, U_3/n_3 is the squared vertical velocity. For an infinite offset, the values of U_3/n_3 are:
(i) P-wave:

$$\operatorname*{Lim}_{x_0 \to \infty} \frac{U_3}{n_3} = \frac{1}{\rho} \left\{ c_{44} + \frac{(c_{13} + c_{44})^2}{c_{11} - c_{44}} \right\}$$

[Hake et al. 1984, Eq. (16)]
(ii) SV-wave:

$$\operatorname*{Lim}_{x_0 \to \infty} \frac{U_3}{n_3} = \frac{1}{\rho} \left\{ c_{33} - \frac{(c_{13} + c_{44})^2}{c_{11} - c_{44}} \right\}$$

[Hake et al. 1984, Eq. (16)]

9.3.1.1
Variation of Velocity V and Intercept Time 't_0' as a Function of Offset

Inhomogeneity and R go to determine the slope of the t^2–x^2 graph at a certain offset. R depends on the angle of incidence.

Hake et al. (1984) studied the effect of transverse isotropy on velocity and intercept time by considering a homogeneous layer with the properties of a k-medium

(Section 9.3), the reflector depth 1 and unit vertical P-wave velocity. The t^2–x^2 graph for various values of λ is shown in Fig. 9.10 a. Variations of V for various values of λ calculated from the slope of curves in Fig. 9.10 a are presented in Fig. 9.10 b. For a zero offset, the value of V is identical for all curves and equal to the vertical velocity

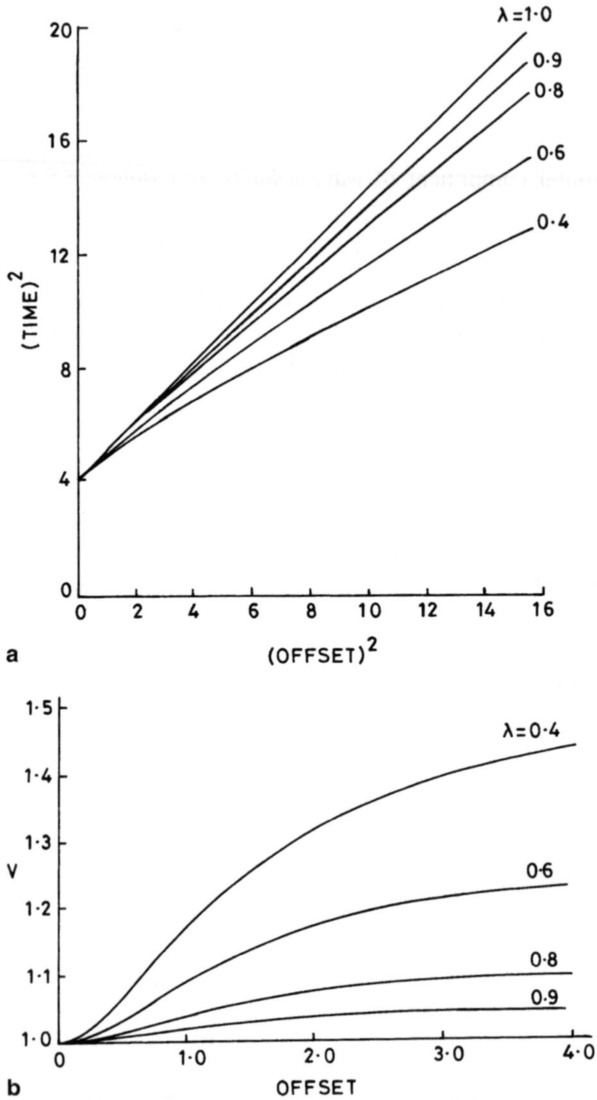

Fig. 9.10 a–c. a (t^2–x^2) curve for P-waves for a homogeneous transversely isotropic K-medium. Reflector depth and P-wave velocity are taken as units, $\tau=0.3$. The curves are calculated for several values of λ; **b** velocity V calculated from the slope of the curves in Fig. 9.10 a as a function of the offset; **c** intercept time t_0 corresponding to curves in Fig. 9.10 a as a function of offset (Source: Hake et al. 1984, from Geophysical Prospecting with permission)

Fig. 9.10c

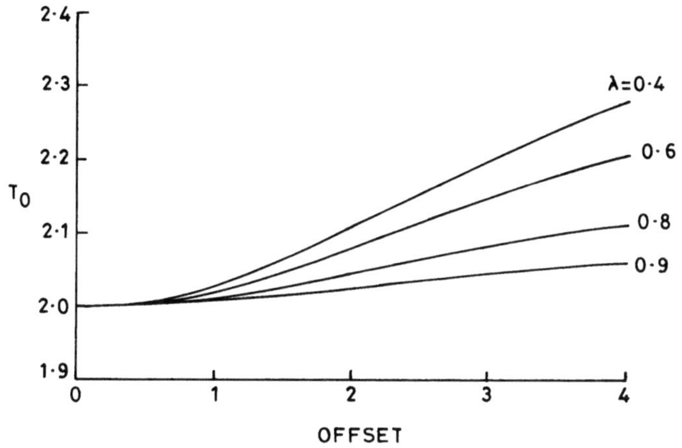

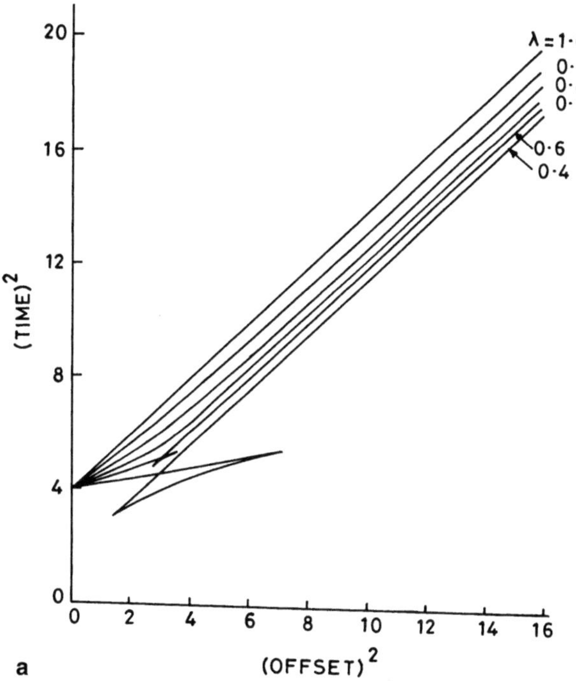

a

Fig. 9.11 a–c. a (t^2–x^2) curves for SV waves for a homogeneous transversely isotropic k-medium. Reflector depths and vertical SV-wave velocity are taken as units, $\tau=0.3$. These curves are calculated for several values of λ; **b** velocity V calculated from the slope of the curves in Fig. 9.11a as a function of offset; **c** intercept time 'τ_0' corresponding to the curves in Fig. 9.11a as a function of offset (Source: Hake et al. 1984, from Geophysical Prospecting with permission)

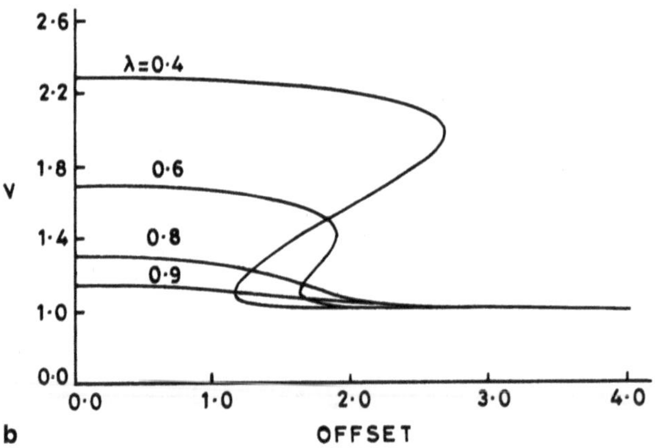

b

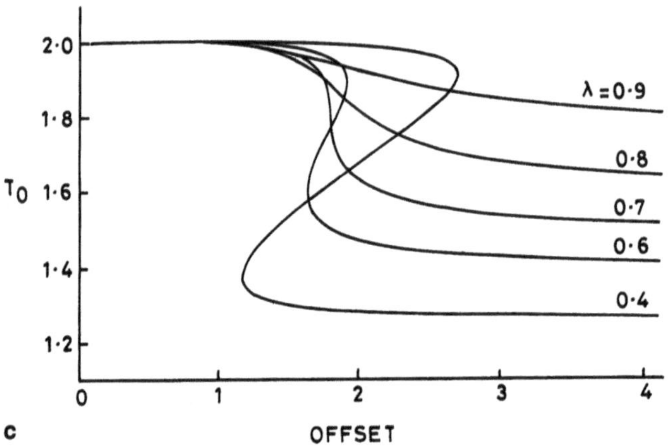

c

Fig. 9.11 b, c

as the layer is a k-medium. With increasing offset, V increases towards the horizontal velocity. Fig. 9.10 c shows the variations of 't_0' against offset.

Corresponding curves for SV waves appear in Fig. 9.11 a–c. The curves have been calculated for a homogeneous layer (k-medium) of unit thickness with vertical SV velocity taken as unity. The curves with $\lambda=0.6$ and $\lambda=0.4$ in Fig. 9.11 a show cusps. In Fig. 9.11 b for small offsets, V increases for increasing anisotropy, and for a large offset V approaches the horizontal velocity. Figure 9.11 c shows the variation of 't_0' with offset for various λ values.

9.3.2
Recommended Combined Use of SH and SV Waves to Probe Anisotropy

Summarizing the results on the t^2–x^2 graph from Hake et al. (1984) studies, the following points emerge:

The t^2–x^2 curves for P-waves are convex upwards, and it is to be noted that this may be due to *transverse isotropy* as well as *inhomogeneity.* In the case of SV waves, transverse isotropy causes the t^2–x^2 curve to be convex downward, something, which may not be caused by vertical inhomogeneity. In the case of SH-waves, the transverse isotropy has no effect on curvature.

Using directly the slope of the t^2–x^2 curve as squared slowness over a transversely isotropic ground would result in erroneous reflector depths.

One of the ways to detect transverse isotropy is to observe both SH- and SV wave reflections. For transverse isotropy due to layering the travel-time curve of the SH and SV wave have the same intercept time. However, the slope of the t^2–x^2 curve at zero-offset for the SV wave is always smaller than for SH-waves. They are equal only in the case of isotropy.

In conclusion, Hake et al. (1984) recommended the use of shear waves over compressional waves to detect transverse isotropy from surface observations.

Estimation of Anisotropy and Layer Parameters 10

Accurate measurement of subsurface anisotropy provides very useful clues for the identification of lithology, structure and the degree of stratification in sedimentary rocks with hydrocarbon potential. With the development of new observational techniques such as Vertical Seismic Profiling (VSP) and Multicomponent Seismology, anisotropy measurements have become more quantitative. The following two sections explain and illustrate how data from Offset Vertical Seismic Profiles and Surface Seismic Surveys may be utilized to determine the stiffness constants for azimuthally isotropic media (Section 10.1); and the elastic stiffness constants and axis of symmetry for azimuthally anisotropic (Section 10.2) media. Section 10.3 deals with *travel time inversion* for determining anisotropy and model parameters.

10.1 Semblance Analysis of Skewed Hyperbolic Time-Distance Curves

We make use of the skewed hyperbolic travel time relation (Eq. 8.16 – renumbered as 10.1) for transverse isotropy and offset VSP geometry (Fig. 8.3 reproduced here as Fig. 10.1):

$$t^2 = \frac{z^2}{V_v^2} + \left[\frac{z^2}{V_\gamma^2} + \frac{x^2}{V_h^2}\right] \frac{x^2}{x^2 + z^2} \tag{10.1}$$

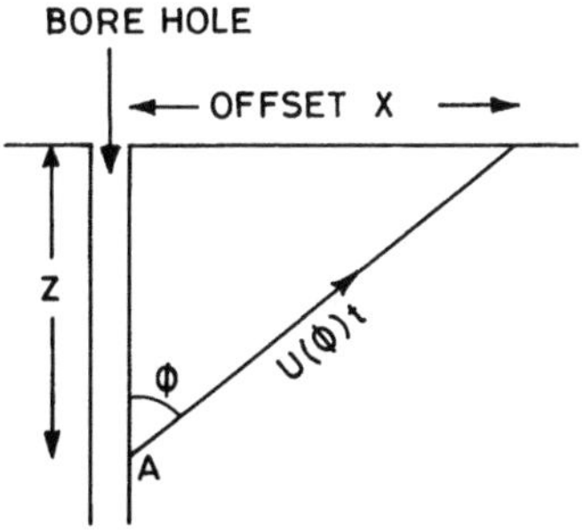

Fig. 10.1. Ray geometry for an offset VSP, showing a raypath from a borehole source to a surface receiver. The subsurface is assumed to be transversely isotropic (Source: Byun and Corrigan 1990)

[Byun and Corrigan 1990, Eq. (7)]

Byun and Corrigan's (1990) method works with five measured quantities: V_v, V_h, V_γ, U_v, and U_h. V_v, V_h, and V_γ are the average vertical, horizontal and skew moveout velocities, respectively for P-wave. U_v and U_h refer to average vertical and horizontal moveout velocities of SH-waves for a horizontal stack of different transversely isotropic layers.

This method is based on an *iterative model based optimization* scheme to determine five *elastic stiffness coefficients* (A, C, F, L and M) for each layer in a *layer-stripping model.* It is approximate in the sense that the measurement parameters are determined through a best fit to the time distance curve. Further, the estimation errors tend to increase with the number of layers, as is the case with any other *layer stripping method.* Byun and Corrigan (1990) found that this inversion process gives the least reliable estimate of *skew moveout velocity.* In Subsection 10.3.2 of this chapter the travel-time inversion algorithm of Lee (1990) is given in which all layer parameters are estimated simultaneously, and therefore the errors in estimates for the upper layers do not propagate to the lower ones. The method of Byun and Corrigan (1990) works in three steps, starting from the surface layer down to the lower layers.

Step 1: In this step, coefficients L_n and M_n for layer 'n' are determined from SH-wave measurements. Equations (8.19 a) and (8.19 b) give the interval properties $U_{v,n}$ and $U_{h,n}$ from measurements of U_v and U_h as well as τ_0. From the measured values of $U_{v,n}$ $U_{h,n}$, stiffness coefficients L_n and M_n of transversely isotropic layers 'n' may be determined from the relations

$$U_{v,n} = \sqrt{\frac{L_n}{\rho_n}} \tag{10.2 a}$$

$$U_{h,n} = \sqrt{\frac{M_n}{\rho_n}} \tag{10.2 b}$$

It is assumed that the density 'ρ' is known from the density log.

Step 2: The elastic stiffness coefficient C_n is determined from P-wave vertical travel time measurements using Eqs. (8.20) and (8.20 a). The average vertical velocity $(V_v)_{aver}$ may be determined from zero-offset VSP data. Using Eq. (8.20 a), the coefficient C_n for layer 'n' may be determined..

Step 3: Under this step the remaining two parameters A_n and F_n are determined from multi offset P-wave 't–x' curves through a least mean-squares (LMS) error analysis. These are obtained as follows:

From the measured parameters V_v, V_h, and V_γ, a real travel-time curve is reconstructed using Eq. (10.1), assuming that this equation reproduces in an accurate way the true 't–x' relationship for a multilayered transversely isotropic medium.

Next model 't–x' curves are generated for a set of (A_n, F_n) trial values and by simulating the acquisition geometry used to obtain the field data.

Finally, those A_n and F_n values are accepted which correspond to the least-mean-squares (LMS) error between the real and model travel time curves.

The method outlined above is discussed with the help of an example under Subsection 10.1.1.

For extending this method to surface reflection surveys, the depths to reflectors should be known independently. Then the borehole source point is to be treated as a reflection point for surface measurements, and times and offsets need be doubled to yield two-way time for use with surface seismic data.

10.1.1 Elastic Constants Determination using Layer Stripping Method

Byun and Corrigan (1990) analyzed the correlation of *anisotropy factor and vertical velocity* with lithology. Confirming a similar result as in Byun et al. (1989) they found that the degree of anisotropy, combined with vertical velocity, might be a valuable *lithology indicator. Vertical velocity* itself is not a good *lithology indicator* to discriminate sands from shales. The anisotropy factor, defined as the ratio of horizontal to vertical velocity of a P wave, i.e. $\sqrt{A/\rho}/\sqrt{C/\rho}$ may be used to distinguish sand from shale. Further, Byun and Corrigan used data from six geophone levels between 61 to 655 m for offset VSP geometry, and also obtained zero-offset SH-wave VSP data to get information on the vertically traveling SH-wave. Multiple offset and zero offset data were measured and analyzed to determine the required parameters. Values of V_h and V_γ for P-waves for six geophone levels were obtained by *semblance analysis* (refer to Fig. 8.6). Values of 't_0' were obtained from zero-offset VSP. The travel time parameters used to invert for five elastic coefficients are given in Table 10.1.

Table 10.1. Traveltime parameters for the Sulfur Springs test site (Source: Byun and Corrigan 1990. From Geophysics with permission)

Layer no	Depth (ft)*	SH-wave		P-wave		
		T_0 (S)	U_h (ft/s)*	T_0 (s)	V_h (ft/s)*	V_γ (ft/s)*
1	1050 (320)	0.4861	-	0.1676	7360 (2243)	6480 (1975)
2	1250 (381)	0.5545	-	0.1928	7520 (2292)	6680 (2036)
3	1450 (442)	0.6170	-	0.2185	7560 (2304)	6920 (2109)
4	1700 (518)	0.7100	-	0.2524	7680 (2341)	7000 (2134)
5	1950 (594)	0.8078	-	0.2865	7800 2377	7120 (2170)
6	2149 (655)	0.8510	-	0.3069	8400 (2560)	7400 (2256)

* Numbers in parentheses represent meters for depths and m/s for velocities

The least-mean-squares (LMS) error analysis for layer 6 is given as an illustrative example in Fig. 10.2.

Figure 10.2 shows the velocity analysis of P-wave travel times after elastic constants of the overlying 5-layers were found in a *layer-stripping mode*. It shows

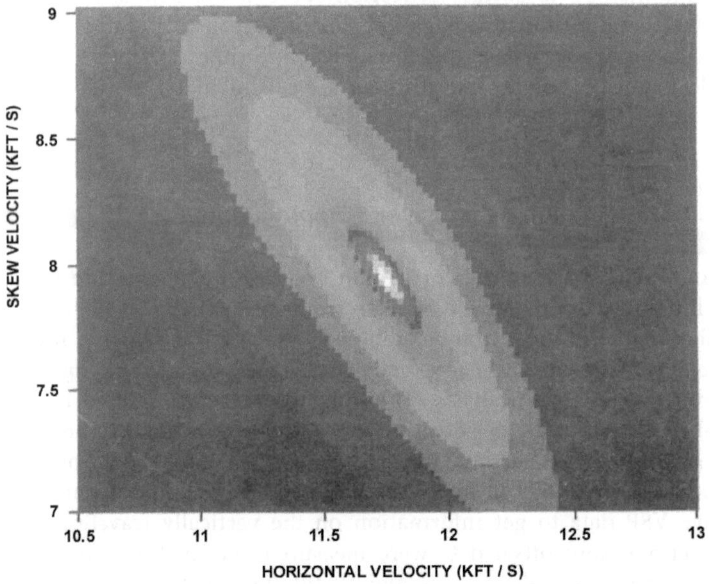

Fig. 10.2. Velocity analysis of P-wave travel times for layer 6 showing the mean squared error between the measured and model t–x curves. *Least dark* is the minimum error and *most dark* is the maximum error (Source: Byun and Corrigan 1990)

Table 10.2. Inversion model for the Sulfur Springs test site (Source: Byun, Corrigan 1990. From Geophysics with permission)

Layer no	Representative material	Depth (ft)*	Elastic constants (ft/s)*				
			$\sqrt{A/\rho}$	$\sqrt{C/\rho}$	$\sqrt{F/\rho}$	$\sqrt{L/\rho}$	$\sqrt{M/\rho}$
1	Shaly soil	1050 (320)	7400 (2256)	6295 (1919)	5575 (1699)	2160 (658)	–
2	Sands and limestone	1250 (381)	8400 (2560)	7693 (2345)	6700 (2042)	2925 (892)	–
3	Sands and shale	1450 (442)	8025 (2446)	7782 (2372)	6825 (2080)	3200 (975)	–
4	Shale	1700 (518)	8400 (2560)	7375 (2248)	6350 (1935)	2688 (815)	–
5	Shale	1950 (594)	8600 (2621)	7331 (2234)	6725 (2050)	2556 (779)	–
6	Chalk	2149 (655)	11750 (3581)	9755 (2973)	7950 (2423)	4606 (1404)	–

* Numbers in parentheses represent meters for depths and m/s for elastic constants

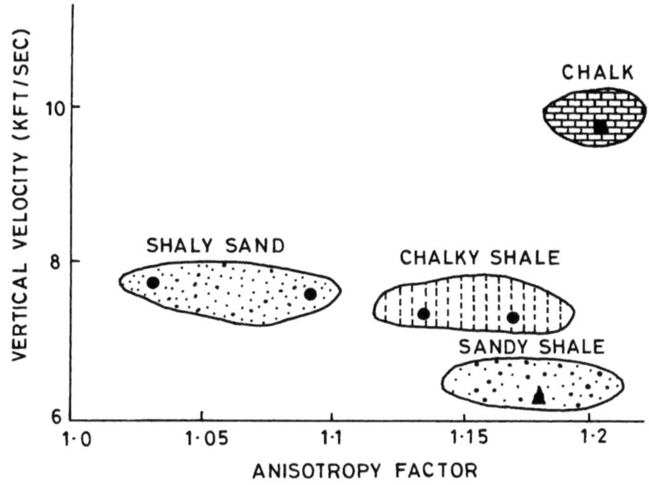

Fig. 10.3. Cross-plot of the vertical velocity and anisotropy factor for layer intervals at the Sulfur Springs test site. Data are grouped according to lithology types (Source: Byun and Corrigan 1990)

the mean-squared error between measured and model 't–x' curves for a 2-D array of horizontal velocity $\sqrt{A/\rho}$ and skew velocity $\sqrt{F/\rho}$ for layer 6. Two velocity parameters V_h and V_γ were found for each layer in Byun and Corrigan (1990). The values of $\sqrt{C/\rho}$ and $\sqrt{L/\rho}$ were obtained from zero-offset P and SH-wave data. The results of the inversion process are given below in Table 10.2.

Figure 10.3 shows a cross-plot of anisotropy factor and vertical velocity for different layer intervals. It shows that the cross-plot may be helpful for identifying lithology.

10.2 Inversion using Non-Linear Fit to t^2-x^2 Curves

This method is based on reflection survey travel time data to known reflector depths or multiple offset VSP data for quasi-P and SH-waves. It also makes use of the density log for retrieving *elastic stiffness coefficients* for *azimuthally isotropic* or *anisotropic media.*

In this method, the initial step is to estimate bulk velocities $V_v(Z)$, $V_\gamma(Z)$ and $V_h(Z)$ to each reflector. Here $V_v(Z)$ is the average vertical velocity; $V_\gamma(Z)$ and $V_h(Z)$ are skew and horizontal moveout velocities, respectively. The average vertical velocity V_v is obtainable from zero-offset information of the travel-time curve. Horizontal moveout velocity V_h and skew moveout velocity V_γ may be obtained either from

- *semblance analysis* along skewed hyperbolic travel time curves as explained in the previous section (refer to Fig. 10.2), or

- by finding a *non-linear fit* of Eq. (8.3) (refer to Subsection 8.1.1) to the observed curves, if the travel time offset curves are available.

Thus, application of the above steps to P and SH waves data for a seismic line yields bulk velocities for each reflector depth. Subsequently, interval velocities may be determined from Eqs. (10.3) to (10.5) as defined below.

If Z_k (k=1,2,...n) denote the reflector depths (assumed known) we have, from Eqs. (8.7a), (8.7b) and (8.7c)

(i) Vertical interval velocities V_{vk} [from Eq. (8.7a)]

$$V_{vk} = \frac{T_0(Z_k)V_v(Z_k) - T_0(Z_{k-1})V_v(Z_{k-1})}{T_0(Z_k) - T_0(Z_{k-1})} \tag{10.3}$$

From Eq. (10.3) vertical interval velocities for both quasi-P and SH wave may be determined, using zero offset information.

(ii) Skew moveout interval velocities $V_{\gamma k}$ [from Eq. (8.7b)]

$$V_{\gamma k}^2 = \frac{T_0(Z_k)V_\gamma^2(Z_k) - T_0(Z_{k-1})V_\gamma^2(Z_{k-1})}{T_0(Z_k) - T_0(Z_{k-1})} \tag{10.4}$$

Equation (10.4) gives skew moveout velocities for P and SH-waves. For the SH wave, the skew moveout velocity $V_{\gamma k}$ equals the horizontal moveout velocity V_{hk}.

(iii) Horizontal moveout interval velocity V_{hk} [from Eq. (8.7c)]

$$\frac{1}{V_{hk}^2} = \frac{1}{V_{\gamma k}^2} + \frac{f_k - f_{k-1}}{S_k} \tag{10.5}$$

where

$$f_k = \frac{T_0(Z_k)V_\gamma^4(Z_k)}{8}\left[\frac{4V_\gamma^4(Z_k)}{V_v^2(Z_k)}\left(\frac{1}{V_h^2(Z_k)} - \frac{1}{V_\gamma^2(Z_k)}\right) - 1\right] + \frac{1}{4}\sum_{m=1}^{k} t_m V_{\gamma m}^4$$

and

$$S_k = t_k \frac{V_{\gamma_k}^8}{V_{v_k}^2}$$

Equations (10.3) to (10.5) were derived in Sena (1991).

Interval velocities V_v, V_γ, and V_h for P and SH waves are related to elastic parameters C_{11}, C_{33}, C_{44}, C_{66} and C_{13}. These are given separately for azimuthal isotropy and anisotropy in the following subsections.

10.2.1
Interval Velocity and Elastic Constants for Azimuthal Isotropy

For a single layer, we have
(i) for quasi-P motion

$$V_v^2 = \frac{C_{33}}{\rho} = \alpha_0^2 \tag{10.6 a}$$

$$V_\gamma^2 = \alpha_0^2(1 + 2\delta) \tag{10.6 b}$$

$$V_h^2 = \frac{C_{11}}{\rho} = \alpha_0^2(1 + 2\varepsilon) \tag{10.6. c}$$

(ii) for quasi SV motion

$$V_v^2 = \beta_0^2 = \frac{c_{44}}{\rho} \tag{10.7 a}$$

$$V_\gamma^2 = \beta_0^2\left[1 + 2\left(\frac{\alpha_0}{\beta_0}\right)^2(\varepsilon - \delta)\right] \tag{10.7 b}$$

$$V_h^2 = \frac{c_{44}}{\rho} \tag{10.7 c}$$

(iii) for SH-motion

$$V_v^2 = \frac{C_{44}}{\rho} = \beta_0^2 \tag{10.8 a}$$

$$V_\gamma^2 = V_h^2 = \frac{C_{66}}{\rho} \tag{10.8 b}$$

ε, γ and δ are Thomsen's (1986) anisotropy parameters for a single layer defined as

$$\varepsilon = \frac{c_{11} - c_{33}}{2c_{33}}$$

$$\gamma = \frac{c_{66} - c_{44}}{2c_{44}}$$

$$\delta = \frac{(c_{13} + c_{44})^2 - (c_{33} - c_{44})^2}{2c_{33}(c_{33} - c_{44})}$$

ρ is the density.

10.2.2
Interval Velocity, Elastic Constants and Direction of Symmetry Axis for Azimuthal Anisotropy

When a layer possesses azimuthal anisotropy, the orientation of the horizontal axis of symmetry with respect to the seismic line needs to be calculated in addition to the five elastic stiffness constants. Sena (1991) showed that from travel time measurements along two seismic lines, it is possible to calculate the orientation of the horizontal axis of symmetry (Fig. 10.4a). This is possible by calculating the value of $a_1^{(k)}$ expressed in terms of $V_{\gamma k}$ and V_{vk} through Eqs. (8.4a) and (8.4b) (refer to Subsection 8.1.1) as:

$$a_1^{(k)} = \frac{1}{V_{\gamma_k}^2} - \frac{1}{V_{v_k}^2} \tag{10.9}$$

In Fig. 10.4a the seismic lines 1 and 2 make angles ϕ' and $\phi' + \Delta$ from the x-axis, respectively. The axis of symmetry shown in this figure depicts the orientation of horizontal axis of symmetry in an azimuthally anisotropic layer (say the k-th layer) at depth. Letting that $a_1^{(k)}(\phi')$ and $a_1^{(k)}(\phi' + \Delta)$ are the estimates of the coefficient $a_1^{(k)}$ for seismic lines 1 and 2 respectively, then the angle between the axis of symmetry and seismic line 1 is given by

$$\phi' - \phi = \text{arc tan}\left\{\frac{1}{\sin\Delta}\left[\cos\Delta - \sqrt{\frac{a_1^{(k)}(\phi' + \Delta)}{a_1^{(k)}(\phi')}}\right]\right\} \tag{10.10}$$

[Sena 1991, Eq. (45)]

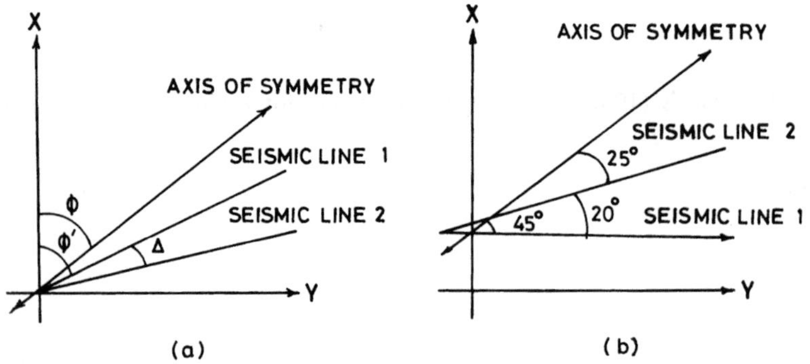

Fig. 10.4a, b. a Plan view of a surface reflection survey. The horizontal axis of symmetry corresponds to a given azimuthally anisotropic layer at depth. Travel time data from *seismic lines 1 and 2* are used to obtain anisotropy parameters of the multilayered anisotropic medium; **b** plan view of the surface reflection survey for the synthetic model data of Table 10.3. The horizontal axis of symmetry shown corresponds to the second layer (Source: Sena 1991)

Equation (10.10) may be utilized to determine the azimuthal orientation of the horizontal axis of symmetry with respect to seismic line 1 for the k-th layer.

The next step in inversion is the determination of interval velocities V_v, V_γ and V_h. Knowing the values of ϕ'–ϕ, these velocities may be calculated from the following formulae for velocities for azimuthal anisotropy (Sena 1991):

(i) Quasi P-wave:

$$V_v^2 = \frac{C_{11}}{\rho} \tag{10.11 a}$$

$$V_\gamma^2 = \alpha_0^2\left[1 + 2\varepsilon\sin^2(\phi\prime - \phi) - 2(\varepsilon - \delta)\cos^2(\phi' - \phi)\right] \tag{10.11 b}$$

$$V_h^2 = \alpha_0^2\left\{1 + 2\sin^2(\phi' - \phi)\left[\varepsilon\sin^2(\phi' - \phi) + \delta\cos^2(\phi' - \phi)\right]\right\} \tag{10.11 c}$$

(ii) Quasi-SV motion:

$$V_v^2 = \beta_0^2 = \frac{C_{44}}{\rho} \tag{10.12 a}$$

$$V_\gamma^2 = \beta_0^2\left[1 + 2\left(\frac{\alpha_0}{\beta_0}\right)^2(\varepsilon - \delta)\cos^2(\phi' - \phi)\right] \tag{10.12 b}$$

$$V_h^2 = \beta_0^2\left[1 + 2\left(\frac{\alpha_0}{\beta_0}\right)^2(\varepsilon - \delta)\sin^2(\phi' - \phi)\cos^2(\phi' - \phi)\right] \tag{10.12 c}$$

(iii) SH motion:

$$V_v^2 = \frac{C_{66}}{\rho} \tag{10.13 a}$$

$$V_\gamma^2 = V_h^2 = \beta_0^2\left[1 + 2\gamma\sin^2(\phi' - \phi)\right] \tag{10.13 b}$$

From Eqs. (10.11) to (10.13) – the expressions for interval velocities, it is possible to calculate the elastic constants for an azimuthally anisotropic layer, knowing 'ρ' from the density log.

The above-described method was applied by Sena (1991) for a 3-layer azimuthally anisotropic/isotropic layer model for testing the error in estimates of inverted anisotropy parameters. The top and bottom layers were taken as azimuthally isotropic and the middle layer as azimuthally anisotropic – the horizontal axis of symmetry making an angle of 45° with respect to seismic line 1 (Fig. 10.4b). Seismic line 2 was used only to estimate the direction of the horizontal axis of symmetry; the five velocities were estimated by travel time measurements on seismic line 1. Table 10.3 gives the velocities and depths used in the analysis.

The travel time curves were generated in Sena (1991) using paraxial ray-tracing code. Two way vertical travel time 'T_0' was determined from vertical incidence ray

Table 10.3. Azimuthally anisotropic/isotropic layer model. Second layer is azimuthally anisotropic with a horizontal axis of symmetry at 45° from the X-axis. Depth in meters and elastic constants (velocities) in m/s (Source: Sena 1991. From Geophysics with permission)

Layer	Material	Depth	$\sqrt{C_{11}/\rho}$	$\sqrt{C_{13}/\rho}$	$\sqrt{C_{33}/\rho}$	$\sqrt{C_{44}/\rho}$	$\sqrt{C_{66}/\rho}$
1	Anisotropic shale	610	3016	2745	1724	1508	1962
2	Limestone/ shale	1220	3824	3306	2074	1819	2259
3	Sandstone/ anisotropic shale	1830	3048	3009	1890	1654	1682

Table 10.4. Relative error (%), E_{ij} for inverted anisotropy parameters for the azimuthally anisotropic/isotropic layer model. Second layer has a true horizontal axis of symmetry at 45° from seismic line 1 (Source: Sena 1991. From Geophysics with permission)

Layer	Depth (m)	E_{11} (%)	E_{33} (%)	E_{13} (%)	E_{44} (%)	E_{66} (%)	Angle (deg) of symm. axis from seismic line 1
1	610	-0.106	0.827	0.124	0.812	0.005	
2	1220	0.012	4.290	-4.393	1.026	0.011	47
3	1830	1.007	-0.012	0.638	-0.026	2.424	

tracing. For a quasi-P wave the horizontal moveout velocity V_h and skew moveout velocities V_γ were determined by a Levenberg-Marquardt non-linear fit to the t^2–x^2 curves (Gill et al. 1988). For SH-waves, horizontal moveout velocity V_h (= Vg) was measured by a linear least square fit to the t^2–x^2 curves. The coefficient a_1 [refer to Eq. (10.9) was calculated for seismic line 1 and 2, and the orientation of the horizontal axis of symmetry with respect to seismic line 1 was found from Eq. (10.10). From the values of the bulk velocities for each reflector calculated for seismic line 1 and the orientation of the horizontal axis of symmetry with respect to seismic line 1, Sena (1991) estimated the interval velocities of the model. Errors in the estimates were subsequently found. This is given here as Table 10.4.

The vertical velocities for quasi P- and SH waves ($\sqrt{C_{11}/\rho}$ and $\sqrt{C_{66}/\rho}$) agree very well, and the direction of the axis of symmetry is also estimated accurately. The maximum error is 4.3% in $\sqrt{C_{13}/\rho}$ for the second layer. The computation of $\sqrt{C_{13}/\rho}$ involves estimation of $\sqrt{C_{33}/\rho}$ and $\sqrt{C_{44}/\rho}$. The errors in their computation accumulate to a larger error in $\sqrt{C_{13}/\rho}$. By interchanging the role of seismic lines, i.e. using seismic line 2 as seismic line 1, Sena (1991) found that the largest error in $\sqrt{C_{13}/\rho}$ was reduced to 1.7%. The orientation of the axis of symmetry was also estimated accurately as before. The reason for less error in the second computation is that the deviation of the rays from the *sagittal plane* is less.

The Sena (1991) method of inversion is given for *weak-anisotropy approximation*. The accuracy of this method depends on the measurement of moveout velocities from travel time curves. Further, this method should have limited use in the presence of strongly dipping layers and lateral heterogeneity because of its intrinsic limitations of ray theory. This method gives a quantitative measure of anisotropy, which should prove very helpful for *lithology discrimination.*

10.3 Inversion using Travel Times

Travel time inversion is a typical *non-linear inversion problem*, which has been discussed extensively (c.f. Backus and Gilbert 1970; Jackson 1972, 1979; Aki and Richards 1980; Novotny 1981; Tarantola and Valette 1982; Lines and Treitel 1984; Tarantola 1987; Lines 1988).

The *linear approximation* of the non-linear inverse problem makes use of Taylor series expansion. We give below two such approximations leading to the application of *Damped Least Square formulation* and *Jacobian Matrix formulation.* The application of these methods is shown on combined VSP and surface reflection data, and on VSP data respectively. The theory and results are based on investigations in Chiu and Stewart (1987) and Lee (1990). Application of the method using a Jacobian matrix of ray velocity (Okoye et al. 1996) for inversion of elastic parameter δ^* and α_0 of a transversely isotropic medium is further discussed.

10.3.1 Damped Least Square Formulation

Let T_i be the observed travel time for a given source-receiver location. Then T_i can be approximated by first-order Taylor-series expansion (Wiggins et al. 1976; Jackson 1979):

$$T_i = t_i\left(x^0\right) + \sum_{j=1}^{m} \frac{\partial t_i}{\partial x_j}\bigg|_{x=x_0} \left(x_j - x_j^0\right) \tag{10.14}$$

[Chiu and Stewart 1987, Eq. (6)]
where T_i is the observed travel time, x_j^0 is the initial estimate of the j-th model parameter, x_j is the updated j-th model parameter and t_i (x^0) is the calculated travel time from the initial model.

Equation (10.14) may be rearranged as,

$$T_i - t_i\left(x^0\right) = \sum_{j=1}^{m} \frac{\partial t_i}{\partial x_j}\bigg|_{x=x_0} \left(x_j - x_j^0\right) = \Delta t_i \tag{10.15}$$

Let N denote the total number of observations and m the total number of model parameters, we may then define

$$\Delta x_j = x_j - x_j^0, \; j = 1, \ldots\ldots\ldots m \tag{10.16 a}$$

$$\Delta t_i = T_i - T_i(x^0), \; i = 1, \ldots\ldots\ldots N \qquad (10.16\,b)$$

and

$$A_{ij} = \frac{\partial t_i}{\partial x_j}, \; i = 1, \ldots\ldots\ldots N; \; j = 1, \ldots\ldots\ldots m \qquad (10.16\,c)$$

Equation (10.15) would form a set of linear equations for the observed data. A_{ij} consists of partial derivatives of travel time function with respect to velocities and model parameters defining the interface for *simultaneous inversion.*

With the notation in Eqs. (10.16a–c), Eq. (10.15) may be written in vector and matrix notation as,

$$\mathbf{\Delta t} = \mathbf{A\Delta x} \qquad (10.17)$$

[Chiu and Stewart 1987, Eq. (10)]

In Eq. (10.17) $\mathbf{\Delta t}$ and $\mathbf{\Delta x}$ are the (N×1) and (m×1) vectors respectively, and $\mathbf{A}$ is the (N×m) matrix.

The above formulation in Chiu and Stewart (1987) is quite general, and provides a method for inverting 2-D and 3-D velocity structure, and model parameters. For example, in Chiu and Stewart (1987), the interfaces separating media with constant velocities were simulated by the equation

$$Z_i = Z_i^0 + F_i(X, Y) \qquad (10.18)$$

where Z_i^0 is the depth of the interface at the origin (X=0, Y=0) and $F_i(X, Y)$ is a function of X, and Y. $F_i(X, Y)$ may represent a sinusoidal or a polynomial surface.

In seismic applications, the number of observations is more than the model parameters, and in this case a solution may be sought through Damped Least Square formulation. In the *Damped Least Square inversion*, the model parameters are updated by minimizing the difference between observed times and calculated times from a guessed Earth structure, and the solution is,

$$\mathbf{\Delta x} = \left(\mathbf{A}^T\mathbf{A} + \mathrm{KI}\right)^{-1}\mathbf{A}^T\mathbf{\Delta t} \qquad (10.19)$$

[Chiu and Stewart 1987, Eq. (11)]

In $(A^TA + KI)$, constant K (damping factor) is added to the main diagonal of the matrix A^TA to make matrix $\mathbf{A}$ non-singular and to stabilize the inversion process. There are a number of ways to solve the system of Eq. (10.19) (see, for example, Lawson, Hanson 1974). One approach is the use of the *Singular Value Decomposition method.* For details of this method, reference may be made to Jackson (1979) and Aki and Richards (1980).

Chiu and Stewart (1987) tested their algorithm of *travel time inversion* through *tomographic determination* of velocity and interfaces. The basic idea in their *tomographic reconstruction* is to guess an initial model of geologic structure, trace the rays through this initial guessed model and calculate the travel time for a given seismic source-receiver position. Chiu and Stewart (1987) used the algorithm of Chiu et al. (1986) for ray tracing. The mismatch between observed and calculated travel times is then used to update the initial model in an iterative manner.

Chiu and Stewart (1987) illustrated the working of their algorithms for surface seismic and VSP data individually and in a combined manner for 2-D and 3-D data. They also tested the effects of the presence of random noise in the data. For illustration, results of the 3-D analysis are presented below.

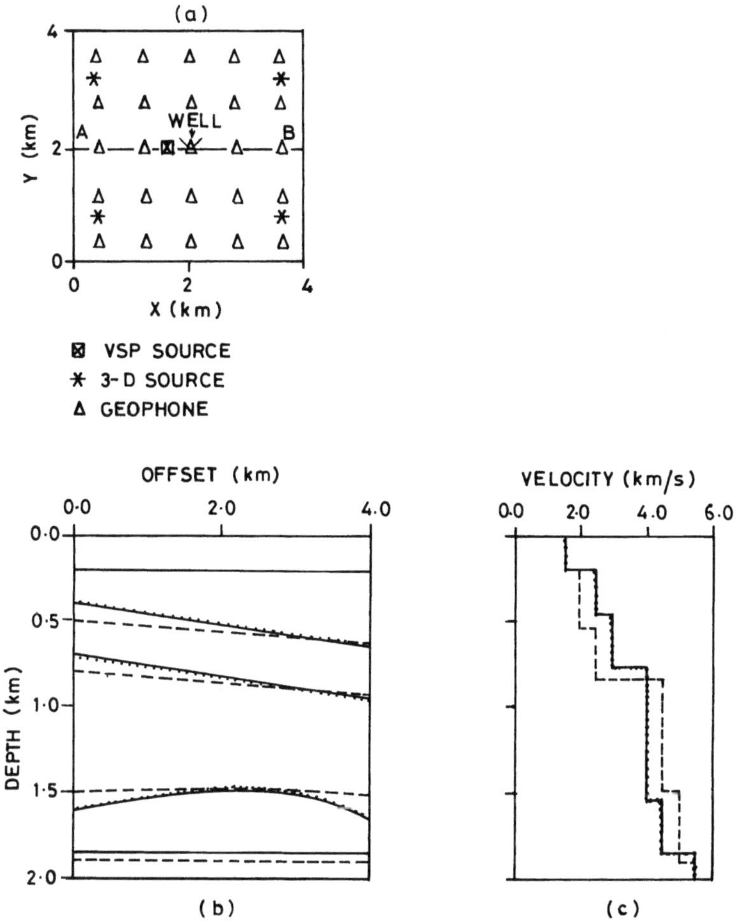

Fig. 10.5 a–c. a Configuration of the sources and receivers for both VSP and 3-D surface seismic data. The receiver spacing for the 3-D surface seismic was 80 m; for the VSP, 40 m; **b**, **c** inversion results for both VSP and 3-D surface seismic data. Random noise of 1, 2 and 10 ms was added to direct and reflected travel times of the VSP, and to the reflected travel times of the surface seismic, respectively. The *continuous line* (–), *dashed* (--) *line*, and *dotted* (...) *lines* are for the model, initial guess and result of the 8th interaction, respectively (Source: Chiu and Stewart 1987)

10.3.1.1
Use on Combined VSP and 3-D Surface Seismic Data

The model tested by Chiu and Stewart (1987) consists of five interfaces: the first one horizontal, the second and third are dipping planes with a 5° dip, the fourth one is a third order polynomial surface and the fifth is a horizontal plane. Figure 10.5a illustrates the source and receiver configurations in both surface and VSP data.

Tomographic inversion was done in this 3-D model. The depth to the first interface was assumed to be fixed. Here the inversion results were obtained in two cases: (i) considering surface data alone, and (ii) considering surface seismic and VSP data together, with or without random noise in travel time data. The results of inversion for combined VSP and 3-D surface seismic data for a cross-section along line AB in Fig. 10.5a is shown in Fig. 10.5b, c.

Chiu and Stewart (1987) made the following general observation about the use of VSP and surface seismic data together for inversion (Chiu and Stewart 1987, p 1095):

VSP observations appear to provide the constraint that determines the interval velocities needed to resolve the ambiguity in surface seismic data. Therefore, inverting both sets of data can result in a more constrained geological model in depth.

They tested their algorithm for a transversely isotropic medium and came to the conclusion that offset VSP surveys can be used to estimate the velocities in transversely isotropic media, and the velocity structure so determined may be valuable inputs for migration and amplitude inversion processes.

10.3.2
Jacobian Matrix Formulation: Simultaneous Determination of All Layer Parameters

Let 't_N' be the travel time from the N-th reflector. Then, this may be expressed in a Taylor series approximation of the form

$$t_N\left(\mathbf{P}^{n+1}\right) = t_N(\mathbf{P}^n) + \sum_{j=1}^{L}\left(\frac{\partial t_N}{\partial P_j}\right)_{\mathbf{P}=\mathbf{P}^n} \delta P_j \qquad (10.20)$$

[Lee 1990, Eq. (1)]
where
t_N: travel time from the Nth reflector,
$\mathbf{P}^{n+1}$: $(n+1)^{th}$ estimation of parameters with dimension L and
$\delta \mathbf{P}_j$: the parameter update

$$= \left(P_j^{n+1} - P_j^n\right) \qquad (10.21)$$

The least square solution of δP_j in Eq. (10.20) for the given observation O_i $(i=1, 2,\ldots.M)$ is found by minimizing the mean squared error E:

$$
\begin{aligned}
E &= \sum_{i=1}^{M} \left[O_i - t_i\left(\mathbf{P}^{n+1}\right)\right]^2 \\
&= \sum_{i=1}^{M} \left[O_i - t_i(\mathbf{P}^n) - \sum_{j=1}^{L} \left(\frac{\partial t_i}{\partial P_j}\right)_{\mathbf{P}=\mathbf{P}^n} \delta P_j\right]^2
\end{aligned}
\tag{10.22}
$$

The solution of Eq. (10.22) from the least square method is

$$
\delta \mathbf{P} = \mathbf{G}^{-1}(\mathbf{O} - \mathbf{t}^n) \tag{10.23}
$$

where $\mathbf{G}^{-1}$ is the generalized inverse of the M×L *Jacobian matrix*, **O** is the M dimensional column vector of observations, $\mathbf{t}^n$ is the M dimensional column vector of the modeled travel time at the n-th iteration, and **P** is the L dimensional column vector of the layer parameters.

Accurate estimation of the Jacobian matrix in Eq. (10.23) is a very important step for solving a *non-linear inversion problem*. Lines et al. (1984), for example, used a finite difference approximation of the Jacobian matrix. Lee (1990) derived explicit expressions of the Jacobian matrix. Using explicit expression for the Jacobian matrix, and by suitable parameterization, the problem reduces to iteratively computing the inverse of a reduced size matrix. Lee (1990) has illustrated his method by considering isotropic layers separated by dipping interfaces (Fig. 10.6). In this figure the i-th layer is characterized by slowness S_i. The dipping layers are defined by using slopes a_i and intercepts b_i as

$$
Z_i = a_i X_i + b_i \tag{10.24}
$$

Equation (10.23) has been put in Lee (1990) in a suitable form for computation.

Considering that the reflections are from N layers and each of these reflected arrivals is recorded by M geophones, Eq. (10.23) may be written as,

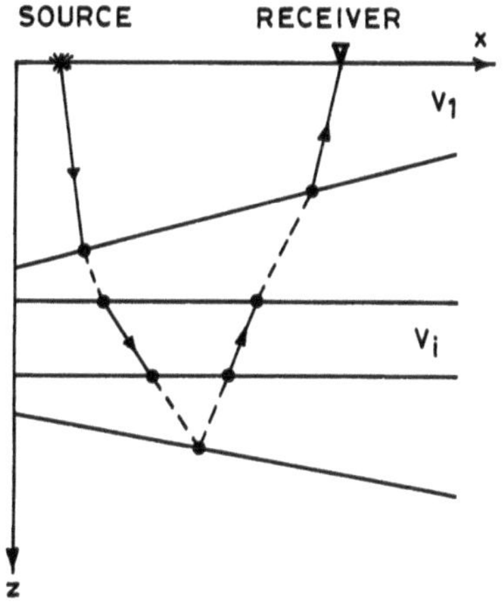

Fig. 10.6. Schematic ray path for a reflection from dipping reflectors. V_i is the interval velocity. *Dots* (.) represent intersections of downgoing and upgoing waves with reflectors

$$\mathbf{G}\delta\mathbf{P} = \delta\mathbf{m} \tag{10.25}$$

[Lee 1990, Eq. (8)]

where $\delta\mathbf{P}$ is a column vector defined as

$$\delta\mathbf{P} = \mathrm{Col}(\delta\mathbf{P}_1, \delta\mathbf{P}_2, ...\delta\mathbf{P}_N) \tag{10.26 a}$$

and $\delta\mathbf{P}_i$ is a three component column vector

$$\delta\mathbf{P}_i = \mathrm{Col}(\delta s_i, \delta a_i, \delta b_i) \tag{10.26 b}$$

Further, $\delta\mathbf{m}$ is a column vector given by

$$\delta\mathbf{m} = \mathrm{Col}(\delta\mathbf{m}_1, \delta\mathbf{m}_2, ...\delta\mathbf{m}_N) \tag{10.27}$$

and $\delta\mathbf{m}_i$ itself is a M dimensional vector given by

$$\delta\mathbf{m}_i = \mathrm{Col}\left(O_i^1 - T_i^1, O_i^2 - T_i^2......O_i^M - T_i^M\right) \tag{10.28}$$

$\delta\mathbf{m}_i$ is the difference between the observed and computed arrival times. $\delta\mathbf{P}$ is a column vector of length 3 N and $\delta\mathbf{m}$ is a column vector of length MN.

Equation (10.25) may be written explicitly as

$$\begin{bmatrix} \mathbf{G}_{11} & & & 0 \\ \mathbf{G}_{12} & \mathbf{G}_{22} & & \\ \cdots\cdots & \cdots\cdots & \cdots\cdots & \cdots\cdots \\ \mathbf{G}_{1N} & \mathbf{G}_{2N} & \cdots\cdots & \mathbf{G}_{NN} \end{bmatrix} \begin{bmatrix} \delta\mathbf{P}_1 \\ \delta\mathbf{P}_2 \\ \cdots\cdots \\ \delta\mathbf{P}_N \end{bmatrix} = \begin{bmatrix} \delta\mathbf{m}_1 \\ \delta\mathbf{m}_2 \\ \cdots\cdots \\ \delta\mathbf{m}_N \end{bmatrix} \tag{10.29}$$

Matrix elements G_{ij} are defined as

$$\mathbf{G}_{ij} = \begin{bmatrix} \frac{\partial T_j^1}{\partial s_i} & \frac{\partial T_j^1}{\partial a_i} & \frac{\partial T_j^1}{\partial b_i} \\ \cdots\cdots & \cdots\cdots & \cdots\cdots \\ \frac{\partial T_j^M}{\partial s_i} & \frac{\partial T_j^M}{\partial a_i} & \frac{\partial T_j^M}{\partial b_i} \end{bmatrix} \tag{10.30}$$

Matrix elements G_{ij} consists of derivatives of the travel time reflected from the j-th interface with respect to i-th model parameters. G in Eq. (10.29) is in lower-triangular form. The iterative solutions of the general inverse are

$$\begin{aligned} \delta\mathbf{P}_1 &= \mathbf{G}_{11}^{-1}\delta\mathbf{m}_1 \\ \delta\mathbf{P}_2 &= \mathbf{G}_{22}^{-1}(\delta\mathbf{m}_2 - \mathbf{G}_{12}\delta\mathbf{P}_1) \\ &\cdots\cdots\cdots\cdots\cdots\cdots \\ &\cdots\cdots\cdots\cdots\cdots\cdots \\ \delta\mathbf{P}_N &= (\mathbf{G}_{NN})^{-1}\left(\delta\mathbf{m}_N - \sum_{i=1}^{N-1}\mathbf{G}_{iN}\delta\mathbf{P}_i\right) \end{aligned} \tag{10.31}$$

Equation (10.31) is in a simple form, suitable for implementation in a personal computer.

The above formulation for a surface reflection is adaptable for transmitted VSP data with slight modifications. Since the arrival times in the first layer give information on velocity only, the column vector **P** may be rearranged as

$$\begin{aligned} \mathbf{P}_1 &= \mathrm{Col}(s_1) \\ \mathbf{P}_2 &= \mathrm{Col}(s_2, a_1, b_1) \\ &\ldots\ldots\ldots\ldots\ldots\ldots \\ \mathbf{P}_N &= \mathrm{Col}(S_N, a_{N-1}, b_{N-1}) \end{aligned} \tag{10.32}$$

Because of the change in parameters, the matrix elements in Eq. (10.29) would change as follows:
G_{11} $(1\times M)$ = derivative with respect to s_1,
G_{ij} $(3\times M)$ = derivative with respect to a_i, s_{i-1} and b_{i-1},
G_{NN} $(3\times M)$ = derivative with respect to s_N, a_{N-1}, b_{N-1}.

In a VSP application, M changes according to current estimates of reflector depth. For inverting surface shot gathers M is fixed.

The method of Lee (1990) for inverting travel times has the following distinct features:

(i) This method estimates all layer parameters simultaneously. This is distinct advantage over the *layer stripping method* (c.f. Gjøystdal and Ursin 1981; Lines et al. 1984) in which the cumulative error resulting from upper layers accumulate.
(ii) The method uses explicit expressions of derivations of travel times with respect to layer parameters and is computationally efficient and accurate.
(iii) Since all layer parameters are estimated simultaneously, the method is suitable for inversion of VSP data with unknown reflector depths.

10.3.2.1
Example of Travel Time Inversion Applied To Offset VSP Data

Lee (1990) tested his travel time inversion algorithm on offset VSP data, considering only transmitted first-arrival times. The VSP data were obtained from two explosive sources located 2 km North and South of the well site. First arrival times are shown in Fig. 10.7 a for the two data sets. There exists a marked difference between the two data sets.

Through the algorithm developed for travel time inversion an explanation for the observed time discrepancy was explored in Lee (1990). For the purpose of inversion, initial estimates of the parameters were made from the two VSP data sets assuming flat layers. Initial model parameters and the ray-tracing results are given in Fig. 10.7 b. The root mean square (rms) travel time error between the observed data and the initial model was approximately 63 ms. With the application of travel time inversion, the root mean square (rms) travel time error was reduced to a much smaller amount of approximately 4 ms. The final model found is shown in Fig. 10.7 c. The observed discrepancy between travel times between two offset VSP may be explained by the dip of the layers. Further, the computed

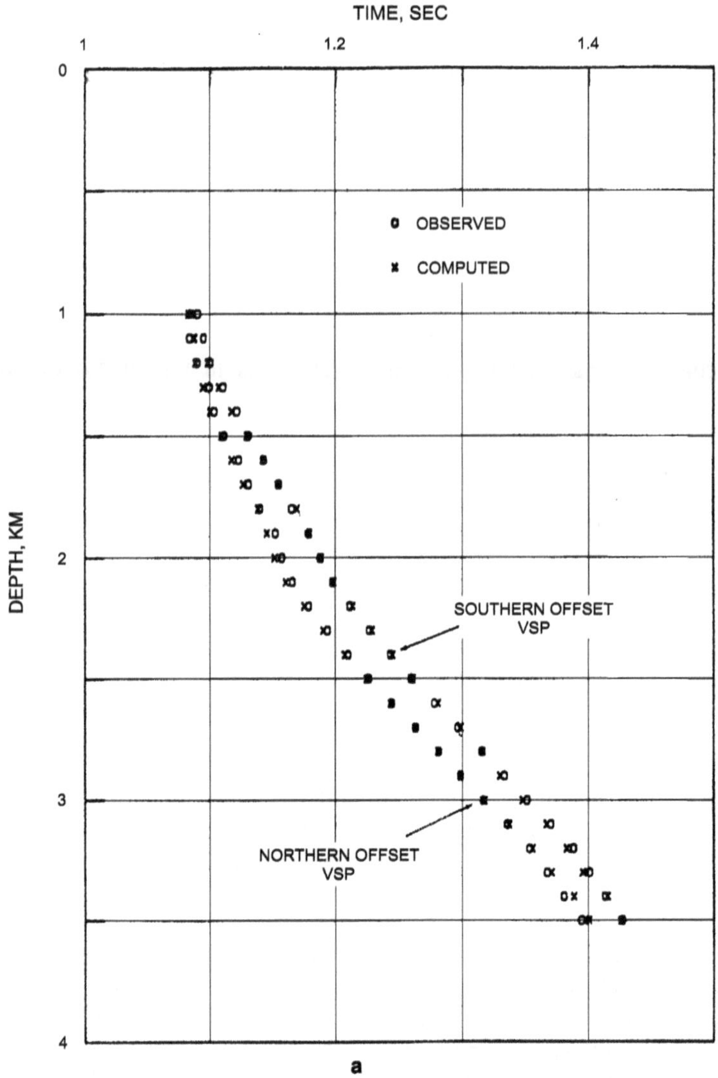

Fig. 10.7 a–c. **a** First arrival times for offset VSP data as acquired at Bekes Basin, Hungary. *Crosses* represent actual arrival times and *circles* represent the arrival times of the final model shown in Fig. 10.7 c; **b, c** ray tracing for the offset VSP data with ±2 km source offset at Bekes Basin, Hungary. *Numbers shown at right* of each panel represent interval velocities in km/s (**b** initial estimate; **c** final estimate) (Source: Lee 1990, from Geophysics with permission)

arrivals for the final model are shown in Fig. 10.7 a as circles. Computed results match very well with the observed arrival time. Although the match is very good, it should be noted that it is one of the many possible models, which can explain the observed time discrepancy, and a feasible geological model at the well site should be found using other geological and geophysical information.

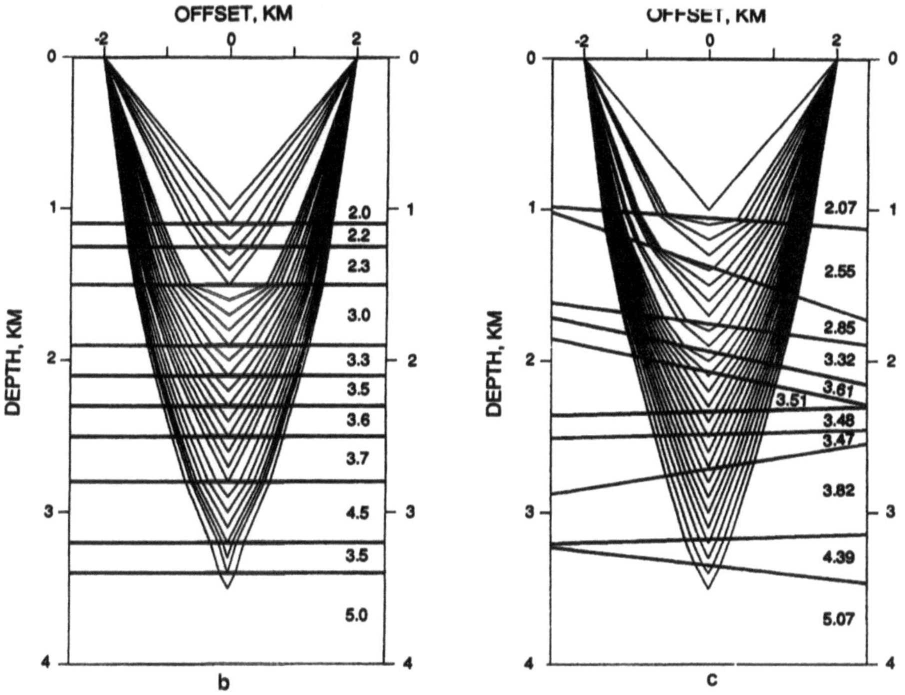

Fig. 10.7 b, c

Lee (1990) concluded from the study of VSP data that, in general, travel time inversion for estimating all three layer parameters (interval velocity, dip and depth) cannot be accomplished from a single offset VSP survey, even if there are no measurement errors. To get reliable results on layer parameters, more than one offset VSP data is necessary.

10.3.2.2
Method to Recover Elastic Constants

Okoye et al. (1996) gave a least square iterative inversion technique for the determination of δ^* and a_0 of a transversely isotropic material in the laboratory. The parameter δ^* is a very important parameter for near vertical P-wave propagation in seismic exploration work. Podio (1968) described a method to determine elastic constant c_{13} and δ^* of a T.I. material by making a measurement of P-wave group velocity at 45° to the *symmetry axis*. This method has its inherent ambiguity. The inversion technique described in Okoye et al. (1996) is an improvement over the method proposed in Podio (1968). Since δ^* is very sensitive to small differences between vertical and oblique travel time measurements, accurate picking of first arrival travel time is essential in the method proposed by Okoye et al. (1996). Also, reasonably wide angular coverage of the data is found necessary for this determination. A computer program is given in Okoye et al. (1996), which yields the values of δ^* and a_0 using observed times as input.

The *anisotropic inversion technique* of Okoye et al. (1996) uses exact expressions for ray and phase velocity [refer to Eqs. (5.54 a) and (5.51 a) renumbered as 10.33 a, b] in a transversely isotropic medium

$$U^2(\phi(i)) = V^2(i) + \left(\frac{dV}{di}\right)^2 \tag{10.33 a}$$

and

$$V^2(i) = \alpha_0^2\left[1 + \varepsilon \sin^2 i + D^*(i)\right] \tag{10.33 b}$$

where

$$D^*(i) \equiv \frac{1}{2}\left(1 - \frac{\beta_0^2}{\alpha_0^2}\right) \times \left\{\left[1 + \frac{4\delta^*}{\left(1 - \beta_0^2/\alpha_0^2\right)^2}\sin^2 i \cos^2 i + \frac{4\left(1 - \frac{\beta_0^2}{\alpha_0^2} + \varepsilon\right)\varepsilon}{\left(1 - \beta_0^2/\alpha_0^2\right)^2}\sin^4 i\right]^{\frac{1}{2}} - 1\right\}$$

Here α_0 and β_0 are vertical P- and SH-wave velocity and δ^* is a combination of elastic constants.

Travel-time measurements at different angles are used to determine the parameters δ^*, α_0 and β_0 by using Eqs. (10.33 a, b) on ray velocity. Ray velocity U is a function of the parameters δ^*, α_0 and β_0:

$$U = U(\alpha_0, \beta_0, \delta^*) \tag{10.34}$$

Equation (10.34), being a non-linear function of α_0, β_0 and δ^*, may be expanded in a Taylor Series with starting values of α_0, β_0 and δ^*. To the first order, Taylor Series expansion gives

$$U = U_0 + \frac{\partial U}{\partial \alpha_0}\Delta\alpha_0 + \frac{\partial U}{\partial \beta_0}\Delta\beta_0 + \frac{\partial U}{\partial \delta^*}\Delta\delta^* \tag{10.35}$$

If there are 'n' observational data, a Jacobian matrix with elements $\partial U^1/\partial\alpha_0$, $\partial U^1/\partial\beta_0$, $\partial U^1/\partial\delta^*$, -------- $\partial U^n/\partial\alpha_0$, $\partial U^n/\partial\beta_0$, $\partial U^n/\partial\delta^*$ is calculated:

$$[G] = \begin{bmatrix} \frac{\partial U^1}{\partial \alpha_0} & \frac{\partial U^1}{\partial \beta_0} & \frac{\partial U^1}{\partial \delta^*} \\ \cdots\cdots & \cdots\cdots & \cdots\cdots \\ \frac{\partial U^n}{\partial \alpha_0} & \frac{\partial U^n}{\partial \beta_0} & \frac{\partial U^n}{\partial \delta^*} \end{bmatrix} \tag{10.36}$$

(Okoye et al. 1996)

The least square inversion method is finally applied to update values of α_0, β_0 and δ^*. This inverse modeling procedure allows for controlled iterative adjustment for determining the required elastic parameters. Okoye et al. (1996) suggest using known 'β_0' values determined from inversion of SH-wave data by fitting an elliptical velocity function to the data. Thereby, the ray velocity becomes a function of only two variables α_0 and δ^*. This improves stability and convergence of the inversion procedure.

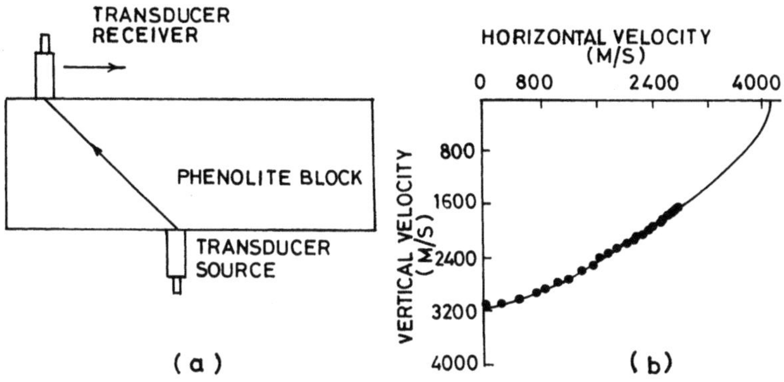

Fig. 10.8 a, b. a Measurement of the P-wave velocity function in a phenolite block. The source is centrally positioned at the bottom while the receiver moves at equal spacing along the profile direction centered over the source transducer; **b** plot of the velocity function for a phenolite generated from inversion results (*solid curve*) together with the associated observed velocity values (•). The degree of fit could be improved by an accurate picking of the first-arrival travel times (Source: Okoye et al. 1996, from Geophysics with permission)

The algorithm was verified by Okoye et al. (1996) by taking different starting parameters. It's application to laboratory data is illustrated through Fig. 10.8 a, b.

10.3.3
Artifacts of Isotropic Travel Time Inversion under Conditions of Anisotropy

Multi offset VSP data has been found useful (c.f. Galperin 1974; Hardage 1985; Oristaglio 1985) to analyze transmitted and reflected seismic waves with the aim to determine the properties and structures of the medium around the borehole. *Travel time inversion* has been applied to arrival times of waves for inverting the seismic velocities of the layered media (Stewart 1984; Pujol et al. 1985), for the structure of interfaces separating the layers (Lines et al. 1984), and both for the velocity and structure (Deplante and Oristaglio 1986) of a layered medium. In all these investigations, a condition of isotropy is assumed for travel-time inversion. It has to be realized that if anisotropy exists and the travel-time data is analyzed assuming isotropy, the interpretation results would be questionable or even misleading. To explore the artifacts generated due to the assumption of isotropy Douma (1990) analyzed synthetic multi-offset VSP transmitted wave travel time data (Fig. 10.9 a) for a layered transversely isotropic (T.I.) medium with either a horizontal or vertical axis of symmetry. Similar studies were conducted by Cerveny and Psencik (1972) and Doyle et al. (1982) in earthquake seismology.

In general for forward modeling two types of analytical results and programs have been published. Under the first category are analysis of travel times and amplitudes of body waves through laterally homogeneous layered anisotropic media (c.f. Booth and Crampin 1983; Fryer and Frazer 1984; Van der Hijden 1987). Under the latter category are investigations for laterally inhomogeneous anisotropic

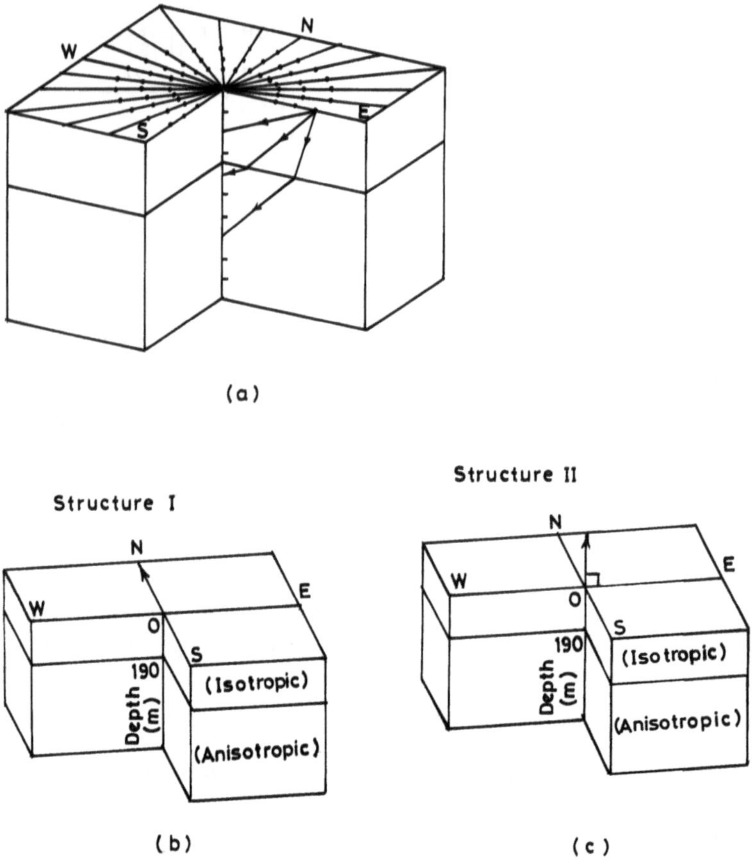

Fig. 10.9 a–c. a The VSP geometry. The sources (*circles*) are located at different azimuths and offsets from the borehole generate waves that are recorded at receivers in the borehole; **b, c** two transversely isotropic structures consisting of an isotropic low velocity layer 1 on the top of a transversely isotropic layer 2. The *arrows* indicate the direction of rotational symmetry of the transversely isotropic layer (Source: Douma 1990)

media (c.f. Vlaar 1968; Cerveny 1972). Gajewski and Psencik (1987, 1988) developed the numerical algorithm ANRAY 86 for ray tracing in 3-D laterally varying anisotropic media; and a description of this program may be found in Gajewski and Psencik (1986).

Douma (1990) used the ANRAY 86 program for forward modeling to calculate travel time for multi offset VSP geometry under the conditions of anisotropy. The travel time inversion was carried out using Schlumberger's travel time inversion program FERMAT to invert synthetic multi offset VSP travel time data. The FERMAT program assumes isotropy. As the synthetic travel time data is for anisotropic model (calculated using ANRAY 86), and inversion is done assuming isotropy, the results reveal the kind of error resulting due to *isotropic travel time inversion*. Velocities are kept constant at values obtained (say, from the well-log), only the interfaces between the layers are inverted.

Two of the models considered by Douma (1990) for analysis are an isotropic low velocity layer on the top of a transversely isotropic layer, with either a horizontal or vertical symmetry axis (Fig. 10.9b, c). Figure 10.10 shows the result of isotropic travel time inversion. The position of the actual interface and the one obtained by inversion are shown (Fig. 10.10a–d) for different source azimuths (source-borehole direction). The inversion results in an anticlinal structure for a horizontal interface (Fig. 10.10a). For the vertical axis of symmetry, the inversion results in a synclinal structure (a 'bowl' structure in 3-D), independent of the source azimuth (Fig. 10.10e).

Douma's result shows that away from the borehole, the error in depth to the interfaces may be very large if anisotropic conditions are interpreted as being isotropic.

A given travel-time data of qP and leading qS wave may be explained by a layered transversely isotropic or a layered isotropic medium, which points to the non-uniqueness of travel time inversion. Once an indication for anisotropy (say, for example, from shear-wave splitting) is detected, multi offset VSP measurements using isotropic travel time inversion should be used with caution.

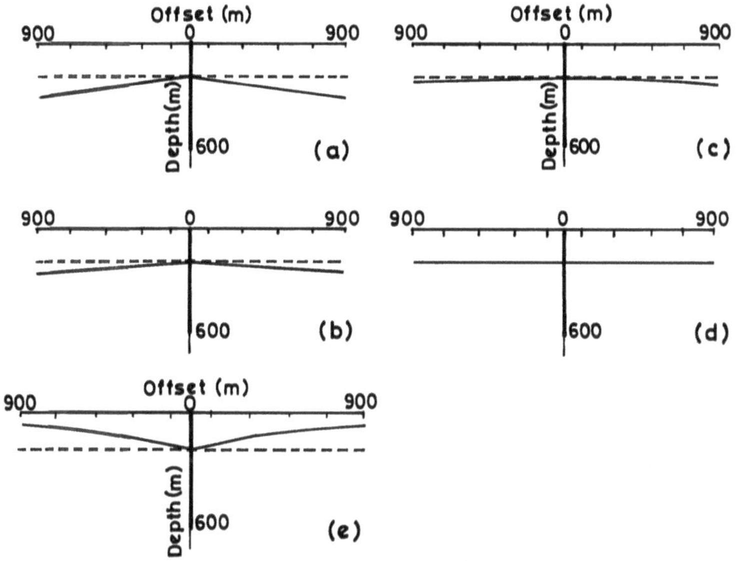

Fig. 10.10a–e. The original interface (*dashed line*) of the structure I and the interface (*solid line*) obtained by isotropic travel time inversion when applied to the q^P- and leading q^s-travel time data calculated for structure I at different source azimuths: **a** N 0°E/N 180°E, **b** N 30°E/N 150°W, **c** N 60°E/N 120°W, **d** N 90°E/N 90°W. The first direction in this notation is along the *right-hand side* of the borehole in **a–d. e** The original interface (*dashed line*) of structure II and the interface (*solid line*) obtained by isotropic travel time inversion when applied to the q^P- and leading q^s-travel time data for structure II (Source: Douma 1990)

It may be mentioned that ignoring anisotropy results in artifacts that are far different from the actual model. Such an example would also be found under tomographic reconstruction where 'X' like velocity inhomogeneity appears if anisotropy conditions are interpreted in terms of isotropy (Carrion et al. 1992). Such pitfalls may also result due to a local anisotropic condition (c.f. Chen and Castagna 2000). These pitfalls (although discussed in a different context) in Chen and Castagna (2000) may be in the form of apparent faults, folds, channels, flat spot, dim spot, bright spot and AVO anomalies due to the assumption of isotropy under actual anisotropic conditions.

10.3.4 Inversion Algorithms and Model Assessment

In inverse problems the goal is not only to obtain a solution but also to estimate the reliability of the solution. Several methods exist for solving an inverse problem and estimating the uniqueness and uncertainty in the solution. In this respect a classical method is *generalized inverse* based on singular value decomposition (SVD) of the sensitivity matrix (c.f. Jackson 1972; Wiggins 1972). Although generalized inverse is a commonly used method, it can only be applied to inverse problems with a modest number of unknowns and observations. The practical limitation is CPU memory requirements of the SVD method.

For problems as in *tomographic velocity estimations* involving hundreds of thousands of unknowns, and millions of observations, methods such as the *simultaneous iterative reconstruction technique* (SIRT) or *conjugate gradient* methods may be used. Both these methods work on one equation of the matrix at a time and therefore the computer memory requirement is small. The LSQR algorithm is a conjugate gradient type of algorithm (c.f. Hestenes Stiefel 1952; Paige and Saunders 1982). LSQR is faster in convergence than SIRT and has become widely used for obtaining solutions to *linear inverse problems.*

The advantages of conjugate gradient (CG) methods are their far less computer memory requirements than for the full singular value decomposition (SVD) method and iteration may be stopped at any time to provide an approximate solution. Therefore, CG methods may be used to obtain the solution of problems that are too large for SVD. However, the disadvantage of CG is that it does not conveniently provide auxiliary information on the quality of the solution (*resolution and covariance matrices*). It is possible to do this by extending the LSQR algorithm, which is one of the family of CG solutions. This was accomplished by Zhang and McMechan (1995). The new method of Zhang and McMechan (1995) gives approximate *resolution* and *covariance* estimates that asymptotically approach SVD solutions as the number of iterations increases.

It may be mentioned that algorithms for *model assessment* have not kept pace with the algorithms for obtaining solutions for large linear geophysical inverse problems. For example, it is now possible to obtain approximate solutions of linear systems of the order of 1 million equations and over 100,000 unknowns (see, for example, Vasco et al. 1995). This has become possible by the development of a robust iterative solver for linear systems, such as the LSQR algorithm. However, large-scale model assessment has been mostly based on inversion of synthetic data sets (c.f. Inoue et al. 1990). Approaches for assessing solutions of large inverse problems have been discussed in Scales (1987), Nolet and Schneider (1990), Jhang and McMechan (1995) and Minkoff (1996).

In a recent publication Vasco et al (1998) discuss the use of the *Sparse Matrix Technique* for model assessment. This investigation uses developments in Scales (1989), Zhang and McMechan (1995) and Minkoff (1996) in presenting the Lanczos algorithm for model assessment. In addition to this, the *Sparse-Matrix Technique* defines steps for the *Subspace Iteration Technique* that has not been applied for model assessment. Practical issues associated with estimation of the resolution and covariance matrices, computer storage memory, obtaining estimates of singular values and maintaining orthogonality between singular values is addressed.

The *Sparse Matrix Technique* enables us to conduct a *full model assessment.* The method is applicable to all large sparse inverse problems, and should cover a wide range of applications. This method is illustrated in Vasco et al. (1998) for studying the distribution of anisotropy in the subsurface. Anisotropy particularly leads to a larger system of equations to be solved, and creates difficulties in model assessment. Problems arising due to anisotropy have been dealt with in Pratt and Chapman (1992). One simple reason for additional complication in anisotropic problems is the requirement to deal with a number of anisotropy coefficients for each cell, which may trade off. The underlying need is to develop algorithms for assessing reliability of the solution obtained.

CHAPTER 11

Dip Moveout Processing and True Amplitude Imaging

11

11.1 Introduction

In Subsection 3.5.8 *normal moveout* and *dip moveout* corrections were defined. It was shown that for stacking *common mid point* (CMP) gather data for a dipping reflector, the correct *stacking velocity* depends on the angle of dip of the reflector [Eq. (3.27) – renumbered as (11.1)] in the following form

$$t^2 = \left(t_0 + \frac{4h^2\cos^2\theta}{V^2}\right) \tag{11.1}$$

It was further shown in Subsection 4.4.3 that for an anisotropic medium having direction dependent velocity, the *stacking velocity* is very sensitive to the angle of dip when the *symmetry axis* is perpendicular to the ground surface.

It may be emphasized that, for a horizontal reflector, NMO correction transforms *non-zero offset data* to *true zero offset data,* i.e. data obtained with a coincident source-receiver position. When the dip is present, this is not true. On the other hand, the application of Dip Moveout (DMO) is able to transform the NMO corrected data on a dipping reflector to true zero offset data. *After the application of DMO, the stacking velocity for reflected events from horizontal and dipping reflectors is the same* (refer to Subsection 3.5.9).

One very significant feature of reflections from a dipping reflector in a common mid-point gather is that the reflection point moves up dip as the source-receiver separation increases. This is referred to as *reflection point dispersal* (refer to Fig. 3.31 b). The dip moveout operator compensates for dispersal of reflection points on a dipping reflector.

In a very significant contribution, Judson et al. (1978) showed the phenomenon of *dip selectivity* using commonly used processing steps–*NMO and CMP stacking. Common-mid-point stacking* and *normal moveout* have the effect of enhancing reflections having a particular *moveout velocity.* Also, they have an attenuating effect on such reflections, which have different *moveout velocities,* i.e. the process acts as a *dip filter.* In conventional NMO and stacking, one choice of V_{nmo} is used, amounting to an optimally stacked dip 'θ' for that CMP and time.

It must be mentioned that DMO algorithms for an isotropic medium are inadequate for a vertically transversely isotropic (VTI) medium (Alkhalifah 1996).

Contributions for formulating suitable dip moveout operators for constant or variable isotropic velocity, and the anisotropic velocity field have progressed mainly for space-time domain data and transformed wave-number frequency domain data. The dip moveout operator plays a very significant role in achieving the ultimate goal of seismic processing, i.e. mapping true reflectivity of the subsurface in their true positions from where the reflections originated.

Liner (1990) gives the general theory and comparative anatomy of dip moveout.

The following sections present the evolution, brief theory and application aspects of the DMO operator.

11.2 Common Mid Point Gather and Dip Moveout Operator

In a *common-mid-point* (CMP) gather, over a dipping reflector, the reflection points move up dip as the source-receiver separation (offset) increases (refer to Fig. 3.31 b). The phenomenon, called *reflection point dispersal*, has been described in Deregowski (1982). The stacking velocity is $V/\cos\alpha$ [Eq. (3.27)] where α is the angle of dip. The CMP gather may therefore be NMO corrected and stacked. But

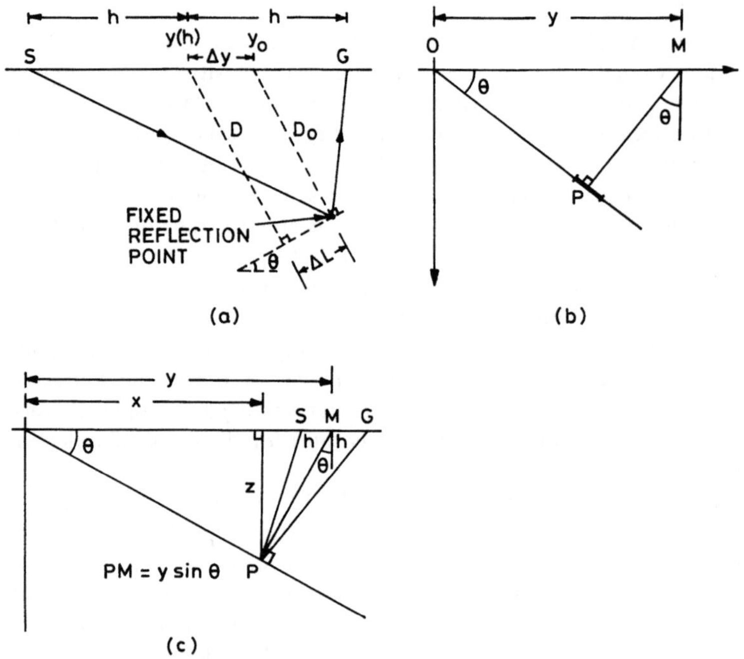

Fig. 11.1 a–c. a Reflection point dispersal; **b** dipping reflector at zero offset; **c** dipping reflector at constant offset (Sources: Fig. 11.1 a from Deregowski 1982; Fig. 11.1 b, c from Deregowski and Rocca 1981; Fig. 11.1 a–c from Geophysical Prospecting with permission)

the stack is *smeared* due to reflector point dispersal, and *DMO is a method to correct for reflection point dispersal.* One DMO algorithm contributed by Forel and Gardner (1988) forms gathers having a common-reflection point (in place of reflection point dispersal) in transformed (t_1–k) space. This will be discussed in Section 11.4.

Levin (1971) showed that reflections in a common mid-point gather over a dipping reflector are distributed over a segment *DL* of the reflector,

$$\Delta L = \frac{h^2}{D} \cos\theta \sin\theta \tag{11.2}$$

where θ is the dip angle, h is the half-offset and D is the perpendicular distance from the reflector to the mid point (Fig. 11.1).

Referring to Fig. 3.31 b, if the mid-point P is kept fixed, the reflection points would be dispersed. On the other hand, if we wish to keep the reflection point fixed, i.e. wish to stack energy from a common reflection point, the corresponding mid-point (Fig. 11.1 a) has to be a function of offset (Deregowski 1982). Thus,

$$y(h) = y_0 + \Delta y(h) \tag{11.3}$$

where

$$\begin{aligned} \Delta y(h) &= \frac{\Delta L}{\cos\theta} \\ &= \frac{h^2 \sin\theta}{D(h)} \end{aligned} \tag{11.3 a}$$

In Eq. (11.3 a), D itself is a function of offset. Let us say that D_0 is the perpendicular distance to the surface from a fixed (target) reflection point, then

$$D - D_0 = \Delta y \sin\theta \tag{11.4}$$

Substituting for D from Eq. (11.3 a) in (11.4) gives

$$\sin\theta \cdot \Delta y^2 + D_0 \Delta y - h^2 \sin\theta = 0 \tag{11.5}$$

From Eq. (11.5) we have

$$\Delta y(h) = \frac{1}{2}\left[\left\{\left(\frac{D_0}{\sin\theta}\right)^2 + 4h^2\right\}^{\frac{1}{2}} - \left(\frac{D_0}{\sin\theta}\right)\right] \tag{11.6}$$

[Deregowski 1982]

Equation (11.6) gives an expression for the amount the mid-point needs to be shifted in order to stack energy from the fixed (target) reflection point. It is a function of offset.

Deregowski (1982) showed that the *dip-moveout operator* moves energy reflected from dipping reflectors on finite offset sections by precisely this amount, thereby insuring that common-mid point gathers are equivalent to common depth point (CDP) gathers. The *dip-moveout* compensates for the dispersal of reflection points such that a subsequent mid point gather is a true common depth point gather.

The mapping from a finite offset section to a zero-offset section includes both NMO and DMO. This is discussed below from Deregowski and Rocca (1981).

Consider the geometry shown in Fig. 11.1b for a zero-offset section. M denotes the coincident shot-geophone position. O is the origin and is the point of emergence of the plane reflector on the ground surface. The planar element P dips at angle θ. The time of normally reflected energy from planar element P has its two way travel time, t_0:

$$t_0 = \left(\frac{2y}{V}\right)\sin\theta \tag{11.7}$$

Differentiation of two way travel time gives

$$\frac{dt_0}{dy} = \left(\frac{2}{V}\right)\sin\theta \tag{11.8}$$

Assuming plane wave *eigen functions* of the form

$$\exp\left[i\left(k_y y + k_z z + \omega t\right)\right],$$

we have

$$FT(\partial_t) = i\omega \text{ and } FT\left(\partial_y\right) = ik_y$$

The time differential

$$\frac{dt}{dy} = \left(\frac{2}{V}\right)\sin\theta$$

becomes

$$\frac{k_y}{\omega} = \left(\frac{2}{V}\right)\sin\theta \text{ in the transfer domain.}$$

$$\text{Or, } \sin\theta = \frac{k_y V}{2\omega} \tag{11.9}$$

Considering now the non-zero offset section. We have from Fig. 11.1c

$$y - x = y\sin^2\theta \tag{11.10}$$

$$z = y\sin\theta\cos\theta \tag{11.11}$$

By taking a small offset and deep reflector approximation the mid point is supposed to be the recipient of the reflected ray from the planar element at distance x along the line (Fig. 11.1 c).

The total travel path SPG is of length

$$\begin{aligned} d &= \overline{PS} + \overline{PG} \\ &= \left[(y - x - h)^2 + z^2\right]^{\frac{1}{2}} + \left[(y - x + h)^2 + z^2\right]^{\frac{1}{2}} \end{aligned} \tag{11.12}$$

As the half-offset is assumed small, the binomial expansion yields,

$$d \cong 2y \sin\theta + \frac{h^2 \cos^2\theta}{y \sin\theta} \tag{11.12 a}$$

and

$$t = \frac{d}{V} \cong t_0 + \frac{2h^2 \cos^2\theta}{V^2 t_0} \tag{11.12 b}$$

Equation (11.12 b) may be written as

$$t = t_0 + \Delta t_N \tag{11.12 c}$$

where t is the observed two way travel time on finite-offset (ray path SP and PG), t_0 is the (required) zero-offset travel time and Δt_N is the *moveout operator* required to map *constant offset sections* to *zero offset.*

We have

$$\Delta t_N \cong \frac{2h^2 \cos^2\theta}{V^2 t_0} = \frac{2h^2}{V^2 t_0}\left(1 - \frac{k_y^2 V^2}{4\omega^2}\right) = \Delta t_n + \Delta t_d \tag{11.13}$$

(Deregowski, Rocca 1981)
where Δt_n is the standard *normal moveout* term, and Δt_d is the *dip moveout.*

Therefore, the operator, which maps the offset section to a zero offset section, consists of both normal moveout and dip moveout.

11.3 Mapping the Constant Offset Section to Zero Offset

Deregowski and Rocca (1981) discussed the transformation from constant offset to zero offset as a mapping problem. We give below the steps from Deregowski and Rocca (1981). Assume a constant offset gather with a single impulsive event. The problem is to find a geometrical structure that would give rise to such an observation. We begin with a constant offset section in the (y–t) plane, where y is the mid point and t is the two way travel time (Fig. 11.2 a). This plane is 'seeded' with a single impulse at $y = y_h$ and $t = t_h$. The equation of the correspond-

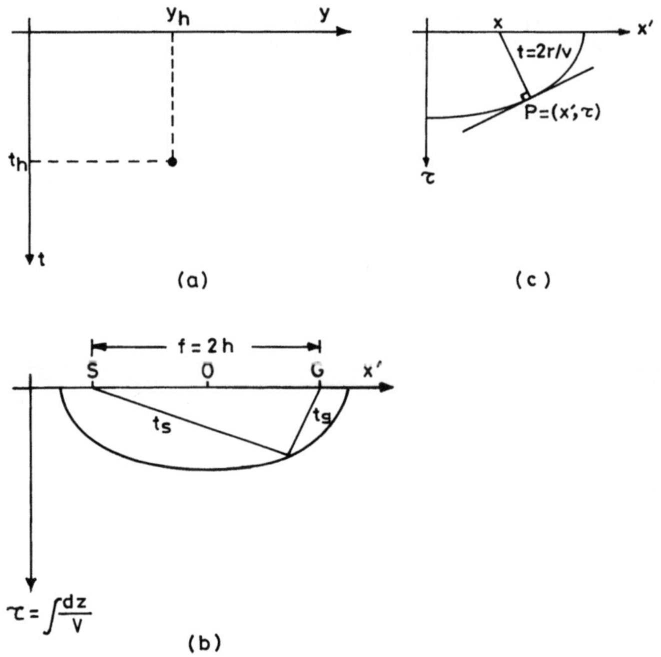

Fig. 11.2 a–c. Mapping from (y, t) to (x', τ). **a** (y, t) plane: constant offset section; **b** (x', τ) plane: the migrated section $t_s + t_g = t_h =$ constant, τ is one way vertical travel time; **c** mapping from (x', τ) to zero offset (Source: Deregowski and Rocca 1981, from Geophysical Prospecting with permission)

ing reflector is to be determined in (x', τ) plane where τ is one way vertical travel time. As the impulse response defines a constant two-way travel time t_h equal to the sum of the one-way travel times over shot and receiver rays, the required reflector is an ellipse (Fig. 11.2 b).

The next step is the simulation of the zero-offset experiment for this elliptical reflector. As in Fig. 11.2 c, each point P(x', τ) on the elliptical reflector maps to an image point (x, t) in the zero-offset section, where (x, 0) is the surface point of intersection with the normal to the ellipse at P and t (=2r/v) is the two way travel time.

The mapping operator smears a point impulse at (y, t_h) on the finite offset section over a "smile" in the (x, t) plane of the zero-offset section, and the smile forms the lower segment of the ellipse (e.g. see Fig. 11.3 a, b):

$$\frac{(x-y)^2}{h^2} + \frac{t^2}{t_0^2} = 1 \tag{11.14}$$

(Deregowski 1982)

The locus of the smear in the (x, t) plane is obtained by solving the equation of the ellipse for t(x) to obtain (Fig. 11.3)

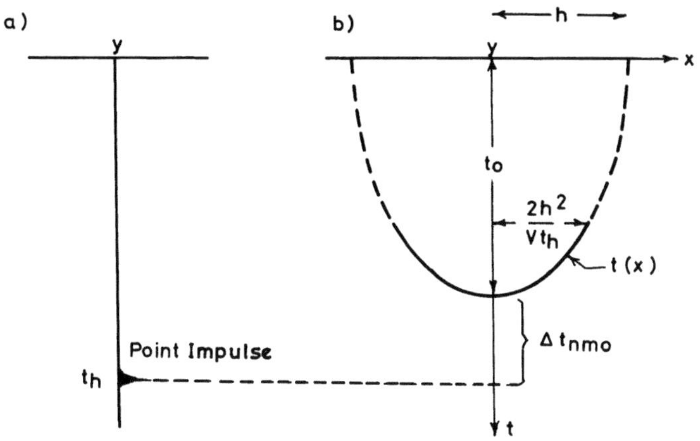

Fig. 11.3 a, b. Single point on finite-offset section maps to a 'smile' on zero-offset section. **a** Trace for finite offset 2 h; **b** result of mapping to zero offset (Source: Deregowski 1982, from Geophysical Prospecting with permission)

$$t(x) = t_0 \left[1 - \frac{(x-y)^2}{h^2} \right]^{\frac{1}{2}} \tag{11.15}$$

Here

$$|x - y| < \frac{2h^2}{Vt_h}$$

The smile is the curve along which energy at any point (y_h, t_h) on a non-zero offset section is to be "smeared" when mapped to zero offset.

11.4 Kinematics of DMO Equations

Consider Fig. 11.4 a. It shows that an event recorded at time 't' with a finite offset 2 h over a medium of constant velocity V may have arisen from any point 'P' on an ellipsoid of revolution with the line joining the source and receiver as the axis of rotation.

This ellipsoidal surface can be thought of as a sum of spherical wave fronts with centers on the line joining shot point and the receiver (Fig. 11.4 b). Figure 11.4 b shows an ellipsoidal wave front is an envelope of circular wave fronts. By considering that circles are tangential to the elliptical surface, the radius R of the circles could be expressed in terms of major and minor axes (A and B) of the ellipse and the distance 'b' of the centers of the circles from the mid-point of S and R (Fig. 11.4 c), and we have

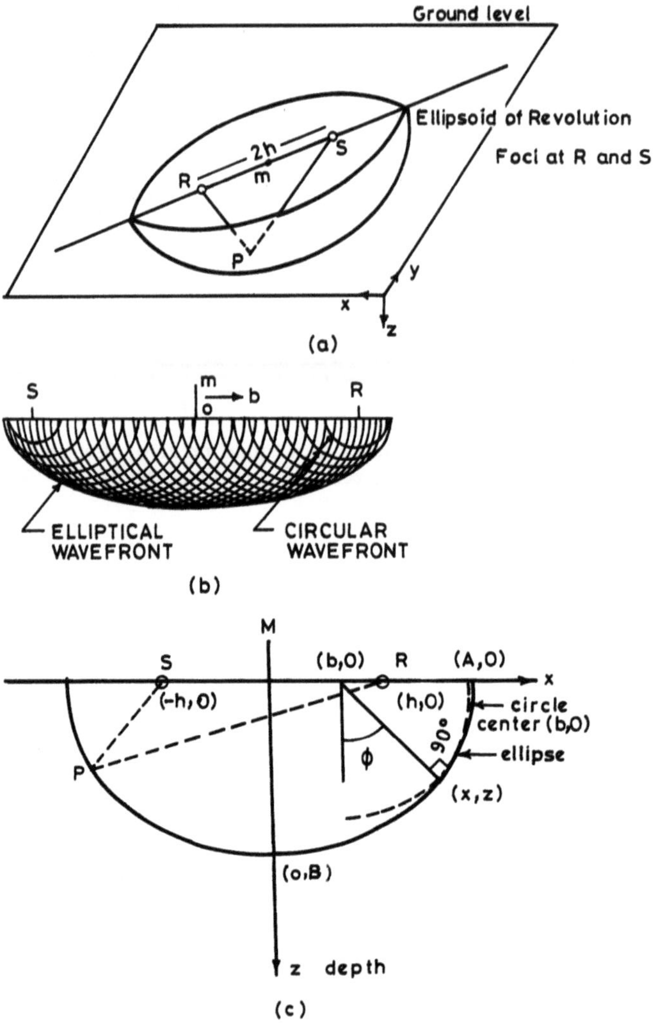

Fig. 11.4a–c. a Loci of points in 3-D from which reflection could arise for a reflected event with time 't'; **b** an ellipsoidal wavefront created as an envelope of circular wavefronts; **c** the ellipsoidal surface may be regarded as the envelope of a family of spheres with centers at (b, 0) along the segment SR, joining source and receiver positions S and R (Source: Forel and Gardner 1988, from Geophysics with permission)

$$R^2 = B^2\left(\frac{A^2 - B^2 - b^2}{A^2 - B^2}\right) \tag{11.16}$$

Forel and Gardner 1988, Eq. A-8)

The idea that the ellipsoidal surface with foci at the source and receiver positions for a finite offset trace may be obtained by migrating a family of spheres corresponding to zero-offset traces placed along the line segment from the source

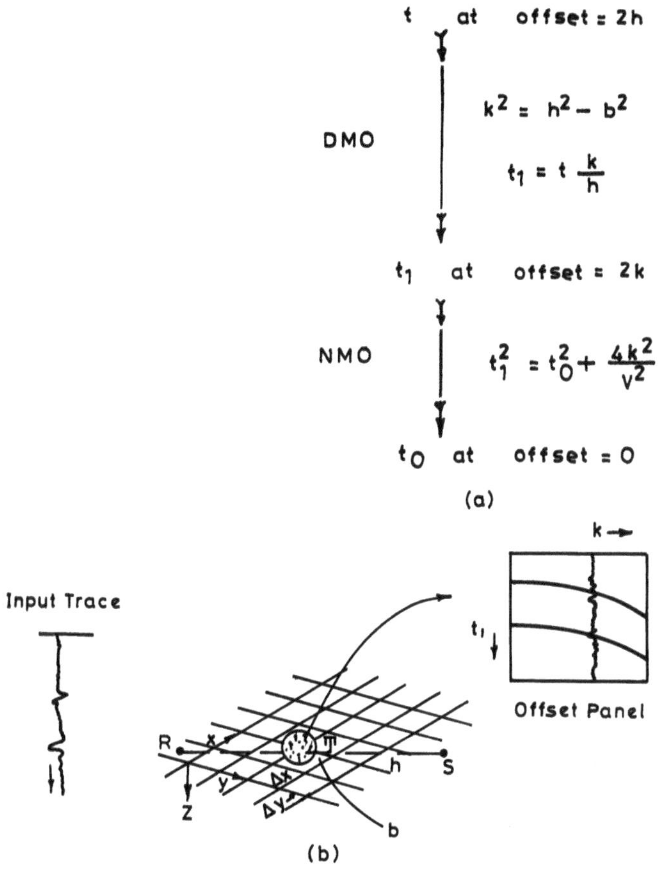

Fig. 11.5 a, b. a The source and receiver have offset 2 h and the data are recorded at time t. To apply DMO to a single trace, replace the trace by a family of traces at distances b from the midpoint along the source-receiver segment and having offset axis K and time axis t_1. After all traces have been replaced, each (K, t_1) matrix can be analyzed with the NMO equation to get the subsurface velocity and stacked to get zero-offset traces; **b** the 3-D survey is conducted in any orientation on a $t=0$ plane over a medium of constant velocity V. The source-receiver offset is 2 h; midpoint is m, and the events have arrival times t. With h fixed, the individual traces are mapped to matrices with axes (k, t_1). The velocity of the subsurface is found by applying standard NMO to the (k, t_1) matrix (Source: Forel and Gardner 1988, from Geophysics with permission)

to the receiver, was given practical form by formulating kinematic equations governing this concept. The aim is to transform finite offset traces to zero-offset traces: the DMO-NMO algorithm of Forel and Gardner (1988) converts any trace to an equivalent set of zero-offset traces. The flow chart incorporating the Gardner DMO is given in Fig. 11.5 a.

To see mathematically how the ellipsoidal surface (finite offset) may be replaced by a family of spheres (zero offset) tangential to the ellipsoid, it is necessary to relate the two times 't' and 't_0' for finite offset and zero offset, respectively.

We may analyze Eq. (11.16) in terms of time. We have

$$R^2 = B^2\left(\frac{A^2 - B^2 - b^2}{A^2 - B^2}\right)$$

Substitution of values of A and B in Eq. (11.16) yields (refer to Forel and Gardner 1988)

$$R^2 = \left[\left(\frac{Vt}{2}\right)^2 - h^2\right]\left(1 - \frac{b^2}{h^2}\right) \tag{11.17}$$

As the medium has constant velocity V, $R^2 = V^2 t_0^2/4$, and therefore

$$t_0^2 = \left(t^2 - \frac{4h^2}{V^2}\right)\left(1 - \frac{b^2}{h^2}\right) \tag{11.18}$$

Equation (11.18) shows that to calculate 't_0' from 't', we should know velocity V. Forel and Gardner (1988) suggested that the solution to this is to construct a trace with offset 2 k instead of zero and with arrival time t_1 where

$$t_1^2 = t_0^2 + \frac{4k^2}{V^2} \tag{11.19}$$

From Eqs. (11.17) to (11.19) we have

$$t_1^2 = t^2\left(1 - \frac{b^2}{h^2}\right) + \frac{4}{V^2}(k^2 + b^2 - h^2) \tag{11.20}$$

[Forel and Gardner 1988, Eq. (A-12)]

Choosing 'k' in such a way that

$$k^2 = h^2 - b^2 \tag{11.21}$$

then

$$t_1^2 = t^2\left(1 - \frac{b^2}{h^2}\right) \tag{11.22}$$

or

$$t_1 = t\frac{k}{h} \tag{11.22 a}$$

Equations (11.21) and (11.22 a) define the process of DMO. It is independent of velocity. Equation (11.19) is the NMO equation for a (k, t_1) matrix and it gives the subsurface velocity V.

The complete DMO-NMO process may be described as a two-step process as in Fig. 11.5 a.

The process may be illustrated with the help of Fig. 11.5 b.

Let R and S denote the receiver and shot point and let 'm' be the mid-point, let the sampling point be at a distance 'b' from 'm', then the corresponding offset k is

$$k^2 = h^2 - b^2$$

The original trace is added to column k after uniform contraction of the time axis. Time 't_1' for column 'k' is related to time 't' on the original trace by the relation [Eq. (11.22 a)]

$$t_1 = t\frac{k}{h}$$

Thus, we obtain the trace shown on the right side of Fig. 11.5 b. For different values of 'b', we would get a set of traces (an array of transformed traces).

Consider Eq. (11.19):

$$t_1^2 = t_0^2 + \frac{4k^2}{V^2}$$

It is a standard NMO equation, and hence each array may be analyzed by standard velocity analysis methods. Stacking velocity may be found and data may be stacked into a zero offset trace in (t_1–k) space.

The method may be applied to an ensemble of traces whatever the azimuth and offset is, and therefore would work with 3-D data. Forel and Gardner (1988) illustrate the use of their technique, described above, with 3-D synthetic data.

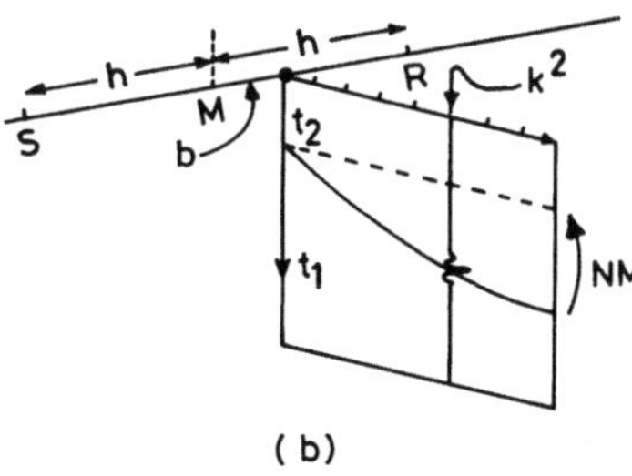

Fig. 11.6 a, b. a A point scatterer is located at a point P at a distance D from ground level location X on a seismic line in a direction which makes an angle θ with the seismic line; **b** a (k^2, t_1) array can be constructed at any location X using a velocity-independent DMO, the first three points of Eq. (11.23) (Source: Gardner and Forel 1990, from Geophysics with permission)

Gardner and Forel (1990) further examined the effect of their algorithm on amplitude, and formulated equations that preserve the physical meaning of the amplitude in the DMO process. These equations were derived for a point single scatterer model (Fig. 11.6 a). It is assumed that the scatterer has the same density and rigidity as the surrounding medium. The generalization from a point scatterer to a thin plane layer is explained, and this involves only changing the multiplication factor to the amplitude. The physical meaning of the equation derived is that the scaled amplitude (B_2) is independent of the offset variable (k^2), and has the same value whatever the depth or dip of the layer may be. Their main results are given below.

Gardner and Forel (1990) gave the following set of amplitude preserving equations:

$$\begin{aligned} k^2 &= h^2 - b^2 \\ t_1 &= k\frac{t}{h} \\ A_1 &= Aht \\ t_2^2 &= t_1^2 - \frac{4k^2}{V^2} \\ &\text{and} \\ B_2 &= B_1 t_2^{\frac{1}{2}} \end{aligned} \tag{11.23}$$

[Gardner and Forel 1990, Eq. (1)]

In Eq. (11.23), A, t and 2 h are true amplitude, time and offset for an input trace; A_1, t_1 and 2 k are the new amplitude, time and offset for a replacement trace at distance b from the mid-point along the shot-receiver segment ($-h<b<h$); t_2 is the NMO corrected time for offset 2 k using NMO velocity V; B_1 is the accumulated amplitude at time t_1 after all data is processed; B_2 is the mapped amplitude at time t_2 after NMO correction. For the meaning of initial and derived quantities, reference should be made to Fig. 11.6 b.

Figure 11.6 b depicts a common mid-point gather after DMO at location X. In this depiction, all input traces lying between S and R contribute to the gather. This is governed by the first three equations in Eqs. (11.23). Contributions form a curved event with amplitude B_1. This may be horizontally aligned with the use of the fourth equation, and scaled for amplitude B_2 using the fifth equation.

For derivation of amplitude preserving Eqs. (11.23) three factors that affect the amplitude are considered: the DMO process, the spherical divergence and the sampling interval of data in the mid point and offset. The effect of each of these factors is expressed as

$$\begin{aligned} F_{DMO} &= \frac{t_2^{3/2}}{k\left\{t_2^2 + \frac{4k^2}{V^2}\right\}^{\frac{1}{2}}} \\ F_{ss} &= V^2\left(t_2^2 + \frac{4k^2}{V^2}\right) \\ F_{data} &= h \end{aligned} \tag{11.24}$$

Combined together, it gives

$$\begin{aligned} F &= F_{DMO} \cdot F_{ss} \cdot F_{data} \\ &= \left(V^2 t_2\right)\left(t_2^{1/2} t\right) \end{aligned} \tag{11.25}$$

[Gardner and Forel 1990, Eqs. (7)]

This is an expression by which the point-source and point-receiver amplitude measurements should be multiplied, so that the resulting amplitude in the (k^2, t_1) array does not depend on direction 'θ', or distance 'D' of the scatterer for which there in no change of density. This result may be extended to a thin plane layer. The summed response increases in the proportion of $V^2 t_2$. Hence, to scale the data for a plane layer, the factor F is to be replaced by $(t_2^{1/2} t)$.

11.4.1 Extension of the NMO-DMO Algorithm of Forel and Gardner (1988) to Elliptically Anisotropic Media

Uren et al. (1990) extended the theory of velocity independent DMO (Forel and Gardner 1988) to triaxially elliptically anisotropic media. In the velocity independent DMO contributed by Forel and Gardner (1988) gathers are formed in t_1–k transform space. Relations used to generate t_1–k data are (refer to Eqs. (11.21) and (11.22a))

$$k^2 = h^2 - b^2$$

and

$$t_1 = t\frac{k}{h}$$

It was shown in the preceding discussions that after Forel and Gardner's DMO correction is applied, gathers in the constant velocity medium may be stacked to zero offset using NMO velocity [Eq. (11.19)] which is independent of *structural dip*. The resulting traces are equivalent to zero offset traces, which may subsequently be migrated using *zero-offset migration algorithms*.

It was shown in Uren et al. (1990) that, for elliptically anisotropic media also, the same procedure works with Eq. (11.19) modified as

$$t_1^2 = t_0^2 + \frac{(2k)^2}{V_{horiz}^2} \tag{11.26}$$

[Uren et al. 1990, Eq. (10)]

where stacking velocity, V_{horiz}, is equal to the horizontal component of the elliptical velocity function in the direction of the survey line. Here, also the stacking velocity is independent of dip. Further, data stacked with this horizontal velocity may be migrated with a zero offset anisotropic migration algorithm, as in case of isotropic medium.

Uren et al. (1990) tested their algorithm for dip moveout for isotropic, elliptically and non-elliptically anisotropic media by considering a point scatterer (Fig. 11.7a).

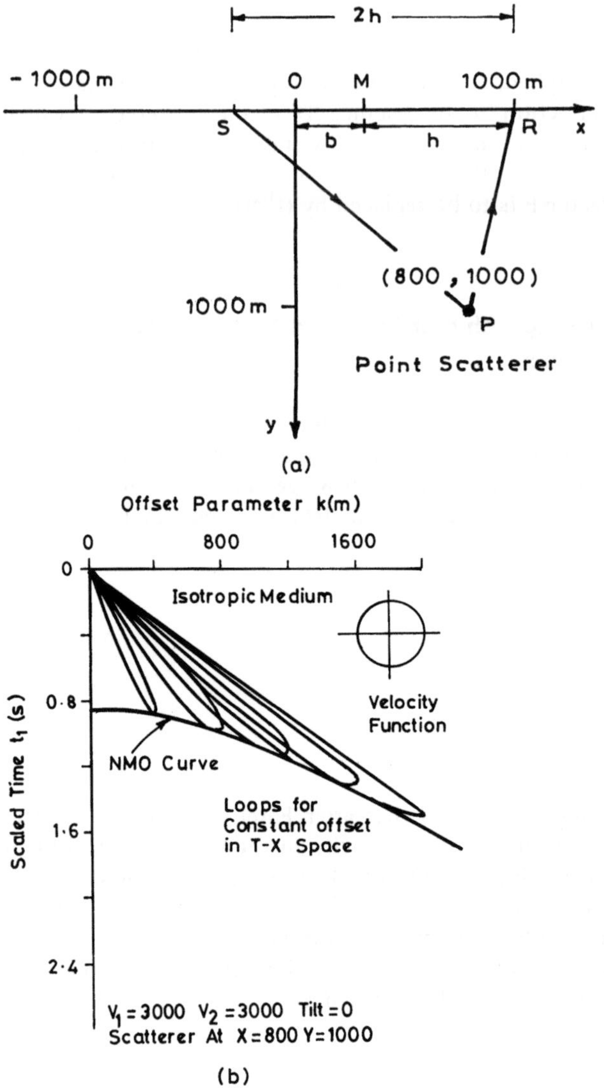

Fig. 11.7a–c. a Point diffractor in a homogeneous medium. M is the mid-point between the source at S and receiver at R; **b** single layer (t_1–k) gather above a point diffractor in an isotropic, homogeneous medium. The stacking velocity after DMO is equal to the true velocity of the medium; **c** single layer (t_1–k) gather above a point diffractor in an elliptically anisotropic homogeneous medium. The stacking velocity after DMO is equal to the horizontal component of the velocity function (Source: Uren et al. 1990, from Geophysics with permission)

Fig. 11.7c

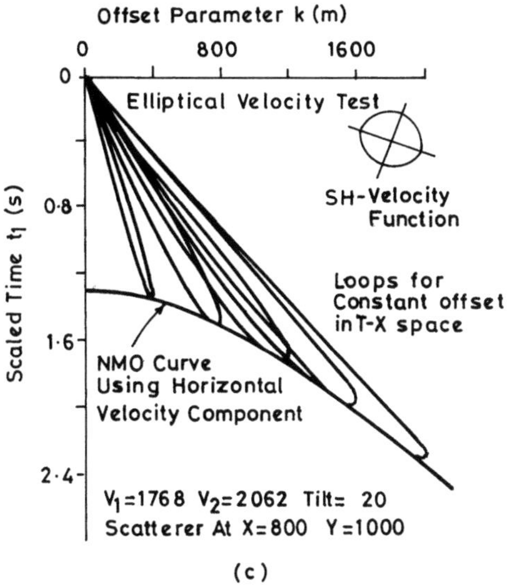

Figures 11.7 b, c shows t_1–k gathers above a point diffractor in isotropic (Fig. 11.7b) and elliptically anisotropic (Fig. 11.7c) homogeneous media. For generating the t_1–k gathers offsets 2 h of 800, 1,600, 2,400, 3,200 and 4,000 m were used for the mid-point location at the origin of coordinate. For five offsets, five loops appear in these figures, which are transformed travel-time loci. In each figure the NMO curve is given. For an isotropic medium, the NMO curve was plotted for a medium velocity of 3,000 m/s. In Fig. 11.7c NMO curves were plotted using a horizontal velocity component.

In Fig. 11.7b (for isotropy) and Fig. 11.7c (for elliptical anisotropy), the NMO curve forms an envelope to the loops in the transformed t_1–k space. Thus, in these cases when the transformed gather is NMO corrected and stacked, wavelets along the NMO curve will stack in, while the remaining data would stack out. The Gardner DMO does not work exactly for non-elliptical anisotropy velocity functions.

11.5 DMO Impulse Response

Before discussing specific cases of impulse response for isotropic and anisotropic media, it is desirable to first present the background theory. This section introduces DMO processing in the frequency wave number domain. Section 11.4 dealt with DMO processing in the space-time domain.

Hale (1984) gave the DMO algorithm that performs a DMO correction in the wave-number-frequency domain. We discuss below Hale's velocity-independent DMO algorithm.

The basis of Hale's algorithm is the NMO equation corrected for dip (refer to Eq. (11.1)):

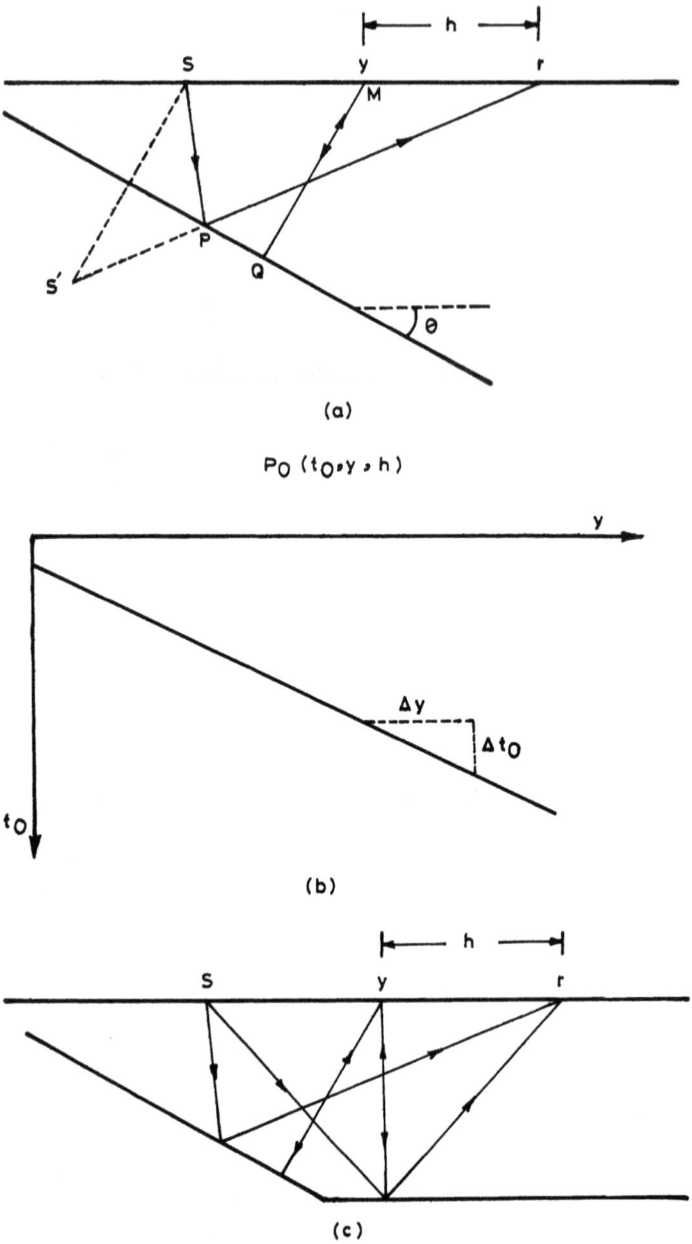

Fig. 11.8 a–d. a Geometry of a dipping reflector, ray path for separated source-receiver, and coincident source-receiver at mid point between source and receiver; **b** zero-offset section (y_i, t_0) corresponding to experiment geometry in **a**; **c** subsurface model with two dips–dipping reflector suddenly becomes horizontal; **d** zero-offset section (y_i, t_0) corresponding to conflicting dip geometry in **c** (Source: Hale 1984, from Geophysics with permission)

Fig. 11.8 d

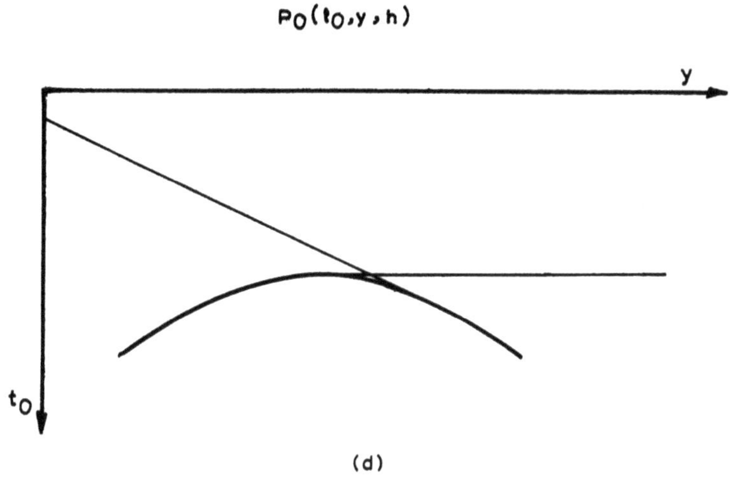

$$t = \left(t_0^2 + \frac{4h^2\cos^2\theta}{V^2}\right)^{\frac{1}{2}}$$

[Hale 1984, Eq. (3)]

This may be obtained by considering the geometry in Fig. 11.8 a. We have

$$V^2t^2 = (s'r)^2 = (ss')^2 + (sr)^2 - 2(ss')(sr)\cos\left(\frac{\pi}{2} + \theta\right) \tag{11.27}$$

If y_i is taken to denote the intersection of the reflector and the surface

$$ss' = 2(s - y_i)\sin\theta$$

The travel-time 't' in terms of source-receiver mid point 'y', and half-offset 'h' is given by

$$t = \frac{2}{V}\left[(y - y_i)^2\sin^2\theta + h^2\cos^2\theta\right]^{\frac{1}{2}} \tag{11.28}$$

Defining 't_0' as the zero-offset ($s = r = y$) travel-time for the path MQM we have

$$t_0 \equiv \frac{2}{V}(y - y_i)\sin\theta \tag{11.29}$$

Therefore travel time 't' for the path SPR may be written as

$$t = \left(t_0^2 + \frac{4h^2\cos^2\theta}{V^2}\right)^{\frac{1}{2}} \tag{11.30}$$

This equation is of special value as it may be used to compensate for travel time differences among seismograms recorded with different source-receiver offsets. If we take p (t, y, h) to denote a pressure seismogram recorded as a function of time 't', mid point 'y', and half offset 'h' it suggests the transformation

$$p_0(t_0, y, h) \equiv p\left(\sqrt{\left(t_0^2 + \frac{4h^2\cos^2\theta}{V^2}\right)}, y, h\right) \tag{11.31}$$

[Hale 1984, Eq. (4)]

We have from Eq. (11.8)

$$\frac{\Delta t_0}{\Delta y} = \frac{2\sin\theta}{V}$$

The zero-offset section for the geometry shown in Fig. 11.8 a is shown in Fig. 11.8 b.

Here $p_0(t_0, y, h)$ is the (computed) zero offset seismogram from a non-zero offset seismogram p (t, y, h).

The true zero offset seismogram is denoted as $p(t_0, y, h=0)$. The reflection $p_0(t_0, y, h)$ occurs at the same time as that of $p(t_0, y, h=0)$. Hale (1984) called the transformation expressed by Eq. (11.31) as the *dip corrected NMO.*

Dip-corrected NMO may be performed as two cascaded processes. For this purpose, Eq. (11.30) is written as

$$t = \left(t_0^2 + \frac{4h^2}{V^2} - \frac{4h^2\sin^2\theta}{V^2}\right)^{\frac{1}{2}} \tag{11.32}$$

NMO time 't_n' (not dip-corrected) is defined as

$$t \equiv \left(t_n^2 + \frac{4h^2}{V^2}\right)^{\frac{1}{2}} \tag{11.33}$$

For these Eqs. (11.32) and (11.33) to be equivalent, one must have

$$t_n = \left(t_0^2 - \frac{4h^2\sin^2\theta}{V^2}\right)^{\frac{1}{2}} \tag{11.34}$$

Equations (11.33) and (11.34) suggest that the dip-corrected NMO may be performed in two steps. The first Eq. (11.33) suggests that NMO may be defined as

$$\text{NMO}\; p_n(t_n, y, h) \equiv p\left(\sqrt{\left(t_n^2 + \frac{4h^2}{V^2}\right)}, y, h\right) \tag{11.35}$$

[Hale 1984, Eq. (8)]

The second Eq. (11.34) suggest DMO to be defined as

$$\mathrm{DMO}\; p_0(t_0, y, h) \equiv p_n\left(\sqrt{t_0^2 - \frac{4h^2 \sin^2\theta}{V^2}}, y, h\right) \tag{11.36}$$

This may be expressed using Eq. (11.8) in terms of the *time dip* ($\Delta t_0/\Delta y$) as

$$\mathrm{DMO}\; p_0(t_0, y, h) \equiv p_n\left(\sqrt{t_0^2 - \left(\frac{\Delta t_0}{\Delta y}\right)^2 h^2}, y, h\right) \tag{11.37}$$

[Hale 1984, Eq. (10)]

Thus, NMO is a transformation from recording time 't' to NMO time 't_n', DMO is then the transformation from NMO time 't_n' to zero-offset time 't_0'.

The splitting of dip-corrected NMO [Eq. (11.1)] into NMO and DMO [Eqs. (11.35) and (11.36) or (11.37)] has an important basis, which is, DMO may be applied through Eq. (11.37) without knowledge of velocity 'V' or the angle of dip 'θ' which are typically unknown.

Incidentally, the DMO algorithm of Hale (1984) compares with the *adjusted NMO* performance proposed by Cressman (1968) [refer to Subsection 3.5.8.2, Eq. (3.30)]. Cressman (1968) observed that NMO applied, assuming a horizontal reflector for a dipping reflector case, overcorrects, which is compensated for by using *adjusted NMO*. Hale (1984) observed the whole process of DMO as two cascaded processes: first the transformation from record time t to t_n, and secondly from t_n to t_0.

If correct values of V and θ are used in NMO and DMO, the computed zero offset section $p_0(t_0, y, h)$ should be equal (considering travel time only) to the time of the true zero-offset section $p(t_0, y, h=0)$ for all values of h. For the model shown in Fig. 11.8a, this zero offset section is as in Fig. 11.8b.

The practical use of Eq. (11.37) creates problems. For using it, one needs to measure the zero offset slope ($\Delta t_0/\Delta y$). This is not done in field practice. A second more important problem with Eq. (11.37) occurs when more than one slope is present at a given (t_0, y), as in Fig. 11.8c.

Referring to Fig. 11.8c, the dipping reflector suddenly becomes horizontal. The zero offset section for this model is shown in Fig. 11.8d. Zero-offset typically involves triplication in a (t_0, y) plot, as opposed to a straight line for a single reflector (refer to Fig. 11.8b).

After NMO, the application of Eq. (11.37) with $\Delta t_0/\Delta y = 0$ to a non-zero offset section would correctly align the horizontal reflection in the CMP stack while incorrectly positioning the dipping reflections. Subsequent CMP stacking (summing over 'h') would attenuate the misaligned dipping reflections as compared to the horizontal reflections. The dipping reflection would be enhanced by taking $\Delta t_0/\Delta y \neq 0$ in Eq. (11.37), but at the cost of attenuating horizontal reflections in the CMP stack. Hale (1984) remarks that the most troublesome is the point where the reflections intersect. Even with knowledge of two distinct slopes at that point, the conventional process of stacking after dip corrected NMO must enhance one slope at the expense of attenuating the other. Hale's (1984) work focuses on treating conflicting *zero-offset slopes.*

Equation (11.37) implies that a different DMO correction is required for each different zero-offset slope $\Delta t_0/\Delta y$, i.e. *DMO is a dip-dependent process.* Two-dimensional F.T. over (t_0, y) is found useful for dip-dependent processing. The reason for this is that all events with a particular slope in the (t_0, y) domain transform to a single radial line in the $(\omega_0\text{–}k)$ domain. This is according to the relation [refer to Eqs. (11.8) and (11.9)]

$$\frac{k}{\omega_0} = \frac{\Delta t_0}{\Delta y} = \frac{2\sin\theta}{V}$$

The two-dimensional F.T. may be written as

$$p_0(\omega_0, k, h) \equiv \int_{-\infty}^{+\infty} dt_0 e^{i\omega_0 t_0} \int_{-\infty}^{+\infty} dy e^{-iky} p_0(t_0, y, h) \qquad (11.38)$$

It is possible to write this transformation in terms of NMO corrected data, by changing the integration variable from 't_0' to 't_n' as suggested by Eq. (11.34):

$$\begin{aligned} t_0 &= \left(t_n^2 + \tfrac{4h^2\sin^2\theta}{V^2}\right)^{\frac{1}{2}} \\ &= \left(t_n^2 + \left(\tfrac{\Delta t_0}{\Delta y}\right)^2 h^2\right)^{\frac{1}{2}} \end{aligned}$$

Defining

$$A \equiv \frac{dt_n}{dt_0} = \frac{t_0}{t_n} = \left[1 + \left(\frac{\Delta t_0}{\Delta y}\right)^2 \frac{h^2}{t_n^2}\right]^{\frac{1}{2}} \qquad (11.39)$$

and replacing

$$p_0\left(\sqrt{t_n^2 + \left(\frac{\Delta t_0}{\Delta y}\right)^2 h^2}, y, h\right) \text{ by } p_n(t_n, y, h)$$

the F.T. [Eq. (11.38)] is written in the form

$$p_0(\omega_0, k, h) = \int dt_n A^{-1} e^{i\omega_0 t_n A} \int dy e^{-iky} p_n(t_n, y, h) \qquad (11.40)$$

[Hale 1984, Eq. (12 a)]

This integral transform depends through A on the unknown zero-offset slope $\Delta t_0/\Delta y$. Since $\Delta t_0/\Delta y$ may be interpreted as a ratio of k/ω_0, 'A' in Eq. (11.39) may therefore be replaced as

$$A = A(t_n, \omega_0, h, k) \equiv \left(1 + \frac{k^2h^2}{\omega_0^2 t_n^2}\right)^{\frac{1}{2}} \tag{11.41}$$

The desired zero-offset section $P_0(t_0, y, h)$ is obtained by taking the inverse F. T. of $p_0(\omega_0, k, h)$:

$$p_0(t_0, y, h) = \frac{1}{4\pi^2} \int d\omega_0 e^{-i\omega_0 t_0} \int dk e^{iky} p_0(\omega_0, k, h) \tag{11.42}$$

[Hale 1984, Eq. (12.c)]

It is to be understood that the replacement of $\Delta t_0/\Delta y$ by k/ω_0 is valid only with respect to the arrival times of reflections. Therefore, computed $p_0(t_0, y, h)$ is to be interpreted as zero offset sections for which only reflection times are the same as those of a true *zero-offset section* $p_0(t_0, y, h=0)$.

Equation (11.42) is versatile in the sense that it gives a method for applying DMO not just for one slope, but also for each slope $\Delta\tau_0/\Delta y = k/\omega_0$ *in the zero offset section.*

In the limiting case of a zero offset $h/t_n \to 0$ and a zero slope ($k/\omega_0 \to 0$), DMO does not do anything [e.g., see Eqs. (11.40) and (11.41)]. DMO correction is most significant for early reflections at large offsets ($h/t_n \to \infty$) and steep slopes ($k/\omega_0 \to \infty$) (Hale 1984).

Hale (1984) proposed a simple implementation of his algorithm (Fig. 11.9). The transform over 'y' is a standard Fourier transform and may be obtained by using any Fourier transform algorithm. The transform over t_n is not a F. T., and may be performed by numerical integration over all t_n for each ω_0 and k.

The DMO algorithm of Hale is for a constant velocity, and accurate for all offset and dips.

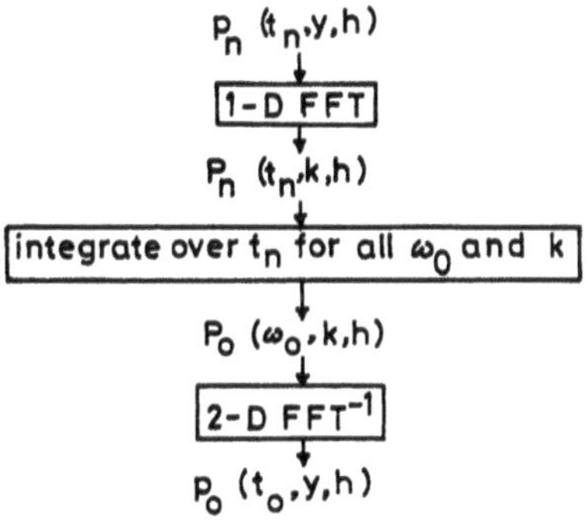

Fig. 11.9. Steps in implementation of dip-move-out. The half-offset h is constant in each step, and the entire sequence is repeated for each offset (Source: Hale 1984, from Geophysics with permission)

11.5.1
DMO Impulse Response for Constant and Varying Isotropic Velocity

It is known (Deregowski and Rocca 1981) that if p (t, y, h) has impulses at (t_p, y_p), then the corresponding subsurface elliptical reflectors are represented by

$$\frac{\left(y - y_p\right)^2}{\left(\frac{V^2 t_p^2}{4}\right)} + \frac{t_m^2}{\left(t_p^2 - \frac{4h^2}{V^2}\right)} = 1 \tag{11.43}$$

(Hale 1984)

Here t_m is migrated (two-way vertical) time defined by $t_m \equiv 2z/V$ where z is depth.

To illustrate correct DMO processing, Hale (1984) considered an example of a time-variable impulse response. It was shown that correct processing (NMO and DMO) resulted in imaging the elliptical reflectors.

The effect of constant velocity DMO on reflection times is summarized by the equation

$$t_0 = t_n\left(1 - \frac{x^2}{h^2}\right)^{\frac{1}{2}} \tag{11.44}$$

(Deregowski and Rocca 1981)

where t_0 is the zero-offset time (after DMO), t_n is the NMO corrected time (before DMO), x is the horizontal distance measured from the source-receiver midpoint, and h is half the source-receiver offset.

Hale and Artley (1993) gave the following squeezed version of *DMO mapping* (Fig. 11.10a) to approximately handle velocity variations with depth

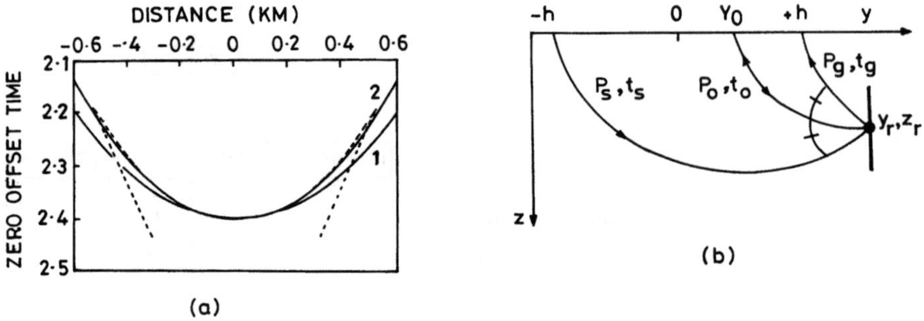

Fig. 11.10 a, b. a Zero-offset time t_0 as a function of distance x for NMO time t_n = 2.4 s and half offset h = 1.5. '1' is the constant velocity DMO, '2' is the squeezed DMO, the dotted curve is the exact DMO. Velocity varies as v(z) = 1.5 + 0.8z km/s.; **b** diagram showing a DMO raypath trio. All three rays must terminate at the reflection point and the zero offset ray bisects the angle formed by the source ray and geophone ray. The zero offset ray is normal to the reflecting surface at the reflection point (Sources: for Fig. 11.10 a – after Hale, Artley 1993; for Fig. 11.10 b – Artley and Hale 1994, from Geophysics with permission)

$$t_0 = t_n(1 - S) + t_n S \left[1 - \frac{x^2}{S\gamma(t_n)h^2}\right]^{\frac{1}{2}} \tag{11.44 a}$$

[Hale and Artley 1993, Eq. (2)]

In Eq. (11.44 a) S is the constant DMO *squeeze* factor and $\gamma(t)$ is the time-variable *squeeze function* (Hale 1983, 1988).

Squeezed DMO takes the form of constant velocity DMO for $\gamma(t) = S = 1$. For details of squeezing DMO and its performance with examples, interested readers may refer to the original paper of Hale and Artley (1993).

In a further development of *dip-moveout processing* for depth-variable velocity, Artley and Hale (1994) derive a method for kinematically computing the exact DMO for arbitrary variations of velocity with depth. This development over the *approximate DMO* in Hale and Artley (1993) does not make any assumption about offset, dip, velocity function, or hyperbolic moveout. This method makes use of *ray-tracing* (Slotnick 1959) to determine the *DMO mapping* for the variable-velocity case. Each ray is characterized by its ray parameter (p) and *two way* travel-time (t) along a particular ray. Computed quantities are the lateral distance traveled by the ray $x(p,t)$, the two-way vertical travel-time $\tau(p,t)$ and the propagation angle $\theta(p,t)$. The interval velocity function $V(\tau)$ is assumed known. The vertical travel-time $\tau(p,t)$ has an analogy to the depth of the ray, stretched according to the relation $d\tau/dz = 2/v(z)$.

To determine the appropriate combination of ray geometry, Artley and Hale (1994) describe in a unique way the several conditions that describe the DMO transformation. One combination described in Artley and Hale (1994) is a trio of rays (Fig. 11.10 b).

Figure 11.10 b shows various ray paths. These paths are from the source to the reflector and from the reflector to the receiver. It also shows the zero offset ray path and its relationship with source and receiver ray paths. The equality of the angle between source ray and zero offset ray, and zero offset ray and receiver ray holds. In addition to this the zero offset ray has to be perpendicular to the reflecting surface. These conditions on ray paths are used to generate a system of *non-linear equations*. By simultaneously solving these equations, the *mapping* for transforming *non zero-offset data* to *zero-offset data* is determined.

For a constant velocity medium, the DMO mapping function $t_n(t_0, p_0)$ is described by the relation

$$t_n^2 = t_0^2 - p_0^2 h^2 \tag{11.45}$$

[Artley and Hale 1994, Eq. (1)]
where t_n is time on normal moveout corrected section, and t_0 is time on the corresponding zero-offset section.

Equation (11.45) does not hold when velocity varies with depth. For a simple velocity function, Dietrich and Cohen (1993) give an analytic expression for the DMO mapping. A very general situation is taken care of in Artley and Hale (1994) in which *DMO mapping* may be found for any velocity function that varies *arbitrarily* with depth. Once the *DMO mapping* is found, its practical use in the *dip-decomposition procedure* of Jacobowicz (1990) – (refer to Section 11.6) is straightforward.

11.5.2 DMO Impulse Response under Conditions of Anisotropy

Hale (1984) showed that the *DMO impulse response* for an isotropic homogeneous medium is an ellipse. For an elliptically anisotropic medium also, its form is elliptical. Elliptical anisotropy is a special case of transverse isotropy, which results when certain relations hold between elastic constants. Anderson and Tsvankin (1995) extended the dip-moveout method developed by Hale (1984) to treat transversely isotropic (T.I.) media. Alkhalifah's (1996) method of transformation to zero-offset is based on ray tracing for anisotropic media, which is a modified version of Artley and Hale's (1994) v(z) ray tracing algorithm.

Alkhalifah (1996) showed that the *DMO impulse response* in typical transversely isotropic (T.I.) models (such as that associated with shales) deviate substantially from the familiar elliptical shape associated with isotropic homogeneous media. It is shown that triplication in the DMO impulse response may arise even in a simple homogeneous medium. Also such deviations may exceed those created by vertical inhomogeneity. This emphasizes the need to take anisotropy into consideration in DMO processing. Aspect of triplication in the DMO impulse response has been investigated in Alkhalifah (1996). We present it through Fig. 11.11.

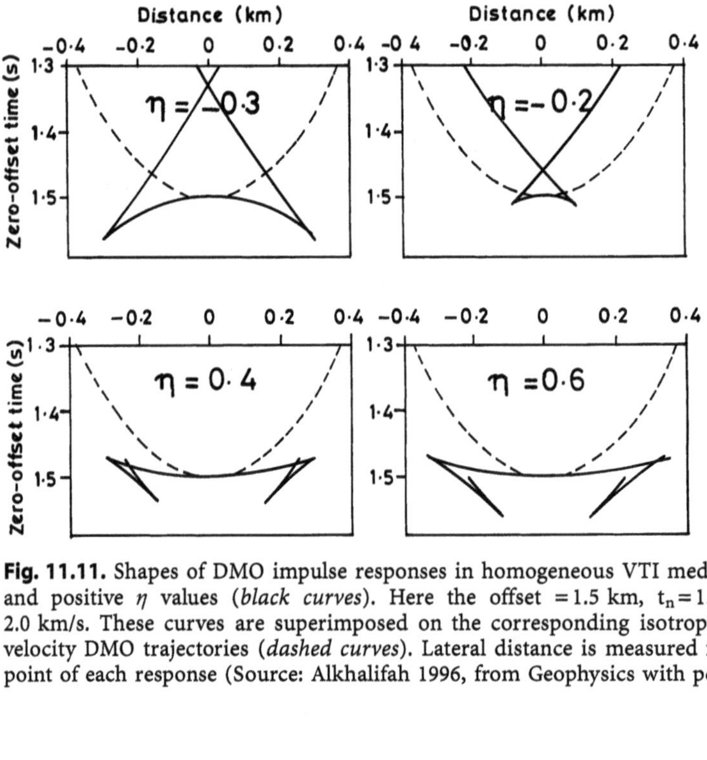

Fig. 11.11. Shapes of DMO impulse responses in homogeneous VTI media for four negative and positive η values (*black curves*). Here the offset = 1.5 km, t_n = 1.5 s and $V_{nmo}(0)$ = 2.0 km/s. These curves are superimposed on the corresponding isotropic (η = 0) constant-velocity DMO trajectories (*dashed curves*). Lateral distance is measured from the symmetry point of each response (Source: Alkhalifah 1996, from Geophysics with permission)

The goal of DMO processing is to correctly image both horizontal and dipping events on the same velocity panel. If the processing DMO algorithm were not suitable, horizontal and dipping events would not be focused with the same velocity panel. It is found in Alkhalifah (1996) that reflectors with small dips are imaged best with a lower stacking velocity; steep reflectors (faults) are imaged better with a much higher velocity using conventional isotropic DMO processing. These results in Alkhalifah (1996) show that conventional isotropic DMO is unable to remove the dip dependence of stacking velocity. The net result is that a poorly-focused image of the fault plane is produced (Alkhalifah 1996, Fig. 1). Conventional isotropic DMO fails here because stacking (moveout) velocity increases with dip much faster than in an isotropic medium assumed in the DMO algorithm. If the shortcomings of the DMO processing were a result of a vertical velocity gradient, steep events would be imaged at a lower stacking velocity than that of events with a small dip (Hale and Artley 1993). Taking that velocity increases here with depth, the inference is that anisotropy introduces an error in conventional (isotropic, homogeneous) DMO which is contrary and stronger than that caused by a vertical velocity gradient (Alkhalifah 1996).

Alkhalifah and Tsvankin (1995) demonstrated that a new representation of anisotropy in terms of just two parameters is sufficient for performing all time-related processing, such as normal moveout correction (including non hyperbolic moveout correction, if necessary), dip moveout removal and prestack and post stack time migrations. These two parameters are normal moveout velocity for a horizontal reflector, $V_{nmo}(0)$ and anisotropy coefficient η, defined as

$$V_{nmo}(0) = V_{PO}\sqrt{1+2\delta} \tag{11.46}$$

and

$$\eta \equiv 0.5\left(\frac{V_h^2}{V_{nmo}^2(0)} - 1\right) = \frac{\varepsilon - \delta}{1+2\delta} \tag{11.47}$$

[Alkhalifah 1996, Eqs. (1) and (2)]
where V_h is the horizontal velocity, and ε and δ are:

$$\varepsilon \equiv \frac{c_{11} - c_{33}}{2c_{33}}$$

$$\delta \equiv \frac{(c_{13} + c_{44})^2 - (c_{33} - c_{44})^2}{2c_{33}(c_{33} - c_{44})}$$

Alkhalifah (1996) studied the dependence on various parameters that enter DMO in VTI media. These are: η, $V_{nmo}(0)$, offset and NMO corrected time t_n.

The dependence on η, of the DMO impulse response shape in a homogeneous VTI medium (transversely isotropic media with a vertical axis of symmetry) is shown in Fig. 11.11. The striking behavior is of triplication that develops as η increases. η is the key anisotropy parameter that controls the impulse response in VTI media because of which the shape of the response differs substantially from elliptical. The amplitude distributed along DMO operators differs significantly from the one for isotropic media (Alkhalifah 1996).

The DMO operator for isotropic homogeneous media is independent of velocity. This is not so for VTI media with $\eta \neq 0$. However, V_{nmo} has a small influence on the DMO impulse response (Alkhalifah 1996).

The algorithm developed in Alkhalifah (1996) gives a method of computing travel-time exactly and uses this information for accurate transformation to zero offset (TZO) for an arbitrary depth-variable anisotropy function. 'η' defined in Eq. (11.47) is the parameter which primarily controls the impulse response in an anisotropic medium. The presence of triplication (Fig. 11.11) shows the large differences between the DMO impulse response for VTI and for isotropic media. This complex shape leads to significantly different amplitude distribution along the DMO operator than in isotropic media. With these large differences the DMO algorithm for isotropic media proves inadequate for typical VTI media. The parameter 'h' is found appropriate for processing data for VTI media and is of potential value in lithological interpretation (refer, for example, to Alkhalifah and Tsvankin 1995).

11.6 DMO in the Dip Domain

A practical form of Hale's (1984) technique is briefly discussed below because of its accuracy and efficiency. The presentation derives from Jakubowicz (1990).

It may be mentioned that *NMO at migration velocity followed by DMO is equivalent to a single application of NMO using the appropriate dip dependent stacking velocity as required by Eq. (11.1).*

From Hale's (1984) DMO algorithm, seismogram p_0 in the ω–k domain is related to the input (fixed) non-zero offset seismogram p in the t–k domain through the relationship

$$P_0(k, \omega_0, h) = \int_0^{\infty} A^{-1} \exp(i\omega_0 t_n A) p(k, t_n, h) dt_n \tag{11.48}$$

[Jakubowicz 1990, Eq. (2)]

Here

$$A = \sqrt{1 + \frac{k^2 h^2}{\omega_0^2 t_n^2}}$$

and

$$t^2 = t_n^2 + \frac{4h^2}{V^2}$$

t_n is the travel-time after NMO, k is the spatial wave number, and ω_0 is the output temporal angular frequency at zero offset. Zero offset data is obtained by transforming $p_0(k, \omega_0, h)$ to the space-time domain.

Equation (11.48) bears a resemblance to the Fourier transform. In fact, for zero offset it is an F–T. This is because DMO does not apply any corrections to data that are already at zero offset. For a non-zero offset if differs from F–T and cannot be calculated with the help of the FFT algorithm.

In fact, Eq. (11.48) may be transformed into a Fourier transform by suitable substitution. This is shown below.

To devise an efficient implementation of the Hale DMO, Jakubowicz (1990) considered Hale's (1984) algorithm for a constant velocity medium having reflectors with same angle of dip. In this case (refer to Section 11.2)

$$\frac{k}{\omega_0} = \frac{dt_0}{dy} = \frac{2\sin\theta}{V}$$

Here dt_0/dy is the *time slope* of the reflections in the zero offset data set [see for example, Hale 1984, Eq. (11)]. Eq. (11.48) reduces to

$$p_0(k = \omega_0 D, h) = \int_0^{\infty} \frac{1}{\sqrt{1 + D^2h^2/t_n^2}} \exp\left(i\omega_0 t_n \sqrt{1 + \frac{D^2h^2}{t_n^2}}\right) p_D(k, t_n, h)\, dt_n \tag{11.49}$$

where

$$D = \frac{k}{\omega_0}$$

Use of P_D emphasizes that the data contains only a single time slope D.

Using the transformation

$$t_0^2 = t_n^2 + D^2h^2 \tag{11.50}$$

in Eq. (11.49) gives

$$p_0(k = \omega_0 D, h) = \int_0^{\infty} e^{(i\omega_0 t_0)} p_D\left(k, t_0 = \sqrt{t_n^2 + D^2h^2}, h\right) dt_0 \tag{11.51}$$

[Jakubowicz 1990, Eq. (9)]

This equation is now in the form of a Fourier transform. Consideration of only a single structural dip makes Eq. (11.48) expressible as a Fourier transform through Eq. (11.51).

The above analysis shows that when a single dip is considered, DMO reduces to a simple change of travel time given by Eq. (11.50). Hale (1984) showed that this form is of *NMO inverse*. The transformation in Eq. (11.51) depends only on offset 'h' and 'D', and is therefore independent of velocity. The original form of the Hale DMO is also independent of velocity. The application of NMO using a medium velocity overcorrects dipping data, DMO exactly compensates for the overcorrection, i.e. DMO takes care of the dip dependent part of the correction to zero offset travel time and the dip independent part is taken care of by NMO.

The above explanation is good for a single structural dip. In field conditions, seismic data in general, contains all types of dips. Therefore, the above simple result would not be applicable. DMO in this case becomes a multi channel process as against a single channel process for one structural dip.

Even if all dips are present, Eq. (11.50) can still be applied, for a given value of D and part of the data set with that particular time slope would be properly processed using DMO. The resulting data could be converted into a 'pure' DMO data set for the dip under consideration by removing all other dipping data by using *dip filters*. Repeating this procedure for all values of D and summing the results is equivalent to applying DMO to all of the data set. Therefore partitioning of data into separate dips (i.e. using the dip domain) DMO can be expressed as

$$p_0(x \text{ or } k, t_0, h) = \int_0^{\infty} p_D\left(x \text{ or } k, t_0 = \sqrt{t_n^2 + D^2h^2}, h\right) dD \tag{11.52}$$

[Jakubowicz 1990, Eq. (10)]
'x' or 'k' stands for spatial coordinate or spatial frequency.

Equation (11.52) is equivalent to the Haley (1984) DMO with the difference that it is expressed in the dip domain and not in the frequency wave number (ω–k) domain. The dip-domain form [Eq. (11.52)] provides an opportunity for introducing the efficiencies of DMO in the Hale algorithm (Jakubowicz 1990).

Examples of numerical analysis of Dip-domain DMO may be found in Jakubowicz (1990).

The Hale DMO and its related practical form demonstrates that the number of dip ranges required to simulate substantially the results of the Hale algorithm depends on offset and travel time. Jakubowicz (1990) finds that for typical seismic data (having an offset up to 3,000 m, a velocity in excess of 1,500 m/s, a frequency up to 100 Hz, dips of up to 90° and typical mute patterns) results equiva-

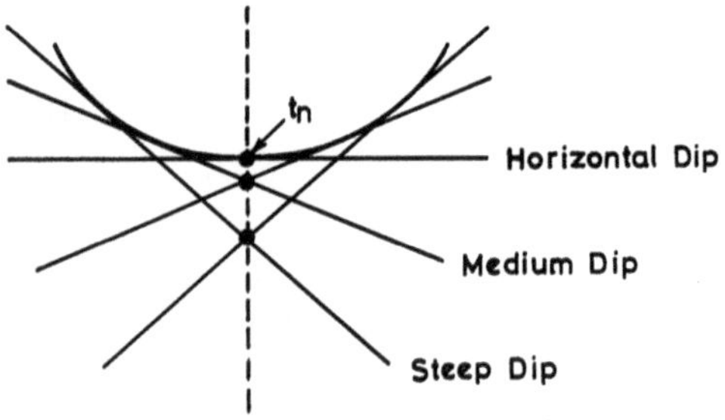

Fig. 11.12. Schematic illustration of how dip-domain DMO generates the DMO impulse response and corrects for reflection point smear. The *dashed line* corresponds to an original non-zero offset trace with t_n, the time of impulse after NMO at medium velocity. The DMO impulse response (i.e. zero offset section corresponding to the original impulse) is shown as a *bold curve* resulting from the constructive interference between the dipping events (Source: Jakubowicz 1990, from Geophysical prospecting with permission)

lent to that of Hale (1984) could be obtained using the dip domain DMO with, at most, up to 30 dip ranges. It is claimed in Jakubowicz (1990) that implementation of dip domain DMO can provide substantial computational cost saving over that of Hale's original algorithm.

How the process of NMO and *dip filtering* in dip domain DMO generate the DMO impulse response may be illustrated with the help of Fig. 11.12.

The original impulse is located at time t_n. This corresponds to the travel time of the event after NMO at a medium velocity. The application of *inverse NMO* [Eq. (11.50)] for non-zero dips relocates the impulse in time using a dip-dependent stacking velocity. The positions are shown for medium dip and steep dips. The subsequent dip filtering spreads the impulse along the corresponding time slope such that it is tangent to the required impulse response shown by the curved line. When this processing of NMO and dip filtering is done for all dips, it results in constructive superposition at all points along the required *impulse response curve*, and destructive superposition elsewhere. This produces the required *DMO impulse response*. The effect of NMO and dip filtering in *dip-domain DMO*, described above, for a given dip is to shift energy upwards and sideward, exactly in accordance with the correction that is required to correct for the *reflection point smear* for that of the dip angle.

11.7 Dip-Moveout and True Amplitude Imaging

The primary goal in seismic prospecting is to generate a true-amplitude estimate of the Earth's reflectivity in its fully migrated position. '*True amplitude*' is meant to specify that a migrated event's peak amplitude is proportional to the reflection coefficient. In an ideal situation, the proportionality constant between the peak amplitude and reflectivity should be the same, and should be independent of depth, dip, or final wavelet.

The standard imaging sequence is shown in Fig. 11.13. The pertinent question is to make the standard imaging sequence such as to produce true amplitude images. The exploding reflector concept, an appealing approach to dealing with the kinematics of zero-offset migration, does not fully address the amplitude question at zero offset. The application of DMO, however, makes it possible to appropriately address the question of amplitude due to arbitrarily dipping events at finite offsets. The *ideal DMO processing* is one that properly treats the amplitudes of reflection events both from horizontal and dipping reflectors. In general, DMO processing algorithms guarantee only the event positioning (kinematics). Black et al. (1993) gave a *revised DMO* technique which preserves peak amplitudes on horizontal and dipping data equally well with correct positioning of events. The *revised DMO* of Black et al. (1993) is an *amplitude preserving DMO*. The concept behind the *revised DMO* of Black et al. (1993) is explained with the help of Fig. 11.14. It shows the ray path geometry for finite offset and zero-offset with the various travel times symbols considered in Black et al. (1993).

Black et al. (1993) follows Hale's (1984) notation (refer to Section 11.5) to the extent possible, with changes as necessary for adopting it to 3 dimensions. For example,

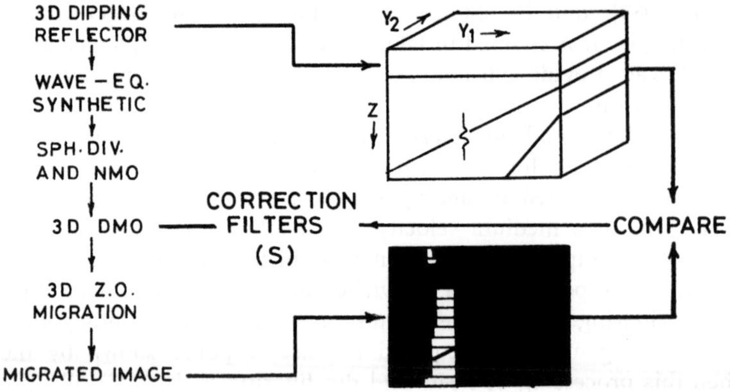

Fig. 11.13. Processing flow diagram for the standard imaging sequence, showing the wave-equation calibration procedure for a dipping reflector (Source: Black et al. 1993, from Geophysics with permission)

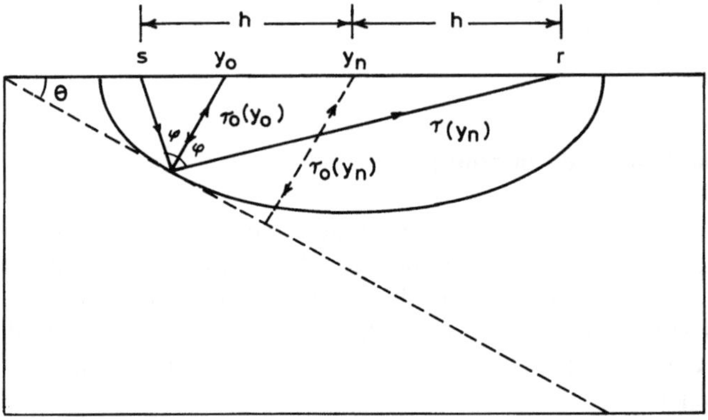

Fig. 11.14. DMO raypath geometry for finite offset and zero-offset, with various travel times annotated (Source: Black et al. 1993, from Geophysics with permission)

the mid-point position on a zero-offset section is y_0, in place of y in Hale's notation; 't' serves as an independent variable (in the argument) and 'τ' [as $\tau(y_0)$] as dependent variables giving the travel times of a dipping event as a function of the mid-point.

Black et al. (1993) mentions a subtle flaw in Hale's derivation of F-K DMO (refer to Section 11.5). The subtle flaw is that the reflection point smear has not been completely taken into account. According to Deregowski (1982) the reflection point smear means that the input event located at y_n (refer to Fig. 11.14) would be repositioned by application of DMO to correct zero-offset location y_0. In Hale's (1984) algorithm, the definition of DMO is

$$p_0(t_0, y_n, h) \equiv p_n(t_n, y_n, h) \tag{11.53}$$

[Black et al. 1993, Eq. (8)]

According to this, each sample of p_n is mapped from time t_n to t_0 without changing its mid point location y_n. Black et al. (1993) mentions that 'the lack of a mid-point change is the subtle flaw in Hale's argument'. *The effect of this flaw is that amplitudes of dipping events are not preserved.* This is proved in Black et al. (1993).

According to Black et al. (1993) the definition of DMO should be

$$p_0(t_0, y_0, h) \equiv p_n(t_n, y_n, h) \tag{11.54}$$

The correct relationship between t_0 and t_n in this case takes the form

$$t_o = \frac{t_n}{A} \tag{11.55}$$

as against $t_0 = At_n$ in Hale (1984). Here

$$A \equiv \left[1 + \left(\frac{Dh}{t_n}\right)^2\right]^{\frac{1}{2}} \tag{11.55 a}$$

[Black et al. 1993, Eq. (14)]
where D is the *time dip* given by

$$D = \frac{\partial \tau_0}{\partial y_0} = \frac{2\sin\theta}{v} = \frac{k}{\omega}$$

The relationship between y_0 and y_n (refer to Fig. 11.14) is

$$y_0 = y_n - \frac{Dh^2}{A\tau_n(y_n)} = y_n - \frac{h^2k}{A\omega t_n} \tag{11.56}$$

[Black et al. 1993, Eq. (17)]

Using Eq. (11.54), Black et al. (1993) define a new amplitude preserving DMO integral. The modified DMO of Black et al. (1993) is *amplitude preserving* for all dips, whereas Hale's DMO is strictly *amplitude preserving* for horizontal reflectors. It is given by

$$\begin{aligned} p_0(\omega, kh) &= \iint p_0(t_0, y_0, h)e^{i(\omega t_0 - ky_0)}dt_0dy_0 \\ &= \iint p_n(t_n, y_n, h)e^{i(\omega t_0 - ky_0)}dt_0dy_0 \\ &= \iint p_n(t_n, y_n, h)J_T e^{i\phi}dt_ndy_n \end{aligned} \tag{11.57}$$

[Black et al. 1993, Eq. (18)]
where J_T is a *Jacobian*, given by the determinant of a 2×2 matrix.

$$J_T = \left[\frac{\partial(t_0, y_0)}{\partial(t_n, y_n)}\right] = \frac{2A^2 - 1}{A^3}$$

The Jacobian for Hale's (1984) algorithm is:

$$J_H = \left|\frac{dt_0}{dt_n}\right| = \frac{1}{A}$$

It was mentioned in the beginning of this section that *true amplitude* is meant to specify that migrated events peak amplitude is proportional to reflectivity. This aspect has been discussed by Black et al. (1993) for a band - limited wavelet. The *image wavelet* depends on factors influencing the ray path such as dip, velocity, depth and offset. Black et al. (1993) explained the effect of ray path obliquity on the *image wavelet.* It was mentioned that two events plotted for same reflectivity and same deconvolved zero-phase *source wavelet* would show different *image wavelets* (e.g., see Fig. 3.9). This example shows that the *zero phase image wavelet* W_I *on migrated section* must vary from event to event.

In the processing sequence, we finally get migrated image p_m. It is convolution of the earth's reflectivity R $(\phi)\delta[t_m - \tau_m(y_m)]$ with a band-limited event-dependent image wavelet $W_I(t, E)$:

$$P_m(t_m, \mathbf{y}_m, \mathbf{h}) = R(\phi)\delta[t_m - \tau_m(\mathbf{y}_m)] \cdot W_I(t, E) = R(\phi)W_I[t_m - \tau_m(\mathbf{y}_m), E] \tag{11.58}$$

[Black et al. 1993, Eq. (1)]

In Eq. (11.58) $R(\phi)$ is the reflectivity; (t_m, y_m) are the migrated time and position. Reflectivity depends on angle ϕ (see Fig. 11.14). E is short notation for (D, τ_m, h), exhibiting the dependence on time-dip D, migrated event time, $\tau_m(y_m)$, and half-offset '**h**'. Because of the dependence of the image wavelet on these factors, extracting reflectivity from the migrated image P_m is not possible, unless the *image wavelet* $W_I(t, E)$ is global.

In the search for finding a global connection between P_m and R_ϕ, Black et al. (1993) adopted the following convention: 'keep the peak value of the zero phase *image wavelet* the same for all events'. Thus, the data can be so processed as to keep the peak value of W_I a global constant W(0), which is independent of **D**, τ_m and **h**. This implies that two events with the same (ϕ-dependent) reflectivity will have the same peak amplitude in the migrated image, i.e.

$$p_m(t_m = \tau_m(\mathbf{y}_m), \mathbf{y}_m, h) = R(\phi)W_I(\phi, \mathbf{D}, \tau_m, \mathbf{h}) = R(\phi)W(0) \tag{11.59}$$

[Black et al. 1993, Eq. (2)]

Black et al. (1993) investigated the method of maintaining the convention of keeping peak amplitudes constant for image wavelets. Their *wave equation analysis* of a dipping reflector establishes a wavelet stretch relationship between W_I and W(0).

$$W_I(t, \mathbf{D}, \tau_m, \mathbf{h}) = W(\lambda t) \tag{11.60}$$

[Black et al. 1993, Eq. (3)]

'λ' is a product of three *stretch* factors corresponding to NMO, DMO and zero-offset migration. The last Eq. (11.60) implies that the peak amplitude of W_I,

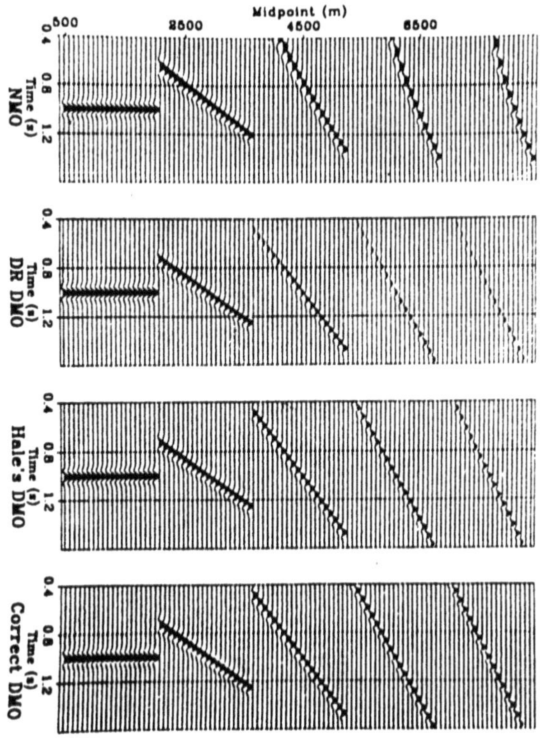

Fig. 11.15. Processing of synthetic-dipping reflector data through NMO and DMO steps of the standard sequence (Source: Black et al. 1993, from Geophysics with permission)

the t = 0 value is globally equal to W(0). Derivations of the stretch factors may be referred to in Black et al. (1993).

We give here an example from Black et al. (1993) to study the relative performance for various DMO algorithms including the revised DMO of Black et al. (1993). Figure 11.15 shows a series of events with dips ranging from 0 to 80° in increments of 20°. Each event is with same source wavelet and the value of R(ϕ) is independent of ϕ.

The top part is the result of applying *spherical divergence* and *NMO corrections*, and is input to subsequent DMO algorithms. The next three panels show the F–K DMO based outputs based on the methods of Deregowski and Rocca (DR), Hale and correct amplitude outputs.

The preservation of *peak amplitude* in true amplitude (correct) DMO is clear as is the loss of amplitudes of steep dips with Hale DMO and DR DMO algorithms (Black et al. 1993).

Another field example from Black et al. (1993) is presented in Fig. 11.16 for a marine survey in the Gulf of Mexico. Steeply dipping salt flanks are nicely imaged with the help of true amplitude integral DMO (for more details, refer to Black et al. 1993). True amplitude DMO produces a *stronger image* at large offsets.

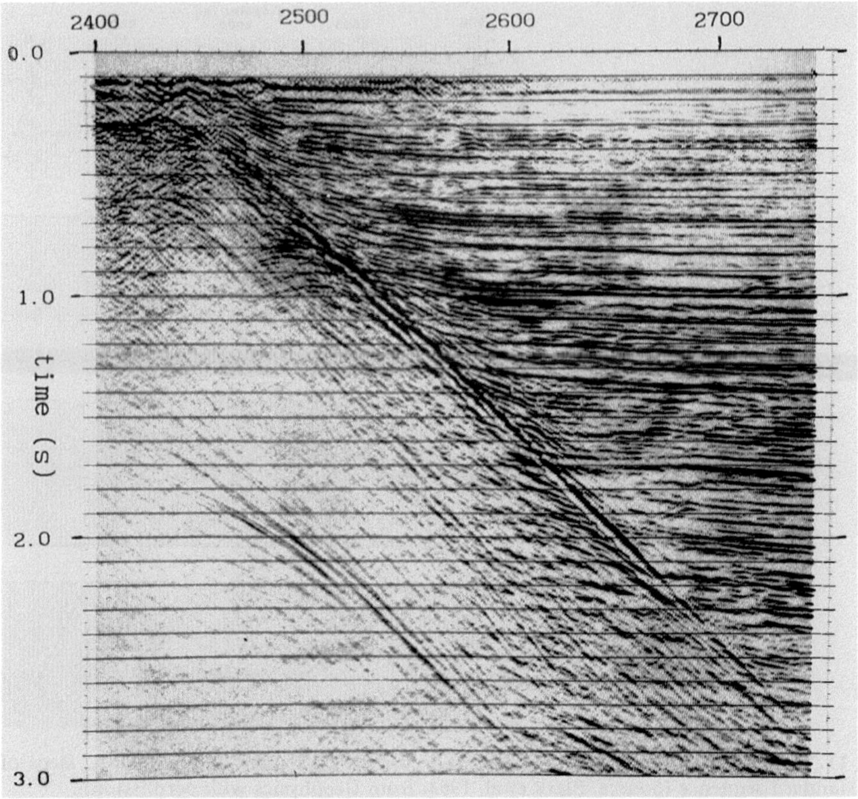

Fig. 11.16. Stack section after application of true-amplitude integral DMO steps of the standard processing sequence. The data is from a marine survey in the Gulf of Mexico. The CDP interval is 41 ft (Source: Black et al. 1993, from Geophysics with permission)

The true amplitude DMO algorithm is governed by its context. It is intimately related to the processing sequence of which it is a part. The revised DMO algorithm of Black et al. (1993) is correct when spherical divergence correction is made prior to DMO.

The true amplitude formulation algorithm has its advantage in not only obtaining the final migrated image of *steeply dipping events*, but also should be useful in DMO-processed amplitudes for amplitude vs. offset (AVO) studies.

Gardner and Canning (1998) propose a new development parallel to Black et al. (1993). It is also a velocity independent DMO. In their method NMO is applied after DMO, and then zero-offset migration is done. The method proposed a connection between offset (h) and the angle of incidence (ϕ) to get the reflectivity function R(ϕ). The final result is that each peak amplitude is a reflection coefficient linked to an angle of incidence.

CHAPTER 12

Basic Reflection Theory for Anisotropic Models 12

Moveout velocity and reflection amplitude have a dominant role in reflection prospecting. Both these quantities change in the presence of anisotropy. A particular type of anisotropy – the transverse isotropy, has been the main subject of study because of its greater relevance in seismic exploration. In this context it may be mentioned that the presence of anisotropy results in an error in calculation of depth if anisotropy is not accounted for in interpretation (e.g., see Banik 1984). Banik (1984) concludes that whenever a subsurface medium contains well-consolidated shales, a large error exists ($\approx$10%) in depth determination from seismic data. The theoretical work of Helbig (1983) and the laboratory experiments of Jones and Wang (1981) suggest that shales are anisotropic. For systematic and accurate interpretation of seismic records, study of both kinematic and dynamic properties in anisotropic media are most relevant. The study of dynamic properties is emphasized for generating synthetic seismograms. In seismic interpretation there has been a growing emphasis on extracting 'lithology' and 'fluid content' from changes in reflection amplitude with shot-geophone group offset. In this context it is very pertinent to know what role the *anisotropy* plays in affecting *reflectivity*. The study of reflection, transmission and surface conversion coefficients in anisotropic models is very important for generating synthetic seismograms to interpret field records for lithology, fluid contents etc. It is also very important to quantify the anisotropy parameters that account for the effects observed in the presence of anisotropy.

In the following, the questions related to the effects of anisotropy on reflection, transmission and surface conversion coefficients are addressed by discussing relevant theory supported by examples. The presentation would mainly relate to transverse isotropy.

12.1 Ray Series and Zero Order Approximation

Asymptotic ray theory provides an efficient method for generating synthetic seismograms. In this context, notable contributions are, for example, Hron and Kanasewich (1971), Hron (1973), Hron, Kanasewich and Alpaslan (1974). Babich (1961), Cerveny (1972), and Daley (1976) have applied asymptotic ray theory to the most general case of an anisotropic media. The application of *asymptotic ray theory* to anisotropic media shows that substitution of an assumed solution into equations of particle motion leads to an *eigenvalue problem*. The resulting *eigenvalues* are the normal velocities of a *quasi-compressional* P wavefront and *quasi-shear* SV and SH wavefronts. Cerveny and Psencik (1972) derived expressions for normal velocities for compressional and shear wavefronts for a transversely isotropic medium. The corresponding *eigenvectors* provide a preferred direction of *displacement vectors* for compressional and shear waves. For an isotropic medium, the *compressional eigenvector* is tangent to the wavefront normal and SV and SH *eigenvectors* are normal and binormal to the compressional eigenvector. For an anisotropic medium, the three eigenvectors form an orthogonal system. Musgrave (1961) showed that the P *eigenvector* is in general not tangent to the wavefront normal, and therefore the prefix *quasi* is used.

The treatment given below is based on asymptotic ray theory applied to anisotropic models for calculation of reflection, transmission and surface conversion coefficients as contributed by Daley and Hron (1977). In this investigation the displacement vector approaching or leaving a boundary is represented by a ray-series:

$$\mathbf{u}_v(x, y, z, t) = \sum_{n=0}^{\infty} \mathbf{U}_{nv}(x, y, z)\frac{e^{i\omega(t-\tau_v)}}{(i\omega)^n} \tag{12.1}$$

(Daley and Hron 1977)

Here ω is the frequency, τ is the *phase function* describing the wavefront, and v specifies the type of the wavefront and the medium according to the following scheme:

v=0 incident wave (either P or SV)
v=1 reflected P wave
v=2 transmitted P wave
v=3 reflected SV wave
v=4 transmitted SV wave

Accordingly, the amplitude terms, $\mathbf{U}_{nv}$, in the ray series have the following interpretation:

$\mathbf{U}_{n0} = \mathbf{P}_{n0}$ or $\mathbf{S}_{n0}$
$\mathbf{U}_{n1} = \mathbf{P}_{n1}$
$\mathbf{U}_{n2} = \mathbf{P}_{n2}$
$\mathbf{U}_{n3} = \mathbf{S}_{n3}$
$\mathbf{U}_{n4} = \mathbf{S}_{n4}$

If we are not very close to the source, and if the geometrical parameters of the layers are much greater than wavelength, taking n=0 may give a reasonable approximation. This is called *zero order* or *plane wave approximation*.

$\mathbf{U}_{nv}$ is not purely in the direction of $\mathbf{U}_{n0}$. Further $\mathbf{U}_{nv}$ may be expressed as a sum of the combination of scalar amplitudes (P_{nv}, p_{nv} and S_{nv}, s_{nv}) and two unit eigenvectors ($\mathbf{g}_v^{(1)}$ and $\mathbf{g}_v^{(2)}$) (Fig. 12.1):

$$\mathbf{P}_{nv} = P_{nv}\mathbf{g}_v^{(1)} + p_{nv}\mathbf{g}_v^{(2)} \tag{12.2 a}$$

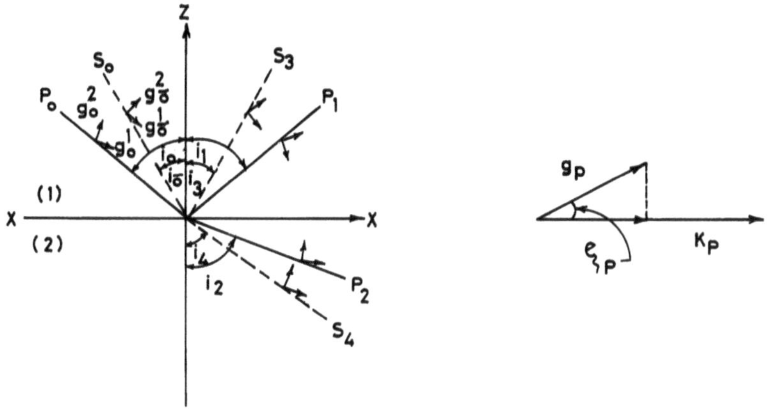

(a) (b)

Fig. 12.1 a, b. a Geometry of incidence for wavefront normals and displacement vectors at the interface of two transversely isotropic layers; **b** orientations of propagation vector ($\mathbf{k}_p$) and eigenvector ($\mathbf{g}_p$) (Source for Fig. 12.1 a: Daley and Hron 1977, from Bull Seism Soc Am, with permission of the author)

$$S_{nv} = S_{nv}\mathbf{g}_v^{(1)} + s_{nv}\mathbf{g}_v^{(2)} \tag{12.2 b}$$

[Daley and Hron 1977, Eqs. (2) and (3)]

Explicit expressions for $\mathbf{g}_v^{(1)}$ and $\mathbf{g}_v^{(2)}$ were found by Daley and Hron (1977). It may be mentioned that particle motion of a quasi-P-wave is polarized in the direction of eigenvector $\mathbf{g}_p$. This direction is not parallel to the direction of the *propagation vector* $\mathbf{K}_p$ and the angle ξ_p between $\mathbf{K}_p$ and $\mathbf{g}_p$ is given by (Fig. 12.1 b)

$$\cos\xi_p = \frac{1}{|\mathbf{K}_p||\mathbf{g}_p|}\left(\mathbf{K}_p \cdot \mathbf{g}_p\right) \tag{12.3}$$

In case of transverse isotropy, a P or SV wavefront incident on an interface produces only reflected and transmitted P and SV wavefronts. This is schematically shown in Fig. 12.1, which shows the wave front normals and displacement vectors. In this figure solid lines are normal to P wavefronts and the dashed line to SV wavefronts.

12.2 Reflecting Interface Separating Two Transversely Isotropic Media

Consider the case of two transversely isotropic media in welded contact (Fig. 12.1). The axis of symmetry for both media is the z-axis. *Eigenvectors* $\mathbf{g}_v^{(1)}$ and $\mathbf{g}_v^{(2)}$ are also shown. $\mathbf{g}_v^{(1)}$ and $\mathbf{g}_v^{(2)}$ have components (Daley and Horn 1977):

$$\mathbf{g}_v^1 = \left(l_v \sin i_v, (-1)^{v+1} m_v \cos i_v\right) \tag{12.4 a}$$

$$g_v^2 = (m_v \cos i_v, (-1)^v l_v \sin i_v) \tag{12.4b}$$

[Daley and Hron 1977, Eq. (6)]

'i_v' is the acute angle between the z-axis and vth wavefront normal. l_v, m_v in Eqs. (12.4a, b) are functions of the elastic constants, density and the angle between the z-axis and the wavefront normal.

At the interface between two media, the continuity of x and z components of displacement and normal and shear stresses are required. From these equations reflection and transmission coefficients may be derived.

Assuming that displacement vectors are plane waves, the boundary conditions of the continuity of displacement and stress give four linear equations in four unknowns, corresponding to amplitudes of reflected and transmitted P and SV waves. From these four equations, ratios of reflected and transmitted P and SV type displacement amplitude to the amplitude of incident disturbance, which correspond to displacement reflection and transmission coefficient normalized with respect to incident wave, may be calculated.

For details of the equations obtained on applying a boundary condition, reference may be made to the publication of Daley and Hron (1977).

Equations relating to reflection and transmission between two transversely isotropic media (Fig. 12.1) may be written in matrix form [Daley and Hron 1977, Eqs. (15) and (16)]:

$$\mathbf{AP} = \mathbf{H} \tag{12.5}$$

where **A** is a 4×4 coefficient matrix, and **P** and **H** are 4×1 column matrices. The elements of matrix **P** correspond to reflection and transmission coefficients.

Banik (1987) visualized the matrices **A** and **H** as a combination of two matrices; the first corresponding to isotropy and the second corresponding to linear terms in anisotropy. In this form, it is possible to analyze the effect of *anisotropy.* This would form the material of the next Subsection 12.2.1.

In the present context, we give below the meaning of *surface conversion coefficients.*

Assume a receiver (G) on the Earth's surface. The Earth's surface is treated as an interface between a T.I. medium and a vacuum. The receiver does not record only the disturbance caused due to the incident wavefront but also the disturbances caused due to two reflected wavefronts. We consider only the case of a P wave incidence. The case of a SV wave may be dealt with in a similar manner. We have:

Displacement at G ($\mathbf{u}$(G)) on the surface of the Earth = contributions to the displacement due to the incident wave front + reflected P wavefront + reflected SV wavefront. Or,

$$\mathbf{u}_P(G) = \mathbf{u}_P^i(G) + \mathbf{u}_P^R p(G) + \mathbf{u}_{PSV}^R(G) \tag{12.6}$$

In the zero-order approximation of the asymptotic ray theory, an arbitrary vector may be defined as

$$\mathbf{u}_\nu(G) = A_\nu(G) \exp i\omega(t - \tau_\nu(G))\mathbf{n}_\nu(G) \tag{12.7}$$

where $\mathbf{n}_\nu$ (G) is a unit vector.

As the value of t in Eq. (12.7) for incident and reflected waves at the interface are the same ($\tau_i = \tau_{RP} = \tau_{RSV} = \tau$, say), we have

$$\mathbf{u}_P(G) = P_i(G) \exp i\omega(t - \tau(G)) \times \{\mathbf{n}_i(G) + R_{PP}\mathbf{n}_{RP}(G) + R_{PSV}\mathbf{n}_{RSV}(G)\} \tag{12.8}$$

[Daley and Hron 1977, Eq. (26)]
where the incident wave is a P wave; subscript 'i' is for the incident wave; R_{PP} is the reflection coefficient corresponding to the reflected P wave and R_{PSV} is the reflection coefficient for the SV wave; $\mathbf{n}_{RP}$ and $\mathbf{n}_{RSV}$ are unit vectors for reflected P and SV waves, respectively.

Define a vector

$$\mathbf{g}_p(G) = \mathbf{n}_i(G) + R_{PP}\mathbf{n}_{RP}(G) + R_{PSV}\mathbf{n}_{RSV}(G) \tag{12.9}$$

$\mathbf{g}_p(G)$ is called the *surface conversion vector*, and its x and z components are called the *surface conversion coefficients.*

A similar expression for the *surface conversion vector and coefficients* may be written for an incident SV wave.

12.2.1 Linearized Reflection and Transmission Coefficients

Thomson (1986) defined an (replacement) anisotropy parameter 'δ' in terms of three elastic constants in the following form [refer to Eq. (5.52d)]:

$$\delta = \frac{\left[(c_{13} + c_{44})^2 - (c_{33} - c_{44})^2\right]}{2c_{33}(c_{33} - c_{44})} \tag{12.10}$$

Banik (1987) assigned physical meaning to 'δ' for a transversely isotropic medium. The elastic properties in such a medium depend on the angle between the axis of symmetry and directions of propagation and polarization. Banik assumed the axis of symmetry to be vertical and analyzed P-SV waves in a vertical plane. In a transversely isotropic medium, the elastic waves separate into two independent groups: SH-waves and P-SV waves. For SH-waves the anisotropy is elliptical and can be described by one anisotropy parameter. Further, as the SH-waves do not convert to P or SV waves, their reflection and transmission properties are similar to acoustic waves in elliptically anisotropic media. The P-SV group of waves requires two anisotropy parameters to describe their anisotropy. Further conversion from P to SV waves or vice-versa is possible. Considering P-SV wave motion in a vertical plane, Banik (1987) defined two anisotropy parameters ε_p and ε_s as

$$\varepsilon_p = \frac{(\alpha_h - \alpha_0)}{\alpha_0} \tag{12.11 a}$$

and

$$\varepsilon_s = \frac{(\beta_{45} - \beta_0)}{\beta_0} \tag{12.11 b}$$

[Banik 1987, Eqs. (4a) and (4b)]
where,
α_h: P-wave phase velocity in the horizontal direction,
α_0: vertical P-wave phase velocity,
β_{45}: SV-wave phase velocity at an angle of 45° to the axis of symmetry,
β_0: vertical SV-wave phase velocity

'ε_p' is approximately equal to the parameter ε in Thomsen [1986, Eq. (18b)], and is the anisotropy parameter for elliptically anisotropic media. ε_s describes the deviation of the phase velocity surface from ellipticity. For an elliptically anisotropic media $\varepsilon_s = 0$ and there is no SV-wave anisotropy. ε_s may be referred to as *SV-wave anisotropy.* The corresponding meaning of ε_p being zero does not imply that P-wave anisotropy is zero. Description of P-wave anisotropy requires both ε_p and ε_s. Banik calls ε_p *P wave anisotropy.* This becomes evident on linearizing the exact phase-velocity expression in Gassman (1964) in terms of ε_p and ε_s. Banik obtained the linearized form of phase velocity for a P and SV-wave in the following forms:

$$\alpha(\mathrm{i}) \cong \alpha_0 \left[\begin{matrix} 1 + (\varepsilon_p - 4\varepsilon_s\beta_0^2/\alpha_0^2)\sin^2 \mathrm{i} \cos^2 \mathrm{i} + \\ \varepsilon_p \sin^4 \mathrm{i} \end{matrix} \right] \tag{12.12 a}$$

$$\beta(\mathrm{i}) \cong \beta_0 \left[1 + 4\varepsilon_s \sin^2 \mathrm{i} \cos^2 \mathrm{i}\right] \tag{12.12 b}$$

[Banik 1987, Eqs. (5 a) and (5 b)]

The dependence of P and SV wave phase velocity surfaces on ε_s are shown in Figs. 12.2 a, b. In Fig. 12.2 a, P-wave (ε_p) anisotropy is the same for all curves, and SV wave anisotropy is different.

In the linearized expressions Eqs. (12.12 a) and (12.12 b), the ray velocities are equal to the phase velocities.

Banik (1987) determined the following relationship between the Thomsen parameter δ and P- and SV-waves anisotropies, ε_p and ε_s:

$$\delta \cong \varepsilon_p - 4\varepsilon_s\beta_0^2/\alpha_0^2 \tag{12.13}$$

The physical meaning to δ in Eq. (12.13) may be assigned (Banik 1987). 'δ' describes the relative competitiveness between *P wave anisotropy* and *SV-wave anisotropy* in affecting properties of a transversely isotropic medium. For many consolidated rocks which have $\alpha_0 \cong 2\beta_0$, $\delta \cong \varepsilon_p - \varepsilon_s$. Thus, depending on whether $\varepsilon_p > \varepsilon_s$ or $< \varepsilon_s$, δ is positive or negative.

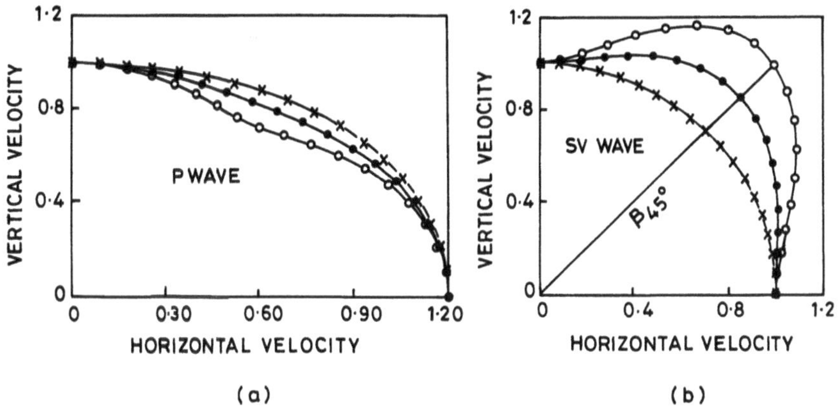

Fig. 12.2 a, b. a P-wave phase velocity surface for shale in model 1 of Table 12.1. All three curves have the same P-wave anisotropy, $\varepsilon_p = 0.20$ but different SV wave anisotropies, ε_s. The curves represented by a cross (x) corresponds to $\varepsilon_s = 0$ and $\delta \cong 0.20$, that represented by dots (●) to $\varepsilon_s = 0.25$ and $\delta = -0.07$, and that by circles to $\varepsilon_s = 0.5$ and $\delta \cong -0.30$. **b** SV wave phase velocity surface for shale. Parameters are the same as in Fig. 12.2a (Source: Banik 1987, from Geophysics with permission)

'δ' happens to be a very important parameter for reflection prospecting. It effectively describes the variation of moveout velocity [Thomsen 1986, Eq. (26a)]. Further changes in δ values at an interface adequately describe the effects of transverse isotropy on P-wave reflection amplitude. The offset-dependent effects in reflection prospecting arise mainly through δ, which is roughly the difference between *P-wave anisotropy* and *SV-wave anisotropy*. Banik (1987) derived the following relation for offset-dependent P-P reflection amplitude at short offsets (subscripts 'a' and 'i' stand for anisotropy and isotropy, respectively):

$$R_{aPP}(i) \cong R_{iPP}(i) + \frac{\Delta\delta}{2}\sin^2 i \tag{12.14}$$

[Banik 1987, Eq. (2)]
where

$R_{aPP}(i)$: PP reflection amplitude at an angle of incidence i at the interface between two transversely isotropic media

$R_{iPP}(i)$: PP reflection amplitude in the absence of anisotropy, and

$\Delta\delta$: $\delta_2 - \delta_1$ where δ_2 and δ_1 are values of δ on the lower and upper side of the interface (Fig. 12.3).

Equation (12.14) is a linearized form of the expression for the reflection coefficient in Daley and Hron (1977).

For isotropic media $R_{iPP}(i)$ is given by

$$R_{iPP}(i) = \frac{1}{2}\left(1 - 4\frac{\beta_0^2}{\alpha_0^2}\sin^2 i\right)\frac{\Delta\rho}{\rho_0} + \frac{1}{2\cos^2 i}\frac{\Delta\alpha}{\alpha_0} - 4\frac{\beta_0^2}{\alpha_0^2}\sin^2 i\frac{\Delta\beta}{\beta_0} \tag{12.15}$$

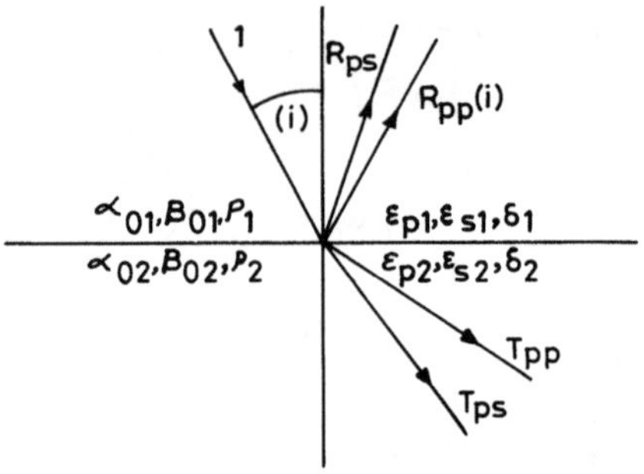

Fig. 12.3. Reflected and transmitted waves at an interface separating two transversely isotropic media (Source: Banik 1987, from Geophysics with permission)

(Aki and Richards 1980, p 153)
where
$\Delta\alpha = \alpha_2 - \alpha_1$
$\Delta\beta = \beta_2 - \beta_1$
$\Delta\rho = \rho_2 - \rho_1$
$\beta_0 = (\beta_2 + \beta_1)/2$
$\alpha_0 = (\alpha_2 + \alpha_1)/2$

and

$$\rho_0 = (\rho_2 + \rho_1)/2 \tag{12.15 a}$$

ρ, α and β in Eq. (12.15) are density and velocities in two media.

For a complete set of *linearized reflection* and *transmission coefficients* for the interface between two transversely isotropic media, we refer to the original contribution by Banik (1987).

Equation (12.14) reveals that knowing the change in δ at an interface, approximate effects of transverse isotropy on the offset

dependence of the P-wave reflection coefficient may be evaluated. From this equation, the following conclusions are drawn:

1. The quadratic in sin i is the same offset dependence as in the isotropic case at a small angle of incidence (Shuey 1985). Therefore, any anisotropy should have an important role in lithology estimation from the offset-dependent amplitude information.
2. The larger the contrast ($\Delta\delta$) in δ values, the stronger is the effect on amplitude. Conversely, when $\delta_2 = \delta_1$, the effect on amplitude is very small.
3. There is competition between *P-wave anisotropy* and *S-wave anisotropy* in influencing the offset dependence. If *S-wave anisotropy* is of the same order as *P wave anisotropy* and $\alpha_0 = 2\beta_0$, there is little effect on amplitude. The S-wave anisotropy balances out the effect of P-wave anisotropy.

Equation (12.14) is derived in Banik (1987) from the general form of Eq. (12.5).

The elements of **P** in Eq. (12.5) are the reflection and transmission coefficients corresponding to anisotropy:

$$\mathbf{P} = \begin{bmatrix} R_{aPP} \\ R_{aPS} \\ T_{aPP} \\ T_{aPS} \end{bmatrix} = \mathbf{A}^{-1}\mathbf{H} \tag{12.16}$$

In Eq. (12.16), R and T stand for reflection and transmission coefficients; the subscript 'a' stands for anisotropy, the second subscript P stands for source type P-wave and the third subscript (P or S) stands for reflected or transmitted wave type (refer to Fig. 12.3). The incident wave has been assumed to have unit amplitude.

To separate the effect of anisotropy, Banik (1987) expressed matrices **A** and **H** as a summation of two matrices:

$$\mathbf{A} = \mathbf{A}_0 + \Delta\mathbf{A} \tag{12.17 a}$$

and

$$\mathbf{H} = \mathbf{H}_0 + \Delta\mathbf{H} \tag{10.17 b}$$

The first term in Eqs. (12.17 a, b) corresponds to isotropy. The second terms are obtained by expanding **A** and **H** in small values of ε_{p1}, ε_{p2}, ε_{s1}, ε_{s2}, and by retaining only linear terms in these quantities. Further, the elements of $\Delta\mathbf{A}$ and $\Delta\mathbf{H}$ are assumed small compared to corresponding elements of $\mathbf{A}_0$ and $\mathbf{H}_0$.

By assuming **A** and **H** in the above form, P may be expressed as [Banik 1987, Eq. (18)]

$$\mathbf{P} = \mathbf{A}_0^{-1}\mathbf{H}_0 - \mathbf{A}_0^{-1}\Delta\mathbf{A}\left(\mathbf{A}_0^{-1}\mathbf{H}_0\right) + \mathbf{A}_0^{-1}\Delta\mathbf{H} \tag{12.18}$$

In Eq. (12.18), the first term $\mathbf{A}_0^{-1}\,\mathbf{H}_0$ is the column matrix $\mathbf{P}_0$ corresponding to reflection and transmission coefficients, R_{iPP}, R_{iPS}, T_{iPP}, T_{iPS}, in isotropic limit:

$$P_0 = \begin{bmatrix} R_{iPP} \\ R_{iPS} \\ T_{iPP} \\ T_{iPS} \end{bmatrix} \tag{12.19}$$

Equation (12.18) may be written as

$$\mathbf{P} = \mathbf{P}_0 - \left(\mathbf{A}_0^{-1}\Delta\mathbf{A}\right)\mathbf{P}_0 + \mathbf{A}_0^{-1}\Delta\mathbf{H} \tag{12.20}$$

Equation (12.20) is further simplified in Banik (1987) under the assumption of small changes in velocity and density at the interface resulting in the form

$$\mathbf{P} = \mathbf{P}_0 + \Delta\mathbf{H}^1 \tag{12.21}$$

where

$$\Delta\mathbf{H}^1 = -\mathbf{A}_0^{-1}\Delta\mathbf{A}\mathbf{P}_0^1 + \mathbf{A}_0^{-1}\Delta\mathbf{H}$$

Equation (12.21) is in a form which reveals the effect of anisotropy. The first term is the isotropic contribution and the second term contains linear terms in anisotropy parameters.

The first element of **P**, i.e. R_{aPP} [refer to Eq. (12.16)]:

$$R_{aPP}(i) \cong R_{iPP}(i) + \frac{\sin^2 i}{2}\left(\Delta\varepsilon_p - 4\frac{\beta_0^2}{\alpha_0^2}\Delta\varepsilon_s\right)$$

The variation of $R_{aPP}(i)$ from $R_{iPP}(i)$ for the isotropic case occurs because of two effects: (1) due to a variation in the phase velocity surface, and (2) due to a variation in the polarization direction of displacement vectors. Thomsen (1986) showed that polarization effects are at least of second order in anisotropy and, therefore, for linear effects only the effects of the phase velocity surface need be considered.

12.2.2 Numerical Results

Results of Wright (1987)

Wright (1987) analyzed the P-P reflection coefficient for isotropic and anisotropic models making use of the theory developed by Daly and Hron (1977, 1979) and Gassmann (1964). Wright's main conclusions are:

(i) In amplitude-offset studies involving shales, anisotropy should be taken into account.

(ii) The anisotropy may affect the amplitude-offset dependence in such a way that the expected trend of reflection coefficient variation for the isotropic case may be reversed in seismic exploration work.

Wright (1987) considered a transversely isotropic medium characterized by five independent elastic constants A_{11}, A_{33}, A_{55}, A_{66}, and A_{13}, containing the volume density ρ where $A_{ij} = c_{ij}/\rho$ (Fig. 12.4), c_{ij} are an elastic stiffness constant of a T.I. medium.

With reference to Fig. 12.4, velocities of propagation in the horizontal plane are

$$V_P^2 = A_{11} = V_{11}^2 \tag{12.22 a}$$

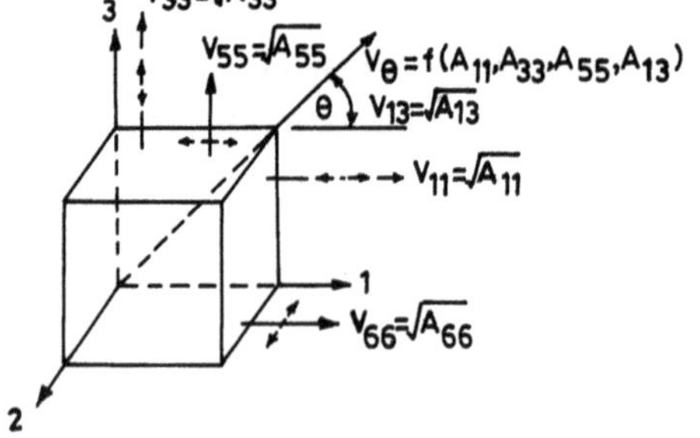

Fig. 12.4. Relations between elastic constants and wave velocities for a transversely isotropic medium (Source: Wright 1987, from Geophysics with permission)

$$V_{SV}^2 = A_{55} = V_{55}^2 \tag{12.22 b}$$

$$V_{SH}^2 = A_{66} = V_{66}^2, \tag{12.22 c}$$

and in the vertical plane are

$$V_P^2 = A_{33} = V_{33}^2 \tag{12.23 a}$$

$$V_{SV}^2 = A_{55} = V_{SH}^2 \tag{12.23 b}$$

At an angle to the axis of symmetry, combinations of A_{11}, A_{33}, A_{55}, and A_{13} control P and SV velocities. For convenience in modeling, Wright (1987) assumed that $V_{13}^2 = A_{13}$.

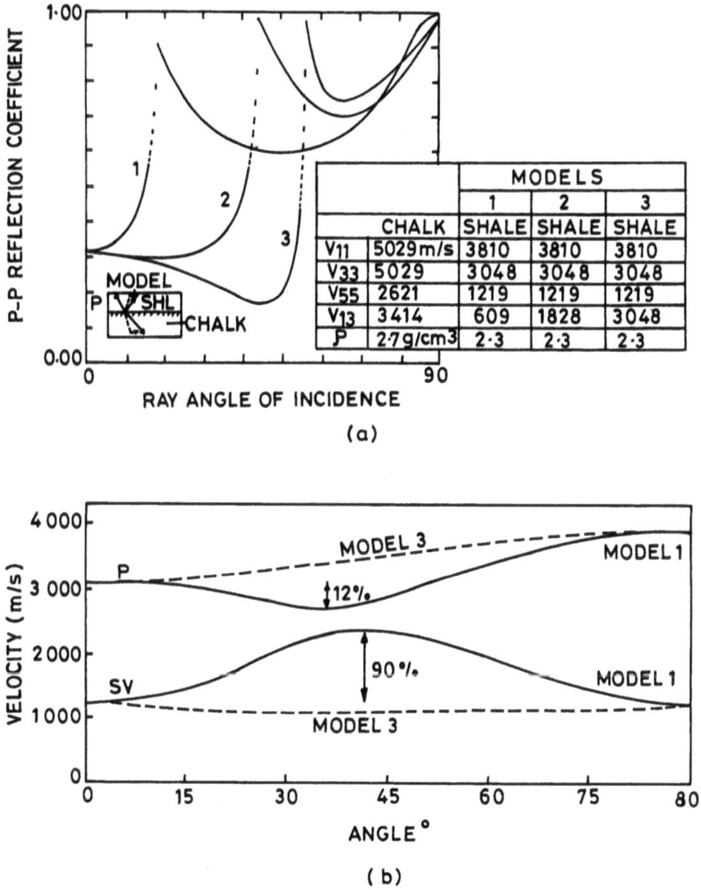

		MODELS		
		1	2	3
	CHALK	SHALE	SHALE	SHALE
V_{11}	5029 m/s	3810	3810	3810
V_{33}	5029	3048	3048	3048
V_{55}	2621	1219	1219	1219
V_{13}	3414	609	1828	3048
ρ	2·7 g/cm3	2·3	2·3	2·3

Fig. 12.5 a, b. a P-wave reflectivity at three shale-chalk interfaces showing the effect of varying only the shale V_{13} parameter. **b** P and SV phase velocities of plane waves in the two extreme shale models of Fig. 12.5 a showing the maximum percentage velocity change from vertical velocity (Source: Wright 1987, from Geophysics with permission)

In isotropic media

$$A_{11} = A_{33} = V_P^2$$

$$A_{55} = A_{66} = V_S^2, \text{ and}$$

$$A_{13} = V_P^2 - 2V_S^2$$

Wright considered anisotropic rock velocities from various shale and chalk models (Fig. 12.5 a). It shows the variation of P-wave *reflectivity* in a shale-chalk model, showing the effect of varying V_{13} for shale. Chalk is assumed isotropic. The V_{13} parameter affects the *critical angle.* As V_{13} decreases, the critical angle is smaller than for a higher V_{13}. For the smallest V_{13}, the reflection coefficient is found to increase immediately, for the largest V_{13} reflection coefficient decreases (curve 3) first and then begins to increase to the critical angle. Wright (1987) describes it as an effect of the variation of *phase velocity* (Fig. 12.5 b). The figure is self-explanatory. *Phase velocity* behavior contrasts the position of the *critical angle.*

In the second example examined by Wright (1987), a shale-sand interface was considered. The result is given here in Fig. 12.6. In the case of isotropic shale the reflection coefficient is found to increase, whereas for anisotropic shale, it is found to decrease with offset.

Banik (1987) analyzed Wright's (1987) result in the light of proposed anisotropy parameter 'δ' [refer to Eq. (12.13)], which is a combination of *P- and SV-wave anisotropies.* He observed that had the effect of SV-wave anisotropy also been considered in Wright's calculation the results would have been different, and, instead of decrease, an enhancement in the increase of amplitude should have occurred. Nevertheless, both investigations emphasize the importance of considering anisotropy in studying the *offset-dependence of reflection amplitude.*

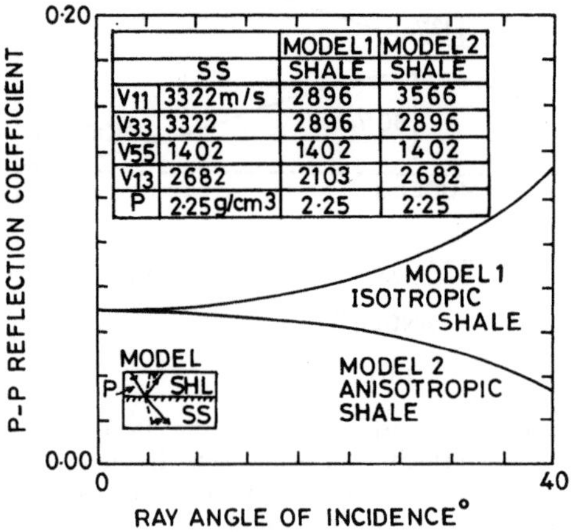

Fig. 12.6. P-wave reflection at two shale-sand interfaces. Model 1 – isotropic shale-sand. Model 2 – anisotropic shale and isotropic sand (Source: Wright 1987, from Geophysics with permission)

12.2.2.1
Results of Banik (1987)

Banik's (1987) approximate form of *reflection amplitude* [Eq. (12.14)) is elegant as it brings out separately the *effect of anisotropy.* In Banik's contribution, the reflection amplitudes are examined with a view to assessing the effect of *anisotropy* as against isotropy, and also how good the approximate equations are when compared with numerical results based on an exact equation. For the exact numerical evaluation, an exact form of reflection amplitude from Daly and Hron (1977) has been considered. The effects of the anisotropy parameters, ε_p and ε_s have been examined. In Banik's investigation, three geologically important models have been considered. These are the sand-shale model (model 1), the shale-gas sand model (model 2) and the shale-chalk model (model 3). Model 1 has been considered in Wright (1987) and model 3 has been considered in Shuey (1985). The isotropic elastic parameters are shown in Table 12.1.

The reflection coefficients were evaluated as a function of incident phase angle. A section of Banik's (1987) results are presented here in Fig. 12.7 a–f.

Each figure gives the model number. There are four curves in each figure. The thin continuous line represents the exact calculations assuming isotropy and the thick continuous curve is for the exact transversely isotropic expression in Daly and Hron (1977). The dash-dot curve is based on the approximate expression Eq. (12.14), and the dashed curve has been drawn to evaluate the error on the reflection coefficient in neglecting the second order polarization effects. The difference between the curves for isotropy (shown by the thin continuous line) and the exact curve for transverse isotropy (shown by the thick line) gives the effect of transverse isotropy. Similarly, the difference between the curve shown by a thick line and the dash-dot line (an approximate expression for transverse isotropy) reveals the accuracy of approximate expression Eq. (12.14). Further, the difference between curves shown by the thick line and a dashed-line brings out the error in neglecting the *polarization effects.*

The conclusions (Banik 1987, p 1,662) are as follows:

(i) the approximate relation [Eqs. (12.14) or (12.24)] is very good for angles of incidence to about 30° for models 1 and 2. However for model 3 (Fig. 12.7 e and f), the approximate relation is not good except for very small angles.
(ii) Except for large changes in the *elastic parameter* (as in model 3), the approximate expression Eq. (12.14) describes adequately the effect of transverse isotropy on the P-wave *reflection coefficient.*

Table 12.1. Interface models [Source: Banik 1987. From Geophysics with permission]

	α_1 (m/s)	β_1 (m/s)	ρ_1 (g/cm^3)	α_2 (m/s)	β_2 (m/s)	ρ_2 (g/cm^3)
Model 1	2,896	1,402	2.25	3,322	1,402	2.25
Model 2	2,307	942	2.15	1,951	1,301	1.95
Model 3	2,307	942	2.15	3,450	1,849	2.25

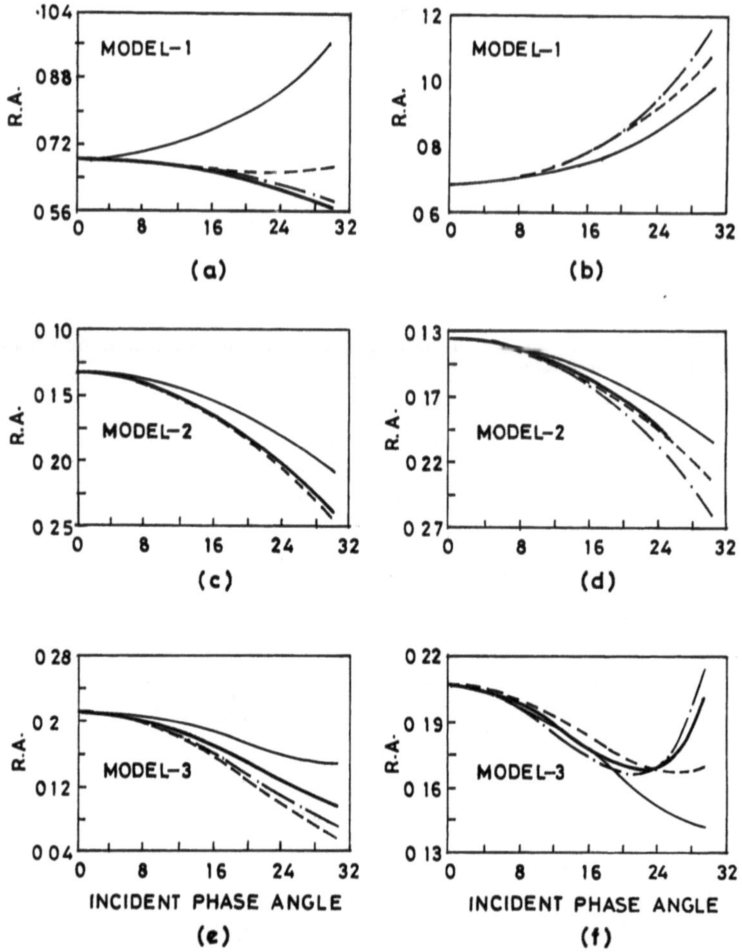

Fig. 12.7. P-wave reflection amplitudes (*R.A.*) for models in Table 12.1. The *thin continuous curve* represents exact calculations assuming isotropy; the *thick continuous curve* is for exact transversely isotropic expressions in Daley and Hron (1977); the *dash-dot curve* represents Eq. (12.14), and the dashed-curve corresponds to Eq. (12.24). P- and S-wave anisotropy parameters for different models are: (model 1: (a) $\varepsilon_{p1}=0.2$; $\varepsilon_{p2}=\varepsilon_{s2}=\varepsilon_{s1}=0.0$; (b) $\varepsilon_{p1}=\varepsilon_{p2}=\varepsilon_{s1}=\varepsilon_{s2}=0.2$); (model 2: (c) $\varepsilon_{p1}=0.2$, $\varepsilon_{p2}=\varepsilon_{s2}=\varepsilon_{s1}=0.0$; (d) $\varepsilon_{p1}=\varepsilon_{p2}=\varepsilon_{s1}=\varepsilon_{s2}=0.2$); and (model 3: (e) $\varepsilon_{p1}=0.2$, $\varepsilon_{p2}=\varepsilon_{s2}=\varepsilon_{s1}=0.0$); (f) $\varepsilon_{p1}=\varepsilon_{p2}=0.2$, $\varepsilon_{s1}=\varepsilon_{s2}=0$). For model 1 (b), the dash-dot and thick curves overlap (Source: Banik 1987, from Geophysics with permission)

More details on reflection amplitude approximations for isotropic, vertically transversely isotropic (VTI) and horizontally transversely isotropic (HTI) media is given in Chapter 13.

12.3
Case of Elliptical Anisotropy

Elliptical anisotropy is a special case of transverse isotropy in which wavefronts are the ellipsoid of revolution. Cholet and Richards (1954) and Richards (1960) used elliptical anisotropy to explain travel times at Berraine in the Sahara and in the foothills of Western Canada. Elliptical anisotropy incorporates many features inherent to wave propagation in a more general anisotropic medium, and can serve as a starting point for analyzing more complicated anisotropic models.

Assumption of ellipsoidal anisotropy simplifies greatly the quantities needed to compute the reflection and transmission coefficients (Daley and Hron 1979). The corresponding expressions in transversely isotropic media are more complicated. In the computation of reflection, transmission and surface conversion coefficients, Daly and Hron (1979) assumed that the axes of rotational invariance of the ellipsoids are perpendicular to the interfaces.

Reflection, transmission and surface conversion coefficients were calculated in Daly and Hron (1979) for varying degrees of anisotropy in medium 1. The degree of anisotropy is defined as

$$\left(\frac{ax_v - az_v}{az_v}\right) \times 100\,\text{percent}$$

(Fig. 12.8). (Daley and Hron 1979, p 36)

The parameters of medium 2 are kept constant at 10% anisotropy. In medium 1, the value of $(ax_1 + az_1)$ is kept constant to eliminate the effects of changing *impedance*. The graphs of the reflection and transmission coefficient therefore show only the effect of *anisotropy*.

Analytical expressions for coefficients in Daley and Hron (1979) are complex quantities. It requires two plots for each coefficient: the first one being the modulus vs. ray angle of incidence, and the second is the argument vs. ray angle.

Four possible cases of incidence exist at an interface between the ellipsoidally anisotropic media corresponding to a P or SV type wave, and an incidence from above or below. The computation of reflection and transmission coefficients along with mathematical expressions is given in detail in Daly and Hron (1979). Here, however, without giving mathematical expressions, we give some selected results only (Fig. 12.9a–d). Figure 12.9a,b shows the reflection and transmission coeffi-

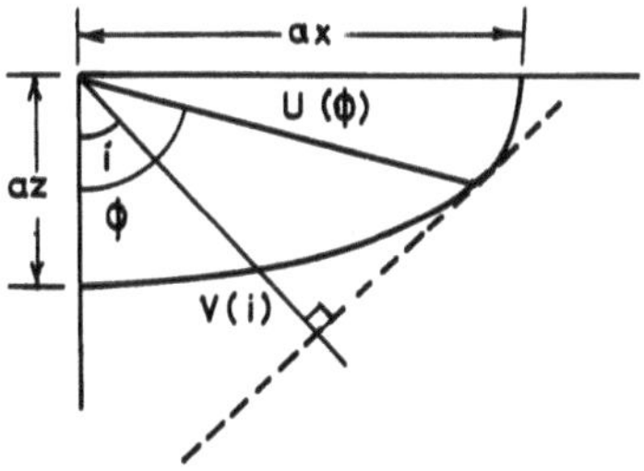

Fig. 12.8. The relation between ray and wavefront normal angles and velocities in an elliptical case (Source: Daley and Hron 1979, from Geophysics with permission)

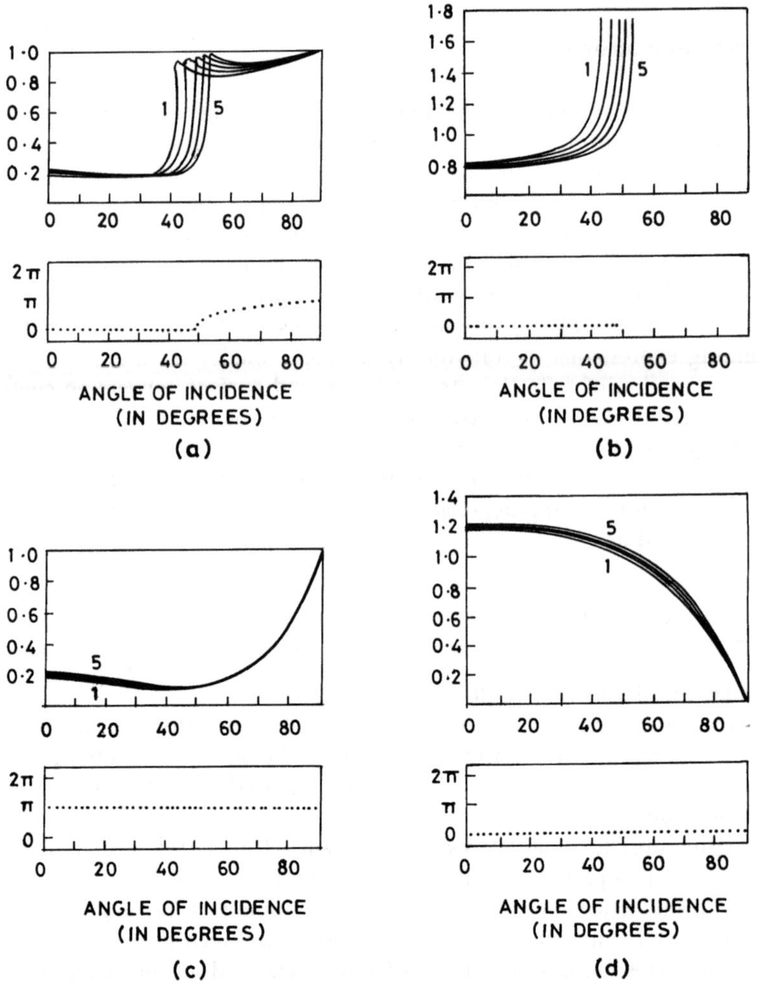

Fig. 12.9 a–d. a Modulus (*above*) and argument (*below*) of P_1P_1 reflection coefficients versus angle of incidence. **b** P_1P_2 transmission coefficients. **c** P_2P_2 reflection coefficients. **d** P_2P_1 transmission coefficients (Source: Daley and Hron 1979, from Geophysics with permission)

cients for P-wave incidences from medium 1 on medium 2. Figure 12.9 c, d depict corresponding quantities for incidence from medium 2. Each graph has five curves 1 to 5. These curves are for varying degrees of anisotropy (in percent) from 0% to 20% in steps of 5% increase. An elaborate set of graphical depictions of reflection and transmission coefficients from medium 1 or 2 may be found in Daley and Hron (1979).

The behaviour of these coefficients is very similar to coefficients for the similar isotropic case (refer also to Young and Braile 1976; McCamy et al. 1962; Tooley et al. 1965). An incremental change in anisotropy brings about incremental increase or decrease in the modulus – the relation is roughly linear.

12.4 Concluding Remarks

An important step in quantifying anisotropy effects on the reflection coefficient is to identify the parameter that produces anomalous results under conditions of anisotropy. Thomsen (1986) found that parameter 'δ' [Eq. (12.10)], defined in terms of three elastic constants' c_{13}, c_{44} and c_{33}, controls most anisotropic phenomenon of importance in exploration geophysics. Banik (1987) assigned physical meaning to 'δ'.

For a transversely isotropic medium with a vertical axis of symmetry, two groups of waves may propagate in the vertical symmetry plane: a P-SV group and a SH-group. The P-SV group of waves requires two anisotropy parameters to describe its anisotropy effects. In this group conversion from P to SV and SV to P on reflection is possible. The two-anisotropy parameters of the P-SV group are ε_p and ε_s (Banik 1987). ε_p and ε_s are referred to as P and SV wave anisotropy, respectively. ε_s describes the deviation of the phase velocity surface from elliptical. For an elliptically anisotropic medium, $\varepsilon_s = 0$. The relationship between δ and ε_p and ε_s is given by Eq. (12.13)

$$\delta \cong \varepsilon_p - 4\varepsilon_s \beta_0^2 / \alpha_0^2$$

The numerical value of δ depends on the value of ε_p and ε_s. ε_p and ε_s are defined in terms of phase velocity values by Eqs. (12.11 a,b). 'δ' is roughly the difference between P wave and SV wave anisotropy. The physical meaning of 'δ' is that it describes the relative competitiveness of P and SV wave anisotropy in influencing properties of a transversely isotropic medium.

Contrast in δ, i.e. $\Delta\ \delta$, across the reflecting boundary, controls offset-dependent reflection coefficient variation in a transversely isotropic medium, given by Eq. (12.14)

$$R_{aPP}(i) \cong R_{iPP}(i) + \frac{\Delta\delta}{2} \sin^2 i$$

'$\Delta\delta$' at an interface is given by

$$\Delta\delta = \delta_1 - \delta_2 \cong \Delta\varepsilon_p - 4\beta_0^2 \Delta\varepsilon_s / \alpha_0^2$$

The larger the contrast in δ, the stronger is the effect on reflection amplitude.

It should be realized that Eq. (12.14) is an approximation for reflection coefficient variation in a transversely isotropic medium. Exact expressions for the reflection and transmission coefficient were derived in Daley and Hron (1977). These expressions, however, are not as explicit as Eq. (12.14) to separate out anisotropy effects from isotropy.

The deviations of the reflection coefficient in the anisotropic case ($R_{aPP}(i)$) from the corresponding isotropic case (R_{iPP} (i)) arise because of two effects: Variations of *phase velocity surface*, and a variation in the polarization direction of *displacement vectors.* For linear effects, only the effect of phase velocity need be considered. Polarization effects are at least of second order in anisotropy parameters and may be neglected. For the sake of completeness in treatment, a brief discussion of the particle motion displacement vector and its relationship with wavefront, and propagation vector (K) in a transversely isotropic medium are given in Sections 12.1 and 12.2. These sections also cover the explanation of reflection, transmission and surface conversion coefficients. For more detailed discussion on particle motion polarization, Section 5.8 may be referred to.

The form of Eq. (12.14) does not explicitly reveal a phase velocity dependence of the reflection coefficient. This dependence may be appreciated if we consider Eq. (12.15) for the reflection coefficient variation for an isotropic medium and assume that velocities α and β vary with angle of incidence in accordance with Eqs. (12.12a) and (12.12b) as in a transversely isotropic medium. Under this assumption, and by taking linearized isotropic expressions in terms of $\Delta\alpha$ (i), $\Delta\beta$ (i), and $\Delta\rho$ at each angle 'i' (refer, for example, to Bortfeld 1961; Aki and Richards 1980), we have in the anisotropic case

$$R_{aPP}(i) \cong \frac{1}{2}\left[1 - 4\frac{\beta^2(i)\sin^2 i}{\alpha^2(i)}\right]\frac{\Delta\rho}{\rho} + \frac{1}{2\cos^2 i}\frac{\Delta\alpha(i)}{\alpha(i)} - 4\frac{\beta^2(i)}{\alpha^2(i)}\sin^2 i\frac{\Delta\beta(i)}{\beta(i)} \tag{12.24}$$

[Banik 1987, Eq. (30)]
where
$\alpha(i) = [\alpha_1(i) + \alpha_2(i)]/2$
$\beta(i) = [\beta_1(i) + \beta_2(i)]/2$
$\Delta\beta(i) = [\beta_2(i) - \beta_1(i)]$, and
$\Delta\alpha(i) = [\alpha_2(i) - \alpha_1(i)]$

Equation (12.24) shows explicitly the phase velocity dependence of the reflection coefficient. It is possible to write the simple form of Eq. (12.14) from Eq. (12.24). This may be achieved by substituting the following forms of $\Delta\alpha(i)/\alpha(i)$ and $\Delta\beta(i)/\beta(i)$ in Eq. (12.24) and retaining only the dominant terms [Banik 1987, Eqs. (31) and (32)]

$$\frac{\Delta\alpha(i)}{\alpha(i)} \cong \frac{\Delta\alpha_0}{\alpha_0} + \left(\frac{\sin^2 i}{\alpha_0}\delta_2\frac{\alpha_2^3}{\alpha_1^2} - \alpha_1\delta_1\right) \tag{12.25 a}$$

$$\frac{\Delta\beta(i)}{\beta(i)} \cong \frac{\Delta\beta_0}{\beta_0} + \frac{4\sin^2 i}{\beta_0}\left(\frac{\beta_2^3\varepsilon_{s2}}{\alpha_1^2} - \varepsilon_{s1}\frac{\beta_1^3}{\alpha_1^2}\right) \tag{12.25 b}$$

Equation (12.16), although approximate, is elegant in its ability to separate out the anisotropy effect from isotropy.

Figures 12.2a,b show the phase velocity surfaces for P and SV waves for shale. Figure 12.2a shows how the phase velocity surface for a P wave is affected both by P wave and SV-waves anisotropies, ε_p and ε_s. Three distinctly different curves have the same P-wave anisotropy, but different SV wave anisotropy. The case of SV anisotropy being zero (i.e. $\varepsilon_s = 0$, a case of elliptical anisotropy) is also shown.

The relationship of reflection coefficient variation including position of critical angle with phase velocity variation is observed more distinctly in Wright's (1987) results (Fig. 12.5 a, b). Figure 12.5 a shows how the elastic parameter A_{13} ($=c_{13}/\rho$; c_{13} is the elastic constant and ρ is density) changes the position of the critical angle for three shale-chalk anisotropic models. For these models all other elastic parameters (A_{11}, A_{33} and A_{55}) are kept equal; only A_{13} is varied. It may be observed in Fig. 12.5 b that, for model 3, the phase velocity increases smoothly from vertical velocity ($V_{33}=\sqrt{A_{33}}$) 3,048 m/s to 3,810 m/s in a horizontal direction. For model 1, the phase velocity decreases to a minimum of 2,682 m/s before rising to 3,810 m/s. This behaviour controls the position of the critical angle at different values for these models. It may be mentioned that Wright's (1987) results are based on an exact expression for the reflection coefficient derived in Daley and Hron (1977). From Figs. 12.5 a, b it may be inferred that c_{13} happens to be an important elastic constant, governing the offset dependence of the reflection coefficient in a transversely isotropic medium.

It may be mentioned that c_{13} is the only elastic constant which has both terms on P and SV wave anisotropy, ε_p and ε_s (or anisotropy quantifying parameter δ) in it. This relation is [Banik 1987, Eq. (7 a)]

$$c_{13} \cong c_{33} - 2c_{44} + c_{33}\left(\varepsilon_p - 4\varepsilon_s \beta_0^2/\alpha_0^2\right) \tag{12.26}$$

[Banik 1987, Eq. (7 a)]
where

$$c_{33} = \rho\alpha_0^2$$
$$c_{44} = \rho\beta_0^2$$

α_0 is the vertical P-wave phase velocity and β_0 is the vertical SV wave phase velocity. c_{11} is given by

$$c_{11} \cong c_{33}\left(1 + 2\varepsilon_p\right)$$

[Banik 1987, Eq. (7 b)]

c_{33} and c_{11} are elastic constants determining the vertical and horizontal P-wave velocities in a transversely isotropic medium with a vertical axis of symmetry.

Conclusively, Wright's (1987) and Banik's (1987) numerical results on reflection coefficient variation in a transversely isotropic medium suggest that *reflection coefficient variation with angle of incidence (or offset) is intimately related with the directional dependence of phase velocity.*

On the basis of numerical analysis, Banik (1987) concludes that the approximate Eq. (12.14) for reflection coefficient variation is good for angles of incidence to about 30° except in the case of large changes in elastic parameters. Further, changes in anisotropy parameter 'δ' describe adequately the small angle variation of the reflection coefficient in a transversely isotropic medium. 'δ' is roughly the difference between *P and SV wave anisotropies*, ε_p and ε_s, respectively.

Reflection coefficient variation for elliptical anisotropy is very similar to the isotropic case (Fig. 12.9). Incremental change gives rise to an incremental increase or decrease in the modulus of the reflection coefficient – the relation is roughly linear.

Chugh (1994) developed computer programs for computing the reflection coefficient for transversely isotropic media as a function of the ray parameter based on the work of Graebner (1992).

CHAPTER 13

Reflection Amplitude and AVO-Interpretation 13

13.1 Introduction

Partitioning of seismic energy at a geologic interface gives rise to changes in reflection amplitude, which varies with the angle of incidence. As the angle of incidence and offsets are related, the variation of amplitude with offset (seismic shot-geophone separation) of a reflected seismic signal would also be there. The first formulation of the problem of reflection and transmission coefficients for plane waves as a function of angle of incidence may be found in Zoeppritz (1919). The important inverse problem is to infer about the geologic boundary (elastic parameters on either side of the boundary) from an observation of reflection amplitudes as a function of the angle of incidence (or offset). The explicit Zoeppritz amplitude expression for calculating reflected P-wave amplitude may be solved, and may be programmed (refer, for example, to Young and Braile 1976). In a classic article, Muskat and Meres (1940) published results on the variation of reflection and transmission coefficients as a function of angles of incidence.

Since the publication of Muskat and Meres's results, various investigations have been attempted to make use of simplified Zoeppritz equations. Notable amongst these are contributions by Koefoed (1955), Tooley et al. (1965), Ostrander (1984), Shuey (1985), Rüger (1997, 1998) and Castagna et al. (1998).

In Koefoed's (1955) pioneering work, the interplay of *velocity contrast*, and the *Poisson ratio* on the *variation of reflection coefficient* vs. angle of incidence is clearly observed (refer, for example, to Shuey 1985, p 609 for Koefoed's rules).

Koefoed's (1955) investigation reveals that the *Poisson ratio* is the elastic property most directly related to the variation of the *reflection coefficient* with the angle of incidence (or offset).

The Poisson ratio for an isotropic elastic material may be defined as:

$$\sigma = \frac{\left(V_P/V_s\right)^2 - 2}{2\left[\left(\frac{V_P}{V_s}\right)^2 - 1\right]} \tag{13.1}$$

[Ostrander 1984, Eq. (1)]

where V_P and V_s are velocity of longitudinal and shear waves, respectively.

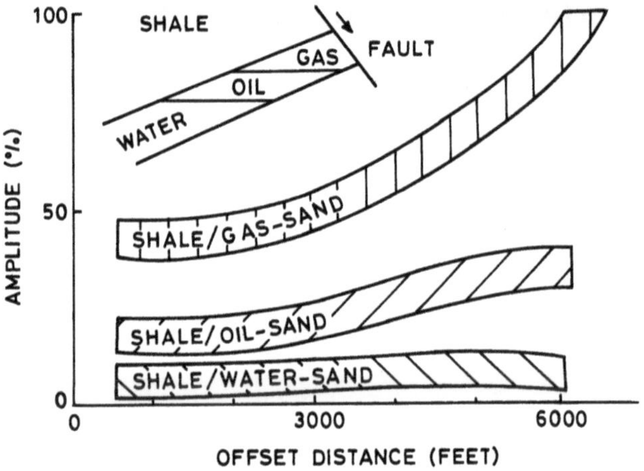

Fig. 13.1. Typical reservoir model and offset-amplitude patterns (Source: Yu 1985, from Geophysics with permission)

Thus, measurement of V_p and V_s may be used to find the value of the Poisson ratio.

The work of Gregory (1976) and Domenico (1976) are of direct interest in studies of the gas/fluid saturation effect in sediments on the reflection amplitude vs. angle of incidence, via the effect of *Poisson ratio.*

The contrast in the *Poisson ratio* between a gas-sand and the encasing medium (shale) is found to be large. This has focused attention on exploring *gas sand,* and using amplitude-vs. -offset anomalies (AVO anomalies) of seismic reflections (Rutherford and Williams 1989). Yu (1985) determined patterns (Fig. 13.1) in AVO anomalies for a *typical sand reservoir.* Results in Yu (1985) show that changes in amplitude-patterns may be used to visually demarcate various fluid, pore-fluid contacts in reservoir sand, and delineate the size of the reservoir.

It must be mentioned here that raw field results on amplitude variation are to be processed to obtain *true-amplitude information,* which is *geologic-in-origin,* free from other causes which affect the amplitude value at a particular offset. These effects are many, the prominent being *spherical divergence, transmission loss, radiation pattern* and *absorption.*

Amplitude variation with offset (AVO) in *gas sands* are of particular interest. This is because the contrast in Poisson ratio between gas sand and the encasing medium is large and this gives rise to detectable changes in amplitudes with offset (Rutherford and Williams 1989). On the basis of AVO response attempts have been made to classify gas sands. This is discussed below.

13.2 Classification of Gas Sands

Rutherford and Williams (1989) classified gas sands in three categories on the basis of *AVO characteristics* defined in terms of normal incidence reflection coefficient R_0 at the top of the gas sand. These categories are (Fig. 13.2)

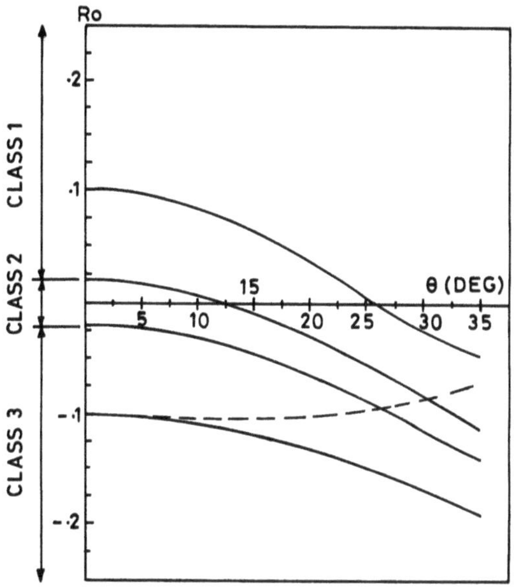

Fig. 13.2. Zoeppritz P-wave reflection coefficients, shown by *continuous curves*, for a shale-gas sand interface for a range of R_0 values. The Poisson's ratio and density of the shale were assumed to be 0.38 and 2.4 g/cm^3, respectively. The Poisson's ratio and density of the gas sand were assumed to be 0.15 and 2.0 gm/cm^3, respectively (Source: Rutherford and Williams 1989, from Geophysics with permission). Class IV sands, not discussed by Rutherford and Williams, have a negative normal-incidence reflection coefficient, but decrease in amplitude magnitude with offset (shown by the *dashed line*) (Source: Castagna et al. 1998)

Class 1: *high impedance sands,*
Class 2: *near-zero impedance contrast sands, and*
Class 3: *low-impedance sands.*

A *polarity change* of the reflection coefficient for *class 1-type gas sand* can have a pronounced effect on *stacked data.* If the *polarity change* is strong, the reflection amplitudes for the class 1 sand is likely to cancel in *CMP stacking*, or can have an opposite polarity to that given by normal incidence modeling.

This underlines the need to accurately model the reflector depth and acquisition geometry for a class 1 sand which may change polarity. Rutherford and William's (1989) results suggest that to enhance detectivity of class 1 sands, instead of a conventional stacked section, a *limited offset* or *constant angle seismic section* should be used.

For a class 2 gas sand, the normal incidence reflectivity R_0 is near zero. *Inversion techniques* using *stacked data* are questionable for *class 2 sands.*

Class 3 sands possess *large reflectivities* at all offsets and no *polarity change.* Therefore, their detectability on a stacked section is good. Class 3 sands have the characteristics that they are *bright* at near offset and brighten with offset. Therefore, they offer the best chance of having an adequate signal to noise ratio as a function of offset for AVO analysis.

Class 1 sands are found in hard rock onshore areas; *class 2 sands* in onshore and offshore areas and *class 3 sands* are found in marine environments (Rutherford and Williams 1989).

The classification of gas-sands by Rutherford and Williams (1989) is based only on the *normal incidence reflection coefficient.*

Castagna et al. (1998) gave a classification of gas sand which is based on the cross-plot of the AVO intercept A, and gradient B. They suggested that a gas-sand should be classified on the basis of their location in the A-B plane, rather than on the basis of the normal incidence reflection coefficient alone.

Castagna et al. (1998) showed that there is fourth category of *gas-sands* for which the reflection coefficient becomes more positive, but decreases in magnitude with increasing offset (refer to Fig. 13.2).

Porous sand overlain by a high-velocity material like hard-shale (e.g. siliceous or calcareous), siltstone, tightly cemented sand, or carbonate, frequently results in *class IV type behavior* of reflection coefficient with offset.

13.3 Approximation to Reflection Coefficient Variation

Since exact formulae for the plane wave reflection coefficient are often too complicated to be interpreted in terms of effects produced by geological models, attempts have been made in the past to derive approximate formulas for the reflection coefficient in forms suitable for quantifying effects of small changes in velocities, densities, and elastic parameters. These aspects would be discussed below.

13.3.1 Condition of Isotropy

The approximate formula for *reflection amplitude,* $R(\theta)$, in an isotropic case with a small percentage change of *elastic properties* may be written in the form

$$R(\theta) \approx \frac{1}{2}\left(1 - 4\frac{V_S^2}{V_P^2}\sin^2\theta\right)\frac{\Delta\rho}{\rho} + \frac{\mathrm{Sec}^2\theta}{2}\frac{\Delta V_P}{V_P} - 4\frac{V_S^2}{V_P^2}\sin^2\theta\frac{\Delta V_s}{V_s} \tag{13.2}$$

(Aki and Richards 1980 p 153)

Here $\Delta\rho/\rho$, $\Delta V_p/V_p$, $\Delta V_s/V_S$ are small fractional changes in density, compressional and shear wave velocity, respectively. Symbols in Eq. (13.2) are defined as

$$\Delta V_P = (V_{P2} - V_{P1})$$
$$V_P = (V_{P2} + V_{P1})/2$$
$$\Delta V_s = (V_{s2} - V_{Vs1})$$
$$V_s = (V_{s2} + V_{s1})/2$$
$$\Delta\rho = (\rho_2 - \rho_1)$$
$$\rho = (\rho_2 + \rho_1)/2$$

The incident wave is on side 1, and the transmitted wave on side 2 of the interface between the isotropic media. 'θ' is the average of incidence and transmission angle given by

$$\theta = \frac{\theta_1 + \theta_2}{2}$$

'p' is the ray parameter defined as

$$p = \frac{\sin\theta_1}{V_{p1}} = \frac{\sin\theta_2}{V_{p2}}$$

Motivated by Koefoed's (1955) observation that the *Poisson ratio* is the most directly related parameter to the angular dependence of reflection coefficient, Shuey (1985) modified Eq. (13.2) in terms of the Poisson ratio σ in place of ΔV_s and V_s. Defining

$$\Delta\sigma = \sigma_2 - \sigma_1,$$
$$\sigma = \frac{\sigma_2 + \sigma_1}{2},$$

and using Eq. (13.1), Shuey (1985) obtained the following form for relative variation of *reflection amplitude* ($R(\theta)/R_0$) with offset

$$\frac{R(\theta)}{R_0} \approx 1 + A\sin^2\theta + B\left(\tan^2\theta - \sin^2\theta\right) \quad (13.3)$$

where

$$R_0 \approx \frac{1}{2}\left(\frac{\Delta V_P}{V_P} + \frac{\Delta\rho}{\rho}\right)$$
$$A = A_0 + \frac{1}{(1-\sigma)^2}\frac{\Delta\sigma}{R_0}$$
$$A_0 = B - 2(1+B)\frac{1-2\sigma}{1-\sigma}, \text{ and}$$
$$B = \frac{\Delta V_P/V_P}{\Delta V_P/V_P + \Delta\rho/\rho}$$

Equation (13.3) is in a form suitable for interpreting which combinations of *elastic properties* are effective in different ranges of incidence angle θ. Equation (13.3) may be written as

$$R(\theta) \approx R_0 + \left[A_0 R_0 + \frac{\Delta\sigma}{(1-\sigma)^2}\right]\sin^2\theta + \frac{1}{2}\frac{\Delta V_P}{V_P}\left[\tan^2\theta - \sin^2\theta\right] \quad (13.4)$$

Equation (13.4) has three terms, which are important in three different angle ranges, and are related with different elastic properties.

The first term R_0 (amplitude at normal incidence) is given by

$$R_0 \approx \frac{1}{2}\left(\frac{\Delta V_P}{V_P} + \frac{\Delta \rho}{\rho}\right) \approx \frac{1}{2}\frac{\Delta \rho V_P}{\rho V_P} = \frac{1}{2}\Delta \ln\left(\rho V_P\right) \qquad (13.4\,a)$$

[Shuey 1985, Eq. (16)]

R_0 is half the change in the natural logarithm of *impedance* ρV_P.

The second term A $\sin^2\theta$ in Eq. (13.3) is important in the intermediate angle range ($0<\theta<30°$). $R(\theta)/R_0$ in the intermediate angle range depends on A, which is a combination of two terms, namely, A_0 and $(1/(1-\rho)^2)(\Delta\sigma/R_0)$. Shuey (1985) showed that A_0 is accurately predictable under certain conditions, and therefore, information on $\Delta\sigma/R_0$ is contained in A.

The third term in Eq. (13.3) is pertinent at wide angles, and the reflection amplitude at wide angles is related to the change in velocity. At these angles, the third term dominates the first two.

Referring to Eq. (13.3), the amplitude increases or decreases depending on whether A is greater than or less than zero. B controls the sign at large angles. Shuey (1985) argues that $B<0$ is rare.

Replacing $\sin\theta$ by θ_1, which is accurate up to $\theta_1<30°$, we have

$$R = R_0\left(1 + A\theta_1^2\right) \qquad (13.5)$$

[Shuey 1985, Eq. (18)]

'θ_1' is the angle of incidence in medium 1. θ is the average of the angle of incidence θ_1 and the angle of transmission θ_2.

Figure 13.3 a, b shows variations of the reflection coefficient for two cases $\Delta V<0$ and $\Delta V>0$, using an exact equation and approximate equations.

Variation of $R(\theta)$ for different values of parameter A, based on Eq. (13.5) [Shuey 1985, Eq. (18)] is shown in Fig. 13.3 c.

Equation (13.5) could be used to extract information on A from the value of $R(\theta)/R_0$. From knowledge of A, $\Delta\sigma/R_0$ could be obtained, provided a very approximate value of the *average Poisson ratio*, and the *velocity/impedance ratio* B is available or hypothesized (Shuey 1985, p 613). $\Delta\sigma/R_0$ could provide useful information on *lithology*. Since R_0 is a change in the natural logarithm of acoustic impedance, $\Delta\sigma/R_0$ is the slope on a cross plot of Poisson ratio *s* vs. impedance ρV_P on a log scale.

Shuey's (1985) results are valid provided the percentage change in *elastic properties* $\Delta V_P/V_P$, $\Delta V_s/V_s$, $\Delta\rho/\rho$ are small compared to unity, which holds good in the majority of cases in which the reflection coefficient R_0 does not exceed 0.2.

Using Biot-Gassman theory on the effects of pore-fluid on rock-velocities for fluid-contact interfaces, Wright (1986) showed that Eq. (13.2) reduces to

$$R(\theta) \approx \frac{\Delta\rho}{2\rho} + \mathrm{Sec}^2\theta\frac{\Delta V_P}{2V_P} \qquad (13.6)$$

[Wright 1986, Eq. (8)]

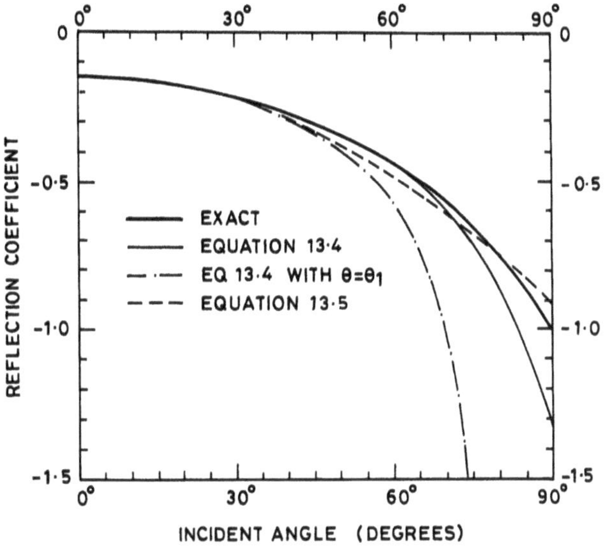

(a)

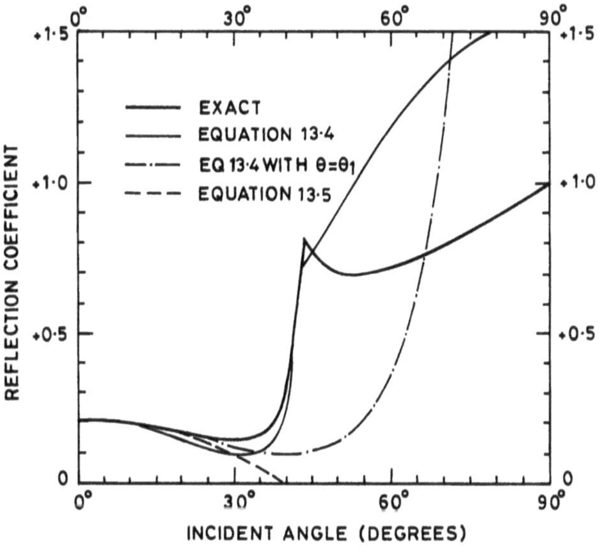

(b)

Fig. 13.3a–c. a Approximations in an instance with $\Delta V < 0$. The elastic properties correspond to an actual gas sand in the Gulf of Mexico: $V_{p1} = 7{,}570$ ft/s, $\rho_1 = 2.15$ and $\sigma_1 = 0.40$ for the overlying shale, and $V_{p2} = 6{,}400$ ft/s, $\rho_2 = 1.95$ and $\sigma_2 = 0.10$ for the gas sand. **b** Approximation in an instance $\Delta V > 0$. The elastic properties on the *upper side* are the same as for Fig. 13.3a, while the underlying medium has $V_{p2} = 11{,}350$ ft/s, $\rho_2 = 2.20$ and $\sigma_2 = 0.30$. These values are appropriate for a chalk with 30% porosity. Beyond the critical angle (42°), the reflection coefficient becomes complex and the absolute value is shown for the exact solution and for Eq. (13.4). **c** Parabolic approximation to $R(\theta)$ for different values of the dimensionless parameter A [Eq. (13.5)] (Source: Shuey 1985)

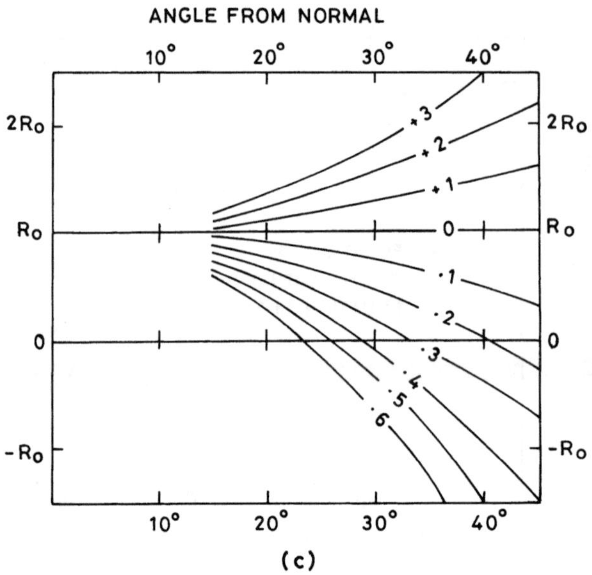

Fig. 13.3c

Equation (13.6) may be further simplified by changing $\mathrm{Sec}^2\theta$ in terms of powers of $\sin^2\theta$ (by expansion) as

$$R(\theta) \approx \frac{\Delta\rho}{2\rho} + \frac{\Delta V_P}{2V_P} + \frac{\Delta V_P}{2V_P}\sin^2\theta + \frac{\Delta V_P}{2V_P}\sin^4\theta \tag{13.7}$$

Further taking

$$R_0 \approx \frac{1}{2}\left(\frac{\Delta\rho}{\rho} + \frac{\Delta V_P}{V_P}\right),$$

and assuming a small angle of incidence Eq. (13.7) reduces to,

$$R(\theta) \approx R_0 + \frac{\Delta V_P}{2V_P}\sin^2\theta \tag{13.7 a}$$

[Wright 1986, Eq. (11)]

Equation (13.7 a) tells us that the reflection coefficient equals the normal-incidence reflection coefficient plus a correction term that depends on offset and P-wave velocity. This form tells us that the reflection coefficient is independent of the Poisson ratio, and therefore may be studied without knowledge of shear wave velocity.

Figure 13.4 depicts *reflection coefficient* behavior at a *gas brine* contact in sandstone (Wright 1986).

The corresponding equation of SH-SH reflection is

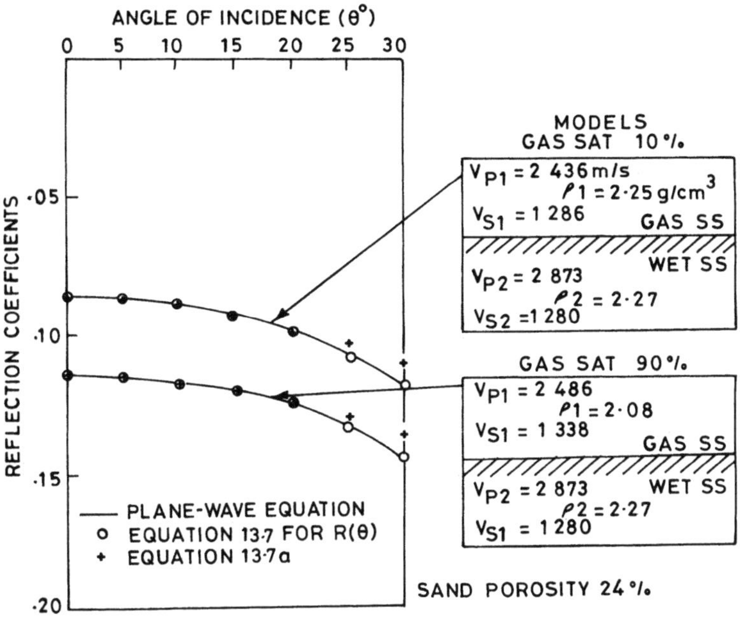

Fig. 13.4. Comparison of exact plane-wave and approximate P-wave reflection coefficients $R(\theta)$ for a gas-brine contact in sandstone (Source: Wright 1986, from Geophysics with permission)

$$S(\theta) \approx S_0\left(1 + \frac{V_2}{V_{s1}}\sin^2\theta_1\right) \tag{13.8}$$

[Wright 1986, Eq. (17)]

For a small change in S-wave velocity, it may be written as

$$S(\theta) \approx S_0\left(1 + \sin^2\theta\right) \tag{13.8 a}$$

The ratio of the P to SH reflection coefficient is related to *gas saturation*. Wright (1986) gives the following relationship:

$$\frac{R_0}{S_0} \approx -2\left(1 + \frac{\rho\Delta V_p}{V_p}\frac{1}{S_g\phi}\right), \tag{13.9}$$

[Wright 1986, Eq. (25)]

where S_g is fractional gas saturation, ϕ is fractional porosity. Wright (1986) relates *gas saturation* with the reflection coefficient ratio. Wright's (1986) result is a confirmation of the findings in Meissner and Hegazy (1981) who showed that the study of the ratio of the normal-incidence P-S reflection coefficient has the potential for discriminating between different *gas concentrations*.

13.3.2
Condition of Anisotropy

Anisotropy presents special problems for studying the reflection coefficient variation with offset. The conventional AVO signature for isotropic conditions may be significantly distorted by the presence of anisotropy. Rüger (1997, 1998) has further extended the approximate solution study for anisotropic conditions – transversely isotropic medium with either a vertical axis of symmetry (VTI), or a horizontal axis of symmetry (HTI). These approximate analytic solutions form the basis of our following discussions.

13.3.2.1
Transversely Isotropic (T.I.) Medium with a Vertical Axis of Symmetry: Azimuthal Isotropy

Consider quasi P-wave reflection at a boundary between two transversely isotropic media with a vertical axis of symmetry (*VTI media*). Further, assume conditions of *weak anisotropy*, and a boundary with small discontinuities in elastic properties (which are geologically reasonable assumptions).

Rüger expresses vertical P-wave velocity, S-wave vertical velocity and density in two media in terms of the average velocity value and the difference in velocity values in two media. Accordingly,

$$V_{PO1} = \overline{V_{PO}}\left(1 - \frac{1}{2}\frac{\Delta V_{PO}}{\overline{V_{PO}}}\right), \text{ and}$$

$$V_{PO2} = \overline{V_{PO}}\left(1 + \frac{1}{2}\frac{\Delta V_{PO}}{\overline{V_{PO}}}\right),$$

[Rüger 1997, Eqs. (1) and (2)]
are vertical P-wave velocities in upper and lower transversely isotropic media. Similar expressions are taken for S-wave velocities and densities (i.e. V_{SO1}, V_{SO2}, ρ_1 and ρ_2). The bar denotes the average value, and Δ – the difference in values for lower and upper medium parameters. The assumption of small discontinuities in elastic parameters and weak anisotropy are expressed as

$$\left|\frac{\Delta V_{PO}}{\overline{V_{PO}}}\right|, \left|\frac{\Delta V_{SO}}{\overline{V_{SO}}}\right|, \left|\frac{\Delta \rho}{\overline{\rho}}\right|, |\delta|, |\varepsilon|, |\gamma| \ll 1$$

Rüger (1997) gives, under assumptions of small discontinuities in elastic parameters and weak anisotropy, the following expression for the P-P reflection coefficient as a function of angle of incidence θ:

$$\begin{aligned} R_P^{VTI}(\theta) = &\frac{1}{2}\frac{\Delta Z}{\overline{Z}} + \frac{1}{2}\left\{\frac{\Delta V_{PO}}{\overline{V_{PO}}} - \left(\frac{2\overline{V_{SO}}}{\overline{V_{PO}}}\right)^2\frac{\Delta G}{\overline{G}} + \Delta\delta\right\}\sin^2\theta \\ &+ \left\{\frac{\Delta V_{PO}}{\overline{V_{PO}}} + \Delta\varepsilon\right\}\sin^2\theta\tan^2\theta \end{aligned} \tag{13.10}$$

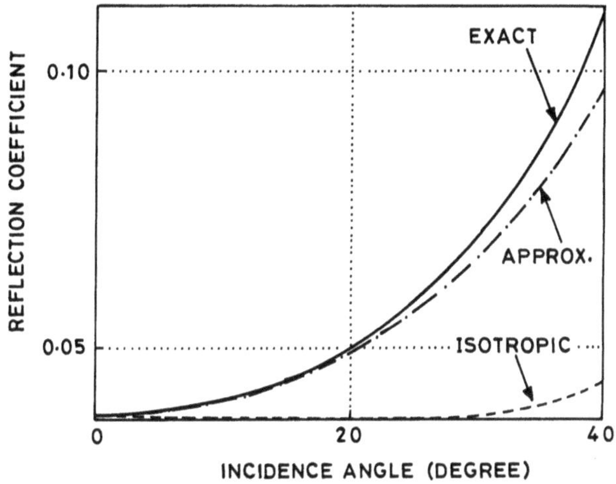

Fig. 13.5. Reflection coefficient for VTI media. Shown are the exact VTI solution (*continuous line*), the approximation based on Eq. (13.10) (*dash-dot line*), and the corresponding isotropic ($\delta_2=0$, $\varepsilon_2=0$) reflection coefficient (*dash line*). Model parameters are $V_{PO1}=2.9$, $V_{SO1}=1.8$, $\rho_1=2.18$, $\varepsilon_1=0$, $\delta_1=0$, and $V_{PO2}=3.1$, $V_{SO2}=1.85$, $\rho_2=2.2$, $\varepsilon_2=0.1$, $\delta_2=0.2$ (Source: Rüger 1997, from Geophysics with permission)

[Rüger 1997, Eq. (4)]
where Z denotes the vertical P-wave impedance ($Z=\rho V_{PO}$), $G=\rho V^2{}_{SO}$ is the vertical shear wave modulus. $\Delta\delta$ and $\Delta\varepsilon$ are equal to $(\delta_2-\delta_1)$ and $(\varepsilon_2-\varepsilon_1)$, respectively.

The validity of Eq. (13.10), i.e. the approximate solution for a VTI medium for weak anisotropy is depicted in Fig. 13.5 vis-a-vis exact values of the reflection coefficient for transversely isotropic and isotrokpic media.

13.3.2.2 Transversely Isotropic Medium with a Horizontal Axis of Symmetry (HTI Medium): Azimuthal Anisotropy

Referring to Fig. 5.7b, an incident P-wave in an isotropy plane would excite S^{11} mode polarized within the fractures. If the incident P-wave is in the symmetry plane, it would generate $S_\perp$. The speeds of S^{11} and $S_\perp$ are different.

Computation of the *exact P-P reflection* in the isotropy plane is straightforward, and may be written from existing solutions for isotropic media [Aki and Richard 1980, e.g., see Eq. (13.3)] by inserting fracture plane velocities.

The most significant outcome of the HTI/VTI analogy is the derivation of new effective "*Thomsen style*" *anisotropy parameters* for HTI media (refer to Subsection 5.3.3 and Section 5.4). For example, the approximate P-wave reflection coefficient in the symmetry plane of HTI media may be written from Eq. (13.10) for VTI media.

Mallick and Frazer (1991) described the sensitivity of P-wave amplitude variation vs. offset to fractures or cracks. They showed that the P-wave AVO response has azimuthal dependence from a fractured layer. Azimuthal dependence of the P-wave response has also been observed on field data in Lynn et al. (1995) and

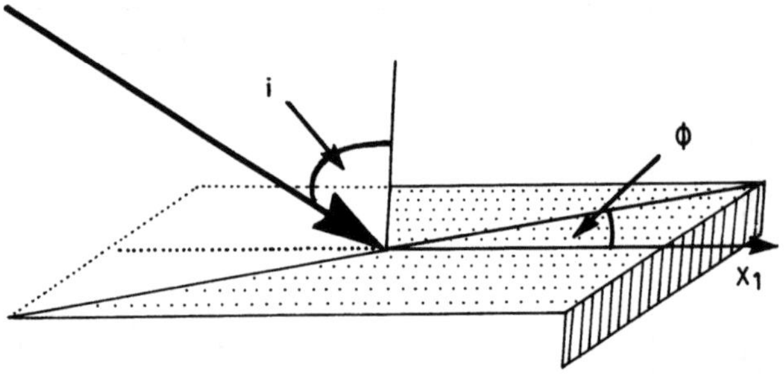

Fig. 13.6. The angle between the slowness vector of the incident wave and vertical is denoted by i. The azimuth angle ϕ is defined with respect to a symmetry axis pointing in the x_1-direction (Source: Rüger 1998, from Geophysics with permission)

Johnson (1995). Rüger (1998) considered azimuthal variation of the P-wave reflection including contributions at large incidence angles, and to lower symmetry systems.

Referring to Fig. 13.6 the expression for the compressional plane-wave reflection coefficient between a HTI/HTI interface is given as

$$R_p(i,\phi) = \frac{1}{2}\frac{\Delta Z}{\overline{Z}} + \frac{1}{2}\left\{\frac{\Delta\alpha}{\overline{\alpha}} - \left(\frac{2\overline{\beta}}{\overline{\alpha}}\right)^2\frac{\Delta G}{\overline{G}} + \left[\Delta\delta^{(V)} + 2\left(\frac{2\overline{\beta}}{\overline{\alpha}}\right)^2\Delta\gamma\right]\cos^2\phi\right\}\sin^2 i$$
$$+\frac{1}{2}\left\{\frac{\Delta\alpha}{\overline{\alpha}} + \Delta\varepsilon^{(V)}\cos^4\phi + \Delta\delta^{(V)}\sin^2\phi\cos^2\phi\right\} \times \sin^2 i \tan^2 i \tag{13.11}$$

[Rüger 1998, Eq. (5)]

In this equation, $Z = \rho\alpha$ is the vertical P-wave impedance, $G = \rho\beta^2$ is the vertical shear modulus. Changes in elastic parameters are expressed as relative differences. The bar denotes the average, i.e. $\alpha = 1/2\ (\alpha_1 + \alpha_2)$ of velocities (α_1) in the upper and lower (α_2) medium. $\Delta\alpha$ is equal to $(\alpha_2 - \alpha_1)$. Corresponding expressions hold for shear modulus, the density, and P-wave impedance. $\Delta\gamma$ is the difference in *shear-wave splitting parameters* of the reflecting medium and overburden.

Equation (13.11) describes the anisotropic contribution as a function of azimuth through the parameters $\Delta\delta^{(V)}$, $\Delta\gamma$ and $\Delta\varepsilon^{(V)}$. Equation (13.11) may be simplified for the special case of $\phi = 0$ m, i.e. for *isotropy plane* of HTI media and $\phi = \pi/2$ i.e. *symmetry axis* plane of HTI media.

Equation (13.11) may be utilized to study the variation of reflection coefficient vs. angle of incidence (i) for a fixed azimuth (ϕ = constant), or it's variation with azimuth (ϕ) for a fixed angle of incidence (i = constant). For computational results reference may be made to Rüger (1998).

The second term in Eq. (13.11) for R_p (i, ϕ), describing the AVO gradient, consists of two parts – i.e. one azimuthally invariant part β^{iso} and the anisotropic contribution β^{ani}. The *AVO gradient* in azimuth ϕ_k may be written from Eq. (13.11) as

$$B(\phi_k) = B^{iso} + B^{ani}\cos^2(\phi_k - \phi_{sym}) \qquad (13.12)$$

where

$$B^{iso} = \frac{1}{2}\left[\frac{\Delta\alpha}{\overline{\alpha}} - \left(\frac{2\overline{\beta}}{\overline{\alpha}}\right)^2 \frac{\Delta G}{\overline{G}}\right] \text{ and}$$

$$B^{ani} = \frac{1}{2}\left[\Delta\delta^{(V)} + 2\left(\frac{2\overline{\beta}}{\overline{\alpha}}\right)^2 \Delta\gamma\right]$$

Equation (13.12) opens up the possibility to invert for the *anisotropy* contribution to the reflection coefficient amplitude. This is discussed in Subsection 13.6.3.

13.4 Factors Affecting Reflection Amplitude

The reflected wave amplitude is affected due to the geometrical spreading of wavefront, and the radiation pattern. Also, the *anelasticity* of the medium would produce its effect to alter reflection amplitudes. To give a meaningful interpretation of measured reflection amplitudes with offset, it is most important to apply corrections to the recorded amplitudes to estimate the "true amplitude" resulting only due to *physical property contrasts* at the interfaces.

13.4.1 Geometrical Spreading

Figure 13.7a shows how energy flowing over a small area near the source gets distributed over a larger area on an emerging reflected wavefront. Corresponding areas are of annular form as shown by the dashed lines, which are bounded by common ray paths. This approach opened up the possibility of relating the *divergence effect* on amplitude as a function of reflection time (Newman 1973). This concept (Newman 1973) was also explained for a layered sequence (Fig. 13.7b). We have energy flowing through a conical shell formed by rotating the ray paths about a vertical axis. Initially, energy flowing on a spherical surface of radius 'r' through an area A_r flows subsequently through an annular area A_x as the *reflected wavefront* surface.

Acoustic intensity is defined as the rate of flow of acoustic energy per unit area normal to the direction of flow (Newman 1973). Therefore, ratio of acoustic intensities in incident and reflected waves due to the effect of divergence is equal to

$$\frac{I_x}{I_r} = \frac{A_r}{A_x} \qquad (13.13)$$

Using this concept, Newman derived the *divergence factor* for the reflection from the base of the n-th layer observed at offset distance x from the source (Fig. 13.7b)

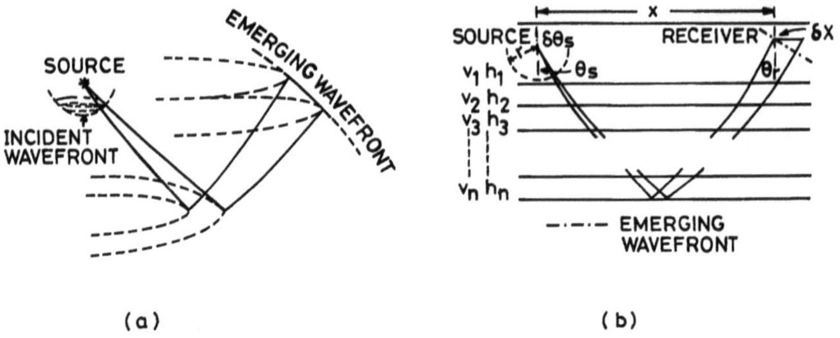

Fig. 13.7 a, b. a Intensity ratios are determined from area ratios. **b** Raypath geometry for determining divergence factors (Source: Newman 1973)

$$D_{x,\theta} = \frac{\left[x^2 + 2x\sum_{i=1}^{n} h_i \tan^3 \theta i\right]^{\frac{1}{2}}}{\tan\theta_1} \tag{13.13 a}$$

[Newman 1973, Eq. (8)]
where

$$x\left(= 2\sum_{i=1}^{n} h_i \tan\theta_i\right) \text{is offset distance}$$

θ_i = incidence angle in the i-th layer,
h_i = thickness of i-th layer.

θ_1 is defined in the following manner: when the source and receiver are both located in the first layer, and all layers are horizontal, $\theta_s = \theta_r = \theta_1$.

It may be mentioned that the offset distance x and incidence angles 'θ_i' in the layers are not independent variables. In fact, Snell's law in terms of the parameters of the layered model relates these. Therefore determination of the divergence factor (D) for a horizontal layer model is simple.

In Ursin (1978) the radius of curvature of the reflected wavefront is expressed as a function of source receiver offset 'x' in the following form (e.g., see Fig. 6.15):

$$R(x)^2 = R(0)^2 + \sum_{k=1}^{\infty} C_k x^{2k} \tag{13.14}$$

[Ursin 1978, Eq. (14)]
where

$$R(0) = \frac{t(0)(\bar{v}_2)^2}{v_1}$$

t(0) is the two way *normal incidence travel time* given by

$$t(0) = 2\sum_{k=1}^{n} \frac{h_k}{v_k} \tag{13.14 a}$$

$\bar{v}_2$ is the time average velocity, written from the general expression [Ursin 1978, Eq. (12)]

$$\overline{v_{2j}} = \left[\frac{1}{t(0)}\sum_{k=1}^{n} v_k^{2j}\Delta t_k\right]^{\frac{1}{2j}} \tag{13.14 b}$$

[Ursin 1978, Eq. (12)]
and $\Delta t_k = 2h_k/v_k$ is the two way travel time in the k-th layer.

A first approximation of the radius of curvature of the reflected wavefront is obtained as,

$$R(x) = \left[R(0)^2 + c_1 x^2\right]^{\frac{1}{2}} \tag{13.15}$$

with

$$c_1 = \left(\frac{3\overline{v}_4^4}{v_1^2\overline{v}_2^2} - 2\right) \tag{13.15 a}$$

C_k's in Eq. (13.14) are expressible in terms of layer velocities, and two-way normal incidence travel times in the layers.

Ursin (1990) showed, with a new approach, that the *geometrical spreading* effect may be expressed as a function of velocity in the first layer, the offset, and the first and second-order travel time derivatives in the form

$$g(x)^2 = \left[\frac{1}{V_0^2}\left(\frac{dT}{dx}\right)^{-1} - \frac{dT}{dx}\right]\left(\frac{d^2T}{dx^2}\right)^{-1} x \tag{13.16}$$

[Ursin 1990, Eq. (12)]

For a standard hyperbolic travel-time approximation in the form

$$T(x) = \left[T(0)^2 + \frac{x^2}{\overline{v}_2^2}\right]^{\frac{1}{2}}$$

Equation (13.16) casts itself in the form

$$g(x)^2 = g(0)^2 + \left[2\left(\frac{\overline{v_2}}{v_0}\right)^2 - 1\right]x^2 + \frac{1}{t(0)^2}\left(\frac{1}{v_0^2} - \frac{1}{\overline{v}_2^2}\right)x^4 \tag{13.17}$$

[Ursin 1990, Eq. (23)]
where,

$$g(0) = \frac{t(0)\overline{v}_2^2}{v_0} \tag{13.17 a}$$

Equation (13.17) is an *offset-dependent geometrical spreading* effect expressed as a function of routinely estimated quantities in standard velocity analysis. These quantities are the two-way *zero offset travel time* and the *root mean square velocity* (rms velocity). This formula has an ease of application and a *geometrical spreading* correction may be applied by multiplying the data with g(x) at the same time as correcting for *normal moveout* (nmo).

The *divergence factor* at normal incidence is of special interest. For a horizontally layered medium Newman showed that, in this case, the divergence factor is given by

$$g^{(0)} = \frac{t(0)(\mathrm{VRMS})^2}{V_1} \tag{13.18}$$

[Newman 1973, Eq. (12)]
where

$$t(0) = \sum_{j=1}^{n} \Delta t_j$$

is the *two-way reflection time* at normal incidence and Δt_j is the two way time in layer j, and VRMS is the rms velocity given by

$$\mathrm{VRMS} = \left[t^{-1}(0) \sum_{j=1}^{n} \Delta t_j v_j^2 \right]^{\frac{1}{2}}$$

Newman (1973) showed that the normal incidence divergence factor is given by the radius of the emerging reflected wavefront in the region of the source. He proved this considering the position of the image of the source in a layered medium (refer to Newman 1973, Fig. 5).

13.4.2 Anisotropy of Incident Medium: Radiation Pattern Effects

Seismic source directivity and *spherical divergence* of amplitude enter the *far field point-source radiation pattern* in an isotropic homogeneous, non-attenuating medium (c.f. Aki and Richards 1980). In addition to these, the radiation pattern of body waves is affected by the presence of anisotropy. The most significant distortions in an anisotropic medium are caused by *focusing* and *defocusing* of seismic energy. Tsvankin and Chesnokov (1990) found that energy increases (focuses) in the part of the wavefront with a high concentration of *group-velocity vectors* of elementary plane waves, which comprise point source radiation. *Defocusing*, on the other hand, corresponds to areas with a low concentration of group velocity vectors. Focusing and defocusing are often associated with velocity maxima and minima, respectively. Tsvankin (1995) examined the P- and S-wave radiation patterns for most common anisotropic model-transversely isotropic media, and concise analytic results were obtained for a weak-anisotropy approximation. The results convey the distortions of point source radiation caused due to Thomsen parameters ε, δ (for P- and SV-waves), and γ (for SH-waves). The primary aim in Tsvankin's (1995) analysis is to compare the influence of transverse isotropy on wave propagation above the reflector, and of the reflection coefficient, i.e. the two principal components deciding the *AVO signatures.* The specific goal of AVO analysis is to obtain the reflection coefficient versus angle, and not to study directly the amplitude variation versus offset. This underlines the need for correction for angular amplitude variation caused by wave phenomenon due to anisotropy above the reflector. In a nutshell, AVO analysis must include *anisotropic correction* to the angular amplitude distribution. To highlight this, we describe a simple anisotropic model considered in Tsvankin (1995) and the wavefront amplitudes associated with it.

13.4.2.1
Radiation Pattern of P-, SV- and SH-Waves

Consider a model as shown in Fig. 13.8 – a horizontal reflector below a transversely isotrokpic medium with a vertical symmetry axis.

Transverse isotropy in Tsvankin's (1995) investigation is described by the vertical velocities, V_{PO} and V_{SO}, of P- and S-waves, and three dimensionless anisotropic parameters ε, δ and γ (Thomsen 1986). P-SV propagation is describable by V_{PO}, V_{SO}, ε and δ, and SH-wave propagation by V_{SO} and γ (SH-wave anisotropy is elliptical).

In Tsvankin's (1995) investigation, *anisotropy-induced distortions* of *wavefront amplitudes* are investigated. The treatment is general which holds for any layer encountered in ray path. The *radiation pattern* is expressed as a function of *phase* or *group angle* with the symmetry axis, and the results are applicable for any orientation of axis of symmetry and not necessarily for the vertical. For a detailed derivation of point source far field radiation for weak-anisotropy, reference may be made to the original contribution of Tsvankin (1995). The final outcome of derivations is that the far field radiation pattern of a P-, SV or SH wave from point force for weak-transverse isotropy ($|\varepsilon|\ll|$, $|\delta|\ll|$ and $|\gamma|<<|$) is given by

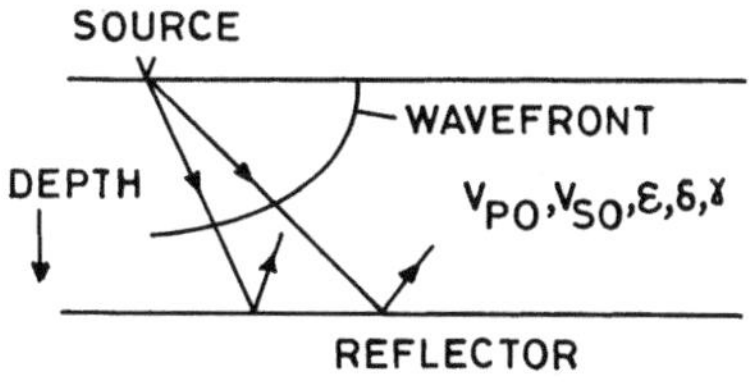

Fig. 13.8. Reflection from the bottom of a transversely isotropic layer. Anisotropy distorts the angular amplitude distribution of the incident wave (Source: Tsvankin 1995, from Geophysics with permission)

$$u(R,\theta) = \frac{F_u}{4\pi\rho V^2(\theta)R} \frac{1}{\sqrt{\frac{\sin\psi}{\sin\theta}\left(1+\frac{1}{V}\frac{d^2V}{d\theta^2}\right)}} \tag{13.19}$$

[Tsvankin 1995, Eq. (4)]
where

u = magnitude of displacement,
θ = phase angle measured from symmetry axis,
V = the phase velocity,
ρ = density,
R = $\sqrt{(z^2+r^2)}$ (z is the receiver depth, and r is the horizontal source-receiver offset), and
F_u = the source term – the projection of the force on the *displacement (polarization) vector.*
ψ = $\tan^{-1}$ (r/z)) is the ray (group velocity) angle of the incident wave.

Displacement u is expressed in terms of phase angle θ, corresponding to a given ray (group-velocity) angle $\psi = \tan^{-1}$ (r/z) of the incident wave. At the velocity maximum and minimum, the phase velocity and group-velocity vectors coincide with each other.

Although the term $F_u/4\pi\rho V^2(\theta)R$ is an expression for far-field point force radiation in an isotropic medium (Aki and Richards 1980), the phase velocity V is angle dependent here because of anisotropy. In addition to this, F_u is also distorted by the anisotropy, as body-wave polarization (refer to Section 5.8 for particle motion polarization due to anisotropy) depends on the elastic constants. Also, the displacement is measured in phase direction θ, which is generally different from the source-receiver direction due to the presence of anisotropy. The term under square root is a pure contribution of anisotropy to the radiation pattern. The second-derivative of the phase-velocity is responsible for '*focusing*' and '*defocusing*' phenomenon (refer to Tsvankin and Chesnokov 1990). *Due to focusing and defocusing of energy in anisotropic media, the distribution of energy along wavefronts is substantially distorted.* This is in complete contrast to the case of isotropy and homogeneity in which the distribution of energy along wavefronts is uniform.

Linearization of Eq. (13.19) in terms of ε, δ and γ leads to following concise expressions for the P-wave radiation pattern:

$$u_P(R,\theta) = \frac{F_u}{4\pi\rho V_{PO}^2 R} \frac{1-2(\varepsilon-\delta)\sin^2 2\theta + \delta\sin^2\theta}{(1+2\delta)} \tag{13.20}$$

$$(|\delta| \ll 1, |\varepsilon| \ll 1)$$

[Tsvankin 1995, Eq. (8)]

For obtaining expression Eq. (13.20) for $u_P(R, \theta)$, terms having higher powers of ε and δ have been dropped.

Single anisotropy parameter 'δ' determines the *focusing* or *defocusing* of P-wave energy at vertical incidence. 'δ' is the anisotropic parameter, responsible for near-vertical P-wave propagation. For $\delta<0$, the velocity function $V_p(\theta)$ is given by (refer to Subsection 5.6.1)

$$V_p(\theta) = V_{PO}\left(1 + \delta \sin^2\theta \cos^2\theta + \varepsilon \sin^4\theta\right)$$

has a maximum at $\theta=0$, and the amplitude at vertical incidence increases as a result of *focusing* of the energy. For $\delta>0$, the velocity minimum at $\theta=0$ gives rise to lower amplitude, because of the defocusing effect.

The lowest order anisotrokpic angular term in $U_p(R, \theta)$ depends on the difference $(\varepsilon-\delta)$. For $(\varepsilon-\delta)>0$, the most common case in crustal rocks, the transverse isotropy, causes *P-wave amplitude to decrease away* from the vertical. In the case of elliptical anisotropy, $\varepsilon=\delta$, the second term in $u_p(R, \theta)$ vanishes, and anisotropic angular correction depends only on $\delta \sin^2\theta$. If $\varepsilon-\delta<0$, transverse isotropy would cause P-wave amplitude to increase with angle.

Further, angular distortion to the radiation pattern is also caused by the source term F_u, dependent on the *polarization vector.* Tsvankin and Chesnokov (1990) argue that the source term F_u for a P-wave is almost *isotropic,* i.e. for a unit force it is close to the absolute value of cosine of angle between the force and the group velocity vector. For $|\varepsilon|\leq 0.2$ and $|\delta|\leq .0.2$, distortion to point force directivity factor F_u in the angular range 0–40° is only a few percent. For dislocation or explosive sources, the dependence of the source term on anisotropy is more complicated, and can contribute significantly to distortions of the angular amplitude distribution.

13.4.2.2
Weak Anisotropy Effects on P-Wave Amplitude

Radiation patterns in Tsvankin (1995) are depicted as a function of group angle ψ which is given by the source-receiver direction: $\psi=\tan^{-1} r/Z$. To bring out the effects of anisotropy parameters, all curves are normalized by the radiation pattern for the corresponding isotropic medium ($\varepsilon=0$, $\delta=0$). The isotropic term for a unit force is given as

$$\frac{\cos\psi}{4\pi\rho V_{PO}^2 R}$$

In the Tsvankin (1995) investigation, systematic results on amplitude variation vs. group angle are given for a suit of transversely isotropic models parameterized in Thomsen anisotropic parameters ε and δ. Fig. 13.9a is for a transversely isotropic olivine model for a point *vertical force.* As a result of transverse isotropy, a decrease in amplitude from the vertical occurs, and a minimum occurs near 45–50°. This matches with the phase-velocity maximum and minimum, maximum at $\theta=0$ and minimum near $\theta=49°$. Amplitude variation occurs as a result of the focusing of energy near vertical incidence and defocusing near 45–50°.

The influence of transverse isotropy on amplitude is mostly determined by the parameters ε and δ for P-waves. Figure 13.9a is given for negative values of δ and ε. ε is almost always positive (Thomsen 1986). Tsvankin (1995) finds that the P-wave radiation pattern is controlled more by the difference $(\varepsilon-\delta)$, than by individual val-

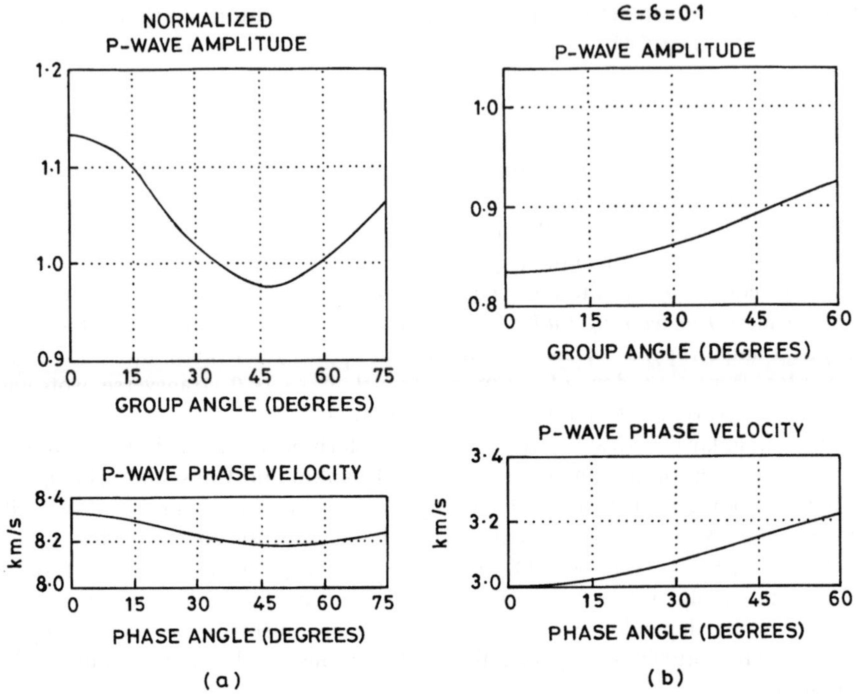

Fig. 13.9 a, b. a P-wave amplitude from a point vertical force in transversely isotropic olivine ($V_{PO}=8.328$ km/s, $V_{SO}=4.606$ km/s, $\varepsilon=-0.008$, $\delta=-0.059$), based on a weak anisotropy approximation of the P wave radiation pattern [Eq. (13.20)]. The values are normalized by the radiation pattern in the corresponding isotropic medium ($\varepsilon=0$, $\delta=0$). The receiver is located at a constant distance R from the source in the far-field ($R/\lambda_p \approx 12$, where λ_p is the average P-wavelength in the isotropy plane). **b** Normalized P-wave amplitude from a vertical force for models $\varepsilon-\delta=0$ (Source: Tsvankin 1995, from Geophysics with permission)

ues of ε and δ. Existing measurements for seismic frequencies show that ε is greater than δ (see, for example, Thomsen 1986; Tsvankin and Thomsen 1994). 'ε' is always greater than δ for transversely isotropic media formed by thin bedding of isotropic layers (Berryman 1979). For this case, the decrease in amplitude with angle may be typical for P-wave propagation through T.I. subsurface layers such as shales.

With increasing ($\varepsilon-\delta$) value, the influence of anisotropy on P-wave amplitude becomes more pronounced. If the ($\varepsilon-\delta$) value is negative, it may produce an amplitude increase with group-angle.

In the case of elliptical anisotropy, $\varepsilon=\delta$, the expression for the P-wave radiation pattern reduces to,

$$u_p(\text{elliptical}) = \frac{F_u}{4\pi\rho V_{PO}^2 R}\frac{1}{\sqrt{(1+2\delta)(1+2\delta\cos^2\psi)}} \tag{13.21}$$

[Tsvankin 1995, Eq. (10)],
without application of weak-anisotropy approximation.

In this case P-wave amplitude variation and phase velocity variation are well correlated (Fig. 13.9b).

13.4.2.3
Radiation Pattern vs. Reflection Coefficient for P-Waves

Tsvankin (1995) compared the distortion produced in radiation pattern due to anisotropy with the influence of transverse isotropy on the reflection coefficient. The following expression derived in Thomsen (1993) was used for the reflection coefficient

$$R(\theta) = R_{isot}(\theta) + R_{anis}(\theta) \tag{13.22a}$$

[Tsvankin 1995, Eq. (11)]

$R_{isot}(\theta)$ stands for the case $\varepsilon=0$, $\delta=0$, and

$$R_{aniso}(\theta) = \frac{1}{2}(\delta_2 - \delta_1)\sin^2\theta + \frac{1}{2}(\varepsilon_2 - \varepsilon_1)\sin^2\theta\tan^2\theta \tag{13.22b}$$

δ_1, δ_2 and ε_1, ε_2 in Eq. (13.22b) stand for Thomsen parameters above and below the reflector.

Comparing Eq. (13.22b) with Eq. (13.20) for the radiation pattern, the reflection coefficient depends on the contrast in anisotropy parameters across the reflector whereas the angular term in the radiation pattern depends entirely on the properties of the incidence medium.

For comparison purposes Tsvankin (1995) takes the shale/sand model (Kim et al. 1993). Sand is taken as isotropic. In this case,

$$R_{anis}(\theta) = -\frac{1}{2}\delta_1\sin^2\theta - \frac{1}{2}\varepsilon_1\sin^2\theta\tan^2\theta \tag{13.23}$$

Two models are analyzed for $\varepsilon=0.1$ and $\delta=0.1$ and $\varepsilon=0.2$ and $\delta=-0.1$. This result is shown in Fig. 13.10.

Analysis of Tsvankin's (1995) results shows that for typical T.I. media for which $\varepsilon-\delta>0$ above the reflector, a decrease of amplitude with angle would be produced which may, therefore, reduce or reverse the influence of anisotropy on reflection coefficient.

The treatment of radiation pattern in an anisotropic medium has also been the subject of intensive study by other investigators (see, for example, Ben Menahem et al. 1991; Samec and Blangy 1992; Gajewski 1993). For a systematic study of seismic sources and the related radiation patterns reference may be made to Ben Menahem and Singh (1981).

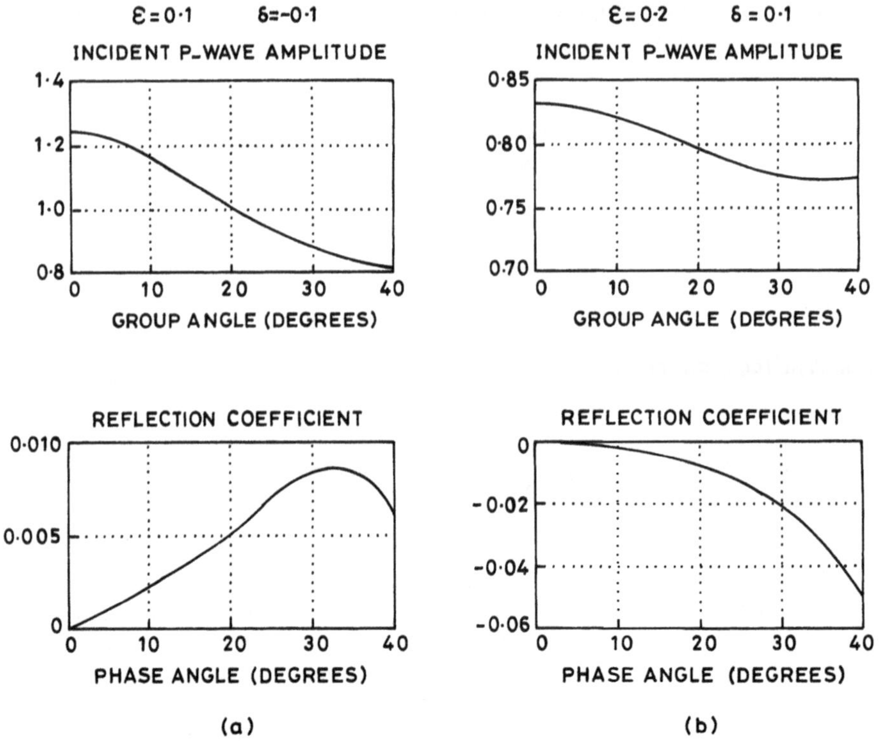

Fig. 13.10a, b. a, b Comparison of the influence of transverse isotropy on the P-wave radiation pattern (from a vertical force) and on the reflection coefficient. (*Top*) The exact far-field amplitude of the incident wave normalized by the amplitude in the corresponding isotropic medium with $\varepsilon=0$, $\delta=0$. (*Bottom*) The angular variation in the reflection coefficient caused solely by the anisotropy [Eq. (13.23)). It is assumed that the transversely isotropic models with ε and δ shown on the plot overlie an isotropic medium (Source: Tsvankin 1995, from Geophysics with permission)

13.4.3
Anelastic Effects and Absorption Loss

When elastic waves propagate in the Earth, energy decreases not only because of *geometrical spreading*, but also as a result of the friction caused by grains of materials sliding past one another along grain boundaries. *Solid friction* is used to denote the non-elastic mechanism, which converts strain energy into heat. The result is damping or attenuation of wave motion in the medium, which eventually accounts for complete disappearance of the wave motion.

Commonly used measures of the *attenuation property* of the medium are

$1/Q$ = specific dissipation constant,
δ = logarithmic decrement,
a = attenuation coefficient,
$\Delta E/E$ = fraction of strain energy lost per cycle.

Specific dissipation constant (1/Q) is related to the rate at which the mechanical energy of vibration is irreversibly converted into heat energy.

Logarithmic decrement (δ) is defined as the natural logarithm of the ratio of amplitudes of two successive maxima or minima in an exponentially decaying free vibration, i.e.

$$\delta = \ln\left(\frac{\text{amplitude}}{\text{amplitude one cycle later}}\right)$$

Amplitude decrease for elastic waves in rocks is found to be exponential. Taking A_1 and A_2 as two values of amplitudes of a wavefront at a distance x apart, we may write it as,

$$A_2 = A_1 e^{-\alpha x}$$

where α is the *attenuation coefficient.*

13.4.3.1
Anelasticity and Refraction

The general form of the expression for *viscoelastic plane waves* is

$$(\cdot)e^{i(\omega t - \mathbf{K}\cdot\mathbf{x})} \text{ where } \mathbf{K} = \mathbf{k} - i\boldsymbol{\alpha}$$

where ω is angular frequency, K is a *complex wave number vector*, with $\mathbf{k}$ and $\boldsymbol{\alpha}$ as real wave number and attenuation vector, respectively. They ($\mathbf{k}$ and α) stand for directions and magnitudes of *propagation* and *attenuation vectors.* When the direction of propagation and attenuation coincide, the wave is referred to as a *homogeneous wave.*

When plane waves propagate in a layered anelastic medium having variation of specific dissipation constant (1/Q) from one layer to the next, the maximum attenuation direction is, in general, different from the direction of propagation. Lochett (1962) first recognized this. Further development of the theory has been done by many investigators (e.g. Buchen 1971; Borcherdt 1973, 1982; Krebes 1984; Wennerberg 1985; Winterstein 1987; Carcione 1990, 1994, 1999; and Carcione et al. 1992). Theoretical investigations reveal unusual plane wave behavior in *isotropic* and *anisotropic, inelastic* media.

Winterstein (1987) investigated the kinematics of the plane wave propagation for two anelastic media with an interface. This simple model illustrates how the angle between propagation and maximum attenuation varies in going from a medium with low Q into a medium with high Q. Q values specify the attenuation characteristic of the media. To illustrate the point, we describe here the important findings of Winterstein's investigation.

Winterstein (1987) considered plane waves W in the form

$$W \propto e^{-\mathbf{k}_I\cdot\mathbf{r}} e^{i(\mathbf{K}_R\cdot\mathbf{r} - \omega t)}$$

[Winterstein 1987, Eq. (1)]

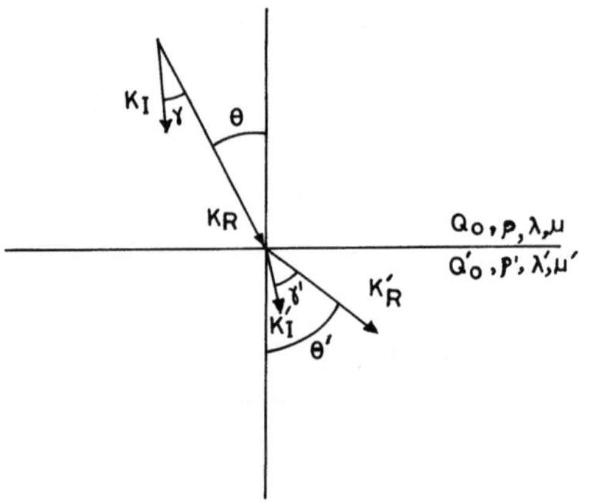

Fig. 13.11. Relation of real and imaginary parts of the wave vector for an incident and a transmitted wave at a boundary between anelastic media. The sign of γ is positive as shown, negative if $\mathbf{K}_I$ is rotated counterclockwise past $\mathbf{K}_R$. Waves are called homogeneous if $\gamma = 0$ (Source: Winterstein 1987)

where $\mathbf{K}_R$ and $\mathbf{K}_I$ denote the real and imaginary parts of the wave vector $\mathbf{K}$, and in general they do not have the same direction (Fig. 13.11).

Borcherdt (1982) gave Snell's law for anelastic media. It has two parts; the first for the real part (K_R) of the wave vector, and the second for the imaginary part (K_I) of wave vector, $\mathbf{K}$. The first part is

$$K_R \sin\theta = K'_R \sin\theta' \tag{13.24 a}$$

[Winterstein 1987, Eq. (5)]
where K_R and K'_R are magnitudes of vectors $\mathbf{K}_R$ and $\mathbf{K}'_R$.

This is identical to the form of Snell's law for elastic media. The point of difference for the elastic and anelastic case is that K_R for a homogeneous, isotropic, anelastic medium is generally not constant for a wave at a given frequency, but depends on γ (Fig. 13.11).

The second part of Snell's law is required for satisfying the boundary condition on imaginary parts of the wave vectors and is given by

$$K_I \sin(\theta - \gamma) = K'_I \sin\left(\theta' - \gamma'\right) \tag{13.24 b}$$

[Winterstein 1987, Eq. (6)]
where K_I and K'_I are magnitudes of the imaginary part of vectors $\mathbf{K}_I$ and $\mathbf{K}'_I$.

Further, Eqs. (13.24 a) and (13.24 b) expressing Snell's law for isotropic, homogeneous, anelastic media, hold good for any combination of incident wave and emergent waves from the interface – for example for an incident P-wave and refracted SV-wave, or vice versa.

An important outcome of Winterstein's analysis is (that concerns us here) that for a given angle θ of the incident wave, ray path of the refracted wave may change considerably for different values of γ (refer to Fig. 13.11) as a consequence of anelasticity.

13.4.3.2
Vector Attenuation and Reflection

Winterstein's (1987) investigation shows that the ray path of the refracted waves is affected due to *anelasticity.* In seismic prospecting, the interest lies withhin the reflected waves arrival times, and with amplitudes for AVO analysis. Carcione (1999) investigated the effects of *vector attenuation* on the AVO of *offshore reflection.* In Carcione, results were derived on the AVO response to an inhomogeneous wave generated at the ocean bottom, which is incident at the lower interface separating two viscoelastic transversely isotropic media (Fig. 13.12 a).

Figure 13.12 a shows an inhomogeneous wave generated at the ocean bottom. The *transmitted attenuation vector* is perpendicular to the ocean bottom. This follows from Snell's law, taking that the attenuation vector of a wave propagating in a water layer is zero, since water is assumed to be lossless. The inhomogeneity angle is equal to the propagation angle θ.

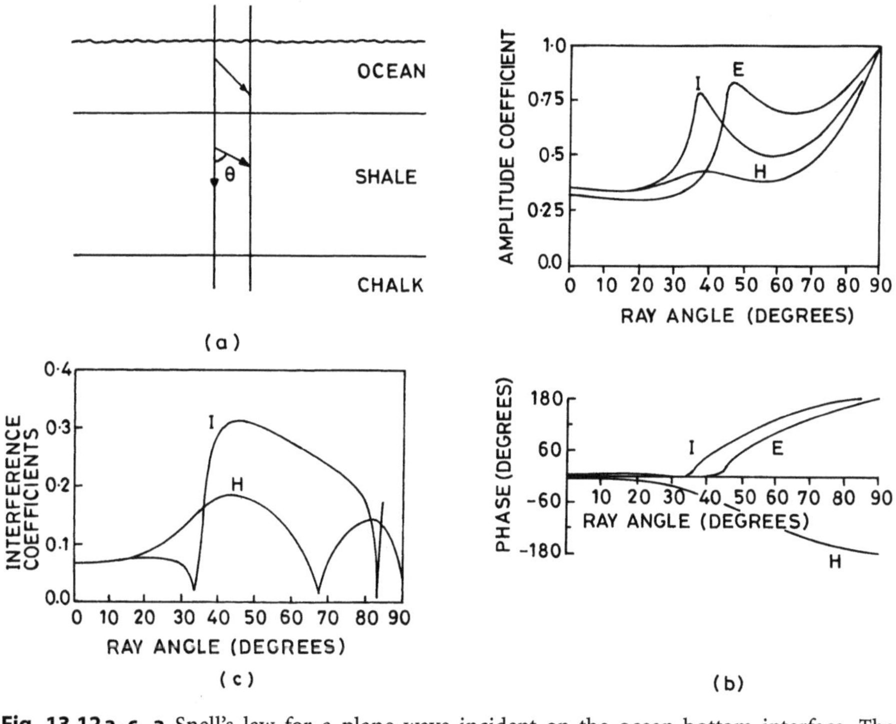

Fig. 13.12 a–c. a Snell's law for a plane wave incident on the ocean bottom interface. The diagram shows the continuity of the horizontal component of the complex slowness vector. In the ocean this vector is real, since water is assumed to be lossless. In the *shale layer* the attenuation vector is perpendicular to the ocean bottom. **b** Comparison between the absolute values of the R_{PP} reflection coefficients. E corresponds to the elastic case (i.e. elastic shale), H to an incident visco-elastic homogeneous wave, and I to an incident inhomogeneous wave with the characteristics indicated in Fig. 13.12 a. **c** Comparison between the interference coefficients of the incident homogeneous (H) and inhomogeneous (I) waves. The coefficients correspond to the interaction of the stress and particle velocity fields of the incident and reflected q^P-waves (Source: Carcione 1999, from Geophysics with permission)

Absolute values of the qP reflection coefficient for the model shown in Fig. 13.12 a are given in Fig. 13.12 b. When shale and chalk are both elastic, the critical angle is between 40–50°.

In near offset (up to 20°), the trend is the same in all three curves. The difference is because of an *inelastic property* of the shale bed. Beyond 30°, the difference is important, mainly for an incident homogeneous wave, which is also observed in the phase plot [Fig. 13.12 b (bottom)].

The *interference coefficient* is a result of the interaction of the stress and particle velocity fields of the incident and reflected qP-wave. Beyond the elastic critical angle much of the energy flow is because of interference. Also for some incidence angles, the interference coefficient can have the same magnitude as the reflection coefficient (Fig. 13.12 c). The *interference effect* does not occur in the case of *perfect elasticity.*

The *conclusion in the Carcione (1999) investigation is that, for media with highly attenuating properties such as unconsolidated sediments, one should consider the vector attenuation character of the seismic pulse. Also, energy flows from the interference of stress and particle velocity, in the case of anelastic media, could be comparable to the energy flux of reflected waves. These effects must be considered in any AVO analysis.*

13.4.3.3
Frequency-Dependent Attenuation: Solid Friction Model

The effect of attenuation may be expressed by assuming the velocity to be a complex quantity (refer for example to Waters 1987).

Taking velocity C in the form

$$C = V + iv$$

If $v^2 \ll V^2$, the plane wave solution is

$$\zeta = \zeta_0 e^{-\alpha x} e^{\left[i\omega\left(t - \frac{x}{V}\right)\right]} \tag{13.25}$$

[Waters 1987, p 165]
where

$$\alpha = \frac{\omega v}{V^2} \quad (\omega = 2\pi f)$$

and

$$\frac{1}{Q} = \frac{2v}{V}$$

Equation (13.25) represents a wave, which attenuates as the first power of frequency. Thus, we find that, as a result of attenuation due to solid friction, the amplitude of the elastic plane wave decreases exponentially with distance.

a, 1/Q, δ and $\Delta E/E$ (defined in the beginning of this section) are related to each other as follows:

$$\frac{2\pi}{Q} = 2\delta = \frac{\Delta E}{E} = \frac{2aV}{f}$$

Absorption in Earth material as a result of solid friction is taken to vary as the first power of frequency. It implies that the decay in *acoustic pressure* over each wavelength is constant. The proportional loss over one wavelength is independent of frequency. This property is used to describe the *absorption characteristics* of the rocks and is expressed in the unit of decibel/wavelength or dB/*wavelength*. Decibel is used to express the ratio of intensity, energy or power. The value in dB is $10 \log_{10}$ (ratio of intensity, energy or power). For amplitudes A_1 and A_2, ratio A_2/A_1 is expressed in decibels as $20 \log_{10} (A_2/A_1)$.

In seismic prospecting, we are mostly concerned with the decibel scale of *amplitudes*. Conversion of the amplitude ratio in decibels is easy, provided some simple rules for conversion are followed (refer to Anstey 1970, pp 16–17).

13.4.3.4
Frequency-Selective Absorption and Seismic Pulse

Absorption is a frequency selective process. Therefore, a seismic pulse consisting of a range of frequencies, and propagating in an absorptive medium would loose progressively higher frequencies. This process is bound to affect the *pulse spectrum*, measured at different times. In the time domain, this accounts for a reduction of *peak amplitude*, and for *pulse broadening*. These points are illustrated in Fig. 13.13 a, b by taking an example from 'Doherty and Anstey (1971).

Consider a rock material having an absorption characteristic of 0.2 dB/wavelength, a velocity of 3,000 m/s, and a path length of 300 m in this material. For a wave of 100 Hz, and a velocity of 3,000 m/s, this corresponds to ten wavelengths. The result of absorption (0.2 dB/wavelength) would be that each compression would be 0.2 dB less in amplitude than the previous one. It means that the second compression amplitude is about 98% of the first one, the third compression amplitude is 98% of second one, and so on. Taking another frequency the decay is similarly exponential. Considering a total distance of 300 m traveled in the rock under consideration, the loss is 2 dB at 100 Hz, 1 dB at 50 Hz, and 0.2 dB at 10 Hz. Using this behavior with frequency, 'Doherty and Anstey (1971) presented a picture of the spectrum of the propagating pulse (Fig. 13.13 a). The effect in the time-domain on an input *spike pulse* is shown in Fig. 13.13 b.

The result presented in Fig. 13.13 a describes the amplitude-frequency response of a sinusoidal component, measured at 1, 2, 3, 4 and 5 seconds, i.e. for larger and larger path lengths in the medium determined by the velocity of the medium.

A pulse may be thought of as composed of a number of sinusoidal components. How the spectrum of the pulse would be affected at different times, would depend on the characteristics of the pulse, which in turn depends on the characteristics of the seismic source.

In Fig. 13.13 b the *spike input* has considerable energy at higher frequencies, which is lost soon as a result of *selective absorption* of high frequencies first. So

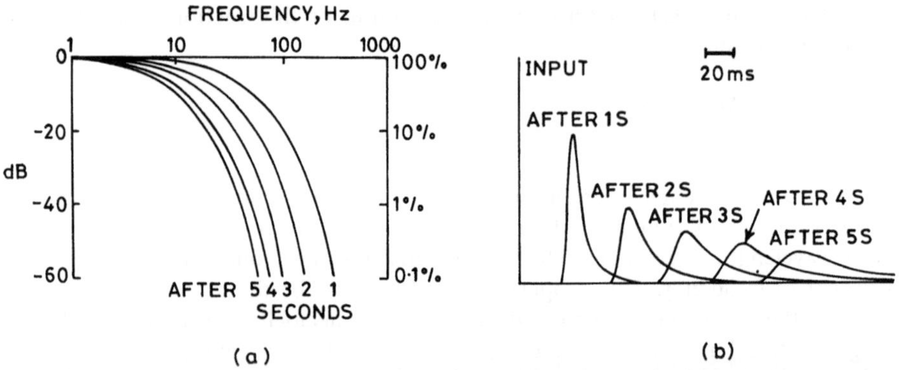

Fig. 13.13 a, b. **a** The progressive high frequency loss with increasing travel time, illustrated for a uniform absorption of 0.2 dB/wavelength. **b** The absorptive effect of Fig. 13.13 a translated into the time domain, on the assumption of minimum phase (Source: O'Doherty and Anstey 1971, from Geophysics Prospecting with permission)

after 1 s, we find a large drop in the *peak-amplitude value.* As the pulse (depicted at 1 s) propagates, absorption of the remaining higher frequencies in it continues, giving rise to further decay in peak amplitude. There is an important role of *dispersion* (frequency-dependent velocity) in lengthening the original *spike pulse*, and in the process decreasing the *peak amplitude*.

Dispersion means the different frequencies have slightly different velocities. So those frequency components that are in phase at the peak of the pulse in its early history are not in phase at later times during pulse propagation. The net result is a lengthening of the original pulse, and a decrease of peak amplitude.

13.4.3.5
Elastic and Intrinsic Attenuation

Richards and Menke (1983) showed that the total attenuation of a medium may be approximately represented by the sum of attenuations due to *elastic scattering* ($1/Q_{sc}$), and *intrinsic attenuation* ($1/Q_i$)

$$\frac{1}{Q_{total}} = \frac{1}{Q_{sc}} + \frac{1}{Q_i} \tag{13.26}$$

Intrinsic attenuation is caused by the viscosity of the fluid filling the pore space (Jones 1986).

The approximate relation that accounts for attenuation relates to velocities V_H and V_L measured at frequencies f_H and f_L ($V_H > V_L$) in the form

$$\frac{V_H}{V_L} = 1 + \frac{1}{\pi Q} \ln\left(\frac{f_H}{f_L}\right) \tag{13.27}$$

[Aki and Richards 1980, Eq. (5.77)]
where subscripts H and L represent measurements at high and low frequencies.

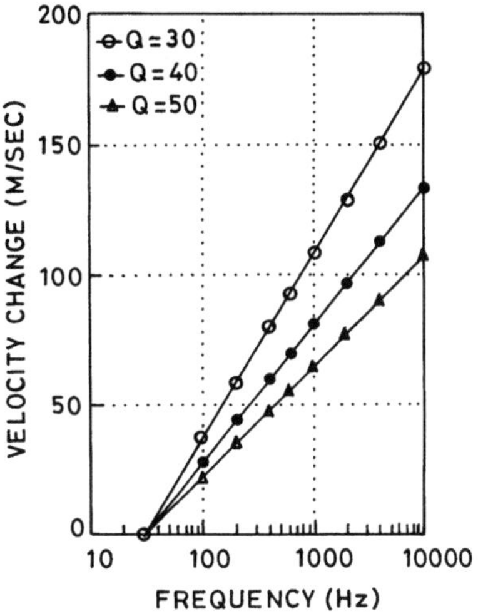

Fig. 13.14. The velocity increase as a function of frequency for a plane wave in a homogeneous-attenuating medium. The computation is based on Eq. (13.27) assuming a reference velocity of 2,896 m/s for V_L at a frequency f_L of 30 Hz (Source: Brown and Seifert 1997, from Geophysics with permission)

Dispersion is described by this equation (Fig. 13.14).

Brown and Seifert (1997) found that dispersion could be a key tool for mapping reservoir properties, using crosswell and surface seismic data. They concluded that velocity-dispersion logging might be a potentially useful method for estimating important reservoir properties. Intrinsic attenuation is indicative of reservoir properties, e.g. the presence of gas, cracks, and permeability; velocity dispersion logging should prove to be an important tool for recognizing and characterizing pay zones, especially gas reservoirs where conventional velocity logging methods have failed.

Toksöz and Johnston (1981) summarize the literature on absorption. Q values for igneous rocks are the largest (75–150), followed by sedimentary rocks (20–150), and rock with gas in the pore spaces (5–50) (Sheriff and Geldart 1982, Table 2.1, p 56). For more detailed coverage on transient waves in viscoelastic media, reference may be made to the works by Ricker (1977), and Christensien (1982). Further, for a detailed description of wave propagation in *visco-elastic anisotropic media*, reference may be made to Carcione (1990, 1992, 1994, 1997).

13.5 Estimating a Target's Reflection Amplitude Behavior with Offset

In Section 13.4 important factors, such as geometrical spreading, radiation pattern and anelasticity that affect the amplitude, were discussed. The recorded reflection seismic trace has to be corrected for these effects to arrive finally at the *target amplitude* behavior with offset. By *target* is meant features of interest such as 'gas sands', 'oil sand', 'coal' etc.

Noteworthy contributions in regard to the conditioning of the recorded reflection trace have been made by Duren (1990, 1991, 1992), Ross and Beale (1994), Ursin and Dahl (1990, 1992) and Ursin and Ekren (1995). Conceptual aspects of these contributions are briefly discussed below.

13.5.1
AVO Correction Factor: Use of the Seismic Range Equation

Duren (1990, 1991) discussed the use of the *Seismic Range Equation*. This formulation integrates the role of the seismic source, subsurface, target and receiving system into a single equation, which represents a systematic approach to AVO analysis. More clearly, the seismic range equation takes into consideration source array geometry and outgoing waveform, the receiving system's (hydrophone array) geometry, and sensitivity and other factors determining the response characteristics of a horizontally layered model. Zimmerman and Fahmy (1990) successfully applied the seismic range equation for a recent gas recovery in the Gulf of Mexico. This has also been applied for lithology prediction for identification of gas sand and coal in offshore Alaska (refer, for example, to Duren 1991).

Duren (1992) gave the following general equation for calculating weights for AVO correction with offset for a marine data gathering system:

$$E_r(x_n, f) = \frac{E(f)D(\theta_s, \phi_s, f)A_c(\theta_h, \phi_h, f)R_T^2}{D_v(x_x)L(x_n, f)} \tag{13.28}$$

[Duren 1992, Eq. (1)]
where f is the frequency, x_n is offset distance to the n-th recording station from the seismic source centroid, and

$E_r(x_n, f)$ = received energy density
$E(f)$ = radiated energy density
$D()$ = source directivity along θ_s and ϕ_s directions
$A(\theta_h, \phi_h, f)$ = capture area along θ_h, ϕ_h direction
$R^2{}_T$ = target's power reflection coefficient
$D_v(x_n)$ = divergence effect
$L(x_n, f)$ = loss factor (includes the product of transmission loss $L_T(x_n)$, anelastic loss $L_Q(x_n, f)$ and interbed multiple loss $L_{ib}(x_n, f)$, respectively)

Spherical coordinates θ_s, ϕ_s are measured in the local coordinate frame at the source array and define the downgoing direction of the primary ray. Vertical direction is defined by $\theta_s = 0$.

Likewise θ_h and ϕ_h define the direction for the upcoming primary pulse at the receiver array. The vertical direction is the acoustical axis for the receiver array defined by $\theta_h = 0$.

Directivity is defined as the ratio of energy density per unit solid angle radiated into a particular direction to the average radiated energy density per unit solid angle taken over all angles (Duren 1987, 1988), assuming an array of seismic sources. *Source directivity* refers to a *directive source* in contrast to an *isotropic source*. The terms, source directivity and capture area, are related to arrays of seismic sources and receivers.

As the amplitudes are corrected relative to near offset, the weight $W(x_n)$ for n-th offset is given by

$$W(x_n) = \sqrt{\frac{\langle AVO(x_1)\rangle}{\langle AVO(x_n)\rangle}} \tag{13.29}$$

[Duren 1992, Eq. (3)]

This factor is a multiplier of a trace with offset x_n to correct the trace's amplitude at the target. The average value of $AVO(x_n)(=\langle AVO(x_n)\rangle)$ includes only those terms that vary with offset. For example E(f) [refer to Eq. (13.28)] is not included in $AVO(x_n)\rangle$ as it does not vary with offset. Although R_T (target's reflection coefficient) may vary with offset, but it is not included in the average value $(\langle AVO(x_n)\rangle)$ to deliberately retain it's behavior in the processed output.

13.5.2 Model-Based Amplitude Balancing

Ross and Beale (1994) proposed a model-based, *offset-dependent amplitude balancing*. This method is based on calibrating the data to a background (brine-charged reservoir) response. The process is initiated by undertaking a detailed petrophysical analysis to determine elastic properties of both brine-charged and gas-charged sands. A petrophysical study may be based on well log data in the area of interest. The theoretical responses of the direct-hydrocarbon indicator and background reflectors are estimated using synthetic seismic modeling algorithms. The seismic response as measured by a carefully selected data window is compared to the theoretical brine-sand/shale (assumed background) data. An offset-dependent scaling function is designed which can be applied to both prospective and non-prospective data within an area for which theoretical responses are appropriate and seismic characteristics are consistent.

Data window selection is a critical part of the procedure. For example, the window length should be large enough to include several reflectors, which have a uniform background signal. Also, the process assumes a high signal-to-noise ratio devoid of strong coherent noise. The data for window selection should have been preprocessed for dip moveout and prestack migration including all other steps aimed at amplitude preservation and restoration.

Ross and Beale (1994) conclude by comparing CMP data with and without *offset balancing*, that offset balancing improves AVO responses over reservoirs.

The final goal is to determine an accurate AVO response for predicting lithology and fluid content of the reservoir rock. Even after completion of offset-dependent processing steps, *residual amplitude* effects remain which represent a significant error when compared with the AVO response due to reservoir. The offset-dependent amplitude balancing procedure of Ross and Beale (1994) aims at correcting these residuals along with other errors due to sub-optimal processing.

13.5.3
Use of Separable Signal Model

The unique aspect of this method, contributed by Ursin and Dahl (1992) and Ursin and Ekren (1995), is that the AVO effect is modeled in the *offset domain* using *time windows*, thus making it more amenable to direct interpretation. Conventionally, exact as well as approximate formulae on reflection coefficient variation are given in terms of angles of incidence. Walden (1991) gives approximate formulas for converting angle of incidence to offset. Since amplitudes may be modeled directly as a polynomial in the offset coordinate using the Ursin and Ekren (1995) method, the need to convert data from offset to angles or slowness is avoided.

Ursin and Ekren (1995) also address the problem of the accurate moveout correction in AVO analysis. AVO methods are very sensitive to residual NMO and erroneous amplitudes (Spratt 1987). Swan (1991) concluded that large errors in estimated AVO parameters might result because of *NMO stretch* (refer, for example, to Buchholtz 1972; Dunkin and Levin 1973).

Walden (1991) suggested that a better estimate of AVO is possible by using time windows. In Ursin and Ekren (1995), *block NMO* is calculated for each time window to correct for the effect of *NMO stretch*. Also, correction for *residual NMO* is performed using an iterative *cross-correlation technique.*

The basis of the Ursin and Ekren (1995) procedure is that data in a *moveout corrected time window* is modeled as a constant pulse multiplied by a polynomial in the offset coordinate. This is expressed as

$$d(t,x) = p(t)a(x) + w(t,x) \tag{13.30}$$

[Ursin and Ekren 1995, Eq. (1)]
where t is the zero offset time, x is the offset, p(t) is the signal pulse and

$$a(x) = \sum_{j=0}^{m} r_j F_j(x)$$

is the amplitude function for a suitable family of functions $F_j(x)$, e.g. polynomials; **r** is the coefficient vector, i.e. r_j is the weight of F_j; and w(t,x) includes noise and modeling errors.

To arrive at a unique model a(0) is set equal to 1. The assumptions in this formulation are that reflections are non-overlapping, and noise is Gaussian with zero mean and is uncorrelated with the data.

The seismic pulse and polynomial coefficients in Eq. (13.30) are found by the least square method. The least square error criterion for the estimation of pulse p(t) and amplitude function a(x) from the data is the minimization of the *objective function* ϕ, given by

$$\phi = \sum_k \sum_n [d(t_k, x_n) - p(t_k)a(x_n)]^2 \tag{13.31}$$

Formulation in the formof Eq. (13.31) permits separate estimations of *optimum polynomial coefficients* and the pulse. It is a *separable least square problem* (refer, for example, to Cadzow 1990).

Ursin and Dahl (1992) took the following form of the amplitude model for the PP reflection coefficient:

$$a(x) = 1.0 + \sum_{j=1}^{m} r_j x^{2j} \quad (13.32)$$

It gives changes in the normalized value of amplitude variation with offset (i.e. $r_0=1$). While dealing with a real data example, Ursin and Ekren (1995) took the following fourth order form of amplitude polynomial in the AVO analysis:

$$a_2(x) = 1.0 + r_1 x^2 + r_2 x^4 \quad (13.33)$$

This gives the r_2 section in addition to the zero offset section $P(t_c)$ and gradient section r_1. r_2 coefficients relate mainly to changes in amplitude at large offsets.

Results in Ursin and Ekren (1995) have been processed for a shallow seismic anomaly believed to arise from *gas-hydrate sediments* trapping underlying gas (Andreassen et al. 1990) in the southern Barents Sea.

The following form of the amplitude function is usually accurate for offsets smaller than the reflector depth (Ursin and Dahl 1992):

$$a_1(x) = 1 + r_1 x^2 \quad (13.34)$$

A positive value of r_1 amounts to an increase in normalized amplitude with offset, and r_1 less than zero indicates a decrease. A product section $|p(t_c)|r_1$ shows where the normalized value (and hence absolute value) of amplitude increases with offset.

Separate formulae have been derived for coefficients r_j (refer to Ursen and Ekren 1995, Eq. (12)) and the *offset-independent pulse* $p(t)$ (refer to Ursen and Ekren 1995, Eq. (15)). Pulse amplitude in the middle of the time-gate, denoted as $p(t_c)$, is used as the zero offset amplitude for building up the *zero offset reflection amplitude section*. Further, r_1 is interpreted as the gradient of the amplitude function with respect to x^2 for $x=0$.

For analysis results (r_1-, r_2-, zero-offset sections, and product section $|p(t_c)|r_1$ with $r_2=0$ and $\neq 0$) on field data interested readers are advised to refer to the original contribution by Ursin and Ekren (1995).

It should be emphasized that, after travel time correction, the data in the time window has to be corrected for offset-dependent geometrical spreading and array directivity. AVO analysis is performed on the three center-most samples of the time window for better results. Further, as AVO analysis requires very accurate alignment, *residual moveout corrections* are applied before performing amplitude analysis.

13.6 AVO Interpretation

13.6.1 AVO Gradient and Intercept for Gas Sands

For analyzing amplitude variation with offset and interpreting them for the presence of hydrocarbons, it is most appropriate to begin with an expression for the reflection coefficient $R(\theta)$ in terms of angular reflection coefficients A, B and C in the form (refer, for example, to Richards and Frasier 1976; Aki and Richards 1980),

$$R(\theta) \approx A + B\sin^2\theta + C\sin^2\theta\tan^2\theta \tag{13.35}$$

where

$$A = \frac{1}{2}\left(\frac{\Delta V_P}{\langle V_P\rangle} + \frac{\Delta\rho}{\langle\rho\rangle}\right)$$

$$B = \frac{1}{2}\frac{\Delta V_P}{\langle V_P\rangle} - 2\left(\frac{\langle V_s\rangle}{\langle V_P\rangle}\right)^2\left(2\frac{\Delta V_s}{\langle V_s\rangle} + \frac{\Delta\rho}{\langle\rho\rangle}\right)$$

and

$$C = \frac{1}{2}\frac{\Delta V_P}{\langle V_P\rangle}$$

Symbol $\langle\rangle$ expresses the average across the reflecting interface, and ΔV_P, ΔV_s and $\Delta\rho$ represent the contrasts. Vp, Vs and ρ refer to compressional and shear wave velocities and density, respectively. Here A (the AVO intercept) is a linearized form of the normal-incidence compressional wave reflection coefficient (R_P); B is the AVO gradient term.

It may be mentioned that the Rutherford and Williams classification of gas sands (refer to Fig. 13.2) is based on the normal-incidence reflection coefficient ($R_P = A$) only. Castagna et al. (1998) reasoned consideration of both intercept A and gradient B for giving a meaningful interpretation of AVO anomalies associated with gas sands. They refer to it as *AVO A vs. B cross plot interpretation*, and elaborate on the physical meaning of B together with A.

The usefulness of cross-plotting seismically extracted AVO intercept (A) and gradient (B) was realized by Smith and Gidlow (1987). As brought out by Smith and Gidlow (1987), the AVO intercept A and gradient B often form a well-defined *background trend* in the absence of hydrocarbon bearing strata.

It is to be realized that the same gas sand may show up very different AVO behavior depending upon the properties of the overlying shale (Castagna et al. 1998). In fact, compressional and shear wave velocities and densities of both media on either side of the reflecting interface decide what would be the total AVO response with offset. To demonstrate this very significant point, the following form of the reflection coefficient approximation (Richards and Frasier 1976) is illustrative:

$$R(\theta) = \frac{1}{2}\left(1 - 4\langle V_s\rangle^2 p^2\right)\frac{\Delta\rho}{\langle\rho\rangle} + \frac{1}{2\cos^2\theta}\frac{\Delta V_p}{\langle V_p\rangle} - 4\langle V_s\rangle^2 p^2\frac{\Delta V_s}{\langle V_s\rangle} \tag{13.36}$$

In Eq. (13.36), p is ray parameter ($=\sin\theta/V_p$) and θ is average of the angle of incidence and refraction.

Based on Eq. (13.36), Fig. 13.15a,b depicts the comparative role of fractional changes in density ρ, compressional velocity V_p, and shear velocity V_s. The summed up total response reveals contrasting AVO behavior in these cases. In the first case, it is shale ($V_p=2{,}900$ m/s, $V_s=1{,}330$ m/s, $\rho=2.29$ gm/cm^3) over gas sand ($V_p=2{,}540$ m/s, $V_s=1{,}620$ m/s, $\rho=2.09$ gm/cm^3). In the second case it is tight calcareous shale ($V_p=3{,}250$ m/s, $V_s=1{,}780$ m/s, $\rho=2.44$ gm/cm^3) overlying the same gas sand. The key parameter is the shear wave velocity contrast (positive or negative), controlling the total AVO responses.

Castagna et al. (1998, Eq. (7)) quantified the general equation for the *background reflections* in the A–B plane in the form

$$B = \frac{A}{1+g}\left[1 - 4\frac{\langle V_s\rangle}{\langle V_p\rangle}\left(\frac{2}{m} + g\frac{\langle V_s\rangle}{\langle V_p\rangle}\right)\right] \tag{13.37}$$

Derivation of this equation assumes that V_p, V_s and ρ are highly correlated for the 'background' trend. The "background" is taken – as brine – saturated rock

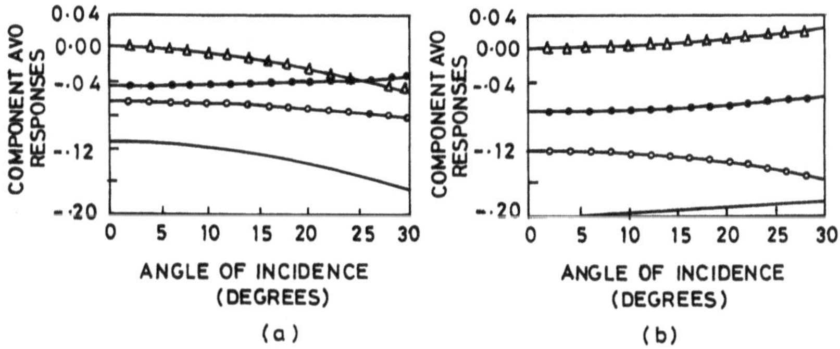

Fig. 13.15 a, b. a Richard and Frasier (1976) decomposition for a class III (quadrant III) shale over gas-sand reflection showing the AVO contribution from fractional changes in density (•), compressional velocity (○), and shear velocity (△). The *continuous line* shows the total response. **b** Richard and Frasier (1976) decomposition for the same gas sand as in Fig. 13.15 **a**, but now overlain with a calcareous tight unit. It's total AVO response is now class IV (quadrant II), due to a reversal of its shear velocity component (Source: Castagna et al. 1998)

(e.g. sandstone or shale). 'g' and 'm' appear in Eq. (13.37) by taking the following reasonable assumptions:

$$\frac{\Delta\rho}{\langle\rho\rangle} \approx g\frac{\Delta V_P}{\langle V_P\rangle}$$

and

$$V_P = mV_s + C$$

The relationship between the zero offset reflection coefficient A vs. AVO gradient B for the 'background' trend may be analyzed using Eq. (13.37). The numerical analysis would reveal that for constant V_P/V_s and $g = 1/4$ (Gardner et al. 1974),

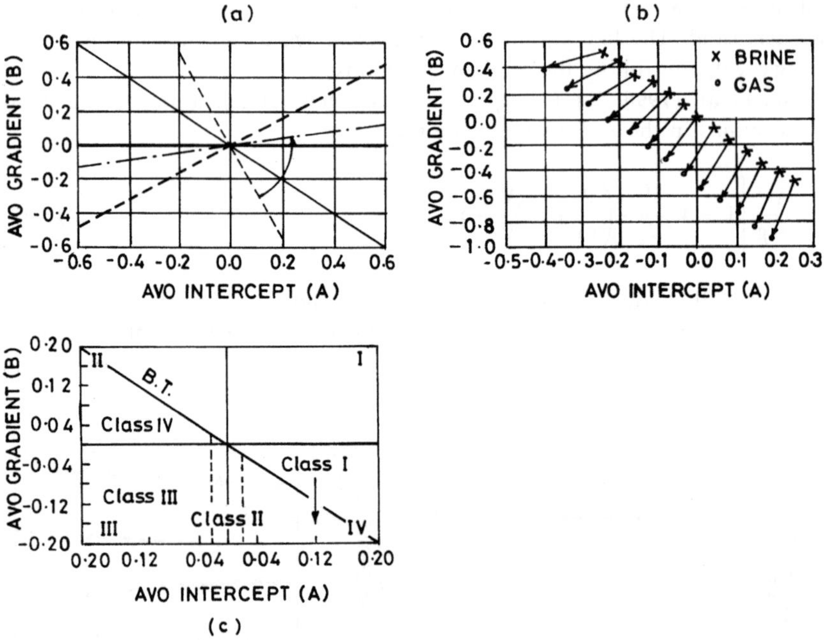

Fig. 13.16 a–c. a Zero-offset reflection coefficient (A) versus the AVO gradient (B) assuming a constant (V_P/V_s) and Gardner's relation. The background trend rotates counterclockwise as (V_P/V_s) increases. The V_P/V_s values are: 3.0 for the *thick continuous line*; 2.0 for the *thin continuous line*; 1.41 for the *thin dashed line*; ∞ for the *thick dashed line*; 3.5 for the *dash – dot line*. **b** Brine sand-gas sand tie lines for shale over brine-sand reflections having an average P-wave velocity of 3 km/s and which conforms to Gardner and mudrock petrophysical trend curves (g = 0.25, m = 1.16, c = 1.36 km/s) and Gassman's equations. **c** AVO intercept A versus gradient B crossplot showing four possible quadrants. For a limited time window, brine-saturated sandstones and shales tend to fall along a well-defined background trend. The top of gas-sand reflections tend to fall below the background trend, whereas the bottom of gas-sand reflections tend to fall above the trend. Augmented Rutherford and Williams (1989) gas sand classes are also indicated for reference (Source: Castagna et al. 1998)

the background trend rotates counterclockwise with an increase of V_p/V_s (Fig. 13.16 a).

It may be mentioned here that deviations from the background in the cross plot A–B may result from the replacement of brine in the pore-space by hydrocarbons, particularly gas. The result of this replacement would be that both A and B become more negative as compared to that for a fully-brine saturated sand. Figure 13.16 b shows the general behavior of A and B for shale/gas sand and shale/brine sand reflections.

The basis of the Smith and Gidlow (1987) hydrocarbon detection methodology is the use of *fluid-factor* (ΔF), which is measure of the deviation from the *background trend* and is given by,

$$\Delta F = \frac{\Delta V_P}{\langle V_P \rangle} - m \frac{\langle V_s \rangle}{\langle V_P \rangle} \frac{\Delta V_s}{\langle V_s \rangle} \tag{13.38}$$

In fact, the interpretation of A and B becoming more negative due to replacement of brine in the pore space by gas is in conformity with Koefoed's (1955) rules describing the variation of reflection coefficient with angle of incidence as a result of a change in the Poisson ratio. The addition of hydrocarbon in the underlying medium decreases the Poisson ratio, making B more negative. A is also made more negative by the addition of hydrocarbon.

The result of the addition of hydrocarbon may be interpreted with respect to Fig. 13.16c. Figure 13.16c shows a *background trend*. Adding hydrocarbon to the layer, otherwise falling on the 'background' trend, would change the trend. For the top of the gas sand, the new trend would plot below the background trend for more negative A and B. For the bottom of the gas sand, the new trend should plot above the background trend for more positive A and B. The tacit assumption is that the overlying and underlying medium above the gas sand has similar properties.

Referring to Fig. 13.2 it shows a class IV gas sand not covered in Rutherford and William's classification. Class III gas sands are classical bright spots and have increasing reflection magnitude with offset. Class IV gas sands are true bright spots, but reflection magnitude decreases with increasing offset. The locations of the 4 classes of gas sands are given in four quadrants of the A–B plot (Fig. 13.16c).

13.6.2
V_p/V_s Ratio, Anisotropy and AVO Response

An important aspect of an AVO study is to determine the magnitude of the effect which transverse isotropy has on seismic data, and to estimate whether it would make the AVO gradient more positive or negative. A situation may arise in which although the AVO response is of interest, accurate measurement of anisotropy parameters is lacking. The only other data that is available is the well log, or a general knowledge of V_p/V_s (the ratio of compressional and shear wave velocity).

Under such a situation, one may make use of the empirical relationship between V_p/V_s and transverse isotropy parameters. By making use of these anisotropy parameters in an interpretation, the chances of correct prediction of the presence of hydrocarbon would be increased. Ryan-Grigor (1997) has found an emkpirical relationship between the anisotropy parameter 'δ' and V_p/V_s. 'δ' is an important parameter for modeling the AVO response in a transversely isotropic medium.

A transversely isotropic medium is characterized by five elastic constants. Four (out of the five elastic constants) control wave propagation along the axes of a transversely isotropic medium. c_{33} and c_{11} control the vertical and horizontal compressional wave (V_p) velocities, respectively. c_{44} determines the velocity of vertically traveling horizontally polarized (SH) and vertically polarized (SV) shear waves. c_{66} controls the velocity of a horizontally traveling SH-wave. The elastic parameter c_{13} controls propagation in the oblique directions for P and SV waves only. The anisotropy parameter 'δ' has c_{13} in it in the form

$$\delta = \frac{(c_{13} + c_{44})^2 - (c_{33} - c_{44})^2}{2c_{33}(c_{33} - c_{44})} \tag{13.39}$$

Out of three Thomsen anisotropy parameters (ε, γ and δ), δ has the largest influence on P-wave reflection amplitude for small offsets, i.e. angles less than 20°–30° (Banik 1987).

Parameter ε gives the fractional difference between horizontal and vertical P-wave velocity squared. It is given by

$$\varepsilon = \frac{c_{11} - c_{33}}{2c_{33}}$$

Parameter γ is given by,

$$\gamma = \frac{c_{66} - c_{44}}{2c_{44}}$$

and approximates the fractional difference between the horizontal (SH) and vertical (vertical SV or SH, or horizontal SV) shear wave velocity squared.

The combination of ε and δ is expressed in the form

$$\eta = \frac{(\varepsilon - \delta)}{(1 + 2\delta)} \tag{13.40}$$

(Alkhalifah and Tsvankin 1994; Tsvankin and Thomsen 1994)
and has a significant role in time-related processing. The deviation $\Delta a(\theta)$ of the P-wavefront from the ellipse is expressible in terms of the difference between ($\delta - \varepsilon$) as

$$\Delta a(\theta) = a_v(\delta - \varepsilon)\sin^2\theta\cos^2\theta \tag{13.41}$$

[Ryan-Grigor 1997, Eq. (8)]

When $\delta=\varepsilon$, the P-wavefront is an ellipse, when $\delta>\varepsilon$, $(\delta-\varepsilon)$ is positive, making $\Delta a(\theta)$ positive. The wavefront bulges outside the ellipse. For $\theta=0°$ and $90°$, there is no deviation from the ellipse (Berge 1991). Deviation of the S-wavefront is determined by $(\varepsilon-\delta)$, and is opposite to that for the P-wave, i.e. when the P-wavefront bulges out, the S-wavefront contracts inwards.

Dividing Eq. (13.39) by c_{44}, δ may be expressed in terms of V_p/V_s and c_{13}/c_{44} as

$$\delta=\frac{[1+(c_{13}/c_{44})]^2-\left((V_p/V_s)^2-1\right)^2}{2(V_p/V_s)^2\left((V_p/V_s)^2-1\right)} \tag{13.42}$$

[Ryan-Grigor 1997, Eq. (9)]

This equation shows that if a relationship between c_{13}/c_{44} and V_p/V_s could be defined, δ would become a function of V_p/V_s only. The following empirical relation is found to give the best fit for measurements on saturated sedimentary rocks (Ryan-Grigor 1997),

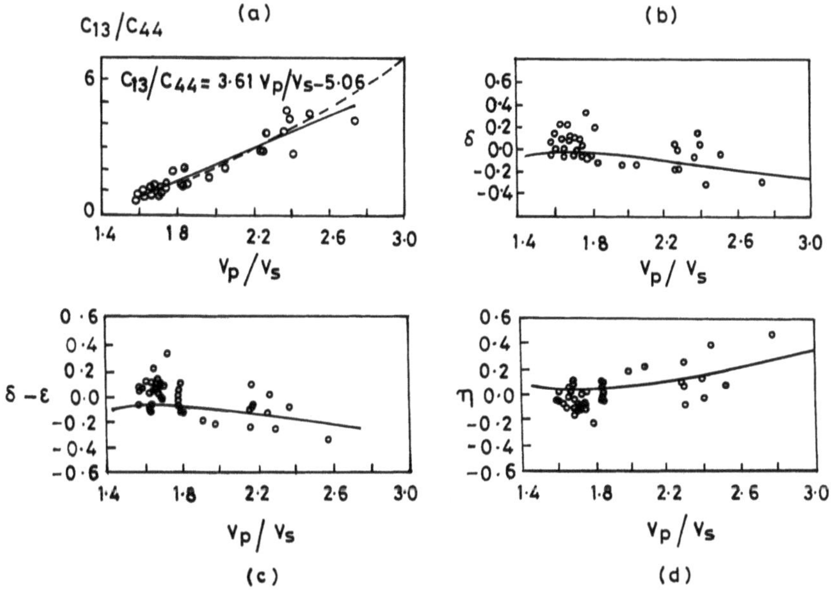

Fig. 13.17 a–d. a Ratio C_{13}/C_{44} versus V_p/V_s. The *solid line* is the line of best fit through the data. The equation $C_{13}/C_{44}=3.61\ V_P/V_S-5.06$ with a correlation coefficient of 0.9. **b** Parameter δ versus V_p/V_s. The *solid line* is the V_P/V_S versus δ empirical relationship, derived from $c_{13}/c_{44}=3.61\ V_p/V_s-5.06$ and Eq. (3.42). **c** Parameter $(\delta-\varepsilon)$ versus V_p/V_s. The *solid line* is the empirical relationship between $(\delta-\varepsilon)$ and V_p/V_s derived from $c_{13}/c_{44}=3.61\ V_p/V_s-5.06$, Eq. (5.42) and $\varepsilon=0.14$. **d** parameter η and V_p/V_s. The *solid line* is the empirical relationship between η and V_p/V_s, derived from $c_{13}/c_{44}=3.61\ V_p/V_s-5.06$, Eq. (13.9) and $\varepsilon=0.14$ (Source: Ryan-Grigor 1997, from Geophysics with permission)

$$\frac{c_{13}}{c_{44}} = 3.61 V_p/V_S - 5.06 \tag{13.43}$$

This relation holds for the condition of weak anisotropy as rocks with ε and δ less than 0.2 were included in Ryan-Grigor's study.

The pattern of variation of c_{13}/c_{44} is shown in Fig. 13.17a. Also, variations of δ, $(\delta-\varepsilon)$, and η with V_p/V_s using the linear relationship Eq. (13.43) and Eq. (13.42) are shown in Fig. 13.17b, c and d.

Figure 13.17b shows that when V_p/V_s is less than 1.8, δ is nearly zero. For V_p/V_s greater than 2.0, the δ value is negative and decreases with increasing V_p/V_s. η is the anellipticity parameter and shows a close correlation with V_p/V_s. This correlation reveals that knowledge of V_p/V_s is sufficient to make an estimate of P and S wavefronts in a rock, irrespective of knowledge of its lithology.

In general shales tend to have higher V_p/V_s than sandstone. However, some shales have V_p/V_s values between 1.6 and 1.8 which is typical range for sandstones. These shales have $(\delta-\varepsilon)$ close to zero, and the P-wavefront approximates an ellipse. Vernik and Nur (1992) attribute low V_p/V_s values to high organic content of shale, and overpressure. These shales would make the AVO gradient more negative, and intensify the AVO bright spot effect. One may infer that an organic rich shale with low V_p/V_s overlying sand should make the AVO gradient more negative whereas shale with high V_p/V_s should make the AVO gradient more positive. A thick shale section, having in it a part with large V_p/V_s lying over another part with low V_p/V_s, would give rise to an AVO anomaly indistinguishable from shale overlying a sandstone (Allen et al. 1993). This happens because of the contrast in δ across low V_p/V_s shale with high V_p/V_s shale in the same shale section.

Rocks tend to be more anelliptic with increasing V_p/V_s. Rocks with V_p/V_s less than about 1.8 (typical sandstone and some shales) are likely to have zero or small positive values of δ; while rocks with larger V_p/V_s than around 2 (tykpical of most shales) tend to have values of δ of zero to large negative values. If the δ of shale is negative (V_p/V_s higher than 2) and overlies sandstone, a positive change in δ across the interface would cause a false negative AVO gas indicator. Similarly if shale with V_p/V_s less than 1.8 (as for organic rich or overpressured shales) overlies a water-filled sandstone, this should cause a false positive AVO gas indicator. These effects underline the necessity for correctly estimating anisotropy effects in interpreting the AVO response due to gas sands.

Kumar (2002) developed computer programs for AVO interpretation for thin or thick gas sand beds.

13.6.3 Perspective for AVO Intercept and Gradient Inversion

Considering firstly the isotropic condition, reflection amplitude variation with offset may be expressed by Eq. (13.3). It gives the AVO intercept – the reflection amplitude at normal incidence, R_0.

The AVO intercept contains information on impedance change and is half the change in the natural logarithm of impedance ρV_p.

The term $A\sin^2\theta$ in Eq. (13.3) is important in the intermediate angle range. A is a combination of two terms, A_0 and $(\Delta\sigma/R_0)\ (1/(1-\sigma)^2)$. A_0 is accurately predictable under certain conditions (Shuey 1985) and therefore the gradient term A has information on $(\Delta\sigma/R_0)$. $\Delta\sigma$ is the change in the Poisson ratio. The change in the Poisson ratio may be helpful in identifying lithology. The third term in Eq. (13.3) is dominant at large angles, and reflection amplitude at large angles is linked with a change in velocity. Therefore, the inversion of reflection amplitude at large angles should provide information on velocity change.

The ratio of AVO intercepts, R_0 (for P-waves) and S_0 (for S-waves) contain information on gas saturation [Eq. (13.9)].

S_g in Eq. (13.9) is the fractional gas saturation and ϕ is *fractional porosity*. This relation holds for a gas-brine contact in sandstone.

Next to the simple case of isotropy is the case of transverse isotropy with a vertical axis of symmetry (VTI). For a VTI medium, the AVO response is independent of azimuth. For small jumps in velocities, densities and anisotropy parameters, the approximate solution for the PP reflection coefficient is given by Eq. (13.10).

Equation (13.10) is a concise Shuey (1985) type approximate reflection coefficient and gives AVO parameters – an intercept, a gradient and higher order terms as a function of change in the Thomsen anisotropy parameters, δ and ε. $\Delta\delta$ appearing with $\sin^2\theta$ term describes the influence of anisotropy on a small angle reflection coefficient and AVO slope. $\Delta\varepsilon$ associated with the $\sin^2\theta\ \tan^2\theta$ term has dominating role at larger incidence angles.

Next in complexity to isotropy and transverse isotropy (with a vertical symmetry axis) is the case of transverse isotropy with a horizontal symmetry axis. Reflection in a HTI medium with a horizontal symmetry axis is distinctly different from isotropy and VTI media in having its dependence on azimuth. The compressional (P-wave) wave reflection coefficient between the HTI/HTI interface is expressed by Eq. (13.11).

In this case, the anisotropic contribution as a function of azimuth ϕ comes through the parameters $\Delta\delta^{(V)}$, $\Delta\gamma$ and $\Delta\varepsilon^{(V)}$. In contrast to a VTI medium, the anisotropy contribution also comes through $\Delta\gamma$. The contrast in *shear wave splitting parameter* γ has an influence on the AVO gradient ($\sin^2 i$ term).

A P-wave is coupled with a $S^{\perp}$ wave in the symmetry plane, which differs from the velocity of the shear wave (S^{II}), confined to the isotropy plane.

The importance of studying reflection amplitude variation with azimuth along with the angle of incidence is in directly detecting and characterizing fractured reservoirs. Naturally fractured reservoirs show azimuthal anisotropy of various types and strengths and the HTI medium is one of them. The HTI medium describes, for example, a system of parallel penny-shaped vertical cracks embedded in an isotropic matrix. The HTI medium is a transversely isotropic medium with a horizontal axis of symmetry.

Returning back to the gradient term in Eq. (13.11), the AVO gradient measured at azimuth $B(\phi_k)$ is expressible as isotropic and anisotropic contributions [Eq. (13.12)] [Rüger 1998, Eqs. (8)–(10)].

The above form of $B(\phi_k)$ [Eq. (13.12)] assumes that the symmetry axis orientation is unknown.

Equation (13.12) is non-linear with three unknowns, B^{iso}, B^{aniso}, and ϕ_{sym}. If directions of symmetry axis are known (as may be obtained from analysis of S-wave data), the equation becomes linear with two unknowns B^{iso}, B^{aniso}. Therefore two independent measurements should suffice to determine the values of B^{iso}, B^{aniso}.

If the symmetry axis direction is unknown, multiazimuthal surveys should be used to find optimum values of the unknowns in Eq. (13.12) using a least square approach. The solution would not be unique and give two possible directions of symmetry axis. These are orthogonal to each other. Under such a situation prior knowledge of the approximate symmetry axis direction should enable us to identify correctly the actual direction of the symmetry axis.

The AVO gradient in a HTI medium may be inverted to get an idea of crack density and discriminate between fluid-filled and dry cracks. This is possible, because for penny-shaped cracks the shear wave splitting parameter (γ) is close to the crack density. The other parameter in the gradient term (δ^{v}) varies significantly as a function of crack filling. Thus accurate estimation of shear wave splitting parameter (γ) and coefficient (δ^{v}) should enable a definition of *crack density*, and discrimination between fluid-filled and dry cracks.

The explicit contribution of the parameters (γ) and (δ^{v}) comes by taking the difference between $B(\phi_{sym}) - B\ (\phi_{sym} + \pi/2)$, given by

$$B(\phi_{sym}) - B\left(\phi_{sym} + \frac{\pi}{2}\right) \cong \left\{ \left(\frac{2\bar{\beta}}{\bar{\alpha}}\right)^2 \gamma_2 + \frac{1}{2}\delta_2^{(v)} \right\} \tag{13.44}$$

[Rüger 1998, Eq. (12)]

The difference in AVO gradients in the above form depends only on the shear wave splitting parameter γ and $\delta^{(v)}$. $\delta^{(v)}$ may be determined, for example, by analysis of P-wave NMO velocities. Knowing $\delta^{(v)}$ and assuming a reasonable value for β/α (say $\approx 1/2$) the difference in AVO gradients may be inverted for the shear wave splitting parameter γ.

For fractured fluid-filled coal layers ($\varepsilon = \varepsilon^{(v)} = 0$) the difference $[B(\phi_{sym}) - B(\phi_{sym} + \pi/2)]$ gives the crack density directly (Tsvankin 1997).

There are some assumptions and limitations on the use of the P-wave AVO gradient. The main assumption is small jumps in elastic parameter across reflecting interface and weak anisotropy. In an exploration context, these assumptions are mostly reasonable. However, the method may fail for large contrasts in elastic parameters and strong anisotropy. If the orientation of the symmetry axis at the interface changes abruptly, the horizontal symmetry axis approximation would prove inaccurate. The limitation of the method comes from an improper processing sequence that does not preserve azimuthally varying amplitude signatures. The method does not cover cases of curved reflectors and dipping symmetry axes.

Appendix 13 A – Computer Program for Approximate Numerical Solutions of the P-P Reflection Coefficient between HTI/HTI Media

This Program is made by Sharma (2002)
Part I: Computation of the variation of the p-wave reflection coefficient with azimuth angle at a constant angle of incidence [Eq. (13.11) for HTI media]

a is the ratio of ΔZ and Z
b is the ratio of $\Delta\alpha$ and α
c is the ratio of β and α
d is the ratio of ΔG and G
e is the $\Delta\gamma$
f is the $\Delta\delta^{(v)}$
g is the $\Delta\varepsilon^{(v)}$
i is the incidence angle
phi is the azimuth angle

****

```
#include⟨iostream.h⟩
#include⟨math.h⟩
#include⟨conio.h⟩
#include⟨iomanip.h⟩
#include⟨fstream.h⟩
void main ()
  {
    float a,b,c,d,e,f,g,i,phi;
    clrscr();
    ifstream input7c("r7c.dat");
    ofstream output7c("r7c.xls");
    input7c>>a>>b>>c>>d>>e>>f>>g;
    output7c<<" value of del Z/Z bar="<<a
      <<"\n value of del alpha/alpha bar ="<<b
      <<"\n value of beta bar/alpha bar ="<<c
      <<"\n value of del G/G bar ="<<d
      <<"\n value of del gamma ="<<e
      <"\n value of del delta(v) ="<<f
      <<"\n del epsilon(v) = "<<g<<endl;
i=10;
  do
  {output7c<<setw(3)<<"\n i="<<i<<endl;
    phi=0;
    output7c<<setw(3)<<"phi"<<setw(10)<<" "<<"Rp(i,phi)"<<endl;
    do
  {
  float C=cos(phi*3.141592/180);
  float s=sin(phi*3.141592/180);
  float S=sin(i*3.141592/180);
  float T=tan(i*3.141592/180);
```

```
    float t1=b-(2*2*c*c*d)+(f+2*2*2*c*c*e)*C*C;
    float t2=b+(g*C*C*C*C)+(f*s*s*C*C);
    float R=(a+(t1*S*S)+(t2*S*S*T*T))/2;
    output7c<<setw(3)<<phi<<setw(15)<<R<<endl;
    phi=phi+10;
    }
  while(phi<90||phi=90);
  i=i+10;
  }
  while(i<40||i=40);
  cout<<"Press any key";
  getch();
  }
```

Part II: Computation of variation of p-wave reflection coefficient with incidence angle at constant azimuth angles [Eq. (13.11) for HTI media]

a is the ratio of ΔZ and Z
b is the ratio of $\Delta \alpha$ and α
c is the ratio of β and α
d is the ratio of ΔG and G
e is the $\Delta\gamma$
f is the $\Delta\delta^{(v)}$
g is the $\Delta\varepsilon^{(v)}$
i is the incidence angle
phi is the azimuth angle
***/

```
  #include⟨iostream.h⟩
  #include⟨math.h⟩
  #include⟨conio.h⟩
  #include⟨iomanip.h⟩
  #include⟨fstream.h⟩
  void main ()
    {
    float a,b,c,d,e,f,g,i,phi;
    clrscr();
    ifstream input7c("r7c.dat");
    ofstream output7c("r7c.xls");
    input7c>>a>>b>>c>>d>>e>>f>>g;
    output7c<<" value of del Z/Z bar ="<<a
      <"\n value of del alpha/alpha bar ="<<b
      <"\n value of beta bar/alpha bar ="<<c
      <"\n value of del G/G bar ="<<d
      <"\n value of del gama ="<<e
      <"\n value of del kde(v) ="<<f
      <"\n del epsilon(v)="<<g<<endl;
    phi=10;
    do
    {
```

```
    output7c<<setw(3)<"\n phi="<<phi<<endl;
    i=0;
    output7c<<setw(3)<"i "<<setw(10)<<" "<<"Rp(i,phi)"<<endl;
    do
    {
    float C=cos(phi*3.141592/180);
    float s=sin(phi*3.141592/180);
    float S=sin(i*3.141592/180);
    float T=tan(i*3.141592/180);
    float t1=b-(2*2*c*c*d)+(f+2*2*2*c*c*e)*C*C;
    float t2=b+(g*C*C*C*C)+(f*s*s*C*C);
    float R=(a+(t1*S*S)+(t2*S*S*T*T))/2;
    output7c<<setw(3)<<phi<<setw(15)<<R<<endl;
    i=i+5;
    }
  while(i<40||i==40);
  phi=phi+15;
  }
  while(phi<90||phii==90);
  cout<<"Press any key";
  getch();
  }
```

CHAPTER 14

Concepts and Methods in Seismic Migration 14

14.1 Definitions and Elementary Considerations

Consider a homogeneous medium, containing only one diffraction point (Fig. 14.1 a). In this case, the time-section (record surface) is a hyperbola (Fig. 14.1 b) whereas the geological inhomogeneity (a diffractor) is a single point. It is evident that the reflection or diffraction events distributed on a hyperbola shown in Fig. 14.1 b are to be brought to their actual (point) position. This step is taken care of in the process of '*migration*'.

The migration process based on the summation of amplitudes on the diffraction hyperbola, and restoring it to the apex of the hyperbola is termed *diffraction stack migration* (Fig. 14.1 c). The diffraction stack migration method was the most popular migration technique in the early 1970s, and has been in use even afterwards. Early developments in the diffraction stack (or summation) process were based on ray tracing concepts and the scalar diffraction theory of Huygen and Fresnel. Later developments (refer, for example, to French 1975; Schneider 1978) provided a firm mathematical foundation by relating the diffraction summation process with Kirchoff integral solutions of the wave equation.

The concept of a point reflector is very useful. A dipping reflector may be simulated by a sequence of points along a straight line on the reflector. Also, a

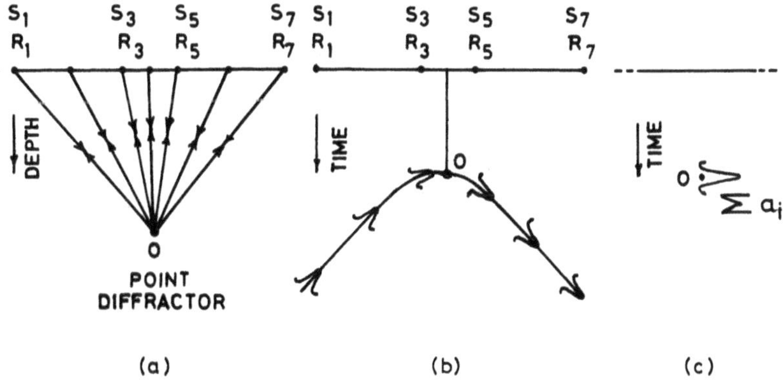

Fig. 14.1 a–c. a Point diffractor; **b** time section; **c** result of diffraction stack migration

curved reflector may be assumed as consisting of small linear segments of varying dips, and each segment may be approximated by point reflectors. In general, the actual *reflector surface* and the *record surface* are different. Nevertheless, the reflector surface is derivable from the record surface.

If the *record surface* is single-valued, the determination of the reflector surface is straightforward. This process referred to as *migration* is shown in Fig. 14.2. Consider a point P on the *record surface*. Its position (P′) on the reflector surface is found by fitting one of the family of curves *of maximum convexity* to the slope of the record surface at P. Point P′ is situated at the minimum depth on the curve of maximum convexity which is tangential to the record surface at P. The reflector surface is tangential to the wavefront at P′, which passes both from P and P′ (Fig. 14.2).

In constructing a *record surface*, the ray paths are assumed to be vertical. The record surface can be constructed in a simple manner knowing the velocity function. For details, reference may be made to Grant and West (1965, pp 156–163). The *record surface* and *reflector surface* would be identical if actual ray paths are vertical, as in the case of a horizontal reflector.

The *wavefront* is the generator of the *reflector surface.* In a similar way, the *surface of maximum convexity*, i.e. the record surface corresponding to reflections from a point reflector, is the generator for the *record surface.* Both *wavefronts* and *record surfaces* may be derived for the assigned velocity function. An example of finding the *diffraction curve* (*curve of maximum convexity*) for a point scatterer and for a linear velocity function may be found in Waters (1987, pp 381–384).

Figure 14.2 shows the relationship between *wavefront*, the *surface of maximum convexity*, and the *ray path* for coincident source-receiver geometry. The wavefront intersects the surface of maximum convexity at two points, one at the actual position of the reflector and the other vertically below the coincident shot-receiver.

The above discussion pertains to a migration concept using amplitude summation along the diffraction hyperbola (Fig. 14.1 c). A contrasting concept of migration is the *wavefront method of migration.* In contrast to summation of ampli-

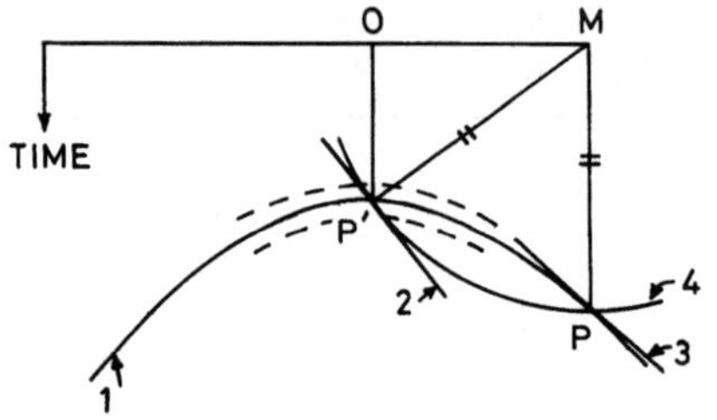

Fig. 14.2. *1* is a curve of maximum convexity (diffraction hyperbola); *2* is a part of the reflector surface; *3* is a part of the record surface; and *4* is the wavefront passing through the point P on record surface and minimum time point P′ at the apex of the diffraction hyperbola. The diffraction curve is tangential to the record surface at P and the wavefront is tangential to the reflector at P′. The dashed curves are different positions of diffraction hyperbola. M is the coincident source-receiver position, and MP′ = MP

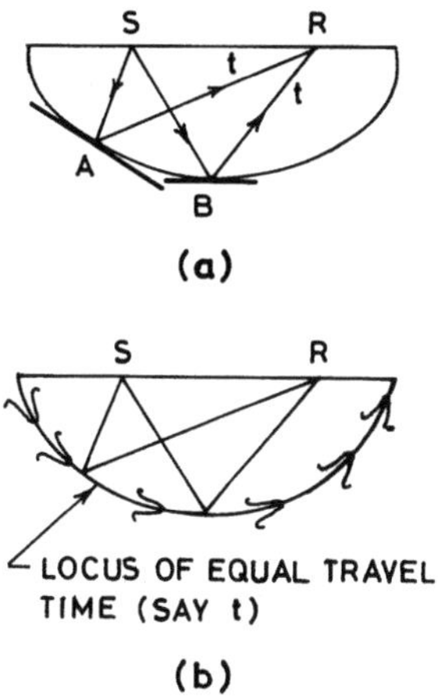

Fig. 14.3 a, b. Schematic diagram illustrating the principle behind the wavefront method of migration. S and R are seismic source and receiver respectively. **a** Reflection with time t could arise from any point, like A or B. The locus of points, from which reflections observed has time t, is elliptical. The reflectors are tangential to the ellipse at the points of reflection. **b** Trace pulse at time t of the record is shown distributed along the wavefront or isochron (Source: after Schneider 1971)

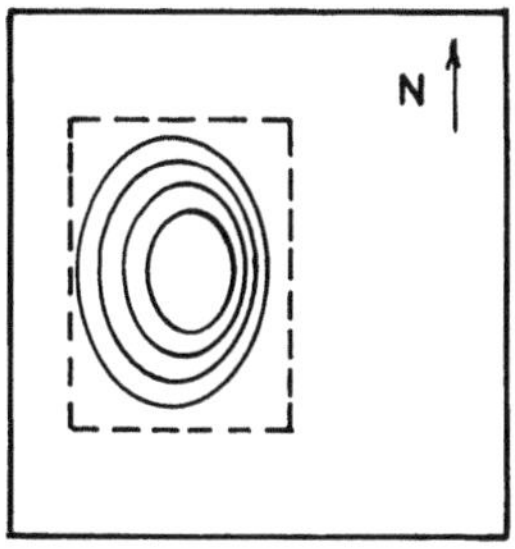

Fig. 14.4. Schematic diagram showing structural contours of a localized subsurface-elevated feature with the eastern flank steeply dipping. The inner rectangle (*dashed line*) shows subsurface lateral extensions; the outer rectangle the larger recording area required to adequately image the subsurface structure

tudes the wavefront method of migration uses the concept of distribution of amplitudes along wavefronts (or isochrons, Fig. 14.3 a, b). For discussion of its practical implementation reference may be made to the original excellent contribution by Schneider (1971).

An important consideration for migration is the concept of *migration aperture*. Referring to Fig. 14.1 we observe that for migration corresponding to a point diffractor, we require reflection data for a number of coincident source-receiver positions along a line for summation of amplitudes. That is to image a point diffractor we need extended data along a line. This concept may be easily extended for a structural feature below the ground surface. Figure 14.4 shows the subsurface extension of the structure, which is steep on the eastern flank. To properly image this structure, we would require reflection data over a larger area on the ground surface. Since the structure is steep on the eastern side, the survey area should be extended more on this side to adequately image the steeply dipping eastern flank. For basic details on migration aperture reference may be made to Chun and Jacewitz (1981), and for elaborate discussions to Berkhout (1982).

14.2 Point Diffractor Model and Application

We now illustrate migration of a specific case applicable to a dipping reflector, using the concept of the point diffractor explained in the previous section. For this purpose we reconsider two point diffractors (A and B) displaced vertically and laterally to each other (Fig. 14.5 a).

Record sections for coincident shot receivers (S_1 R_1, S_2 R_2,... S_N R_N) for point diffractors A and B are shown by two dash and dash-dot curves in Fig. 14.5 a. Assume that the recorded data for each coincident source receiver is of unit amplitude. Further assume that the position of the (sampling) diffraction hyperbola is shown by a continuous curve. It cuts the record section only at points A′ and B′. Therefore, the summed up values of amplitudes along the continuous sampling curves would be of amplitude 2. In other downward positions also (shown by short length continuous curves) the summed up amplitudes would be 2 (Fig. 14.5 b) provided the continuous curve cuts the diffraction hyperbolas at two points. The summed up values are restored at the apex of the sampling hyperbola.

Next consider the position (S_j R_j) and (S_k R_k), when the sampling curve coincides completely with the diffraction curve of point diffractor A. All amplitudes values of the diffraction curve plus the unit amplitude at Q would be added up to

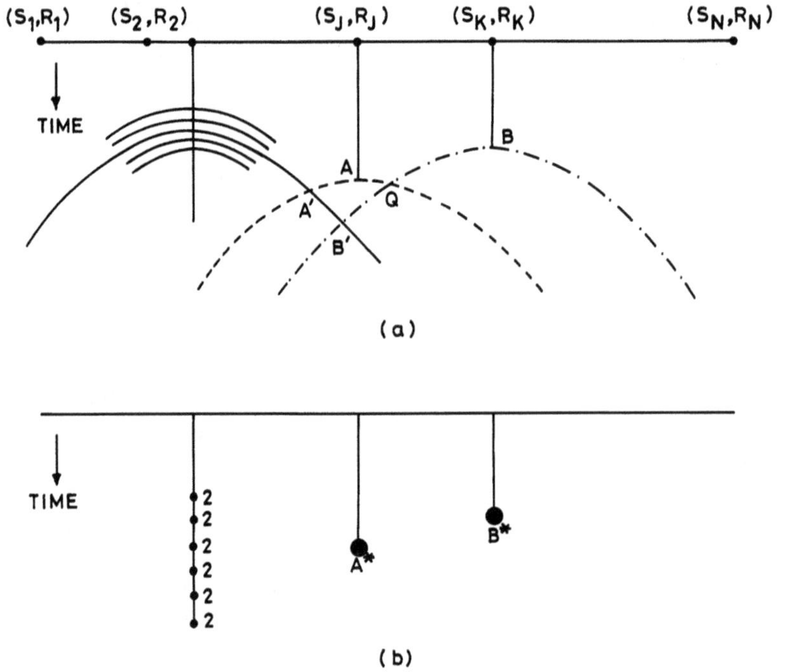

Fig. 14.5 a, b. a Two point diffractors A and B, their time sections shown by *dashed-* and *dot-dash curves*; and the sampling hyperbola shown by the *continuous curve*. **b** Summed up amplitudes for vertically and laterally shifted positions of sampling hyperbola

produce a large amplitude at A* (Fig. 14.5b). Similarly, for point diffractor B a large amplitude would result when the sampling hyperbola coincides with the record section of point diffractor B. These are shown by A* and B* in Fig. 14.5b. A* and B* are the migrated positions of the point diffractors A and B. It is obvious that summed up amplitude values at A* and B* would be more than at any other point.

Points A and B may be the end points of a line segment of a reflector, producing diffraction hyperbolas as shown in Fig. 14.5a.

The procedure of migrating data due to a point diffractor may be easily adapted for a dipping reflector also. The dipping reflector may be simulated by a sequence of point diffractors along the reflector.

Problem 14.1

Given the velocity function in the form $V_z = V_0(1 + az)$, where V_0 is 3,000 m/s and $a = 0.0005$, generate diffraction curves for a point diffractor at depths of 700 m, 1,000 m and 3,000 m.

Hint: For a linear velocity function of the given form, the following relation holds:

$$T = \frac{1}{V_0 a} \ln\left\{\frac{V_z(1 + \cos\theta_0)}{V_0(1 + \cos\theta_z)}\right\}$$

$$X = \frac{1}{a \sin\theta_0}(\cos\theta_0 - \cos\theta_z)$$

[Water 1987, Eq. (9A.6)]
where T is the ray travel time from the point diffractor to the detector on the ground surface and X is the distance from a point vertically above the diffractor to the detector. V_0 and V_z are velocities at the surface and at depth Z. θ_0 is the emergence angle of the ray on the ground surface measured with respect to the vertical; θ_z is the angle which the ray makes with the vertical at any point (depth z) on the ray. θ_0, θ_z, V_0 and V_z are related by Snell's law:

$$\frac{\sin\theta_0}{V_0} = \frac{\sin\theta_z}{V_z}$$

By increasing θ_0 in steps of $\Delta\theta$ ($\theta_0 = n\Delta\theta$), calculate T and X for various θ_0 values. Interpolate T for equal increments of X. Plot T versus X to obtain the diffraction curve. Repeat the procedure for various depths of the diffractor.

14.3 Geometric Techniques of Migration

There is an intimate geometric relation between migration of *dipping events* in the space domain to their counterpart migration in the wave number domain. To illustrate the point we consider simple examples from Chun and Jacewitz (1981).

First consider that there is a vertical reflector (Fig. 14.6 a). Let C denote the position of the seismic source as well as the receiver. The ray paths are shown. Only the ray incident horizontally would be received after reflection. For different positions of C on the x-axis, the record section would be a line inclined at an angle of 45°, corresponding to CO equal to CP. We conclude that in the record section a vertical reflector appears as if it is inclined at 45°.

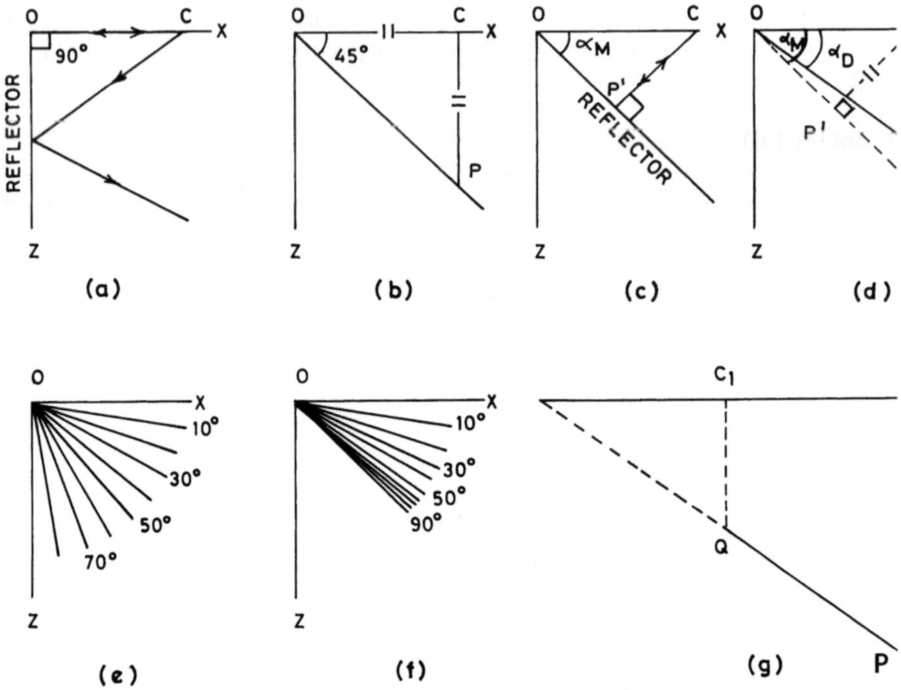

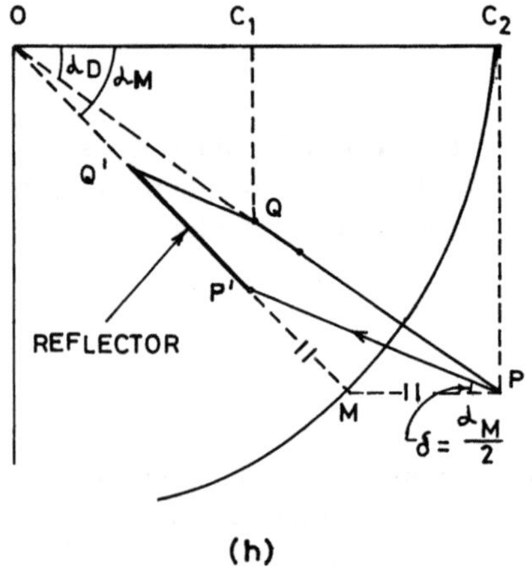

Fig. 14.6 a–h. a A 90° reflector model; **b** record section for the 90° reflector model in **a**; **c** a dipping reflector model (Earth section); **d** record section for a dipping model; **e** the fan Earth model with some representative dips (open fan); **f** corresponding record section (folded fan); **g** a bounded dipping reflector (PQ) in the record section; **h** construction for migration, P′Q′ is the migrated section (the reflector) of the record section PQ of the dipping segment (Source: Chun and Jacewitz 1981)

Next consider Fig. 14.6c with reflector dipping at an angle α_M. 'C' denotes the position both for seismic source and the receiver. The incident and reflected ray-paths would be coincident and perpendicular to the reflector. CP′ shows this. This distance, CP′, would be mapped vertically in the x–z plane to P where CP′ = CP (Fig. 14.6d) for generating the record section. The position of the reflector is shown by the solid line in Fig. 14.6c and by the dashed line in Fig. 14.6d. The following relationship is easily verifiable:

$$\sin \alpha_M = \frac{CP'}{OC} = \frac{CP}{OC} = \tan \alpha_D \tag{14.1}$$

Equation (14.1) relates the migrated dip, α_M, to the recorded dip angle α_D. As point P on the record section moves to point P′ after migration, we may say that the *migration process moves the data up dip.* Migration Eq. (14.1) holds under condition of isotropy. Under an anisotropic condition, there are two velocities associated with a particular wave front – the phase velocity (V), and the ray velocity (U) and the corresponding migration equation takes the form (refer to Subsection 4.4.7 for details):

$$\sin \alpha_M = \frac{V}{U} \tan \alpha_D$$

It may be mentioned that reflections may be considered as a superposition of diffractions. The process of mapping a reflector or a diffractor to the *record section* is referred to as the *diffraction process.* The process of *migration* proceeds in the opposite direction and transforms the *record section* to a *real Earth model.* To further elaborate this point we consider a model (Fig. 14.6e and f) combining and generalizing the results in Fig. 14.6a,b and Fig. 14.6c,d. Figure 14.6e is an Earth model with varying dip angles – a *fan model.* In this model velocity is taken as constant for each layer between the reflectors or ribs of the fan. However, each layer has a different density.

The seismic source and receivers are taken coincident along the x-axis. Following the results on the record sections in Fig. 14.6a,b and 14.6c,d, it is immediately clear that the recorded seismic section of Fig. 14.6e would be a folded fan as in Fig. 14.6f, i.e. the *diffraction model.* The migration process would unfold the folded fan and make it a fully extended fan, i.e. the Earth model as in Fig. 14.6e. It may be noted that the steeper the dip in the unfolded fan (Earth model), the more densely it is folded on the diffraction model and vice-versa.

A simple geometrical method of migration evolves from Eq. (14.1) (Chun and Jacewitz 1981). Referring to Fig. 14.6g, PQ is the reflected (or diffracted) segment, which is to be migrated. The following procedure may be followed for migration:

- extend the segment PQ to the surface to the point O,
- draw a vertical line PC_2 through P,
- take a radius OC_2 and draw an arc of the circle as shown,
- draw a horizontal line from P to intersect the circle at M,
- join M to O. Line OM gives the migrated dip angle (α_M),
- construct the point P′ by taking MP = MP′. Angle (δ) of line PP′ is equal to $\alpha_M/2$.

The projection of PQ at angle $\delta = (\alpha_M/2)$ on the line OM, the migrated segment P′Q′ (the actual reflector) corresponding to the diffracted segment PQ (the record section) is achieved, and the migration process is complete.

Thus, in the migration process there are two steps – first finding the correct line which gives *migrated dip angle* (α_M) and the second is mapping the reflection points linearly.

The *linear mapping* has an important role for explaining *wave number domain migration.* An idealized description follows from Chun and Jacewitz (1981).

14.3.1
Corresponding Wave Number Domain Migration

In the above a *fan model* was described, and it was shown that the *unmigrated section* resembled a partially opened fan (Fig. 14.6f). A real seismic section may be imagined as being a set of scattered diffraction fans.

The migration in the spatial frequency (or wave number) domain makes use of an important property of the 2-D Fourier transform. According to this property, howsoever-scattered events are in (x–z) domain, in the Fourier-transformed domain data will become well organized, and become a nicely gathered fan. Therefore, as a result of subsequent application of the migration process in the frequency domain to the gathered fan, all scattered dipping events in the (x–z) plane (as in Fig. 14.13a) may be handled easily.

It may be emphasized here that the 2-D Fourier transform maps an event in a direction normal to the dip direction of event in the (x–z) plane. Accordingly, horizontal events are mapped along the k_z-axis and vertical events along the k_x-axis where k_x and k_z are Fourier-transformed quantities. The width of the mapped event is inversely proportional to the extension of the event in the x–z plane. This point is further illustrated through the Figs. 14.7a–c and 14.8a–c.

Figure 14.7a shows three bounded horizontal reflectors. It shows a line spike model in the depth domain. Figure 14.7b shows its Fourier transform in the x-direction and Fig. 14.7c shows the typical amplitude contours in the 2-D frequency

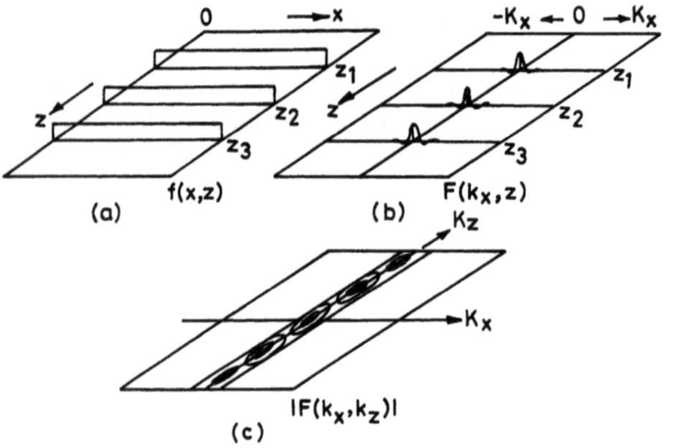

Fig. 14.7a–c. A 3-D sketch of three bounded horizontal reflectors. **a** Line spike model in the depth domain; **b** Fourier transform in the x-direction; **c** typical amplitude contours in the 2-D frequency domain (Source: Chun and Jacewitz 1981)

domain. Figure 14.8a–c shows a similar depiction for vertical reflectors. It is observed that horizontal events are mapped along the k_z-axis. Similarly, vertical events are mapped along the k_x-axis.

By similar arguments as for horizontal and vertical reflectors, a dipping reflector in the x–z plane (Fig. 14.9a) would be mapped along the normal direction to the dip angle (ϕ) (Fig. 14.9b) in the frequency (k_x–k_z plane) domain. The width would be inversely proportional to the extension of the event in the x–z plane.

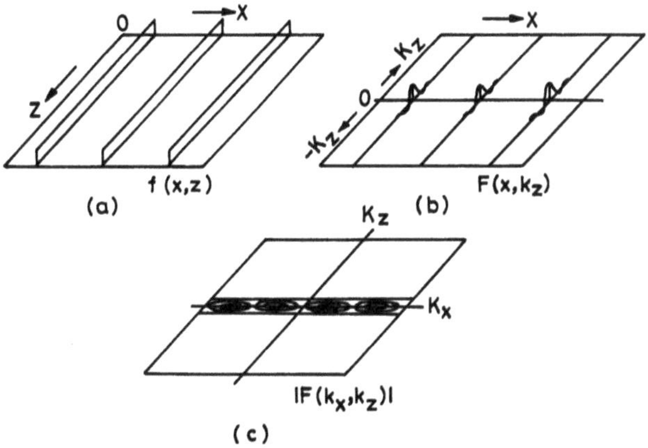

Fig. 14.8a–c. A 3-D sketch of three bounded vertical reflectors. **a** Line spike model in the depth domain; **b** Fourier transform in the depth direction; **c** typical amplitude contours in the 2-D frequency domain (Source: Chun and Jacewitz 1981)

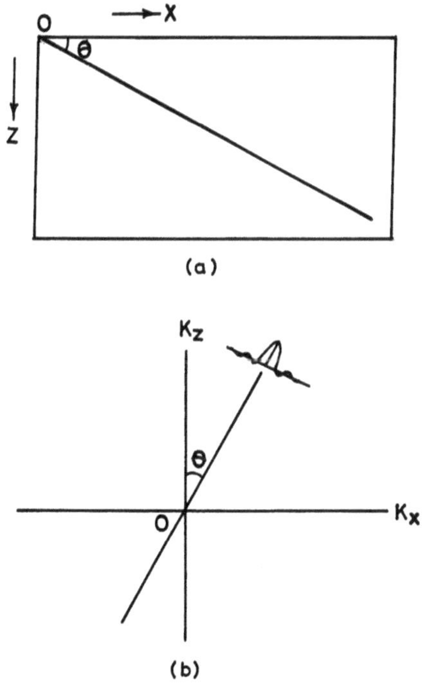

Fig. 14.9a, b. Bounded dipping reflection segment. **a** Reflection in the depth domain; **b** frequency domain representation (Source: Chun and Jacewitz 1981)

In fact, any event in the (x–z) plane is mapped normal to its dip angle in the (k_x–k_z) plane. This fact ensures that the dip angle of both *diffracted* and *migrated* events in the (x–z) plane must be preserved in the (k_x–k_z) plane. Therefore, *all events of the same dip in the (x–z) plane are gathered together along a single line in the (k_x–k_z) plane in the frequency domain.* This is about preservation along a single line in the (k_x–k_z) plane of all reflection events with the same dip in the (x–z) plane. A question still remains concerning the position on the line in the (k_x-.k_z) plane of different (scattered) and equally dipping events in the (x–z) plane. This may be understood by taking an example of a spike.

Consider a spike located at time t=0 (Fig. 14.10 a). Its Fourier transform representation is given in the lower part of this figure. The amplitude part is constant for all frequencies, and the phase is zero. Now consider the spike located not at zero time but at some other time (Fig. 14.10 b). Its frequency domain representation is given in the figure's lower part. It is observed that the amplitude spectrum is the same whereas the phase spectrum is changed. *The information on the position of the spike in time is preserved in its phase spectrum.*

With similar arguments as for the spike, it may be concluded that information on spatial locations of dipping reflections in the (x–z) domain are preserved in phase relation within a complex number in the frequency domain.

The process of migration has been explained for a single dip on a record section. The seismic section may be thought of as a combination of dips of all types. Here the principle of superposition applies, and the migration of all events on a seismic section may be realized as if it has been applied to a vast sum of simple events. Curved events on a seismic section may be thought of as containing short line segments of rapidly changing dips. In this way, the principle of superposition allows even a complicated seismic section to be migrated easily in the frequency domain.

Following the above argument (refer to Fig. 14.9 for dipping reflector transform sets) the diffracted event OP and migrated event OP′ (Fig. 14.11a) would be

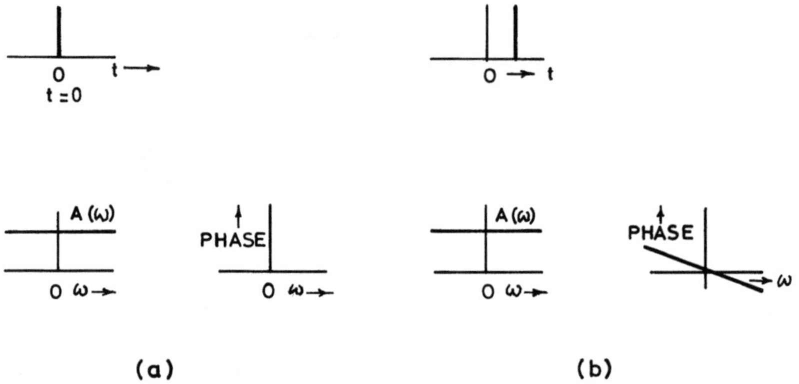

Fig. 14.10 a, b. a Spike at zero time (*upper part*) and its Fourier transform (*lower part*); **b** spike shifted in time (*upper part*) and its Fourier transform (*lower part*) (Source: Chun and Jacewitz 1981)

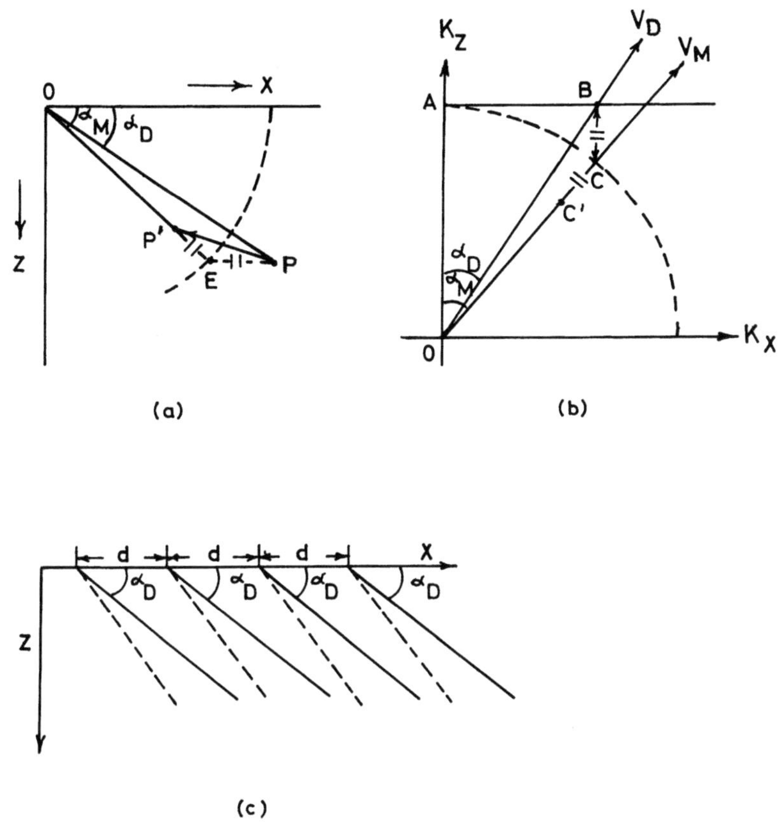

Fig. 14.11 a–c. Migration mapping. **a** Depth domain mapping; **b** construction of frequency domain mapping; **c** preservation of spatial frequency (Source: Chun and Jacewitz 1981)

mapped along normal vectors V_D and V_M of Fig. 14.11b. Given the vector V_D corresponding to dipping event OP, the vector for the migrated event (V_M) may be easily determined following a procedure similar to that for the depth domain (refer to Fig. 14.6h). This procedure (refer to Fig. 14.11b) is as follows:

- draw a circle with radius OA (A is an arbitrary point on the k_z-axis) with O as the center,
- draw a horizontal line from A, and find the point of intersection (point B) with vector V_D,
- draw vertical line from point B and find the intersection point (point C) with the circle,
- draw a line passing from O to C.

O C gives the *migrated dip line* V_M in frequency domain.

It is reasonably obvious that after migration the frequency components along V_D must be mapped along V_M.

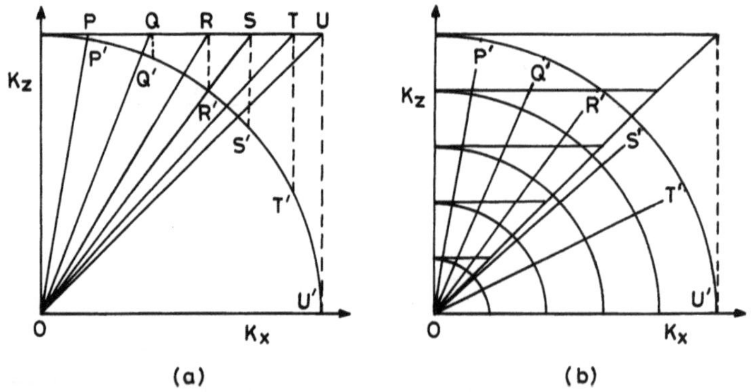

Fig. 14.12 a, b. Migration mapping in the frequency domain. **a** A line of constant k_z frequency and its mapping; **b** a grid of curves of constant k_z (Source: Chun and Jacewitz 1981)

It may be noted that point B is directly mapped to C. It is a vertical mapping. It was a slanted projection with angle ($\alpha_M/2$) in the (x–z) domain (refer to Fig. 14.6 h).

For a seismic section having dipping events with a single dip, the spacing in the x-direction is unchanged after migration (Fig. 14.11 c). It implies that k_x cannot change under migration mapping. Therefore, this mapping must be vertical in the k_x–k_z plane.

Figure 14.12 a shows all dip lines. The horizontal line intersects dip lines in P, Q, R, S, T, U. Projecting the frequency components on this line to the circle, all points on these lines are migrated to P′, Q′, R′, S′, T′, U′. Repeating this procedure of projecting the frequency components on a series of horizontal lines to their corresponding circles vertically (as in Fig. 14.12 a), the migration process in the frequency domain is completed. The mathematical expression for this process is written as

$$F^*(k_x, k_z) = \frac{k_z}{\sqrt{k_x^2 + k_z^2}} F\left(k_x, \sqrt{k_x^2 + k_z^2}\right) \tag{14.2}$$

[Chun and Jacewitz 1981, Eq. (2)]
where F is the 2-D Fourier transform (F.T.) of the original section. Therefore F* (the transform of the migrated section) may be calculated directly from the transform of the original section. Taking the inverse transform of F*, the migrated section in the (x, z) domain would be obtained.

Consider a synthetic data record section (Fig. 14.13 a) resembling a scattered fan. It may be noted that using the depth domain migration procedure as depicted by Fig. 14.6 h would be cumbersome. As the Fourier transform gathers together the events of the same dip, the transform corresponding to scattered dips of all angles (including 10° and 25° shown in Fig. 14.13 a) would be like Fig. 14.13 b. Application of the migration process to the frequency domain model

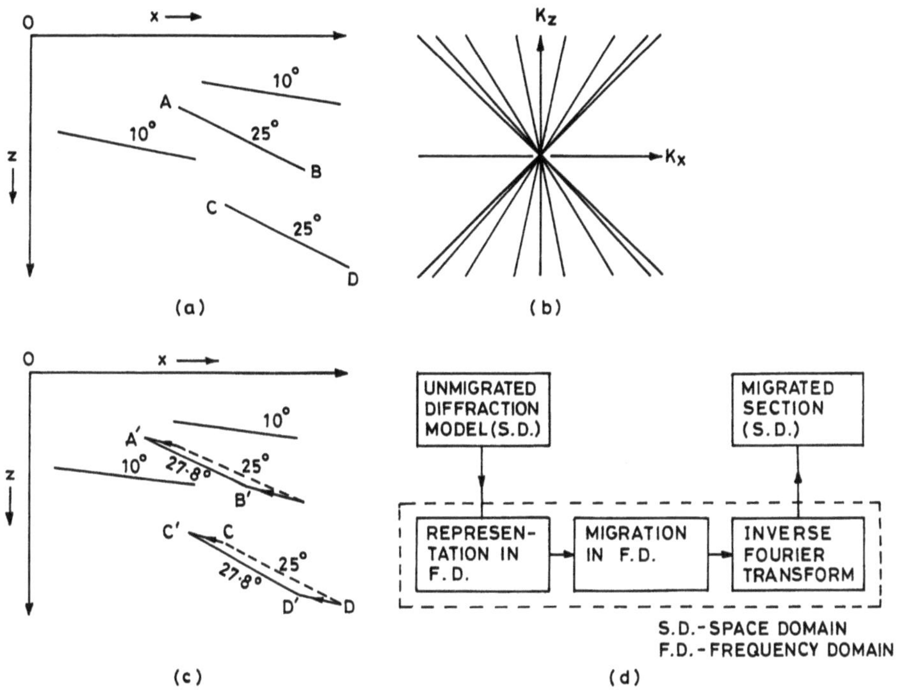

Fig. 14.13 a–d. Synthetic data: **a** Record section in the space domain. **b** Frequency domain representation of scattered fan model data (record section) of all angles including those shown in **a**. **c** Migrated section obtained by migration (Eq. (14.2)) in the frequency domain, and taking the inverse Fourier transform (scattered dip events of 10° and 25° are only shown). **d** Integrated depiction of migration process via Fourier domain (Source for Figs. 14.13 a–c: Chun and Jacewitz 1981)

of Fig. 14.13 b in the frequency domain, and then performing an inverse F.T., a migrated scattered fan model of Fig. 14.13 c (corresponding to the unmigrated section of Fig. 14.13 a) in the (x–z) plane would be obtained. As events of the same dip are migrated together in frequency domain migration, a complex section in the x–z plane before migration, should appear as smooth curves after frequency domain migration. Integration of steps is shown in Fig. 14.13 d.

Sections 14.1 and 14.2 were devoted to generating a feeling for the necessity to migrate seismic data, by using simple graphical and geometric concepts. Subsequent sections deal with achieving the goal of migration by using solutions of the wave equations. In this respect the following discussions are organized to provide basic equations for use of solutions of wave equations for migration, for stacked and unstacked seismic data. Claerbout (1970, 1971) introduced the concept of wave equation migration in a series of conceptual and foundational papers. We initiate our discussion with Finite Difference Migration. Additionally, Section 14.8 deals with *slant stack migration.*

Exercise 14.1

(i) Consider Fig. 14.6 h. Assume a_D (angle of diffracted event) as 30°. Determine a_M (angle of migrated event) by construction. Draw the migrated reflector segment.

(ii) Considering Fig. 14.11 (the construction for frequency domain mapping) justifies the vertical mapping in the frequency domain.

Solution:

(i) Follow the construction procedure as outlined in Section 14.3. Value of a_M is 35·26°

(ii) Referring to Fig. 14.11 c, the migration process moves each line about its intersection point with the surface. However, the point of intersection remains fixed. It amounts to saying that the frequency in x, the spatial frequency, does not change by migration. Therefore, the migration mapping defined as

$$(\bar{k}_x, \bar{k}_z) \rightarrow (k_x, k_z)$$

implies

$$k_x = \bar{k}_x$$

Since a point (k_x, k_z) in the frequency domain corresponds to an event of dip a with $\tan a = k_x/k_z$, we have from migration equation

$\tan a_D = \sin a_M$ applied to one point $(\bar{k}_x, \bar{k}_z)$ to yield

$$\frac{\bar{k}_x}{\bar{k}_z} = \frac{k_x}{\sqrt{k_x^2 + k_z^2}}$$

(Chun and Jacewitz 1981, p 731)

As we have under migration mapping

$$\bar{k}_x = k_x$$

we may write

$$\bar{k}_x^2 + k_z^2 = \bar{k}_z^2$$

$$k_z^2 = \bar{k}_z^2 - \bar{k}_x^2$$

or

$$k_z = \sqrt{\bar{k}_z^2 - \bar{k}_x^2}$$

This implies vertical mapping in Fig. 14.11 b.

14.4 Finite Difference Migration

Every point (x, z) in the Earth is able to transform a downgoing wave into an upgoing wave, and this characteristic is expressed by reflection strength R (x, z) of the point. Measurements are made by placing seismic source and geophone on the surface of the Earth. The record of measurement is supposed to be the reflection from some point in the medium, which is conveniently expressed as a function of coordinates of seismic source (x_s, z_s), and geophone (x_g, z_g), and two way reflection time 't', denoted by $\psi(x_s, z_s, x_g, z_g, t)$. In the migration procedure, one evaluates R (x, z) from the measured quantity $\Psi(x_s, 0, x_g, 0, t)$ on the Earth surface (z = 0). This is made possible by evaluating ψ for assigned positions of sources and receivers in the Earth. As the positions of sources and receivers are assumed at larger and larger depths inside the Earth, two things happen. First, the separation between source and receiver decrease, and second the reflection time 't' reduces. In the continued process when the level of seismic source and receiver reaches the reflection point, the coordinates of source, receiver and the reflection point all coincide. The value of ψ at this point evaluates well the reflection strength of the point at time t = 0. This is mathematically expressed as (Claerbout 1971)

$$R(x, z) \approx \psi(x, z, x, z, 0) \tag{14.3}$$

The process is essentially an extrapolation of wavefield at lowered positions of sources and receivers. The procedure is schematically shown in Fig. 14.14.

To describe the result of this theoretical experiment, an equation is required which governs ψ values for small changes in the position of source and receiver. We know that for wave motion resulting in seismic experiments, changes are described by the wave equation, whose general form in 2-D is

$$\frac{\partial^2 \psi}{\partial x^2} + \frac{\partial^2 \psi}{\partial z^2} - \frac{1}{V^2}\frac{\partial^2 \psi}{\partial t^2} = 0 \tag{14.4}$$

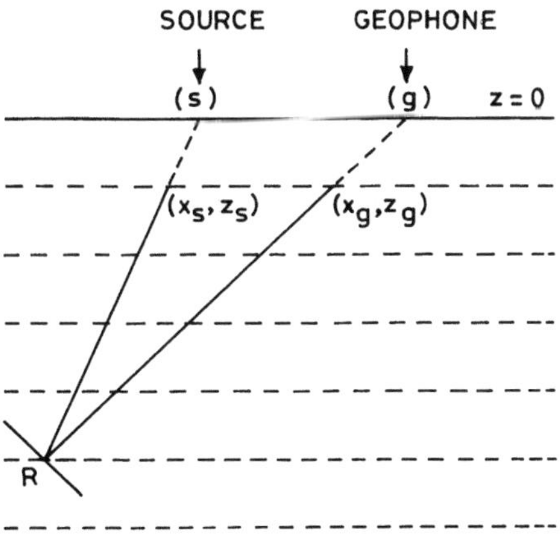

Fig. 14.14. Migration process viewed as a prediction of changes in the seismic field as sources and receivers are moved into the Earth (Source: Stolt 1978)

The role of the wave equation in migration may be described by saying that it is useable under the condition when ψ is a slowly varying function both in space and time to allow prediction of values in the neighboring locations from a set of known values at nearby points; or, more specifically, for projecting values downwards from the values measured (and corrected) on a horizontal plane.

Equation (14.4) describing the wave motion is not useable directly as the change of ψ in space (x, z) and time (t) is too rapid. For putting Eq. (14.4) in a practical form for use in migration, Claerbout defined a new coordinate that moves with the wave. With this new coordinate, the variation is much slower, and projection of wavefield to lower levels becomes possible even in larger steps to accomplish the process of migration.

Writing wave Eq. (14.4), for small changes in source and receiver locations, we have for all points in space-time

$$\frac{\partial^2 \psi}{\partial x_s^2} + \frac{\partial^2 \psi}{\partial z_s^2} - \frac{1}{V^2(x_s, z_s)} \frac{\partial^2 \psi}{\partial t^2} = 0 \tag{14.5}$$

and

$$\frac{\partial^2 \psi}{\partial x_g^2} + \frac{\partial^2 \psi}{\partial z_g^2} - \frac{1}{V^2\left(x_g, z_g\right)} \frac{\partial^2 \psi}{\partial t^2} = 0 \tag{14.6}$$

It becomes possible to write Eqs. (14.5) and (14.6) in more compact form for a coincident source-receiver position, taken as the mid point between source and receiver. *Stacked sections* of seismic data correspond to coincident source-receiver, and, for this situation, the incident and reflected ray paths are identical. The stacked sections correspond to a normal-incidence section. The new coordinate, defined by the mid-point, and offset are

$$X = \frac{(x_s + x_g)}{2}, \ Z = \frac{(z_s + z_g)}{2} \tag{14.7 a}$$

and

$$x = \left(x_s - x_g\right)/2, \ z = \left(z_s - z_g\right)/2 \tag{14.7 b}$$

Under this transformation, the *stacked section* would be written as ψ (X, 0, 0, 0, t) and *migrated section* as ψ (X, 0, Z, 0, 0).

Equations (14.5) and (14.6) may be written in the transformed coordinates, and if the first derivatives with respect to x and z are ignored, a single equation in place of two equations results:

$$\frac{\partial^2 \psi}{\partial X^2} + \frac{\partial^2 \psi}{\partial Z^2} - \frac{4}{V^2(X, Z)} \frac{\partial^2 \psi}{\partial t^2} = 0 \tag{14.8}$$

[Stolt 1978, Eq. (8)]

The form of Eq. (14.8) differs from Eqs. (14.5) and (14.6) in having a factor 4 in the term having a derivative of time. This form may be made identical to Eqs. (14.5) and (14.6) by redefining 't' to be one-way travel time.

Claerbout converted partial differential Eq. (14.8) to an algebraic difference equation by yet another choice of a coordinate system in which ψ varies slowly. This is defined as

$$D = \frac{Vt}{2} + Z, \; d = Z \tag{14.9}$$

In Eq. (14.9), 'D' is the depth of some reflection point and 'd' is the depth of the source-receiver mid-point at the lowered levels of sources and receivers.

Taking

$$\phi(x, d, D) = \psi(X, 0, Z, 0, t), \tag{14.10}$$

the wave equation takes the form

$$\frac{\partial^2 \phi}{\partial X^2} + \frac{\partial^2 \phi}{\partial d^2} + 2\frac{\partial^2 \phi}{\partial d \partial D} = 0 \tag{14.11}$$

[Stolt 1978, Eq. (11)]

The *stacked* and *migrated sections* in the new depth coordinates (D and d) are

$$\phi(X, 0, D) = \phi\left(X, 0, \frac{Vt}{2}\right) = \psi(X, 0, 0, 0, t) \tag{14.12 a}$$

(at $d = z = 0$)

and

$$\phi(X, D, D) = \phi(X, Z, Z) = \psi(X, 0, Z, 0, 0) \tag{14.12 b}$$

(at $d = D$ and $t = 0$)

'ϕ_{dd}' is negligible compared to ϕ_{XX} for an upward traveling wave (which is of concern in the migration process) near the vertical. Also ϕ_d (X, 0, D) is not known a priori, therefore ϕ_{dd} in Eq. (14.17) is ignored. This gives a *15° approximation* to the wave equation (Claerbout 1976, p 211)

$$\frac{\partial^2 \phi}{\partial X^2} + 2\frac{\partial^2 \phi}{\partial d \partial D} = 0 \tag{14.13}$$

Equation (14.13) is cast into a difference equation, by defining discrete variables j and k as

$$D_j = j\Delta D \quad \text{and} \quad d_k = k\Delta d \tag{14.13 a}$$

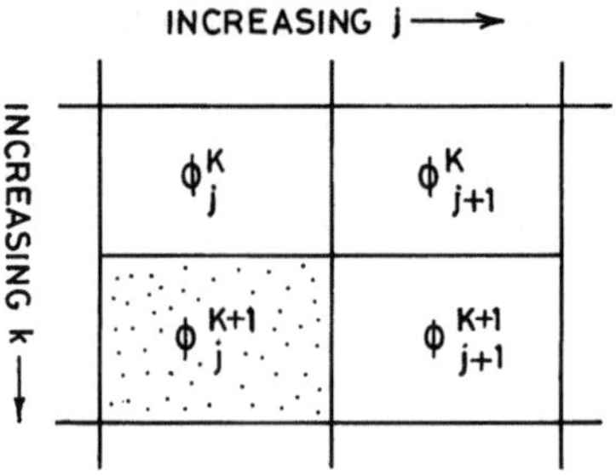

Fig. 14.15. Adjacent grids formed by increasing j (for D) and K (for d). Value (k+1) in the *dotted grid* may be found from values in the adjoining grids of a (d, D) array (Source: after Waters 1987)

In Eq. (14.13 a) ΔD and Δd are increments in reflector depth (transformed travel time) and source-receiver midpoint depth, respectively. Adopting the notation,

$$\phi(X)_j^k \equiv \phi\left(X, d_k, D_j\right),$$

two possible forms of difference equation corresponding to Eq. (14.13) result [refer, for example, to Stolt 1978, Eqs. (19 a) and (19 b)]:

$$(1 - T)\phi_j^{k+1} = -(1 - T)\phi_{j+1}^k + (1 + T)\left(\phi_{j+1}^{k+1} + \phi_j^k\right) \tag{14.14 a}$$

and

$$\phi_j^{k+1} = -\phi_{j+1}^k + (1 + 2T)\left(\phi_{j+1}^{k+1} + \phi_j^k\right) \tag{14.14 b}$$

$$\text{where } T = \frac{\Delta D \Delta d}{8}\frac{d^2}{dX^2}$$

is an operator in X.

The difference form [Eq. (14.14 a)] is referred to as the *implicit solution* for ϕ_j^{k+1}, because to solve for ϕ_j^{k+1} one needs to invert the operator (1−T) also. This is not required in Eq. (14.14 b), and for this reason Eq. (14.14 b) is referred to as the *explicit solution.*

It may be mentioned that although Eq. (14.14 a) is a more expensive algorithm compared to Eq. (14.14 b) it has greater accuracy at *steep dips.*

Figure 14.15 illustrates the working of the 15° approximate solution [Eq. (14.14 b)]. Knowing the three neighboring values of ϕ, i.e. ϕ_j^k, ϕ_{j+1}^k and ϕ_{j+1}^{k+1}, the value of unknown ϕ_j^{k+1} may be determined.

14.5 Frequency Domain Migration

Frequency domain migration is also called Fourier-transform migration or frequency wave number migration. The algorithm for *frequency domain migration* is based on the *acoustic wave equation* (Stolt 1978). Similar independent approaches in frequency domain migration as in Stolt were contributed by Bolondi et al. (1978), Hood (1978), and Whiltlesey and Quay (1977).

Stolt's (1978) approach uses original data in (X, D) space [refer to Eqs. (14.7 a) and (14.9)]. (X, D) space data is Fourier transformed into frequency domain (p, ω) space. The data in (p, ω) space is operated in the manner as prescribed by the wave equation, and finally a 2-D inverse Fourier transform is taken to generate the *migrated section.*

The surface field is ϕ(X,0,D) [refer to Eq. (14.12 a)]. The following Fourier transform pairs may easily be written

$$A(p,\omega) = \frac{1}{2\pi}\int dX \int dD e^{i}_{(pX-2D/c)}\phi(X,0,D), \tag{14.15 a}$$

and

$$\phi(X,0,D) = \frac{1}{2\pi}\int dp \int d\omega e^{-i(pX-2\omega D/c)} A(p,\omega) \tag{14.15 b}$$

[Stolt 1978, Eqs. (46) and (47)]

Equation (14.15 b), for an upcoming wave, at depth d, may be generalized to,

$$\phi(X,d,D) = \frac{1}{2\pi}\int dp \int d\omega e^{-i(pX+qd-2\omega D/c)} A(p,\omega) \tag{14.16 a}$$

and to satisfy the wave Eq. (14.11) q would be defined by

$$q = \frac{2\omega}{c} - \sqrt{\frac{4\omega^2}{c^2} - p^2} \tag{14.16 b}$$

As the *migrated section* is ϕ(X,D,D) [refer to Eq. (14.12 b)] it's form is

$$\phi(X,D,D) = \frac{1}{2\pi}\int dp \int d\omega A(p,\omega) e^{-i\left(pX-\sqrt{\left(\frac{4\omega^2}{c^2}-p^2\right)}D\right)} \tag{14.17}$$

To put Eq. (14.17) in the standard form of a Fourier integral, one additional substitution,

$$k = \sqrt{\frac{4\omega^2}{c^2} - p^2}, \tag{14.17 a}$$

is required.

This gives

$$\phi(X,D,D) = \frac{1}{2\pi}\int dp \int dk\, B(p,k) e^{-i(pX-kD)} \tag{14.17 b}$$

[Stolt 1978, Eq. (51)]

with

$$B(p,k) = \frac{1}{\sqrt{1+\frac{p^2}{k^2}}} A\left(p, \frac{kc}{2}\sqrt{1+\frac{p^2}{k^2}}\right) \tag{14.17 c}$$

Stolt visualized this transformation as shifting data from frequency ω to a lower frequency ω' given by,

$$\omega' = \sqrt{\omega^2 - p^2c^2/4} \tag{14.18 a}$$

and a change of scale represented by

$$\frac{k}{\sqrt{k^2+p^2}} = \frac{\omega'}{\omega} \tag{14.18 b}$$

Equation (14.18 a) amounts to a *moveout correction* with ω taking the place of time and 'p' of offset.

The frequency shift effect is shown in Fig. 14.16. The process of obtaining *migrated dip* α_M from *apparent dip* α_D (before migration) [refer to Eq. (14.1)] is inherent in the process of frequency shift in frequency domain migration.

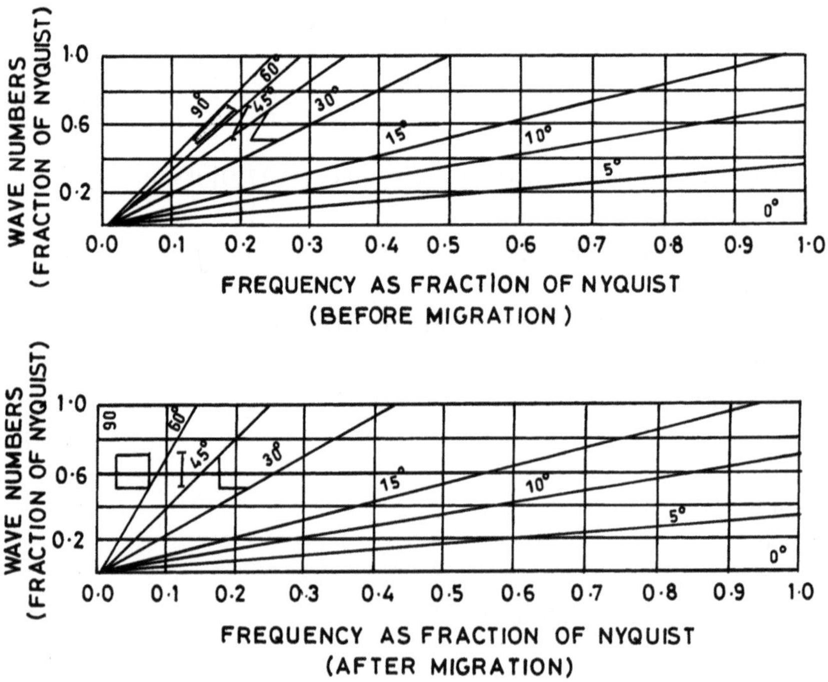

Fig. 14.16. Migration in frequency-wave number (F–K) space (Source: Stolt 1978)

The above procedure holds good for stacked data in which source and receiver are taken coincident. Unstacked data involves source-receiver offset. Therefore, the algorithm for *prestack migration* should be developed using both wave Eqs. (14.5) and (14.6) for seismic source and receiver positions. For mathematical details reference may be made to Stolt (1978).

Stolt showed that a simple transformation is able to complete the *prestack migration* process also. In fact, moveout correction, stack and migration of data are accomplished in a single process.

In the frequency domain migration of *stacked* and *prestack* data, an assumption of constant velocity was taken. For the case of variable velocity, suitable coordinate transformation is required in which both the wave equation and boundary conditions are velocity independent.

For extension of the migration scheme to three dimensions, it is required to consider a three-dimensional form of the wave equation that should be satisfied by the Fourier-transformed quantities. The final result, i.e. $\psi(x, y, z, o)$, is expressible again as a Fourier transform. Extensive discussions of various situations including those discussed above may be found in the original contribution of Stolt (1978).

The advantages of frequency domain migration is fast computing, good performance with data having low signal to noise (S/N) ratio, and excellent performance for steep dips. Difficulties arise with widely varying velocity functions (Chun and Jacewitz 1981).

14.6 Exploding Reflector Model (ERM) and the One Way Wave Equation

The Exploding Reflector Model (ERM) (refer to Loewenthal et al. 1976) is a theoretical experiment for justifying the use of the wave equation for processing wave motion data on the stacked record section. The stacked record section assumes coincident source-receiver geometry (Fig. 14.17a). For this geometry, the downward and upward ray paths are identical, and the ray strikes the reflector surface at right angles (even though the ray may pass through complex geological structure above the reflecting surface). It may be mentioned that under very special situations (Fig. 14.17b) the incident and reflected ray paths may not be identical.

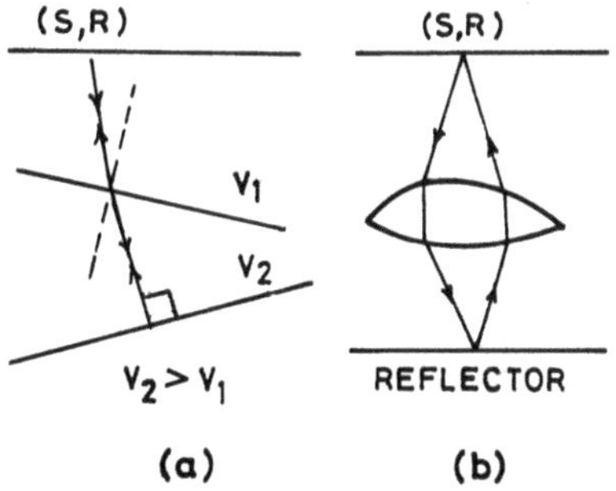

Fig. 14.17a, b. a Normal incidence ray path from a dipping reflector with an identical incident and reflected ray for a coincident source-receiver position. **b** Different incident and reflected raypaths under a special situation (after Claerbout 1985)

The wave equation describes wave motion generated by a physical experiment. On the other hand, the stacked record section is not a result of a single experiment. In fact, many shots are taken in sequence, but the record section gives the appearance as if all the shots were fired simultaneously. In view of this, the theoretical physical experiment is termed – 'Exploding Reflector Model'.

In this model, the seismic sources are not at ground surface. They are taken positioned along every reflector surface, and are fired simultaneously at time $t=0$. The strength of seismic sources is proportional to the reflection coefficients of the reflectors. Multiple reflections and transmission losses at interfaces are ignored. The receivers are taken on the ground surface.

The record section involves two way travel time from the source on the surface to the reflector and back to the receiver. To work with the exploding reflector model we need one-way time, i.e. from the source on the reflector to the receiver on the ground surface. This is achieved by dividing the stacked record section time scale by 2 or by dividing the velocity value by 2.

In working with exploding reflector model, the record section can be taken as boundary condition (i.e. the surface measurements) for a wavefield governed by the wave equation. The migration process involves *depropagating* the wavefield measured on the ground surface back to their subsurface reflector position, corresponding to time $t=0$.

Migration is a two-step process, i.e. *field reconstruction* and *imaging*. Field reconstruction involves determination of the wavefield for the region of interest. Imaging is the generation of the geometrical distribution of true reflecting surfaces in the medium. Implementation of *field reconstruction* and *imaging* produces the *migrated section*. Field reconstruction is accomplished by determining the position-dependent phase-shift for each surface measurement, and superposition of resultants.

An equation describing the spatial and temporal evolution of the pressure field may be described by the equation

$$\nabla^2 \phi_P = (1/V^2)(\partial^2 \phi_P / \partial t^2) \tag{14.19}$$

where $\phi_P(x,y,z,t)$ is pressure amplitude at (x, y, z) and time t; and $V(x,y,z)$ is the propagation speed of the acoustic wave. This equation is referred to as the *scalar wave equation.*

The idea of separating the wavefield into upgoing and downgoing waves is found useful. Downgoing waves propagate in a downward direction in the Earth. They either originate directly from the seismic shot (Fig. 14.18 a) or are generated by multiple reflections (Fig. 14.18 b). The *upgoing wave* is generated from upward reflection and travel towards the surface of the Earth (Fig. 14.18 a, b). In both cases, significant contributions to the energy spectrum come from the waves within a small angle from the vertical (z-axis).

At the reflector, the first arrival of the downgoing wavefield and reflected upgoing wavefield would be time as well as space coincident. This is referred to as the Claerbout *imaging principle* (Claerbout 1976, Fig. 14.18 c). In practical application the form of the imaging condition depends entirely on the model. The *exploding reflector* model (Loewenthal et al. 1976) is one such widely used model.

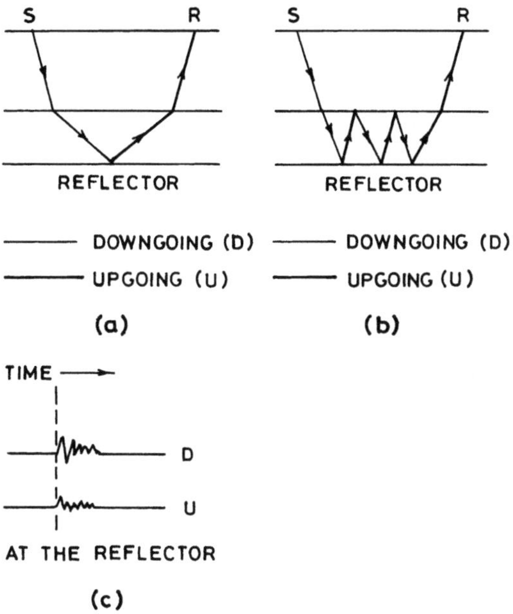

Fig. 14.18a–c. a, b Downgoing and upgoing waves; **c** downgoing and upgoing waves would be time (as well as space) coincident (after Claerbout 1976)

In this model, sources are taken distributed along reflectors which get activated simultaneously at time $t=0$. The energy propagates upwards, and is recorded on the ground surface at time $t=t'$ (say). For this model, the surface-recorded wavefield is continued downwards and the reflector is imaged at time $t=0$, i.e. at the time when the reflectors were activated.

14.6.1 Downward Depropagation of a Surface-Recorded Reflected Wavefield

From the wavefield recorded on the surface of the Earth, it is possible to reconstruct the picture of the wavefield at different depth levels in the Earth. One way of achieving this is by use of Huygen's principle. The other is by using the solution of the wave equation. The process of reconstructing a wavefield from one level to the next lower level is known as *downward continuation. Datumning*, a term used by Berryhill (1979), describes the determination of the wavefield from data recorded on an irregular surface $z=z(x)$ to a datum level which may also be an irregular surface.

Downward propagation of the surface wavefield may be carried out in two ways. In the first method, known as '*downward continuation*', the wavefield is propagated from the surface, taken as $z=0$, to the next surface $z=\Delta z$, and so on (Fig. 14.19a). In the second method, referred to as '*wave tracking*', the wavefield at a given time instant $(t_0+\Delta t)$ propagates to the time instant t_0 (Fig. 14.19b). Each plane in this case represents a *snapshot* of the wave at a given instant of

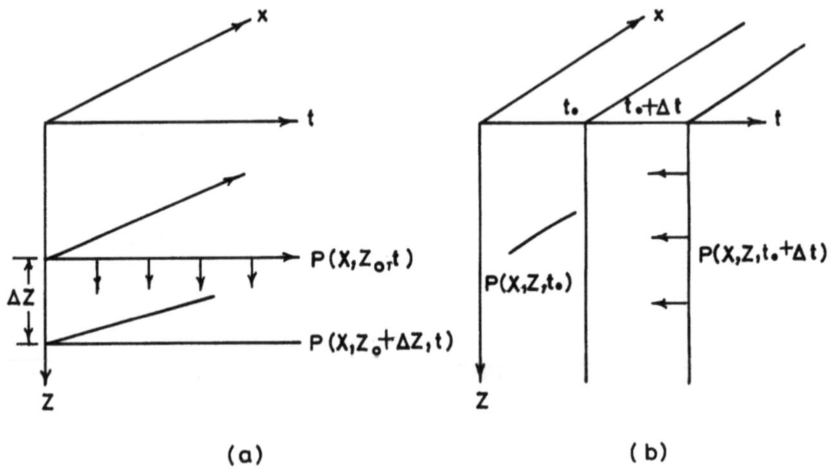

Fig. 14.19 a, b. Migration may be achieved by propagation of energy; **a** in depth – known as downward continuation, or **b** temporarily – known as wave tracking (Source: Hood 1981, p. 169. © Applied Science Publishers Ltd. 1981)

time. Both procedures are equivalent if the velocity is assumed constant. The discussion in Subsections 14.6.1.1 and 14.6.1.2 uses equations and figures from the review article of Hood (1981).

14.6.1.1
Downward Continuation

In solving migration problems, specialized forms of wave equation are developed that account for energy propagation within a small angle about a given axis, usually upwards or downwards. However, the full wave equation has also been used for modeling (Alford et al. 1974) and migration (Deregowski 1978). The one-way wave equation has been derived in different ways (refer, for example, to Claerbout 1970; Claerbout and Johnson 1971; Buchanan 1978). We initiate the discussion using the method in Claerbout (1970).

Taking a Fourier transform over the time of Eq. (14.19) we have

$$\frac{\partial^2 \phi_p^*}{\partial x^2} + \frac{\partial^2 \phi_p^*}{\partial z^2} = -\frac{\omega^2}{V^2} \tag{14.20}$$

where $\phi_p^* = \int_{-\infty}^{+\infty} \phi_p \exp(-i\omega t) dt$

The one way wave equation in Claerbout is derived by writing $\partial\phi_p^*/\partial z$ from Eq. (14.20) in the form,

$$\frac{\partial \phi_p^*}{\partial z} = i\left(\frac{\omega^2}{V^2} + \frac{\partial^2}{\partial x^2}\right)^{\frac{1}{2}} \phi_p^* \tag{14.20 a}$$

Equation (14.20 a) is an exact one-way wave equation. This equation propagates energy exactly in all directions up to the limit of 90°. Validity of this equation requires that velocity V is constant.

For writing a solution of the one-way wave Eq. (14.20 a), it is required to make an approximation of the square root term. This may be done using continued fraction expansions or binomial expansion of the square root term. Gazdag (1980) made use of binomial expansion of the square root term in Eq. (14.20 a) and obtained the following form of the *one-way wave equation*:

$$\frac{\partial \phi_{\mathrm{P}}^{*}}{\partial z} = \frac{i\omega}{V}\phi_{\mathrm{P}}^{*} + \frac{V}{2\omega}\frac{\partial^2 \phi_{\mathrm{P}}^{*}}{\partial x^2} - \frac{V^3}{8\omega^3}\frac{\partial^4 \phi_{\mathrm{P}}^{*}}{\partial x^4} \tag{14.20 b}$$

Kjartansson (1978) obtained a pair of two equations for one-way wave propagation. These equations describe the '*diffracting part*' and '*phase shifting*' part of the one-way wave equation. To present these equations, we consider continued fraction expansion of the square root term in Eq. (14.20 a) using

$$(1+X^2)^{\frac{1}{2}} = 1 + \frac{X^2}{2+\frac{X^2}{2+X^2\ldots\ldots}} \tag{14.21}$$

where $X = \frac{V}{\omega}\frac{\partial}{\partial X}$

R. H. S. of Eq. (14.21) equals (after truncation)

$$1 + \frac{2\left[(V/\omega)\frac{\partial}{\partial X}\right]^2}{4+\left[(V/\omega)\frac{\partial}{\partial X}\right]^2}$$

Substitution of the above expression in Eq. (14.20 a) gives

$$\frac{\partial \phi_{\mathrm{P}}^{*}}{\partial Z} = \frac{i\omega}{V}\left[1 + \frac{2\left[(V/\omega)\frac{\partial}{\partial X}\right]^2}{4+\left[(V/\omega)\frac{\partial}{\partial X}\right]^2}\right]\phi_{\mathrm{P}}^{*} \tag{14.22}$$

Equation (14.22) is transformed to a *moving coordinate frame* by the substitution

$$\phi_{\mathrm{P}}^{*} = \left(\phi_{\mathrm{P}}^{*}\right)' \exp\left(i\omega\frac{Z}{V^{f}}\right) \tag{14.22 a}$$

where V^{f} is constant *frame velocity.*

This gives

$$\frac{\partial\left(\phi_{\mathrm{P}}^{*}\right)'}{\partial Z} = i\left(\frac{\omega}{V} - \frac{\omega}{V^{f}}\right)\left(\phi_{\mathrm{P}}^{*}\right)' + \left(\frac{i(2V/\omega)\frac{\partial^2}{\partial x^2}}{4+(V/\omega)^2\frac{\partial^2}{\partial x^2}}\right)\left(\phi_{\mathrm{P}}^{*}\right)' \tag{14.23}$$

Equation (14.23) is split into two equations as

$$\frac{\partial\left(\phi_P^*\right)'}{\partial Z} = i\left(\frac{\omega}{V} - \frac{\omega}{V^f}\right)\left(\phi_P^*\right)' \tag{14.24 a}$$

and

$$\frac{\partial\left(\phi_P^*\right)'}{\partial Z} = \left(\frac{i(2V/\omega)\frac{\partial^2}{\partial X^2}}{4 + (V/\omega)^2\frac{\partial^2}{\partial X^2}}\right)\left(\phi_P^*\right)' \tag{14.24 b}$$

Equation (14.24 b) is the '*diffracting part*' of the one-way wave equation, and Eq. (14.24 a) is the '*phase shifting*' part.

The solution of Eq. (14.24 a) is given as

$$\left(\phi_P^*\right)'_{(x,z+\Delta z,\omega)} = \left(\phi_P^*\right)'_{(x,z,\omega)} \exp\left[i\omega \int_z^{z+\Delta z} \left(\frac{1}{V} - \frac{1}{V^f}\right)\right] dz' \tag{14.25}$$

In the time domain, this become

$$\left(\phi_P\right)'_{(x,z+\Delta z,t)} = \left(\phi_P\right)'_{(x,z,t+T)} \tag{14.26}$$

where

$$T = \int_z^{z+\Delta z} \left(\frac{1}{V} - \frac{1}{V^f}\right) dz' \tag{14.26 a}$$

The *diffraction part* [Eq. (14.24 b)) in the space-time domain is

$$\frac{\partial^3\left(\phi_P\right)'}{\partial z \partial t^2} - \frac{V^2}{4}\frac{\partial^3\left(\phi_P\right)'}{\partial x^2 \partial z} + \frac{V}{2}\frac{\partial^3\left(\phi_P\right)'}{\partial x^2 \partial t} = 0 \tag{14.27}$$

Equations (14.27) and (14.26) provide the basis for downward continuation. Initial application of Eq. (14.27) moves the data through a distance ΔZ. The subsequent application of Eq. (14.26) is a correction that takes care of velocity variation over Δz from the value of reference frame velocity V^f.

Steps in downward continuation using Kjartansson's method are the following (Fig. 14.20):

- Fourier transformation of surface data to frequency

$$\phi_P(x, z = 0, t) \rightarrow \phi_P^*(x, z = 0, \omega)$$

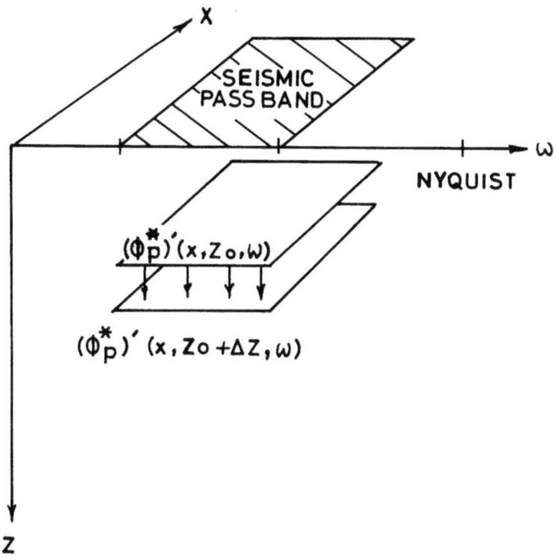

Fig. 14.20. Data are projected from the plane at level Z_0 to $Z_0+\Delta Z$; each frequency component is treated separately (Source: Hood 1981, p. 207. © Applied Science Publishers Ltd. 1981)

- Downward continuation using a diffraction equation to propagate energy from level z to level $z+\Delta z$; and by using a phase shifting equation to account for propagation differences in a variable velocity medium.
- Synthesization of data by summation over all frequencies

$$\phi_p(x, z+\Delta z, t=0) = \sum_{\omega} \phi_p^*(x, z+\Delta z, \omega) \qquad (14.28)$$

The migrated output [L.H.S. of Eq. (14.28)] on the plane $t=0$ is found through an inverse Fourier transform. Since the term $e^{i\omega t}$ in the inverse Fourier transformation at $t=0$ is unity, the integration reduces to a summation.

- Subtraction of the d.c. term

$$\phi_p(x, z+\Delta z, t=0)$$

- Going to the second step

After each step Δz, a part of the data will have reached zero time. This data is then fully migrated.

14.6.1.2 Wave Tracking

The wave equation is utilized to propagate data at a fixed time to an earlier time. Wave tracking is an exact counter part of downward continuation. To illustrate the point we consider a suitable form of wave Eq. (14.19) for a CMP stack appropriate for an exploding reflector model:

$$\nabla^2\phi_p = \frac{4}{V^2}\left(\frac{\partial^2\phi_p}{\partial t^2}\right) \qquad (14.29)$$

Equation (14.29) is transformed by defining a coordinate transformation

$$x' = x$$
$$t' = t$$
$$\tau = t + \int_0^z \frac{2}{V^f} dz \qquad (14.30)$$

where V^f is *frame velocity* independent of x. The new coordinate defines a coordinate υ in place of depth, which moves with *frame velocity*, V^f.

Substituting Eq. (14.30) into Eq. (14.29) and dropping the primes gives

$$\frac{4}{V^2}\frac{\partial^2 \phi_p}{\partial t^2} = \frac{\partial^2 \phi_p}{\partial x^2} + 4\left(\frac{1}{(V^f)^2} - \frac{1}{V^2}\right)\frac{\partial^2 \phi_p}{\partial \tau^2} - \frac{8}{V^2}\frac{\partial^2 \phi_p}{\partial t \partial \tau} + \frac{2\partial}{\partial z}\left(\frac{1}{V^f}\right)\frac{\partial \phi_p}{\partial \tau} \qquad (14.31)$$

[Hood 1981, Eq. (3.10)]

If V^f is constant, Eq. (14.31) may be split into two equations

$$\frac{\partial^2 \phi_p}{\partial t \partial \tau} = -\frac{1}{2}\left(\frac{\partial^2 \phi_p}{\partial t^2}\right) + \frac{V^2}{8}\left(\frac{\partial^2 \phi_p}{\partial x^2}\right) \qquad (14.32\,a)$$

$$\frac{\partial \phi_p}{\partial t} = \frac{1}{2}\left[\left(\frac{V^2}{(V^f)^2}\right) - 1\right]\frac{\partial \phi_p}{\partial \tau} \qquad (14.32\,b)$$

[Hood 1981, Eqs. (3.11a) and (3.11b)]

Solving Eq. (14.32 a) using conventional methods would involve dropping or approximating $\partial^2\phi_p/\partial t^2$ to obtain a one-way 15° or 45° approximation. Eq. (14.32 a) may be solved by the finite difference method in the usual manner. Similarly Eq. (14.32 b) could be integrated to yield a solution.

Migration by means of wave tracking may be carried out in a manner precisely analogous to downward continuation.

In Section 14.6, the mathematical basis for downward continuation and wave tracking of the wavefield was presented. For this purpose a Fourier-transformed pressure field over time (ϕ_p^*) was used in the wave Eq. (14.19), and the one-way wave equation was shown to consist of two parts – the diffracting part and phase shifting part (or refracting part). These equations may be recast for the exploding reflector model (refer to Hood 1981). The equation representing the diffracting part is solvable by the usual finite difference scheme. This permits writing numerical solutions for *downward continuation* or *wave tracking* of the surface-recorded wavefield, and thus accomplishing the goal of seismic migration.

As treatment in the frequency domain becomes easier, a Fourier-transformed wave equation instead of a scalar wave equation in the space-time domain has been adopted to implement seismic migration. This forms the basis of our discussions in the next Section 14.7.

14.7 Migration using a Frequency Domain Form of the Wave Equation

14.7.1 Approximation of the Exact Dispersion Equation

Wave equation in space-time domain or dispersion equation in the frequency domain has in it both upgoing and downgoing waves.

For the purpose of wavefield extrapolation and imaging we really need only upgoing waves. The choice of upgoing wave is easily done using the dispersion relation. From dispersion Eq. (3.6), we have

$$k_z = \pm\sqrt{\frac{\omega^2}{V^2} - k_x^2} = \pm\frac{\omega}{V}\sqrt{\left(1 - \frac{V^2}{\omega^2}k_x^2\right)} = \pm\frac{\omega}{V}\sqrt{(1 - X^2)} = \pm\frac{\omega}{V}R \qquad (14.33)$$

Successive rational approximations of R (refer to Berkhout 1982) enables us to write Eq. (14.33) in the following forms (taking +ve sign):

$$k_z = \frac{\omega}{V} \qquad (14.34\,a)$$

$$k_z = \frac{\omega}{V} - \frac{Vk_x^2}{2\omega} \qquad (14.34\,b)$$

$$k_z = \frac{\omega}{V} - \frac{k_x^2}{\left(\frac{2\omega}{V} - \frac{Vk_x^2}{2\omega}\right)} \qquad (14.34\,c)$$

14.7.2 Wavefield Extrapolation

The first exercise in wavefield extrapolation is to find an expression for the derivative of a wavefield (usually in the depth-z direction). If P(Z) represents the wavefield at depth z, the extrapolated value at $(z+\Delta z)$, i.e. $P\,(z+\Delta z)$ is

$$P(z) + \Delta z\frac{dP}{dz}$$

Thus, we need an expression for dP/dz to achieve extrapolation. The dispersion equation written in the form of Eq. (14.33) helps to achieve this objective.

By using $\partial/\partial z$ in place of ik_z in Eq. (14.33), we get an expression for wavefield extrapolation

$$\frac{\partial P}{\partial z} = i\frac{\omega}{V}\sqrt{1 - \frac{V^2k_x^2}{\omega^2}}P \qquad (14.35)$$

By using the successive rational approximation of the square root term in Eq. (14.35) given by Eqs. (14.34 a–c), we get the following approximate wavefield extrapolation equations valid within ±5, ±15 and ±45° with vertical, respectively:

$$\frac{dP}{dz} = i\left\{\frac{\omega}{V}\right\}P \tag{14.36 a}$$

$$\frac{dP}{dz} = i\left\{\frac{\omega}{V} - \frac{Vk_x^2}{2\omega}\right\}P \tag{14.36 b}$$

$$\frac{dP}{dz} = i\left\{\frac{\omega}{V} - \frac{k_x^2}{2\frac{\omega}{v} - \frac{vk_x^2}{2\omega}}\right\}P \tag{14.36 c}$$

These are *extrapolation equations* when velocity depends only on depth (Claerbout 1985, p 86).

Migration methods based on the use of Eq. (14.35) or its approximation are referred to as *phase-shift methods.*

14.7.3 Application of the Paraxial Equation under a Condition of Anisotropy

An application of the paraxial equation for migrating seismic data in elliptically anisotropic media (Verwest 1989) is given below.

For the case of elliptical anisotropy the expression for k_z is given by (compare with Eq. (14.33) for the isotropic case)

$$k_z = \frac{\omega}{V_z}\sqrt{1 - \frac{a^2V_z^2k_x^2}{\omega^2}} \tag{14.37}$$

[Verwest 1989, Eq. (6)]

In this case rational approximation of the square root in Eq. (14.37) for a 45° approximation (Claerbout 1985) is given as

$$k_z = \frac{\omega}{V_z}\frac{1 - c_1a^2\frac{V^2k_x^2}{\omega^2}}{1 - c_2a^2\frac{V^2k_x^2}{\omega^2}} \tag{14.38}$$

[Verwest 1989, Eq. (7)]
where c_1 and c_2 are constants.

The exact dispersion equation for SH waves

$$\frac{\omega^2}{V^2} = a^2k_x^2 + k_z^2 \tag{14.38 a}$$

Fig. 14.21. Dispersion relation for elliptically anisotropic medium with $a = V_H/V_z = 1.2$. The exact dispersion relation is given by the *continuous curve*. The 45° approximation to the dispersion relation is given by the *dashed curve* (Source: Verwest 1989)

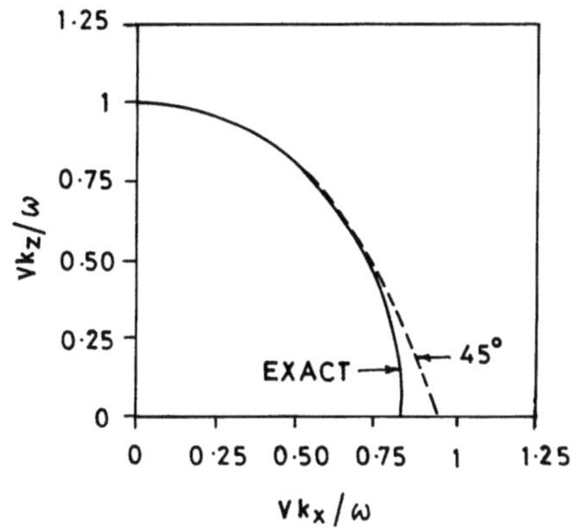

and its approximate form Eq. (14.38) for "a" $(= V_H/V_v = 1.2)$ are diagrammatically depicted in Fig. 14.21.

From Fig. 14.21, the closeness of the exact and approximate form of dispersion relations – for energy propagation up to 45° from the vertical is well observed.

By using the conversions between the space and time derivatives $(\partial/\partial x, \partial/\partial z, \partial/\partial t)$, and Fourier-transformed quantities (k_x, k_z and ω) given by

$$\frac{\partial}{\partial t} \equiv -i\omega$$

$$\frac{\partial}{\partial x} \equiv ik_x$$

$$\frac{\partial}{\partial z} \equiv ik_z$$

the approximate dispersion relation Eq. (14.38) may be converted to an extrapolation operator in x and z, yielding the following form of *paraxial equation*:

$$-\omega^2 P_z - c_2 a^2 V^2 P_{xxz} + i\frac{\omega^3}{V} P + i\omega c_1 a^2 V P_{xx} = 0 \tag{14.39}$$

[Verwest 1989, Eq. (8)]

The approximate dispersion relation Eq. (14.38) has now been converted to a differential equation [the subscripts denote partial derivative(s)]. This differential equation enables a solution both *for modeling* and *migration*. For *modeling* purposes V (in place of V_z) and 'a' (V_H/V_Z) values are assumed; and for *migration* these have to be determined to perform *imaging*.

Verwest (1989) explains how to separate the problem of *focusing* (controlled by a horizontal velocity component) and depth conversion (controlled by a vertical velocity component) into two separate steps. To achieve this a change of variable

$$dz' = a(x, z)dz \tag{14.40 a}$$

is taken which gives new form to the *paraxial equation*

$$-\omega^2 P_{z'} - c_2 V_M^2 P_{xxz'} + i\frac{\omega^3}{V_M} P + i\omega c_1 V_M P_{xx} = 0 \tag{14.40 b}$$

[Verwest 1989, Eq. (10)]
where $V_M = aV_z = V_H$

The *paraxial equation* in the form of Eq. (14.40 b) enables downward continuing of the surface data using only one velocity, i.e. the horizontal velocity; the velocity field being expressed in terms of stretched variable z', and downward continuation being done in equal steps of z'. As a consequence of this, the image is defined in terms of z', which can be restored to true depth by the stretching operation $dz = (1/a)dz'$. This part can be achieved in two parts, by first stretching the image back to vertical two-way travel-time $dt = (2/(aV_z))dz'$, and in the next step stretching it to true depth using $dz = (V_z/2)dt$.

This process, involving depth migration using horizontal velocity, and then stretching to the final depth image, has been termed Split Anisotropic Depth Migration (SADM) by Verwest. This splitting was first suggested by Jacob (1979), and is considered exact when 'a' depends only on z, i.e. only a vertical variation, and not lateral.

Migration in this method is considered a two-step process of *focusing* and *depth conversion.* The surface seismic data may define an apparent horizontal velocity to implement focusing. This apparent horizontal velocity is true horizontal velocity for elliptical SH-waves; and for a pseudo-P-wave this apparent horizontal velocity lies between true vertical and horizontal P-wave velocities. As the surface seismic data does not provide the true vertical component of the velocity field, borehole data or a priori geologic information should be used to determine the vertical velocity component, and consequently the degree of elliptical anisotropy. It should be emphasized that in the case of significant apparent elliptical anisotropy, direct use of the horizontal field for migration would result in significant depth errors in the final depth image.

The effectiveness of this method (SADM) is that it can handle *lateral velocity variation* to the same extent as normal depth migration.

The result of numerical analysis reveal that, in the imaging procedure, provision for correct handling of anisotropy should be made. A further conclusion is that the SADM method gives an accurate depth image in the presence of lateral velocity changes, steep dips, and a slow lateral change in the anisotropy parameter. The procedure is most suited to imaging SH-wave data. It may also be useful for P-wave imaging when anisotropy is small.

Further, development of the theory for application to any type of wave (not only SH-waves) was done by Uren et al. (1990 a). The surface-recorded data (for a exploding reflector model) is shown to be migrated as a one step process in the frequency domain for anisotropic media. It is also explained that only when anisotropic migration is used, is it possible to recover the original structure of the anisotropic model considered. Finally it is concluded that in the presence of significant anisotropy in the field, anisotropy consideration must be done when giving the final depth section. For details we refer to original contributions by Verwest (1989) and Uren et al. (1990).

14.8 Slant Stack Migration

14.8.1 Wave Stack vs. CMP Stack

Wave stack synthesizes a particular wavefront by superposing many spherical wavefronts. Using Huygen's principle, superposition permits synthesis of an arbitrarily shaped wavefront. A simple example is the generation of a downgoing plane wave (Fig. 14.22 a). Figure 14.22 a depicts a vertically traveling downgoing plane wave as a result of superposition of spherical waves from many shots (S_i) that are fired simultaneously. Figure 14.22 b shows the synthesis of a downgoing plane wave at an angle 'θ' to the vertical as a result of firing shots (S_i) with time lags.

The wave stack possesses distinct advantages over the Common Mid Point (CMP) stack (refer to Subsection 3.5.9). Firstly, the wave stack bears the characteristics of being the sampling of a wavefield. CMP stack data does not possess this property. As a result formulation of wave equation migration (which assumes seismic data as sampling in space and time of a single upcoming wavefield) may be accomplished more rigorously using wave stack. Further, as CMP stack uses summation over all angles, identification of angle-dependent features, like the angle-dependent reflection coefficient, is not possible. This aspect is particularly focused in the plane wave stack. For example, using the plane wave stack one may study coherent reflected energy that is produced from a single illumination angle. Results of a *partial coherency stack* (from one direction) are an enhanced signal to noise ratio, which facilitates reliable amplitude measurements. In fact, angle selectivity of a slanted plane wave stack opens up the possibility of detect-

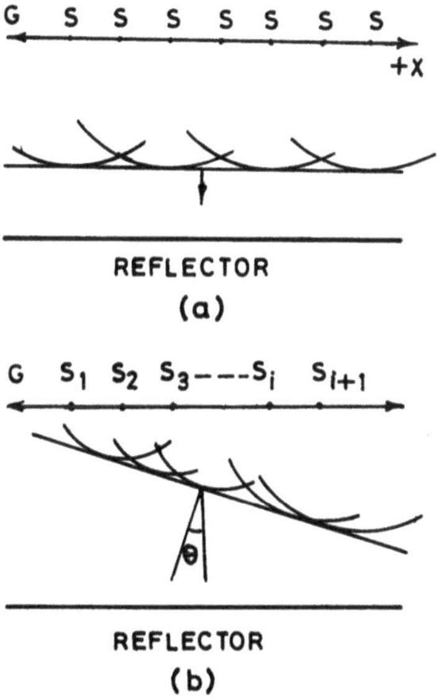

Fig. 14.22 a, b. a If we have a single geophone and many shots, we can imagine firing them all at the same time to generate a downgoing plane wave in a single physical experiment. For a line source in three dimensions, the wavefront is actually cylindrical. **b** We can create a downgoing plane wave at some angle θ by firing the shots with time lags. For a line source in three dimensions, the wavefront is actually conical (Source: Schultz and Claerbout 1978)

ing a reflector that may not be visible on either raw data mid point gathers or on the CMP stack. Further, a slant plane wave stack permits velocity estimation in regions with strong lateral velocity inhomogeneity.

Schultz and Claerbout (1978) originated the basic theory of the wave stack or *slanted plane wave stack* in their excellent contribution. We discuss the relevant part for the consideration of seismic migration.

14.8.2 Plane Wave Stack and p-Gather

In order to develop an expression for plane wave stack, we consider Fig. 14.22 a, b. Figure 14.22 a shows the result of an experiment in which (single) geophone position G is fixed and many single-shot positions vary. Assume 'f_i' as offset between the fixed geophone and shot S_i. Now letting $P(f,t)$ denote the data in the (f,t) coordinate, the operation

$$P'(t) = \sum_{i=1}^{n} P(f_i, t) \tag{14.41}$$

[Schultz and Claerbout 1978, Eq. (3)]
specifies summation over the offset coordinate to produce a single output trace P′. This operation is equivalent to the superposition operation shown in Fig. 14.22 a.

Generalization of Eq. (14.41) to a plane wave traveling at an angle 'θ' to the vertical (Fig. 14.22 b) may be easily determined. Figure 14.22 b shows a time lag of Δt between the shots S_i and S_{i+1}. Assuming

$$\Delta S = S_{i+1} - S_i$$

and

$$\Delta f = f_{i+1} - f_i ,$$

we have

$$\Delta t = \frac{-\Delta s}{V_h} = \frac{\Delta f}{V_h} \tag{14.42}$$

where V_h is the horizontal phase velocity of a propagating wavefront; V_h is positive for a velocity in the +x direction. According to the Snell's law, for a vertically stratified Earth,

$$p \equiv \frac{\sin\theta(z)}{V(z)} = \frac{1}{V_h} = \text{constant} \tag{14.43}$$

where 'p' is called the ray parameter. 'p' is positive when V_h is positive.

From Eqs. (14.42) and (14.43), we may write

$$\Delta t = p\Delta f \tag{14.44}$$

Equation (14.44) provides a hint to synthesize the downward propagating plane wavefront with a particular *ray parameter* p. Equation (14.41) generalizes to

$$P'(p,t') = \sum_{i=1}^{n} P(f_i, t = t' + pf_i) \tag{14.45}$$

[Schultz and Claerbout 1978, Eq. (7)]

Equation (14.45) is a transformation of data from (f, t) coordinates (with common geophone and shot positions varying) to data in the (p, t′) coordinate. The summation operation in Eq. (14.45) is referred to as *slant plane-wave stack* or *slant stack*. This equation is valid both for positive and negative values of 'p' for a wavefront traveling in the positive or negative direction, respectively.

It may be mentioned here that, using reciprocity, slant stack may also be done on common shot gathers as explained above for a common geophone gather.

Figure 14.23 a, b shows schematics for a common geophone gather and its p-gather. According to Eq. (14.45), the summation trajectories for slant stack with ray parameter 'p' are straight lines with slope $\Delta t/\Delta f = p$. The dashed lines in Fig. 14.23 a shows these summation trajectories tangential to hyperbolic reflection events. It is clear that a significant contribution to the slant sum from the reflection event results only from the region of tangency. Figure 14.23 b shows the corresponding p-gather at its appropriate position as an output of coherency summations along the dashed lines. For the dashed lines with different slopes (or p-values), coherency summations would produce the solid curve showing the p-gather picture of the common geophone gather of Fig. 14.23 a.

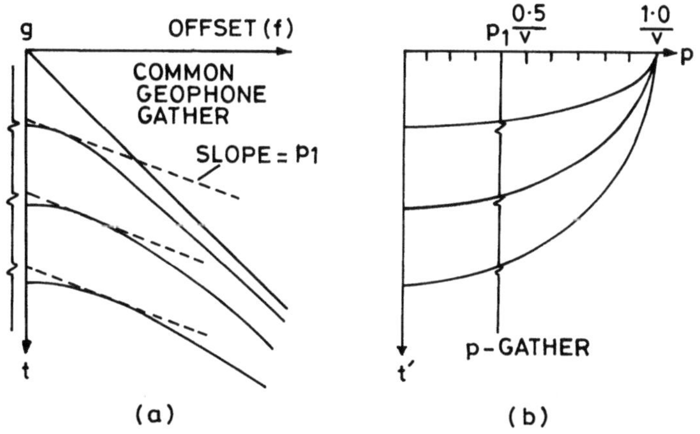

Fig. 14.23 a, b. The slant plane-wave stack transforms a common geophone (or shot) gather into a p-gather. **a** A sample slant stack is implied by the *dashed lines* showing the coherency summation. **b** The output trace is one of those comprising the p-gather and is indicated in its proper position in the p-gather *on the right* (Source: Schultz and Claerbout 1978)

14.8.3 Relationship between Common Geophone Gather and P-Gather

Figure 14.24a shows the relationship of the coordinate (f) and time (t) in a common geophone gather and transformed time (t′) and stacking trajectory (identified by p-value) in the p-gather. The relationship

$$t = t' + \Delta t = t' + fp \tag{14.46}$$

is clearly revealed by this figure. t′ is the intercept on time-axis of the straight tangential summation trajectory to the hyperbolic event at offset ‘f’ (shot-geophone separation).

Δt or fp is the time difference between shot-to-geophone time (t) and slant frame arrival time (t′). This difference is equal to the time it takes for the plane wave to travel the offset ‘f’ at horizontal phase velocity (as p is the inverse of horizontal phase velocity).

Figure 14.24b shows the slant frame geometry in physical space. It depicts the snapshot of a propagating wavefront at time t′ = 0. It also shows the raypath reaching the geophone. It is clear that, in spite of the propagating wavefront, most of the energy reaching the geophone would originate from a shot point at distance S. If ‘f’ is taken as the distance between the geophone and the most effective shot point, the relation between shot-geophone time t and arrival time t′ in slant frame will be

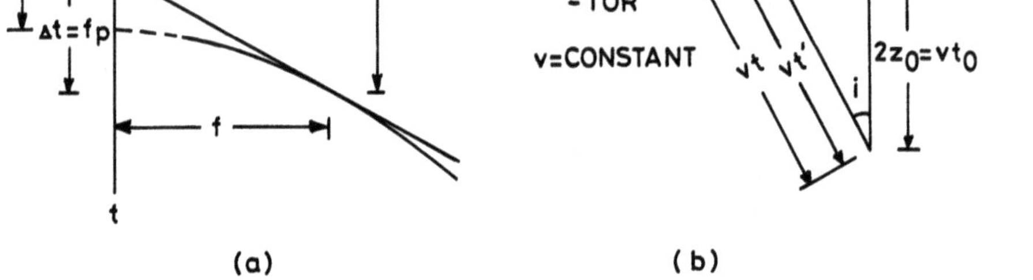

Fig. 14.24a, b. a A single hyperbolic event shown in a common geophone gather. The line labeled *p* shows the stacking trajectory, and its intercept with the time-axis defines *t′* for this particular event and stacking parameter p. The intervals marked f and Δt can also be seen in Fig. 14.24b in physical space. **b** The slant frame geometry. The source is a downgoing plane wavefront at angle i to the horizontal, but for a geophone at distance g, the reflected energy from the reflector at Z_0 emanates mainly from the shot position at distance S. The wavefront is propagating in the +x direction, depicting the geometry for a positive value of p (Source: Schultz and Claerbout 1978)

$$t = t' + \Delta t = t' + fp \tag{14.47}$$

Taking 't_0' as two way vertical travel time to the reflector and V as the constant velocity to the reflector, we have from Fig. 14.24 b

$$\frac{t'}{t_0} = \cos i \tag{14.48 a}$$

and

$$\frac{t_0}{t} = \cos i \tag{14.48 b}$$

Using Eq. (14.43) with Eq. (14.48), we have

$$\frac{t'^2}{t_0^2} + p^2V^2 = 1 \tag{14.49 a}$$

and

$$\frac{t_0^2}{t^2} + p^2V^2 = 1 \tag{14.49 b}$$

[Schultz and Claerbout 1978, Eqs. (10 a, b)]

Equation (14.49 a) is an equation of ellipse in (p, t′) – as shown in Fig. 14.23 b.

It may be mentioned that, for performing slant stack, no velocity information is required prior to stack. Equation (14.49) may be used to estimate velocity in the manner as for equation

$$t^2 = t_0^2 + \frac{f^2}{V^2}$$

in (t, f) space. In this equation t is the time from shot to reflector to geophone, and f is the shot-geophone separation.

14.8.4 Basis of Slant Stack Migration

Figure 14.25 shows the geometry of a propagating wavefront in real time. (x_0, z_0) is a 'point scatterer' scattering energy to the surface. Let 't' represent the time measured from the instant a plane wave at an angle 'i' to the horizontal arrives at the origin O, and x_g represent the horizontal coordinate of the common geophone location. The Snell parameter 'p' for the geometry of the plane wave in Fig. 14.25 is given by

$$p = \frac{\sin i}{V}$$

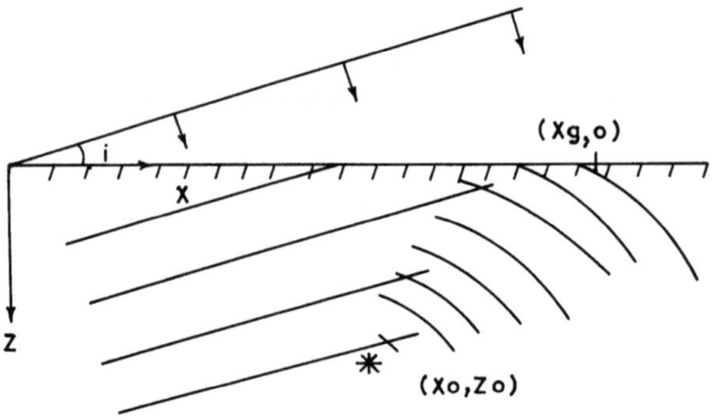

Fig. 14.25. Wavefront propagating into the Earth, shown in real time. The slant stack section transforms real time to a new coordinate in which horizontal beds will all appear at the same time on each trace. Also shown is a point scatterer at (x_0, z_0) which reflects energy back to the surface diffusely (Source: Hood 1981. © Applied Science Publishers Ltd. 1981)

It represents the effective horizontal slowness of the plane wave. Changing 'i' may simulate plane waves of different arrival angles.

In slant stack, the real time 't' is transformed to t′ in the slanted frame by the transformation

$$t = t' + x_g p \tag{14.50}$$

Time coordinate, t′, is the output coordinate in the slanted frame for each value of p, i.e. for each direction of the plane wave (refer to Fig. 14.25). Using Eq. (14.50), the original data in (x_g, t) is transformed in the slant coordinate frame to (p, t'). We refer to the resulting data in (p, t') as *slant stacked data* (Fig. 14.23 b).

To migrate slant-stacked data, the first necessary step is to construct the imaging condition in the slant coordinate frame. Derivation of the imaging condition is given using the approach in Hood (1981).

Referring to Fig. 14.25, the travel time t for a plane wave to reach the point scatterer is given by

$$t = \int_0^{z_0} \frac{\cos i(z)}{V(z)} dz + x_0 p \tag{14.51}$$

[Hood 1981, Eq. (3.14)]

In Eq. (14.51), the medium is assumed to have velocity variation in depth (z) only. Total time t is the combined time for vertical distance z_0, and horizontal distance x_0 with corresponding velocities.

Taking t^u as the time, measured after the plane wave strikes the scatterer, to the common geophone x_g, the total time is given by

$$t = \int_0^{z_0} \frac{\cos i(z)}{V(z)} dz + x_0 p + t^u \tag{14.52}$$

Writing total real time t in terms of slant frame time we have

$$t' = \int_0^{z_0} \frac{\cos i(z)}{V(z)} dz + (x_0 - x_g)p + t^u \tag{14.53}$$

[Hood 1981, Eq. (3.16)]

The *imaging time* is given by $t^u = 0$. By performing the downward continuation of slant stacked data to the time $t^u = 0$, the geophone and scatterer would be coincident, i.e. $x_0 = x_g$ in Eq. (14.53). Therefore, the *imaging conditions* in the *slant frame coordinate* are

$$x_0 = x_g$$

and

$$t' = \int_0^{z_0} \frac{\cos i(z)}{V(z)} dz \tag{14.54}$$

Writing Eq. (14.54) in terms of Snell parameter p, we have

$$t' = \int_0^{z_0} \frac{(1 - p^2V^2(z))^{\frac{1}{2}}}{V(z)} dz \tag{14.55}$$

Equation (14.55) may be evaluated by numerical integration. The transformation from the image at the scatterer to a surface-related coordinate may be obtained by

$$x = x_0 - \int_0^{z_0} \tan i(z) dz = x_0 - \int_0^{z_0} \frac{pV(z)dz}{(1 - p^2V^2(z))^{\frac{1}{2}}} \tag{14.56}$$

[Hood 1981, Eq. (3.19)]

which amounts to a skewing of the data.

After this transformation, migrated sections with different values of 'p' may be superimposed.

For a detailed derivation of Eq. (14.56) the original contribution of Schultz and Claerbout (1978) may be referred.

At this stage it is relevant to explain the meaning of a *slant plane wave section or p-section*. The p-section resembles very closely the cross section of the Earth. We explain it through a simple example of horizontal reflectors (Fig. 14.26).

Consider slant frame data for some fixed value of $p = p_0$. Therefore, data: $P'(x', p = p_0, t')$ would be referred to as the *p-section*. It is clear that a p-section with the particular angle i would have many traces with a similar ray path geom-

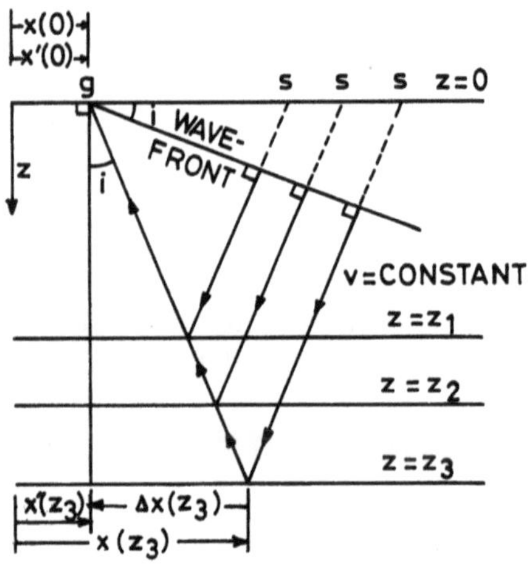

Fig. 14.26. Raypath in the slant frame for three horizontal reflectors in a constant velocity medium. Here the value of p is negative. The energy received at the geophone is displayed as a single trace by the slant stack process, but as seen in the figure really represent subsurface reflection points that lie along a raypath. Points along this raypath all have the same value of the coordinate x′ (Source: Schultz and Claerbout 1978)

etry as shown in Fig. 14.26 for one geophone. All ray paths in this figure contribute a single trace at the surface at $x' = g$. However the data trace does not represent reflection points directly beneath the surface geophone. Our requirement is that reflection points for horizontal reflectors be brought below the geophone. This would involve uniform shearing of the data, and the amount of shearing would depend upon the depth of reflectors of interest. The transformation in Eq. (14.56) aims at forming the required shearing. A new Earth-based coordinate x is defined, which is fixed for a horizontal position in the Earth, for example $x(z_1)$, $x(z_2)$, $x(z_3)$. x is an Earth-based coordinate. The relationship between x′ and x is

$$x = x' - \Delta x(z)$$

Here Δx is negative.

The slant stack migration procedure has been explained assuming that the medium is depth stratified, i.e. $V = V(z)$. However, this procedure is valid in a medium with lateral velocity variations as well. The only requirement is that appropriate imaging conditions be determined, taking into account lateral velocity variations. Slant stack migration is a potential technique for multiple removal in a depth-stratified medium (refer, for example, to Estevez 1977).

14.9 Migration using Explicit Filters for Depth Migration

Numerical wavefield extrapolation is a key aspect of depth migration. Ideally, the extrapolation algorithm should be such as to propagate waves with the minimum of numerically induced distortions over a range of frequencies and angles of propagation. A significant contribution in the field of wavefield extrapolation has been made by Claerbout (1971), Gazdag (1978), Schultz and Sherwood (1980), Holberg

(1987, 1988), Blacquiere et al. (1989), Hale (1991), Kitchenside (1991), Nautiyal et al. (1993), Maeland (1993), Dietrich and Cohen (1993), and Uzcategui (1995). These contributions could be very broadly categorized on the basis of the type of seismic data (*stacked* or *unstacked*) used, and the suitability of the algorithm for different types of media (isotropic, homogeneous or inhomogeneous, or anisotropic). Our concern here is to present the basis of *explicit filter operators* for depth migration. We initiate our presentation by first considering both upgoing and downgoing waves, and then confine our discussions to one-way wave propagation.

Ursin (1983, 1984) gave the following equation for 2D scalar (two-way) wave propagation in a horizontally layered medium:

$$\frac{\partial}{\partial z}\begin{bmatrix} U \\ D \end{bmatrix} = \begin{bmatrix} -ik_z & \\ & +ik_z \end{bmatrix}\begin{bmatrix} U \\ D \end{bmatrix} + r(k_z, z)\begin{bmatrix} 1 & -1 \\ -1 & 1 \end{bmatrix}\begin{bmatrix} U \\ D \end{bmatrix} \tag{14.57}$$

where $k_z = (\omega^2/c^2 - k_x^2)^{1/2}$, $U = U(\omega, k_x, z)$ and $D = D(\omega, k_x, z)$. U and D are the upgoing and downgoing waves, respectively; ω is the temporal frequency; k_x is the horizontal wave number and c is the velocity of propagation, ρ is the density; and $r(k_z, z)$ is the reflectively function equal to

$$\frac{\partial}{\partial z}\left[\log\left(\frac{\rho\omega}{k_z}\right)\right] \Big/ 2$$

The first term in Eq. (14.57) represents the undisturbed one-way wave propagation through a homogeneous medium, and the second term takes care of transmission losses and coupling effects between upgoing and downgoing waves at plane horizontal interfaces.

Neglect of transmission losses and coupling effect would make Eq. (14.57) read as

$$\frac{\partial P}{\partial z} = \pm ik_z P \tag{14.58}$$

where P represents either upgoing (U) or downgoing (D) wave.

The solution of Eq. (14.58) (refer to Holberg 1988)

$$P(\omega, k_x, z + \Delta z) = e^{\pm ik_z \Delta z} P(\omega, k_x, z) \tag{14.59}$$

represents the desired form for downward extrapolation of up- and downgoing waves, respectively. Equation (14.59) incorporates in it the *phase-shift method* of wavefield extrapolation contributed by Gazdag (1978).

14.9.1
Isotropic Laterally Varying Earth Model

Equation (14.59) holds strictly for a perfectly homogeneous medium. Holberg (1988) reviewed finite difference approximations to the one-way wave equation and proposed an algorithm, based on a generalization of the conventional finite difference formulation, which enables downward continuation of the wavefield with *space variant symmetric convolutional operators.* These operators satisfy the desired dispersion relation over a range of frequencies and angle of propagation leading to a minimum of numerical distortions. Also, these are unconditionally stable and can be explicitly applied. The outcome of Holberg's (1988) analysis is that for a general inhomogeneous medium with significant lateral velocity variations downward continuation may be carried out in the frequency-space domain by an x-dependent convolution. This is given by

$$P(\omega, x, z + \Delta z) = \sum_{l=0}^{L} w_l [P(\omega, x + l\Delta x, z) + P(\omega, x - l\Delta x, z)] \qquad (14.60)$$

[Holberg 1988, Eq. (11)]
and is diagrammatically illustrated in Fig. 14.27. The coefficient w in Eq. (14.60) is a function of the ratio of the temporal frequency (ω) and local velocity $c(x, z)$. Details on the method of finding the appropriate coefficients may be found in Holberg (1988).

The determination of coefficient w should be so constrained that Eq. (14.60) represents a best possible solution of the wave equation. In Holberg's analysis, the exact phase shift (which does not change amplitude) is approximated by the operator w $(\omega/c, x)$ having spatial frequency response $W(\omega/c, k_x)$ with the constraint

$$|W| < 1 \quad \text{for} \quad k_{x,\max} < k_x < \pi/\Delta x$$

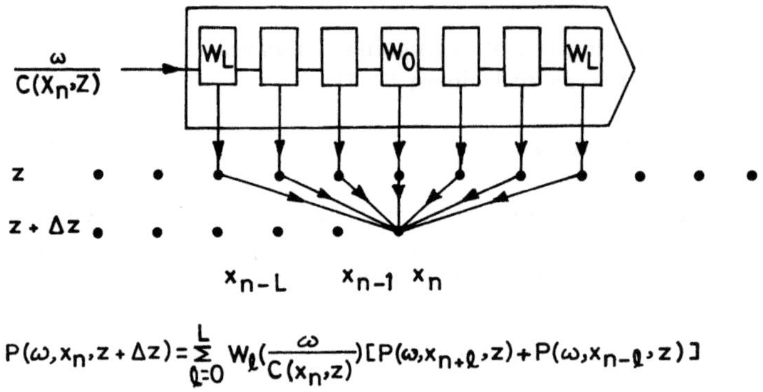

Fig. 14.27. Downward continuation by reference to tables and spatial convolutions. The number of multiplications can be halved by utilizing the operator symmetry (Source: Holberg 1988, from Geophysical Prospecting with permission)

Fig. 14.28. Computational domain for the design of optimum wave propagators (Source: Holberg 1988, from Geophysical Prospecting with permission)

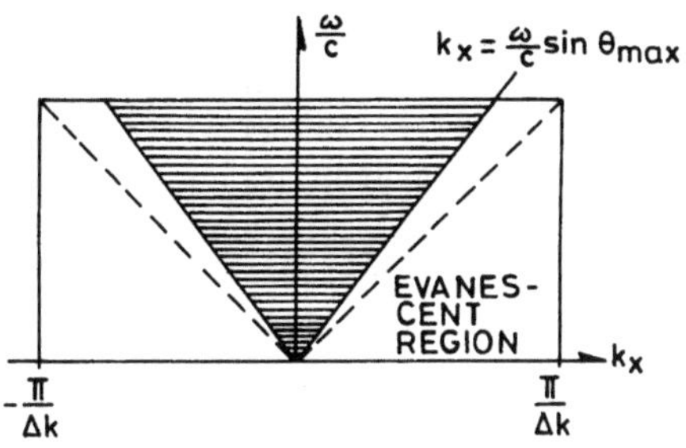

where $k_{x,max} = (\omega/c)\sin\theta_{max}$ and θ_{max} is the maximum design angle (Fig. 14.28), and $k_{\omega,max}$ is the highest ω/c ratio of interest.

For wave energy propagation in inhomogeneous media, the spatial frequency spectra of the wavelets are shrunk and stretched according to local velocity variations (Holberg 1988). As a consequence of this, in $(\omega/c, k_x)$ space (Fig. 14.28) energy moves around during downward continuation in such a medium. In Holberg's (1988) work, the optimization is constrained to ensure the magnitude of the operator's response is less than unity for an angle of propagation higher than the maximum design angle and in the evanescent region. This makes operators unconditionally stable for explicit application. Further operator coefficients can be pre-computed and tabulated in a fashion that would allow us to determine the correct operator at each grid point for downward continuation. The operator coefficients may be determined for the ratio between the temporal frequency and the local velocity.

Hale (1991), Nautiyal et al. (1993) and Uzcategui (1995) further contributed towards designing finite length extrapolation operators for the 2-D Earth. The basis of these investigations is the phase-shift operator $(e^{\pm ik_z\Delta z})$. For a 2-D Earth, the phase-shift operator in the frequency-wave number domain is given by

$$D(k) \equiv e^{i\frac{\Delta z}{\Delta x}\sqrt{\left(\frac{\omega\Delta x}{v}\right)^2 - k^2}} \tag{14.61}$$

(Hale 1991; Nautiyal et al. 1993)

In Eq. (14.61), ω is the frequency, v is velocity, Δz and Δx are vertical and horizontal sampling intervals, respectively, and k is the normalized wave number. 'k' is the product of actual horizontal wave number and Δx. Two quantities $\Delta z/\Delta x$ and $\omega\Delta x/v$ determine the extrapolation filter D(k). Values of k less than $\omega\Delta x/v$ represent a propagating wavefield; and values greater than $\omega\Delta x/v$ are the evanescent waves. In the migration of reflection data, the evanescent wavefield is exponentially damped during extrapolation (Nautiyal et al. 1993) to prevent the amplification of noise at each extrapolation step.

Hale (1991) discussed the problem of finite length filter design, consisting of N coefficients h_n with a Fourier transform H(k), which approximates the desired response D(k) expressed by Eq. (14.61). The filter design is based on using the unconstrained least square, Taylor series and modified Taylor series methods. For details of these methods Hale's publication may be referred. The working of the modified Taylor series method is found most suitable for wavefield extrapolation as it yields extrapolators with flat spectra, and ensures that no spectral component in the wavefield are amplified by a factor greater than one. It may be mentioned (Hale 1991) that amplitude errors for *explicit extrapolators* indicate stability or instability, and phase errors indicate how correctly the explicit extrapolators will position reflectors in depth migration. Further, it is observed that by increasing the length of an extrapolation operator the phase error is reduced.

The Taylor series method yields an unstable extrapolation filter, whereas the modified Taylor series method yields filters that are stable.

Examples of amplitude spectra, amplitude and phase errors are presented in Fig. 14.29 a–c.

The question of *stability* versus *accuracy* for an explicit wavefield extrapolation operator was thoroughly investigated by Nautiyal et al. (1993), although a trade-off between stability and accuracy were first noted by Hale (1991).

Nautiyal et al. (1993) observed that *discretizing* the operator, by sampling on a finite-size grid of spatial locations, produces less effect on accuracy and stability as long as the spatial Nyquist criterion is observed. They further explained that perfect accuracy for wavefield extrapolation, through explicit wavefield extrapolation, can never be attained since the small amplitude tail of the operator D(k) [Eq. (14.61)] in the wave number domain (the evanescent wavefield components) are not accounted for.

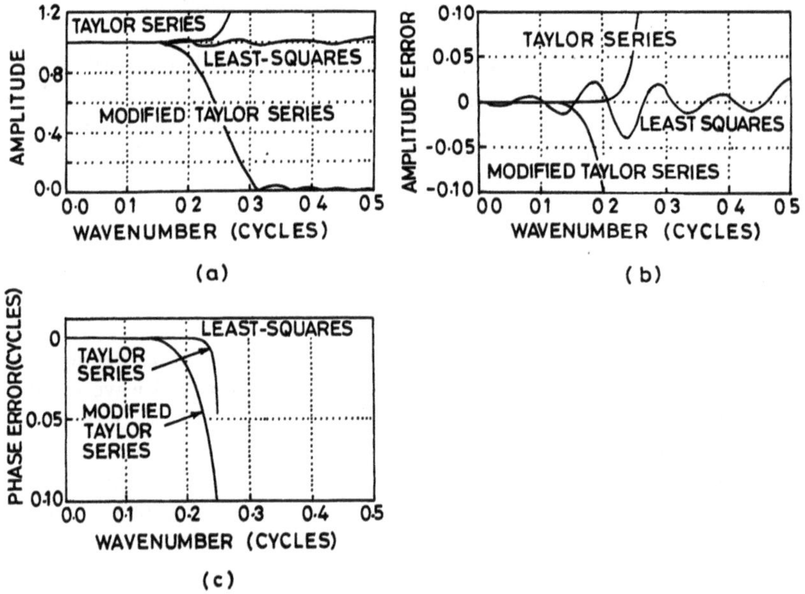

Fig. 14.29 a–c. a Amplitude spectra; **b** amplitude errors; **c** phase errors for 19 coefficient explicit extrapolators designed by an unconstrained least squares method, the conventional Taylor series method and the modified Taylor series method. Normalized wave numbers greater than 0.25 correspond to evanescent waves (Source: Hale 1991)

They also suggested a method to tailor the operator for each frequency using extrapolation steps ΔZ. Suitable details on the design of extrapolators for optimum stability and accuracy for depth migration may be found in Nautiyal et al. (1993).

14.9.2
Anisotropic Earth Model

An implicit and explicit scheme of downward continuation in 2-D transversely isotropic media was developed by Kitchenside (1991, 1993). Explicit filters for extrapolation of seismic wavefields in an isotropic medium proposed by Holberg (1988) and Hale (1991) were extended by Uzcategui (1995) for a transversely isotropic medium with a vertical symmetry axis. The exact downward continuation operator, $D[k_x(\theta),\omega]$ for a transversely isotropic media is given by:

$$D[k_x(\theta),\omega] = \exp\left\{ i\frac{\Delta z}{\Delta x}\left[\left(\frac{\omega\Delta x}{V_p(\theta)}\right)^2 - k_x^2(\theta)\right]^{\frac{1}{2}}\right\} \tag{14.62}$$

[Uzcategui 1995, Eq. (1)]
where Δz and Δx are the vertical and horizontal sampling intervals, and $V_p(\theta)$ and $k_x(\theta)$ are the phase velocity and the horizontal wave number for the q^P or q^{SV} wave evaluated at phase angle θ. $k_x(\theta)$ and phase velocity $V_p(\theta)$ are related as

$$\frac{k_x(\theta)}{\Delta x} = \frac{\omega \sin(\theta)}{V_p(\theta)} \tag{14.63}$$

[Uzcategui 1995, Eq. (2)]
For transversely isotropic media, k_x also depends on 'θ', in contrast to the case of isotropic media.

Forms of the exact downward continuation operators [Eq. (14.61) for isotropic and Eq. (14.63) for transversely isotropic media] are similar, except that for transversely isotropic media k_x and V_p depend on (θ). The design of an explicit filter operator for transversely isotropic media (Uzcategui 1995) follows very similar arguments as in the case of isotropy (Holberg 1988; Hale 1991), i.e. the extrapolation operator should match the theoretical operator in the propagation region and should attenuate the energy in the evanescent region. Denoting $W(k_x,\omega)$ as the Fourier transform of the extrapolation filter, the required condition is

$$W(k_x,\omega) \approx D[k_x(\theta),\omega] \tag{14.64}$$

Uzcategui (1995) obtained explicit depth extrapolation filters by minimizing the squared error of amplitude and phase of the filter in a given range of propagation angles with the vertical ($0 \leq \theta \leq \theta_{max}$), as was earlier accomplished by Holberg (1988) for isotropic media. This is expressed as

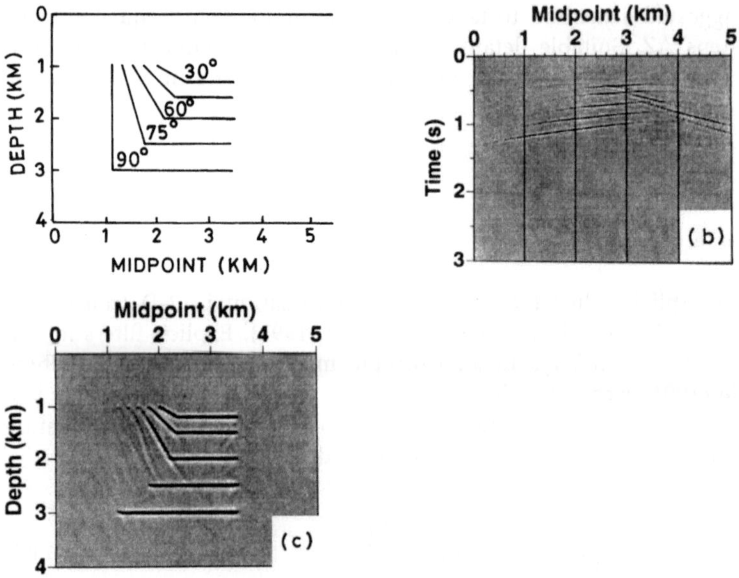

Fig. 14.30 a–c. a Reflector model consisting of horizontal reflectors and segments with dip ranging from 30° to 90°. **b** Zero-offset synthetic section for a model with the anisotropy properties of Mesa Verde Clayshale, corresponding to the reflector model in Fig. 14.30 a and vertical velocity $\alpha_0(x,z) = 3.794 + 0.5x + 0.5z$ km/s. **c** Depth migration obtained using a LS table of extrapolators (Source: Uzcategui 1995)

$$J(\omega) = \int_{\theta=0}^{\theta=\theta_{max}} \left(AE^2 + PE^2\right) dk_x \tag{14.65}$$

[Uzcategui 1995, Eq. (10)]
subject to the constraint

$$|W(k_x, \omega)| < 1.0$$

$$\text{for } |k_x| > \frac{\omega \sin(\theta_{max})}{V_p(\theta_{max})}$$

[Uzcategui 1995, Eq. (11)]

In Eq. (14.65), AE stands for amplitude error, and PE for phase error of the filter.

The analysis in Uzcategui (1995) shows that the accuracy of the extrapolators (for a constant filter length and normalized frequency) is a function of the elastic properties in the medium. It is found that weakly anisotropic media need fewer filter coefficients than do strongly anisotropic media, to generate a good representation of the downward-continuation operator.

As an illustration of the success of depth migration, a model with horizontal reflectors with dipping segments (Uzcategui 1995) is depth-migrated (Fig. 14.30 a–c). As the maximum angle for which the least square extrapolators were designed was 60°, only reflectors up to the dip of 60° are recovered. Larger dips of 75° or 90° are either attenuated and misplaced or lost because of amplitude and phase errors of filters for dips greater than 60°.

CHAPTER 15

Imaging using Integral Solutions of Wave Equations

15

The scalar wave equation is written in the form

$$\nabla^2 U - \frac{1}{c^2} U_{tt} = -4\pi q(\mathbf{r}, t) \tag{15.1}$$

(Schneider 1978)

Here $U(\mathbf{r},t)$ stands for the wavefield; c is the wave velocity and $q(\mathbf{r},t)$ represents the distribution and time dependence of sources at each point in space.

The solution of Eq. (15.1) over an arbitrary volume V_0 is given by a surface integral S_0 enclosing volume V_0 which involves the boundary values; and a volume integral over V_0 which takes into account sources and initial values.

For application in seismic prospecting, contribution from the volume integral may be taken as zero as there are no seismic sources in the subsurface except scatterers and reflectors. Therefore, the contribution to the wavefield comes only from the surface integral, and is of the form

$$U(\mathbf{r}, t) = \frac{1}{4\pi} \int dt_0 \int ds_0 \left[G \frac{\partial U(\mathbf{r}_0, t_0)}{\partial n} - U(\mathbf{r}_0, t_0) \frac{\partial G}{\partial n} \right] \tag{15.2}$$

[Schneider 1978, Eq. (2)]

In Eq. (15.2), **n** is the outward normal vector to the closed surface S_0, which includes the recording surface (say the ground surface), and a hemisphere extending to infinity in the subsurface. **r** and $\mathbf{r}_0$ are vectors from the origin to the observer point and to the source point respectively (Fig. 15.1). t_0 is the time at which the wavefield is observed on the recording surface. G is the Green's function corresponding to a solution of the wave equation for a point source located at $\mathbf{r}_0$. This equation relates the wavefield in the interior of the closed surface to the observation of the wavefield on the recording surface (Morse and Feshbach 1953).

It may be mentioned that, if certain conditions – particularly if the Sommerfeld radiation condition is met, integration only over a limited portion of the closed surface S_0 is sufficient to yield the desired result (Goodman 1968, p 39). From a practical point of view, the seismic observations must extend over that portion of the surface that intercepts reflected waves emanating from reflectors to be imaged.

Equation (15.2) represents wavefield $U(\mathbf{r},t)$ in the image space in terms of the observed wavefield $U(\mathbf{r}_0,t_0)$, its gradient $\partial/\partial n\{U(\mathbf{r}_0,t_0)\}$ and a suitable Green's

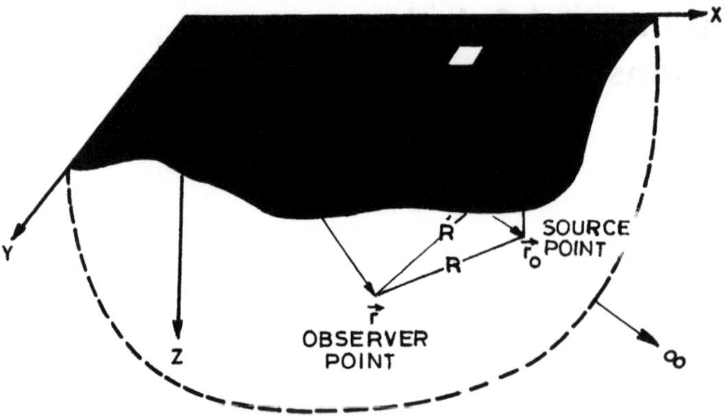

Fig. 15.1. Geometry for boundary value solution (Source: Schneider 1978, from Geophysics with permission)

function G. As the gradient $\partial/\partial n\{U(\mathbf{r}_0, t_0)\}$ is not measured in seismic practice, in order to eliminate the contribution from the gradient term in Eq. (15.2), a suitable form of G is chosen according to the geometry of the measurement surface, such that $G=0$ on this surface.

Equation (15.2) is a basic equation for *wavefield extrapolation* and *imaging* and can be adopted in various forms, some of which are illustrated below.

To begin with, a simple and brief description of the Green's function is given.

15.1 Green's Function and Kirchoff's Integral Representation of a Wavefield in Image Space

The displacement field from a unidirectional, unit impulse localized precisely in space and time is called Green's function. A combination of such simple sources may be used to synthesize displacements due to realistic source models.

Suppose that the unit impulse is applied at $\mathbf{x}=\xi$ and at time $t=\tau$ in the direction $\mathbf{n}$, then the ith component of the displacement at any general value of $(\mathbf{x},t)$ is denoted by $G_{in}(\mathbf{x}, t; \xi, \tau)$. It is a tensor and depends on the position of the source, ξ, and the position of the receiver, $\mathbf{x}$. Useful properties of Green's function are:
(i) Spatial reciprocity of sources and receivers:

$$G_{nm}(\xi_2, \tau; \xi_1, 0) = G_{mn}(\xi_1, \tau; \xi_2, 0) \tag{15.3 a}$$

(ii) Space-time reciprocity:

$$G_{nm}(\xi_2, \tau_2; \xi_1, \tau_1) = G_{mn}(\xi_1, -\tau_1; \xi_2, -\tau_2) \tag{15.3 b}$$

This reciprocal relation is obtained by reversing the direction of flow of time. Taking $\tau_1=0$, we have

$$G_{nm}(\xi_2, \tau_2; \xi_1, 0) = G_{mn}(\xi_1, 0; \xi_2, -\tau_2)$$

Physically, it means that the effect at ξ_2 at time τ_2 later than an impulse started at ξ_1, equals the effect, at ξ_1 at time 0, of an impulse started at ξ_2 at time $-\tau_2$, that is τ_2 earlier. As we would see this concept is useful in wavefield extrapolation.

The simplest case of specifying Green's function G is by considering a unit impulse in an infinite medium. Green's function may be determined from the equation

$$\nabla^2 G - \frac{1}{c^2}\frac{\partial^2 G}{\partial t^2} = -4\pi\delta(\mathbf{r} - \mathbf{r}_0)\delta(t - t_0) \tag{15.4}$$

[Morse and Feshbach 1953, Eq. (7.3.2)]

In Eq. (15.4), the source is an impulse applied at time $t = t_0$ located at $\mathbf{r} = \mathbf{r}_0$; G is the resulting displacement field due to the application of this impulse, as it propagates in the course of time. A detailed discussion on obtaining a solution to this equation may be found in Morse and Feshbach (1953, pp 834–838).

An acceptable solution of G is of the form

$$\frac{\delta\left[+\frac{R}{c} - (t - t_0)\right]}{R}, \quad R(t - t_0) > 0$$

and it represents a spherical shell, expanding about the source with radial velocity 'c'.

For our purpose, for application in seismic theory, we need a suitable form of Green's function for a medium bounded by a surface like the Earth's surface. Under a simplified assumption, the boundary surface may be taken as a plane. In real field conditions it is uneven. To write a complete Green's function under the conditions of a boundary surface, the contribution of image source from the image formed due to the boundary should also be taken into account (Fig. 15.2).

In order to make use of Eq. (15.2) for wavefield extrapolation and imaging Green's function should have a value of zero ($G = 0$) on the observation surface, which is assumed flat. It is given by Green's function for free space (G_{fs}), minus its reflection in the boundary plane. Solution $U(r, t)$ in Eq. (15.2) would correspond to the case where the plane has a reflection coefficient of –1. Green's function, which is zero on a flat surface (seismic data acquisition surface), has consequently the following form (see Fig. 15.2):

$$G(\mathbf{r}, t/\mathbf{r}_0, t_0) = \frac{\delta\left(t - t_0 - \frac{R}{c}\right)}{R} - \frac{\delta\left(t - t_0 - \frac{R'}{c}\right)}{R'} \tag{15.5}$$

[Schneider 1978, Eq. (3)]

where,

$$R = \sqrt{(z - z_0)^2 + (x - x_0)^2 + (y - y_0)^2}$$

and

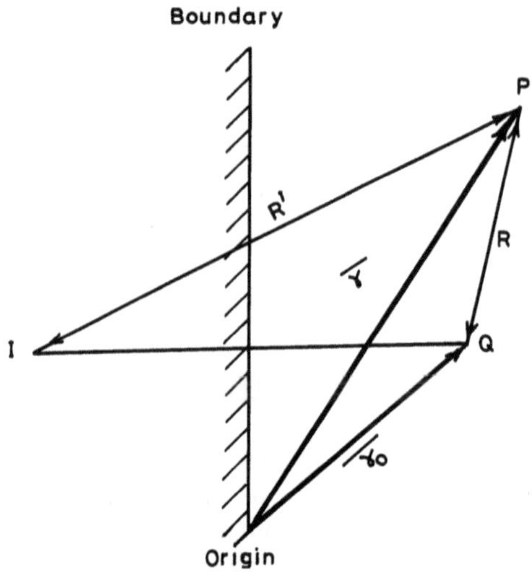

Fig. 15.2. Image at I of an impulsive wave source of Q (Source: after Morse and Feshback 1953. © 1953 by The McGraw-Hill Book Company, Inc., New York)

$$R' = \sqrt{(z + z_0)^2 + (x - x_0)^2 + (y - y_0)^2},$$

Equation (15.5) consists of contributions from a point source at r_0 and its negative image $r_0{}'$.

Substitution of Eq. (15.5) into Eq. (15.2) yields

$$U(\mathbf{r}, t) = \frac{1}{2\pi} \int dt_0 \int dA_0 \cdot U(\mathbf{r}_0, t_0) \frac{\partial}{\partial Z_0} \left[\frac{\delta\left(t - t_0 - \frac{R}{c}\right)}{R} \right] \tag{15.6}$$

[Schneider 1978, Eq. (4)]

This is the integral representation for the wavefield, $U(\mathbf{r}, t)$, at any point in the image space in terms of observed values of wavefield $U(\mathbf{r}_0, t_0)$ on the ground surface. This equation is a rigorous mathematical statement of Huygen's principle, and according to this – wavefield at an earlier time, $U(\mathbf{r}, t)$, may be calculated with the help of measured values of wavefield, $U(\mathbf{r}_0, t_0)$, at subsequent time t_0 at positions $\mathbf{r}_0$ on the measurement surface. This is in complete conformity with the well-known Huygen's principle (refer, for example, to Båth 1968, p 198).

Equation (15.6) is commonly referred to as the *Kirchoff integral.*

Coming to discuss the case of Green's function for a non-planar surface, the resulting Green's function G_{np} is defined as (Wiggins 1984, p 1240)

$$G_{np} = 0$$

on all points on S_0.

$$\frac{\partial G_{np}}{\partial n} = 2 \frac{\partial G_{fs}}{\partial n}$$

$$G_{fs} = \frac{\delta\left(t_0 - t - \frac{R}{c}\right)}{R}$$

and

$$R = |\mathbf{r} - \mathbf{r}_0|$$

c is the speed of sound in the medium.

The above choice of Green's function is referred to as the *Kirchoff approximation.* It is not an exact solution to the wave equation, but represents a good approximation so long as undulations in the observation non-planar surface are small over a wavelength (Beckmann and Spizzichino 1963, pp 17–33).

Accordingly, the Kirchoff integral in the case of a non-planar surface takes the following form:

$$U(\mathbf{r}, t) = -\frac{1}{2\pi}\int dt_0 \int dS_0 \cdot U(\mathbf{r}_0, t_0)\frac{\partial}{\partial n}\left[\frac{\delta\left(t_0 - t - \frac{R}{c}\right)}{R}\right] \tag{15.7}$$

(Wiggins 1984)

Forms of the Kirchoff integral for a flat surface and non-planar surface are strikingly similar, the difference lies in the derivative term. In the case of a flat surface, the direction of the normal to the surface is the z direction or vertical. In the case of a non-planar surface, it is the general direction denoted by n.

In the following section, a detailed discussion of the above forms of Kirchoff integrals would be given for migration of seismic data.

15.1.1 Convolutional Form of the Solution of the Scalar Wave Equation for Wavefield Extrapolation

We begin with Eq. (15.6), and describe concepts taken from Schneider (1978) for using it for migration of seismic data:

$$U(\mathbf{r}, t) = \frac{1}{2\pi}\int dt_0 \int dA_0 \cdot U(\mathbf{r}_0, t_0)\frac{\partial}{\partial Z_0}\left[\frac{\delta\left(t - t_0 - \frac{R}{c}\right)}{R}\right]$$

This equation may also be written as a three-dimensional convolution of the observed wavefield $U(r_0, t_0)$ with a space-time operator related to the point source solution to the wave equation

$$U(x, y, z, t) = U(x, y, z_0, t) * \frac{1}{2\pi}\frac{\partial}{\partial z_0}\left[\frac{\delta\left(t \pm \frac{r}{c}\right)}{r}\right] \tag{15.8}$$

[Schneider 1978, Eq. (7)]
where $r^2 = \Delta z^2 + x^2 + y^2$

This form of the Kirchoff integral is particularly useful, as it provides a way of translating the observed wavefield from one depth to another.

Fourier transformation of Eq. (15.8) over x, y and t gives another form in the frequency wave number domain:

$$U_f(k_x, k_y, z, \omega) = U_f(k_x, k_y, z_0, \omega) \cdot H(k_x, k_y, \Delta z, \omega) \tag{15.9}$$

[Schneider 1978, Eq. (8)]
where subscript 'f' stands for the Fourier transform of the quantity, and

$$H = e^{+i|\Delta z|} \sqrt{\frac{\omega^2}{c^2} - k_x^2 - k_y^2}$$

is a pure phase-shift operator.

Either of the equivalent forms Eq. (15.8) or Eq. (15.9) would enable us to extrapolate waves in space. Use of Eq. (15.9) to extrapolate converging and diverging waves is illustrated in Fig. 15.3.

The sign of H in the transfer function in Eq. (15.9) is important as it determines the direction of extrapolation. A negative sign reflects a phase delay in propagation through a slab of thickness Δz. A positive sign reflects a phase advance in moving a distance Δz towards the source. In the migration of seismic data, a positive sign should be used to extrapolate converging waves towards their source of origin.

For explicit wavefield operators, recent significant contributions in this field may be found, amongst others, in the investigations of Gazdag (1979), Holberg (1988), Hale (1991), Nautiyal et al. (1993), and Uzcategui (1995). Uzcategui developed explicit operators for migration in anisotropic media.

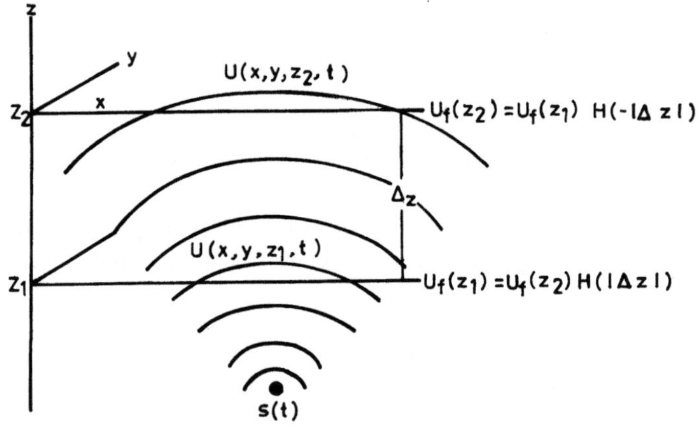

Fig. 15.3. Extrapolation of converging and diverging waves. Migration uses H (+|Δz|) to extrapolate converging waves (Source: Schneider 1978, from Geophysics with permission)

15.1.2
Imaging of Stacked Seismic Data

Although Eqs. (15.8) and (15.9) provide an excellent mathematical formulation for wavefield extrapolation, they are not in a form directly adoptable for application to seismic field records obtained for finite seismic source-receiver offset.

Schneider (1978) gives another form of Eq. (15.6)

$$U(r,t) = -\frac{1}{\pi}\frac{\partial}{\partial z}\int dA_0 \frac{U\left(r_0, t-\frac{R}{c}\right)}{R} \tag{15.10}$$

[Schneider 1978, Eq. (6)]

This form is particularly suitable as it shows that integral transformation is a solution to the wave equation because of the form of the Kernel f(t–R/c)/R. Its application in visualizing migration in seismic data processing has been further illustrated by Schneider (1978).

The assumptions are that:

- The seismic data is Common Depth Point (CDP) stack data for a coincident source-receiver, and one-way time from the reflector to the receiver is considered (Fig. 15.4a).

  CDP one-way time seismic data is amenable to wave equation manipulation and therefore useable for downward extrapolation and imaging of seismic data (Fig. 15.4b).
- Wave sources are located along the reflecting interface(s), their strengths being proportional to reflection coefficient(s). They are fired in unison at time t=0.

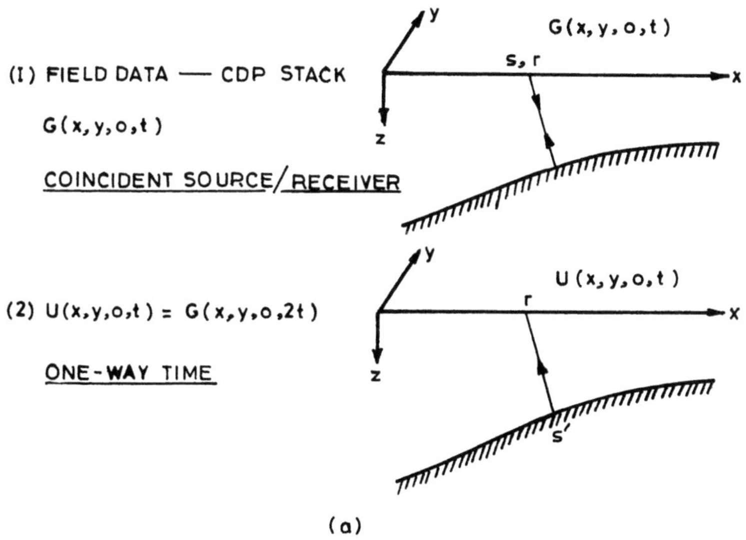

Fig. 15.4a–c. Migration principle: **a** steps 1 and 2

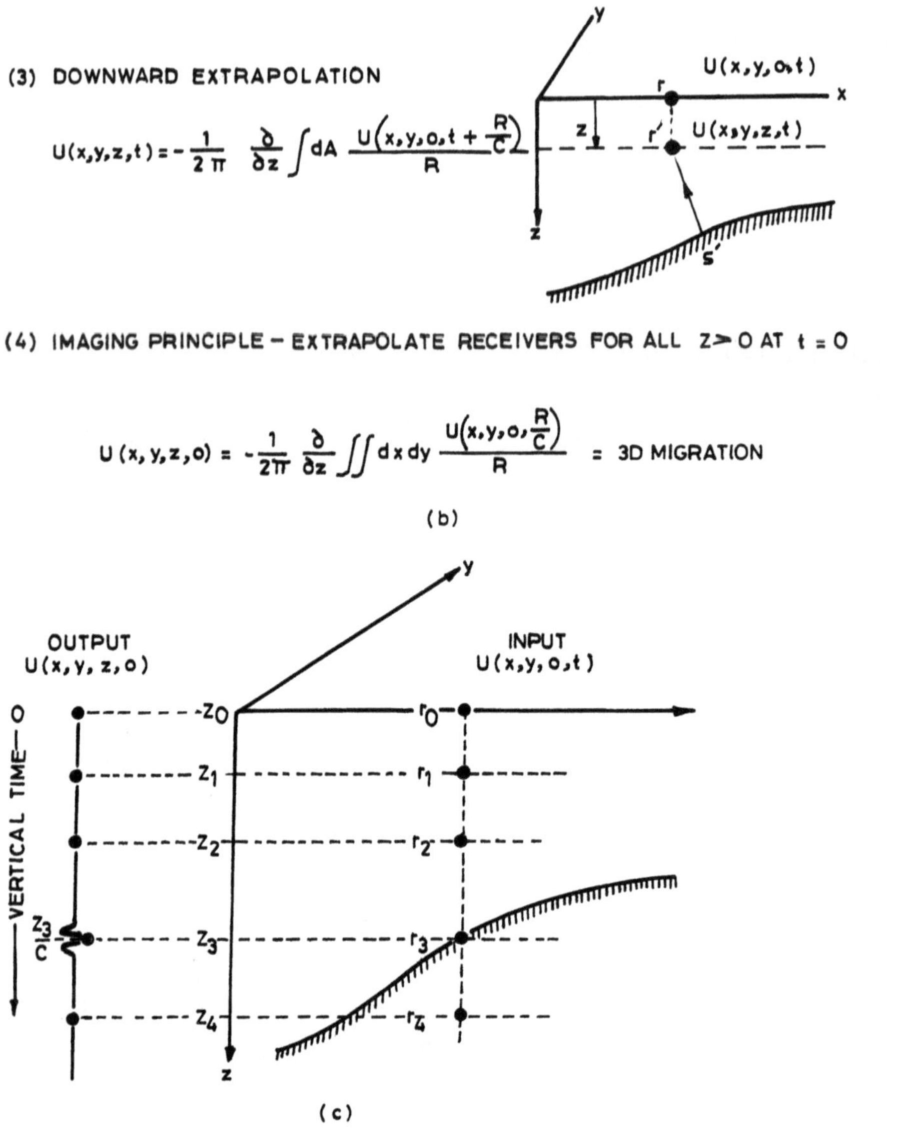

Fig. 15.4b,c. b Steps 3 and 4. **c** Input-output relationship (Source: Schneider 1978, from Geophysics with permission)

Use of Eq. (15.10) for migration (downward extrapolation and imaging) of CDP stacked data is illustrated through Fig. 15.4a,b.

Further, the migration process is visualized in Schneider (1978) as an input-output relationship. In this depiction (Fig. 15.4c), input is the stacked data on the surface at time t, U(x, y, 0, t), and output is U(x, y, z, 0) and corresponds to the depth of the reflection point and time zero when the sources are fired. The output contribution is zero at all depths above the depth of the reflection point. The output is finite at the depth of the reflecting point at vertical time z/c. This concept makes use of the exploding reflector model (refer, for example, to Loewenthal et al. 1976; Robinson et al. 1986, p 222). In Fig. 15.4c, the output value at z_1 and z_2 are zero because the reflections have not arrived. When the reflector position is reached, the reflection wavelet is mapped at the vertical travel time below the coincident source-receiver position, and produces a migrated image.

The excellent ideas and concept of migration using the solution of the scalar wave equation has been further widely illustrated in 3-D and 2-D with field and model examples in Schneider (1978).

15.1.3 Imaging of Unstacked Seismic Data

We return to the discussion of Eq. (15.7), that is, the Kirchoff integral in the case of a non-planar observation surface S_0:

$$U(\mathbf{r},t) = -\frac{1}{2\pi}\int dt_0 \int dS_0 \cdot U(\mathbf{r}_0,t_0)\frac{\partial}{\partial n}\left[\frac{\delta\left(t_0 - t - \frac{R}{c}\right)}{R}\right] \tag{15.7}$$

This equation does not reflect very clearly the effect of a non-planar observation surface. To bring to such a form, the observed data $U(r_0, t_0)$ is Fourier transformed with respect to t_0, and the order of t_0 and ω integrations are interchanged, and an inverse Fourier transform is taken (Wiggins 1984). The resulting wavefield, $U(r,t)$, comes in a form which reflects the effect of non-planar observation geometry:

$$U(r,t) = -\frac{1}{c}\int dS_0 \frac{1}{R}\frac{\partial R}{\partial n}\frac{\partial U}{\partial t}\left(\mathbf{r}_0, t+\frac{R}{c}\right) \tag{15.7 a}$$

[Wiggins 1984, Eq. (1)]

In obtaining Eq. (15.7a), the $1/R^2$ term is neglected. In this form of the Kirchoff integral for a non-planar observation surface S_0, the effect of surface geometry is observable in delayed time $(t+R/c)$ through R. $1/R$ and $\partial R/\partial n$ modify the amplitude part of the integral. Further, with observed field values having sources and receivers on a non-planar surface S_0, Wiggins (1984) derives the wavefield value on plane S_1, i.e. the field resulting from sources on S_0 and receivers on S_1 (Fig. 15.5). The following form is given in Wiggins [1984, Eq. (2)]:

$$U\left(\mathbf{r}_1, \mathbf{r}_0', t\right) = -\frac{1}{c}\int dS_0 \frac{\partial R}{\partial n}\frac{1}{R}\frac{\partial U}{\partial t}\left(\mathbf{r}_0, \mathbf{r}_0', t+\frac{R}{c}\right) \tag{15.7 b}$$

where

$$R = |\mathbf{r}_1 - \mathbf{r}_0|$$

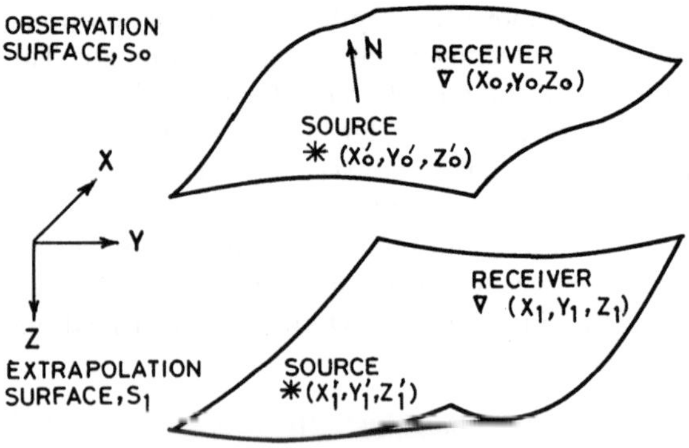

Fig. 15.5. Geometry of the extrapolation problem. Observations made with the sources and receivers on S_0 are used to calculate the results that would be obtained if the sources and receivers were on S_1 (Source: Wiggings 1984, from Geophysics with permission)

The primed variable r_0' in $U(r_1, r_0', t)$ shows that the source is on S_0, and the receiver is on S_1 (Fig. 15.5).

As the integral in Eq. (15.7b) represents a *backward time extrapolation of the received signal* on S_0, the source from S_0 cannot be brought directly to S_1 by application of Eq. (15.7b). For this purpose, Wiggins makes use of the reciprocity principle: a signal received at $\mathbf{r}_1$ from a source at r_0' is equal to the signal that would be received at $\mathbf{r}_0'$ due to an identical source at $\mathbf{r}_1$. With this, $U(\mathbf{r}_1, \mathbf{r}_0', t)$ is thought of as the received signal on S_0 due to the source on S_1 (Schultz and Sherwood 1980). Equation (15.7b) is substituted in Eq. (15.7a) to yield (refer to Fig. 15.5 to distinguish shot, receivers and surfaces)

$$
\begin{aligned}
U(\mathbf{r}_1, \mathbf{r}_1', t) &= -\frac{1}{c}\int dS_0' \frac{1}{R'}\frac{\partial R'}{\partial n}\frac{\partial U\left(\mathbf{r}_1, \mathbf{r}_0', t + \frac{R'}{c}\right)}{\partial t} \\
&= \frac{1}{c^2}\int dS_0 \int dS_0' \frac{1}{RR'}\frac{\partial R}{\partial n}\frac{\partial R'}{\partial n} \cdot \frac{\partial^2 U\left(\mathbf{r}_0 \mathbf{r}_0', t + \frac{R+R'}{c}\right)}{\partial t^2}
\end{aligned}
\tag{15.7 c}
$$

[Wiggins 1984, Eq. (3)]
where

$$R = |\mathbf{r}_1' - \mathbf{r}_0'|$$

This is the required form. It shows that from wavefield $U(\mathbf{r}_0, \mathbf{r}_0', t)$ with the receiver and source on surface S_0, it is possible to write the wavefield on surface S_1 with the receiver and source at $(\mathbf{r}_1, \mathbf{r}_1')$.

Equation (15.7 c) has been assigned the name *KRK integral*, because of the Kirchoff integration, a reciprocal interchange of source and receiver, and a second Kirchoff integration using the principle of reciprocity of source and receiver. This equation is completely symmetrical, and can therefore be used to perform wavefield extrapolation and migration both. It may be mentioned here that by making use of the exploding reflector model, Eq. (15.10) (Schneider 1978) permits single wavefield extrapolation. In contrast, the KRK integral [Eq. (15.7c)] is an extrapolation of both sources and receivers, and is therefore applicable to *unstacked seismic data.*

Kirchoff migration (refer, for example, to Schneider 1978) is achieved by evaluating the receiver extrapolation integral Eq. (15.10) at the time when the source pulse reaches the reflector and the result for each reflector is summed to generate the *migrated image.* The KRK integral [Eq. (15.7c)] makes use of movement of both sources and receivers simultaneously, and in this case a reflector exists at $\mathbf{r}_1$ provided $U(\mathbf{r}_1, \mathbf{r}_{1,0})$ indicates a reflection (Claerbout 1971). Thus, the migrated image $I(r_1)$ is obtained by taking $\mathbf{r}_1 = \mathbf{r}_1'$ and $t=0$ in $U(\mathbf{r}_1, \mathbf{r}_1', t)$ in Eq. (15.7c):

$$I(\mathbf{r}_1) = U(\mathbf{r}_1, \mathbf{r}_1', 0) = -\frac{1}{c^2}\int dS_0 \int dS_0' \frac{1}{RR'}\frac{\partial R}{\partial n}\frac{\partial R'}{\partial n} \cdot \frac{\partial^2 U}{\partial t^2}\left(\mathbf{r}_0, \mathbf{r}_0', t + \frac{R+R'}{c}\right) \tag{15.7d}$$

[Wiggins 1984, Eq. (4)]

It is clear that, because of simultaneous downward movement of both sources and receivers, Eq. (15.7d) would produce a *before-stack migrated picture* in contrast to the *stacked migrated picture* produced by use of Eq. (15.10). Also, the before-stack migrated image is compensated for spherical spreading and for survey geometry. The analysis of Wiggins (1984) assumes independence of the reflection coefficient with the angle of incidence. Also, it is assumed that the portion of the Earth to be imaged is several wavelengths away from the surface of observation, i.e. $R, R' \gg \lambda$.

Equation (15.7d) is completely general. It assumes that extensive observations have been made on a non-planar surface. Its simpler forms (in cases in which observations are available along a line) can also be obtained for extrapolation and imaging. These are further discussed with examples in Wiggins (1984).

15.2 Simultaneous Migration of P- and S-Waves: Elastic Wave Migration

Shear wave and converted shear wave presence in both land and marine seismic surveys provides the incentive that both be used with P-waves for elastic wave migration to achieve a better-migrated picture. Kuo and Dai (1984) developed a method for simultaneously migrating both P- and S-waves, using *offset seismic data.* It is based on Kirchoff-Helmholtz type integrals for elastic waves, and referred to as *Kirchoff Elastic Wave migration.*

The Kirchoff-Helmholtz type integral for elastic waves was derived by Pao and Varthrajulu (1976).

The displacement vector field $\mathbf{u}(\mathbf{x})$ in the interior of a region bounded by closed surface S′ (Fig. 15.6), neglecting body force effects, can be expressed as the surface integral

$$\mathbf{u}(\mathbf{x}) = \int_{S'} \left\{ \mathbf{t}(\mathbf{x}') \cdot \mathbf{G}(\mathbf{x}|\mathbf{x}') - \mathbf{u}(\mathbf{x}') \cdot \left[\mathbf{n}' \cdot \sum (\mathbf{x}|\mathbf{x}')\right]\right\} ds' \tag{15.11}$$

[Kuo and Dai 1984, Eq. (1)]

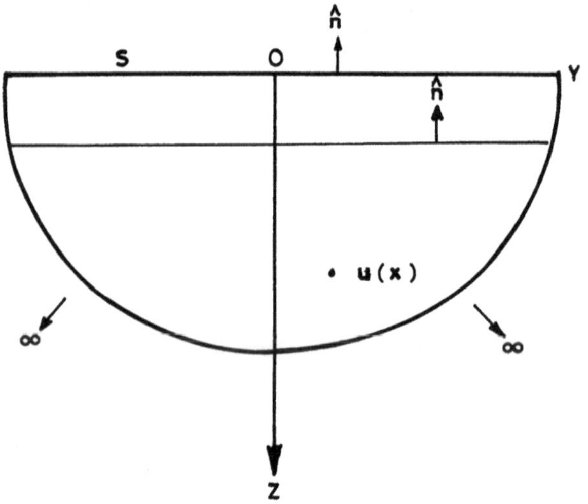

Fig. 15.6. The displacement vector **u**(**x**) interior to a region bounded by a closed surface S can be expressed in terms of the surface integral (Source: Kuo and Dai 1984, from Geophysics with permission)

where **t** is the traction vector, $\mathbf{G}(\mathbf{x}|\mathbf{x}')$ and $\Sigma(\mathbf{x}|\mathbf{x}')$ are *Green's displacement dyadic* and *Green's stress tensor*, respectively.

It may be mentioned here that Green's displacement dyadic and Green's stress tensor have terms, which are contributions of both P and S-waves. These are expressed in the frequency domain and far-field approximation, i.e. $r_k \gg 1$ is taken. When these values of Green's function are substituted in Eq. (15.11), an expression for $\mathbf{u}(\mathbf{x},t)$ is obtained, which is again simplified assuming free surface boundary condition $\mathbf{t}(\mathbf{x}')=0$. Thus, the far-field approximation of $\mathbf{u}(\mathbf{x})$ of Eq. (15.11) is obtained.

Using the far-field approximation value of $\mathbf{u}(\mathbf{x})$, downward continuation becomes possible. The principle of elastic wave migration is explained (Fig. 15.7 a, b) from Kuo and Dai (1984).

Figure 15.7 a shows the generation of P-wave and converted wave PS from a dipping reflector. At a point like R_2 both the reflected P-wave (P_1P_1) and converted $S(P_2S_2)$ wave arrive. This also happens with point R_3. Reflected P wave (P_2P_2) and converted S-wave (P_3S_3) arrive at different times at R_3 as shown in the top part of the Fig. 15.7 a. Figure 15.7 b shows that the travel times of upcoming S-wave $(t_s^u)_{22}$ and P-wave $(t_p^u)_{22}$ recorded at lower level Z (at points R_2' and R_3') will decrease and tend to the travel time of the upcoming wave, $(t_p^u)_2=(SO_2/V_p)$, at reflector point O_2. When downward continuation is done to the reflector point O_2, the traveltimes of upcoming waves approach those of the downgoing wave $(t_p^u)_2=(t_s^u)_2=(t_p^d)_2$ and the reflected P and S-waves coincide. At the instant $(t^u)=(t^d)$ enhanced migration images are formed at the location of reflectors through the P- and S-wave simultaneous migration.

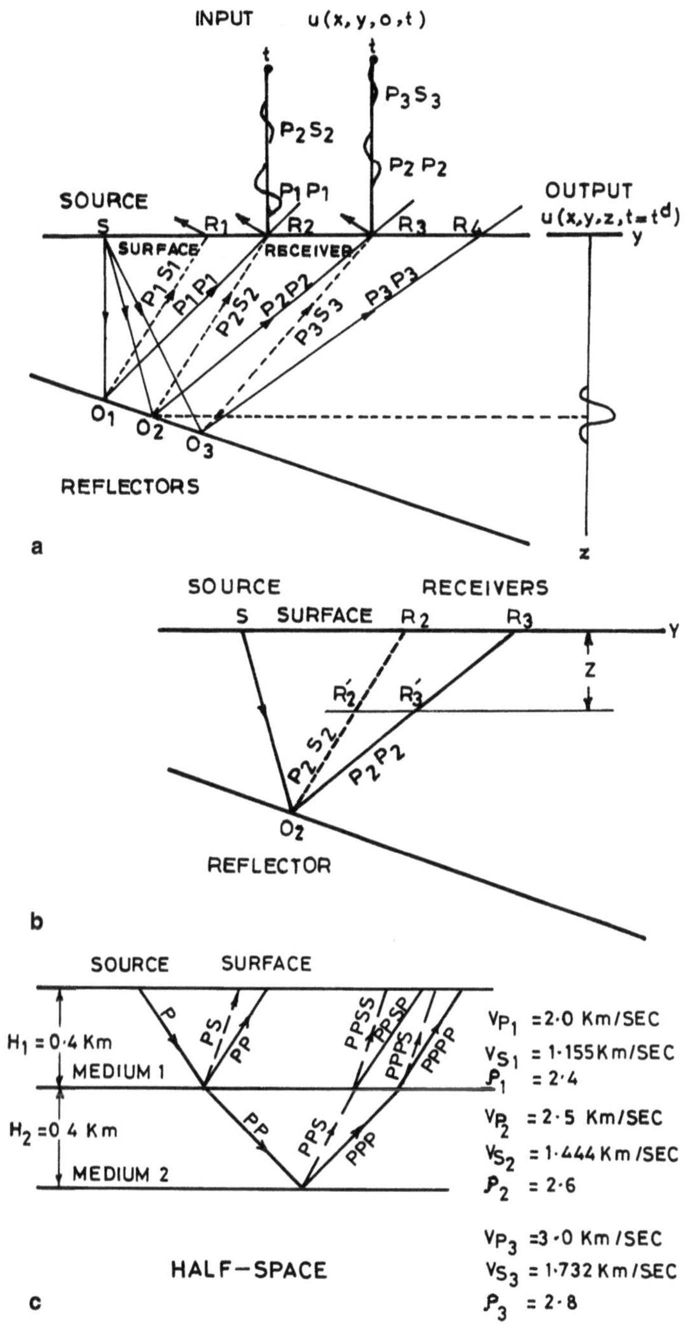

Fig. 15.7 a–c. **a** Principle of P and S wave simultaneous migration. **b** The reflected P and S waves coincide when downward continuation points R_2' and R_3' tend towards reflector point O_2. Criterion of P and S wave simultaneous migration: $t_p^u = t_s^u = t^d$ if R' is at reflector. **c** Two elastic layers over an elastic half space (Source: Kuo and Dai 1984, from Geophysics with permission)

Kuo and Dai (1984) have tested the versatility of their formulation for complex geological situations including a multi layer model. First, synthetic reflection data was generated which depicts the presence of reflected P-waves and converted P-S waves for different types of models. A model consisting of two horizontal layers over a half space (Fig. 15.7 c) is taken here as an example from Kuo and Dai (1984). To illustrate the effectiveness of the elastic wave migration including P, S, and converted waves, Kuo and Dai prepared migrated depth sections from the synthetic data using *Kirchoff acoustic* (P wave only) and *Kirchoff elastic wave* (P- and S-waves both) *migration*. These are illustrated in Fig. 15.8 a, b. It is clear that an *enhanced migrated picture* results using elastic wave migration.

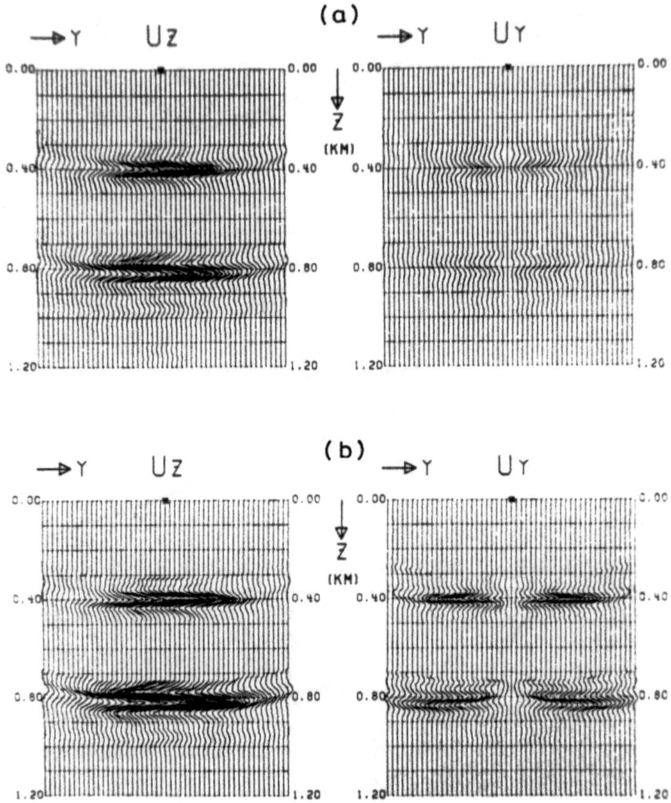

Fig. 15.8 a, b. a Migrated depth section from the synthetic data of Fig. 15.7 c, using Kirchoff P wave migration. **b** Migrated depth section for synthetic data of Fig. 15.7 c, using Kirchoff elastic wave migration (Source: Kuo and Dai 1984, from Geophysics with permission)

15.3 Migration in Inhomogeneous, Anisotropic Media for Converted and Non-converted Waves

Sena and Tohsöz (1993) developed the Kirchoff migration for azimuthally isotropic media for converted as well as non-converted waves. In their investigation, *a new anisotropy velocity analysis* scheme was developed which enables the *generation of realistic anisotropy models* in terms of five elastic constants of azimuthal isotropy to be used in the migration process. The procedure gives accurate images of the subsurface, which is not possible with an isotropic migration algorithm. *Anisotropic Kirchoff migration* enables us to use *converted waves as signals* for migration for *enhanced migrated images.* Converted waves are treated as noise resulting in poorer images in conventional *Kirchoff Acoustic Migration.* The method of Sena and Toksöz (1993) is based on a new suitable *Green's tensor representation* using analytical forms for ray-amplitudes derived in Ben-Menahem and Sena (1990), and traveltimes in Sena (1991).

As in other migration methods described earlier, *Kirchoff migration for an anisotropic inhomogeneous medium* is a two-step process, namely, first the downward extrapolation of the wavefield observed on the surface, and second the imaging.

Analytical steps in the formulation of the method proposed by Sena and Toksöz (1993) for Kirchoff migration in an anisotropic inhomogeneous medium is described below.

Scattered displacement field (u^s_m) emitted by secondary sources inside the volume V is given by

$$u^s_m(\mathbf{r}) = \int_s \left[t_l(\mathbf{r}')G_{lm}(\mathbf{r},\mathbf{r}') - u^s_l(\mathbf{r}') \sum{}_{lm}(\mathbf{r},\mathbf{r}') \right] ds' \tag{15.12}$$

(Pao and Varathrajulu 1976)
where $\mathbf{r} \in V$
$\mathbf{n}'$: Unit vector in the direction of outward normal to S at $\mathbf{r}'$
$G_{lm}(\mathbf{r},\mathbf{r}')$: Green's displacement tensor
t_i: traction vector
Σ_{lm}: Green's traction tensor

For measurements done on free-surface S, t(r) is taken as zero for $r \in S$, to account for a traction-free boundary condition. Equation (15.12) reduces to:

$$u^s_m(\mathbf{r}) = -\int_s u^s_j(\mathbf{r}') \sum{}_{lm}(\mathbf{r},\mathbf{r}') ds' \tag{15.13 a}$$

[Sena and Toksöz 1993, Eq. (5)]
for $Z > 0$

This form of the *Kirchoff-Helmholtz integral* may be used for forward extrapolation of the displacement field in the region $Z > 0$ in the Earth from measured values of the displacement field on the Earth's surface $S(Z=0)$, under the condition that a suitable expression for Green's tensor is known.

To adopt Eq. (15.13 a) for back propagation, needed in the migration process, Green's tensor $\Sigma_{lm}(\mathbf{r},\mathbf{r}')$ is replaced by its complex conjugate $\Sigma_{lm}{}^{*}(\mathbf{r},\mathbf{r}')$ (Keho and Beydoun 1988). We have therefore

$$u_m^s(\mathbf{r}) = -\int_s u_l^s(\mathbf{r}') \sum{}^{*}_{lm}(\mathbf{r},\mathbf{r}')ds' \qquad (15.13\,b)$$

[Sena and Toksöz 1993, Eq. (6)]
for $Z > 0$

This is referred to as the *back-projected Kirchoff field.*

To develop a fast migration scheme, an asymptotic high frequency representation of Green's tensor in elastic anisotropic media developed in Ben-Menahem and Sena (1990) is utilized. This approximation has the form

$$G_{lm}(\mathbf{r},\mathbf{r}') = \sum_{L=1}^{3} A_{lm}^{L}(\mathbf{r},\mathbf{r}')e_{L}^{i\omega\tau}(\mathbf{r},\mathbf{r}') \qquad (15.14)$$

where
A_{lm}^{L} $(\mathbf{r},\mathbf{r}')$: ray amplitude
$\tau_L(\mathbf{r},\mathbf{r}')$: travel time
for mode L. Analysis in Sena and Toksöz (1993) takes into consideration all possible modes (qP, qS_1, qS_2).

Since Green's traction tensor Σ_{lm} is expressed as

$$\sum{}_{lm} = n_j c_{ijkl}\frac{\partial G_{lm}}{\partial x_k}$$

where c_{ijkl} is a fourth rank stiffness tensor, we may use the value of $G_{lm}(\mathbf{r},\mathbf{r}')$ from Eq. (15.14) and determine the Green's traction tensor Σ_{lm}^{*} (r, r′). Substituting this value in Eq. (15.13 b), the expression for the back-propagated Kirchoff field is obtained:

$$u_m^s(\mathbf{r},\omega) = i\omega\sum_{L=1}^{3}\int_s u_l^s(\mathbf{r}',\omega)c_{ijkl}(\mathbf{r}') \times A_{lm}^{L*}(\mathbf{r},\mathbf{r}')p_k^L(\mathbf{r},\mathbf{r}')e^{-i\omega\tau_L(\mathbf{r},\mathbf{r}')}n_j'ds \qquad (15.15)$$

[Sena and Toksöz 1993, Eq. (10)]

Here $p_k^l(\mathbf{r},\mathbf{r}')$ is the kth component of slowness vector corresponding to mode L.

The pseudo-reflectivity function of the subsurface R(r, ω) is defined as the ratio of the back propagated field from receivers, u_m^s(r, ω) to the forward propagated field from the source u_m^I(r, ω). It is given by

$$R(\mathbf{r},\omega) = \begin{pmatrix} R_{11}(\mathbf{r},\omega) & R_{12}(\mathbf{r},\omega) & R_{13}(\mathbf{r},\omega) \\ R_{21}(\mathbf{r},\omega) & R_{22}(\mathbf{r},\omega) & R_{23}(\mathbf{r},\omega) \\ R_{31}(\mathbf{r},\omega) & R_{32}(\mathbf{r},\omega) & R_{33}(\mathbf{r},\omega) \end{pmatrix}$$

[Sena and Toksöz 1993, Eq. (11)]
where 1 stands for q^P, 2 for q^{S1}, and 3 for q^{S2}. Under this notation the q^P to q^P pseudo reflection coefficient is

$$R_{11}(\mathbf{r},\omega) = \frac{\left|U^S_{qp}(\mathbf{r},\omega)\right|}{\left|U^I_{qp}(\mathbf{r},\omega)\right|}$$

The value of the incident field is also computed for its high frequency form

$$U^I_i(\mathbf{r},\omega) = \sum_{M=1}^{3} B^M_{iq}(\mathbf{r},\mathbf{r}_0) e^{i\omega\tau}_M(\mathbf{r},\mathbf{r}_0) \tag{15.16}$$

An incident wave is considered to be generated by a point force in the q-direction at position $\mathbf{r}_0$.

Therefore, the following form is obtained for the M to L pseudo-reflection coefficient:

$$R_{ML}(\mathbf{r},\omega) = \frac{i\omega}{Q^M_q(\mathbf{r},\mathbf{r}_0)} \int_s u^{SL}_i(\mathbf{r}',\omega) C_{ijkl}(\mathbf{r}') \\ P^L_l(\mathbf{r},\mathbf{r}') p^L_k(\mathbf{r},\mathbf{r}') e^{-i\omega[\tau_L(\mathbf{r},\mathbf{r}')+\tau_M(\mathbf{r},\mathbf{r}_0)]} n'_j ds' \tag{15.17 a}$$

[Sena and Toksöz 1993, Eq. (14)]
where

$$P^L_l(\mathbf{r},\mathbf{r}') \equiv \sqrt{\sum_{m=1}^{3} \left|A^L_{lm}(\mathbf{r},\mathbf{r}')\right|^2}$$

and

$$Q^M_q(\mathbf{r},\mathbf{r}_0) \equiv \sqrt{\sum_{i=1}^{3} \left|B^M_{iq}(\mathbf{r},\mathbf{r}_0)\right|^2}$$

$R(r,\omega)$ is transformed back into the time domain by the inverse Fourier transformation.

$$R_{ML}(\mathbf{r},t) = \frac{1}{2\pi} \int_{-\infty}^{+\infty} d\omega e^{-i\omega t} R_{ML}(\mathbf{r},\omega) \tag{15.17 b}$$

Image $I_{ML}(\mathbf{r})$ of the medium is obtained by setting $t=0$ in Eq. (15.17b), i.e. by applying the imaging condition. Thus

$\text{Image} = I_{ML}(\mathbf{r}) = R_{ML}(\mathbf{r}, t = 0)$

The final outcome of the formulation developed by Sena and Toksöz (1993) and described above is the following:

The subsurface image would be created by stacking the observed data along the diffraction curves defined by $t = (\tau_L(\mathbf{r}, \mathbf{r}') + \tau_M(\mathbf{r}, \mathbf{r}_0))$ [Eq. (15.17a)]. The remaining factors in Eq. (15.17a) are weighting functions that depend on elastic properties of the medium (Sena and Toksöz 1993, p 267).

This analytical formulation is completely general, taking into consideration a generalized azimuthal isotropy property of the medium, and all possible combinations of incident and reflected wavefields. In principle, this formulation allows imaging with up to nine different wavefield components. It is also self-contained and self- sustained, as it allows the estimation of "stacking" elastic constants independently for use in the migration process. Best estimation of elastic constants is the most important initial input required. For the application of Kirchoff migration on the q^P–q^P seismic section, the method depends on surface reflection or VSP data for both quasi-P and SH-waves.

Sena and Toksöz (1993) begin with the travel-time relation of Byun et al. (1989) for their velocity analysis and determining *stacking elastic constants* for migration. Referred to as a skewed hyperbolic expression, it is given in the form (Byun et al. 1989).

$$t^2 = T_0^2 + \left[\left(\frac{Z}{V\gamma}\right)^2 + \left(\frac{X}{2V_h}\right)^2\right]\frac{X^2}{Z^2 + \left(\frac{X}{2}\right)^2} \tag{15.18}$$

It involves the determination of three velocities, i.e. V_v the average vertical velocity; V_h and V_γ, horizontal and skew movement velocities, respectively. T_o is the vertical two way travel time given by $T_0 = 2Z/V_v$.

Sena (1991) obtained explicit expressions for V_v, V_h, V_γ for a single azimuthally isotropic layer in terms of elastic constants. For a multi-layered medium, V_v, V_h, and V_γ are cumulative average properties created due to anisotropy and heterogeneity.

Steps in the estimation of bulk velocities to each reflector depth is therefore necessitated for the determination of elastic constants to be used in the Kirchoff migration process. This is depicted in Fig. 15.9.

The final output of this velocity analysis is the estimate of $(C_{11}/\rho)^{1/2}$, $(C_{33}/\rho)^{1/2}$, $(C_{13}/\rho)^{1/2}$ and $(C_{44}/\rho)^{1/2}$ for one fixed offset for each depth of each reflector, at a time. This estimate is repeatedly done for each new offset (refer, for example, to Table 2 of Sena and Toksöz 1993, p 272).

While recording 3-component surface reflection data, converted waves are generally recorded, and initiatives have already started to process converted wave data. Sena and Toksöz (1993) derived a travel time equation for converted q^P–q^{SV} waves

$$t = \sqrt{(Z^2 + X_1^2)(C_0^P + C_1^P D + C_2^P D^2)} + \sqrt{[Z^2 + (X - X_1)^2](C_0^{SV} + C_1^{SV}\bar{D} + C_2^{SV}\bar{D}^2)} \tag{15.19}$$

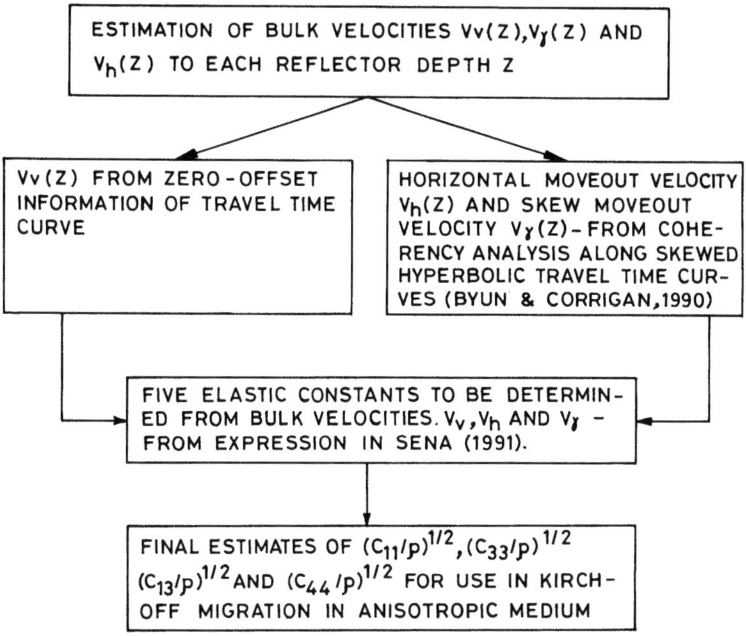

Fig. 15.9. Illustrating steps in the determination of the velocity model (elastic constants) for an anisotropic medium for use in Kirchoff migration, for the algorithm contributed by Sena and Toksöz (1993)

[Sena and Toksöz 1993, Eq. (32)]
where

$$D = \frac{X_1^2}{Z^2 + X_1^2}$$

$$\bar{D} = \frac{(X - X_1)^2}{Z^2 + (X - X_1)^2}$$

and coefficients C_i (i=0, 1, 2) relate to the velocity for each type of wave (refer to Sena 1991).

X is the receiver offset, Z is reflector depth and X_1 is the horizontal distance of the conversion point from source point (Fig. 15.10).

The conversion point offset X_1 is related to X by the formula

$$X_1 = X \frac{\sqrt{C_0^P (C_0^{SV} + C_1^{SV})}}{\sqrt{C_0^P (C_0^{SV} + C_1^{SV})} + \sqrt{C_0^{SV} (C_0^P + C_1^P)}}$$

[Sena and Toksöz 1993, Eq. (38)]

In isotropic limit, this reduces to

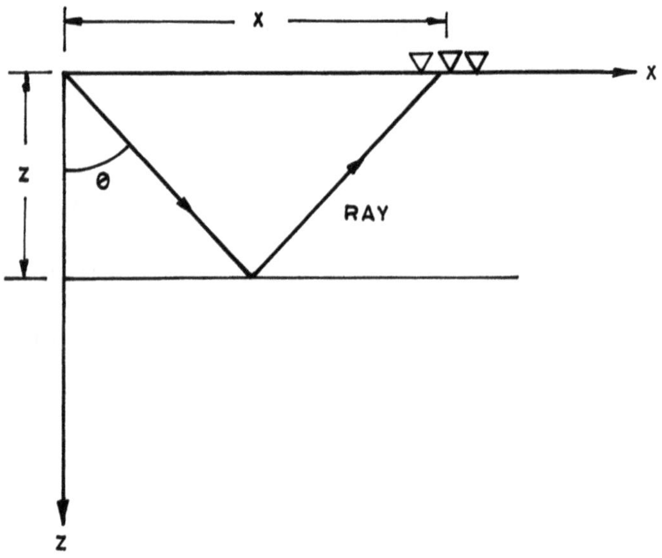

Fig. 15.10. Ray geometry for a surface reflection survey. The layer is considered as azimuthally isotropic (Source: Sena and Toksöz 1993, from Geophysics with permission)

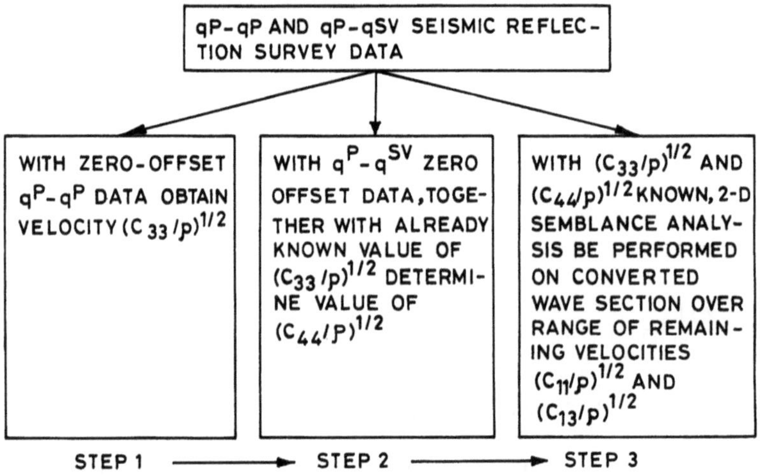

Fig. 15.11. Steps in the determination of velocities (elastic constants) using q^P–q^P and q^P–q^{sv} converted wave data

$$X_1 = X\frac{1}{1+\frac{V_S}{V_P}}$$

which agrees with the result of Fromm et al. (1985).

For performing anisotropic velocity analysis over a converted wave seismic section, q^P–q^P and q^P–q^{SV} data are required from the reflection surveys. Figure 15.11 outlines the steps for determining the velocity model for subsequent use in Kirchoff migration. These four velocities suffice fully for the Kirchoff migration algorithm for q^P–q^{sv} converted waves.

Errors in depth estimates would appear if vertical velocities were not known correctly. Stacking analysis does not provide a correct estimate of this velocity component. Therefore, knowledge of subsurface information from VSP or check shots is essential to generate a complete elastic constant model.

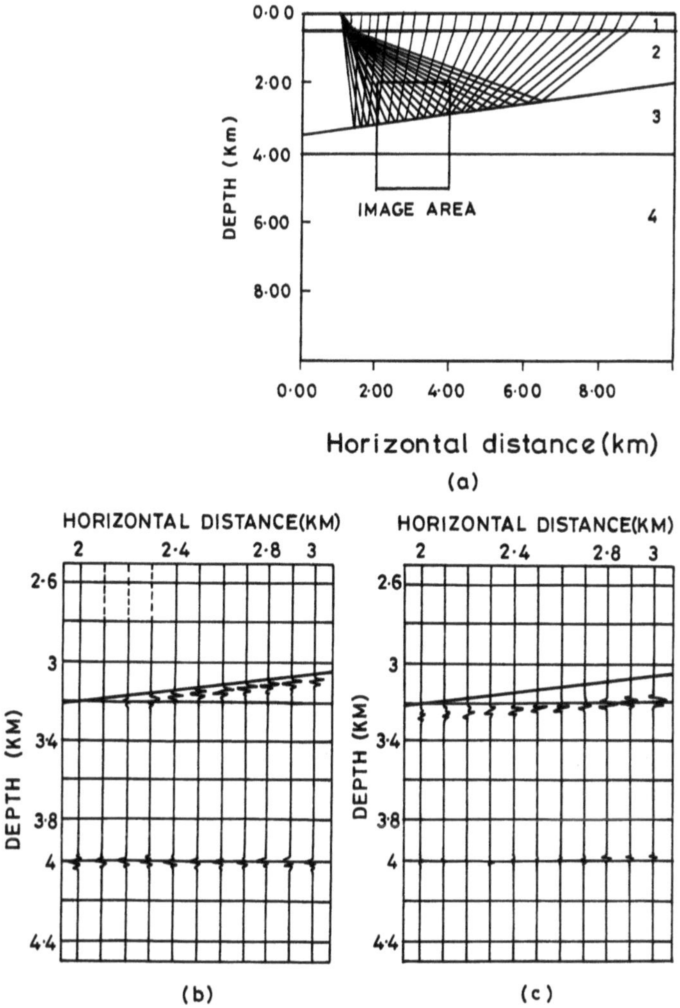

Fig. 15.12a–c. a Image area considered for the migration of converted q^P–q^{sv} waves. The image area is narrower than the one used for q^P–q^P migration to avoid the post critical reflection region. **b** Image created by migrating the q^P–q^{sv} reflected wavefield data using the anisotropic Kirchoff migration technique. **c** Migrated image created by using the q^P–q^{sv} reflected wavefield data using an isotropic Kirchoff migration scheme (Source: Sena and Toksöz 1993, from Geophysics with permission)

Once the anisotropic velocity model is found Green's tensor may be determined with use of the formula for ray amplitude and travel times (refer to Ben Menahem and Sena 1990; Sena 1991). The next step is the implementation of Kirchoff migration. An educative and illustrative model described in Sena and Toksöz (1993) is taken to illustrate the effectiveness of the algorithm.

Sena and Toksöz (1993) analyzed a four-layer model with horizontal and dipping interfaces (Fig. 15.12 a). The image area is shown in the figure. The choice of image area is a very important consideration keeping in view using converted waves. In this respect, analytical results in Tessmer and Behle (1988, 1991), Tsvankin and Thomsen (1994), and Thomsen (1999) should find very useful applications. Apart from this, other basic results exist in these publications, which will help develop the exploration potential of converted waves in reflection seismology. In the model considered (Fig. 15.12 a) the top layer 1 is isotropic, this is followed by layer 2 which is an anisotropic shale, followed in turn by sand/anisotropic shale cyclic layering (layer 3) and the 4th layer which is a cyclic layering of limestone/shale. With the help of the velocity analysis, velocity values $(c_{11}/\rho)^{1/2}$, $(c_{33}/\rho)^{1/2}$, $(c_{13}/\rho)^{1/2}$, and $(c_{44}/\rho)^{1/2}$, were determined in Sena and Toksöz (1993) for various offsets for each reflector. Images were generated using anisotropic and isotropic migration schemes. Results are depicted in Figs. 15.12 b, c.

It is obvious that the use of the isotropic migration scheme for an anisotropic medium produces misplacement of events. The numerical analysis was also conducted in Sena and Toksöz (1993) with converted waves, which points to a similar conclusion concerning the inadequacy of isotropic migration in this case also.

Conclusively, the isotropic migration algorithm cannot properly image the geologic boundaries even with only a weak to moderate percentage of anisotropy. The method developed in Sena and Toksöz (1993), in which a new non-conventional velocity analysis technique was proposed for use in the migration of seismic data, should provide more geologic information on *lithology discrimination*, compared to conventional migration algorithms. The Sena and Toksöz (1993) method has limitations of ray theory and is expected to be of limited use for strong lateral heterogeneities.

Algorithm 15.1: Stepwise Development of an Imaging Scheme using Non-conventional Velocity Analysis

The steps in Kirchoff migration in anisotropic inhomogeneous media (Sena and Toksöz 1993) are the following:

Steps:	Description
0	To begin with the general form of the Kirchoff-Helmholtz integral for elastic waves in an inhomogeneous anisotropic medium involving Green's displacement tensor and Green's traction tensor.
1	Taking a specific form of the Kirchoff-Helmholtz integral for a traction-free surface for forward extrapolation of the wavefield.

2 Writing the Kirchhoff-Helmholtz integral for the *back propagated field* required in the migration process.

3 Using an asymptotic high frequency representation of Green's displacement tensor, expressed in terms of ray-amplitude and travel times for all possible modes (q^P, q^{s1}, q^{s2}).

4 Writing Green's traction tensor expressed in terms of the elastic stiffness tensor, ray amplitude and travel time.

5 Writing the back propagated Kirchhoff-field expressed in terms of elastic stiffness, ray-amplitude, travel-time, and slowness vector.

6 Defining a pseudo-reflectivity function as a ratio of the *back propagated field* from receivers to the *forward propagated field* from the source for all combinations of incident and back propagated q^P, q_1^S, and q_2^S reflection coefficients matrix $\mathbf{R}(\mathbf{r},\omega)$, defined as function of $\mathbf{r}$ and ω for all combinations of incident and back propagated waves.

7 Taking the Inverse Fourier transform to define the pseudo reflectivity matrix in time domain $\mathbf{R}(\mathbf{r},t)$.

8 Obtaining images of the medium by taking time, t, to be zero.

Appendix 15A: Removal of Distortions in the Migrated Image, Caused due to Static Shifting, by using Wavefield Extrapolation

In the early stages of seismic data processing, seismic data obtained on the Earth's surface is transferred to a floating datum by using *static correction*. The floating datum is usually taken close to the Earth's surface. Static correction involves the time shifting of every seismic trace independently by an amount defined by the difference in elevations, and a known or estimated velocity model between the Earth's surface and the floating datum.

In the final stage of processing, migration is performed. The input data for migration is the *zero-offset time section* representing the pressure field along a horizontal datum, fixed for the whole area of investigation. Therefore, to perform migration seismic data should be transferred to this horizontal datum. This final datum correction is normally accomplished by static shift. Again, for small elevation differences between the floating and fixed datum and for slow velocity variations between these levels, static shift works sufficiently well as an accurate *datum correction* procedure. However, in seismic surveys there could arise situations in which the Earth's topography changes are very large amounting to elevation differences of hundreds of meters and significant velocity variations in the shallow part of the sedimentary sections. In such specialized field conditions, the static shift method would not be a suitable choice, and a more exact method of data extrapolation based on the wave equation should be used (Shtivelman and Canning 1988).

To illustrate our point we take two examples from Shtivelman and Canning to emphasize the failure of the static shift method (Figs. 15A.1 and 15A.2). These figures illustrate that static shifting can smear seismic energy, and thereby deteriorate

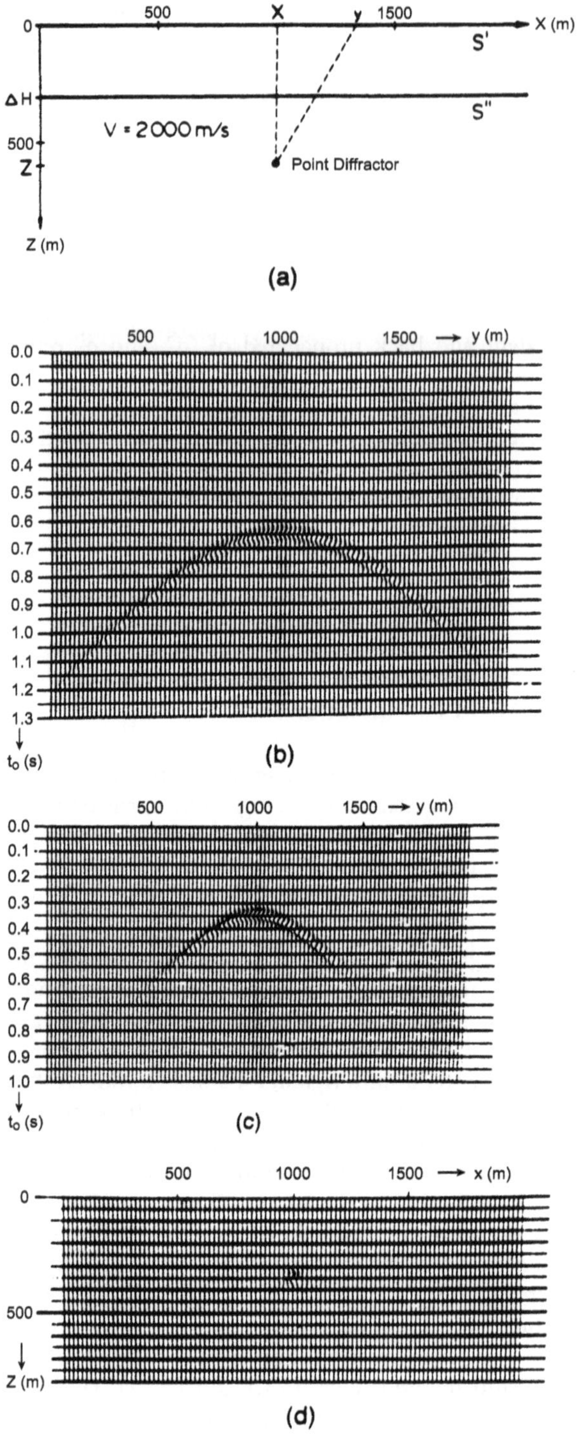

Fig. 15A.1 a–d. a Point diffractor in a homogeneous model containing two horizontal reference surfaces S′ and S″. **b** Zero-offset time section computed at S′. **c** Zero-offset section computed at S″. **d** Depth migration of **c**. The diffraction hyperbola is collapsed to a point

Fig. 15A.1 e, f. e Datum correction of **b** to S″ by static shifting. **f** Datum migration of **e**. The diffraction hyperbola, shifted from S′ to S″, is transformed into a semi-circle with radius ΔH (Source: Shtivelman and Canning 1988, from Geophysics with permission)

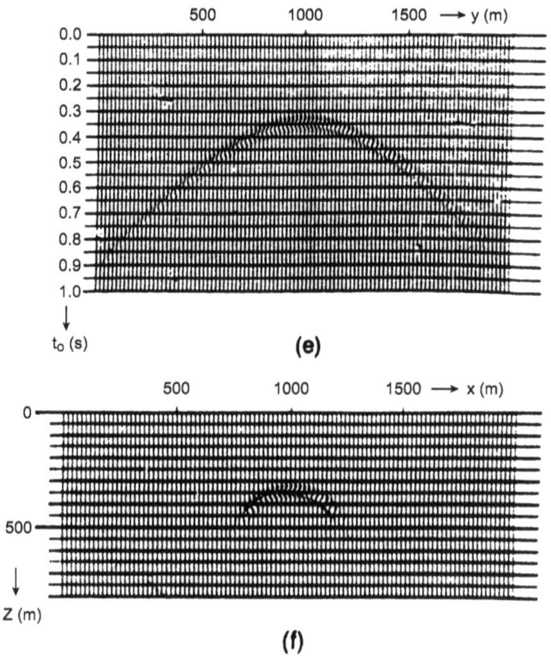

the resolution of the *depth sections.* Figure 15A.1 pertains to distortion in the migrated image for a *point diffractor* whereas Fig. 15A.2 to a *non-planar datum,* and velocity variations. These figures (studied with their captions) explain the inadequacy of the static shift method under the specialized field conditions assumed.

Accurate datum correction may be formulated in terms of wavefield extrapolation defined as

$$E\{P'[S'(x,z),t]\} = P''[S''(x,z),t] \tag{15A.1}$$

[Shtivelman and Canning 1988, Eq. (13)]

Here P′ is a given wavefield specified along a given input datum S′(x, z), P″ is the output (extrapolated) wavefield along a given output datum S″(x, z), and E is an extrapolation operator.

Before discussing the wavefield extrapolation scheme, let us first analyze using a simple example, why a wavefield extrapolation approach should be more effective. Figure 15A.3 a shows two ray paths arriving at one detector b and datum D_1. These raypaths emerge from different points on reflector R. Assuming datum position D_2, two different detectors (c and d) on it would sense the motion of slant and vertical raypaths. It is clear that if we want to predict what should be observed on datum D_2 from a knowledge of what is recorded on D_1, it should not be a straightforward procedure of assigning trace 'b' to location 'd'. It is to be noted that not only time shifts $2L_1/V_1$ and $2L_2/V_2$ are different, but additionally a slanting ray path needs to be assigned to detector c. This reveals that any procedure for changing the datum of a seismic section from D_1 to D_2 must be sensitive to the *time-dip* of reflections in

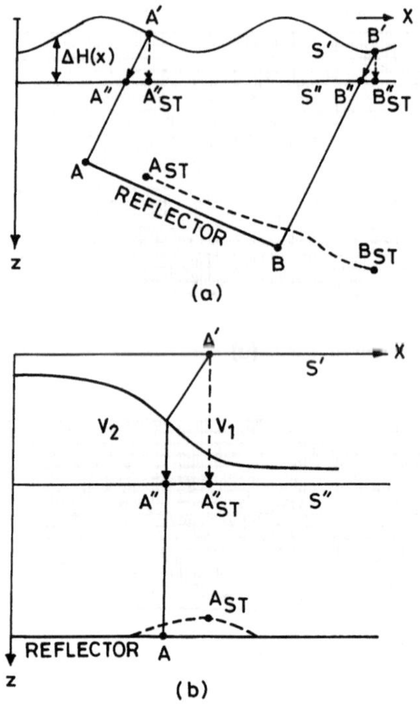

Fig. 15A.2 a, b. Schematic illustration of the cases of non-planar datum and velocity variations. **a** Segment of a dipping reflector (limited by points A and B) in a homogeneous model including non-planar datum S′. The corresponding zero-offset section at S′ is to be corrected to a horizontal datum S″. The static correction shifts the traces of the section along vertical rays (A′ A''_{ST} and B′ B''_{ST}) and results in a non-planar event A_{ST} B_{ST}, shown by the dashed line. **b** Velocity variations between S′ and S″ specified by two layers with velocities V_1 and V_2 separated by a curved interface. The corresponding zero-offset at S′ containing the reflection from the horizontal interface is to be corrected to S″. The static correction shifts the traces along vertical rays (A′ A''_{ST}) and results in a non-planar event shown by the dashed line (Source: Shtivelman and Canning 1988, from Geophysics with permission)

the section (the evidence of a *non-vertical ray path*), and care should be taken to correspondingly shift time and lateral locations. The wavefield extrapolation, based on the wave equation, has the capability of operating simultaneously on a multiplicity of traces to move laterally coherent signals in a manner that is consistent with wave propagation theory (Berryhill 1979).

Extrapolation of the wavefield involves creating from a given seismic section a new section. The given seismic section may be referred to as the input datum and the resulting section as the output datum. Referring to Fig. 15A.3 b, the mathematical statement of extrapolation (a form of Huygen's principle) is:

$$U_j(t) = \frac{1}{\pi} \sum_i \Delta x_i \cos\theta_i \frac{t_i}{r_i} [U_i(t - t_i) * F_i] \qquad (15A.2)$$

(Berryhill 1979, 1984)

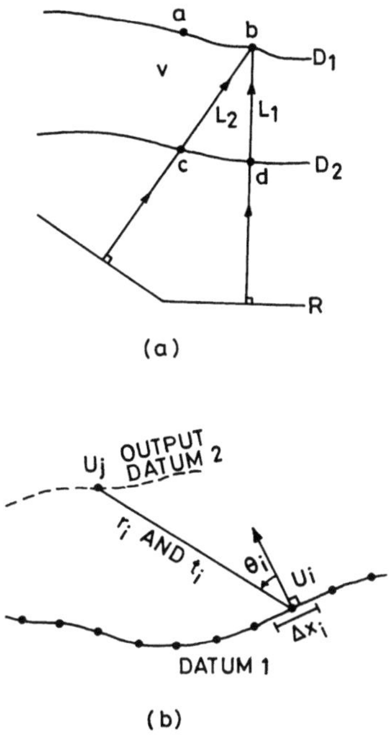

Fig. 15A.3 a, b. a Two raypaths arriving at one point on the surface D_1 cross surface D_2 at separate locations (Source: Berryhill 1979). **b** Definition of the geometrical quantities appearing in the Eq. (15A.2) (Source: Berryhill 1984)

Distance r_i, travel time t_i, angle θ_i and trace spacing Δx_i are defined in Fig. 15-A.3 b. The argument $(t-t_i)$ stands for shifting of U_i in time. F_i is a filter operator of 5 to 10 points length, and is a function of the subscripted variables. This is convolved with the trace to prevent *waveform distortion*. No assumption is made on the geometrical form of the datum; also there is no requirement to treat velocity as a constant.

The formulation of Eq. (15A.2) was given by Berryhill (1979) for stacked seismic data for a zero-offset geometry. It was shown in Berryhill (1984) that an extension to the case of unstacked seismic data required no change in mathematical algorithm. In fact, using the same algorithm, it is possible to move both receivers and sources from one datum to another in two steps: in the first step geophones are moved, and in the second the shots are moved. The two steps are referred to as common-source computation and common receiver computations. It was shown in Berryhill (1984) that common source and common receiver trace groups must take the form of symmetric split spreads for treating equally both directions of dip. This is illustrated by Fig. 15A.4. The ray paths are shown with the help of an image source in common-shot geometry. In the case of a horizontal reflector the minimum of the reflection hyperbola occurs vertically above zero source-receiver distance. From the figure pertaining to the case of a dipping reflector, the vertical upward raypath (the apex of the reflection hyperbola) occurs at some non-zero source-receiver distance. The shift of the reflection hyperbola (rightward in this case, study with Fig. 3.30 a) is determined by the direction of dip. Therefore, to account for dip-direction equally in a common-source geometry, detectors must be present on both sides of the source, corresponding to split-

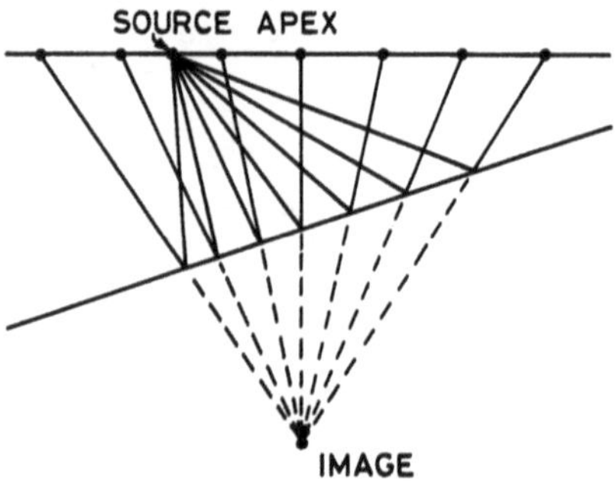

Fig. 15A.4. An illustration of common-source raypath geometry (Source: Berryhill 1984)

spread geometry. It was further explained in Berryhill (1984) that it is possible to construct synthetic split spreads using reciprocity if the data were not actually recorded in the required form of split-spread geometry.

The wave equation extrapolation scheme contributed by Shtivelman (1984, 1985) and Shtivelman and Canning (1988) is based on a Kirchoff integral solution of the 2-D scalar wave equation

$$\frac{\partial^2 u}{\partial x^2} + \frac{\partial^2 u}{\partial z^2} = \frac{1}{V^2}\frac{\partial^2 u}{\partial t^2} - f \tag{15A.3}$$

where $u = u(x, z, t)$ is displacement, V is velocity and $f = f(x, z, t)$ is a source function.

The general form of the extrapolated field P'' in a homogeneous medium along a given datum $S''(x, z)$ is the sum of two integrals

$$P'' = P_F + P_R \tag{15A.4}$$

where

$$P_F = \frac{1}{2\pi}\int_{S'} \left(\frac{\cos\theta}{r} \cdot \frac{\partial P'}{\partial t} * g_F\right) ds' \tag{15A.4 a}$$

and

$$P_R = \frac{1}{2\pi}\int_{S'} \left(\frac{\partial P'}{\partial n} * g_R\right) ds' \tag{15A.4 b}$$

[Shtivelman and Canning 1988, Eqs. (14), (15) and (16)]
where r is the length of the ray connecting output point (x'', z'') on s'' and (x', z') on the input datum s'. θ is the angle subtended by the ray and the normal at s' at point (x', z'). $\partial P'/\partial n$ is the normal derivative of the input field P' along s'. The asterisk (*) stands for convolution. g_F and g_R are Green's function defined by

$$g_F = ts(t - t_r)\Big/\sqrt{t^2 - t_r^2} \tag{15A.4 c}$$

and

$$g_R = s(t - t_r) \Big/ \sqrt{t^2 - t_r^2} \tag{15A.4 d}$$

In Eqs. (15A.4 c and d), $s(t) = 0$ for $t < 0$, and $s(t) = 1$ for $t \geq 0$; $t_r = r/V$ stands for propagation time along the ray and V is the velocity.

In principle, Eqs. (15A.4 a–d) should enable extrapolation of seismic data from s' to s''. However, their numerical implementation happens to be very time-consuming. This is for the reason that for computing a single output trace it is necessary to evaluate time convolution for every input trace within the aperture of spatial integration. The Berryhill (1979) approach is found to be efficient and is based on using a short convolutional operator. The Shtivelman and Canning (1988) method of wavefield extrapolation makes use of a high-frequency asymptotic approximation of *Green's function* Eq. (15A.4 c and d). Using this method time convolution is required to be computed only once for each output trace. The good performance of the wave equation extrapolation scheme is illustrated in Shtivelman and Canning (1988) with various models and examples, and compared with distortions observed by static shift. The interested reader may refer for details to this publication.

Kumar (1999) developed a computer program to evaluate comparative performance of static shift and wavefield extrapolation for simple sources.

CHAPTER 16

Miscellaneous Interpretation Tools 16

16.1 Introduction

Sometimes routine two-dimensional seismic reflection data may not provide adequate details of the underlying medium and it may be necessary either to process this data using specialized mathematical operations, or alternatively to analyze independent measurements based on laboratory or specially designed field experiments. All these aspects are included under the theme miscellaneous interpretation tools. Salient features of the various supporting interpretation tools are the basis of discussions in this chapter. These are broadly categorized as mathematical operations and geophysical applications, the use of tomographic reconstruction, seismic rock physics, 3-D data analysis, and objective oriented interpretation tools. Discussion of each of these follows with examples of laboratory or field survey results and their interpretations.

16.2 Mathematical Operations and Geophysical Applications

16.2.1 Convolution and Deconvolution

The convolution operation may be described for *discrete-time* and *continuous-time data sequences*. We first consider discrete-time sequences, x(t) and h(t). Sampling intervals for x(t) and h(t) are uniform and equal. The convolution of data sequences is represented by the symbol (*). Thus, x(t) convolved with h(t) is written as

$$x(t) * h(t) = y(t) \tag{16.1}$$

where

$$y(t) = \sum_{n=-\infty}^{+\infty} x(n)h(t-n) \tag{16.1 a}$$

Convolution involves the following steps:

(i) Obtain h(−n) by taking the mirror image of h(n) about the time origin.

(ii) Shift h(−n) by 0, 1, 2, 3...intervals, respectively. This would yield h(0−n), h(1−n), h(2−n)...etc.

(iii) For each shift, multiply the shifted sequence h(t−n) with the corresponding input sequence x(n).

(iv) For each shift, add the multiplied values to obtain convolved outputs y(t).

A simpler calculation may be done as follows: Suppose that x(0), x(1), x(2)...is one data sequence; and h(0), h(1), h(2)...is the second. Then, the convolved output may be written as

$$\begin{aligned}
y(0) &= x(0)h(0)\\
y(1) &= x(0)h(1) + x(1)h(0)\\
y(2) &= x(0)h(2) + x(1)h(1) + x(2)h(0)\\
y(3) &= x(0)h(3) + x(1)h(2) + x(2)h(1) + x(3)h(0)\\
&\cdots\cdots\cdots\cdots\cdots\cdots\cdots\cdots\cdots\cdots\\
y(n) &= x(0)h(n) + x(1)h(n-1) + \ldots\ldots\ldots + x(n)h(0)
\end{aligned} \tag{16.2}$$

The following rules on convolution are found useful in seismic data processing:

(i) The convolution operation is commutative:

$$x(t) * h(t) = h(t) * x(t) \tag{16.3}$$

(ii) If two data sets are convolved in the time domain, the effect in the frequency domain is to multiply their amplitude spectra, and add their phase spectra. This may be expressed as

$$x(t) * h(t) \leftrightarrow |X(f)| \cdot |H(f)| e^{i[\phi_x(f)+\phi_h(f)]}$$

where x(t) and h(t) are data sets in the time domain; |X(f)| and |H(f)| are amplitude spectra; and $\phi_x(f)$ and $\phi_h(f)$ are their phase spectra, respectively.

This property sometimes makes the processing of seismic data easier in the frequency domain, since convolution is replaced by the simpler algebraic operations of multiplication and addition.

(iii) Conversely, if x(t) and X(f), and h(t) and H(f) are Fourier transform pairs, then multiplication in the time domain corresponds to convolution in the frequency domain (refer, for example, to Fig. 3.12)

$$x(t)h(t) \leftrightarrow X(f) * H(f)$$

(iv) Superposition and convolution are equivalent operations (refer to Section 3.1).

Next, let us assume now that x(t) and h(t) represent two *continuous-time data sequences*. Then, convolution of x(t) and h(t) would be defined by the integral (replacing the summation for discrete-time sequences)

$$y(t) = x(t) * h(t) = \int_{-\infty}^{+\infty} x(\tau)h(t-\tau)d\tau \tag{16.4}$$

As an example, suppose that we wish to convolve

$$x(t) = e^{-t} \quad \text{with} \quad h(t) = ae^{-at}$$

each defined for $t \geq 0$.

By Eq. (16.4), the convolved output would be

$$y(t) = x(t) * h(t) = \int_{-\infty}^{+\infty} x(\tau)h(t-\tau)d\tau = \int_{0}^{t} e^{-\tau}ae^{-a(t-\tau)}d\tau \tag{16.5}$$

Integral in Eq. (16.5) may be easily evaluated to yield the convolved output $y(t)$.

Further, using Eq. (16.4) it is obvious that convolution of an arbitrary function $f(t)$ with delta function $\delta(t)$ would yield the same function $f(t)$. Mathematically

$$y(t) = \int_{-\infty}^{+\infty} f(\tau)\delta(t-\tau)d\tau = f(t) \tag{16.6}$$

We may summarize our brief discussion on convolution by mentioning that the seismic reflection record is a result of convolution of the appropriate seismic pulse-shape with the reflection coefficient log (Anstey 1970, p 52).

The inverse operation of convolution is *deconvolution. Deconvolution* is also referred to as *inverse filtering.*

To firm up our ideas on deconvolution, we may take it that $x(t)$ represents the digitized seismic pulse, and $h(t)$ the sequence of reflection coefficients. In this example, it is assumed that $y(t)$ – the digitized form of the reflection record and $x(t)$ – the seismic pulse are known. Our aim in deconvolution is to find $h(t)$, the reflection coefficient log. We may use Eq. (16.2) to find out these values, i.e.

$$\begin{aligned} h_0 &= \frac{y(0)}{x(0)}, \\ h_1 &= \{y(1) - x(1)h(0)\}/x(0) \\ h_2 &= \{y(2) - x(1)h(1) - x(2)h(0)\}/x(0) \\ &\ldots\ldots\ldots\ldots\ldots\ldots\ldots\ldots \text{etc.} \end{aligned} \tag{16.7}$$

In this example, we have taken an idealized situation to describe what is meant by deconvolution. However, the presence of noise in the seismic record compounded with inexact knowledge of the seismic pulse requires that sophisticated and suitable deconvolution processing be formulated for a meaningful interpretation of real seismic data. The purpose of deconvolution was explained in Sections

3.5.4 and 3.5.10. For a thorough understanding of deconvolution processing reference may be made to Robinson (1984).

In the context of convolution and deconvolution it is worthwhile to make mention of *homomorphic deconvolution*. Homomorphic deconvolution deals with the transformation from a space in which functions are convolved to a space in which they are added. The domain in which functions are added is called the *cepstrum domain*. Suppose s_t is the convolution of two functions w_t and r_t in the time domain, represented by

$$s_t = w_t * r_t ,$$

then its cepstrum domain representation is

$$\hat{s}(\xi) = \hat{w}(\xi) + \hat{r}(\xi)$$

Transformation from the time domain to the cepstrum domain involves three steps

$$s_t \leftrightarrow S(\omega) = |S(\omega)|e^{i\phi(\omega)}$$

$$\hat{S}(\omega) = \ln\{S(\omega)\} = \ln|S(\omega)| + i\phi(\omega)$$

$$\hat{S}(\omega) \leftrightarrow \hat{s}(\xi) = \frac{1}{2\pi}\int_{-\infty}^{+\infty} \{\ln|S(\omega)| + i\phi(\omega)\}e^{i\omega\xi}d\omega$$

The last step tells that cepstrum $\hat{s}(\xi)$ is obtained by taking the inverse transform of the log of the frequency spectrum.

If s_t is a discrete function, the corresponding steps through Z-transform are

$$s(t) \leftrightarrow S(Z)$$

$$\ln S(Z) = \hat{S}(Z)$$

$$\hat{S}(Z) \leftrightarrow \hat{s}(\xi)$$

It may be noted that convolution in the time domain corresponds to multiplication in the frequency domain, and to an even simpler process of addition in the cepstrum domain.

To return to the time domain from the cepstrum domain the steps are reversed:

$$\hat{s}(\xi) \leftrightarrow \hat{S}(\omega) = \int_{-\infty}^{+\infty} \hat{s}(\xi)e^{-i\omega\xi}d\xi$$

$$S(\omega) = \exp\hat{S}(\omega)$$

$$S(\omega) \leftrightarrow s_t = \frac{1}{2\pi} \int_{-\infty}^{+\infty} \hat{S}(\omega) e^{i\omega t} d\omega$$

Under certain circumstances, frequencies that overlap in the frequency domain are well separated in the cepstrum domain. Therefore, this allows more efficient filtering in the cepstrum domain compared to that in the frequency domain.

Homomorphic deconvolution has found application in extracting the reflectivity of the layered Earth. For the application of homomorphic deconvolution to exploration seismology, reference may be made to the original contribution of Ulrych (1971).

Exercise 16.1

(i) Assuming a linear system, and taking that the form of the impulse response I(t) for a linear system is as shown below in Fig. 16.1 a.

Depict graphically the output when input is kδ(t), and when the system is time-invariant.

Solution: By linearity of the system we have (Fig. 16.1 b)

Input: k δ(t) → Output: k I(t).

For a time-invariant system, we have a shifted version of k I(t) (Fig. 16.1 c).

(ii) Show schematically the convolution of continuous-time data sequences.

Refer to Fig. 16.2. Continuous data sets h(τ) and x(τ) are shown in the top. As given in Eq. (16.4), convolution requires h(t−τ). Function h(−τ) which is a mirror image of h(τ) about the line $\tau=0$ is shown on the left in the middle part. The shifted version of h(τ), i.e. h(t−τ) is shown on the right. Various positions of h(t−τ) are shown by the bottom left part of the figure. Also shown in the bottom left part of the figure is x(τ). The integral in Eq. (16.4) requires the product x(τ) h(t−τ) for each t. The product curve is shown by the bottom right of the figure for t = 1. To determine the convolved output for this time shift the product has to be integrated. Therefore, the area, shown by the slant lines, under the product curve is the convolved output y(1).

We have obtained a single value by this process. To prepare a graph of y(t) for all t, the variable t in h(t−τ) must be allowed to take all values from −∞ to +∞.

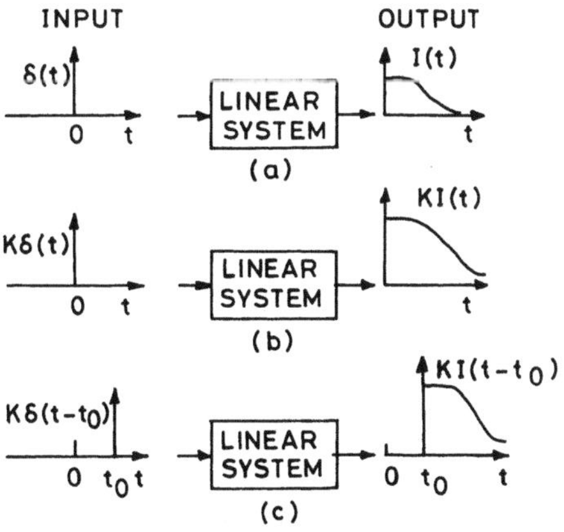

Fig. 16.1 a–c. a Given linear system and its impulse response; **b** output when input is k δ(t); **c** output for a time-invariant system

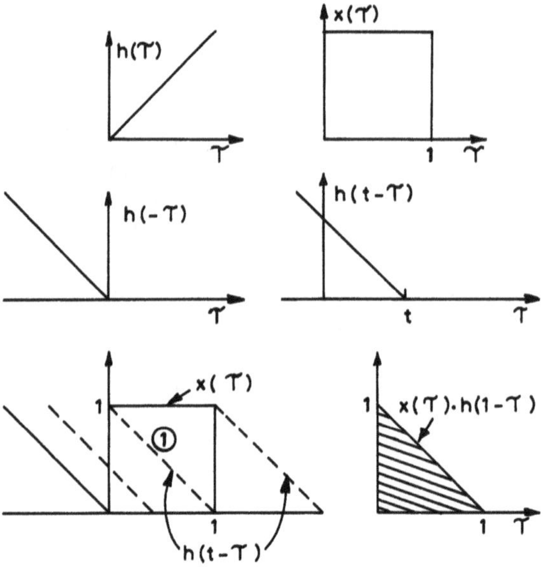

Fig. 16.2. Schematic illustration of convolution of continuous-time data sets

This would produce a graph of convolved output $y(\tau)$ as a result of convolution of x(t) with h(t).

16.2.2
Cross-correlation and Auto-correlation

The cross-correlation function determines the similarity between two data sets. Let us assume the two data sets are x_t and z_t. The cross-correlation function is expressed as

$$\phi_{xz}(\tau) = \sum_k x_{k+\tau} z_k \tag{16.8}$$

where τ stands for displacement of x_t relative to z_t. Conventionally, the shift is taken as positive if the first function x_t is moved to the left with respect to second function z_t.

The following properties of the cross-correlation function find useful applications:

(i)

$$\phi_{xz}(\tau) = \phi_{zx}(-\tau) \tag{16.9}$$

This means that we get the same cross-correlation value when x_t is shifted a given number of time units to the left as z_t is shifted by the same units to the right.

(ii)

$$\phi_{xz}(\tau) = \phi_{zx}(-\tau) = x_\tau * z_{-\tau} \tag{16.10}$$

Therefore, cross-correlation may be found by reversing the second data set and convolving.
(iii)

$$\begin{aligned}\phi_{xz}(\tau) &\leftrightarrow X(f)\overline{Z(f)} \\ &= |X(f)||Z(f)|e^{i(\phi_x-\phi_z)}\end{aligned} \tag{16.11}$$

where

$$x_t \leftrightarrow X(f) = |X(f)|e^{i\phi_x(f)}$$

$$z_t \leftrightarrow Z(f) = |Z(f)|e^{i\phi_z(f)}$$

$$z_{-t} \leftrightarrow \overline{Z(f)} = |Z(f)|e^{-i\phi_z(f)}$$

$$\phi_{xz}(\tau) \leftrightarrow X(f)\overline{Z(f)} = |X(f)||Z(f)|e^{i(\phi_x-\phi_z)}$$

The double arrow ($\leftrightarrow$) stands for Fourier transform pairs.
(iv) For continuous functions x(t) and z(t), cross correlation is defined in integral form as

$$\phi_{xz}(\tau) = \int_{-\infty}^{+\infty} x(t+\tau)z(t)dt \tag{16.12}$$

When the data set is being correlated with itself, the correlation is referred to as *auto-correlation*. This is expressed in the following form:

$$\phi_{zz}(\tau) = \sum_k z_{k+\tau}z_k \tag{16.13}$$

Auto correlation functions are symmetrical i.e. a right shift is equivalent to a left shift:

$$\phi_{zz}(\tau) = \phi_{zz}(-\tau) \tag{16.14}$$

Since the data would always be most similar to itself without a time-shift, the auto correlation has a maximum value for zero time shifts. If the auto correlation has a large value for some time shift Δt, not equal to zero, it is indicative of the fact that the data set is periodic with period Δt. Therefore, the auto correlation function may be used to measure the repetitiveness of the function.

For a continuous function z(t), the auto correlation function takes the following integral form:

$$\phi_{zz}(\tau) = \int_{-\infty}^{+\infty} z(t+\tau)z(t)dt \tag{16.15}$$

The value of the auto correlation function for zero time shift, $\phi_{zz}(0)$, is called the energy of the trace, expressed as

$$\phi_{zz}(0) = \sum_k x_k^2 \tag{16.16}$$

The normalized value of the auto correlation function for any time shift τ is obtained by dividing by the energy of the trace, i.e.

$$[\phi_{zz}(\tau)]_{normalized} = \frac{\phi_{zz}(\tau)}{\phi_{zz}(0)} \tag{16.17}$$

The normalized value of the cross-correlation function is obtained by dividing by the geometric mean of the energy of the two data sets:

$$[\phi_{xz}(\tau)]_{normalized} = \frac{\phi_{xz}(\tau)}{[\phi_{xx}(0)\phi_{zz}(0)]^{\frac{1}{2}}} \tag{16.18}$$

Normalized correlation values may vary between +1 and −1.

Correlation analysis is a very powerful tool for applying accurate corrections to seismic reflection data. For example, cross-correlation helps to determine the amount of time shift required for optimum alignment of two seismic traces, and the determination of residual static correction (refer to Subsection 3.5.5). It is also used to determine the amount of normal moveout required to align traces from different offsets before stacking. For details on the use of the cross-correlation technique for seismic corrections, contributions by Hileman et al. (1968) and Disher and Naquin (1970) may be referred.

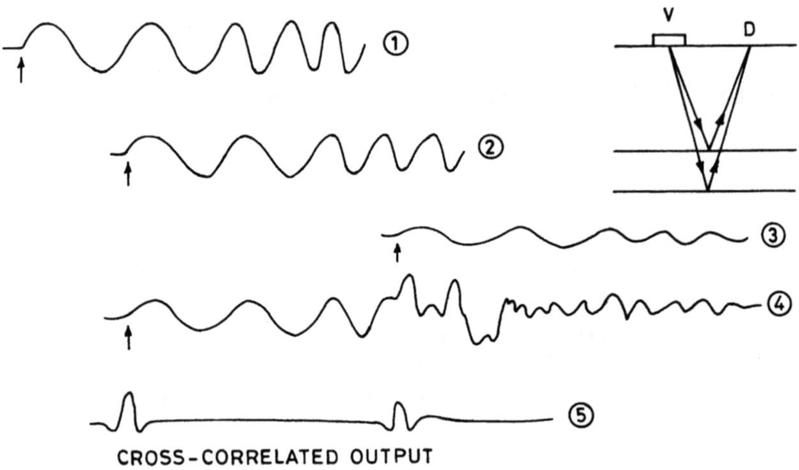

Fig. 16.3. Schematic illustration depicting: *1* Vibroseis upsweep signal; *2* reflection from first reflector; *3* reflection from second reflector; *4* vibroseis record; *5* cross-correlated output as a result of cross-correlations of vibroseis source signal *1* with vibroseis record *4*

The correlation technique is used to process the vibroseis records. For a vibroseis survey, a long duration oscillatory sweep signal (refer to Subsection 2.1.1.1) is sent into the Earth, which after reflections from reflecting boundaries are recorded on the ground surface. Under common field situations, the reflected sweep signals overlap each other producing a complex record which would not permit the picking up of reflected events in the form of sharp wavelets as is possible with impulsive sources.

Correlation processing of the source sweep signal with the vibroseis record brings out the reflections in the form of wavelets for time-shifts equal to two-way travel times to reflectors. This is illustrated in Fig. 16.3. This happens because when the time shift is equal to the two-way reflection time, the source sweep signal and reflected sweep signal are similar.

Exercise 16.2

Determine the cross-correlation values for both positive and negative time shifts for the data sets x_t: (2, 5, −2, 1) and $z_t = (6, -1, -1)$.

Solution: The first data x_t is shifted one time unit each time, starting from zero shift. The product, shown by the cross (x), of corresponding entries in x_t and z_t is obtained and products summed to obtain the value for a particular time-shift. The procedure is repeated both for left shifts (positive) and right shifts (negative) of x. This is illustrated below:

Time shifts	Data sets	Cross-correlation values
0	2, 5, −2, 1 × × × 6, −1, −1	$\phi_{xz}(0) = 12-5+2+0 = 9$
+1	2, 5, −2, 1 × × × 6, −1, −1	$\phi_{xz}(+1) = 0+30+2-1 = 31$
+2	2, 5, −2, 1 × × 6, −1, −1	$\phi_{xz}(+2) = 0+0-12-1+0 = -13$
+3	2, 5, −2, 1 × 6, −1, −1	$\phi_{xz}(+3) = 0+0+0+6+0+0 = 6$
+4	2, 5, −2, 1 6, −1, −1	$\phi_{xz}(+4) = 0$
-1	2, 5, −2, 1 × × 6, −1, −1	$\phi_{xz}(-1) = 0-2-5+0 = -7$
-2	2, 5, −2, 1 × 6, −1, −1	$\phi_{xz}(-2) = 0+0-2+0+0+0 = -2$
-3	2, 5, −2, 1 6, −1, −1	$\phi_{xz}(-3) = 0$

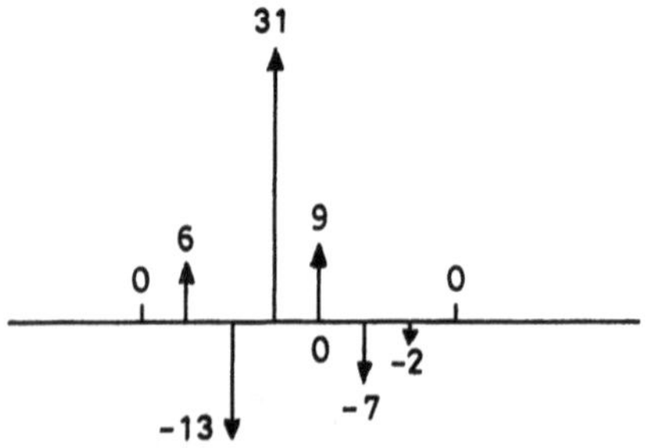

Fig. 16.4. Cross-correlation $\phi_{xz}(\tau)$

Cross correlation values for time shifts are shown in Fig. 16.4. Since $\phi_{xz}(+1)$ has the largest value in comparison to data sets x_t and z_t, this means that the two data sets are most similar to each other for time-shift of +1.

Many times, it is required in seismic data processing to assess the degree of similarity between two traces, and the amount of time shift required for making them match better. For this purpose, cross-correlation is the most commonly used technique.

Normalized cross-correlation values are independent of magnitudes of entries in the data sets and vary between +1 and −1. Normalized cross-correlation value of +1 means traces (or data sets) are identical. Similarly the −1 value means that traces are identical except that the polarity is reversed. A value near zero means little similarity between the data sets.

16.2.3
Digital Filtering

Here we restrict our brief presentation to explain the general principle of digital filtering. Specific purpose filter design is a specialized topic, and we do not intend to go into its details. Readers interested to learn techniques of designing digital filters may refer, for example, to Shanks (1967), Treitel and Robinson (1969), Kulhanek (1976), Sheriff and Geldart (1982), and Waters (1987).

Let us introduce filter function or a filter operator. The filter operator has in it a set of numbers, which are referred to as *filter coefficients.* These coefficients may be viewed as *weighting factors.*

Let the filter function consists of coefficients $f_1, f_2, f_3 \ldots f_m$. Assume further that a digital seismogram is described by the digits $a_1, a_2, a_3 \ldots a_n$. Let the filter function operate on the digital seismogram. Figure 16.5 depicts it. The sequential outputs of the filtering action would yield $C_1, C_2, C_3 \ldots C_i$ where

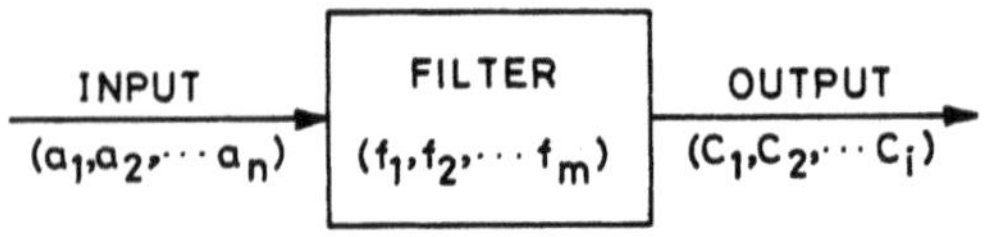

Fig. 16.5. Input (digital seismogram), digital filter, and the output

$$\begin{aligned}
C_1 &= \{f_1a_1 + f_2a_2 + f_3a_3 + \ldots\ldots + f_m a_m\} \\
C_2 &= \{f_1a_2 + f_2a_3 + f_3a_4 + \ldots\ldots + f_m a_{m+1}\} \\
C_3 &= \{f_1a_3 + f_2a_4 + f_3a_5 + \ldots\ldots + f_m a_{m+2}\} \\
C_j &= \left\{f_1a_j + f_2a_{j+1} + f_3a_{j+2} + \ldots\ldots + f_m a_{j+m-1}\right\}
\end{aligned} \tag{16.19}$$

It is clear that a different choice of filter coefficients would produce different outputs. In the following exercise, we take an example of a seven coefficient filter that produces seven point averaging of adjacent values sequentially, i.e. from 1 to 7, 2 to 8, 3 to 9...etc.

Exercise 16.3

Assume a seismogram (Fig. 16.6) which has been digitized at equal spacing at points 1, 2, 3, and 4....25. The digital values of the amplitudes are: (0, 0.5, 0.6, 0.9, 1.1, 1.05, 1, 0.95, 0.9, 0.6, 0, –0.8, –0.7, –0.75, –0.8, –.75, –0.7, –0.65, –0.6, –0.3, +0.3, +0.5, +0.55, 0, –0.5)

A seven-point filter operator is applied on this data set, each filter coefficient is 1/7.

Find out the output, and explain the action of this filter.

Solution: Refer to Fig. 16.6. Using Eq. (16.19), the output is: (0.73, 0.87, 0.92, 0.92, 0.8, 0.52, 0.27, 0.02, –0.22, –0.45, –0.64, –0.735, –0.70, –0.68, –0.62, –0.46, –0.65, –0.5, –0.31, –0.2, –0.028, –0.007).

Both input and output graphs are shown in Fig. 16.6. The output graph shows only a (smooth) low frequency waveform. The high frequency oscillations present in the input waveform have been trapped by the filtering action. Thus, the given filter has worked *as low pass filter.* This happens because, in the seven point averaging process, positive and negative amplitude values of high frequency components cancel each other and the result is a smooth output consisting only of the low frequency component.

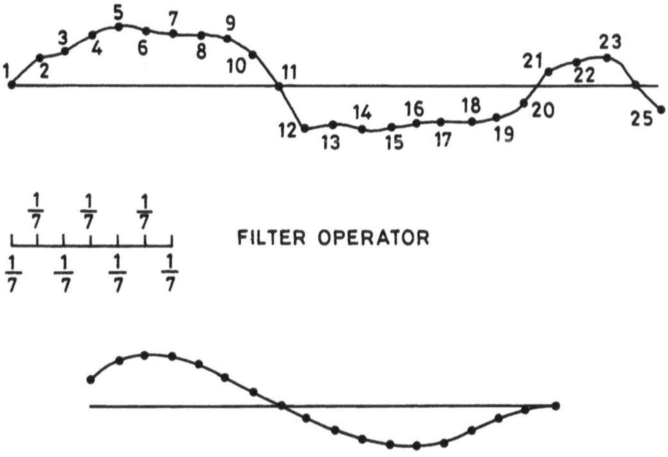

Fig. 16.6. An example of digital filtering in the form of a running average

16.2.4 Z-Transform

Z- transform is a very useful tool for computations involving sampled functions. Suppose a continuous function f(t) is sampled with sampling interval Δ, and the sampled value is f_t: {0, 3, -3/2, 5, 7}. Then its z-transform description would be:

$$F(Z) = 0Z^0 + 3Z^1 - 3/2Z^2 + 5Z^3 + 7Z^4 \tag{16.20}$$

It would be observed that the Z-transform of a sampled function is a polynomial in Z. The coefficient of each term in the polynomial is the sample value of the function, and the exponent of Z is the time of that sample. In the above example, the first sample (zero) is obtained at time t = 0, and subsequent samples at one, two and three times units.....etc., later in the polynomial. Thus, -3/2 is the value of the second sample, and power 2 of Z represents the 2-unit time delay of this sample. As positive powers of Z denote values after zero time, negative powers of Z stand for values in the past. In a general case, therefore, both positive and negative powers of Z are assigned. For example, for the sample function

$$g_t = \left[-1, 0, 3, \overset{\downarrow}{5}, -3, 1, 7\right]$$

in which the superscribed arrow denotes zero time, the Z-transfer description would be:

$$G(Z) = -1Z^{-3} + 0Z^{-2} + 3Z^{-1} + 5Z^0 - 3Z^1 + 1Z^2 + 7Z^3 \tag{16.21}$$

g_t and G(Z) form a transform pair and are written as

$$g_t \leftrightarrow G(Z)$$

It is to be understood that Z is a *unit delay operator.*

This delay operator is very useful for analyzing waves for the reason that waves take a certain amount of time to move from one location to the other.

The Z-transform plays a significant role in the theory of *linear systems.* Some useful deductions in terms of the Z-transform are:

1. If I(Z) is the impulse response of a linear system, then for an input X(Z) the output Y(Z) would be

$$Y(Z) = X(Z)I(Z)$$

2. Since the product of polynomials is commutative, the filtering system depicted in Fig. 16.7 is equivalent. This may be proved using the Z-transform. Thus we have

$$Y_1(Z) = [A(Z)B(Z)]C(Z) = ABC$$

and

Fig. 16.7. Linear systems with characteristics B(Z) and C(Z), and two equivalent filtering systems (*top* and *bottom*)

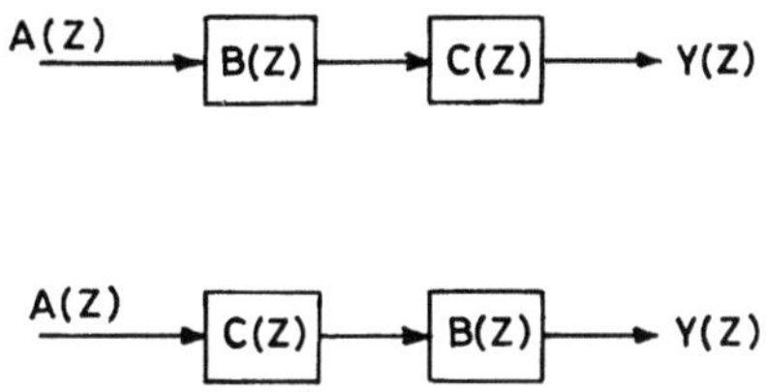

$$Y_2(Z) = [A(Z)C(Z)]B(Z) = ACB = ABC$$

Therefore, $Y_1(Z) = Y_2(Z) = Y(Z)$ and two filtering systems are equivalent.

3. The product of two polynomials in Z is another polynomial, whose coefficients are the result of convolving two corresponding functions in time. Therefore, using Z transform representation of sampled functions, the result of convolution of the two functions may be obtained through multiplication of two polynomials. As convolution may be performed by multiplication in the Z-transform domain, deconvolution may be performed by division. Further, since it is possible to express cross-correlation as convolution [refer to (Eq. 16.10)], the cross-correlation may be easily evaluated using Z-transform. Additionally, as there is an intimate relationship between the Fourier transform and Z-transform of a sampled function, it is possible to write values of amplitude and phase spectra.
The following exercise illustrates simple applications of Z-transform.

Exercise 16.4

(i) Suppose that the digitized impulse response of a linear system is i_t: (5, –3, 2, 1), and the digitized form of the pulse is p_t: (3, 1, –1). The sampling interval is the same both for i_t and p_t. Obtain the filtered output for the linear system with the given impulse response.

Solution: The filtered output in the time domain would be obtained by convolution of p_t with i_t. Using Z-transform descriptions, the result of convolution may be obtained by simple multiplication. For example,

$$I(Z) = 5Z^0 - 3Z^1 + 2Z^2 + 1Z^3$$

$$P(Z) = 3Z^0 + 1Z^1 - 1Z^2$$

Let us now perform the multiplication:

$$I(Z) = 5 - 3Z + 2Z^2 + Z^3$$

$$P(Z) = 3 + Z - Z^2$$

$$\begin{array}{llllll}
3 \times I(Z) \rightarrow & 15 - 9Z + 6Z^2 + 3Z^3 & & & & \\
Z \times I(Z) \rightarrow & \quad\; 5Z - 3Z^2 + 2Z^3 + Z^4 & & & & \\
-Z^2 \times I(Z) \rightarrow & \quad\quad\; - 5Z^2 + 3Z^3 - 2Z^4 - Z^5 & & & & \\
\hline
 & 15 - 4Z - 2Z^2 + 8Z^3 - Z^4 - Z^5 & & & &
\end{array}$$

The filtered output would be (15, −4, −2, 8, −1, −1), since coefficients in the product polynomial represent the output values at the same sampling rate as for i_t and p_t.

It is to be understood that the scale and shift operation involved in the convolution process in the time domain are implicit in the Z-transform multiplication. Convolution in the time domain is equivalent to multiplication in the Z-transform domain or in the frequency domain. The relationship between the Z-transform and Fourier transform is explained in part (iv).

(ii) Assuming that a water layer reverberation filter is given by f_t: $(1, -2r, 3r^2, -4r^3, 5r^4, \ldots\ldots)$ meaning that if the input is a spike of unit amplitude, the returned pulse train is $(1, -2r, 3r^2, -4r^3, 5r^4, \ldots\ldots)$ as a result of reflection from a deep perfect reflector and water layer reverberations. A deconvolution filter or an inverse filter is to be designed which would remove the effect of reverberations and produce an output of 1 as a result of reflection from the deep perfect reflector. 'r' is the reflection coefficient from the bottom of the water layer.

Solution: If i_t represents the inverse filter in the time domain, its action on the filtered record should produce a spike. Mathematically

$$i_t * f_t = 1$$

In Z-transform domain,

$$I(Z) \cdot \left(1 - 2rZ + 3r^2Z^2 - 4r^3Z^3 + 5r^4Z^4 \ldots\right) = 1$$

or

$$I(Z) = \frac{1}{\left(1 - 2rZ + 3r^2Z^2 - 4r^3Z^3 + 5r^4Z^4 \ldots\right)}$$

By division it may be verified that

$$I(Z) = \left(1 + 2rZ + r^2Z^3\right)$$

Therefore, the deconvolution filter or the inverse filter is $(1, 2r, r^2)$.

(iii) Cross-correlate (1, −2, ½) with (1, −1, −1/2)

Solution: We have the relation (16.10) for cross-correlation, i.e.

$$\phi_{xy}(\tau) = x_\tau * y_{-\tau}$$

Writing

$$x_\tau \leftrightarrow \left(1 - 2Z + \frac{1}{2}Z^2\right)$$

$$y_{-\tau} \leftrightarrow \left(-\frac{1}{2}Z^{-2} - Z^{-1} + 1\right)$$

we have

$$x_\tau * y_{-\tau} \leftrightarrow \begin{pmatrix} -\frac{1}{2}z^{-2} - 0z^{-1} + \\ \frac{11}{4}z^0 - \frac{5}{2}z + \frac{1}{2}z^2 \end{pmatrix}$$

Therefore,

$$\phi_{xy}(\tau) = \left(-\frac{1}{2}, 0, \frac{11}{4}, -\frac{5}{2}, \frac{1}{2}\right)$$

The result may be verified for positive and negative shifts for different τ using Eq. (16.8) for cross-correlation of two data sets.
(iv) Determine the frequency and phase characteristic of the digital function a_t given by

$$a_t = \left\{0, 0, 0, -1, 0, \overset{\downarrow}{1}, +1, 0, 0, 0\right\}$$

Solution: A sampled function a_t may be treated as products of continuous function a(t) by a comb function with spacing Δ

$$a_t = \sum_{n=-\infty}^{n=+\infty} a(n\Delta)\delta(t - n\Delta) \tag{16.22}$$

Since $\delta(t \pm t_0) \leftrightarrow e^{\pm j\omega t_0}$, we have

$$a_t \leftrightarrow A(\omega) = \sum_{n=-\infty}^{+\infty} a(n\Delta)e^{-jn\omega\Delta} \tag{16.23}$$

Writing $z = e^{-j\omega\Delta}$, we have

$$A(\omega) = \sum_{n=-\infty}^{n=+\infty} a(n\Delta)Z^{-n} = A(Z) \tag{16.24}$$

Thus, we have $a_t \leftrightarrow A(Z)$
Now the Z-transform of a_t is

$$\begin{aligned} A(Z) &= \left\{0, 0, 0, -1, 0, \overset{\downarrow}{1}, +1, 0, 0, 0\right\} \\ &= -1Z^{-2} + 0Z^{-1} + 1Z^0 + 1Z^1 \\ &= -Z^{-2} + 1 + Z \\ &= \left(e^{-j\omega\Delta}\right)^2 + 1 + \left(e^{-j\omega\Delta}\right) \\ &= e^{-2j\omega\Delta} + 1 + e^{-j\omega\Delta} \end{aligned}$$

Table 16.1. Computation of amplitude and phase spectra of a sampled function

$\omega\Delta$	A(Z)	Amplitude	Phase
0°	3	3	0.0
45°	1.707–1.707 i	2.414	–45°
90°	–i	1	-90°
135°	0.293 + 0.293 i	0.414	45°
180°	1	1	0°
225°	0.293 – 0.293 i	0.41	–45°
270°	i	1	90°
315°	1.707 + 1.707 i	2.414	45°

By substituting values of ω we may write A(Z). A(Z) would in general be of complex form, from which it is possible to calculate the amplitude and phase spectra.

Values of A(Z), the amplitude and phase spectra for various values of $\omega\Delta$ are tabulated in Table 16.1.

16.2.5 Fourier Analysis and Synthesis

An analog form reflection seismogram is a continuous graph of some variable (particle velocity or pressure) as a function of record time. A digital seismogram consists of a sequence of numbers representing the amplitude values at equal intervals of time – called the *sampling interval.* A continuous graph or sampled forms of the data are the *time-domain* presentation of the variable of the record. Records in analog form help in the picking up of reflection events, and their arrival times. From the point of view of fine processing, time-domain form data may not always be suitable. For such processing, there is an alternate and equivalent way of visualizing the given seismic waveform as a combination of sinusoidal waveforms of different amplitudes and frequencies. The presentation in this form is referred to as the *frequency-domain* description of the given waveform.

The sinusoidal break up of the given waveform is based on the *Fourier theorem* which states: 'a single value time function (with only a finite number of discontinuities) within a time interval T may be obtained by combining a constant with a series of sinusoidal signals which have frequencies equal to f_0, $2f_0$, $3f_0$..., and N f_0. Frequency f_0 is called the *fundamental frequency,* and its value, i.e. $f_0 = 1/T$, is uniquely determined by the reciprocal of time interval T for which the equivalence between the time function and combined sinusoidal components is being studied. Frequencies $2f_0$, $3f_0$..., and N f_0 are called the *harmonics.* It is to be understood that the match between the given time function, and approximations obtained by the addition of sinusoidal waveforms increases as N increases. It is clear that the fundamental frequency, all harmonic frequencies and the interval between consecutive frequencies (i.e. between f_0 and $2f_0$, $2f_0$ and $3f_0$,... (N–1) f_0 and N f_0) is determined by the choice of time interval T.

Fourier synthesis is the reverse process of *Fourier analysis*. In Fourier synthesis, a number of constituent sinusoidal waveforms must be combined together. A mathematical statement of Fourier synthesis is

$$a_0 + \sum_{1}^{N} a_j \cos 2\pi j f_0 t + \sum_{1}^{N} b_j \sin 2\pi j f_0 t = s(t) \tag{16.25}$$

[Waters 1987, Eq. (4.1)]
where a_0, a_1, $a_2 \ldots a_N$ and b_1, $b_2 \ldots b_N$ and the fundamental frequency (f_0) are assumed known.

For the Fourier analysis purpose, if we know the time-function f(t) and $f_0(=1/T)$, it is possible to find the values of a_0, a_1, $a_2 \ldots a_N$ and b_1, $b_2 \ldots b_N$, using the relations:

$$\begin{aligned} a_0 &= \frac{1}{T}\int_0^T f(t)dt \\ a_j &= \frac{1}{T}\int_0^T f(t)\cos 2\pi j f_0 t dt \\ b_j &= \frac{1}{T}\int_0^T f(t)\sin 2\pi j f_0 t dt \end{aligned} \tag{16.26}$$

[Waters 1987, Eq. (4.2)]

Fourier analysis or synthesis really describes a repetitive waveform that repeats itself for every time interval T.

Assume that the time-function f(t) is a continuous graph and there is no exact mathematical form for f(t). In this case the evaluation of integral expressions in Eq. (16.26) is not possible. However, if f(t) is in the form of a digital seismogram, the values of a_0, a_1, $a_2 \ldots a_N$ and b_1, $b_2 \ldots b_N$ may be determined, but they need to be redefined as a summation as

$$\begin{aligned} a_j &= \frac{1}{N}\sum_{k=0}^{2N} f_k \cos\frac{2\pi jk}{N} \\ b_j &= \frac{1}{N}\sum_{k=0}^{2N} f_k \sin\frac{2\pi jk}{N} \end{aligned} \tag{16.27}$$

[Waters 1987, Eq. (4.7)]
valid for all j except for 0 and N; and for these values we have

$$a_0 = \frac{1}{2N}\sum_{k=0}^{2N} f_k$$

$$a_N = \frac{1}{2N}\sum_{k=0}^{2N} f_k \cos \pi k$$
$$b_N = \frac{1}{2N}\sum_{k=0}^{2N} f_k \sin \pi k \qquad (16.28)$$

[Waters 1987, Eq. (4.8)]

These coefficients make use of (2N+1) sampled amplitude values and 2N intervals of data points of the waveform.

As mentioned earlier, a tacit assumption for Fourier series representation of a time function is that the function is repetitive and its fundamental period is equal to the time interval used for analysis. The fundamental period may be made theoretically as large as we desire. Or, alternatively the fundamental frequency may be made infinitesimally small by choosing an infinitely large time interval. This would imply that, by choosing an infinitely small fundamental frequency, the waveform could repeat after infinite time – a concept inherent in a *transient waveform*. Thus, a *transient waveform*, which occurs only once, is a result of combining sinusoidal waveforms with an infinitesimally small fundamental frequency and harmonics spaced by this infinitesimally small frequency. This consideration leads to the concept of the Fourier integral as a generalization of the Fourier series. This is expressed as

$$\frac{1}{\sqrt{2\pi}}\int_{-\infty}^{+\infty} F(\omega)e^{i\omega t}d\omega = f(t) \qquad (16.29\,a)$$

This equation may be thought of as expressing the Fourier synthesis of continuous frequencies ω. The inverse form of Eq. (16.29a) is given as

$$F(\omega)\frac{1}{\sqrt{2\pi}}\int_{-\infty}^{+\infty} f(t)e^{-i\omega t}dt \qquad (16.29\,b)$$

f(t) and F(ω) are called Fourier-transform pairs. These are a piecewise continuous function of time and frequency. The transform pairs (f(t) and F(ω)) are symbolically represented by a double arrow

$$f(t) \leftrightarrow F(\omega)$$

Fourier series and Fourier integral representations are an expression of the fact that either domain presentation of a given waveform is equivalent.

In general, the time domain analog form presentation is useful for visual inspection and some basic analysis of the reflection record. On the other hand, the frequency-domain specifications are adopted for illustrating seismic instruments (like the geophone, hydrophone, amplifiers, filters, etc.) response, and for detailed seismic data processing.

We are well familiar with the wiggle form of the reflection record as a function of record time, treating shot time as zero time. However, up to this point of our discussion we have not introduced the method of presentation of frequency domain data (except for a brief mention through Fig. 3.6 of line spectra).

If we look to the statement in Fourier series, it is obvious that we should graph amplitudes for different frequencies, and also show their phase relationships. These are conventionally referred to as *amplitude* and *phase spectra*. The time domain presentation is a single graph f(t) whereas the frequency domain presentation is a pair of graphs of amplitude and phase spectra.

The question arises that if we combine the sinusoids with different frequencies, whether the combined waveform would also be repetitive. The answer is 'yes'. In fact, the frequency of repetition of the combined waveform is the highest common factor of the frequencies of individual sinusoids. Suppose that we are combining sinusoids with frequencies 2, 4, and 10 Hz. Since the highest common factor of these frequencies is 2, the combined waveform would have a frequency of 2, i.e. it would repeat two times in one second. Likewise a combination of frequencies 1, 3, 4, and 7 would produce a waveform that would repeat only once a second, as the highest common factor of these frequencies is 1.

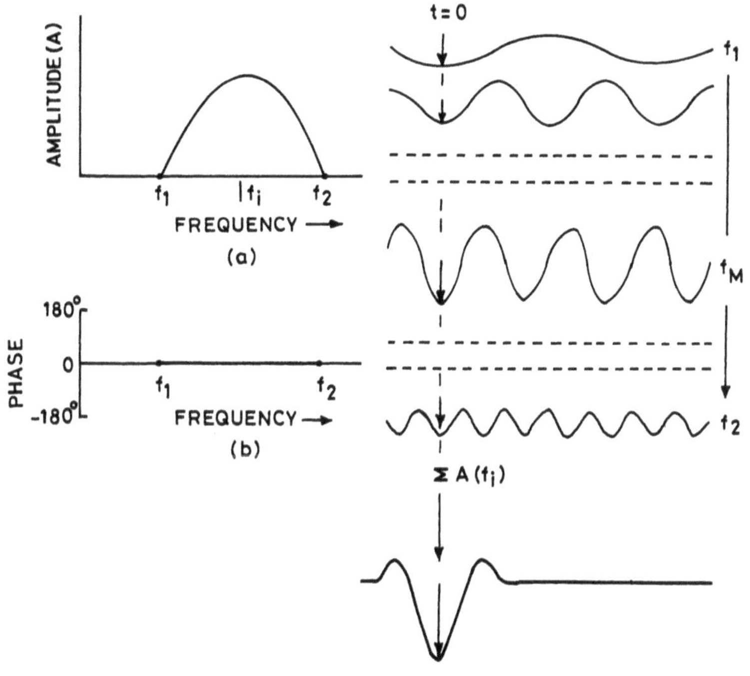

Fig. 16.8 a–c. a Amplitude spectra; **b** zero-phase spectra; **c** individual cosine waveforms in the time domain, and their combined waveform resulting from the synthesis of individual waveforms (after Anstey 1970)

The above paragraph tells us about the relationship between the repetition frequency of a combined waveform with those of individual sinusoidal waveforms which have been combined. The geometrical shape of a combined waveform would depend on the *phase relationships* of the individual waveforms.

First, we consider cosine wave oscillations with zero phase for each frequency. Also, we assume that the frequencies are so close to each other that they represent a continuous coverage. Further, the amplitude of these oscillations first increases and then decreases smoothly from a frequency f_1 to f_2; and frequencies less than f_1 or greater than f_2 have zero amplitudes. The diagrammatic representations of amplitude and zero phase spectra are shown in Fig. 16.8 a, b.

Figure 16.8 c shows individual cosine waveforms in the time domain and also their combined waveform resulting from the synthesis of individual waveforms. We observe that the resulting waveform (or pulse) is *symmetrical*, and its peak amplitude is larger than the amplitude of any of the individual waveforms. Since all cosine waveforms are in phase at $t=0$, the resulting pulse amplitude is simply the summed up amplitude of individual waveforms, and is positioned at $t=0$. This pulse has the characteristic of being *symmetrical*, with a peak amplitude at $t=0$.

To understand what would happen if individual waveforms are not of zero phase, we return back to the sinusoidal waveform, i.e. sin ωt where $\omega=2\pi f$. Consider time-domain depictions in Fig. 16.9 of 'a sin ωt', 'a sin $(\omega t+\theta)$' and 'a sin $(\omega t-\theta)$'. The first sinusoid is of zero phase angle, the second of phase angle $+\theta$, and the third of $-\theta$; amplitude is 'a' for both sinusoids. Depending on whether 'θ' is positive or negative, the waveform would be shifted to the left (a lead) or to the right (a lag). With this convention, a +90° phase shift would produce a cosine wave from an initial sine wave with zero phase, with maximum amplitude at $t=0$. The question is what is the amount of shift on the time axis for a

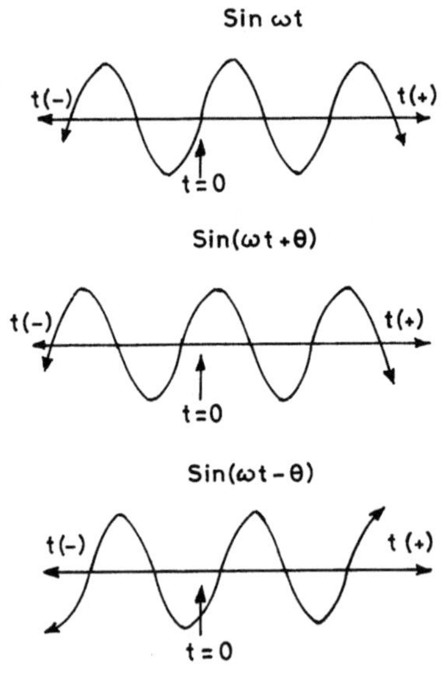

Fig. 16.9. Sinusoidal waveform with zero phase (*top*), and phase angles $+\theta$ and $-\theta$ (*middle and bottom*)

phase shift of θ. Since a phase shift of 2π corresponds to time period T of the waveform, a phase shift of θ would produce a shift equal to θ T/2π or $\theta/2\pi f$. With this argument, it may easily be conjectured that individual cosine waveforms with different phases (other than a zero phase) should produce a resultant (combined) pulse of different characteristics compared to the zero phase case of Fig. 16.8.

Let us analyze the situation more specifically. Two types of phase characteristics for frequencies f_1 to f_2 of Fig. 16.8 are immediately relevant. First, when the *phase spectrum* is *linear* (Fig. 16.10b) and second when it is *curved* (Fig. 16.10c). Without going into details, the net effect may be summarized (for more detailed discussion refer to Anstey 1970, pp 27–31).

A rectilinear phase spectrum, intersecting the phase axis at 0 or at an integral multiple of 2π (Fig. 16.10b), produces an unchanged resultant pulse shape (i.e. the pulse shape is the same as that given by a zero phase) but produces a delay on the pulse proportional to the slope (θ/f) of the phase spectrum line (Fig. 16.10b′ upper).

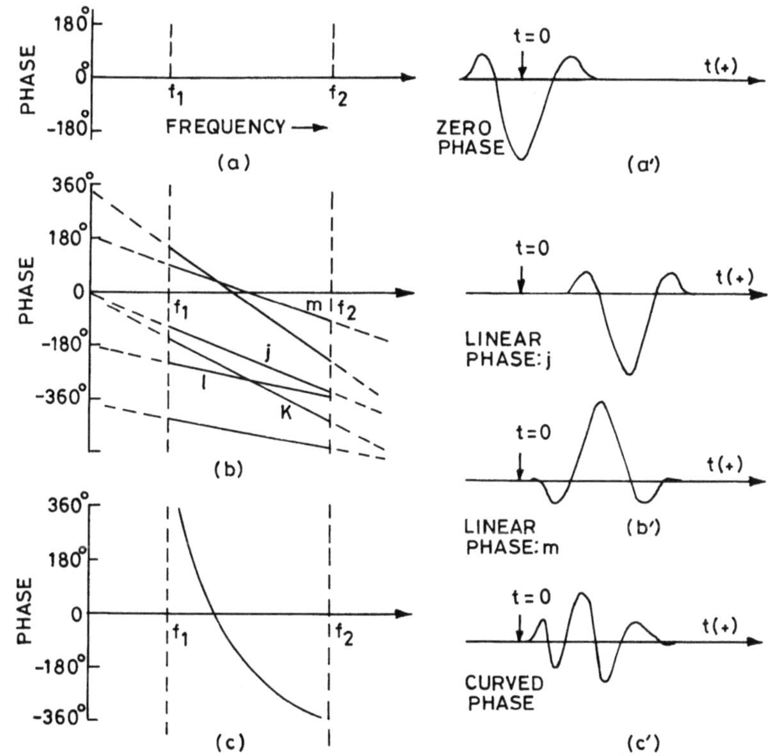

Fig. 16.10a–c. a′–c′ (**a, a′**) Zero phase and corresponding symmetrical pulse form. (**b, b′**) Linear phase spectra passing through 0°, ±180° and ±360° on phase axis and corresponding symmetrical pulse forms. (**c, c′**) Curved phase spectrum and resulting asymmetrical pulse form

This happens because each individual frequency component is delayed by equal amount (decided by the slope θ/f of the phase line). Therefore, the resultant pulse form is unchanged in shape but is delayed in time (Fig. 16.10b′ upper).

Extending the above argument for a linear phase spectrum, passing through 180°, it is clear that this would have the effect of delaying and inverting all frequency components. As a consequence of this, the resultant waveform is both delayed and inverted without change of shape (Fig. 16.10b′ lower).

If the frequency spectrum is curved in the frequency range f_1 to f_2, it is possible to construct individual waveforms for different frequencies between f_1 and f_2 (Fig. 16.10c), with appropriate time shifts for each of them. As the phase spectrum is curved, different time shifts would be there for different frequencies. This would give rise to a combined resultant waveform with the characteristics of reduced peak amplitude, asymmetry and occupation of greater time by the pulse (Fig. 16.10c′).

The above description exemplifies that if the amplitude spectrum and phase spectrum for a range of frequencies is known, it is possible to create an equivalent time domain representation of the waveform. Further, for a common amplitude spectrum, and different phase spectra, the synthesis of individual waveforms would, in general, produce different pulse forms.

16.2.5.1 Phase Characteristics and Wavelets

In the above discussion, we have introduced zero-phase, linear-phase and curved-phase characteristics and explained how these affect the resultant pulse on Fourier synthesis of individual sinusoids. In the following, we would analyze this topic in more detail.

We initiate with a digitized waveform or a pulse. Let its sampled values be:

$$(a_0, a_1, a_2, \dots\dots\dots a_i, \dots\dots\dots a_{N-1}, a_N)$$

Its z-transform representation is

$$\begin{pmatrix} a_0 + a_1 z + a_2 z^2 + \dots\dots a_i z^i \\ + \dots\dots a_{N-1} z^{N-1}, a_N z^N \end{pmatrix}$$

This polynomial in z may be written as a product of its factors:

$$a_0(z - p_1)(z - p_2)\dots\dots(z - p_n)$$

where p_1, p_2, $p_3 \dots p_N$ are *roots* of the polynomial, and these may be *complex*.

When a polynomial in z is expressed in terms of products of its factor, the analysis of amplitude and phases of each of the product factors may be done and the results generalized for the overall characteristic of the polynomial. We first explore what is meant by a *minimum phase wavelet*.

A minimum-delay function is defined as a physically realizable function whose transform has an inverse, that is $a_t \leftrightarrow A(z)$ and $1/A(z)$ is finite. Then, A(z) is said to be a *minimum phase*.

Let us take an example of a simple 2-term wavelet:

$$a_t = (b, -c)$$

It has as its z-transform and inverse given by

$$A(z) = (b - cz),$$
$$\frac{1}{A(z)} = \frac{1}{(b - cz)} = (b - cz)^{-1} = \frac{1}{b}\left\{1 + \frac{c}{b}z + \left(\frac{c}{b}z\right)^2 + \cdots\right\} \tag{16.30}$$

= a convergent series for $c/b < 1$, and a divergent series for $c/b > 1$.

Thus, using the criterion for minimum phase, we should have $|b| > |c|$. By the same argument the wavelet (c, b) is *maximum phase.*

Let us now analyze *minimum phase* and *maximum phase wavelets* diagrammatically. Consider a minimum phase wavelet $P(z) \equiv (5 - z) = 5(1 - z/5)$. Replacing z by $e^{-i\omega\Delta}$, we have

$$P(z) = 5\left\{1 - \frac{1}{5}\cos\omega\Delta + i\left(\frac{1}{5}\right)\sin\omega\Delta\right\} \tag{16.31}$$

Its phase is given by

$$\theta = \tan^{-1}\{\sin\omega\Delta/(5 - \cos\omega\Delta)\}$$

Since the denominator is never zero as the first term (i.e. 5) is greater than the second term (i.e. $\cos(\omega\Delta)$ which may vary between +1 and −1). This means that $|\theta| < \frac{1}{2}\pi$. Further, if we plot Im P(z) vs. Re P(z), it would not enclose the origin.

Thus, for a minimum phase we have a criterion that phase is never more than $\frac{1}{2}\pi$. Also, since *phase* is *periodic,* every time $\omega\Delta$ increases by 2π, phase values repeat producing a variation as shown in Fig. 16.11 a.

Now let us analyze the *phase curve* for a *maximum phase wavelet.* For a maximum phase wavelet, the first term, which is greater than 1, i.e. 5 for the example considered for minimum phase, is less than 1. In this case as $\omega\Delta$ increases by 2π, θ also increases by 2π.

Conclusively, while the *phase* of the *minimum phase wavelet* varies between $-\frac{1}{2}\pi$ and $+\frac{1}{2}\pi$, that of the *maximum phase wavelet* increases without limit (Fig. 16.11 a).

For a minimum phase system, it is possible to compute phase-frequency response uniquely from the amplitude-frequency response.

We have explained through Fig. 16.10 the difference (in the form of delay) between a zero phase wavelet and a linear phase wavelet. It may be guessed intuitively that a zero phase wavelet (symmetrical about time origin, t = 0) can not be the impulse response of a natural system as this would mean that the system gives an output at negative times, or an output before an input is received by it. In the case of a minimum phase, phase shifts are such that no signal can appear before zero time. For this reason, minimum phase wavelets are realistic for a nat-

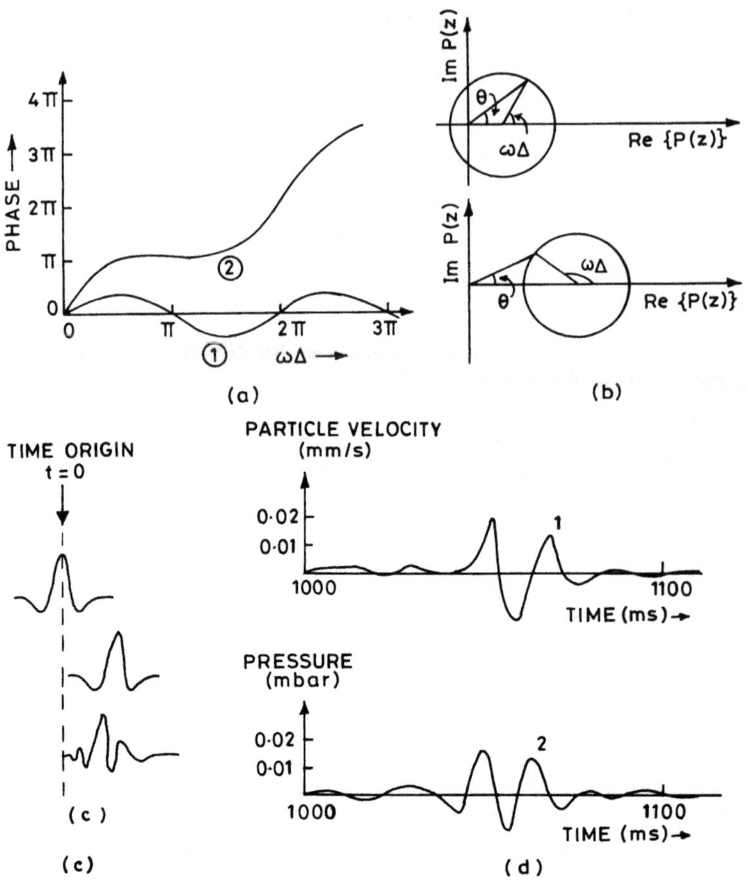

Fig. 16.11 a–d. Schematic diagram showing: **a** Phase variations of minimum phase wavelet (*curve 1*), and of the maximum phase wavelet (*curve 2*); **b** complex plane representation of the minimum phase (*bottom*) and maximum phase wavelets (*top*) (after Sheriff and Geldart 1983). **c** Waveforms with common amplitude spectrum but different phase spectrum: (*top*) zero phase, (*middle*) linear phase, (*bottom*) minimum phase (Source: after Anstey 1970:47). **d** Seismic pulse recorded by reflection upon a 1 km deep reflector: *1* onshore, *2* offshore (Source: Lavergne 1989. © 1989 Editions Technip, Paris. With permission from Editions Technip, Paris)

ural system. Also, in the case of a minimum phase wavelet, as the phase shifts are minimum allowed by the given amplitude spectrum, the waveform has the property that the energy is concentrated as far towards the front as the amplitude spectrum allows.

Figure 16.11 c depicts schematically the time-domain forms of the zero phase wavelet (symmetrical about time origin $t=0$), the linear phase wavelet (with a delay in arrival time), and a minimum phase wavelet – all corresponding to one common amplitude spectrum. The symmetrical wavelet is referred to as the *Ricker wavelet* (for details, refer to Ricker 1940, 1953, 1977).

Figure 16.11 d is an example of a seismic pulse recorded by the reflection from a 1 km deep reflector on shore (1) and offshore (2). These pulse forms have resemblance with a Ricker wavelet.

Let us now have a look at the complex plane representation of minimum phase and maximum phase wavelets. The plot of $|z|$ for various values of $\omega\Delta$ would produce a circle of radius 1, called a *unit circle*. In a graphical plot of imaginary part vs. real part (Fig. 16.11 b), the unit circle for a minimum phase wavelet would not enclose the origin whereas for a *maximum phase wavelet* it would enclose the origin. Angles $\omega\Delta$ and phase angle θ are also depicted in the graph. For a *maximum phase wavelet* it may be seen that $\omega\Delta$ increases every time by 2π, phase angle θ also increases by 2π.

To evolve a criterion about the location of roots of the factor of the polynomial in z in the complex plane, both for minimum phase and maximum phase wavelets, we return to the product representation of the z-polynomial considered in the beginning:

$$a_0(z - p_1)(z - p_2)\ldots\ldots(z - p_n)$$

where $p_1, p_2 \ldots p_n$ are polynomial roots which may be complex. The criterion based on the location of roots in the complex plane is as follows:

If all the roots of the polynomial are plotted in the complex plane with the unit circle centered at origin, all of these would be *outside* the unit circle for a *minimum phase wavelet.*

The above discussion has centered on a two-term wavelet and its classification as minimum phase or maximum phase. When we have a product of such two term wavelets in the z-polynomial, how do we decide the characteristics of the wavelet represented by the polynomial itself. The following rules are useful guidelines:
(i) If $(b_i - c_i z)$, $i = 1, 2, 3 \ldots n$ are all minimum phase, the product

$$\prod_1^n (b_i - c_i z)$$

is always a *minimum phase.*

Also if all individual terms $(b_i - c_i z)$ are maximum phase, the product is also maximum phase.

If some of the terms in the product are minimum phase and some are maximum phase, the product is *mixed phase.*

Similarly,

$$\frac{\prod_1^n (a_i - b_i z)}{\prod_1^n (c_i - d_i z)}$$

is *minimum phase,* if all the factors in the product in the numerator and denominator are minimum phase.

Before closing our discussion let us take an example of a *minimum phase wavelet.*

Suppose that the polynomial in z is factored as follows:

$$[z + (4 - i)] \times [z + (4 + i)] \times$$
$$[z + (3 + 3i)] \times [z + (3 - 3i)]$$

Here,

$$p_1 = -(4 - i), \ |p_1| = \sqrt{4^2 + 1^2} = \sqrt{17}$$

$$p_2 = -(4 - i), \ |p_2| = \sqrt{4^2 + 1^2} = \sqrt{17}$$

$$p_3 = -(3 + 3i), \ |p_3| = \sqrt{3^2 + 3^2} = \sqrt{18}$$

$$p_4 = -(3 - 3i), \ |p_4| = \sqrt{3^2 + 3^2} = \sqrt{18}$$

Since the roots lie outside the unit circle (Fig. 16.12), the pulse represented by the above product is a minimum phase.

The above discussion was confined to properties of minimum phase, maximum phase and mixed phase wavelets using the z-transform. We may also adopt the z-transform description for *zero phase* and *linear phase wavelets.*

For zero phase, the imaginary part should be zero. This implies that the form of the zero phase wavelet in the z-representation should be such that there is only the real part involved. Therefore it should be of the type

$$(z^n + z^{-n}) = 2 \cos n\omega\Delta \qquad (16.32)$$

(i.e. only real part)

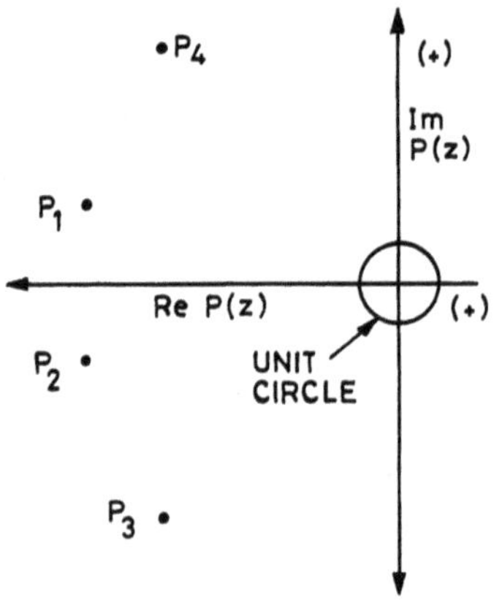

Fig. 16.12. Representation of roots of a Z-polynomial and criterion of minimum phase

Multiplying a pair of elementary wavelets of the form $(bz-1)(bz^{-1}-1)$ would give zero phase wavelets. The roots of zero phase wavelets occur in pairs as $(a, 1/a)$. A linear phase wavelet is obtained by multiplying by z^n.

16.3 Use of Tomographic Reconstruction

For conducting crosshole seismic surveys, seismic sources are placed in one borehole and receivers in the other. The properties of the medium in between two boreholes are explored using data on travel times of seismic waves. The investigation technique is termed *Cross Borehole Tomography* (c.f. Ivansson 1987; Bording et al. 1987; Wang et al. 1987; Macrides et al. 1988; Justice et al. 1989). The theory for borehole tomography was initially developed for exploring isotropic velocity variations, which has subsequently been extended to anisotropic mediums.

Without going into the details of the various techniques of tomographic reconstruction, it is desirable to give an outline of some of the well known tomographic techniques as this should serve to clarify the results that follow, specifically for the anisotropic medium. Amongst the initial developments in tomography are: Back Projection Technique (BPT), Algebraic Reconstruction Technique (ART), and Simultaneous Iterative Reconstruction Technique (SIRT).

The BPT is basically a weighted averaging process and is a non-iterative algorithm (see, for example, Kuhl and Edwards 1963; Humphreys and Clayton 1988). ART involves iterative processing and is appropriate for solving a large number of linear equations with many unknowns. For extensive use of ART for tomographic reconstruction reference may be made to Herman et al. (1973), Peterson et al. (1985) Carrion (1987), and Nolet (1987). The SIRT technique was developed by Gilbert (1972). This technique is similar to ART and differs only in the method of implementation. For example, SIRT operates on all paths passing through a given cell simultaneously whereas ART operates on one path at a time. Computer programs were developed for different tomographic reconstruction techniques by Joseph (1986), Kumar (1989) and Singh (1987).

Seismic tomography has its basis in the Fermat integral relating measured travel time t(l) to unknown slowness $s(\mathbf{x})$,

$$t(l) = \int_{l(s)} s(\mathbf{x})\,dl \tag{16.33}$$

[Carrion 1991, Eq. (4)]
where l(s) is the arc length of the ray. Equation (16.33) tells that tomographic estimation of seismic velocity is highly underdetermined. From a finite number of travel time measurements, one tries to determine slowness at an infinite number of points along the rays. As slowness along the ray may be changed in many ways to keep t(l) constant, it may be inferred that tomography has an infinite degree of freedom.

Equation (16.33) is a non-linear integral. If the ray geometry is known, a set of linear equations of the following form are used for analysis

$$\mathbf{\delta t} = \mathbf{A\delta s} \qquad (16.34)$$

where $\mathbf{\delta t}$ is the vector of travel time residuals, i.e. recorded travel time minus computed for the chosen model.

With the use of Eq. (16.34), the problem reduces to finding slowness residuals $[\mathbf{\delta s} = (\delta s_1, \delta s_2, \delta s_3 \ldots \delta s_n)^T]$ from travel time residuals $[\mathbf{\delta t} = (\delta t_1, \delta t_2, \ldots \delta t_N)^T]$. The residual of slowness is the difference between computed slowness and initial model slowness. The travel time residuals are related to the residual to slowness by the linear map **A**.

The conventional tomographic reconstruction starts from the initial guess model of the geological structure, which is an approximation to the subsurface. The next step is to trace the rays through the initial guess model. The travel times are calculated for the traced rays through the initial model. The travel-time residual – i.e., the mismatch between the actual observed travel time and calculated time is used to update the initial model. An iterative process is continued until some *convergence criterion* is met. The convergence criterion could be, for example, that the residual times become small enough, or that model parameters change insignificantly. The procedure is illustrated in the following subsection for a weakly anisotropic medium.

16.3.1 ART for a Weakly Anisotropic Medium

Consider a discretized medium possessing *weak velocity anisotropy* as in Thomsen (1986) (refer to Section 5.6 on *weak anisotropy*). Further assume that the raypath is straight. A raypath traversing a pixel is shown in Fig. 16.13.

The P-wave slowness s, for the ray path at some angle θ (for the weak anisotropic medium under consideration) is expressed in terms of vertical slowness, s_0 and anisotropy parameters δ and ε as,

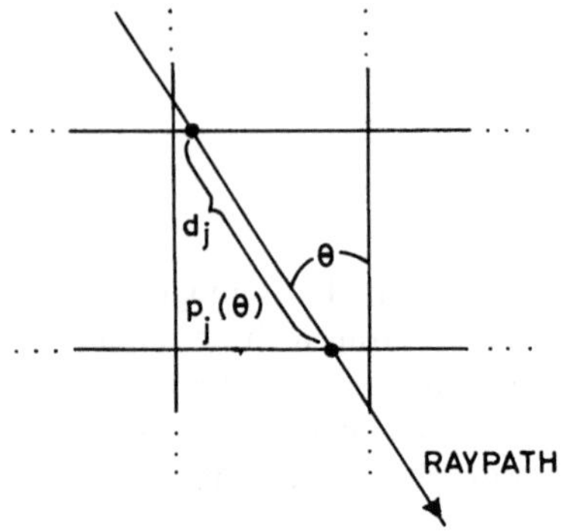

Fig. 16.13. Schematic diagram of the jth pixel, traversed by a straight ray at an angle θ to the vertical. The length of the ray in the pixel is d_j (Source: Stewart 1988, from Geophysics with permission)

$$s(\theta) = s_0\{1 - \delta \sin^2 \theta \cos^2 \theta - \varepsilon \sin^4 \theta\} \tag{16.35}$$

[Stewart 1988, Eq. (1)]

For a small angle of θ, this may be written as

$$s(\theta) = s_0 + ag(\theta)$$

where $a = s_0\,\delta$ and $g(\theta) = -\sin^2\theta\,\cos^2\theta$. Also, for θ near 90° (as in the case of cross-hole seismic), $g(\theta)$ and 'a' may be assumed to be $-\sin^4\theta$ and $s_0\varepsilon$, respectively.

Assume that the ray path at angle θ traverses N pixels; the path length and slowness in direction 'θ' in the jth pixel being d_j and s_j. The total travel time for a single straight ray is given by

$$t = \sum_{j=1}^{N} s_j(\theta) d_j \tag{16.36}$$

In the Algebraic Reconstruction Technique, a guess is made of the slowness, and the travel time is calculated using Eq. (16.36). The difference between the calculated time and observed time is used for updating slowness along each ray path. Al-Yahya (1985) used the method of Lagrange-multipliers for deriving an ART update for an isotropic medium, which has been extended by Stewart (1988) for a *weakly anisotropic transversely isotropic medium.*

The vertical slowness (s_0) and anisotropy factor δ_j are updated by adding Δs_{0j} and $\Delta a_j/s_{0j}$ to the previous estimates of slowness and anisotropy factor, respectively. The slowness and anisotropy factors are updated ray by ray and for the ith ray and the jth pixel. The corrections are

$$\Delta s^i_{0j} = \frac{\Delta t^i d^i_j}{[1 + g^2(\theta^i)] \sum_j \left(d^i_j\right)^2} \tag{16.37 a}$$

and

$$\Delta a^i_j = \frac{\Delta t^i d^i_j g(\theta^i)}{[1 + g^2(\theta^i)] \sum_j \left(d^i_j\right)^2} \tag{16.37 b}$$

[Stewart 1988, Eq. (11)]

Here Δt^i is the time residual for the ith ray, d^i_j is the distance traveled in the jth pixel by the ith ray, and θ^i is the angle the ith ray makes with the vertical in the j-th pixel.

One complete iteration would be a sequential application of time residuals to relevant pixels, ray by ray, for all rays.

It may be mentioned here that geotomography is an information-deficient system. This is because of the fact that, in seismic experiments, complete angular coverage is not achievable, which is very much required for generating an accurate image of the medium. Methods have to be found to compensate for incomplete recording. Carrion (1991) showed that incomplete angular coverage might be balanced by *imposed constraints* in tomographic analysis. He proposed *constrained dual tomography*, which permits us to incorporate any constraints in the inversion and update them in the course of computations. A brief outline of *constrained dual tomography* would be given below.

16.3.2 Constrained Dual Tomography

Carrion (1991) remarks that seismic tomography is usually overdetermined, i.e. the number of rows of **A** [refer to Eq. (16.34)] is greater than the number of columns. Also the rank R(**A**) of A of tomographic matrix **A** is usually less than N (the number of columns of **A**). This implies that rank-deficient geotomography is underdetermined, and that there exists a non-empty null space of matrix **A**, N(**A**) whose dimension is governed by the equation

$$R(\mathbf{A}) + \dim\{N(\mathbf{A})\} = N \tag{16.38}$$

[Carrian 1991, Eq. (2)]

The presence of a non-empty null space causes *image blurring*, and steps need be taken to reduce its effects. Carrion (1991) introduced solution vectors in the form of $\mathbf{A}^{T}\lambda$ where λ is the vector of *Lagrangian multipliers.* These vectors are always orthogonal to the null space of **A**. Dual tomography (Carrion 1991) works on optimization with respect to vectors $\mathbf{A}^{T}\lambda$. For details of the optimization procedure, we refer to the original contribution by Carrion (1991).

Dual tomography does not require accurate knowledge of the initial model. The initiation is done with an arbitrary homogeneous medium, and updating of the medium takes place in the course of iterations. Also, this method of tomography does not require *inversion* of large matrices. Dual tomography yields a better image compared to the Algebraic Reconstruction Technique (ART), Simultaneous Iterative Reconstruction Technique (SIRT), and other conventional techniques especially for *limited angular aperture experiments* typically used in seismic exploration work.

Steps in *dual tomography* (c.f. Carrion 1991) are depicted in the following flow chart (Fig. 16.14).

The important thing to note in *dual tomography* is that, in the course of iterations, constraints may also be updated. Carrion (1991) introduced the concept of *tightening constraints.*

Dual tomography may incorporate any constraints in travel time inversion. *Constraints* work as a *matched filter* for the object. Posing the filter correctly produces accurate images. The constraints quash *image blurring* associated with *incomplete angular recording.*

Working with *dual tomography* for the complicated model of an isotropic salt dome in an anisotropic host medium was analyzed by Carrion et al. (1992). It

Fig. 16.14. Steps in dual tomography for imaging complex structures

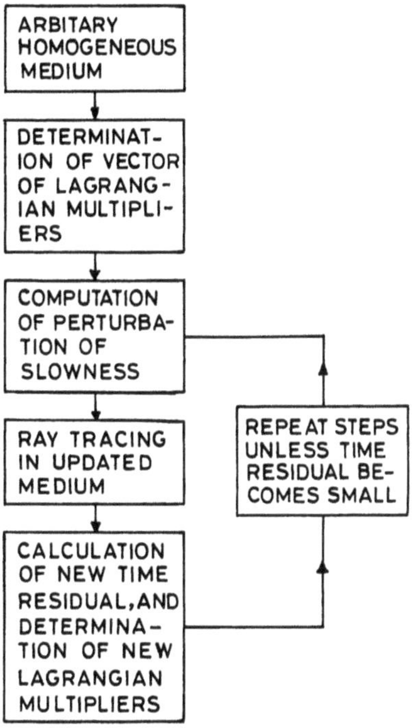

was shown that salt dome image might be fairly well constructed even in the presence of anisotropy by the use of dual tomography.

An interesting result in Carrion et al. (1992, Figs. 2, 3) is, if isotropic travel time tomography is applied to numerical anisotropic model data, the resulting tomogram manifests itself as X-like inhomogeneity. Unrealistic X-like inhomogeneity is a consequence of ignoring anisotropy in the inversion procedure. Carrion et al. (1992) concluded that even relatively *weak anisotropy* couldn't be ignored if an accurate estimate of velocity is desired.

Recent investigations (c.f. Saito 1991; Lines et al. 1992; Qin et al. 1992; Pratt and Chapman 1992; Williamsen et al. 1993) in tomography in anisotropic media deal with estimating parameters that describe the variation of velocity with direction. However, these do not formulate a method for transforming the parameters into five elastic constants that describe a general transversely isotropic (T.I.) medium. Michelana (1994) and Michelana et al. (1995) deal with tomographic determination of elastic constants of T.I. homogeneous or heterogeneous medium. This is discussed below.

16.3.3
Tomographic Estimation of Elastic Constants of a T.I. Medium

Michelana (1995) described a procedure for obtaining elastic constants of a transversely isotropic (T.I.) medium from P-, SV- and SH wave travel times, around the horizontal axis, assuming that the axis of symmetry is vertical. It is a two-step process. In the first step, direct and normal moveout velocities are obtained by separately fitting travel times for P-, SV- and SH wave types with *ray velocity functions* of the form

$$t^2 = \Delta x^2 s_x^2 + \Delta z^2 s_{z,nmo}^2 \tag{16.39}$$

[Michelana et al. 1995, Eq. (1)]
where Δx and Δz determine the distance d equal to $\sqrt{(\Delta x^2 + \Delta z^2)}$ between two points of a ray that travels in a homogeneous medium. The second step is mapping these elliptical velocities into elastic constants of the form:

$$W_{11} = W_{P,X} \tag{16.40 a}$$

$$W_{44} = W_{SV,X} \ldots \text{etc.} \tag{16.40 b}$$

where W_{ij} are elastic moduli (c_{ij}) divided by density. ($W_{P,X}$, $W_{P,ZNMO}$), ($W_{SV,X}$, $W_{SV,ZNMO}$) and ($W_{SH,X}$,$W_{SH,ZNMO}$) are direct and NMO phase velocity squared in each bracket of P, SV and SH waves, respectively.

Michelana et al. (1995) generalized Eq. (16.39) for a heterogeneous anisotropic medium. The heterogeneous model is described as the superposition of homogeneous blocks (Fig. 16.15). It consists of homogeneous elliptically anisotropic blocks with the axis of symmetry making an angle γ_j with the vertical. The blocks are separated by straight interfaces of variable dips 'a_j' and intercepts 'b_j'. The travel time of the i-th ray in the j-th cell '$t_{i,j}$' is given as

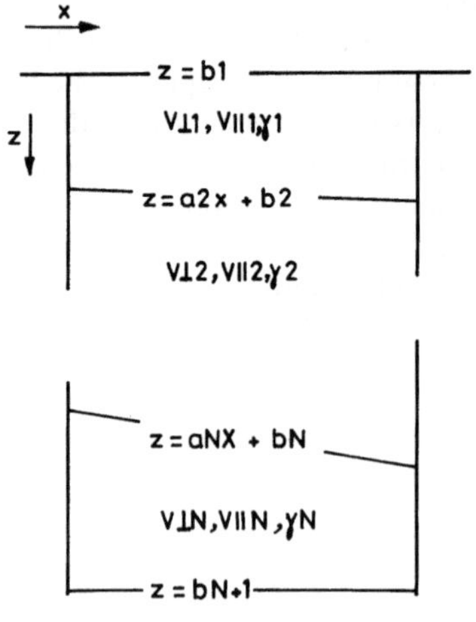

Fig. 16.15. Model of velocities and heterogeneities. *Top and bottom* interfaces are horizontal ($a_1 = a_{N+1} = 0$) and located at known depths (Source: Michelana et al. 1995, from Geophysics with permission)

$$t_{i,j} = \sqrt{\Delta X_{i,j}^2 s_{\perp j}^2 + \Delta Z_{i,j}^2 s_{11j}^2} \qquad (16.41)$$

[Michelana et al. 1995, Eq. (8)]
where $s_{\wedge j}$ and s_{11j} are values of slowness in the directions perpendicular and parallel respectively to the axis of symmetry; $\Delta X_{i,j}$ and $\Delta Z_{i,j}$ are expressible in terms of coordinates of points of intersection ($x_{i,j}$, $z_{i,j}$) between the ith ray and jth interface, and a_j, b_j and γ_j.

The total travel time for a ray traveling from source to receiver is

$$t_i(\mathbf{m}) = \sum_{j=1}^{N} t_{i,j}(\mathbf{m}) \; (i = 1, 2, \ldots M) \qquad (16.42)$$

[Michelana et al. 1995, Eq. (9)]
where **m** is a vector of model parameters of 5N elements.

$$\mathbf{m} = (s_{\perp 1} \ldots s_{\perp N};\ s_{111}, \ldots s_{11N};\ \gamma_1 \ldots \gamma_N;\ b_1 \ldots . b_N;\ a_1 \ldots a_N) \qquad (16.42\,a)$$

Equation (16.42) is a system of non-linear equations relating model parameters to measured travel times. Its linearized version is used to solve the inverse problem.

For successful application of the method, the data aperture should be constrained to satisfy the requirement of elliptical velocity approximation, and tomographic estimation of elliptical velocities. If the data aperture is too large, the elliptical approximation may not be adequate. If it were too small, tomographic estimation of elliptical velocity would not work. For synthetic and field applications of the method, reference may be made to Michelana et al. (1995).

16.4 Use of Seismic Rock Physics

Use of seismic waves to explore rock properties like lithology, porosity, pore fluid and degree of saturation apart from detailing the structural and stratigraphic picture of the subsurface medium, has now become commonplace. Further, *4-D seismic reservoir monitoring*, *seismic lithology discrimination*, and *direct detection of hydrocarbons* using *bright spot* and angle-dependent reflectivity analysis (also called *AVO analysis*) are being applied successfully. It is therefore very pertinent to understand what are the major factors that influence the seismic properties of sedimentary rocks. Table 16.2 lists these factors.

Since a detailed discussion of each of the factors listed in Table 16.2 is not possible here, our presentation below is mainly derived from a recent tutorial publication by Wang (2001), to which readers interested in the finer details may refer.

Table 16.2. Factors influencing seismic properties of sedimentary rocks (with increasing importance from top to bottom) (Source: Wang 2001)

Rock properties	Fluid properties	Environment
Compaction	Viscosity	Frequency
Consolidation history	Density	Stress history
Age	Wettability	Depositional environment
Cementation	Fluid composition	Temperature
Texture	Phase	Reservoir process
Bulk density	Fluid type	Production history
Clay content	Gas-oil, gas-water ratio	Layer geometry
Anisotropy	Saturation	Net reservoir pressure
Fractures		
Porosity		
Lithology		
Pore shape		

16.4.1 Porous Fluid-Saturated Reservoir Rocks: Laws Governing Elastic and Physical Properties

Unconsolidated clean sands and sandstones form ideal reservoir rocks. However, these may also contain clays that affect the seismic properties of these rocks. By seismic properties is meant the P-wave velocity (V_p), S-wave velocity (V_s), or (V_p/V_s) ratio and *impedances*. We list below some useful laws governing elastic and physical properties of reservoir rocks.

The reservoir rock may be thought of as having a solid matrix (consisting of rock-forming minerals), the frame (skeleton of the rock sample), and the pore fluid which could be gas, oil, water or a mixture of all these. The Gassmann (1951) equation may be used for the calculation of the bulk modulus (k*) of a fluid-saturated porous medium, using the known bulk moduli (k_m, k_d and k_f respectively) of the solid matrix, the frame and the pore fluid [refer to Wang 2001, Eq. (1)]

$$k* = k_d + \frac{(1 - k_d/k_m)^2}{\frac{\phi}{k_f} + \frac{1-\phi}{k_m} - \frac{k_d}{k_m^2}} \tag{16.43}$$

where 'ϕ' is the porosity.

As the shear modulus is not affected by fluid saturation, we have

$$\mu^* = \mu_d \tag{16.44}$$

where μ_d is the frame shear modulus of the rock.

The following relationship exists between fluid-saturated density (ρ*), dry density (ρ_d), and pore fluid density (ρ_f) of the rock:

$$\rho^* = \rho_d + \phi\rho_f \tag{16.45}$$

Also

$$\rho_d = (1 - \phi)\rho_m \tag{16.45 a}$$

where ρ_m is the matrix (grain) density.

Wood's (1941) equation gives bulk modulus (k_f) of a fluid mixture [refer to Wang 2001, Eq. (6)]:

$$\frac{1}{k_f} = \frac{s_w}{k_w} + \frac{s_O}{k_O} + \frac{s_g}{k_g} \tag{16.46}$$

where k_w, k_O and k_g are bulk moduli of water, oil and gas respectively. s_w, s_O and s_g are saturations of water, oil and gas, respectively. Saturations are expressed as volume fractions of pore space, with $s_w + s_O + s_g = 1$. Equation (16.46) assumes that pore fluid is uniformly distributed in the pores.

The bulk density (ρ_f) of the fluid mixture is calculated using the equation [refer, for example, to Wang 2001, Eq. (7)]

$$\rho_f = S_w\rho_w + S_o\rho_o + S_g\rho_g \tag{16.47}$$

where ρ_w, ρ_O and ρ_g are bulk densities of water, oil and gas, respectively.

Murphy et al. (1993) gave the following relationships between frame bulk and shear modulus to porosity:
(i)

$$\begin{aligned} k_d &= 38.18\left(1 - 3.39\phi + 1.95\phi^2\right) \\ \mu_d &= 42.65\left(1 - 3.48\phi + 2.19\phi^2\right) \end{aligned} \tag{16.48 a}$$

for porosity less than or equal to 0.35, and
(ii)

$$\begin{aligned} k_d &= \exp\{-62.60\phi + 22.58\} \\ \mu_d &= \exp\{-62.69\phi + 22.73\} \end{aligned} \tag{16.48 b}$$

for porosity greater than 0.35,
where k_d and μ_d are the bulk and shear moduli, respectively of the frame rock (i.e. rocks with empty pores) in Gp_a. ϕ represents porosity in the volume fraction.

Shear modulus may be calculated using Eq. (16.48 a) or Eq. (16.48 b) or using the relationship between shear modulus, density and shear velocity data, i.e.

$$\mu_d = \rho V_s^2 \tag{16.49}$$

Values of μ_d obtained alternately may be compared. If the two values are close to each other, the rock is sand. In case the calculated frame shear modulus using Eq. (16.49) has a much lower value compared to Eq. (16.48 a) or Eq. (16.48 b), the rock is shale. Wang (2000) classified granular materials – sands and sandstones on the basis of the ratio of frame shear modulus to bulk modulus. When frame shear modulus is plotted against frame bulk modulus for clastic granular rocks, the correlation coefficient (R) is found to be 0.9639. This corresponds to a V_P/V_s ratio of about 1.54, and a Poisson ratio of 0.135. The near equality of frame bulk and shear moduli for clastic rocks was also reported by Castagna et al. (1985).

16.4.2 Effects on Seismic Velocities of Reservoir Rock Properties

Amongst the various rock properties, lithology and porosity affect the compressional wave velocity. Domenico (1984) categorized rock lithology and porosity using shear and compressional wave velocities and porosities. In his study, he examined previously published data on water-saturated sandstones, calcareous sandstones, and dolomite and limestone cores. Porosity of sandstone was found to range from 0.092 to 0.299, and of limestone from 0.006 to 0.229, respectively. Laboratory measurements of porosity simulated approximate burial depths of 290 to 3,460 m. Domenico's studies showed that sandstones, limestones, and dolomites are effectively separated by Poisson ratio σ, or by the ratio of P- and S-wave velocities. Differences of Poisson ratios for sandstones and limestones were related to the corresponding difference of the matrix materials of these rocks. For quartz, σ value is 0.056, and for calcite 0.316. A schematic variation of Poisson ratio vs. V_P/V_s is shown in Fig. 16.16.

Amongst the reservoir rocks, for a given porosity and *pore aspect ratio* spectrum, dolomite has the highest P-wave velocity followed by limestone, sandstone and unconsolidated sand (Wang 2001). In terms of the V_P/V_s ratio limestone ranks first followed by dolomite, deeply buried unconsolidated sand, and sandstones. Shales (which form reservoir cap rock) possess a wide range of velocities and impedances. The V_P/V_s ratio for shale is higher than reservoir sands. It may happen that cap shale and reservoir rock have similar P-wave impedances, but a large contrast in S-wave impedance. For such a situation, P to S converted waves should prove a successful *imaging tool* (MacLeod et al. 1999).

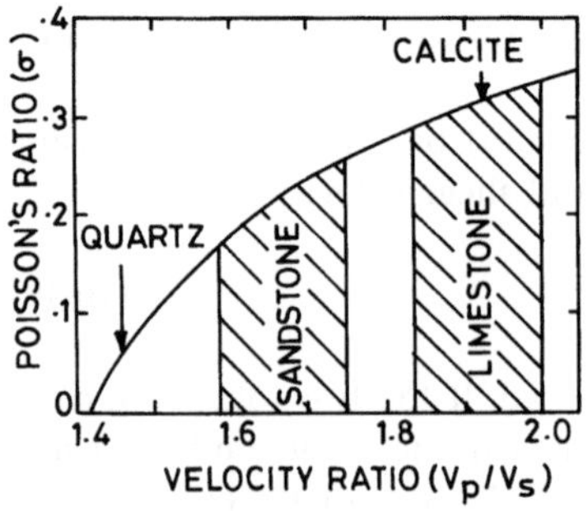

Fig. 16.16. Poisson's ratio versus velocity ratio graph on which quartz and calcite values are indicated as well as the range of sandstone and limestone values (Source: Domenico 1984)

Gardner et al. (1974) gave the following empirical relationship between density and P-wave velocity:

$$\rho = 0.23V_P^{0.25} \tag{16.50}$$

where V_P is the P-wave velocity in ft/s, and ρ is the bulk density in gm/cm^3. This relation is useful for estimating P-wave velocities from bulk-densities in water-saturated sedimentary rocks.

Texture properties, like grain-to-grain contacts, roundness, sorting, cementation, etc. also control seismic properties. Poor grain-to-grain contacts are much more likely to bring down both seismic velocities and impedances. For the reason that large grain-size sands have larger contact areas amongst grains, the seismic velocities are higher for large grain sands compared to fine-grain sands.

Wyllie et al. (1956, 1958) proposed an empirical relation relating P-wave velocity to porosity. They showed that, in *clean* water-saturated sands and sandstones at appreciable depth, the following relationship holds:

$$\frac{1}{V_P} = \frac{\phi}{V_\omega} + \frac{(1-\phi)}{V_m} \tag{16.51}$$

where ϕ is porosity, V_m is the velocity of the matrix (solid) material, and V_ω is the water velocity.

Equation (16.51) is also called the "*time-average equation*".

Pickett (1963) proposed the following relationship between velocity and porosity:

$$\frac{1}{V} = A + B(\phi) \tag{16.52}$$

where V is either P or S-wave velocity, and A and B are constants for a given rock type.

The following empirical relationship between velocity, porosity, and clay content were proposed by Han et al. (1986):

$$V_P = V_{PO} - a_1\phi - a_2C \tag{16.53}$$

and

$$V_s = V_{SO} - b_1\phi - b_2C \tag{16.54}$$

where ϕ and C are porosity and clay content in the volume fraction, respectively. These relations show that both P and S-wave velocity decrease (statistically) with increasing porosity and clay content.

A pore fluid's properties and degree of saturation have a significant role in influencing the seismic response of the reservoir, and 4-D seismic reservoir monitoring. Rocks having less compressible fluids in their pores possess higher velocity ratios (V_P/V_s); also their rocks have higher P-wave velocity and impedances.

On the other hand S-wave velocities and impedances are little affected by the pore fluids. Fluid saturation effects are primarily controlled by the compressibility of pores (or pore shape). For this reason, rocks with cracks or fractures always show a large effect of *fluid saturation* on seismic velocity, independent of how low a value the porosity is (Wang 2001).

Anisotropy may significantly affect the seismic properties of rocks. It may be present in sedimentary rocks as *intrinsic* or *induced* anisotropy. Intrinsic anisotropy in sedimentary rocks is normally present as transverse isotropy, and results due to the effect of the preferential alignment of elongate grains or pores and fine laminations. Induced anisotropy in rocks is a result of stress anisotropy and fracturing. The role of stress anisotropy is to preferentially align pores, grains, cracks, and fractures resulting in a seismically anisotropic rock from an isotropic rock.

Most shales are intrinsically anisotropic. Measurements show that seismic anisotropy in shale may range from a few percent to as high as 50% (refer, for example, to Kaarsberg 1959; Jones and Wang 1981; Tosaya and Nur 1982; Lo et al. 1986; Vernik and Nur 1992; Hornby et al. 1994; Johnston and Christensen 1995; Vernik and Liu 1997).

Due to the possibility of high magnitudes (as much as 50%) of seismic anisotropy in shales (Wang 2001), seismic processing and interpretations would yield reliable results only by orientating these to account suitably for the presence of anisotropy.

16.4.3 The Influence of Environmental Factors and Geological History

Environmental factors include pressure and temperature. These affect the current or present state of the reservoir rock. Geological history refers to the depositional environment of the rock, and changes that took place after the rock was deposited.

To describe the effect of pressure, we should consider two types of pressure conditions that exist in the reservoir rock. First is the *overburden pressure*, which is also referred to as the *confining pressure*. It is the pressure exerted by overburden rock mass. A second type of pressure – the reservoir pressure (referred to also as the *fluid pressure* or *pore pressure*) is exerted by the fluid mass. The *effective pressure* is the net overburden pressure and equals the difference between overburden pressure and reservoir pressure. This happens because pore fluid pressure supports a part of the overburden pressure. The seismic properties of the reservoir rocks are controlled by *net overburden pressure*.

Both seismic velocities and impedances of P and S waves increase with net increase in overburden pressure. The relationship between changes in velocity with a change in overburden pressure is non-linear. Figure 16.17a (Wang 2001) shows this property, i.e. the compressional velocity increase is larger at low net overburden pressure. Both the reservoir pressure and fluid saturation change during production and enhanced oil recovery (EOR) process. The effects of changes in reservoir pressure and fluid saturation on the change in seismic properties (velocity and impedance) may be either additive or they may oppose each other and cancel

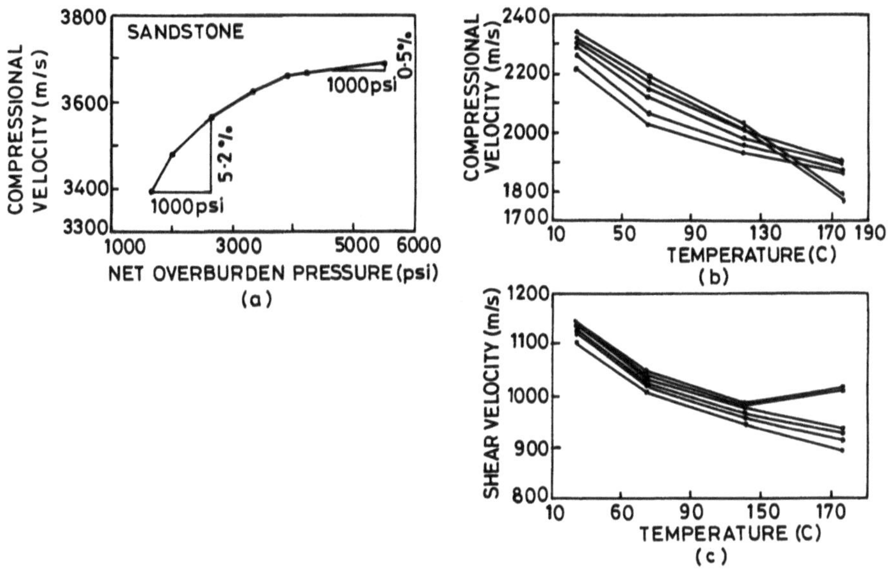

Fig. 16.17 a–c. a V_p versus net overburden pressure in sandstone. Change in velocity with change in overburden pressure is non-linear. **b, c** V_p and V_s versus temperature in heavy oil sand at an overburden pressure of 350 psi. Curves from the *top to bottom* relate to a pore pressure of 50, 100, 150, 200, 250 and 300 psi. Both V_p and V_s decrease by about 15% as temperature increases from 22 to 177°C. V_p drops further between 120° and 177 °C at low pressures of 50 and 100 psi – *the top two curves* in **b** as the water inside the rock transforms to steam, adding another 10% or so V_p decrease. In contrast, V_s increases by about 5% as the results of steam (Source: Wang 2001, from Geophysics with permission)

out each other's effect. Thus, the link between the reservoir process and rock property change must be clearly understood for *4-D seismic surveys.*

Results in Timur (1977) and Wang and Nur (1990 b) show that in rocks, saturated with gas or water, the seismic velocities and impedances decrease slightly with the increase in temperature. However, oil-saturated rocks show a large decrease in seismic properties with increasing temperature. This effect is found enhanced in unconsolidated sands with heavy oils. This effect is the underlying basis for seismic monitoring of thermal EOR. A substantial decrease in compressional wave velocity in heavy oil sands was first brought out in the studies of Tosaya et al. (1987). A variety of results in Wang and Nur (1990) for heavy-oil sands and hydrocarbon-saturated rocks reveal that both V_p and V_s decrease with increasing temperature: Figure 16.17 b, c depicts the variations of compressional and shear wave velocities with temperature for a given overburden and different pore-pressure changes.

The depositional environment may also affect the seismic properties. This includes the rate and source of deposition, and available depositional energy. For example, sand/shale sequences are the result of alternating depositional energy, rate and source.

The stress history of the rock may help to plan the measurement and interpretation of the data.

Older rocks are subjected to better consolidation and compaction, which in turn influences the seismic properties. More compacted rocks have higher values of compressional and shear velocities and impedances.

16.4.4 V_p/V_s as an Indicator of Lithology and Porosity

Tatham (1982) and Domenico (1984) recognized that the V_p/V_s ratio may be used as a *lithology indicator.*

For example, shale (if assumed isotropic) always possesses a higher V_p/V_s ratio compared to reservoir sands. Rafavich et al. (1984) showed that limestone could be discriminated from dolomites on the basis of the V_p/V_s ratio.

The V_p/V_s ratio has also found an application in direct hydrocarbon detection, particularly with AVO techniques. Since shear waves are insensitive to fluid changes but compressional waves are, changes in fluid type and saturation would give rise to a V_p/V_s ratio change (refer, for example, to Castagna and Backus 1993).

16.4.5 Velocity-Porosity Cross Plot as a Measure of the Kerogene Contents of Shales

The organic richness or kerogene content of shales influence the velocity and anisotropy of shales (Meissner 1984; Vernik and Liu 1997). Vernik and Liu (1997) specifically analyzed the variation of bedding normal velocity, $V_P(O)$ versus porosity at high confining pressure for different degrees of richness of kerogene content in the source rock (shale). The aim of this study was to improve upon the accuracy of porosity and/or lithology prediction from seismic wave velocities in shales. They concluded that the cross plot of bedding normal velocity $V_P(O)$ vs. porosity should enable us to map, with reasonable accuracy, the richness of kerogene content.

The cross plot (Fig. 16.18) of the shale data points form a triangle with the upper boundary shown by the linear regression line corresponding to organic free sand rock with volumetric clay content C = 15–35% and the lower boundary corresponds to linear-regressive for the richest-in-kerogene (k = 0.4) of shales.

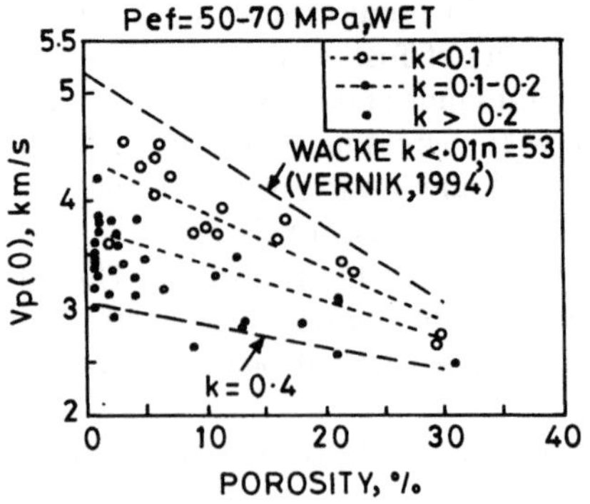

Fig. 16.18. P-wave velocity (at 70 MPa and room temperature) normal to bedding versus porosity in source rocks. Parameter K is the volume fraction of kerogene in the rock (Source: Vernik and Liu 1997)

The entire data could be divided into three groups: $k<0.1$, $k=0.1-0.2$ and $k>0.2$ respectively. With increasing kerogene content, the slope $d(V_p/d\phi)$ flattens.

16.4.6 Practical Guidelines Relating some Significant Rock/Fluid Properties with Seismic Properties

In the following we discuss briefly only those points that were either not covered or mentioned only briefly. These are derived from the tutorial article by Wang (2001), which deals with this particular aspect along with others in detail.

16.4.6.1
Rock Properties

Bulk density: Statistical results show that rocks with higher bulk density have higher seismic properties.
Fractures: Aligned fractures cause seismic velocity anisotropy.
Porosity: Seismic properties decrease as porosity increases. V_p/V_s is not found to be strongly linked to porosity.
Pore shape: This is the dominant factor that influences seismic properties. It is difficult to quantify its effect. In general, rocks with flat pores bear lower seismic properties.

16.4.6.2
Fluid Properties

Viscosity: Rocks with more viscous oil tend to bear higher seismic properties.
Density: Higher density oils are associated with higher compressional properties, and shear impedance, but shear velocity may be lower.
Saturation: Full saturation of a liquid in a rock results in an increase of V_p/V_s ratio. When gas is introduced to a fully liquid-saturated rock, it results in a lowering of the V_p/V_s ratio.

16.5 Use of 3-D Data Analysis

Data presented in a linear seismic cross-section has its own limitation in dealing with following types of situations:

(i) Events, which are recorded from outside the vertical plane of the cross-section and may provide useful clues about the properties of the medium, are not accounted for in conventional 2-D interpretations.
To understand and account for the origin of such events some form of three-dimensional control of seismic data is necessary.

(ii) In a number of field situations of hydrocarbon occurrence, it is essential that the details of seismic reflection data along a line be matched with similar details across the line.
An example is the association of oil or gas fields with fluvial sands in meandering ancient stream channels. A vertical plane through a single seismic line may cross the underground channel at an angle, and may cut at one or more

sporadic locations. Interpretation of seismic data from one or a few such locations would not give definite clues about the meandering stream. On the other hand, if in-line and cross-line data were interpreted together it would be easy to define the meander.

The 2-D seismic section display can at best give subsurface details along a vertical section. The 3-D data display, on the other hand, brings out immensely more details through a volume display of the data (Fig. 16.19). Therefore, even very intricate interpretation problems may be handled successfully, by choosing the appropriate processing of initial 3-D data.

In the early stages (1970s and 1980s) of 3-D seismic technology, the use of 3-D seismic data was rather limited. It was utilized to appraise a discovery, and was limited to small areas. With realization of its consistently growing success, 3-D seismic

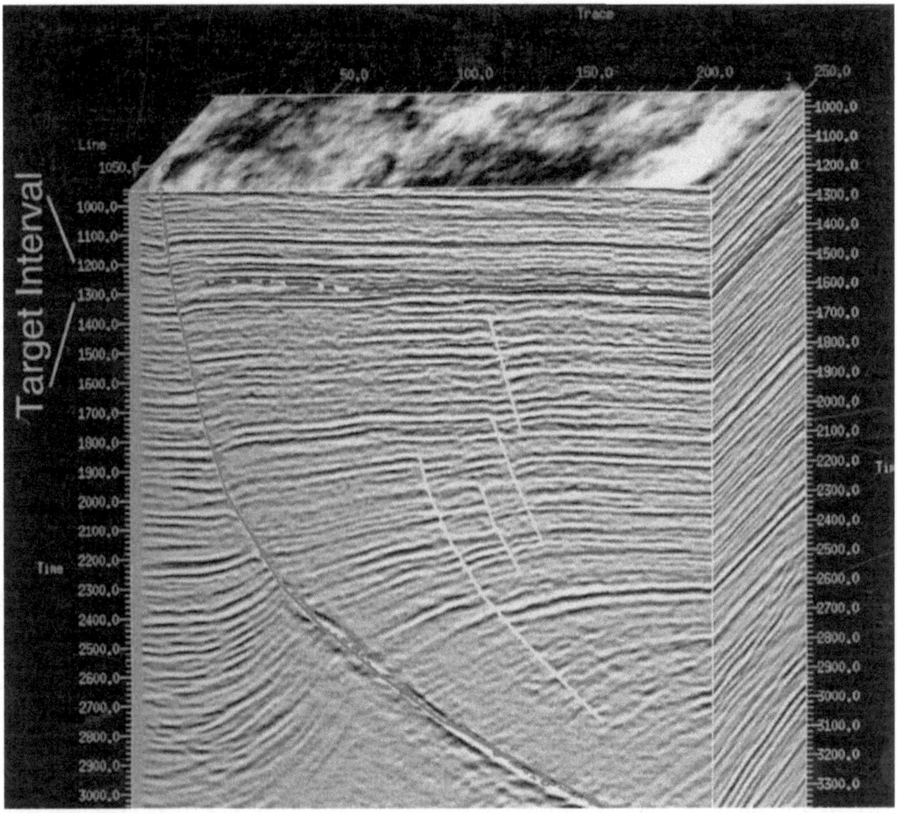

Fig. 16.19. 3-D seismic survey display covering a migrated area of 18 mi^2. The 3-D survey covers two fields associated with the faulted, rollover anticline shown in the 3-D seismic data cube. The variable density amplitude display is normal polarity, and trace spacing is 100 ft (33 m) [Source: Horkowitz and Davis 1996. In: Weimer and Davis (eds) Application of 3D seismic data to exploration and production. AAPG © 1996, reprinted by permission of the AAPG whose permission is required for further use, p 38]

technology today is being greatly embraced by oil companies for wide scale exploration, delineation, and planning optimum production from newly discovered or already producing oil fields. It is now realized that true appreciation of reservoir complexities from 2-D seismic data alone is simply not possible. In fact 3-D seismic imaging is now the most vital or only tool to correctly see into the Earth, and map the subsurface in its finest details. Summarizing, it may be said that 3-D seismic data has tremendous potential for exploration, field development and production.

3-D seismic technology is applied in the rejuvenation of producing oil fields. As the oil production continues, the reservoir model needs to be updated continuously during the life history of each oil field. This may be fulfilled by repeat 3-D seismic surveys.

As the 3-D seismic image has received resounding success in the oil industries, its use is expanding particularly to reduce exploration and production costs.

In the last 2–3 decades, 3-D seismic exploration has been used to solve a variety of reservoir-related problems. Weimer and Davis (1996) organize excellent

Table 16.3. Specific problems solved through 3-D Seismic Technologies (From Davis and Weimer 1996)

S N	Problem	Technique used	Contributors*
1	To track change of seismic attributes with oil pool drainage (fluvial - deltaic)	4-D seismic monitoring	Anderson et al.
2	Interpretation of fluvially deposited thin bed reservoirs (fluvial - deltaic)	3-D seismic imaging with VSP data	Hardage et al.
3	Delineation of thin-sandstone reservoirs in a shale-rich sequence (fluvial - deltaic)	Instantaneous frequency and reflection amplitude attributes of 3-D seismic data	Horkowitz and Davis
4	Reservoir management (fluvial - deltaic)	3-D seismic interpretation	Gutierrez et al.
5	3-D evaluation of oil fields (fluvial - deltaic)	Integrated 3-D seismic interpretations for structural mapping, detailed fault interpretations, and analysis of seismic amplitudes	Silverman et al.
6	Reservoir property characterization (carbonates)	3-D, 3-C seismology	Arestad et al.
7	To increase ultimate recovery (carbonates)	Practical application of amplitude information in field development	Skaloud DK
8	Fault-displacement distributions (structure)	Interpretation of high resolution 3-D seismic data	Mansfield CS
9	Remediation of contamination (near surface geophysics)	3-D seismic interpretation	House et al.

* Complete references given in reference list.

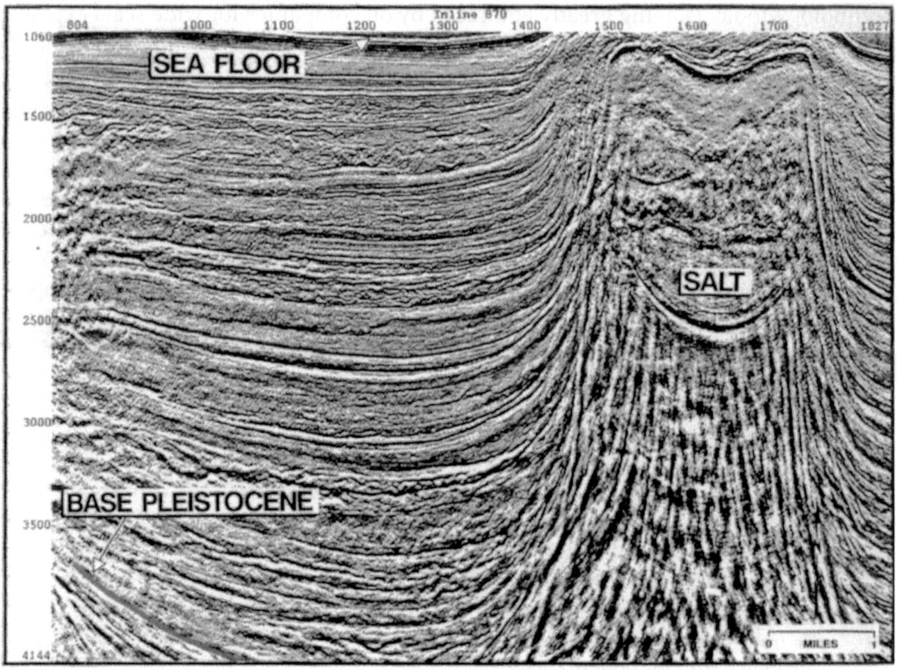

Fig. 16.20. Deep-water facies analysis using 3-D seismic data from the Gulf of Mexico [Source: Risch et al. 1996. In: Weimer and Davis (eds) Application of 3D seismic data to exploration and production. AAPG © 1996, reprinted by permission of the AAPG whose permission is required for further use, p 145]

case studies for 3-D seismic. Table 16.3 lists a few of these, which are considered of characteristics type.

To conclude we reproduce an example (Fig. 16.20) of deep-water *facies analysis* using 3-D seismic data from the Gulf of Mexico (Weimer and Davis 1984, p 145). It shows the *structure and cyclic stratigraphy.* The sediments are 3,050 m thick. The structure is a trough flanked by shallow salt on the east and buried salt towards the west. The high spatial resolution of data is of significance.

16.6 Objective-Oriented Interpretation Tools

16.6.1 Detection of a Coal Seam

Detection and detailed mapping of coal seams has important differences from similar exercises for oil and gas. The shallow occurrence of coal necessitates special considerations for high quality data acquisition and processing. Some of these may be mentioned:

1. If there are large variations in surface topography or large changes in near surface velocity, accuracy in the application of static and dynamic corrections play a significant role for the success of investigations.
2. The action of the coal seam on the incident seismic pulse is like a low-pass filter. It is found that the reflected pulse spectrum is rich in high frequencies and of the transmitted pulse in low frequencies (refer, for example, to Waters 1987, Figs. 11.25 and 11.26). Because of this effect, the detection of the top seam would be easy; however, this would bring in the problem of detection of lower seams. The reason for this is the following. Each deeper seam would reflect prominently higher frequencies from the frequencies incident on it, and pass lower frequencies. Thus, the pulses reflected upward towards the surface from deeper seams would contain lower and lower frequencies. Hence, the overall poor resolution of the record (due to the lower frequencies) would affect detection of deeper seams. It is obvious that the effect just described would be repeated both on downward as well as on upward paths of the pulse. It is assumed that the coal seam is a low velocity (2,134 m/s), low density (1,500 kg/m^3) medium embedded in rock of much higher velocity (4,268 m/s), and density (2,500 kg/m^3) (refer, for example, to Waters 1987, p 459).
3. Choice of the appropriate size of explosive, as the source of seismic energy, has an important role for coal seam detection. For a high resolution of reflected events, it is desired that the charge size be optimum. With a large charge size there is the problem of an increase in the level of ground roll; at the same time a lower charge size results in poor signal-to-noise ratio. Figure 16.21 a shows an example of the effect on a seismogram of varying charge size in exploration for coal. Charge size M/4 (mass M of charge is 0.45 kg) produces the best resolution and alignment of reflected events at 0.33 s and 0.5 s. For clarity of record, short duration wavelets would be desirable. One of the means of controlling it is by using a small quantity charge at the bottom of the shot hole at the base of the weathered zone. The nature of the far field pulse (either in terms of pressure or particle velocity) is linked with mass M of the explosive charge in the following manner (refer to Ziolkowski and Lerwill 1979):
   (i) both the pulse duration and amplitudes are proportional to $M^{1/3}$,
   (ii) the absolute spectral bandwidth of the pulse is inversely proportional to $M^{1/3}$,
   (iii) the amplitude of the spectrum of the pulse is proportional to $M^{2/3}$.

   The assumptions behind these deductions are:
   (i) the radiation generated by the explosion is spherically symmetric;
   (ii) for a given type of explosive in a given medium, the fraction of total explosive energy that is converted into seismic energy is constant, independent of the charge size.

   Figure 16.21 b shows the effect on signal period and amplitude as the mass of the explosive source is scaled down.
   Jain (1982, unpublished) showed, through numerical computations, that by using a vibroseis type seismic sweep signal the thickness of the coal seam may be directly determined. Under suitable conditions it may dispense with the need for drilling and coring to find the thickness of a coal seam in the subsurface. The analysis is based on considering the results of interference for reflections from

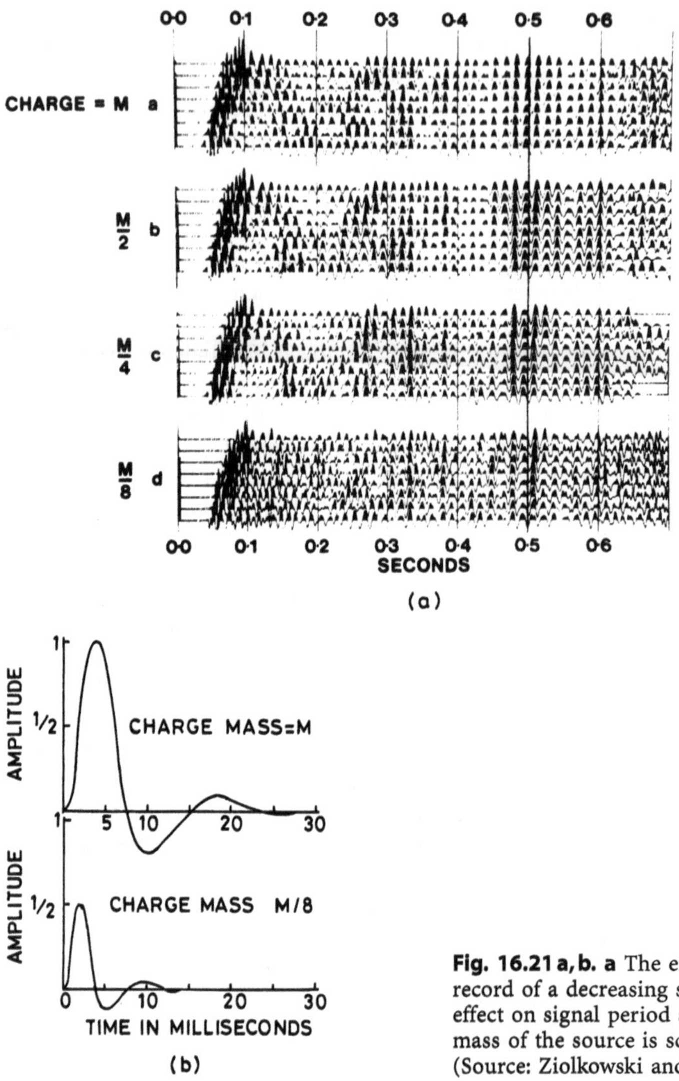

Fig. 16.21 a, b. a The effect on a monitor record of a decreasing source mass. **b** The effect on signal period and amplitude as the mass of the source is scaled down (Source: Ziolkowski and Lerwill 1979)

the top and bottom of the seam, and makes use of the condition of constructive interference related with the ratio of wavelength and thickness of the seam.
The effect of the lowering of signal-to-noise ratio from small quantity charges may be obviated to some extent by *vertical stacking.*
Stacked records, although useful for focusing weak events, produce distortions of relative amplitudes of the events. In view of this prestack migration for shallow investigations is preferred. Although post-stack migration is less costly than prestack migration, the latter has distinct advantages in terms of generating more accurate migrated results.

4. For shallow coal seams, there is a limit on offset traces that may be combined for stacking. Large offset traces may be questionable for stacking with short

offset traces because of the possible inclusion of converted (PS or SP) waves. To resolve this aspect, synthetic seismogram forward modeling proves helpful. In the Subsection 16.6.1.1 that follows, the *reflectivity method* of generating a synthetic seismogram is briefly described. This method is chosen, as it was specifically applied (Fertig and Müller 1978) for coal seam field models. Also, this method deals with important generalizations in terms of model parameters, and wave types by including both P and S waves, mode converted (PS or SP) waves, and multiples. Additionally it provides the foundation for mathematical formulation to treat wave propagation and reflections in the frequency domain, and finally view the results in the form of a synthetic seismogram in the usual space-time domain.

16.6.1.1 Generation of a Synthetic Seismogram: The Reflectivity Method

A synthetic seismogram is a very useful tool for improving interpretations of field records. Various approaches have been published to find a link between model parameters and their seismic (seismogram) response for a given seismic source. Some of these may be mentioned:

(i) Methods using basic considerations on the evolution of a seismogram (Peterson et al. 1955; Berryman et al. 1958; Wuenschel 1960).
(ii) Reflectivity method (Fuchs 1968; Fuchs and Müller 1971; Kind 1976; Fertig and Müller 1978).
(iii) Methods resembling the reflectivity method (Kennet 1974, 1975, 1978).
(iv) Numerical simulations including finite difference and finite element methods (Cherry and Hurdlow 1966; Boore 1972; Alterman and Loewenthal 1972; Kelly et al. 1976).

Each of these methods have their own merits under the assumptions made for the generation of a synthetic seismogram. For getting close to real form seismograms, it is essential to simulate realistically the wave propagation by including reflection, diffractions, mode conversions, multiples generation and transmission, and take into consideration the characteristics (elastic, anelastic, or anisotropic) of the medium.

Early basic formulations for generating synthetic seismograms were done in the time domain considering a spike as the source and considering reflection, with or without transmission losses, for a layered medium. The generalization for different types of sources (point source and vertical force) is accomplished through a frequency-domain treatment of the problem. Stepwise implementation of the reflectivity method is as follows:

- Assumes a point source located on the surface of a plane-parallel layered medium; and each layer's parameters (P-wave velocity (α), S-wave velocity (β), density (ρ), and thickness (h)) are assumed known.
- Spherical waves at constant frequency generated by a source are treated through the equivalent set of plane waves.
- Both real and complex angles of incidence are taken. Complex angles simulate mathematically inhomogeneous waves that decay with depth. Flexibility of using only real angles within a limit, corresponding to waves starting from the source and reaching the geophone, exists.

- Treatments of each plane wave for reflection from a plane-layered medium is done, using the theory developed by Thomsen and Haskell (for more detailed discussion on the use of the Thomson Haskell matrix formulation refer to Fuchs and Müller 1971; and for making the reflectivity computational procedure considerably faster refer to Kind (1976).
- Calculation of the response of each plane wave, and summing up the responses for all real angles of incidence.
- Response on the surface is found in the time domain by inverse FFT.

Final results yield vertical and horizontal component displacement seismograms in the usual wiggle form display.

The following Fig. 16.22 gives the main mathematical steps of the reflectivity method (Fertig and Müller 1978):

Steps		Comments
$\phi_0(r, z, t)$	→	Compressional potential from an explosive point source
↓ F. T.		
$\phi_0(r, z, \omega)$	→	Includes F. T., $F(\omega)$, of the excitation function $F(t)$
↓ Use of Sommerfeld integral	→	Expansion of spherical waves into plane waves

$$\phi_0(r,z,\omega) = F(\omega)\int_0^\infty \frac{k}{j\nu_1} J_0(kr)\exp(-j\nu_1 z)dk$$

(k is horizontal wave number; and ν_1 the vertical wave number)

↓ Simulation of reflection process	→	Reflection of plane waves from reflecting zone
$\phi_1(r, z, \omega)$	→	Reflected P-wave potential in top layer; includes complex reflectivity function $R_{PP}(\omega, k)$ for plane waves (general algorithm includes incident and reflected S-wave, conversions P to S and vice-versa, and multiples of any order)
↓ Inclusion of effect of reflection from free surface		
$\phi_r(r, z, \omega)$, $\psi(r, z, \omega)$	→	Potentials of reflected P- and S-wave in layer 1
↓ From potentials to spectral displacements	→	From P and S wave potentials in layer 1
$U(r, o, \omega)$, $W(r, o, \omega)$	→	Spectral horizontal and vertical displacements at the free surface ($z = 0$)
↓ Conversion of horizontal wave number to angle γ	→	γ real interpreted as angle of incidence at top of reflecting zone
$U(r, o, \omega)$, $W(r, o, \omega)$	→	Integral variable k (horizontal wave number) changed to γ and integral evaluated for a range of angles of incidence, γ_1 to γ_2. Reflectivity $R_{PP}(\omega, k)$ is now $R_{PP}(\omega, \gamma)$.
↓ Use of inverse FFT		
Synthetic seismogram in terms of distance and time.		

Fig. 16.22. Mathematical steps in reflectivity method

An example of the use of the reflectivity method for generating synthetic seismograms for coal seams is given in Fig. 16.23 a, b. Figure 16.23 a shows a two coal seam model separated from each other in an overburden material. Figure 16.22 b shows the synthetic seismogram (vertical component) using the reflectivity method just described.

In Fig. 16.23 b, the P reflections (PP) from two seams arrive in the beginning followed by multiples in between the seams. Two PS pulses (P to S reflected wave) are observed in later parts of the synthetic seismogram. For distances larger than 700 m, the PS reflections are almost as strong as the PP reflections. Also the PS reflections have better time resolution, because of their lower velocities.

16.6.2 Detection of an Overpressured Zone

Overpressured zones are found in the exploration for hydrocarbons. Whenever such zones are likely, care is necessary in drilling when approaching the depth of these zones. Overpressure zones may arise, for example, when a thick-section of clastics may have been laid down quickly, and might have retained water within an impermeable matrix. The pressure in the overpressured zone might be considerably greater than the normal pressure expected at that depth.

Oil sands may well be associated with overpressured shale zones, and there may be more than one prolific oil sand horizon with overpressured shale horizons in between. Therefore, as far as possible accurate predictions of overpressured zones becomes necessary. Interpretations of seismic average or interval velocity, or reflection coefficient may give some clue to the depths of overpressured zones within certain depth limits.

An accurate estimate of average or interval velocity with depth may help in detecting an overpressured zone. Figure 16.24 shows the *knee* form of the velocity depth curve for a location in which overpressured shales are present. However, it is to be ascertained that the decrease of average velocity is due to the low P-wave velocity of shale, and not an artifact of using multiple events or diffractions for velocity determinations.

The overpressured zone gives rise to a gradual change of acoustic impedance, i.e. it acts like a transition zone. Berryman, Goupillad and Water (1958) showed that a characteristic property of transition zones is that they have a high reflection coefficient at low frequencies.

Sometimes a fault may displace overpressured zones on either side by a large amount. In such cases, the contoured interval velocity section superimposed on the normal seismic section may provide important clues to their depths.

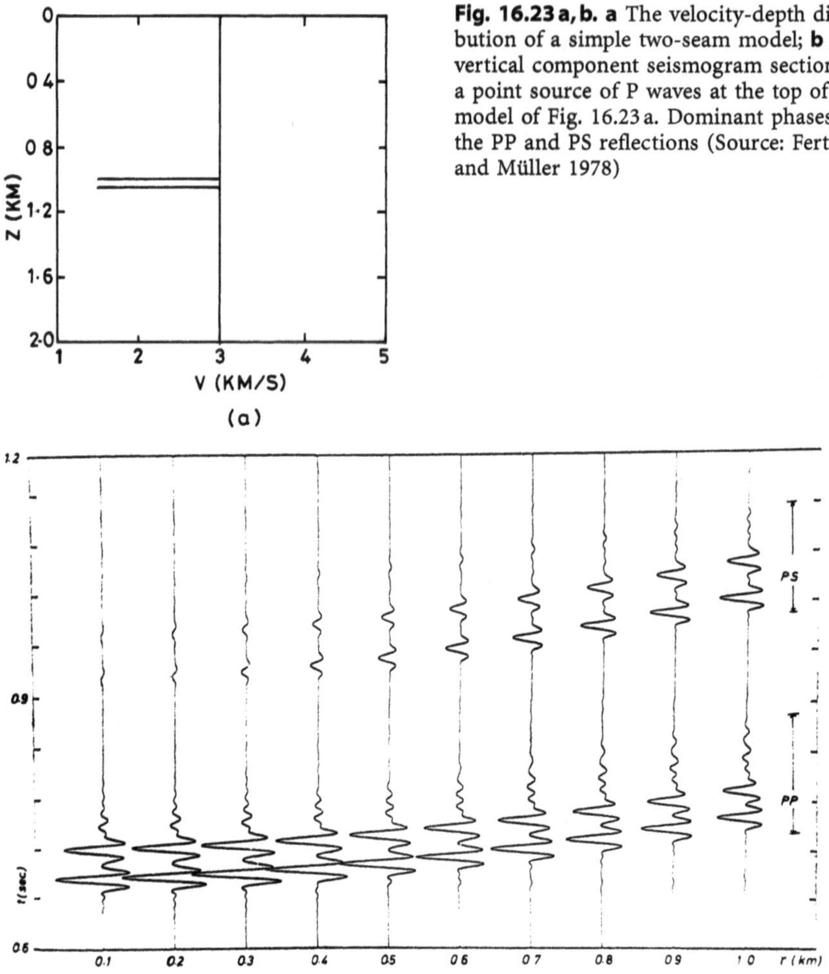

Fig. 16.23 a, b. a The velocity-depth distribution of a simple two-seam model; **b** the vertical component seismogram section for a point source of P waves at the top of the model of Fig. 16.23 a. Dominant phases are the PP and PS reflections (Source: Fertig and Müller 1978)

16.6.3
The Anisotropic Property of a Shale Formation

Sometimes it may be required to explore in detail the anisotropic properties of a shale formation, in order to explain some discrepancies in interpretations. Winterstein and Paulsson (1990) investigated a combined use of *crosshole seismic and VSP data* for better definition of elastic and anisotropic properties for such a case.

Figure 16.24 shows the recording geometry used by Winterstein and Paulsson. The crosshole source was a *hydraulic axial vibrator* (Paulsson 1988) designed to produce S-waves. The vibrator swept frequencies from 10 to 360 Hz. A weight drop was used as a source for the VSP survey. Different types of three-component downhole receivers were employed for crosshole (refer to Wuenschel 1976) and

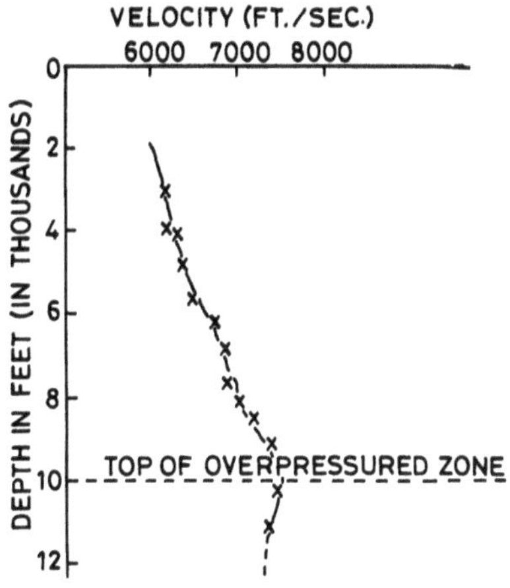

Fig. 16.24. A velocity depth curve, showing the effect of an overpressured zone (Source: Waters 1987 Reflection seismology, 3rd edn. © 1987 by John Wiley and Sons. This material is used by permission of John Wiley and Sons)

VSP recordings. For VSP data, the receiver in well I was lowered from 50′ to 750′ in increments of 50′ for recording wave data at each level from weight drops on the surface. Two sets of crosshole data were generated. In one set, the vibrator source was fixed at a depth of 450′ in well I, and receivers were lowered successively in well II by 25′ from a depth of 25 to 725 ft. The second crosshole data set was obtained by lowering simultaneously both source and receiver each by 25 ft.

The well axis was very nearly vertical and three components (V, H1, and H2) were recorded for each level of the receivers in the VSP or crosswell experiments.

In the case of the crosshole survey, raw data were found interpretable as some key features, like separated arrivals of the P wave, and two S waves could be marked. P-wave arrival times were from 40–90 ms, which were separated from two S-wave arrivals between 100–250 ms. To optimize the two S-wave (characteristic of the anisotropy of the medium) arrivals, processing steps like subtracting *tube waves*, reversing some polarities, forcing codas to be zero, and rebalancing wavelet amplitudes were adopted for the crosshole raw data. Also, in the optimization process, two S-wave arrivals were first identified, and receiver coordinate frames were rotated in order to minimize the energy on one of the components over a time window that included only the faster of the two waves. The purpose of the rotation was to minimize the energy on one of the two components over a chosen time window. Figure 16.25 shows different S-wave arrivals from a fixed source after *data conditioning*. Data conditioning was found more stringent for the VSP record. It was realized that rotating VSP S-wave data was a necessary interpretation step. In the final stage of rotation of the S-wave data, use was made of cross-correlation criterion. The angle that separates best the two S-waves is defined by that angle which makes the peak value of their cross-correlation function a global maximum (Winterstein and Paulsson 1990, p 474).

Preliminary modeling assumed that shale is transversely isotropic with a vertical symmetry axis. Based on this assumption, values of stiffness constants c_{33}, c_{11}, c_{44} and c_{66} were found by analyzing P, S_1 and S_2 VSP and crosshole records. The value of c_{33} was determined from vertically traveling P-wave arrivals in the

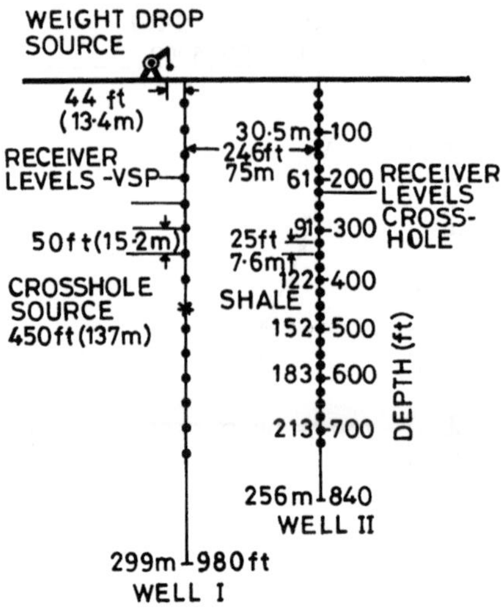

Fig. 16.25. Recording geometry at Chevron-experimental well site. Three separate data sets were recorded. The two indicated here are the crosshole with source at fixed depth and the VSP with weight drop source (Source: Winterstein and Paulsson 1990)

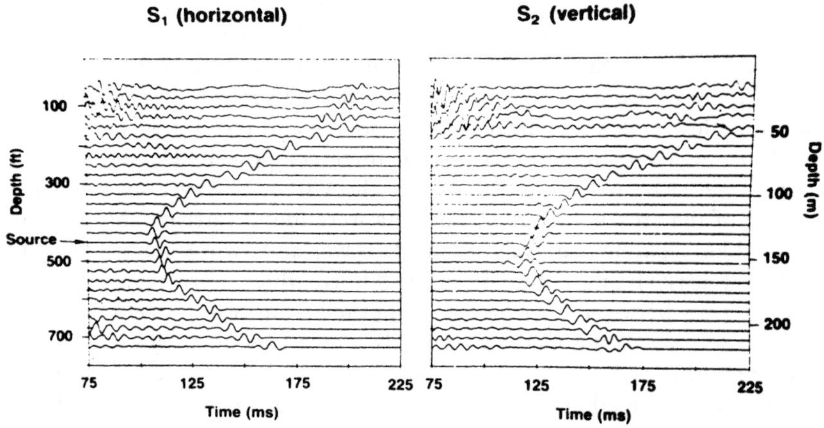

Fig. 16.26. Crosshole S-wave arrivals from the fixed source after data conditioning. Data conditioning involved subtracting a tube wave and a P-wave reflection, rotating signal energy onto separate components, zeroing codas, and equalizing amplitudes (Source: Winterstein and Paulsson 1990)

VSP record. Values of c_{11}, c_{44} and c_{66} were found from horizontally traveling P-wave and S-wave crosshole arrivals. For determining the value of c_{13} non-horizontally traveling crosshole arrivals were used.

S-wave birefringence, both on the crosshole and VSP record, were analyzed to evaluate the tilt of the *symmetry axis*. Detailed analysis of the crosshole arrival time data at depths near the source depth revealed that the transversely isotropic medium was isotropic for two of the waves (P and S_2) out of three waves (P, S_1 and S_2) (Fig. 16.26). The S_1 wave exhibited ellipsoidal anisotropy.

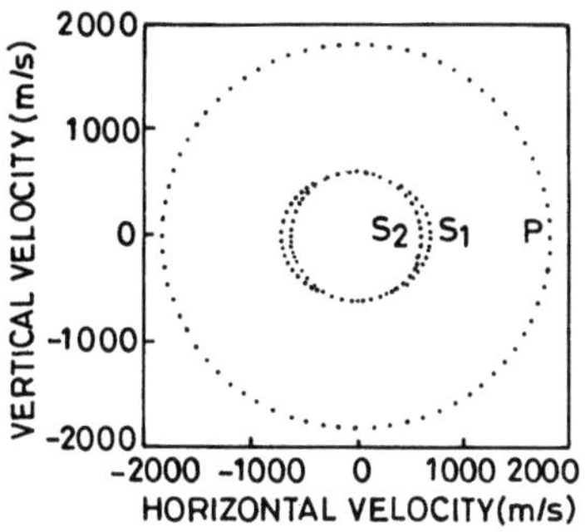

Fig. 16.27. Group velocity surfaces for the TI model that best fit crosshole data at depths near the source depth. The *P* and S_2 wave surfaces are almost perfect circles. The S_1 surface is an ellipse with the ratio of major to minor axis of 1.15 (Source: Winterstein and Paulsson 1990)

It may be mentioned that as the measurements taken for crosshole and VSP data are direct, there is a greatly reduced chance of misinterpretation. The measurements are in situ, and provide a guideline for planning exploration of a given depth interval in detail.

CHAPTER 17

Exploration Value of Fracture-Induced Anisotropy 17

17.1
The Objective and Need for Specific Development

Backus (1962) showed that a heterogeneous mixture of isotropic elements would reveal itself as homogeneous and anisotropic when the scale of its fabric is smaller than the seismic wavelengths that measure it (refer to Subsection 5.3.1 for details). The fabric may originate from thin layering or from oriented micro-cracks or fractures. Further, in some situations, anisotropy may be quite complex resulting from fractures in a previously anisotropic medium. Under such situations, anisotropy may combine and produce a combined effect. Therefore their effects have to be separated to study anisotropy due to fractures or pre-existing anisotropy.

An important objective in seismic investigations is to determine the cause of anisotropy, and locate and map the zone of anisotropy on a fine scale. In the case of fracture-induced anisotropy, the need for specific development is to clearly understand the relationships between crack or fracture characteristics and their various seismic responses, with the aim to use them effectively for seismic imaging. Knowledge of fracture-characteristics and their connectivity helps to understand fluid transport through the rock mass. The fracture characteristics include orientation of fractures, their density and aspect ratio, and the fluid contents filling the fractured zone.

17.2
Observable Signatures of Seismic Anisotropy

The most common form of observable signature of seismic anisotropy is the directional dependence of velocity. Such results are well known, both from field and laboratory measurements. Attempts to quantify fracture characteristics from measured seismic parameters have recently been emphasized. In the following, we discuss these relationships.

When a shear wave propagates through vertically aligned fractures, *shear wave splitting* is observed; split shear waves are denoted by $s_\perp$ and s_{11}. $s_\perp$ denotes the shear mode with polarization perpendicular to the fracture strike plane. It propagates at the velocity of fractured rock, and is referred to as a *slow shear wave.*

Similarly, S_{11} denotes the shear mode with polarization parallel to the strike plane of the fractures and is called the *fast shear wave*. It propagates at the velocity of unfractured rock. In Crampin et al. (1986) it was observed that, to the first order, the group velocity of quasi-P(q^P) and slow shear waves ($s_\perp$) vary with *fracture density* alone and have 2θ variations, where θ is defined as the azimuth angle measured from the crack normal direction. Further, in water-saturated fractures – in addition to *fracture density* – the *crack aspect ratios* (crack shape) has an important influence on quasi-P(q^P) and shear wave ($s_\perp$) group velocities. Thomsen (1995) found the angular dependence on aspect ratio of (q^P) and $s_\perp$ phase velocities. For a more complete coverage of the effective properties of materials containing cracks, reference may be made to recent publications by Thomsen (1995) and Liu et al. (2000).

Field as well as analytical results provide proof of observable anomalies in stacking and P-wave *NMO velocities*, due to aligned fractures. For example, Paul (1993) noted anomalously low stacking velocity that was attributed to localized fractures. In another study, Lynn et al. (1995) attributed azimuthally dependent stacking velocities to the presence of fractures. Corrigan et al. (1996) found substantial azimuthal variations in NMO velocities in fractured reservoirs. Theoretical investigations by Thomsen (1988), Tsvankin (1997) and Sayers and Ebrom (1997) open up new possibilities for finding crack orientation, and in some cases to estimate the crack density in fractured reservoirs by making use of P-wave NMO velocities.

Fracture-induced anisotropy has angle and azimuth-dependent effects on amplitudes. Mallick et al. (1998) found periodic azimuthal dependence of P-wave reflection amplitudes as characteristic of anisotropy induced by fractures. Sayers and Rickett (1997) investigated the effect of natural fractures on the azimuthal AVO response from a gas sandstone reservoir encased within shale. Ramos and Davis (1997) used a 3-D AVO technique to map crack density in a coal-bed methane reservoir. In general, both numerical simulations and physical experiments point towards azimuthal amplitude versus offset (AVO) variations that may be interpreted for determining the principal direction of fractures (Mallick and Frazer 1991; Chang and Gardner 1993; Lynn et al. 1996; Sayers and Richett 1997). Shen et al. (2002) conclude that for quantitative characterization of a fractured reservoir P-wave azimuthal AVO variation alone is not sufficient and should be combined with other types of data for accurate results. Shen et al. have studied closely what causes variations in anisotropy parameters ($\delta^{(v)}$, $\varepsilon^{(v)}$, $\gamma^{(v)}$) for fractured reservoirs.

$\delta^{(v)}$ and $\varepsilon^{(v)}$ show differing behavior for gas-saturated and water-saturated fractures. Their analysis shows that for gas-saturated fractures, the absolute value of anisotropy parameters $\delta^{(v)}$ and $\varepsilon^{(v)}$ increase with increasing fracture density alone (Shen et al. 2002, Fig. 1). However, in water-saturated fractures, $\delta^{(v)}$ and $\varepsilon^{(v)}$ depend both on fracture density and crack aspect ratio. It is only anisotropy parameter $\gamma^{(v)}$ that is most related to crack density only, and is insensitive to saturating fluid and aspect ratio (refer to Shen et al. 2002, Figs. 1 and 2).

It should be noted that the expected AVO response over a fracture reservoir may be changed due to anisotropic (refer to Tsvankin 1995) overburden rock. Further, in the case of a thin fractured reservoir, the *tuning effect* (depending on wavelength and reservoir thickness) can also produce changes in expected AVO gradient and azimuthal AVO variations. Unless these effects are not corrected, AVO response results may be misleading.

Castoro et al. (2001) have discussed remedial measures to remove the effects of *offset-dependent tuning* and *NMO stretch.* They claim that by using their technique the correct AVO gradient relative to normal incidence amplitude may be recovered.

The common occurrence of anisotropy complicates AVO analysis. As a means of understanding anisotropic behavior on an AVO cross plot, Chen et al. (2001) suggest three parameter cross plots of AVO intercept (A), gradient (B), and curvature (C) terms.

17.3 Elastic and Seismic Properties of Fractured Rocks

Consider an isotropic medium containing a single set of vertically aligned fractures. The effective compliance tensor of fractured rock is the sum of the compliance tensor of the unfractured background rock, s^b_{ijkl} and the excess compliance tensor, s^f_{ijkl}, induced by fractures (Liu et al. 2000)

$$s_{ijkl} = s^b_{ijkl} + s^f_{ijkl} \tag{17.1}$$

By assuming that fractures are invariant with respect to rotation about an axis along fracture normal direction, the fracture compliance tensor set reduces to only two i.e. the fracture normal compliance tensor and the tangential compliance tensor (Schoenberg and Sayers 1995). The fracture compliance tensor may be expressed as (refer, for example, to Liu et al. 2000; Shen et al. 2002)

$$s^f_{ijkl} = \frac{\eta}{4\mu}\left[U_{11}\left(\delta_{jk}n_in_j + \delta_{jk}n_ln_i + \delta_{il}n_kn_j + \delta_{jl}n_kn_i\right) + 4(U_{33} - U_{11})n_in_jn_kn_l\right] \tag{17.2}$$

Here μ is the Lame' parameter of the unfractured rock; n_i and n_j are components of the local unit normal to the surface S; η is the *fracture density* – equal to the number of fractures per unit volume (i.e. number density) times the cube of the average radius of circular cracks. U_{11} and U_{33} are values of the displacement discontinuities resulting due to the response of a circular crack to shear and normal stresses, respectively. Hudson (1981) gave expressions for U_{11} and U_{33} for fluid-filled penny-shaped cracks (refer to Shen et al. 2002):

$$U_{11} = \frac{16}{3}\left[\frac{\lambda + 2\mu}{3\lambda + 4\mu}\right][1 + M]^{-1} \tag{17.2 a}$$

where

$$M = \frac{4}{\pi}\left[\frac{\mu'}{\frac{c}{a}\mu}\right]\left[\frac{\lambda + 2\mu}{3\lambda + 4\mu}\right] \tag{17.2 b}$$

and

$$U_{33} = \frac{4}{3}\left[\frac{\lambda + 2\mu}{\lambda + \mu}\right][1 + K]^{-1} \tag{17.2 c}$$

where

$$K = \frac{1}{\pi}\frac{\left[k' + \frac{4}{3}\mu\right]}{\frac{c}{a}\mu}\left[\frac{\lambda + 2\mu}{\lambda + \mu}\right] \tag{17.2 d}$$

'c/a' in Eqs. (17.2 b) and (17.2 d) is the *crack aspect ratio* (crack shape); λ and μ are the Lame′ parameters of unfractured rock, and λ' and μ' are Lame parameters of the fluid content.

The *elastic stiffness tensor* may be obtained by inverting the compliance tensor calculated using Eqs. (17.2) to (17.2 d). Fracture density (η), saturated fluid content, and crack aspect ratio are referred to as *fracture parameters*. It is clear, from the above step, that seismic anisotropy is a function of fracture parameters. For example, an increase of fracture density should lead to a decrease in the elastic stiffness tensor of a fractured medium. Further, considering fluid-saturated fractures, M in Eq. (17.2 b) would vanish because of the zero shear modulus of the fluid. Therefore, the discontinuity in shear displacement U_{11} is a function of Lame′ parameter of unfractured rock only. However, discontinuity of normal displacement, U_{33}, is affected (through K) by the bulk modulus. These features account for differences between the seismic anisotropic characteristics of P and S waves in fluid-saturated fractured rocks.

Equations (17.1) to (17.2 d) describe the relationship between stiffness coefficients and *fracture parameters*. Since anisotropy parameters (δ, ε, γ) are expressible in terms of five independent elastic stiffness coefficients, it is clear that anisotropy parameters are linked to the fracture parameters, and the relationship between these must be understood to better interpret NMO velocity and amplitude in terms of anisotropy parameters. The anisotropy parameter ($\delta^{(v)}$, $\varepsilon^{(v)}$, $\gamma^{(v)}$) for a medium containing vertically aligned cracks, referred to as a HTI medium (refer to Subsection 5.3.2) are defined by Eqs. (5.46 a–c).

It may be mentioned that the parameter $\delta^{(v)}$ is responsible for near vertical P-wave velocity variations in the crack normal direction. Parameter $\varepsilon^{(v)}$ defines P-wave anisotropy, i.e. the difference between P-wave velocities along the crack normal and vertical direction. It is a measure of magnitude of P-wave anisotropy. For HTI media, $\delta^{(v)}$ and $\varepsilon^{(v)}$ are typically negative (refer to Shen et al. 2002). $\gamma^{(v)}$ is the parameter governing the degree of shear-wave splitting at near normal (vertical) incidence.

$\delta^{(v)}$ may be expressed as

$$\delta^{(v)} = \frac{\left[\frac{c_{13}}{c_{55}} + 1\right]^2 - \left[\left[\frac{V_{Pvert}}{V_{s\perp}}\right]^2 - 1\right]^2}{2\left[\frac{V_{Pvert}}{V_{s\perp}}\right]^2\left[\left[\frac{V_{Pvert}}{V_{s\perp}}\right]^2 - 1\right]} \tag{17.3}$$

[Shen et al. 2002, Eq. (10)]

where V_{Pvert} and $V_{s\perp}$ are given by $\sqrt{(c_{33}/\rho)}$ and $\sqrt{(c_{55}/\rho}$, respectively. It may be mentioned that V_{S11} is given by $\sqrt{(c_{44}/r)}$ and is also equal to $V_{s\perp}/\sqrt{(1+2\gamma^{(v)})}$. This equation shows that $\delta^{(v)}$ is a function of the velocity ratio $V_{Pvert}/V_{s\perp}$ and stiffness constants ratio c_{13}/c_{55}.

Because of the ratio c_{13}/c_{55}, which is significantly affected by the fluid content, $\delta^{(v)}$ shows differing characteristics in gas- and water-saturated fractured rocks (Shen et al. 2002).

NMO velocities are routinely measured from surface seismic data. It is possible to give the principal orientations of fractures from the azimuthal variation picture of NMO velocity. The method is based on contributions by Tsvankin (1997) and Grechka and Tsvankin (1998). Tsvankin (1997) showed that exact NMO velocity from horizontal reflectors in HTI media is a simple function of three parameters: the vertical velocity, the azimuthal angle (a) between the survey line and symmetry axis, and the anisotropic term A:

$$V_{nmo}^2 = V_{Pvert}^2 \frac{1+A}{1+A\sin^2 a} \tag{17.4}$$

[Tsvankin 1997, Eq. (21)]
where A [P-wave] $= 2\delta^{(v)}$.

Grechka and Tsvankin (1998) showed that azimuthal NMO velocity variation represents an ellipse in the horizontal plane, and the orientation of the axis is determined by the properties of the medium and the direction of reflection normal. $\delta^{(v)}$ for a HTI medium is typically negative. This implies that, in any other direction different from the fracture strike direction, the NMO velocity is smaller than the vertical P-wave velocity. In the fracture strike direction, the NMO velocity is the same as vertical P-wave velocity since there is no influence of fracture. This implies that the long axis orientation of the NMO velocity ellipse is the indicator of principal orientation of fractures. It may be mentioned that at least three NMO velocity measurements are needed to generate the best-fit velocity ellipse of NMO velocity.

The discussions in this section are mainly based on the very recent publication by Shen et al. (2002).

17.4 Stress Anisotropy and Fracturing

Dilatancy is a term used to signify an increase in volume due to the opening of cracks in rocks subjected to shear stress (Doyle et al. 1982). Aligned stress-induced cracks in the dilatancy zone produce anisotropic effects on seismic velocities and elastic parameters. This effect was noted in laboratory experiments with rock samples under loading (refer, for example, to Brace, Paulding and Scholz 1966; Gupta 1973; Hadlay 1975). Brace et al. (1966) showed that such cracks open parallel to the directions of maximum compressive stress with faces perpendicular to the direction of least compressive stress (Fig. 17.1). Further, such dilatancy cracks are effectively anisotropic to seismic waves with wavelengths longer than the crack di-

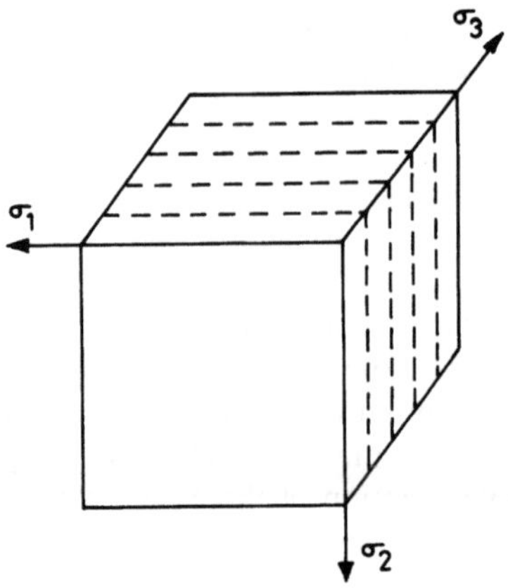

Fig. 17.1. σ_1 – the maximum, and σ_3 – the minimum compressive stress directions. Cracks open parallel to maximum stress direction, and their faces are perpendicular to the minimum stress direction (after Crampin 1987)

mensions and shorter than the dimensions of the dilatancy zone (Crampin and McGonigle 1981). The term extensive-dilatancy anisotropy (EDA) has been assigned for anisotropy resulting from the alignment of stress-aligned fluid-filled cracks, microcracks, and preferentially oriented pore space pervading most rocks in the upper most 10 to 20 km of the crust (refer, for example, to Crampin and Lovell 1990).

We would briefly discuss some salient features of rock failures by fractures under the action of stress based on laboratory experiments and theoretical deductions. Under the action of stress, most rocks deform elastically and failure by rupture occurs at the elastic limit. Ruptures may be classified as *tension* or *shear fractures*. Tension fractures are the result of stresses that tend to pull the specimen apart. Shear fractures result due to stresses that act to slide one part of the rock sample with respect to the adjacent part. It may be mentioned that tension fractures may result not only from tension, but also from couples and even from compression (Billings 1999, p 154). Also, shear fractures may develop not only from compression, but may also form under the actions of couples or tension.

With the above very brief qualitative introduction on the relationship between ruptures and applied stress, we would now explore its basis quantitatively.

Stress at a point is referred to three axes perpendicular to each other. These axes are called the principal stress axes. Figure 17.2 a shows these axes. These axes coincide with the maximum, intermediate and minimum stress directions at the point. All three stresses may be of one type, or a combination of different types. Figure 17.2 a is an example of all three compressive stresses. ξ_1 is the greatest, ξ_2 the intermediate and ξ_3 the least stress. Geological sign convention of stresses is the following: compression is taken as positive, and tension as negative. The *stress difference* is defined as the algebraic difference of the greatest and the least principal stresses. Suppose that a cylindrical rock sample is subjected on its top and bottom to a tensile stress equal to $-1{,}500$ kg/cm^2, and the curved surface of the sample is under a confining compressive stress equal to 200 kg/cm^2. In this case, the intermediate and greatest stresses are each equal to $+200$ kg/cm^2 and the least stress is $-1{,}500$ kg/cm^2. Therefore, the stress difference (greatest minus the least) is equal to 1,700 [i.e. $200-(-1{,}500)$] kg/cm^2.

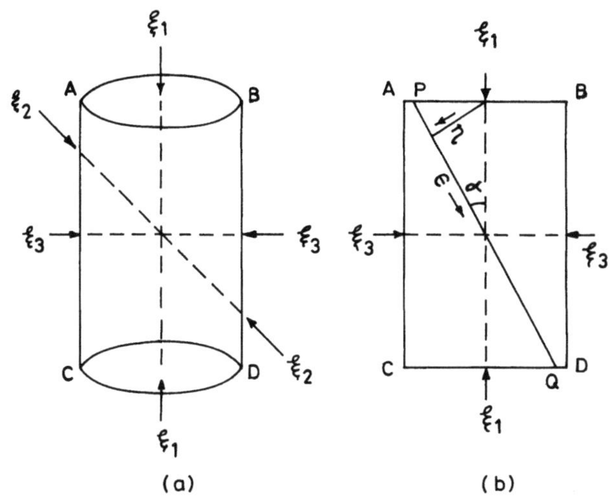

Fig. 17.2 a, b. a Compressive stresses: ξ_1 (greatest), ξ_2 (intermediate), ξ_3 (least); **b** section through the cylinder: ε is shearing stress, η the normal stress and a is the angle between ξ_1 and the plane

In order to find a relationship between the fracture plane and the direction of maximum stress, we consider that a cylindrical sample (Fig. 17.2 a) is subjected to compressive stresses: ξ_1 (the greatest), ξ_2 (the intermediate), and ξ_3 (the least). To illustrate our point we consider a section (Fig. 17.2 b) of the cylindrical sample defined by the plane passing through ξ_1 and ξ_3. The intermediate stress (ξ_2) is perpendicular to the plane of the paper and is neglected in the analysis.

Prior to the fracture, the fracture plane is under the action of tangential, and normal stresses acting on it and working to shear rupture the sample along this plane. At the same time, the sample gives resistance (to shearing) that is the sum of the *internal friction*, and *cohesive strength* of the sample. The meaning of these and the method of determining their values can be best explained using the *Mohr stress circle.*

First referring back to Fig. 17.2 b, the shearing (τ) and normal (η) stress along the plane (a line PQ in the section) is

$$\varepsilon = \frac{\zeta_1 - \zeta_2}{2}(2 \sin a \cos a) = \frac{\zeta_1 - \zeta_2}{2} \sin 2a \qquad (17.5\,a)$$

and

$$\eta = \frac{\zeta_1 + \zeta_2}{2} - \frac{\zeta_1 - \zeta_2}{2} \cos 2a \qquad (17.5\,b)$$

where a is the angle the plane PQ makes with the greatest principal stress axis.

From Eq. (17.5 a) the shearing stress (ε) is maximum when $a = 45°$, also normal stress (η) is maximum when $a = 90°$ (because $\cos 2a = -1$) and minimum when $a = 0°$.

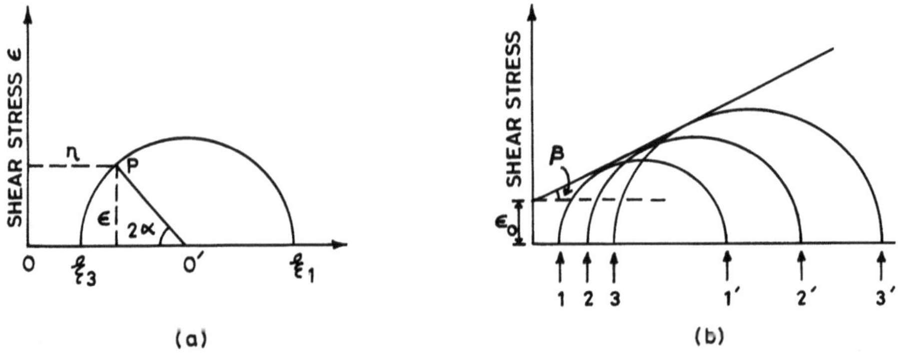

Fig. 17.3 a, b. a Mohr stress circle. **b** Mohr stress envelope

Shear fracture takes place when the combined effect of shear stress and normal stress (η) becomes most effective.

Mohr stress circle provides a graphical method to depict the relationship between the stresses and the rupture. In Fig. 17.3 a, the Mohr stress circle is shown based on a single triaxial compression test run on a cylindrical rock sample. Maximum stress is taken along the axis of the cylinder, and intermediate (ξ_2) and least (ξ_3) stresses are taken as equal. In this figure, O is taken as the origin, and the vertical axis depicts the shearing stress (ε). The values of confining pressure (ξ_3) and compressive stress ξ_1 (causing rupture) are plotted along the horizontal axis. A circle is drawn with the center at mid-point O between ξ_1 and ξ_3 [having its coordinate value as $(\xi_1+\xi_3)/2$] and a radius equal to $(\xi_1-\xi_3)/2$ where $(\xi_1-\xi_3)$ is the stress difference, i.e. the greatest stress minus the least stress].

Considering a plane making an angle a with the greatest principal stress ξ_1, the shearing and normal stresses on this plane are determined by drawing a line, at an angle $2a$ taken in the clockwise direction from the horizontal axis. The coordinates of this point, P, give the values of shear and normal stress on this plane. These are given by expressions Eqs. (17.5 a) and (17.5 b).

In order to determine the values of the cohesive strength and angle of internal friction, a series of experiments are run for different values of confining pressure. Each experiment is done until rupture takes place. Figure 17.3 b depicts the graphical plot of Mohr circles for each experiment. The Mohr circle cuts the horizontal axis at two places, the left side value is the confining pressure and right side value is the stress at which the rupture occurs in each experiment. Experimental results through Mohr circle diagrams reveal that with an increase in confining pressure, both the stress and stress difference must be increased to cause rupture.

The angle of the line, drawn as the envelope to the Mohr circles, with the horizontal axis, gives the angle of internal friction β. The equation of this line is

$$\varepsilon = \varepsilon_0 + \eta \tan\beta$$

[Billings 1999, Eq. (5)]
where ε_0 is the *cohesive strength* or *adhesive strength* of the rock.

The angle which the fracture plane should make with the greatest principal stress is $\alpha = \pm 45 \pm \beta/2$.

If the angle of friction is 34°, the fracture should make an angle of 28° with the direction of the greatest principle stress axis.

17.5 Model of Vertically Fractured and Horizontally Stratified Media

Fine-scale layering in the Earth's crust is prevalent. Common examples are sedimentary basins, and regions of basaltic volcanism. Fine scale layering produces transverse isotropy, having a symmetry axis perpendicular to the layering. After the formation of rock mass, it may get subjected to regional stresses which gives rise to micro-cracks normal to minimum compressive stress (Hubbert and Willis 1957; Crampin 1985). The opening of aligned cracks results in anisotropy. The total seismic response of fine layering and aligned cracks causes compound anisotropy, which is combined anisotropy due to fine layering (formational anisotropy) and aligned micro-cracks.

In a similar way, horizontal layering from lava flows, breccia zones and pillow units results in anisotropy, which gets enhanced by horizontal fractures (Fryer et al. 1989). Horizontal fracturing occurs at the time of crustal formation (Newmark et al. 1985). Vertical fractures are produced in the crust at the time of uplift during the formation of a rift valley (Stephen 1985). Layering and fracturing thus produce compound anisotropy.

Schoenberg (1986) showed that horizontal bedding and a set of parallel vertical fractures generate orthorhombic symmetry in long wavelength limit. When the single set of parallel fractures is neither perpendicular nor parallel to the bedding, the equivalent medium is monoclinic (Schoenberg 1985). These types of situations as well as those involving triclinic symmetries are observed in the Earth, which may be dealt with by matrix algebra and ideas from group theory.

Hood and Schoenberg (1989) dealt with the problem of separating anisotropic effects of fractures from pre-existing anisotropy in formation or the background anisotropy. Specifically they considered a transversely isotropic formation, which develops vertical fracturing, and developed a method of separating transverse isotropy from fracture-induced *azimuthal anisotropy* (refer to Subsection 5.3.2). Attempts to model anisotropic effects of fractures and separate it from the background anisotropy effect would also be found in the works of Schoenberg and Douma (1988), Nicholas et al. (1989), Hood (1991), Schoenberg and Sayers (1995), and Sayers and Ebrom (1997).

In the case of compound anisotropy, the effects of both layering and fracturing in producing anisotropic effect is to be analyzed. In the pioneering work of Schoenberg and Muir (1989), calculus was developed that simplifies in a big way the combined effects of layering and fracturing. Hood and Schoenberg (1989) make use of Schoenberg and Muir calculus for the estimation of vertical fracturing from measured elastic moduli from seismic velocity measurement. They derived explicit formulae for the fracture compliance, and background transversely isotropic elastic moduli.

17.5.1
Schoenberg and Muir Calculus: Quantification of Fracture Compliance

In Schoenberg and Muir calculus, each constituent in a finely layered medium, and each set of fractures can be represented by a separate group of elements (G-elements) in a transform domain. This permits fractures and anisotropy of rocks to be dealt with in a consistent and uniform manner. In the transform domain, G-elements may be combined using simple addition (assuming non-linear interaction between constituent layers and/or fracture sets). The result may be transformed back to the physical domain to generate elastic properties of the more complicated system. Complex anisotropic media may be decomposed in the transform domain, and simple subtraction would bring out properties of the separate anisotropic constituents of the total system-like sets of parallel fractures, aligned micro-cracks, etc. Multiple fractures may also be accounted for by rotating a rock appropriate to each fracture system, and adding the appropriate fracture system group element. A brief discussion of Schoenberg and Muir (1989) calculus is given below from Hood and Schoenberg (1989).

Schoenberg and Muir (1989) make use of matrix algebra and group theory to map properties of fine layers and/or fracture, to an element of a commutative group.

Assume a region of thickness H composed of layers with welded interfaces, where the x_1-axis is taken as perpendicular to layering. Let the total thickness of the ith constituent be H_i, and the density be ρ_i. Further, let three elastic moduli submatrices be C_{NN}, C_{TN} and C_{TT} which are defined as

$$C_{NN} = \begin{bmatrix} c_{11} & c_{15} & c_{16} \\ c_{15} & c_{55} & c_{56} \\ c_{16} & c_{56} & c_{66} \end{bmatrix}$$

$$C_{TT} = \begin{bmatrix} c_{22} & c_{23} & c_{24} \\ c_{23} & c_{33} & c_{34} \\ c_{4} & c_{34} & c_{44} \end{bmatrix}$$

$$C_{TN} = \begin{bmatrix} c_{12} & c_{25} & c_{26} \\ c_{13} & c_{35} & c_{36} \\ c_{14} & c_{45} & c_{46} \end{bmatrix} \mathbf{C_{NT}} = \mathbf{C^{t}_{TN}} \tag{17.6}$$

These submatrices are formed from an expression for Hook's law for a general elastic medium:

$$\begin{bmatrix} \sigma_1 \\ \sigma_2 \\ \cdot\cdot \\ \cdot \\ \cdot \\ \sigma_6 \end{bmatrix} = \begin{bmatrix} c_{11}c_{12}\ldots\ldots\ldots\ldots c_{16} \\ c_{12}c_{22}\ldots\ldots\ldots\ldots c_{26} \\ \ldots\ldots\ldots\ldots\ldots\ldots\ldots\ldots \\ \ldots\ldots\ldots\ldots\ldots\ldots\ldots\ldots \\ \ldots\ldots\ldots\ldots\ldots\ldots\ldots\ldots \\ c_{16}c_{26}\ldots\ldots\ldots\ldots\ldots c_{66} \end{bmatrix} \begin{bmatrix} \varepsilon_1 \\ \varepsilon_2 \\ \varepsilon_3 \\ \\ \\ \varepsilon_6 \end{bmatrix} \tag{17.7}$$

In terms of submatrices, Eq. (17.7) may be written as

$$\begin{bmatrix} \sigma_1 \\ \sigma_5 \\ \sigma_6 \end{bmatrix} = C_{NN} \begin{bmatrix} \varepsilon_1 \\ \varepsilon_5 \\ \varepsilon_6 \end{bmatrix} + C_{NT} \begin{bmatrix} \varepsilon_2 \\ \varepsilon_3 \\ \varepsilon_4 \end{bmatrix} \tag{17.8 a}$$

and

$$\begin{bmatrix} \sigma_2 \\ \sigma_3 \\ \sigma_4 \end{bmatrix} = C_{TN} \begin{bmatrix} \varepsilon_1 \\ \varepsilon_5 \\ \varepsilon_6 \end{bmatrix} + C_{TT}\varepsilon_2 \begin{bmatrix} \varepsilon_2 \\ \varepsilon_3 \\ \varepsilon_4 \end{bmatrix} \tag{17.8 b}$$

NN relates stress and strain components acting on plane normal to x_1 (i.e. 1, 5, 6 components); TN acting in the plane normal to x_1 (i.e. 2, 3, 4 components); NT is the transpose of the TN submatrix and relates normal stress components to tangential strain components.

H_i, ρ_i, C_{NNi}, C_{TNi} and C_{TTi} are the physical model parameters of the ith constituent. These five physical model parameters map into a 5-vector $\mathbf{G}_i$. The first two components of **G** are scalars; and the third, fourth and fifth components are 3×3 matrices. Mapping is given by

$$\begin{bmatrix} H_i \\ H_i\rho_i \\ H_i C_{NNi}^{-1} \\ H_i C_{TNi} C_{NNi}^{-1} \\ H_i\left[C_{TTi} - C_{TNi} C_{NNi}^{-1} C_{NTi}\right] \end{bmatrix} = \begin{bmatrix} g_i(1) \\ g_i(2) \\ g_i(3) \\ g_i(4) \\ g_i(5) \end{bmatrix} = G_i \tag{17.9}$$

For any G_i with $g_i(1) \neq 0$, and $g_i(3)$ invertible, the set of physical model parameters are found by inverse mapping:

$$\begin{bmatrix} g_i(1) \\ g_i(2)/g_i(1) \\ g_i(1) g_i(3)^{-1} \\ g_i(4) g_i(3)^{-1} \\ \left[g_i(5) + g_i(4) g_i(3)^{-1} g_i(4)^t\right]/g_i(1) \end{bmatrix} = \begin{bmatrix} H_i \\ \rho_i \\ C_{NNi} \\ C_{TNi} \\ C_{TTi} \end{bmatrix} \tag{17.10}$$

[Hood and Schoenberg 1989, Eq. (A7)]

This formulation tells that 5-vector **G** of a homogeneous medium that is equivalent to a layered medium composed of n constituents is simply the sum of the 5-vector G of the n-constituent media, i.e.

$$G = \sum_{i=1}^{n} G_i \tag{17.11}$$

[Hood and Schoenberg 1989, Eq. (A8)]

In the domain of the G-vector, addition corresponds to combining layers. The set of physical model parameters of the equivalent medium is found by inverse mapping [Eq. (17.10)] to G.

Thus, G of a vertically fractured transversely isotropic medium may be found by adding G_f of the set of fractures to the G_b of the background, i.e.

$$G = G_f + G_b \tag{17.12}$$

This forms the basis in the investigation of Hood and Schoenberg (1989). The background medium is transversely isotropic and embedded in this medium is a set of large vertical fractures perpendicular to the x_1 direction. This results in a model with orthorhombic symmetry.

5-G elements of the vertically fractured, horizontally stratified media (orthorhombic media) is written using Schoenberg and Muir calculus.

G_b, i.e. of the background rock, is written as

$$G_b = G - G_f, \tag{17.13}$$

expressing

Background rock fractured rock – vertical fractures

Equation (17.13) gives $g_b(1)$ to $g_b(5)$ for the background medium.

Inverse mapping Eq. (17.10) is used to retrieve the elastic moduli of the background medium.

The unknown fracture compliances are evaluated by satisfying 4 conditions for the background medium to be transversely isotropic,
i.e.

$$c_{44b} = c_{55b}, \qquad c_{13b} = c_{23b}$$

$$c_{11b} = c_{22b}, \text{ and } c_{12} = c_{11} - 2c_{66b} \tag{17.14}$$

This gives fracture compliances from measurement parameters of a vertically fractured horizontally stratified (orthorhombic) medium.

17.6
An Inverse Method to Separate Fracture Anisotropy from Background Anisotropy

Hood (1991) describes fractures in terms of their effect on compliance moduli of the background medium and proposes an inverse method to separate fracture-induced anisotropy. Hood's method does not require use of Schoenberg and Muir calculus. If $\mathbf{S}_b$ represents the compliance matrix for the unfractured background medium, the fracture system introduces additional compliance **F**, giving

$$\mathbf{S} = \mathbf{S}_b + \mathbf{F} \tag{17.15}$$

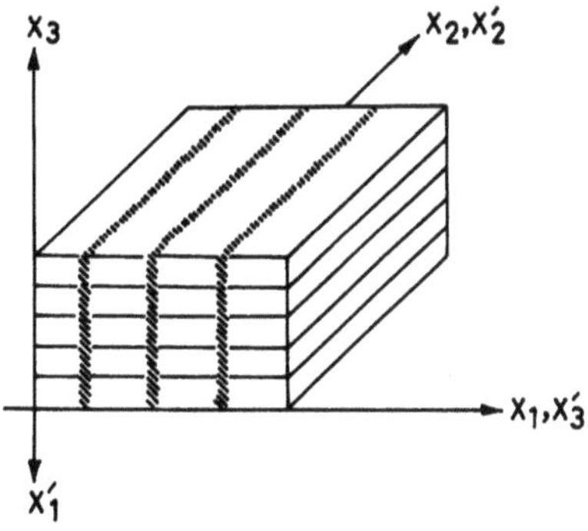

Fig. 17.4. Vertical fractures in a layered medium: fracture compliance Z_N is normal to fracture planes along x_1 (=x_3'); fracture compliance Z_2 parallels the fracture strike along x_2 (=x_2'); fracture compliance Z_3 along x_3 (=$-x_1'$) (Source: Hood 1991)

Hood (1991) chooses two coordinate systems: one is the observation coordinate system (x_1, x_2, x_3) and the other is the fracture-oriented coordinate system (x_1', x_2', x_3') and Hood refers to it as the fracture *eigen-coordinate system* (Fig. 17.4). Eigen coordinates and compliance of different fracture sets may be transformed individually to those of background compliances S_b and any number of sets may be included provided no non-linear interaction occurs between constituent fracture systems.

In the fracture coordinate system, it may be expressed as

$$S' = S_b + F' \tag{17.16}$$

Transformations back from the fracture coordinate system to the observation coordinate gives

$$S = S_b + \Delta S$$

ΔS is the fracture contribution to compliance S and is rotated from eigen coordinates to observation coordinates. Different fracture sets have different eigen coordinates. Assuming that no non-linear interaction occurs, compliance matrix for the background medium and rotated fractures combine, giving for n fracture sets, the general expression (Nichols et al. 1989)

$$S = S_b + \sum_{i=1}^{n} \Delta S_i \tag{17.17}$$

17.6.1 Example

Consider a horizontally layered medium (transversely isotropic) with vertical fractures. Such a system has orthorhombic symmetry (Hood and Schoenberg 1989). The compliance matrix has the form in the observation coordinate (x_1, x_2, x_3)

$$S + \Delta S = S = \begin{bmatrix} S_{11} & S_{12} & S_{13} & 0 & 0 & 0 \\ S_{12} & S_{22} & S_{23} & 0 & 0 & 0 \\ S_{13} & S_{23} & S_{33} & 0 & 0 & 0 \\ 0 & 0 & 0 & S_{44} & 0 & 0 \\ 0 & 0 & 0 & 0 & S_{55} & 0 \\ 0 & 0 & 0 & 0 & 0 & S_{66} \end{bmatrix} \tag{17.18}$$

The fracture contribution ΔS, to the elastic compliance moduli is to be separated.

Vertical fractures in their eigen-coordinates (x_1', x_2', x_3') may be represented as (refer to Fig. 17.4)

$$Z = \begin{bmatrix} Z_N & 0 & 0 \\ 0 & Z_2 & 0 \\ 0 & 0 & Z_3 \end{bmatrix} \tag{17.19}$$

The most general elastic response of any fracture set embedded in an elastic medium may be represented by a 3×3 matrix in the form

$$Z = \begin{bmatrix} Z_{11} & Z_{12} & Z_{13} \\ Z_{12} & Z_{22} & Z_{23} \\ Z_{13} & Z_{23} & Z_{33} \end{bmatrix} \tag{17.20}$$

It requires only 6 parameters which are components of the fracture compliance matrix. For details reference may be made to Schoenberg and Douma (1988).

With reference to Fig. 17.4, ΔS in the observation coordinate (x_1, x_2, x_3) is obtained by rotation $(\phi = -\pi/2)$ about the x_2-axis in the form

$$\Delta S = \begin{bmatrix} Z_N & 0 & 0 & 0 & 0 & 0 \\ 0 & 0 & 0 & 0 & 0 & 0 \\ 0 & 0 & 0 & 0 & 0 & 0 \\ 0 & 0 & 0 & 0 & 0 & 0 \\ 0 & 0 & 0 & 0 & Z_3 & 0 \\ 0 & 0 & 0 & 0 & 0 & Z_2 \end{bmatrix} \tag{17.21}$$

$S = S_b + \Delta S$ gives the long-wavelength equivalent of a transversely isotropic background medium embedded with a set of vertical fractures. Removal of the vertical fractures is simple, and the background medium has

$$S_b = S - \Delta S \tag{17.22}$$

From this relation, the compliance matrix for the background medium is

$$S_b = \begin{bmatrix} S_{11} - Z_N & S_{12} & S_{13} & 0 & 0 & 0 \\ S_{12} & S_{22} & S_{23} & 0 & 0 & 0 \\ S_{13} & S_{23} & S_{33} & 0 & 0 & 0 \\ 0 & 0 & 0 & S_{44} & 0 & 0 \\ 0 & 0 & 0 & 0 & S_{55} - Z_3 & 0 \\ 0 & 0 & 0 & 0 & 0 & S_{66} - Z_2 \end{bmatrix} \qquad (17.23)$$

[Hood 1991, Eq. (20)]

S_{ij} may be determined from seismic velocity measurements (see, for example, Hood and Schoenberg 1989). Above is the compliance matrix for the background medium, which is transversely isotropic. For a transversely isotropic medium the constraints are

$$S_{44b} = S_{55b},\ S_{13b} = S_{23b},\ S_{11b} = S_{22b},$$

$$S_{12b} = S_{11b} - \frac{1}{2} S_{66b}$$

This gives: $Z_3 = (S_{55} - S_{44})$,

$Z_N = (S_{11} - S_{22})$, and $Z_2 = 2(S_{11} - S_{22}) + S_{66}$.

Thus, fracture compliance Z_N, Z_3, Z_2 may be determined from measurements of compliance for a fractured medium (orthorhombic medium). As compliance matrix **S** is the inverse of stiffness matrix **C**, it is possible to write Z_N, Z_3, Z_2 in terms of stiffness moduli in the form

$$Z_3 = \frac{1}{c_{55}} - \frac{1}{c_{44}}$$
$$Z_N = \frac{c_{23} - c_{13}}{c_{11}c_{23} - c_{12}c_{13}}$$
$$Z_2 = \frac{1}{c_{66}} - \frac{2c_{13}}{c_{11}c_{23} - c_{12}c_{13}}$$

(Hood and Schoenberg 1989; Hood 1991)

The technique developed by Hood, which has been explained above, facilitate the separation of different contributions to anisotropy in a complex system without using Schoenberg and Muir calculus.

17.7 Numerical Method to Model Fracture Anisotropy: Use of a Synthetic Seismogram

We take an example from Kelner et al. (1999) to demonstrate the use of numerical simulation of P and SV waves to model anisotropy arising due to fractures in granite. Conclusions on anisotropy derived from the study of shear-wave splitting are verified by calculations, which take into account multiple scattering and crack diffraction. It is found that results on the time delays between fast and slow S-waves may be reproduced by different models of fractured media. Numerical calculations and measurement of attenuation of transmitted S-waves help to detail the picture of the fractured medium.

In earlier studies Archuleta et al. (1992) and Coutant (1996) observed anisotropy from the analysis of data from a vertical array of accelerometers at Garner valley in Southern California. The occurrence of shallow anisotropy in these investigations was modeled by Kelner et al. (1999 a) by computing synthetic seismograms.

In Kelner et al. (1999 b), the calculation of diffraction of seismic waves by empty or fluid-filled cracks of arbitrary shape are simulated, using the *boundary integral equation method* (refer to Bouchon 1987; Coutant 1989). It uses Huygen's principle, which assumes crack surfaces as diffracting boundaries. Basic details on the methods of calculation of the wavefield diffracted by cracks may be found in Kelner et al. (1999 a).

The model considered is an unbounded homogeneous and isotropic elastic medium with two-dimensional, vertical, empty planar cracks. The thickness of the cracks is taken as much smaller than their length. With this model Kelner et al. (1999 b) consider the presence of seismic anisotropy for propagating seismic waves, based on the observation by Grechka and McMechan (1995) that *any heterogeneous medium must behave like an anisotropic medium to elastic waves at some frequencies.* Kelner et al. (1999 b) evaluate anisotropy manifesting itself in the form of the time delay between a vertically incident plane S-wave, representing the fast S-wave, and the transmitted wave, which has been diffracted by cracks, and which corresponds to the slow S-wave.

The model considered in Kelner et al. (1999 b) contains long vertical fractures, representing long near-continuous fractures in reality [Fig. 17.5 (top)]. This models the granite at the Garner valley site, as if it were fractured from 220 m depth to 22 m depth – the depth range over which anisotropy is observed. The zone acts like a wave-guide since waves are trapped between the adjacent empty cracks. Wavefields (incident and total wavefield) are calculated and synthetic seismograms generated for various spacings of vertical cracks. An illustrative example of the synthetic seismogram for a 100 m crack spacing is shown in Fig. 17.5 (bottom). This figure also explains the method to measure anisotropy.

Synthetic seismogram pictures for various crack spacings show that the configuration of cracks with 85 m spacing produces anisotropy of the order of 8% – the anisotropy observed at the Garner site.

Kelner et al. (1999 b) analyze other models of fractured media that would produce 8% anisotropy. For this purpose they make use of results from the theory of an equivalent homogeneous medium (Hudson 1975, 1981). It states that a frac-

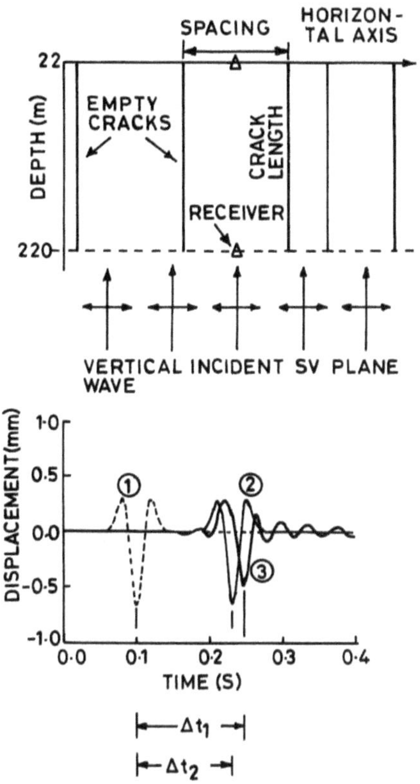

Fig. 17.5. (*Upper*) Model of a fractured medium with vertical empty 198 m long cracks. The crack thickness is 1 cm. The positions of receivers (Δ) used to measure S-wave anisotropy are shown. These positions exactly correspond to the difference of depth between the two receivers used to measure S-wave anisotropy at the Garnar Valley Downhole Array (GVDA) by Coutant (1996). The incident wavefield is a vertically propagating plane SV wave. (*Lower*) Example of synthetic seismograms in the case of vertical cracks with a spacing of 100 m, for crack and receiver configurations as shown in the *upper part*, and with a Ricker central frequency of 20 Hz. 1 is incident wavefield at 220 m depth; 2 is incident wavefield at 22 m depth, and 3 the total wavefield at 22 m depth; Δt_1 is slow S-wave travel time, Δt_2 is fast S-wave travel time. The magnitude of anisotropy is $(\Delta t_1 - \Delta t_2)/\Delta t_1$ (Source: Kelner et al. 1999b. From Bull Seism. Soc. Am. with permission)

tured material acts mechanically as a homogeneous medium if excitation wavelengths are large compared to size of the cracks or inclusions. The speed of the S-wave, in a material containing 2D parallel empty cracks, is to first order in (νa^2), where ν is number-density of cracks per unit area, and a is the average crack half-length (refer to Kelner et al. 1999b, p 710):

$$v^2 = \beta^2 \left[\begin{array}{l} 1 - \nu a^2 \frac{\pi(\lambda+2\mu)}{4(\lambda+\mu)} \times \\ \left(1 + 4\sin^2\theta\cos^2\theta\right) \end{array} \right] \tag{17.24}$$

(Kelner 1999b)

In this equation, λ, μ are Lame' parameter constants, β is shear wave velocity of a solid unfractured material; θ is the angle between direction of propagation and normals to the cracks. This is applicable for a low concentration of cracks.

For numerical evaluation β is taken as 1,520 m/s, $\lambda=\mu$, $\theta=\pi/2$.

For introducing 8% anisotropy, the relationship between β and v is

$$1-\frac{v}{\beta}=0.08$$

For numerical evaluation in Kelner et al. (1999b) all parameters are fixed except ν and a, the crack density and crack half-length.

Kelner et al. (1999b) find that models with different *crack half-length* and *crack density* can explain the 8% value, because of the trade-off between these two parameters.

In order to resolve between these two parameters, Kelner et al. (1999b) also considered modeling the effect of attenuation of the medium. For this purpose, they studied S-wave attenuation for both the data recorded at Garner Valley Downhole Array (GVDA), and synthetics obtained for different models of fractured media. From comparison of attenuation curves, they concluded that the cracks have to be nearly vertical and crack length distribution need be *fractal* to explain the observed shear-wave splitting.

It is found that attenuation factor Q^{-1} of the fractured media peaks around kd=2 (high attenuation) for P, SV and SH wave modes, where k is the incident wave number and d is the mean lengths of the cracks or cavities (Kawahara and Yamashita 1992; Kawahara 1992; Yomogide and Benites 1995; Murai et al. 1995; Kelner et al. 1999a). Therefore, measurement of Q^{-1} provides a way to estimate the mean crack length of the fractured medium.

In order to measure attenuation factor Q^{-1} of the S-wave in Kelner (1999b), S-wave signals are corrected so that the attenuation of direct S-waves comes only from the diffracting effect of the fractures of the granite. To calculate the attenuation, Kelner et al. (1999b) used the following relationship:

$$A(\omega)=A_0\exp\left(\frac{-\omega h}{2\beta Q}\right) \tag{17.25}$$

[Kelner 1999b, Eq. (4)]

This relates peak amplitude of $A(\omega)$ of the direct arrival that has propagated through the scattering zone (S-wave signal recorded at 22 m depth) to the fracture zone thickness h (here h=220–22=198 m). A_0 is the peak amplitude of the incident wave (S-wave at 22 m depth), β is S-wave velocity (1,520 m/s) and ω is angular frequency.

The attenuation factor Q^{-1} is measured on NS and EW polarized S-waves by Kelner et al. (1999b). It is found that attenuation is weaker on the N-S component than on the EW component. This confirms that the fracturing of the granite is parallel to the N-S direction. The absence of a peak in the attenuation curve reveals that there is no characteristic crack length in the medium – indicating fractal crack length distribution in the shallow granite at Garner valley.

For investigating the influence of crack orientation on S-wave attenuation, Kelner et al. (1999 b) consider the same crack density and crack-length distribution. However, in one case cracks are vertical and in other case they are horizontal. For horizontal cracks and a vertical incident S-wave, no shear wave splitting due to anisotropy can be observed as the cracks have the same direction as the incident shear wave polarization. In conclusion, it is inferred in Kelner et al. (1999 b) that any time delay in the case of horizontal cracks and for a vertical incident S-wave is due to heterogeneity and not due to anisotropy.

Kelner et al. (1999) found a frequency dependence of anisotropy in modeling the fractured medium. This was earlier noticed in works of Marion-Pidgeon and Savage (1997) on teleseismic S, $S_C S$ and $S_K S$ waveforms recorded in New Zealand, concluding that time delay increases with period (or wavelengths). Further Kelner's work illustrates the application of a semi-analytical, semi-numerical method to simulate the diffraction of seismic waves by empty or fluid-filled cracks of arbitrary shape. Also, it illustrates the integrated use of attenuation and anisotropy to model characteristics of the fractured zone in a precise manner.

17.8 Field System Examples of Reservoir Characterization

17.8.1 Determination of Bulk Density and Strike of Fractures in a Geothermal Reservoir: Use of Shear Wave Splitting

Observations on shear-wave splitting due to the alignment of vertical cracks in the Earth's crust has been made in earthquake recordings as well as in controlled source-seismic data (see Lou and Rial 1997). It has also been recognized that the polarization of the fast split shear wave is usually parallel to the local strike of the cracks (direction of maximum horizontal stress) (refer to Fig. 17.1). The time delay between the fast and slow shear wave is related to the intensity of crack-induced anisotropy in the medium (Crampin and Lovell 1991; Crampin 1993). The time delay also depends on the length of the propagation path in the anisotropic medium, and on the direction of propagation with respect to the orientation of anisotropy.

Hudson (1981) gave a formula to measure crack-induced anisotropy in the medium. The velocity of split shear waves in the symmetry plane is (Crampin 1981; Hudson 1981)

$$\beta_1^2 = \beta_0^2\{1 - 8C_D\alpha_0^2[\cos(4\theta) + 1]/[3(3\alpha_0^2 - 2\beta_0^2)]\} \tag{17.26 a}$$

$$\beta_2^2 = \beta_0^2\{1 - 8C_D\alpha_0^2[\cos(2\theta) + 1]/[3(3\alpha_0^2 - 2\beta_0^2)]\} \tag{17.26 b}$$

[Lou and Rial 1997, Eqs. (1) and (2)]
where β_1 and β_2 are velocities of two shear waves polarized parallel and perpendicular to the plane of symmetry, respectively; θ is the angle of propagation from

one symmetry axis (the strike of aligned fractures); α_0 and β_0 are compressional and shear-wave velocities in isotropic rock matrix; C_D is the crack density given by,

$$C_D = \frac{Na^3}{V} \tag{17.27}$$

where N is the number of cracks of average radius 'a' in volume V.

The anisotropy is assumed to be caused by aligned water-filled cracks in an isotropic homogeneous medium. Each crack is a thin disk-like opening with a radius much smaller than the typical wavelength of shear waves.

Time delay τ of split shear waves is given by

$$\tau = \left| \frac{1}{\beta_1} - \frac{1}{\beta_2} \right| L \tag{17.28}$$

[Lou and Rial 1997, Eq. (3)]
where L is the propagation distance between source and receiver.

Assuming the Poisson ratio of isotropic rock matrix as 0.25, and $C_D < 1$, Sato et al. (1991) gave the following formula (to the first order):

$$\tau = 4C_D[\cos 4\theta - \cos 2\theta]L/7\beta_0 \tag{17.29}$$

[Lou and Rial 1997, Eq. (4)]

Lou and Rial (1997) used microearthquake recordings from a downhole three-component seismic network around the Coso, geothermal reservoir, California, showing clear shear-wave splitting. Polarization provides subsurface fracture directions, and time delays of the split shear waves give information on 3-D fracture density distribution in the reservoir. Salient points of these determinations in Lou and Rial (1997) are described below:

The Coso geothermal area is an active seismic zone, generating on the average 20 microearthquakes per day. The microearthquake activity is monitored by a three-component downhole seismographic station network (Alvarez 1992). Seismometers are 100 m or more deep. Background noise is thus minimized. Shear wave interference with the free surface is much reduced. The records thus obtained are well suited for studying shear-wave splitting and anisotropy in the reservoir area.

The results in Lou and Rial (1997) demonstrate the potential usefulness of using the shear-wave splitting technique for surveying other geothermal areas or hydrocarbon deposits. Two aspects in shear-wave splitting, namely *polarization* and *time delays*, are analyzed in Lou and Rial (1997). From the 'polarization' direction of the 'fast shear wave' and rose diagrams for frequency of occurrence of polarization, three predominant subsurface fracture directions (crack-orientations) are identified. In Lou and Rial (1997), the 3-D crack density distribution is found. The assumption in this investigation are: (i) the formulation on the relationship of crack density and anisotropy given by Hudson (1981) holds, (ii) the isotropic velocity model of the investigation area (Coso, California) is known. With these assumptions, 450 *time delays* of split shear waves are utilized for *to-*

mographic inversion proposed by Shalev and Lou (1995) to generate horizontal sections at various depths of crack density in a 3-D volume of 21 (km)× 21 (km)×6 (km). It is thought important to describe the technique in its illustrative details with steps in interpretation.

An example (Lou and Rial 1997) of a record showing shear-wave splitting is shown in Fig. 17.6.

Splitting of shear waves is well demonstrated on the polarization diagram. The two split shear waves depict dominant orthogonal polarization directions.

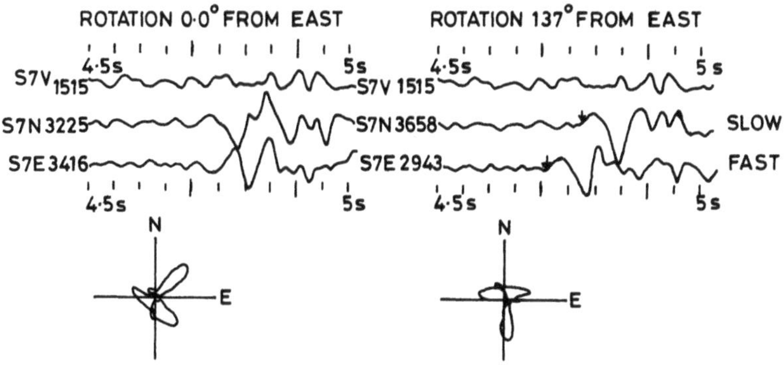

Fig. 17.6. Shear wave splitting recorded on station S7. The shear wave splitting is best detected from the polarization diagram of the horizontal component (S7N, S7E) over a windowed shear-wave train. The polarization direction of the leading shear wave is about 137° measured counterclockwise from an assumed east direction. The two split shear waves (fast and slow) are clearly identified after the seismograms are rotated into the polarization directions corresponding to fast and slow shear waves (Source: Lou and Rial 1997)

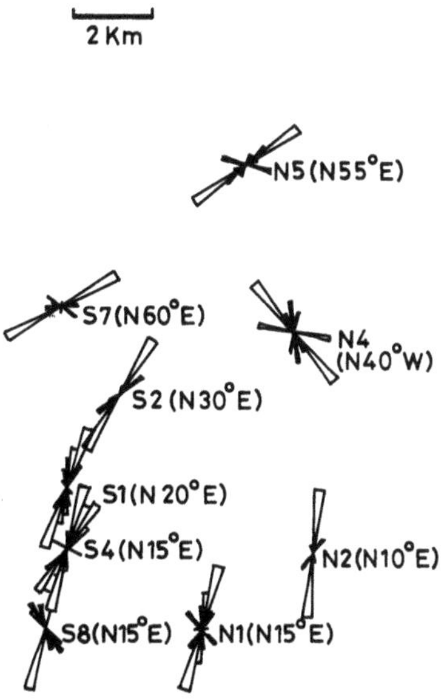

Fig. 17.7. Map view of the crack strikes inferred from the polarizations of the leading shear waves for the Coso geothermal field (Source: Lou and Rial 1997)

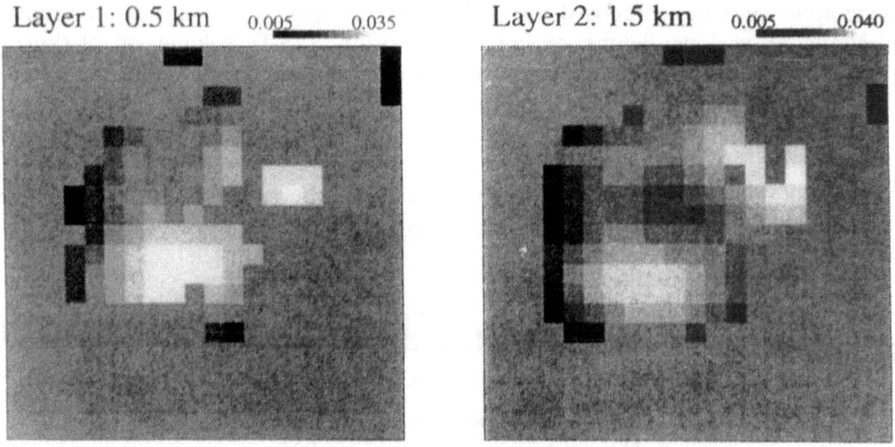

Fig. 17.8. Tomographic inversion results for the crack density determined from the time delays of the split shear waves for the Coso geothermal field. A total of 450 data points (time delays) were used for a target volume of 21 (km)×21 (km)×6 (km). The horizontal sections of crack density distributions are displayed at depths of 0.5 and 1.5 km. The *light shading* indicates high crack density and *dark shading* low crack density. The inversions are limited to those areas covered by split shear wave raypaths (Source: Lou and Rial 1997)

For determining the local strikes of cracks in the Coso area, rose diagrams of the fast-shear wave polarizations directions vs. their relative frequency have been plotted in Fig. 17.7. Most rose diagrams possess a dominant polarization direction, which is interpreted as the local strike of fracturing or the direction of maximum horizontal stress. The three predominant directions of fracturing in the rose diagram are – N40°–60°E, N0°–25°E and N25°–35°W. These trends match with lineament maps, the surface geology map, and field tests conducted in the area.

Figure 17.8 shows the result of tomographic inversion of the time delays of split shear waves for two depths only of 0.5 and 1.5 km; results for larger depths up to 5.5 km appear in Lou and Rial (1995). The results of the tomographic inversion have the limitation of an uneven distribution of microearthquake source locations.

17.8.2 Delineation of Cracked Rock Regime around Boreholes: Use of Velocity Anomalies

This method relies on high-resolution travel time measurement, between a borehole sensor and an array of shot points distributed radially and azimuthally about the well (Fig. 17.9 a). The directional dependence of velocity in the vicinity of the well is determined by comparison of travel times from different azimuths.

The theoretical basis of this method is the velocity variation in a fractured rock, which largely depends on crack density, the degree of preferred orientation, and the degree of fluid saturation (see, for example, Garbin and Knopoff 1973; Oconnel and Budiansky 1974; Crampin et al. 1980; Hudson 1981). Parameters

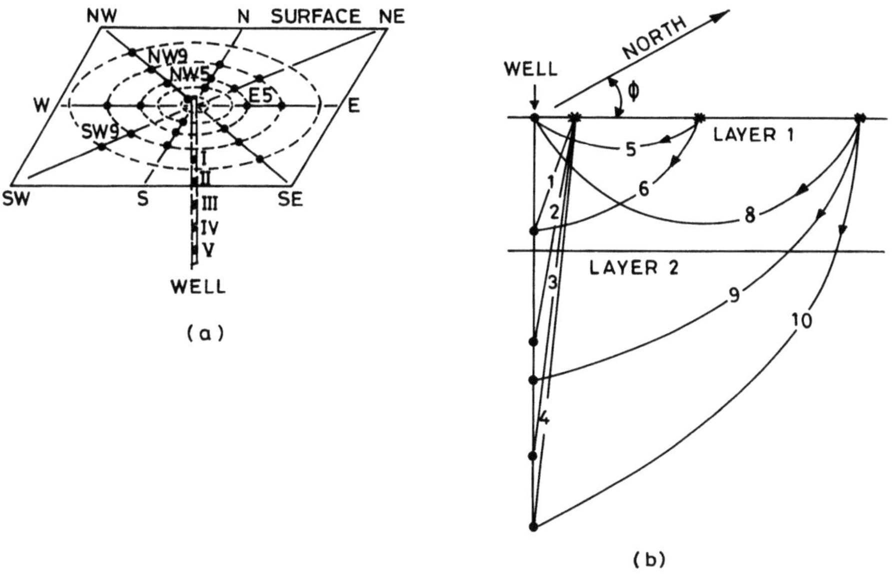

Fig. 17.9 a, b. a Schematic of shot and sensor configurations for areal well shot. Shots are radially distributed along a number of azimuths. N, NE, etc., refer to the azimuths; NW5, SW9 etc. refer to the shot points; I, II etc. are downhole sensor positions; **b** classification of raypaths. Rays *1–4* provide near vertical travel paths, rays *5* and *8* provide data on alluvium (*layer 1*), and ray *9* contains a large portion of horizontal travel path in crystalline rock (*layer 2*) (Source: Leary and Henyey 1985)

controlling velocity variation in fractured rocks may be found from observations of seismic wave behavior (Crampin et al. 1980), and in principle may be used for inferring porosity.

Garbin and Knopoff (1973) showed that P-wave velocity in a medium with a low concentration of parallel penny shaped cracks is given for "dry cracks" (either truly dry or those in which fluid is mobile at seismic frequencies) and saturated cracks (immobile fluid inclusions) as

$$V_p^2 \cong \frac{V_{PO}^2}{\left(1 + 8\big/3\varepsilon\left\{\frac{8}{7}(c^2 - c^4) + \left[(1 + 2c^2)^2\big/4\right]\right\}\right)} \tag{17.30}$$

[Leary and Henyey 1985, Eq. (1)]
where V_{PO} is the velocity in an uncracked medium; ε is the crack density assumed $\ll 1$; $c = \cos\theta$, where θ is the angle of incidence relative to crack plane normal. Crack density ε is given by

$$\varepsilon = \frac{Nr^3}{V}$$

where N is the number of cracks of radius r in a volume V.

According to Crampin et al. (1980) the dry crack case includes all the terms in Eq. (17.30) and in the saturated case the terms in square brackets are omitted.

Crampin (1978) and Crampin et al. (1980) found that the velocity variation of Eq. (17.30) above may be modeled by an equivalent perfectly elastic anisotropic solid.

Expanding Eq. (17.30) and neglecting the terms in ε^2, the velocity variation for a dry crack may be written as

$$V_p^2 = V_{PO}^2(1 - 71/21\varepsilon - 8/3\,\varepsilon\cos 2\,\theta + \varepsilon/21\cos 4\,\theta) \qquad (17.31\,a)$$

and for saturated cracks as

$$V_p^2 = V_{PO}^2(1 - 8/21\varepsilon + 8/21\,\varepsilon\cos 4\,\theta) \qquad (17.31\,b)$$

Leary and Henyey (1985) assumed the dominant crack structure to be vertical. For this θ may be considered as equivalent to the azimuth of a distant shot point from the borehole measured from normal to the crack planes. Assuming a dry crack density of 0.1, Eq. (17.31 a, b) gives the velocity variation ranging from V_{PO} to $0.63V_O$ for rays parallel to the crack system to the rays perpendicular to the crack system, i.e. a reduction of 37% in velocity, taking $V_{PO}=5$ km/s. This amount of velocity reduction would correspond to 12 ms delay in travel times for a 100 m travel path. It is possible to easily detect anomalies in travel time of this magnitude.

On the basis of the theoretical formulation described very briefly above, Leary and Henyey (1985) interpreted the travel time delays in terms of the fracture zone in crystalline rock in the basement. The analysis is educative for modeling the fracture zone from high-resolution travel time measurements in a borehole for areal well shoot (Fig. 17.9 a) in a number of azimuths. Velocity modeling of the structure is done for two types of ray paths: those that have nearly vertical travel paths and the others that have largely horizontal travel paths.

The first step in velocity modeling in Leary and Henyey's work is to estimate the vertical velocity structure, which is considered as an approximation to the isotropic velocity structure.

The second step in velocity modeling is to find the lateral (radial) continuation of this structure to determine an isotropic and 3-D velocity model for raypaths originating at shot points distant from the wellhead, which accounts to first order structural variation in travel times. As a final step obvious travel time deviations from the 3-D isotropic velocity model are isolated, and these travel paths are examined for evidence of anisotropic velocity conditions.

Standard, smoothly varying velocity functions of the type

$$V(z) = V_0(1 + z/l)^a$$

are fitted to different segments of the data. V_0, l, and a are determined for selected depth intervals corresponding to the upper most weathering layer, deeper alluvium, and underlying crystalline rock. The determination of the complete velocity structure from the surface downwards is the most important step in this method, and an excellent description of the details may be found in the publication by Leary and Henyey (1985).

Once the velocity structure is found, the travel times for the final class of rays with significant horizontal path lengths are examined (refer to Fig. 17.9 b).

The rays with a large horizontal path length in the basement on the north and northwest azimuth show large travel time anomalies, as evidence of the failure of the isotropic 3-D velocity model.

In Leary and Henyey's work, the observed travel time deviations (anomalies) from those computed is considered to be evidence for the presence of oriented cracks.

As the cracks are assumed predominantly vertical, a comparison between observed and model travel time is made for rays having large segments of horizontal travel paths. Southwest segments show no significant travel time disparity between model and observation travel time. For western and northwest azimuths, similar paths are delayed by 24 and 45 ms, respectively.

The velocity anomaly present for a horizontal ray path for one azimuth and absent for similar rays on an azimuth 90° away, may be interpreted in terms of the presence of oriented cracks with velocity anisotropy governed by Eq. (17.31 a). From the velocity analysis an average isotropic velocity V_{PO} of 6.0 km/s is assigned in the depth range of 1,200′–2,000′. Horizontal path length is about 550 m over which anomalous travel time develops. Over this path length, Eq. (17.31 a) gives a delay of 25 to 45 ms if the crack density parameter is in the range $0.04 \leq e \leq 0.06$ for an azimuth between -20° to +20° for normal to strike of recent faulting.

The method of Leary and Henyey (1985) described very briefly above enables us to map crack orientation and crack density in basement rock, from the knowledge of velocity/travel time anomalies inferred from the areal well shoot data. This method works successively by first finding the vertical velocity structure, then their lateral continuations in all azimuths, and finally studying travel time anomalies for selected azimuths for defining cracks characteristics.

17.8.3 Determination of Fracture Orientation in Marine Seismic Studies: Use of P-Wave AVO

In Mallick et al. (1998) it was shown that quantitative estimates of subsurface fracture properties might be made by analyzing azimuthal P-wave reflection amplitudes. To illustrate the method, they considered a model consisting of isotropic shale over fractured chalk (Mallick et al. 1998, Table 1). Normalized amplitude variation with angle (AVA) for various orientations of seismic lines to vertical fractures in anisotropic chalk is shown in Fig. 17.10. This figure shows that AVO/AVA for angles greater than 15° could be an effective tool in fracture detection. Also, the rate of decrease of reflection amplitude with angle/offset is largest for an azimuthal angle perpendicular to the fractures and least for an azimuthal angle parallel to fractures.

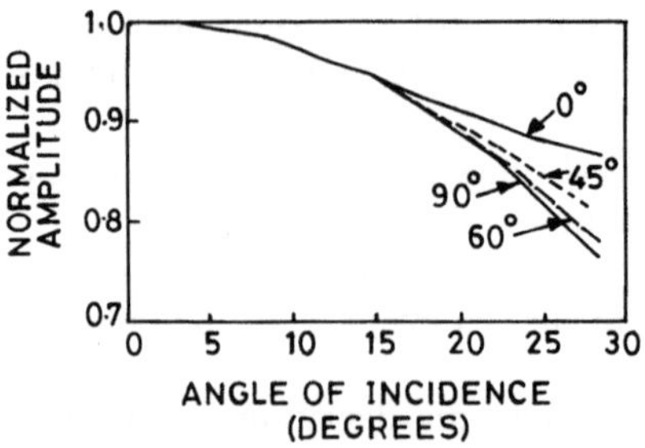

Fig. 17.10. Amplitude variation with the angle for reflector between an isotropic Taylor shale and an azimuthally anisotropic fractured Austin Chalk along different azimuthal directions with respect to the fracture strike (Source: Mallick et al. 1998, from Geophysics with permission)

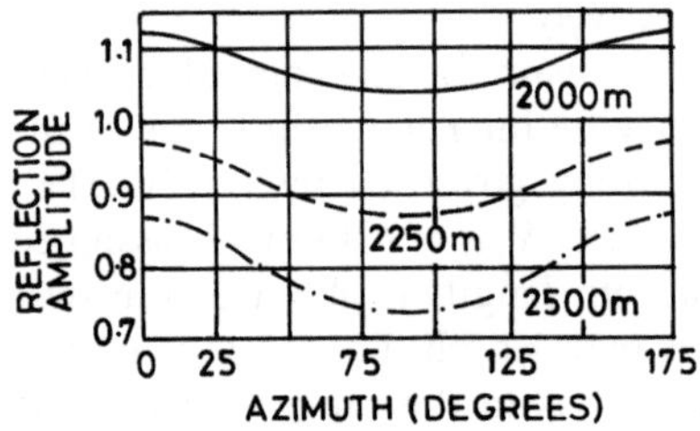

Fig. 17.11. Amplitude variation with azimuth angle of the shooting direction with respect to the fracture strike for the reflection between an isotropic Taylor shale and an azimuthally anisotropic fractured Austin Chalk for three fixed offsets of 2,000, 2,250 and 2,520 m (Source: Mallick et al. 1998, from Geophysics with permission)

Mallick et al. examined reflection amplitude behavior over a fractured medium (after applying geometrical spreading correction) at a fixed offset for different azimuth directions. They found that the P-wave reflection amplitude response, $(\mathrm{RAR})_{\text{P-waves}}$ at a fixed offset may be expressed as

$$(\mathrm{RAR})_{\text{P-wave}} = \mathrm{R}_{\phi} = \mathrm{A} + \mathrm{B}\cos 2\phi \tag{17.32}$$

where ϕ is the angle the shooting direction makes with respect to the strike of fractures, 'A' is an offset-dependent biasing factor, 'B' is the offset and fracture-property-dependent amplitude modulation factor.

Figure 17.11 shows plots of reflection amplitude vs. azimuth angle ϕ of the shooting direction with respect to fracture strike for three different offsets.

Equation (17.32) may be used to determine fracture orientation from P-wave data. Two cases are identified in Mallick et al. (1998). The first case is where we have observations from three azimuth lines for every common mid-point (CMP) location. This case is exactly determined one. The second case is overdetermined where observations for more than three azimuth lines are available, for a given CMP location.

Considering the first case. Assume that, for every CMP, we have data for three azimuthal angles, namely ϕ, $\phi + a$, and $\phi + \beta$. ϕ is the azimuth of the first line with respect to the fracture strike; a and β are known. Using Eq. (17.32) we have

$$R_\phi = A + \cos 2\,\phi \tag{17.33 a}$$

$$R_{\phi+a} = A + B\cos 2\,(\phi + a) \tag{17.33 b}$$

$$R_{\phi+\beta} = A + B\cos 2\,(\phi + \beta) \tag{17.33 c}$$

From 1Eqs. (.33 a–c) we have

$$\phi = \frac{1}{2}\arctan\left[\frac{(V_\phi - V_{\phi+\beta})\sin^2 a - (V_\phi - V_{\phi+a})\sin^2\beta}{(V_\phi - V_{\phi+a})\sin\beta\cos\beta - (V_\phi - V_{\phi+\beta})\sin a\cos a}\right] \pm n\pi \tag{17.34}$$

(Mallick et al. 1998)
for n=0, 1, 2.......

This Eq. (17.34) gives a unique solution to directions parallel and perpendicular to fractures using data for three azimuthal angles for every CMP location. Once ϕ is determined, Eqs. (17.33 a–c) may be used to get A and B.

Mallick et al. (1998) showed that if the upper layer is also azimuthally anisotropic, it is not possible to find fractures by applying a constant geometrical spreading compensation to all the azimuths. In this case, the data is to be corrected for *residual geometrical spreading* effects due to propagation through an azimuthally anisotropic (AA) medium.

Mallick et al. (1998) showed that a higher fracture density generates higher values of B/A [refer to Eq. (17.32)]. B/A is also offset dependent. This offset dependence may be eliminated by using stacked data over a range of fixed offsets, preferably over those offsets for which angles of incidences are greater than 15°.

The other case considered by Mallick et al. (1998) is the one for which, in a given CMP location, data from many azimuthal angles exists. In this case, the determination of fracture orientation is an overdetermined problem. Suppose that fractures are oriented at an angle ϕ, measured clockwise from north. Assume that we have seismic observations at many azimuthal angles $a_i (i = 1, \ldots N)$, also measured clockwise from north.

From Eq. (17.32), we have

$$(RAR)^i_{P-wave} = R_{(a_i - \phi)} = A + B\cos 2(a_i - \phi) \tag{17.35}$$

A, B and ϕ may be estimated from a least square fit of Eq. (17.35) over many azimuthal observations. Mallick et al. (1998) present the results of such an analysis in a least-squares (LS) *fractograms display*, showing the orientation angle ϕ, average amplitude A, and the amplitude modulation B as a function of time (Fig. 17.12). *The time zones in the LS fractogram where ϕ appears to be stable, and A and B are sufficiently high are considered to be fractured zones of economic interest.*

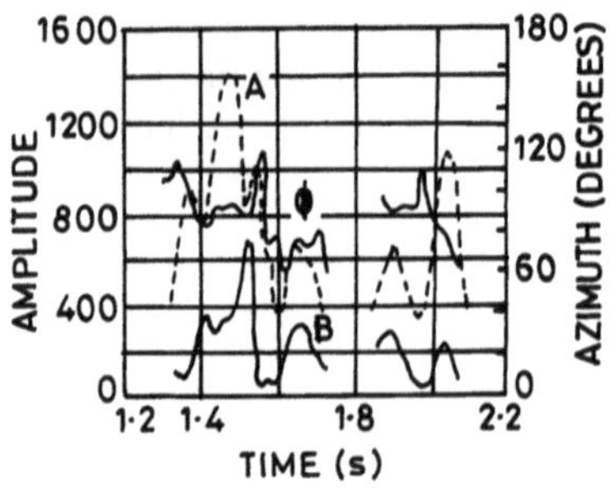

Fig. 17.12. LS (least squares) fractogram display computed from the normalized stacks. Fracture azimuths are angles in degrees, measured clockwise from north (Source: Mallick et al. 1998, from Geophysics with permission)

These zones for example in Fig. 17.12 correspond to (i) around 1.5 s with $\phi \approx$ N 105°, (ii) 1.7 s with $\phi \approx$ N 80° and (iii) 1.9 s with $\phi \approx$ N 105°. As an alternative to the least square method, Mallick et al. also outlines the multiple exact solutions (MES) method on normalized stacked data.

The work by Mallick, Craft, Meister and Chambers (1998) provides the basis for using conventional P-wave data, in a planned survey, to find the characteristics of fractures in the subsurface.

17.8.4
Spatial Location of Cracks and Crack Density Variations in a Coal Bed Methane Reservoir: Use of 3-D AVO

Amplitude vs. offset (AVO) analysis has been used successfully to predict reservoir properties and fluid contents, allowing for the spatial location of gas-water and gas-oil contacts. Ramos and Davis (1997) illustrate the use of the *3-D AVO technique* for characterizing fractured reservoirs and for determining spatial locations of *crack density variations*. Their analysis pertains to a fractured coal bed reservoir, critical to methane production, from the Cedar Hill Field in San Juan Basin, New Mexico. Zones of large Poisson ratio contrast (or high crack densities) in the coal bed methane reservoir are identified using prestack P-wave amplitude data. Also, source-receiver azimuth sorting is used to detect the preferential direction of azimuthal anisotropy generated due to fracturing of the coal. Further, their study illustrates that *integration of 3-D, three-component seismic data and AVO analysis* forms a new approach for characterizing fractured reservoirs. However, the essential requirement of this approach is careful data processing, and sophisticated forward wave-propagation modeling.

Figure 17.13 shows the data processing sequence to get AVO gathers from 3-D, three component seismic data. In this sequence, two major processes are: *true amplitude signal recovery* and *surface-consistent amplitude balancing*. Also, careful editing is required to take care of coherent noise contamination. AVO measurements cannot be interpreted successfully without precision, which accounts for these processes.

The data should be sampled sufficiently well in the offset and azimuth domains. In case the data includes unbalanced offset and azimuth distributions, the

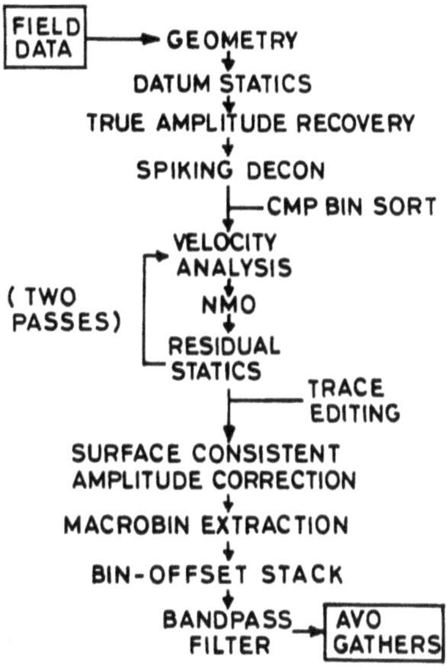

Fig. 17.13. Data processing sequence used to obtain the AVO gathers from the 3-D, three-component seismic data of the Cedar Hill Field (Source: Ramos and Davis 1997)

amplitude variations determined with this type of data may be biased. The controlling factors are fold per offset range and fold per azimuth range.

Figure 17.14 shows the field layout of the 3-D, three component survey area (3.4 km^2). Each swath consists of five receiver lines spaced at 600 ft, with 50 three-component receivers. Each of the square CMP bins (60 ft×60 ft) was defined (not shown in the figure). A total of 102 EW (cross-line) and 104 NS (in-line) CMP tracks were created, generating 10608 CMP-bin locations. Nine adjacent CMP gathers (3×3 microbins) were extracted from the CMP sorted volume for *pre-stack amplitude analysis.* These microbins have better offset and azimuth distribution than single CMP gather contributing to improved velocities and amplitude analysis.

17.8.4.1
AVO Modeling

AVO modeling is an important step in AVO analysis. For the modeling, Ramos and Danis (1997) used data on sonic and density logs as well as data from laboratory measurements. The measurements reveal a large increase in the Poisson ratio of the coal beds with the degree of fracturing. Therefore, large Poisson ratio contrasts are expected between sandstones and coals wherein coals are considerably fractured. Figure 17.15 shows the results for Poisson ratio for little, moderately and considerably fractured coal. Ramos and Davis (1997, p 1684) list Poisson ration ranges for representative rock types.

Gas or water is expected in the Cedar Hill Field to fill the crack system of coals. The presence of gas in cracks in coal lowers P-wave velocity to a large extent. This result is not as much for S-wave velocity. The net result for lower confining pressure is a lowering of the V_p/V_s ratio in gas-saturated conditions. Labo-

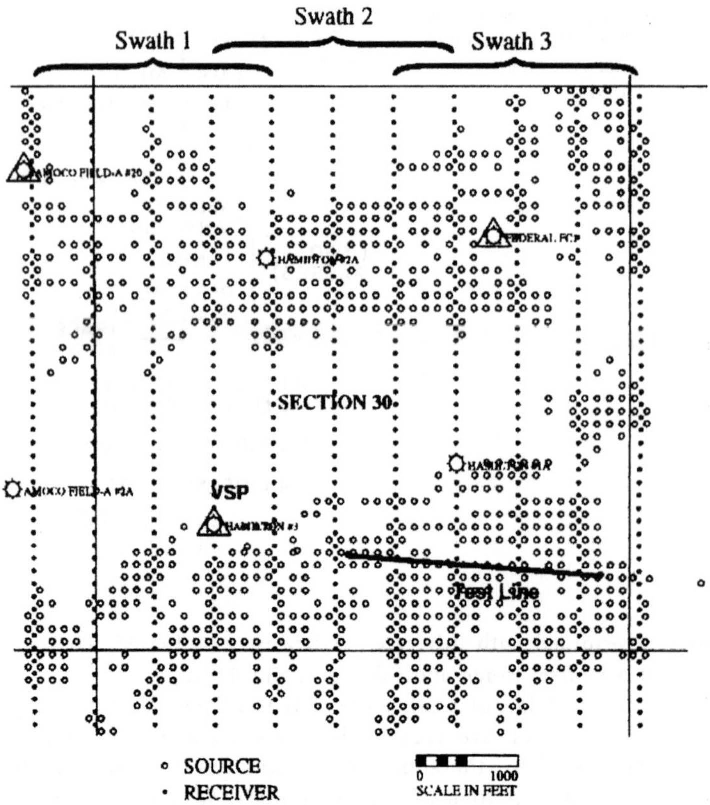

Fig. 17.14. Field layout of 3-D, three-component seismic survey showing the location of the Hamilton 3 well, where a multi-component VSP was acquired (from Shuck 1993). There were five receiver lines per swath. Shots are shown as *small open circles*, and receivers are shown as *black dots*. Because of logistical problems, source locations were not possible in a large portion of the center of the survey area. Wells providing from the Fruitland coal beds are designated by a *triangle around the well symbol* (Source: Ramos and Davis 1997)

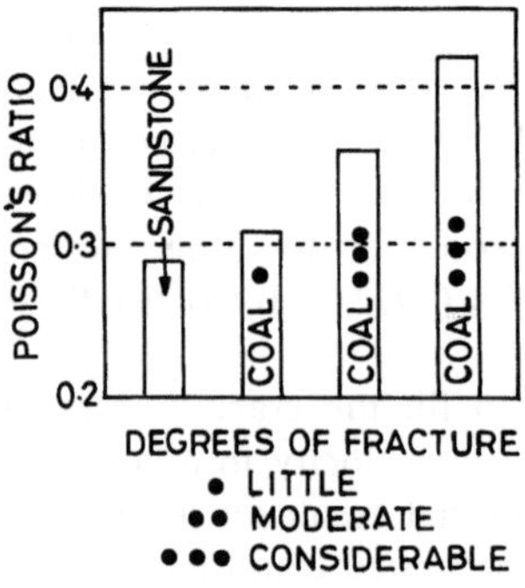

Fig. 17.15. Laboratory measurements for Poisson's ratio in vertical cores (Source: Ramos and Davis 1997)

ratory measurements on coal cores show that the V_p/V_s ratio increases with fracture density. The average V_p/V_s ratios in vertical coal cores for brine-saturated samples were 1.97, 2.15 and 2.34 for small, moderate and large fracture densities. Corresponding V_p/V_s ratios for air-saturated cores were 1.85, 1.89, and 1.95 (Ramos and Davis 1997, p 1687). The V_p/V_s ratio (and hence the Poisson ratio) for coal increases with the degree of fracturing and decreases with the presence of gas in the fracture system.

Figure 17.16 illustrates the effect on P-wave reflectivity of increasing crack density of coal for the coal-sandstone model. This result shows that, near vertical incidence, P-wave reflectivity is little affected due to crack density variation. For intermediate incidence angles up to say 50°, the AVO gradient increases as the crack density increases. For purely isotropic models, variations of the crack density of the coal are simulated by reducing the Poisson ratio in the coal, and therefore increasing the Poisson ratio contrast at the interface, and increasing the AVO gradient as a large contrast in Poisson's ratio corresponds to a large AVO gradient. The effect of changing the velocity contrast at the interface in the model (upper) sandstone and (lower) coal, but keeping $\Delta\sigma$ constant has the effect of an overall increase in reflectivity response (and in the AVO intercept) as the velocity contrast is increased.

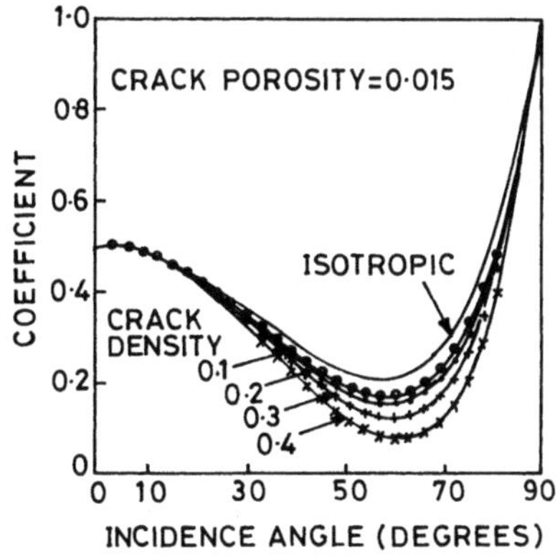

Fig. 17.16. Effects of increasing the crack density of the coals on P-wave reflectivity. Anisotropic ray tracing modeling was used to generate the responses (Source: Ramos and Davis 1997)

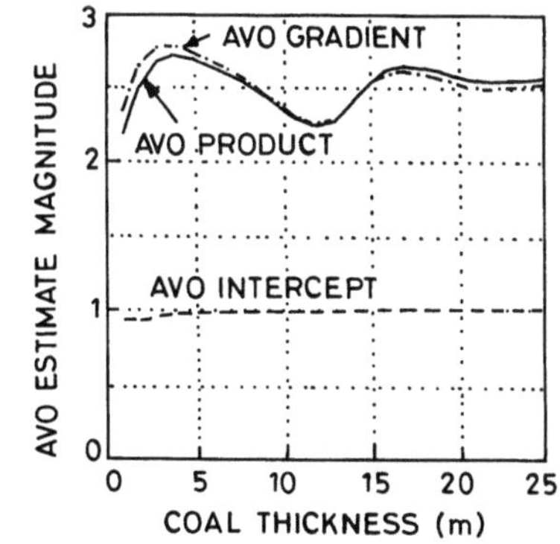

Fig. 17.17. Tuning effect on AVO estimates. Note the effect on the AVO gradient (and therefore in AVO product) as the coal thickness is changed (Source: Ramos and Davis 1997)

Coal thickness variations may have an effect on prestack amplitude analysis. Figure 17.17 shows the effects on AVO gradient, intercept (and AVO product) due to a change of coal thickness. Ramos and Davis (1997) conclude that for large coal thickness variations in an area, a correction should be made and results carefully interpreted.

Modeling results show that converted waves have a strong effect on AVO response.

17.8.4.2
AVO Analysis

For *AVO analysis*, the AVO response is taken as (refer to Shuey 1985; Ramos and Davis 1997, p 1689)

$$R_a(\theta) = R_0 + G\sin^2\theta \tag{17.36}$$

where R_0 is the AVO intercept and G is the AVO gradient; θ is the average angle of incidence and transmission, determined by ray tracing through the Earth model.

A plot of normalized amplitude versus $\sin^2\theta$, and finding a linear regression fit to the data gives estimates of intercept (R_0) and gradient (G) for the microbins selected.

Figure 17.18 shows the Poisson ratio variation for the top of the lower producing coal in the area computed from AVO estimates. This estimate is also known as *Poisson reflectivity.* It is determined from the AVO gradient and intercept estimates at each microbin location. It is based on the Hilterman approximation as (refer to Ramos and Davis 1997, p 1691)

$$R_p(\theta) \approx R_0\cos^2\theta + 2.25\Delta\sigma\sin^2\theta \tag{17.37}$$

This approximation assumes that the average Poisson ratio of the two media involved is equal to 1/3, which is nearly the value for the sandstone-coal interface. Equation (17.36) may be adopted in a modified form as Eq. (17.36) for finding $\Delta\sigma$ from the gradient

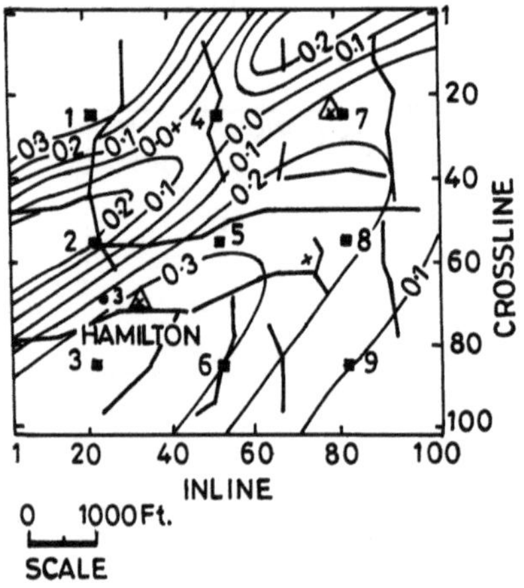

Fig. 17.18. Poisson's ratio variation map computed from AVO estimates at microbin locations using a linearized model for the reflection coefficient (Source: Ramos and Davis 1997)

$$\Delta\sigma = \frac{R_0 + G}{2.25} \tag{17.38}$$

[Ramos and Davis 1997, Eq. (3)]

Large variations in Poisson ratio are found along a north-east trend with high values in the center of the area, in agreement with the AVO product map. It is believed that strike slip movement along an EW-trending fault near crossline 50 produced the stress relief, which may have increased the opening of fractures in the central part of the area, and therefore giving rise to large *Poisson reflectivity anomalies.*

Figure 17.19 shows the bin-azimuth traces for microbin 5. The traces in Fig. 17.19 were obtained by partially stacking the data in approximately 5° source-receiver azimuth ranges. There are 37 traces each one for a particular source-receiver azimuth range. For example, trace 1 is for the west azimuth, trace 19 is for the north-south direction. Assuming that offset and azimuth distributions are perfect (i.e. the number of traces in all offset and azimuth ranges), and amplitudes are perfectly balanced, i.e. corrected for array effects and surface-con-

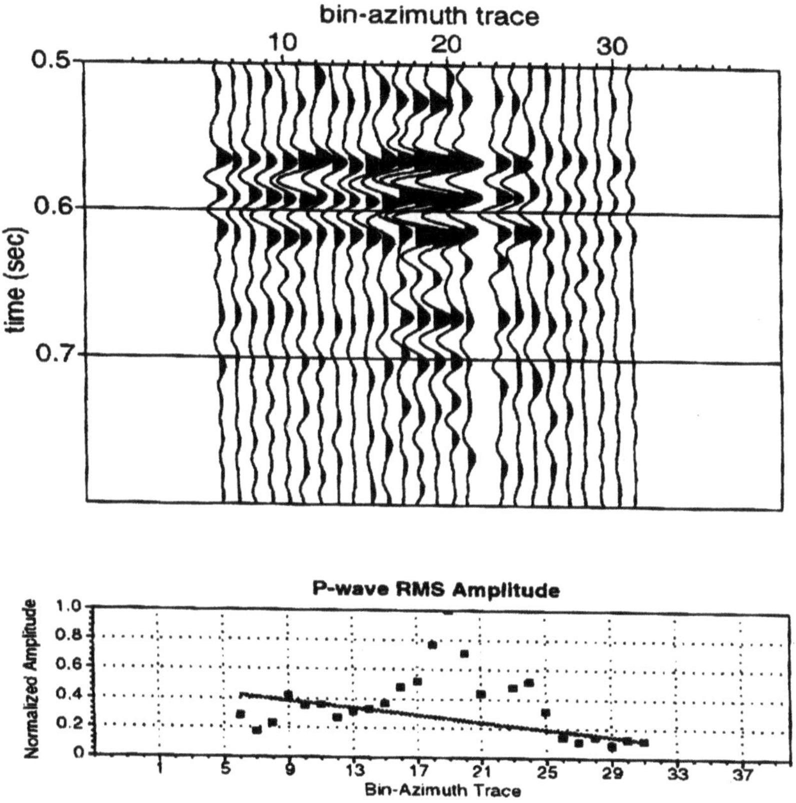

Fig. 17.19. P-wave data from macrobin stacked in 5° azimuth ranges. Trace *1 (left)* represents a source-receiver azimuth towards the west. Note the strong amplitudes in traces *17* to *20*, which parallel the north-northwest direction (Source: Ramos and Davis 1997)

sistent amplitude variations, the display in Fig. 17.19 gives an indication of the characteristic direction of azimuthal anisotropy in the medium. The larger trace amplitude for traces 17 to 20 in Fig. 17.19 shows that the fracture system is oriented between northwest-southeast and north-south directions. This result is in agreement with the polarization of the fast shear wave in the area (Shuck 1993).

In Ramos and Davis (1997) amplitude anomalies, Poisson ratio variations, fracture orientation, preferential trends of P-wave anisotropy and many related aspects are examined in enough detail which are extremely educative from the point of view of interpreting the 3-D data. It is difficult to discuss all of these in detail. Therefore, only as an illustration, we have taken examples of Poisson ratio variations, and bin azimuth traces.

It is felt here that for many educative interpretation diagrams related to 3-D interpretations, readers would find the Ramos and Davis (1997) publication to have very distinctive features for adopting these in routine analysis.

Appendix 17A: Combined Application of v_{nmo} and v_H for Fracture Characterization in an Azimuthally Anisotropic Medium

This method was contributed by Sayers and Ebrom (1997) and is based on a non-hyperbolic travel time equation for an azimuthally anisotropic layered medium, derived by using the expansion of the inverse squared ray-velocity in spherical harmonics. The method permits the determination of the strike of fractures (from travel time P-wave data from multiple azimuths) using the multiple offset VSP technique. A brief description of the method, based on the results derived by Sayers and Ebrom (1997), is given below.

Consider an azimuthally anisotropic homogeneous medium and a coordinate system as in Fig. 17A.1 a. Travel time squared from source S to receiver R is given by

$$t^2(x) = \frac{Z^2}{v^2(\theta, \phi)\cos^2\theta} \tag{17A.1}$$

Here t(x) is the travel time at horizontal offset x; z is the vertical distance traveled. As the medium is azimuthally anisotropic, ray velocity v is a function of both θ and ϕ.

To obtain spherical harmonic expansion, Sayers and Ebrom (1997) introduced a function $r(\theta, \phi)$ defined as

$$r(\theta, \phi) = \frac{1}{v^2(\theta, \phi)} \tag{17A.2}$$

Details of spherical harmonic expansion of $r(\theta, \phi)$ may be found in Sayers (1988, 1995) and its application to the present case may be found in Sayers and Ebrom (1997). For the present case $r(\xi, \phi)$ where $\xi = \cos\theta$ is expressible as

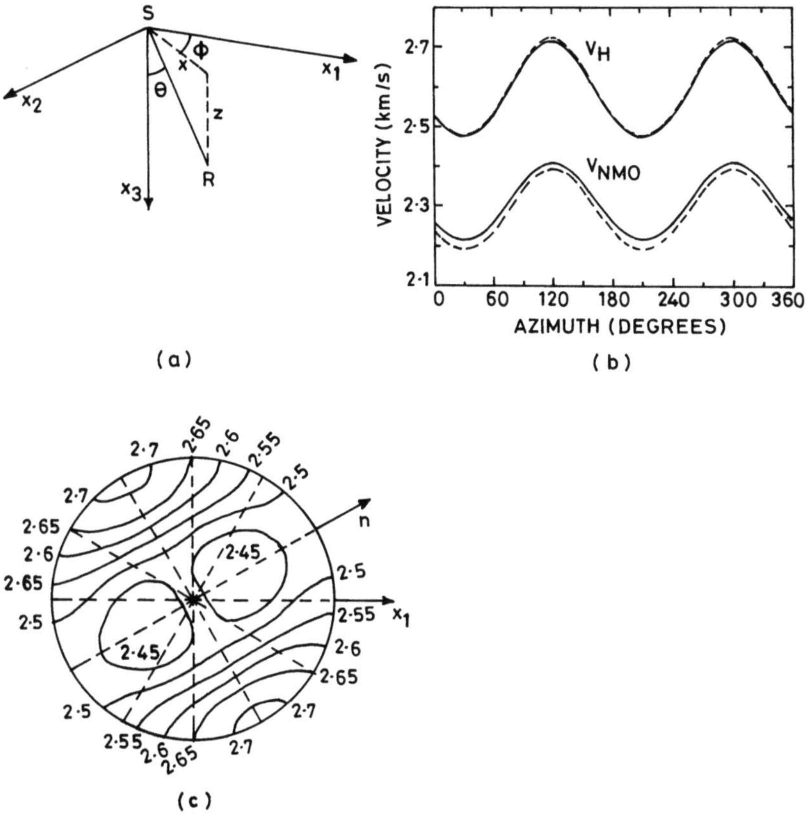

Fig. 17A.1 a–c. **a** Ray geometry in an azimuthally anisotropic homogeneous medium. **b** Variation in small-offset NMO velocity v_{nmo}, and horizontal velocity v_H for model 1 plotted as a function of azimuth ϕ obtained from Eq. (15A.5) by fitting Eq. (17A.3) to the travel time squared vs. offset squared data for various azimuths (refer to Fig. 3 of Sayers and Ebrom (1997). **c** Stereographic projection of ray velocity obtained by fitting the travel time data to Eq. (17A.3) for model 1 (Sayers and Ebrom 1997)

$$r(\xi, \phi) = A_0 + A_1 \cos^2\theta + A_2 \cos^4\theta \tag{17A.3}$$

[Sayers and Ebrom 1997, Eq. (8)]

This equation has the same form as that of Byun et al. (1989) for a transversely isotropic medium with a vertical axis of symmetry, with the difference that A_0, A_1 and A_2 are functions of azimuth ϕ.

Substitution of Eq. (17A.3) in (17A.1) gives

$$t^2(x) = t^2(0) + \frac{x^2}{v_{nmo}^2} - \frac{Ax^4}{(x^2 + z^2)} \tag{17A.4}$$

[Sayers and Ebrom 1997, Eq. (12)]

where t(0) is the vertical one-way travel time given by z/v_v(v_v is vertical velocity), v_{nmo} is the normal moveout velocity, A is a measure of anellipticity of the medium. We also have (refer to Sayers and Ebrom 1999)

$$\begin{aligned} v_H &= A_0^{-\frac{1}{2}} \\ v_v &= (A_0 + A_1 + A_2)^{-\frac{1}{2}} \\ v_{nmo} &= (A_0 - A_2)^{-\frac{1}{2}} \\ A &= -A_2 \\ v_H &= \frac{v_{nmo}}{\sqrt{1 - Av_{nmo}^2}} \end{aligned} \tag{17A.5}$$

In Sayers and Ebrom's (1997) work, the parameters in Eq. (17A.4) are obtained from the fit of travel time data.

The travel time in Eq. (17A.4) has azimuthal dependence; also velocities v_H and v_{nmo} have azimuthal dependence.

To illustrate the use of the algorithm for the determination of the strike of fractures, Sayers and Ebrom (1997) consider a horizontal shale layer containing a set of vertical fractures with normal at 30° to the x_1 direction. Fractures sets are described by the elastic parameters Z_N and Z_T (Schoenberg and Sayers 1995). For gas-filled open fractures $Z_N/Z_T \approx 1$. A lower value of Z_N/Z_T is likely to result from the presence of cement or clay within the fractures. Two models are considered in Sayers and Ebrom: one with gas-saturated cracks, and the other with cracks filled with cement or clay.

Figure 17A.1 b shows variations in small offset NMO velocity, v_{nmo}, and horizontal velocity v_H for model 1 plotted as a function of azimuth ϕ, and computed from the Eqs. (17A.5). For this model, the normal to strike direction of the fractures is found from the minimum of either v_{nmo} or v_H.

For model 2, the normal to fractures ($\phi = 30°$) may be obtained as the angle corresponding to the minimum velocity value from v_{nmo}; variation of v_H shows that Z_N/Z_T is small for this case.

Stereographic projection of ray velocity [obtained using Eq. (17A.4)] is shown in Fig. 17A.1 c. In this depiction the vertical propagation corresponds to the center of the figure and horizontal propagation to the outer edge. In Fig. 17A.1c, the orientation of the fracture set is denoted by normal vector **n**.

Error analysis shows that the determination of the orientation of fractures using v_{nmo} is preferable. A comparison of the variation of v_H and v_{nmo} gives important information on the value of Z_N/Z_T. Also accurate determination of the azimuthal variation of v_{nmo} requires reasonably large source-receiver offsets, and therefore non-hyperbolic moveout to be taken into account.

In the work by Sayers and Ebroms (1997), a non-hyperbolic travel-time equation is derived which can be used for velocity analysis and subsequent determination of the orientation of fractures and for inference of filling materials, like gas/clay, filling the fractures of the azimuthally anisotropic medium.

CHAPTER 18

Future Vision

18

18.1 Enhanced Scope for Applications in Engineering Geophysics Problems

Exploration seismology has been almost exclusively devoted to the search for hydrocarbons, occurring in relatively deeper horizons. However, if suitably adopted, seismic investigations may be utilized for very detailed shallow investigations as well. We consider here two examples with very encouraging results. Detailed shallow investigations, used as high-resolution methods, require specific choice of the source-receivers survey plan, and of processing methods. The first example discussed below relates to a *very high resolution (VHR) 3-D surface marine seismic investigation.*

18.1.1 VHR 3-D Shallow Seismic Investigations

Figure 18.1 a is a schematic diagram of a small-scale 3-D marine seismic data acquisition system. It consists of 12 dual channel towed micro-streamers, the seismic source, the hydrophones and the component (prism crown) of the navigation system. The boomer seismic source provides a frequency range of a few hundred hertz to well over 2,000 Hz. The streamer spacing and channel spacing are both 1 m. In all, seismic records from 20,000 shots were generated; each shot generating 24 seismic traces. Digital recording was done at the sampling rate of 0.25 ms; and an anti-alias filter of 1,400 Hz was applied during the survey. The shooting rate was 1 s.

For the purpose of advanced processing very accurate positioning of the seismic source and receiver array is required. This was achieved by an auto-tracking laser ranging theodolite, which continuously tracked the crown of a reflector prism mounted on the top of the catamaran. With this positioning system, a horizontal and vertical positioning accuracy of a few decimeters could be achieved.

The aim in the investigation was to accurately image a small clay diapir of an apparent diameter of 60 m, and concretions in a clay bed with a diameter of 0.5 to 1 m. With this in view, 3-D acquisition on the River Schelde was planned with a bin size of 1 m×1 m (Henriet et al. 1992), for a very shallow target depth (<25 m), this bin size would provide moderate oversampling of the first Fresnel zone of 4 m for an average sampling frequency of 1 kHz (Marsset et al. 1998).

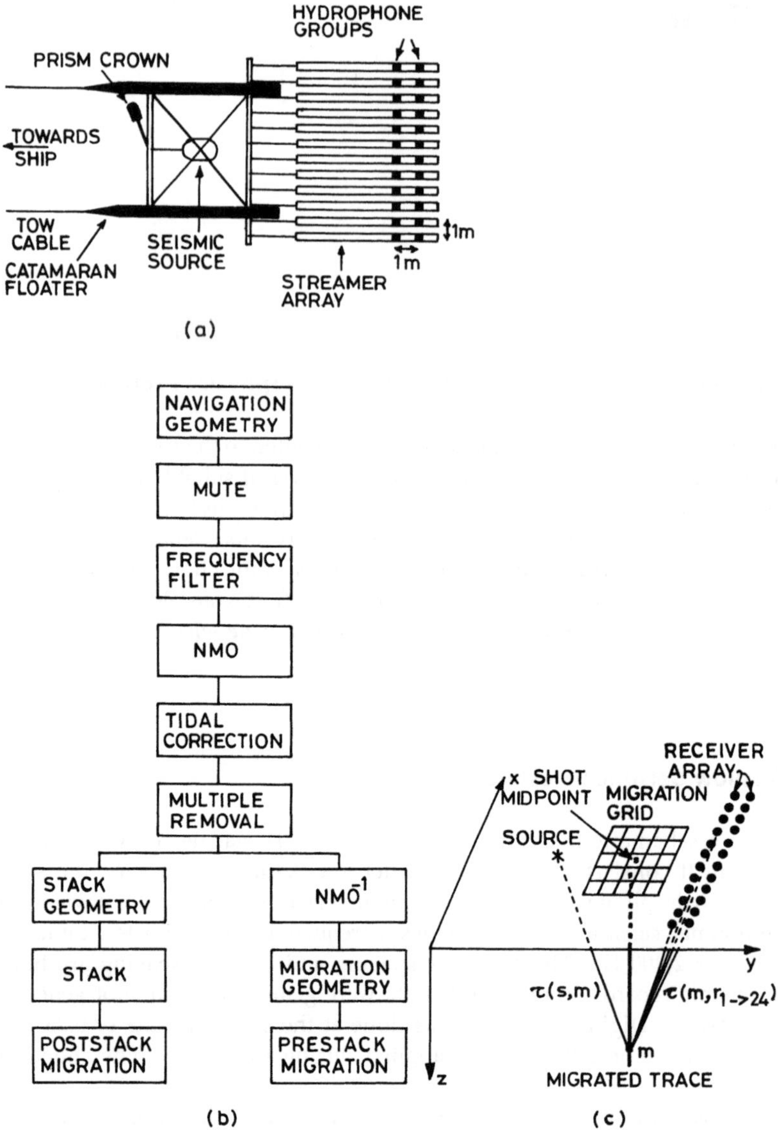

Fig. 18.1 a–c. a Schematic overview of the 3-D seismic acquisition system SEISCAT; **b** different processing steps applied to the VHR 3-D shallow marine data set; **c** geometry of the prestack migration (t = travel-time) (Source: Marsset et al. 1998, from Geophysical Prospecting with permission)

Figure 18.1 b shows the preliminary data processing steps and the major steps of migration – *poststack* and *prestack*.

Since the initial part of each trace contained direct wave arrivals *mute* served the purpose of removing these before arrival of the river-bed reflections. A band pass filter was applied with a 40 dB/decade slope. No amplitude correction was carried out.

Tide amplitudes could reach up to 5.5 m. A correction was made on the basis of the vertical coordinate of the positioning prism crown, and the arrival times were corrected to a common level.

The river-bed multiples were suppressed using deconvolution.

NMO and stack were implemented using exact locations of each source-receiver pair. After the computation of the mid point position for each trace, a stack-grid was formed in the bins of 1 m×1 m. Because of the irregular grid, the stack coverage could not be kept uniform.

The choice of migration method depends mainly on the type of data to be migrated. For this case of VHR small-scale 3-D work, the *Kirchoff summation* approach of migration was adopted (refer to Marsset et al. 1998).

The basis of Kirchoff summation migration is that every point in the depth section is considered a possible diffractor, producing diffracted events along diffraction hyperbolas. In the migration summation the events along hyperbolas are summed up and stored at the diffraction point from where they originated. For migration processing, the velocity model of the subsurface must be known or assumed with a reasonable basis.

The first step in prestack migration is the application of *inverse NMO* to the zero-offset data (Fig. 18.1 b). This step prepares the data for prestack migration. The following steps are involved in the overall prestack migration operation (Fig. 18.1 c):

(i) computation of geometry parameters
(ii) computation of first arrival times, and
(iii) summation migration

Marsset et al. (1998) emphasize the use of Kirchoff migration for the small scale 3-D seismic survey for shallow investigations as it is able to handle in an easy way both irregular shots and line spacing.

Before migration, a grid of 1 m×1 m bins was created. For an assigned or assumed aperture, the maximum offset (source-migrated trace or receiver-migrated trace) was calculated to be able to calculate the travel time grid. The 2-D time grid uses 0.1 m grid cells; the size of the cell is chosen on the basis of the wavelengths of the seismic signal (less than one tenth).

Migration processing is done on a shot-by-shot basis. Again for each trace, migrated traces are calculated for the bins lying within areas centered on the corresponding CMP and extending up to the aperture size taken. Each migrated depth sample is the sum of 24 different trace values taken at the travel times corresponding to the source-migrated trace-receiver triplet. Migrated traces from different shots and located within the same bin are summed together to yield the final migrated volume.

The final results after prestack migration processing is shown in the common display form of depth sections as well as in the form of time slices at an interval of 0.5 m (Fig. 18.2). The migrated picture of the diapir is very clear.

The purpose of this example is to illustrate the degree of fine accuracy that is possible to achieve by conducting very high-resolution 3-D seismic surveys. Many

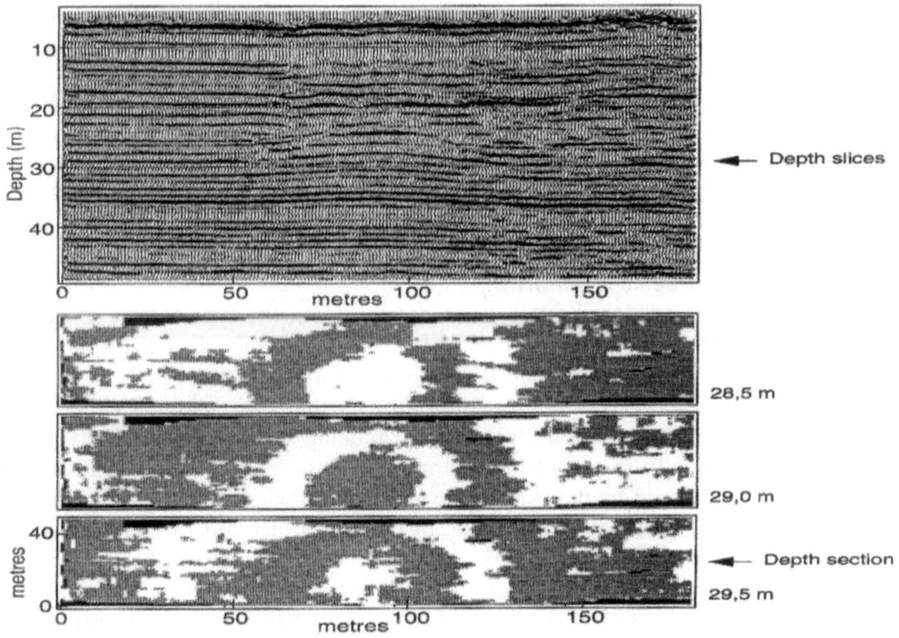

Fig. 18.2. Results after prestack migration. The depth section is taken from a central row of bins. The depth slices are separated by 0.5 m (Source: Marsset et al. 1998, from Geophysical Prospecting with permission)

such geological or geotechnical investigations would require detailed knowledge of the shallow subsurface, and the full potential of 3-D seismic for shallow investigations must be utilized.

18.1.2
Necessity for Combined Application of High Resolution Approaches

Some times a difficult problem with the desired precision cannot be solved by a single method. An example may be the exact location of a fracture, after the general picture of fractures and their orientations is known from conventional low-resolution methods. Majer et al. (1997) showed by an experiment at a field test facility that a combined interpretation of high-resolution (1 to 10 kHz) cross-well and single well seismic data is able to locate fractures within 1 m of its precise occurrence. Results on fracture orientation, width of fracture, their precise locations and connectivity should help greatly successful drilling for enhanced production.

The primary aim in the test investigation carried out by Majer et al. (1997) was to develop an effective method for imaging fractures important in controlling fluid transport. The results reveal that, to achieve the goal, a combination of approaches (crosswell and single well) should be adopted. The investigation is based on high-resolution cross-well and single well seismic data in a shallow (15 to 35 m) water-saturated fractured limestone sequence. Further, the investigations are part of a project to develop equipment, field techniques, and interpretational methods for characterizing fractured heterogeneous reservoirs. The focus is on integrating geology, rock physics, geophysics and hydrology into a unified method for predicting fluid migration. The objective in this investigation is to develop seismic methodologies for imaging gas-filled fractures in naturally fractured gas reservoirs. Although the investigations have been performed at shallow depths, the sizes of the aperture (1 mm) of fractures detected are representative of fractures, and the results are considered to be of significance for fracture detection at much greater depths. Adoption of the high resolution (1 to 10 kHz) seismic technique used in this investigation is important enabling accurate location of fractured zones, which are difficult for conventional methods like VSP and surface reflection to locate.

Figure 18.3 gives a plan view of the region being investigated. It is a test facility for conducting seismic and well-test field experiments. This facility (Queen et al. 1992) has six deep and five shallow wells for geophysical and hydrologic tests. The tests in Majer et al. were conducted on five shallow ground-water wells, GW1 to GW5. These wells penetrate a fractured shale and limestone sequence (Queen and Rizer 1990). A large velocity contrast exists between shale and limestone (Harlan 1990; Lines et al. 1992). Two orthogonal sets of vertical fractures have been mapped from a near surface exposure of limestone: a systematic set striking N70 °E and a non-systematic set at N25 °W (Queen and Rizer 1990). Strong evidence exists for conductive fractures trending N-70 °E (Datta-Gupta et al. 1994). Pump tests reveal that GW-5 and GW-2 appear to be connected by a "fast path". This is not so for GW-3, 2 and 1. These wells are considered unconnected to each other, or to wells GW-5 and GW-2. A seismic investigation undertaken earlier to the one by Majer et al. (1997) in the form of VSP and crosshole in the GW wells revealed seismic anisotropy consistent with the mapped fracture direction of N70 °E.

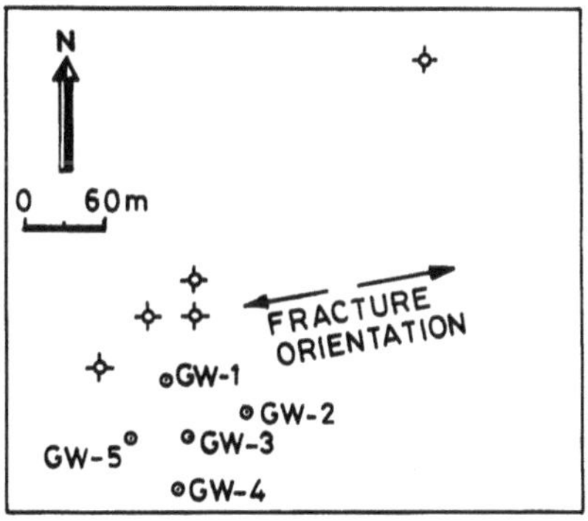

Fig. 18.3. Plan view of the Conoco borehole test facility located near Newkirk, Oklahoma, showing the geometry of wells used (*GW wells*) and the predominant fracture directions as inferred from mapping nearby outcrops of the limestone formation in which the seismic imaging was performed (Source: Majer et al. 1997, from Geophysics with permission)

Based on the pump test data, the character of fractures mapped in the nearby outcrops, and seismic anisotropy, it was inferred that there exists strong evidence for a fracture-controlled transport system, but the individual "target" fracture or fracture system was not clearly identified which was responsible for a "fast path" between GW-5 and GW-2, as found from hydrologic results. For enhancing the seismic visibility of the suspected fracture system, an experiment on air injection into the formation was undertaken by Majer et al. on the assumption that air traveling in the permeable feature should increase the reflectivity and/or attenuation properties of the fractures, and the existence of the fast fracture path from GW-5 to GW-2 should provide an improved seismic target.

Air injection was carried out to displace water from the fractures, and increase the seismic visibility of conductive fractures. Measurement utilizes a high-frequency imaging system (Majer et al. 1991): a 16 bit, 12-channel system with a sampling rate of 100,000/s per channel. The source was a piezoelectric source (cylindrical bender) with swept sinusoids from 100 to 10,000 Hz over a 50 ms time window and a recording time of 80 ms at 50,000 samples/s. Single well reflection surveys were undertaken in GW-1 and GW-3. It used an eight-element hydrophone string as receivers. Both source and receivers were used in a single well. The net outcome of the single well survey was a multifold imaging data set using a split spread.

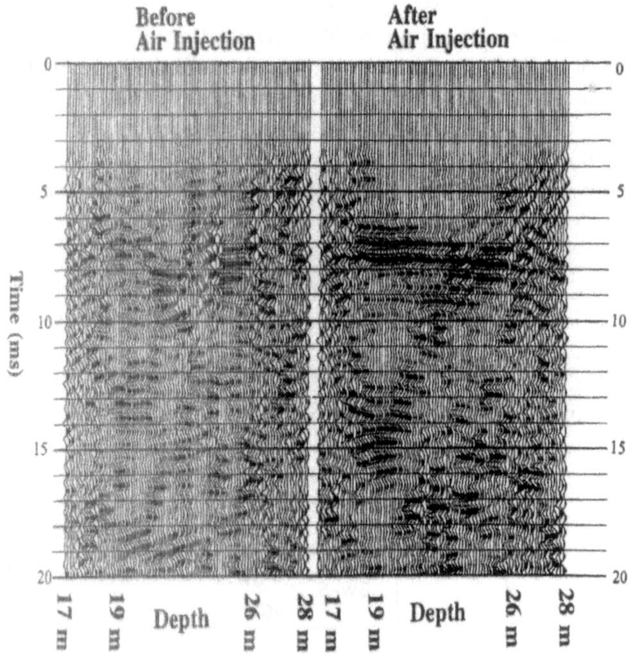

Fig. 18.4. A CDP stack of the single well reflection data from GW-3 before the air-injection (*left*) and after air-injection (*right*). Note the increase in the reflection energy at 7 ms (Source: Majer et al. 1997, from Geophysics with permission)

Details of the field experiment may be found in Majer et al. (1997).

Results of seismic monitoring during air-injection showed a general drop of amplitude over time. There had been small effects on travel time. Air-injection reduced significantly the signal amplitude and decreased the travel time of the initial arrival.

After the air-injection, a crosswell experiment was performed between GW3/GW1 and GW-3/GW-4. A sharp decrease of amplitude of the first arrival is observed for almost all traces. For GW3 to GW4, the crosswell first arrival result looks almost identical before and after the air injection.

The conclusion from the crosswell data is that there is a feature between GW3 and GW1 giving rise to the amplitude anomaly.

Single well survey data was processed as a common CDP reflection survey. Single well surveys have been conducted for GW3 and GW1 before and after the air injection. The before and after injection traces were sorted in CDP gathers. Some processing of the data was necessary to remove strong secondary S-wave arrivals. The resulting CDP stack before and after the air-injection in GW-3 is shown in Fig. 18.4. The results for before and after air injection look identical except for a strong reflector at 7 ms after injection. Reflection is strong for a 19–26 m-deep region of high velocity. Similar analysis for GW-1 does not show any strong reflector.

The drilling target was set based on the results of the crosswell and single well surveys at a depth of the maximum amplitude change and at a distance of half the 7-ms two-way travel time of the average velocity. This was a target at a depth of 24–25 m from the surface and 13.54 m from GW-3 on a line connecting GW3 and GW1.

18.2 Research Areas Requiring Special Emphasis

The following research areas may be identified towards better achievement of exploration goals:

(i) Understanding interpretational links between 3-D seismic data and anomalous medium properties.
(ii) Understanding the possible role of converted waves as an interpretation tool.
(iii) Recovery of true amplitudes.

Recently, phenomenal success has been achieved in oil and gas exploration and production programs using 3-D seismic data. This has been made possible by clearly understanding wave propagation and reflection parameters that characterize properties of the concerned media. This aspect has been well understood for isotropic, homogeneous media. Also, in the case of anisotropic media, valuable theoretical advancements made in recent years have facilitated correct interpretations. Comparatively, attempts to study the influence of a viscoelastic medium on wave propagation and reflection parameters have been much less. It may be mentioned that conditions of anisotropy and/or viscoelasticity are associated with hydrocarbon reservoirs, and these conditions create interpretation gaps as algorithms for routine interpretations assume simpler medium properties.

A 3-D seismic data record contains all types of information, like structural and stratigraphic settings, and medium property influences in the form of wave characteristics of the record. Structural and stratigraphic settings may be visually appreciated in the 3-D record. However, details of the medium properties are hidden. The need is to understand the retrieval of anomalous medium properties by appropriate processing of the record. Since anisotropy has a 3-D character, it is expected that 3-D seismic data would be ideally suited to studying anisotropy of the medium from its characteristic signatures in the record.

Both anisotropy and viscoelasticity are key factors in the interpretation of AVO trends. Both these properties of the medium affect the reflection coefficient as well as the wave propagation. Elastic anisotropy causes energy focusing and defocusing, whereas viscoelastic dissipation distorts energy and phase distribution of incident and reflected wavefronts. As these effects may be of the same order of magnitude as variations in reflection amplitudes, the reflection data character is grossly altered due to these effects, making AVO interpretations questionable. In view of this, attempts to accurately model and interpret reservoir character must account for anisotropy and viscoelastic attenuations. The initial attempt would be in the form of a 3-D mathematical formulation of anomalous (anisotropic or viscoelastic or both) behavior of the reservoir medium, supported by responses in the form of *3-D synthetic seismograms* to serve as the basis for interpretation of recorded 3-D seismic data. The task may be formidable to achieve, but only such attempts would serve to realize the unique role of 3-D seismic data analysis.

Converted wave reflection seismology is a new subject, still in the stage of development for applications in hydrocarbon-related problems. However, it is receiving attention because of the possibility of obtaining high quality multicomponent data. The initial theoretical foundation to the process of converted waves was provided in the classic work of Tessmer and Behle (1988). The initial model taken in Tessmer and Behle was isotropic and homogeneous. Further advancements to more complicated models of anisotropic, homogeneous or inhomogeneous media were done in the contributions by Tsvankin and Thomsen (1994) and Thomsen (1999). Attempts to make converted wave seismology practically adoptable should prove very valuable in reflection seismology as a new exploration tool. It is an important area that needs to be developed and strengthened to its full potential to support, supplement or function as an alternative to conventional PP or SS reflection interpretations.

AVO interpretations rely on accurate corrections for propagation and other factors that influence reflection amplitude behavior arising only due to a contrast in elastic properties at an interface. The recovery of *true reflection amplitude* requires very sophisticated analytical developments and processing to attach meaningful interpretations, solely to reflection amplitude variations. Notable contributions in this respect have been made in very recent years by Black et al. (1993), Canning and Gardner (1996), Padhi and Holley (1997), Gardner and Canning (1998) and Gesbert (2002). *True reflection amplitude* recovery is a challenging task because of many unknown interfering parameters in the recovery process. Therefore putting its formulation on a firm foundation would make AVO interpretations of immediate direct significance to the exploration of hydrocarbons.

References

Aki K, Richard PG (1980) Quantitative seismology - theory and methods vol 1, 2. WH Freeman, San Francisco

Al-Chalabi M (1973) Series approximation in velocity and travel time computations. Geophys Prospect 21:783–795

Al-Chalabi M (1974) An analysis of stacking, rms, average and interval velocities over a horizontally layered ground. Geophys Prospect 22:458–475

Alford RM (1986) Shear data in the presence of azimuthal anisotropy. Dilley, Texas, 56th Ann Intern Mtg Soc Expl Geophys Expanded abstracts, pp 476–479

Alford RM, Kelly KR, Boore DM (1974) Accuracy of finite difference modeling of the acoustic wave equation. Geophysics 39:834–842

Alkhalifah T (1995) Efficient synthetic seismogram generation in transversely isotropic, inhomogeneous media. Geophysics 60:1139–1150

Alkhalifah T (1996) Transformation to zero offset in transversely isotropic media. Geophysics 61:947–963

Alkhalifah TA, Tsvankin I (1994) Velocity analysis for transversely isotropic media. 64th Ann Intern Mtg Soc Expl Geophys. Expanded abstracts, pp 1000–1003

Alkhalifah T, Tsvankin I (1995) Velocity analysis for transversely isotropic media. Geophysics 60:1550–1566

Allen JL, Peddy CP, Fasnacht TL (1993) Some AVO failures and what (we think) we have learned. The Leading Edge 12:162–171

Altherman Z, Loewenthal D (1972) Computer-generated seismoforms. Methods in Computational Physics 12:35–164

Alvarez MG (1992) The seismotectonics of the Southern Coso range observed with a new borehole seismographic network. MSc Thesis, Duke University, Durham, NC

Al-Yahya K (1985) An iterative solution to seismic tomography. Stanford Exploration Project, 42 pp

Anderson J, Tsvankin I (1995) Dip moveout processing from Fourier transform in anisotropic media. Center for Wave Phenomena, Colorado School of Mines (CWP-146)

Anderson RN, Boulanger A, He Wei Sun YF, Xu L, Hart B (1996) 4-D seismic monitoring of reservoir production in the Eugene Island 330 Field, Gulf of Mexico. In: Weimer P, Davis TL (eds) Applications of 3-D seismic data to exploration and production. American Association of Petroleum Geologists and Society of Exploration Geophysicists, Tulsa, Oklahoma, USA, pp 9–20

Ando MY, Ishikawa Y, Yamazaki F (1983) Shear wave polarization anisotropy in upper mantle beneath Honshu, Japan. J Geophys Res 88:5850–5864

Andreassen K, Hogstad K, Berteussen KA (1990) Gas hydrate in the Southern Barents Sea indicated by a shallow seismic anomaly. First Break 8:235–245

Anstey NA (1970) Seismic prospecting instruments, vol 1: Signal characteristics and instrument specifications. Gebrüder Borntraeger, Berlin

Archunlete RJ, Scale SH, Somgas PV, Baber LM, Sesain T (1992) Garner valley downhole array of accelerometers - instrumentation and preliminary data analysis. BSSA 82:1592–1621

Arestad JF, Davis TL, Benson RD (1996) Utilizing 3-D, 3-C seismology for reservoir property characterization at Joffre Field, Alberta, Canada. In: Weimer P, Davis TL (eds) Applications of 3-D seismic data to exploration and production. American Association of Petroleum Geologists and Society of Exploration Geophysicists, Tulsa, Oklahoma, USA, pp 171–178

Artley C, Hale D (1994) Dip moveout processing for depth-variable velocity. Geophysics 59:610–622

Auld BA (1973) Acoustic fields and waves in solids, vol 1. John Wiley & Sons, Inc, New York

Babich VM (1961) Ray method for the computation of the intensity of wavefronts in elastic inhomogeneous media. In: Problems of the dynamic theory of propagation of seismic waves, vol 5. Leningrad University Press, Leningrad (in Russian)

Backus G, Gilbert F (1970) Uniqueness in the inversion of inaccurate gross earth data. Phil Trans, Roy Soc London 266a:123–192

Backus GE (1962) Long wave elastic anisotropy produced by horizontal layering. J Geophys Res 67:4427–4440

Backus MM (1959) Water reverberations – their nature and elimination. Geophysics 24:233–261

Banik NC (1984) Velocity anisotropy in shales and depth estimation in the North Sea Basin. Geophysics 49:1411–1419

Banik NC (1987) An effective anisotropy parameter in transversely isotropic media. Geophysics 52:1654–1664

Båth M (1968) Mathematical aspects of seismology. Elsevier, Amsterdam

Bazelaire E de (1988) Normal moveout revisited: inhomogeneous media and curved interfaces. Geophysics 53:143–157

Beckmann P, Spizzichino A (1963) The scattering of electromagnetic waves from rough surfaces: International series of monographs in electromagnetic waves 4. Pergamon Press, Oxford

Ben-Menahem A, Gibson RL Jr, Sena AG (1991) Green's tensor and radiation patterns of point sources in general anisotropic inhomogeneous elastic media. Geophysics J Int 107:297–308

Ben-Menahem A, Sena AG (1990) Seismic wave theory in stratified anisotropic media. J Geophys Res 95:15395–15427

Ben-Menahem A, Singh SJ (1981) Seismic waves and sources. Springer, Berlin Heidelberg New York

Berge PA (1991) Estimating SV-wave stacking velocities for TI solids. Geophysics 56:1596–1602

Berkhout AJ (1982) Seismic migration: imaging of acoustic energy by wave field extrapolation. A. Theoretical aspects, 2nd edn. Developments in solid earth geophysics 14a. Elsevier, Amsterdam

Berryhill JR (1979) Wave equation datuming. Geophysics 44:1329–1344

Berryhill JR (1984) Wave-equation datuming before stack. Geophysics 49:2064–2066

Berryman JG (1979) Long-wave elastic anisotropy in transversely isotropic media. Geophysics 44:896–917

Berryman LH, Goupilland PL, Waters KH (1958) Reflections from multiple transition layers – theoretical results. Geophysics 23:223–243

Billings MP (1999) Structural geology, 3rd edn. Prentice Hall, Englewood Cliffs, New Jersey

Biondi B (1992) Velocity estimation by beam stack. Geophysics 57:1034–1047

Black JL, Schleicher KL, Zhang L (1993) True amplitude imaging and dip moveout. Geophysics 58:47–66

Blacquiere G, Debeye HWJ, Wapenaar CPA, Berkhout AJ (1989) 3D table driven migration. Geophys Prospect 37:925–958

Blair JM, Korringa J (1987) Aberration-free image for SH-reflection in transversely isotropic media. Geophysics 52:1563–1565

Boise S (1978) Calculation of velocity from seismic reflection amplitude. Geophysics Prospecting 26:163–174

Bolondi G, Rocca F, Savelli S (1978) A frequency domain approach to two dimensional migration. Geophysics Prospecting 26:750–772

Boore DM (1972) Finite difference method for seismic wave propagation in heterogeneous materials. In: Methods in computational physics. Academic Press, New York, pp 1–37
Booth DC, Grampin S (1983) The anisotropic reflectivity technique theory. Geophys J Roy Astr Soc 72:755–766
Borcherdt RD (1973) Energy and plane waves in linear viscoelastic media. J Geophys Res 78:2442–2453
Borcherdt RD (1982) Reflection-refraction of general P- and type-I S-waves in elastic and anelastic solids. Geophys J Roy Astr Soc 70:621–638
Bording R, Gerszten Corn A, Lines L, Scales J, Treitel S (1987) Application of seismic travel time tomography. Geophys J Roy Astr Soc 90:285–303
Bortfeld R (1961) Approximation to reflection and transmission coefficients of plane longitudinal and transverse waves. Geophysics 9:485–502
Brace WF, Paulding BW, Scholz C (1966) Dilatancy in the fracture of crystalline rocks. J Geophys Res 71:3939–3953
Brown RJ, Stewart RR, Lawfon DC (2002) A proposed polarity standard for multicomponent seismic data. Geophysics 67:1028–1037
Brown RL, Seifert D (1997) Velocity dispersion: a tool for characterizing reservoir rocks. Geophysics 62:477–486
Buchanan DJ (1978) An exact one way wave equation. Presented at the 48th Ann Mtg of the SEG, San Francisco
Buchen PW (1971) Plane waves in linear viscoelastic media. Geophys J Roy Astr Soc 23:531–542
Buchholtz H (1972) A note on signal distortion due to dynamic (NMO) corrections. Geophys Prospect 7:395–402
Byun BS (1982) Seismic parameters for media with elliptical velocity dependencies. Geophysics 47:1621–1626
Byun BS (1984) Seismic parameters for transversely isotropic media. Geophysics 49:1908–1914
Byun BS, Corrigan D (1990) Seismic travel time inversion for transverse isotropy. Geophysics 55:192–200
Byun BS, Corrigan D, Gaiser JE (1989) Anisotropic velocity analysis for lithology discrimination. Geophysics 54:1564–1574
Cadzow JA (1990) Signal processing via least squares error modeling. IEEE ASSP Magazine, No 10:12–31
Canning A, Gardner GHF (1996) Another look at the question of azimuth. The Leading Edge 15:1177–1183
Carcione JM (1990) Wave propagation in anisotropic linear viscoelastic media: theory and simulated wavefields. Geophys J Int 101:739–750
Carcione JM (1994) Wavefronts in dissipative anisotropic media. Geophysics 59:644–657
Carcione JM (1999) Effects of vector attenuation on AVO of offshore reflections. Geophysics 64:815–819
Carcione JM, Kosloff D, Behle A (1991) Long-wave anisotropy in stratified media: A numerical test. Geophysics 56:245–254
Carcione JM, Seriani G, Priolo E (1992) Wave simulation in three-dimensional anisotropic viscoelastic media. 62nd Ann Int Mtg Soc Expl Geophys. Expanded abstracts, pp 1251–1254
Carrion P (1987) Inverse problems and tomography in acoustics and seismology. Penn Publ, Atlanta, Georgia
Carrion P (1991) Dual tomography for imaging complex structures. Geophysics 56:1395–1404
Carrion P, Costa J, Jose E, Ferrer P, Schoenberg M (1992) Cross borehole tomography in anisotropic media. Geophysics 57:1194–1198
Castagna J (1993) AVO analysis – tutorial and review. In: Castagne J, Backus M (eds) Offset dependent reflectivity – Theory and practice of AVO analysis. Soc Expl Geophysics, pp 3–36

Castagna JP, Backus MM (1993) Offset-dependent reflectivity – Theory and practice of AVO analysis. Soc Expl Geophysics, 348 pp
Castagna JP, Herbert WS, Douglas JF (1998) Framework for AVO gradient and intercept interpretation. Geophysics 63:948–956
Castoro A, White RE, Thomas RD (2001) Thin-bed AVO: compensating for the effects of NMO on reflectivity sequences. Geophysics 66:1714–1720
Cerveny V (1972) Seismic rays and ray intensities in inhomogeneous anisotropic media. Geophys J Roy Astr Soc 29:1–13
Cerveny V, Psencik I (1972) Rays and travel-time curves in inhomogeneous anisotropic media. Z Geophysik 38:565–577
Chang CH, Gardner GHF (1993) Effects of vertically aligned fractures on reflection amplitude: an amplitude-versus-offset study. 63rd Ann Int Mtg Soc Expl Geophys. Expanded abstracts, pp 769–771
Chen HE, Castagna JP (2000) Anisotropic effects on full and partial stacks. Geophysics 65:1021–1031
Chen He, Castagna JP, Brown RL, Ramos ACB (2001) Three-parameter AVO cross plotting in anisotropic media. Geophysics 66:1359–1363
Cherry JJ, Hurdlow WR (1966) Numerical simulations of seismic disturbances. Geophysics 31:33–49
Chiu SK, Kanasewich ER, Phadke S (1986) 3-D determination of structure and velocity by seismic tomography. Geophysics 51:1559–1571
Chiu SKL, Stewart RR (1987) Tomographic determination of three-dimensional seismic velocity structures, using well logs, vertical seismic profiles, and surface seismic data. Geophysics 52:1085–1098
Cholet J, Richards H (1954) A test on elastic anisotropy measurements at Berraine. Geophys Prosp 2:232–246
Chugh S (1994) Reflection study in anisotropic medium. MTech Dissertation (under supervision of the author), University of Roorkee, Roorkee, India
Chun JH, Jacewitz CA (1981) Fundamentals of frequency domain migration. Geophysics 46:717–733
Claerbout JF (1985) Imaging the Earth's interior. Blackwell Scientific Publishing, Oxford, 414 pp
Claerbout JF (1970) Numerical holography, acoustical holography, ed by AF Metherell. Plenum Press, New York 3:273–283
Claerbout JF (1971) Towards a unified theory of reflector mapping. Geophysics 36:467–481
Claerbout JF (1976) Fundamentals of geophysical data processing with applications to petroleum prospecting. McGraw-Hill, New York
Claerbout JF, Doherty SM (1972) Downward continuation of moveout corrected seismograms. Geophysics 37:741–768
Claerbout JF, Johnson AG (1971) Extrapolation of time dependent waveforms along their path of propagation. Geophys J 26:285–294
Cook EE, Taner MT (1969) Velocity spectra and their use in stratigraphic and lithologic differentiation. Geophys Prospect 17:433–448
Copson CT (1935) Theory of functions of complex variables. Oxford University Press, New York
Cordier JP (1985) Velocities in reflection seismology. Reidl Publishing, Dordrecht, Netherlands, 203 pp
Corrigan D, Withers R, Darnall J, Skopinski T (1996) Fracture mapping from azimuthal velocity analyses using 3-D surface seismic data. 66th Ann Int Mtg Soc Expl Geophys. Expanded abstracts, pp 1834–1837
Coutant O (1996) Observation of shallow anisotropy on local earthquake records at the Garner Valley, Southern California, downhole array. Bull Seism Soc Am 86:477–488
Crampin S (1977) A review of the effects of anisotropic layering on the propagation of seismic waves. Geophys J Roy Astr Soc 49:9–27
Crampin S (1978) Seismic wave propagation through a cracked solid: polarization as a possible dilatancy diagnostic. Geophys J Roy Astr Soc 53:467–496

Crampin S (1981) A review of wave motion in anisotropic and cracked elastic media. Wave Motion 3:343–391
Crampin S (1984) An introduction to wave propagation in anisotropic media. Geophys J Roy Astr Soc 76:17–28
Crampin S (1985) Evaluation of anisotropy by shear wave splitting. Geophysics 50:142–152
Crampin S (1993) A review of the effects of the crack geometry on wave propagation through aligned cracks. Can J Expl Geophys 29:3–17
Crampin S, McGonigle R, Bomford D (1980) Estimating crack parameters from observations of P-wave velocity anisotropy. Geophysics 45:345–360
Crampin S, Kirkwood SC (1981) Velocity variation in systems of anisotropic symmetry. J Geophys 49:35–42
Crampin S, Lovell JH (1991) A decade of shear wave splitting. Earth's crust: what does it mean? What use can we make of it? And what should we do next? Geophys J Int 107:387–407
Crampin S, McGonigle R (1981) The variation of delays in stress-induced anisotropic polarization anomalies. Geophys J Roy Astr Soc 64:115–131
Crampin S, McGonigle R, Ando M (1986) Extensive dilatancy anisotropy beneath Mount Hood: origin and the effect of aspect ratio on seismic velocities through aligned cracks. J Geophys Res 91:12703–12710
Crampin S, Radovich BJ (1982) Interpretation of synthetic common depth-point gathers for a single anisotropic layer (P-wave, SH-wave and SV-wave velocity). Geophysics 47:323–335
Crawford JM, Doty WEND, Lee MR (1960) Continuous signal seismograph. Geophysics 25:95–105
Cressman KS (1968) How velocity layering and steep dip affect CDP. Geophysics 33:399–411
Daley PF (1976) Reflection and transmission coefficients for transversely isotropic media. Master's Thesis, Department of Physics, University of Alberta, Edmonton
Daley PF, Hron F (1977) Reflection and transmission coefficients for transversely isotropic media. Bull Seism Soc Am 67:661–675
Daley PF, Hron F (1979) Reflection and transmission coefficients for seismic waves in ellipsoidally anisotropic media. Geophysics 44:27–38
Davis JL (1975) US patent no 3890 593, June 17, 1975
Dellinger J, Muir F (1988) Imaging reflections in elliptically anisotropic media. Geophysics 53:1616–1618
Dennison AT (1953) The design of electromagnetic geophones. Geophys Prospect 1:3–28
Denoyer de Segonzac P, Laherrere J (1959) Application of the continuous velocity log to anisotropy measurements in Northern Sahara: results and consequences. Geophys Prospect 7:202–217
Deplante C, Oristaglio M (1986) Traveltime inversion of offset vertical profile, using an application of Fermat's principle. 56th Ann Int Mtg Soc Expl Geophys. Expanded abstracts, pp 577–579
Deregowski SM (1978) A finite difference method for CDP stacked section migration. Presented at 40th Mtg of EAEG, Dublin
Deregowski SM (1982) Dip moveout and reflector point dispersal. Geophys Prospect 30:318–322
Deregowski SM, Rocca F (1981) Geometrical optics and wave theory of constant offset section in layered media. Geophys Prospect 29:374–406
Dietrich M, Cohen JK (1993) Migration to zero offset (DMO) for a constant velocity gradient: an analytical formulation. Geophys Prospect 41:621–643
Disher DA, Naquin PJ (1970) Statistical automatic statics analysis. Geophysics 35:574–585
Dix CH (1955) Seismic velocities from surface measurements. Geophysics 20:68–86
Dobrin MB (1976) Introduction to geophysical prospecting, 3rd edn. McGraw-Hill, New York, 630 pp
Dobrin MB, Savit CH (1988) Introduction to geophysical prospecting. McGraw-Hill, New York, pp 867
Domenico SN (1974) Effects of water saturation on seismic reflectivity of sand reservoirs encased in shales. Geophysics 39:759–769

Domenico SN (1976) Effect of brine-gas mixture on velocity in an unconsolidated sand reservoir. Geophysics 41:882–894

Domenico SN (1977) Elastic properties of unconsolidated porous sand reservoir. Geophysics 42:1339–1368

Domenico SN (1984) Rock lithology and porosity determination from shear and compressional wave velocity. Geophysics 49:1188–1195

Douma J (1990) The effect of transverse isotropy on isotropic travel time inversion of vertical seismic profile data – a modeling study. Geophysics 55:1235–1241

Doyle M, McGonigle R, Crampin S (1982) The effect of crack anisotropy on the hypo central locations of local earthquakes. Geophys J Roy Astr Soc 69:137–157

Dunkin S, Levin FK (1973) Effect of normal moveout on a seismic pulse. Geophysics 38:635–642

Duren RE (1987) Theory of marine source arrays. 57th Ann Int Mtg Soc Expl Geophys. Expanded abstracts, pp 204–207

Duren RE (1988) A theory of marine source arrays. Geophysics 53:650–658

Duren RE (1990) Seismic range equation. 60th Ann Int Mtg, Soc Expl Geophys. Expanded abstracts, pp 1495–1498

Duren RE (1991) Seismic range equation. Geophysics 56:1015–1026

Duren RE (1992) Range-equation weights for AVO. Geophysics 57:1203–1208

Dutta-Gupta A, Vasco DW, Long JCS (1994) Detailed characterization of a fractured limestone formation using stochastic inverse approaches. SPE/DOE paper 27744:71–86

Embree P, Burg JP, Backus MM (1963) Wide band velocity filtering: The pie-slice process. Geophysics 28:948–974

Ensley RA (1985) Evaluation of direct hydrocarbon indicators through comparison of compressional and shear wave data: A case study of the Myrnam Gas Field, Alberta. Geophysics 50:37–48

Estevez R (1977) Wide angle diffracted multiple reflections. PhD Thesis, Geophysics Department, Stanford University, California

Evenden BS, Stone DR, Anstey NA (1971) Seismic prospecting instruments, vol 2: Instrument performance and testing. Gebrüder Borntraeger, Berlin

Fertig J, Müller G (1978) Computations of synthetic seismograms for coal seams with the reflectivity method. Geophys Prospect 26:868–883

Forel D, Gardner GHF (1988) A three-dimensional perspective on two-dimensional dip moveout. Geophysics 53:604–610

French WS (1975) Computer migration of oblique seismic reflection profiles. Geophysics 40:961–980

Fryer GJ, Frazer LN (1984) Seismic waves in stratified anisotropic media. Geophy J Roy Astr Soc 78:691–710

Fryer GJ, Miller DJ, Berge PA (1989) Seismic anisotropy and age-dependent structure of the upper oceanic crust. In: Sinton JM (ed) The evolution of mid-oceanic ridges. AGU, Washington, DC

Fuchs K (1968) Das Reflexions- und Transmissionsvermögen eines geschichteten Mediums mit beliebiger Tiefenverteilung der elastischen Module und der Dichte für schrägen Einfall ebener Wellen. Z Geophysics 34:389–413

Fuchs K, Müller G (1971) Computation of synthetic seismograms with the reflectivity method and comparison with observation. Geophys J Astr Soc 23:417–433

Gajewski D (1993) Radiation from point sources in general anisotropic media. Geophys J Int 113:299–317

Gajewski D, Psencik I (1986) Numerical modeling of seismic wavefields in 3-D laterally varying layered anisotropic structures – Program ANRAY86. Internal Report, University of Alberta

Gajewski D, Psencik I (1987) Computation of high frequency seismic wavefields in 3-D laterally inhomogeneous anisotropic media. Geophys J Roy Astr Soc 91:383–411

Gajewski D, Psencik I (1988) Ray synthetic seismograms for a 3-D anisotropic lithospheric structure. Phys Earth Planet Int 51:1–23

Galperin EI (1973) Vertical seismic profiling. Society of Exploration Geophysicists, Tulsa, Oklahoma, USA
Galperin EI (1974) Vertical Seismic Profiling. Society of Exploration Geophysicists, Tulsa, Oklahoma, USA
Garbin HD, Knopoff HL (1973) The compressional modulus of a material permeated by a random distribution of circular cracks. Quart App Math 30:453–464
Gardner GHF, Canning A (1998) AVA analysis after velocity-independent DMO and imaging. Geophysics 63:686–691
Gardner GHF, Forel D (1990) Amplitude preservation equations for dip moveout. Geophysics 55:485–487
Gardner GHF, Gardner LW, Gregory AR (1974) Formation velocity and density - The diagnostic basis for stratigraphic traps. Geophysics 39:770–780
Gassmann F (1951) Über die Elastizität poröser Medien. Vierteljahr Naturforsch Gs Zuerich 96:1–23
Gassmann F (1964) Introduction to seismic travel time methods in anisotropic media. Pure and Applied Geophys 58:63–112
Gazdag J (1978) Wave equation migration with the phase-shift method. Geophysics 43:1342–1351
Gazdag J (1979) Wave equation migration with phase shift method. Geophysics 43:176–185
Gazdag J (1980) Wave equation migration with the accurate space derivative method. Geophys Prospect 28:60–70
Geertsma J (1961) Velocity-log interpretation: the effect of rock bulk compressibility. Soc Petr Eng J 1:235–248
Gesbert S (2002) From acquisition footprints to true amplitude. Geophysics 67:830–839
Gilbert P (1972) Iterative methods for the three dimensional reconstruction of an object from projections. J Theor Biol 36:105–117
Gill PE, Murray W, Wright MH (1988) Practical optimization. Academic Press, New York
Gjøystdal H, Ursin B (1981) Inversion of reflection times in three dimensions. Geophysics 46:972–983
Gonzalez-Serrano A, Claerbout JF (1984) Wave equation velocity analysis. Geophysics 49:1432–1456
Goodman JW (1968) Introduction to Fourier Optics. McGraw-Hill, New York, 287 pp
Graebner M (1992) Plane wave reflection and transmission coefficients for a transversely isotropic solid. Geophysics 57:1512–1519
Grant FS, West GF (1965) Interpretation theory in applied geophysics. McGraw-Hill, New York
Grechka V, Tsvankin I (1998) 3-D description of normal moveout in anisotropic inhomogeneous media. Geophysics 63:1079–1092
Grechka V, Tsvankin I (1998) Feasibility of non-hyperbolic moveout inversion, in transversely isotropic media. Geophysics 63:957–967
Grechka V, McMechan GA (1995) Anisotropy and non-linear polarization of body waves in exponentially heterogeneous media. Geophysics J Int 123:959–965
Gregory AR (1976) Fluid saturation effects on dynamic elastic properties of sedimentary rocks. Geophysics 41:895–921
Gupta IN (1973) Seismic velocity in rock subjected to axial loading up to shear fracture. J Geophys Res 78:6936–6942
Gutierrez M, Lozano E, Sroczynski M (1996) Impact of 3-D seismic interpretation on reservoir management in the Apiay-Ariari oil fields, Llanos Basin, Colombia. In: Weimer P, Davis TL (eds) Applications of 3-D seismic data to exploration and production. American Association of Petroleum Geologists and Society of Exploration Geophysicists, Tulsa, Oklahoma, USA, pp 57–68
Hadley K (1975) Azimuthal variations of dilatancy. J Geophys Res 80:4845–4850
Hagedoorn JG (1954) A process of seismic reflection interpretation. Geophys Prospect 2:85–127
Hake H, Helbig K, Mesdag CS (1984) Three term Taylor series for t^2–x^2 curves of P and S-waves over layered transversely isotropic ground. Geophys Prospect 32:828–850

Halbouty MT (1982) The time is now for all explorationists to purposefully search for the subtle trap. AAPG Memoir 32:1–10
Hale D (1984) Dip moveout by Fourier transform. Geophysics 49:741–757
Hale D (1991) Stable explicit depth extrapolation of seismic wavefields. Geophysics 56:1770–1777
Hale D, Artley C (1993) Squeezing dip moveout for depth-variable velocity. Geophysics 58:257–264
Hale D, Hill NR, Stefani J (1992) Imaging salt with turning seismic waves. Geophysics 57:1453–1462
Han DH, Nur A, Morgan D (1986) Effects of porosity and clay content on wave velocities in sandstones. Geophysics 51:2093–2107
Hardage BA (1985) Vertical seismic profiling part A: principles, 2nd enlaarged edn. Pergamon Press, Oxford
Hardage BA, Levey RA, Pendleton V, Simmons J, Edson R (1996) 3-D seismic imaging and interpretation of fluvially deposited thin-bed reservoirs. In: Weimer P, Davis TL (eds) Applications of 3-D seismic data to exploration and production. American Association of Petroleum Geologists and Society of Exploration Geophysicists, Tulsa, Oklahoma, USA, pp 27–34
Hardage BA, Toksöz MN (1983) Vertical seismic profiling, Part A, B. Geophysical Press, London
Harlam WS (1990) Tomographic estimation of shear velocities from shallow crosswell seismic data. 60th Ann Int Mtg Soc Expl Geophys Expanded abstracts, pp 86–89
Helbig K (1981) Systematic classification of layer induced transverse isotropy. Geophys Prospect 29:550–577
Helbig K (1983) Elliptical anisotropy, its significance and meaning. Geophysics 48:825–832
Helbig K (1984) Anisotropy and dispersion in periodic layered medium. Geophysics 49:364–373
Henriet JP, Verschuren M, Versteeg W (1992) Very high resolution 3-D seismic reflection imaging of small-scale structural deformation. First Break 10:81–88
Herman GT, Lent A, Rowland SW (1973) ART: mathematics and applications. J Theor Biol 42:1–32
Hestens MR, Steifel E (1952) Methods of conjugate gradient for solving linear systems. J Res Natl Bur Stand 49:409–436
Hilemam JA, Embree P, Pflueger JC (1968) Automated static corrections. Geophys Prospect 16:326–358
Holberg O (1987) Computational aspects of the choice of operator and sampling interval for numerical differentiation in large-scale simulation of wave-phenomenon. Geophys Prospect 35:629–655
Holberg O (1988) Towards optimum one way wave propagation. Geophys Prospect 36:99–114
Hood JA (1991) A simple method for decomposing fracture-induced anisotropy. Geophysics 56:1275–1279
Hood JA, Schoenberg M (1989) Estimation of vertical fracturing from measured elastic moduli. J Geophys Res 94:15611–15618
Hood P (1978) Finite difference and wavenumber migration. Geophys Prospect 26:773–789
Hood P (1981) Migration. In: Fitch AA (ed) Development in geophysical exploration methods-2. Applied Science Publisher, London, England, pp 151–230
Horkowitz KO, Davis DR (1996) Seismic delineation of thin sandstone reservoirs in a shale-rich sequence using instantaneous frequency and reflection amplitude attributes from 3-D data, Texas Gulf Coast. In: Weimer P, Davis TL (eds) Applications of 3-D seismic data to exploration and production. American Association of Petroleum Geologists and Society of Exploration Geophysicists, Tulsa, Oklahoma, USA, pp 35–44
Hornby BE, Schwartz LM, Hudson JA (1994) Anisotropic effective medium modeling of elastic properties of shale. Geophysics 10:1570–1583
House JR, Boyd TM, Haeni FP (1996) Haddam meadows, Connecticut: a case study for the acquisition, processing and relevance of 3-D seismic data as applied to the remediation

of DNAPL contamination. In: Weimer P, Davis TL (eds) Applications of 3-D seismic data to exploration and production. American Association of Petroleum Geologists and Society of Exploration Geophysicists, Tulsa, Oklahoma, USA, pp 257–266
Hron F (1973) A numerical ray generation and its application to the computation of synthetic seismograms for complex layered media. Geophys J 35:345–349
Hron F, Kanasewich ER (1971) Synthetic seismogram for deep seismic sounding studies using asymptotic ray theory. Bull Seism Soc Am 61:1169–1200
Hron F, Kanasewich ER, Alpaslan T (1974) Partial ray expansion required to suitably approximate the exact wave solution. Geophys J 36:607–625
Hubbert MK, Willis DG (1957) Mechanics of hydraulic fracturing. Trans Am Inst Min Metall Pet Engg 210:153–170
Hubral P (1978) Discussion on "Wavefront curvature in a layered medium" by B. Ursin (Geophysics-August 1978, 1011–1013). Geophysics 48:1051–1062
Hubral P, Krey T (1980) Interval velocities from seismic reflection measurements. Soc Expl Geophys
Hudsen JA (1975) Overall properties of a cracked solid. Math Proc Camb Phil Soc 88:371–384
Hudson JA (1981) Wave speeds and attenuation of elastic waves in materials continuing cracks. Geophys J Roy Astr Soc 64:133–150
Humphreys E, Clayton RW (1988) Adaptation of back projection tomography to seismic travel time problems. J Geophys Res 93:1073–1085
Inoue H, Fukao Y, Tanabe K, Ogata Y (1990) Whole mantle P-wave traveltime tomography. Phys Earth Planet Int 59:294–328
Ivansson S (1987) Crosshole transmission tomography. In: Nolet G (ed) Seismic tomography with application to global seismology and exploration geophysics. Reidl Publ, Dordrecht, Netherlands, pp 159–188
Jackson DD (1972) Interpretation of inaccurate, insufficient and inconsistent data. Geophys J Roy Astr Soc 28:97–110
Jackson DD (1979) The use of a priori data to resolve non-uniqueness in linear inversion. Geophys J Roy Astr Soc 57:137–157
Jacob B (1979) Velocity anisotropy. Stanford exploration project report 16:918–936
Jakubowicz H (1990) A simple efficient method of dip moveout correction. Geophys Prospect 38:221–245
Johnson WE (1995) Direct detection of gas in pre-Tertiary sediments. The Leading Edge 14:119–122
Johnston JE, Christensen NI (1995) Seismic anisotropy of shales. J Geophys Res B4:5991–6003
Jones LEA, Wang HF (1981) Ultrasonic velocities in Cretaceous shales from the Williston Basin. Geophysics 46:288–297
Jones T (1986) Pore fluids and frequency dependent wave propagation in rocks. Geophysics 51:1939–1953
Joseph J (1986) A study of geotomographic techniques for velocity estimation using crosshole seismic data. MTech Dissertation (under supervision of the author), University of Roorkee, Roorkee, India
Joseph J, Shastry RGS, Upadhyay SK (1987) Comparative study of geotomographic techniques for velocity estimations. J Assoc Expl Geophys VIII:207–217
Judson DR, Schultz PS, Sherwood JWC (1978) Equalizing the stacking velocities of dipping events via DEVILISH. 48th SEG Meeting, San Francisco, USA; brochure published by Digicon Geophysical Corp
Justice J, Vassilion A, Singh S, Logel J, Hansen P, Hall B, Hutt P, Solanky J (1989) Acoustic tomography for monitoring enhanced oil recovery. Leading Edge 8:12–19
Kaarsberg EA (1959) Introductory studies of natural and artificial aggregates by sound propagation and X-ray diffraction methods. J Geol 67:447–477
Kawahara J (1992) Scattering of P, SV-waves by random distribution of aligned open cracks. J Phys Earth 40:517–524

Kawahara J, Yamashita T (1992) Scattering of elastic waves by a fracture zone containing randomly distributed cracks. Geophysics 139:121–144

Keho TH, Beydoun WB (1988) Paraxial ray Kirchoff Migration. Geophysics 53:1540–1546

Kelly KR, Ward RW, Treitel S, Alford RM (1976) Synthetic seismograms: a finite difference approach. Geophysics 41:2–27

Kelner S, Bouchon M, Coutant O (1999a) Numerical simulation of the propagation of P-waves in fractured media. Geophys J Int 137:197–206

Kelner S et al. (1999b) Characterization of fractures in shallow Granites from modeling of anisotropy and attenuation of seismic waves. BSSA 89:706–717

Kennet BLN (1974) Reflections rays, and reverberations. Bull Seism Soc Am 64:1685–1696

Kennet BLN (1975) The effect of attenuation on seismograms. Bull Seism Soc Am 65:1643–1651

Kennet BLN (1978) Theoretical reflection seismograms for elastic media. Geophys Prospect 27:301–321

Kind R (1976) Computation of reflection coefficient for layered media. J Geophys 42:191–200

Kirlin RL (1992) The relationship between semblance and eigenstructure velocity estimators. Geophysics 57:1027–1033

Kitechenside PW (1991) Phase shift-based migration for transverse isotropy. 61st Ann Int Mtg Soc Expl Geophys. Expanded abstracts, pp 993–996

Kitechenside PW (1993) 2-D anisotropic migration in space-frequency domain. J Seiso Expl 2:7–22

Kjartansson E (1978) The Effect of Q on Bright Spots. Presented at 48th Ann Mtg SEG, San Francisco

Koefoed O (1955) On effects of Poisson's ratio of rock strata on reflection coefficients of plane waves. Geophys Prospect 3:381–387

Krebes ES (1984) On the reflection and transmission of viscoelastic waves – some numerical results. Geophysics 49:1374–1380

Krey T, Helbig K (1956) A theorem concerning anisotropy of stratified media and its significance for reflection seismics. Geophys Prospect 4:294–302

Kuhl DE, Edwards RQ (1963) Image separation radioisotope scanning. Radiology 80:653–662

Kulhanek O (1976) Introduction to digital filtering in geophysics. Elsevier, Amsterdam

Kumar A (1999) Datum correction by wave field extrapolation. MTech Dissertation (under supervision of the author), University of Roorkee, Roorkee, India

Kumar J (2000) Wavefield extrapolation for migration of seismic data. MTech Dissertation (under supervision of the author), Indian Institute of Technology Roorkee, Roorkee, India

Kumar J (2000) Wavefield-extrapolation for migration of seismic data. MTech Dissertation (under supervision of the author), University of Roorkee, Roorkee, India

Kumar S (1989) Picture of synthetic seismic offset VSP and forward modeling for geotomographic velocity imaging. MTech Dissertation (under supervision of the author), University of Roorkee, Roorkee, India

Kumar S (2002) AVO interpretation for gas sand. MTech Dissertation (under supervision of the author), University of Roorkee, Roorkee, India

Kuo JT, Dai Ting fan (1984) Kirchoff elastic wave migration for the case of non-coincident source and receivers. Geophysics 49:1223–1238

Landraum RA et al. (1994) Polarity convention for vibratory source recording system. Geophysics 59:315–322

Lapidus L (1962) Digital computation for chemical engineers. McGraw-Hill, New York

Larner KL, Hatten L, Gibson BS, Hsu IC (1981) Depth migration of imaged time section. Geophysics 46:734–750

Lavergne M (1989) Seismic methods. Editions Technip, Paris

Lawson CL, Hanson RJ (1974) Solving least-squares problems. Prentice Hall, Englewood Cliffs, New Jersey

Leary PC, Henyey TL (1985) Anisotropy and fracture zones about a geothermal well from P-wave velocity profiles. Geophysics 50:25–36
Lee MW (1990) Travel time inversion using transmitted waves of offset VSP data. Geophysics 55:1089–1097
Lefeuvre F, Nicoletis L, Ansel V, Cliet C (1992) Detection and measure of the shear wave birefringence from vertical seismic profile data: theory and applications. Geophysics 57:1463–1481
Levin FK (1971) Apparent velocity from dipping interface reflections. Geophysics 36:510–516
Levin FK (1978) The reflection, refraction, and diffraction of waves in media with elliptical velocity dependence. Geophysics 43:528–537
Levin FK (1979) Seismic velocities in transversely isotropic media. Geophysics 44:918–936
Levin FK (1980) Velocities for transversely isotropic media II. Geophysics 45:3–17
Levin FK (1990) Reflection from a dipping plane-transversely isotropic solid. Geophysics 55:851–855
Levin FK (1994) Reflection point behaviour in transversely isotropic solids. Geophysics 59:309–314
Liner CL (1990) General theory and comparative anatomy of dip moveout. Geophysics 55:595–607
Lines LR, Bourgeois A, Covey JD (1984) Travel time inversion of offset vertical seismic profiles: a feasibility study. Geophysics 49:250–264
Lines LR, Kelly KR, Queen JH (1992) Channel waves in cross-borehole data. Geophysics 57:334–342
Lines RL (1988) Inversion of geophysical data. Soc Expl Geophys
Lines RL, Treitel S (1984) Tutorial: A review of least square inversion and its application to geophysical problems. Geophys Prospect 32:159–186
Liu E, Hudson JA, Pointer T (2000) Equivalent medium representation of fractured rock. J Geophys Res 105:2981–3000
Lo TW, Cayner KB, Toksöz MN (1986) Experimental determination of elastic anisotropy of Berea Sandstone, Chicoper Shale and Chelmsford Granite. Geophysics 51:164–171
Lockett FJ (1962) The reflection and refraction of waves at an interface between viscoelastic materials. J Mech Phys Solids 10:53–64
Loewenthal D, Lu L, Robertson R, Sherwood J (1976) The wave equation applied to migration. Geophys Prospect 24:380–399
Lou M, Rial JA (1997) Characterization of geothermal reservoir crack patterns using shear-wave splitting. Geophysics 62:487–494
Love AEH (1944) A treatise on mathematical theory of elasticity, 4th edn. Dover, New York
Lynn HB, Bates CR, Simon KM, Van Dok R (1995) The effect of azimuthal anisotropy in P-wave 3-D Seismic. 65th Ann Int Mtg, Soc Expl Geophys. Expanded abstracts, pp 723–730
Lynn HB, Siman KM, Bates CR (1996) Correlation between P-wave AVO and S-wave traveltime anisotropy in a naturally fractured gas reservoir. The Leading Edge 15:931–935
Macleod MK, Hansen RA, Bell CR, McHugo S (1999) The Alka field ocean bottom cable seismic survey: Impact on field development. The Leading Edge 18:1306–1312
Macrides C, Kanawich E, Baratha S (1988) Multiborehole seismic imaging in steam injection heavy oil recovery projects. Geophysics 53:65–75
Maeland E (1993) On the construction of the 3-D band-limited extrapolation operator in the space-frequency domain. Geophys Prospect 41:645–648
Majer EL, Peterson JE, Benson SM, Long JCS (1991) High frequency seismic tomographic imaging for hydrologic properties of near-surface materials. Earth Sciences Division Annual Report (1990). Lawrence Berkeley Laboratory, Berkeley, CA
Majer EL, Peterson JE, Daley T, Kaelin B, Myer L, Queen J, D'Onfro P, Rizer W (1997) Fracture detection using crosswell and single well surveys. Geophysics 62:495–504
Mallick S, Craft KL, Meister LJ, Chambers RE (1998) Determination of the principal directions of azimuthal anisotropy from P-wave seismic data. Geophysics 63:692–706
Mallick S, Frazer LN (1991) Reflection/transmission coefficients and azimuthal anisotropy in marine seismic studies. Geophys J 105:241–252

Mansfield CS (1996) Fault displacement distributions: insights using high resolution 3-D seismic data. In: Weimer P, Davis TL (eds) Applications of 3-D seismic data to exploration and production. American Association of Petroleum Geologists and Society of Exploration Geophysicists, Tulsa, Oklahoma, USA, pp 231–236

Marion-Pidgeon K, Sawage MK (1997) Frequency dependent anisotropy in Wellington, New Zealand. J Geophys Res 24:3297–3300

Marsset B, Missiaen T, De Roeck YH, Noble M, Versteeg W, Henriet JP (1998) Very high resolution 3-D marine seismic data processing for geotechnical applications. Geophys Prospect 46:105–120

McCamy K, Meyer RP, Smith TJ (1962) Generally applicable solutions of Zoeppritz amplitude equations. Bull Seism Soc Am 52:923–955

McCormack MD, Dunbar JA, Sharp WW (1982) A case study of stratigraphic interpretation using shear and compressional seismic data. Expanded abstracts. 52nd Int Soc Explor Geophys Meeting, pp 21–22

McQuillin R, Bacon M, Barclay W (1979) An introduction to seismic interpretation. Graham and Trotman, UK

Meckel LD Jr, Nath AK (1977) Geologic considerations for stratigraphic modeling and interpretations. In: Seismic stratigraphy – applications to hydrocarbon explorations. American Association of Petroleum Geologists Memoir 26, pp 417–423

Meissner F (1984) Petroleum geology of the Bakken Formation, Williston Basin, North Dakota and Montana. AAPG Memoir 35, pp 159–179

Meissner R, Hegazy MA (1981) The ratio of the PP to the SS reflection coefficient as a possible future method to estimate oil and gas reservoirs. Geophys Prospect 29:533–540

Michelena RJ (1994) Elastic constants of transversely isotropic media from constrained aperture travel time. Geophysics 59:658–667

Michelena RJ, Harris JM, Muir F (1995) Crosswell tomographic estimation of elastic constants in heterogeneous transversely isotropic media. Geophysics 60:774–783

Minkoff SE (1996) A computationally feasible approximate resolution matrix for seismic inverse problems. Geophys J Int 126:345–359

Morgan NA (1970) Wavelet maps – a new analysis tool for reflection seismogram. Geophysics 35:447–460

Morse PM, Feshbach H (1953) Methods of theoretical physics, Part I. McGraw-Hill Book Company, Inc, New York

Muir F (1976) Stanford Exploration Project 8:54

Murai Y, Kawahara J, Yameshita T (1995) Multiple scattering of SH-waves in 2-D elastic media with distributed cracks. Geophysics J Int 122:925–937

Murphy W, Ruscher A, Hsu K (1993) Modulus decomposition of compressional and shear velocities in sand bodies. Geophysics 58:227–239

Musgrave MJP (1961) Elastic waves in anisotropic media. Progr Solid Mech 2:64–85

Musgrave MJP (1970) Crystal acoustics: introduction to the study of elastic waves and vibrations in crystals. Holden-Day, San Francisco

Muskat M, Meres MW (1940) Reflection and transmission coefficients for plane waves in elastic media. Geophysics 5:115–155

Nautiyal A, Samuel HG, Whitmore ND, Garing JD (1993) Stability versus accuracy for an explicit wavefield extrapolation operator. Geophysics 58:277–283

Naville C (1986) Detection of anisotropy using shear wave splitting in VSP surveys: Requirements and applications. 56th Ann Int Mtg Soc Expl Geophys. Expanded abstracts, pp 391–394

Negi JG, Upadhyay SK (1968a) Effect of anisotropy on Love wave propagation. Bull Seism Soc Am 58:259–266

Negi JG, Upadhyay SK (1968b) Effect of anisotropy and inhomogeneity on Love wave dispersion. PAGEOPH 70:28–38

Negi JG, Upadhyay SK (1971) Cutoff-frequency clustering due to anisotropy. Bull Seism Soc Am 61:217–219

Neidell NS, Taner MT (1971) Semblance and other coherency measures for multichannel data. Geophysics 36:482–497

Newman P (1973) Divergence effects in a layered earth. Geophysics 38:481–488

Newmark RL, Anderson RN, Moos D, Zoback MD (1985) Sonic and ultrasonic logging of hole 504B and its implications for structure, porosity and stress regime of upper 1 km of the oceanic crust, initial rep. Deep Sea Drilling Project 83, pp 479–510

Nichols D, Muir F, Schoenberg M (1989) Elastic properties of rocks with multiple set of fractures. 59th Ann Int Mtg Soc Expl Geophys. Expanded abstracts, pp 471–474

Nolet G (1987) Seismic wave propagation and seismic tomography. In: Nolet G (ed) Seismic tomography. Reidl Publishing, Dordrecht, Netherlands, pp 1–23

Nolet G, Schneider R (1990) Solving large linear inverse problems by projection. Geophys J Int 103:565–568

Novotny M (1981) Two methods of solving the linearized two dimensional inverse seismic kinematic problem. J Geophys:7–15

Nowroozi AA (1989) Generalized form of Dix equation for the calculation of interval velocities and layer thicknesses. Geophysics 54:659–661

O'Doherty RF, Anstey NA (1971) Reflections on amplitudes. Geophys Prospect 19:430–458

Okoye PN, Zhao P, Uren NF (1996) Inversion techniques for recovering the elastic constants of transversely isotropic materials. Geophysics 61:1247–1257

Oristaglio ML (1985) A guide to current uses of vertical seismic profiles. Geophysics 50:2473–2479

Ostrander WJ (1984) Plane wave reflection coefficients for gas sands at non-normal angles of incidence. Geophysics 49:1637–1648

Padhi T, Holley TK (1997) Wide azimuths - why not? The Leading Edge 16:175–177

Paige CC, Saunders MA (1982) LSQR: An algorithm for sparse linear equations and sparse linear systems. ACM Trans Math Software 8(2):195–209

Pao YH, Varatharajulu V (1976) Huygen's principle, radiation condition and integral formulas for scattering of elastic waves. J Acoustical Soc Am 58:1361–1371

Parasnis DS (1986) Principles of applied geophysics, 4th edn. Chapman and Hall, London

Paul RJ (1993) Seismic detection of overpressure and fracturing: an example from the Oaidam Basin, Peoples Republic of China. Geophysics 58:1532–1543

Paulsson BNP (1988) Three component downhole seismic vibrator. 58th Ann Int Mtg Soc Expl Geophysics. Expanded abstracts, pp 139–141

Petersen IE, Paulsson BNP, McEvilly TV (1985) Application of algebraic reconstruction techniques to crosshole seismic data. Geophysics 50:1566–1580

Petersen RA, Raymond A, Dobrin MB (1966) A pictorial digital atlas. United Geophysical Corporation, Pasadena, California

Peterson RA, Fillipone WR, Coker FB (1955) The synthesis of seismograms from well log data. Geophysics 20(4):516–638

Pickett GR (1963) Acoustic character logs and their applications in formation evaluation. J Petr Tech 15:659–667

Plesinger A, Hellweg M, Seidl D (1986) Interactive high-resolution polarization analysis of broad-band seismograms. J Geophys 59:129–139

Podio AL (1968) Experimental determination of the dynamic elastic properties of anisotropic rocks, ultrasonic pulse method. PhD Thesis, University of Texas

Podio AL, Gregory AR, Gray KE (1968) Dynamic properties of dry and water-saturated Green-River Shale under stress. Soc Petr Cong J 8:389–404

Postma GW (1955) Wave propagating in a stratified medium. Geophysics 20(4):780–806

Pratt RG, Chapman CH (1992) Travel time tomography in anisotropic media-II, application. Geophys J Int 109:20–37

Pritchett WC (1990) Acquiring better seismic data. Chapman and Hall, London

Pujol J, Burridge R, Smithson SB (1985) Velocity determination from offset vertical seismic profiling data. J Geophys Res 90:1871–1880

Qin F, Luo Y, Cai W (1992) Travel time inversion for anisotropic seismic data. 62nd Ann Int Mtg Soc Expl Geophys. Expanded abstracts, pp 762–765

Queen JH, Rizer WD, Liu E, Crampin S, Lines LR (1992) A review of downhole source effects at the Conoco Borehole Test Facility. In: Lines L, Queen J (eds) Symposium of the Geophysical Society of Tulsa, pp 62–63

Queen JH, Rizer WD (1990) An integrated study of seismic anisotropy and natural fracture system at Conoco Borehole Test Facility, Kay County, Oklahoma. J Geophys Res 95:11255–11273

Radovich BJ, Levin FK (1982) Instantaneous velocities and reflection times for transversely isotropic solids. Geophysics 47:316–322

Rafavich F, Kendau CHSC, Todd TP (1984) The relationship between acoustic properties and petrographic character of carbonate rocks. Geophysics 49:1622–1636

Ramos, Antonio CB, Davis TL (1997) 3-D AVO analysis and modeling applied to fracture detection in coalbed methane reservoirs. Geophysics 62:1683–1695

Richards PG, Frasier CW (1976) Scattering of elastic waves from depth-dependent inhomogeneities. Geophysics 41:441–458

Richards PG, Menke W (1983) The apparent attenuation of a scattering medium. Bull Seism Soc Am 73:1005–1021

Richards TC (1960) Wide angle reflections and their application to finding limestone structures in the foothills of Western Canada. Geophysics 25:385–407

Ricker N (1940) The form and nature of seismic wavelets and the structure of the seismogram. Geophysics 5:348–366

Ricker N (1953) The form and laws of propagation of seismic wavelets. Geophysics 18:10–40

Ricker N (1977) Transient waves in viscoelastic media. Elsevier, Amsterdam, 278 pp

Risch DL, Donaldson BE, Taylor CK (1996) Deep water facies analysis using 3-D seismic sequence stratigraphy and workstation techniques: an example from Plio-Pleistocene strata, Northern Gulf of Mexico. In: Weimer and Davis (eds) (1996), pp143–148

Robertson JD, Pritchett WC (1985) Direct hydrocarbon detection using comparative P-wave and S-wave seismic sections. Geophysics 50:383–393

Robinson EA (1967) Statistical communication and detection: with special reference to digital data processing of radar and seismic signals. Hafner, New York

Robinson EA (1981) Predictive deconvolution. In: Fitch AA (ed) Developments in geophysical exploration methods-2. Applied Science Publishers, London, pp 77–106

Robinson EA (1984) Seismic inversion and deconvolution part A. Geophysical Press, London

Robinson EA, Durrani TS, Peardon LG (1986) Geophysical signal processing. Prentice/Hall International, Englewood Cliffs, NJ

Robinson ES, Coruh C (1988) Basic exploration geophysics. Wiley, New York

Roden R, Sepulvede H (1999) The significance of phase to interpretation: practical guide lines for phase analysis. The Leading Edge 18:774–777

Roksandic MM (1978) Seismic facies analysis concepts. Geophys Prospect 26:383–398

Ross CP, Beale PL (1994) Seismic offset balancing. Geophysics 59:93–101

Rüger A (1997) P-wave reflection coefficient for transversely isotropic models with vertical and horizontal axes of symmetry. Geophysics 62:713–722

Rüger A (1998) Variation of P-wave reflectivity with offset and azimuth in anisotropic media. Geophysics 63:935–947

Rüger A, Alkhalifah T (1996) Efficient two-dimensional ray tracing. In: Proc 6th Int Workshop on Seismic Anisotropy, SEG, Tulsa, pp 556–600

Rutherford SR, Williams RH (1989) Amplitude-versus-offset variations in gas sands. Geophysics 54:680–688

Ryan-Grigor S (1997) Empirical relationships between transverse isotropy parameters and V_P/V_S: Implications for AVO. Geophysics 62:1359–1364

Saito H (1991) Anisotropic travel time tomography at the Buckhorn Test Facility in Illinois. 62nd Ann Int Mtg Soc Expl Geophys. Expanded abstracts, pp 123–126

Samec P, Blangy JP (1992) Viscoelastic attenuation, anisotropy, and AVO. Geophysics 57:441–450

Sangree JB, Widmier JM (1979) Interpretation of depositional facies from seismic data. Geophysics 44:131–160

Sato R, Lapwood ER (1968) SH-waves in a transversely isotropic medium-I. Geophys J Roy Astr Soc 14:463–470

Savage MK, Peppin WA, Vetter UR (1990) Shear wave anisotropy and stress direction in and near Long Valley Caldera, California, 1979-1988. J Geophys Res 95(B7):11165–11177
Savage MK, Shih XR, Meyer RP, Aster RC (1989) Shear wave anisotropy of active tectonic regions via automated S-wave polarization analysis. Tectonophysics 165:279–292
Sayers CM (1988) Inversion of ultrasonic wave velocity measurements to obtain the microcrack orientation distribution function in rocks. Ultrasonics 26:73–77
Sayers CM (1995) Anisotropy velocity analysis. Geophys Prospect 43:541–568
Sayers CM, Ebrom DA (1997) Seismic travel time analysis for azimuthally anisotropic media: theory and experiment. Geophysics 62:1570–1582
Sayers CM, Rickett J (1997) Azimuthal variations in AVO response for fractured gas sands. Geophys Prospect 45:165–182
Scales JA (1987) Tomographic inversion via the conjugate gradient method. Geophysics 52:179–185
Scales JA (1989) On the use of conjugate gradient to calculate the eigen-values and singular values of large sparse matrices. Geophys J 97:179–183
Schneider WA (1971) Developments in seismic data processing and analysis. Geophysics 36:1043–1073
Schneider WA (1978) Integral formulation for migration in two and three dimension. Geophysics 43:49–76
Schneider WA, Backus M (1968) Dynamic correlation analysis. Geophysics 33:105–126
Schoenberg M, Douma J (1988) Elastic wave propagation in media with parallel fractures and aligned cracks. Geophys Prospect 36:571–590
Schoenberg M (1985) An anisotropic model for a fractured thinly layered elastic medium. Presented at 47th EAEG Mtg, Budapest, Hungary
Schoenberg M, Muir F (1989) A calculus for finely layered anisotropic media. Geophysics 54:581–589
Schoenberg M, Sayers CM (1995) Seismic anisotropy of fractured rock. Geophysics 60:204–211
Schoenberger M, Levin FK (1978) Apparent attenuation due to intrabed multiples. Geophysics 43:730–737
Schoenberg M (1986) Wave propagation in an anisotropic medium with a system of aligned cracks. Presented at Int Workshop on Seismic Anisotropy, Moscow
Schultz PS, Claerbout JF (1978) Velocity estimation and downward continuation by wavefront synthesis. Geophysics 43:691–714
Schultz PS, Sherwood JWC (1980) Depth migration before stack. Geophysics 45:376–393
Sena AG (1991) Seismic travel time equations for azimuthally anisotropic and isotropic media: estimation of interval elastic properties. Geophysics 56:2090–2101
Sena AG, Toksöz MN (1993) Kirchoff migration and velocity analysis for converted and non-converted waves in anisotropic media. Geophysics 58:265–276
Shah PM (1973) Use of wavefront curvature to relate seismic data with subsurface parameters. Geophysics 38:812–825
Shalev E, Lou M (1995) A preliminary tomographic inversion of crack density in the Coso geothermal field. EOS (Transactions) 46:351
Shanks JL (1967) Recussion filters for digital processing. Geophysics 32:33–51
Sharma BM (2002) Study of reflection coefficients in anisotropic media. MTech Dissertation (under supervision of the author), Indian Institute of Technology Roorkee, Roorkee, India
Shen F, Ziu X, Toksöz MN (2002) Effects of fractures on NMO velocities and P-wave azimuthal AVO response. Geophysics 67:711–726
Sheriff RE, Geldert LP (1982) Exploration seismology vol I: History, theory and data acquisition. Cambridge University Press, Cambridge
Sheriff RE (1975) Factors affecting seismic amplitudes. Geophys Prospect 23:125–138
Sheriff RE (1978) A first course in geophysical exploration and interpretation. International Human Resources Development Co, Boston
Sheriff RE (1989) Geophysical methods. Prentice Hall, Englewood Cliffs, NJ
Sheriff RE, Geldart LP (1983) Exploration seismology, vol 2. Cambridge University Press, Cambridge

Shih XR, Meyer RP, Schneider JF (1985) An automated analytical method to determine shear-wave splitting. Tectonophysics 165:271–278
Shtivelman V (1984) A hybrid method for wavefield computation. Geophys Prospect 32:236–257
Shtivelman V (1985) Two dimensional acoustic modeling by a hybrid method. Geophysics 50:1273–1284
Shtivelman V, Canning A (1988) Datum correction by wave-equation extrapolation. Geophysics 53:1311–1322
Shuck E (1993) Multicomponent three-dimensional seismic characterizations of a coal bed methane reservoir, Cedar Hill Field, San Juan Country, New Mexico. PhD Thesis, Colorado School of Mines
Shuey RT (1985) A simplification of the Zoeppritz equations. Geophysics 50:609–614
Silverman MR, Wang Q, Byrd BL, Kun ZJ, Ying ZH, Ming C, Weimer P (1996) 3-D evaluation of the Ping Hu Field, East China Sea. In: Weimer P, Davis TL (eds) Applications of 3-D seismic data to exploration and production. American Association of Petroleum Geologists and Society of Exploration Geophysicists, Tulsa, Oklahoma, USA, pp 91–102
Simpson JK (1967) Traveling signal-to-noise ratio and signal power estimates. Geophysics 32:485–493
Singh KP (1995) Travel time analysis for common midpoint gather in transversely isotropic medium. MTech Dissertation (under supervision of the author), University of Roorkee, Roorkee, India
Singh Y (1987) Solution of forward problem in EM Geotomography. MTech Dissertation (under supervision of the author), University of Roorkee, Roorkee, India
Skaloud DK (1996) The use of 3-D seismic data as a tool to increase ultimate recovery from the Yibal Shuaiba Reservoir, Oman in applications of 3-D seismic data to exploration and production. In: Weimer P, Davis TL (eds) American Association of Petroleum Geologists and Society of Exploration Geophysicists, Tulsa, Oklahoma, USA, pp 211–218
Slotnick MM (1959) Lessons in seismic computing. Society of Exploration Geophysics
Smith G, Gidlow PM (1987) Weighted stacking for rock property estimation and detection of gas. Geophys Prospect 35:993–1014
Spratt S (1987) Effect of normal moveout errors on amplitude. Int Mtg Soc Expl Geophys. Expanded abstracts, pp 634–637
Sri Niwas (1973) Theoretical treatment of some problems in interpretation of electrical behaviour of layered rock system. PhD Thesis (under the guidance of the author), Banaras Hindu University, Varanasi, India
Stephen RA (1985) Seismic anisotropy in the upper oceanic crust. J Geophys Res 90:11383–11396
Stewart RR (1984) VSP Interval velocities from travel time inversion. Geophys Prospect 32:608–628
Stewart RR (1988) An algebraic reconstruction technique for weakly anisotropic velocity. Geophysics 53:1613–1615
Stewart RR, Lawton DC (1998) 3C–3D Seismic polarity deformation. In: Stewart RR, Gaise GE (eds) Converted wave seismic exploration. Society of Exploration Geophysics
Stolt RH (1978) Migration by Fourier Transform. Geophysics 43:23–48
Stoneley R (1949) The seismological implications of allotropy in continental structure. Monthly Notices Roy Astro Soc Geophys Suppl 5:222–232
Swan HW (1991) Amplitude-versus-offset measurements errors in a finely layered medium. Geophysics 56:41–49
Smith G, Gidlow PM (1987) Weighted stacking for rock property estimation and detection of gas. Geophys Prospect 35:993–1014
Taner MT, Koehler F, Sheriff RE (1979) Complex seismic trace analysis. Geophysics 44:1041–1063
Taner MT, Sheriff RE (1977) Application of amplitude, frequency, and other attributes to stratigraphic and hydrocarbon determination. AAPG memoir 26, pp 301–328
Taner MT, Koehler F (1969) Velocity spectra-digital computer derivation and applications of velocity functions. Geophysics 34:859–881

Tarantola AT (1987) Inverse problem theory. Elsevier, Amsterdam
Tarantola A, Valette B (1982) Generalized non linear inverse problems solved using the least square criterion. Rev Geophys Space Phys 20:219–232
Tatham RH (1982) V_P/V_S and lithology. Geophysics 47:336–344
Telford WM, Geldart LP, Sheriff RE, Keys A (1976) Applied geophysics. Cambridge University Press, London
Tessmer G, Behle A (1988) Common reflection point data-stacking technique for converted waves. Geophys Prospect 36:671–688
Tessmer G, Behle A (1991) Conversion points and travel times of converted waves in parallel dipping layers. Geophys Prospect 39:387–405
Thomsen L (1986) Weak elastic anisotropy. Geophysics 51:1954–1966
Thomsen L (1988) Reflection seismology over azimuthally anisotropic media. Geophysics 53:304–313
Thomsen L (1995) Elastic anisotropy due to aligned cracks in porous rock. Geophys Prospect 43:805–829
Thomsen L (1999) Converted wave reflection seismology over inhomogeneous, anisotropic media. Geophysics 64:678–690
Timur A (1977) Temperature dependence of compressional and shear wave velocities in rocks. Geophysics 42:950–956
Toksöz MN, Johnston DH (1981) Seismic wave attenuation. SEG (Geophysical Reprint Series No 2), Tulsa
Tooley RD, Spencer TW, Sagoci HF (1965) Reflection and transmission of plane compressional waves. Geophysics 33:552–570
Tosaya C, Nur A (1982) Effects of diagenesis and clays on compressional velocities in rocks. Geophys Res Lett 9:5–8
Tosaya C, Nur A, Vo-Thanh D, Da Prat G (1987) Laboratory seismic method for remote monitoring of thermal EOR. SPE reservoir engineering 2:238–242
Treutel S, Robinson EA (1969) Optimum digital filters for signal to noise ratio enhancement. Geophys Prospect 17:248–293
Trorey AW (1961) The information content of a Rieber sonogram. Geophysics 26:761–764
Trorey AW (1970) A simple theory for seismic diffractions. Geophysics 35:762–784
Tsvanhin I (1995 a) Normal moveout from dipping reflectors in anisotropic media. Geophysics 60:268–284
Tsvankin I (1995 b) Body wave radiation patterns and AVO in transversely isotropic media. Geophysics 60:1409–1425
Tsvankin I (1996) P-wave signatures and notation for transversely isotropic media: an overview. Geophysics 61:1290–1304
Tsvankin I (1997 a) Moveout analysis for transversely isotropic media with a tilted symmetry axis. Geophys Prospect 45:479–512
Tsvankin I (1997 b) Reflection moveout and parameter estimation for horizontal transverse isotropy. Geophysics 62:614–629
Tsvankin I, Chesnokov E (1990) Synthesis of body wave seismograms from point sources in anisotropic media. J Geophys Res 95(B7):11317–11331
Tsvankin I, Thomsen L (1994) Nonhyperbolic reflection moveout in anisotropic media. Geophysics 59:1290–1304
Tucker PM, Yorston HJ (1973) Pitfalls in seismic interpretations. SEG monograph series no 2, Tulsa, 50 pp
Ulrych TJ (1971) Application of homomorphic deconvolution to seismology. Geophysics 36:650–660
Upadhyay SK (1972) Equivalent propagation characteristics of acoustic waves in a fluid layer and SH-waves in a solid layer on a rigid half-space. Bollettino Di Geofisica Teorica Ed Applicata XIV:350–355
Upadhyay SK, Gupta OP (1972-III) SH-wave propagation in anisotropic inhomogeneous crustal layer. PAGEOPH 95:67–74
Uren NF, Gardner GHF, McDonald JA (1990) Dip moveout in anisotropic media. Geophysics 55:863–867

Uren NF, Gardner GHF, McDonald JA (1990) Normal moveout in anisotropic media. Geophysics 55:1634–1636
Uren NF, Gardner GHF, McDonald JA (1990 a) The migrator's equations for anisotropic media. Geophysics 55:1429–1434
Ursin B (1978) Wavefront curvature in layered medium. Geophysics 43:1011–1013
Ursin B (1983) Review of elastic and electromagnetic wave propagation in horizontally layered media. Geophysics 48:1063–1081
Ursin B (1984) Seismic migration using the WKB approximation. Geophys JL Roy Astr Soc 79:339–352
Ursin B (1990) Offset-dependent geometrical spreading in a layered medium. Geophysics 55:492–496
Ursin B, Dahl T (1990) Least-squares estimation of reflectivity polynomial. 60th Ann Int Mtg Soc Expl Geophys. Expanded abstracts, pp 1069–1071
Ursin B, Dahl T (1992) Seismic reflection amplitudes. Geophys Prospect 40:483–512
Ursin B, Ekren BO (1995) Robust AVO analysis. Geophysics 60:317–326
Uzcategui O (1995) 2-D depth migration in transversely isotropic media using explicit operators. Geophysics 60:1819–1829
Van der Hijden (1987) Propagation of transient elastic waves in stratified anisotropic media. North Holland Publishing, Amsterdam
Van der Stoep DM (1966) Velocity anisotropy measurements in wells. Geophysics 31:900–916
Vasco DW, Petersen JE Jr, Majer EL (1998) Resolving seismic anisotropy: sparse matrix methods for geophysical inverse problems. Geophysics 63:970–983
Vasco DW, Pulliam RJ, Johnson LR (1995) Lateral variations in mantle velocity structure and discontinuities determined from P, PP, S, SS, and SS–SdS traveltime residuals. J Geophys Res 100:24037–24059
Verma P (2001) Dip moveout analysis. MTech Dissertation (under supervision of the author), University of Roorkee, Roorkee, India
Verma RK (assisted by Mallick K) (1986) Offshore seismic exploration. Gulf Publishing Company, Houston
Vernik L, Liu Xungzhou (1997) Velocity anisotropy in shales: a petrophysical study. Geophysics 62:521–532
Vernik L, Nur A (1992) Ultrasonic velocity and anisotropy of hydrocarbon source-rocks. Geophysics 57:727–735
Verwest BJ (1989) Seismic migration in elliptically anisotropic media. Geophys Prospect 37:149–166
Vlaar NJ (1968) Ray theory for an anisotropic inhomogeneous elastic medium. Bull Seism Soc Am 58:2053–2072
Walden AT (1991) Making AVO sections more robust. Geophys Prospect 39:915–942
Wang J, Bregman N, Wut G, Hurtley P (1987) Crosshole seismic scanning and tomography. Leading Edge 6:36–41
Wang Z (2001) Fundamentals of seismic rock physics. Geophysics 66:398–412
Wang Z, Nur A (1990) Wave velocities in hydrocarbon saturated rocks: experimental results. Geophysics 55:723–733
Wang Z, Nur F (1990) Wave velocities in hydrocarbon saturated rocks: experimental results. Geophysics 55:723–733
Wang Z (2000) Seismic and acoustic velocities in reservoir rocks 3: Recent developments. Soc Expl Geophysics
Waters KH (1987) Reflection seismology: a tool for energy resource exploration. Wiley, New York
Weimer P, Davis TL (1996) Application of 3-D seismic data to exploration and production. AAPG and SEG
Wennerberg L (1985) Snell's law for viscoelastic materials. Geophys J Roy Astr Soc 81:13–18
White JE (1965) Seismic waves radiation, transmission and attenuation. McGraw-Hill, New York
White JE (1982) Computed waveforms in transversely isotropic media. Geophysics 47:771–783

Whittlesey JR, Quay RG (1977) Wave equation migration operators using 2-D Z-transform theory. Presented at 47th Ann Mtg of SEG, Calgary

Wiggins JW (1984) Kirchoff integral extrapolation and migration of nonplanar data. Geophysics 49:1239–1248

Wiggins RA (1972) The general linear inverse problem. Rev Geophys Space Phys 10(1):251–285

Wiggins RA, Larner KL, Wisecup RP (1976) Residual static analysis as a general linear inverse problem. Geophysics 41:922–938

Wills HA, Rethford GL, Bielanski F (1986) Azimuthal anisotropy: occurrence and effect on shear wave data quality. 56th Ann Int Mtg Soc Exploration Geophysics. Expanded abstracts, pp 479–481

Winterstein DF (1987) Vector attenuation: some implications for plane waves in anelastic layered media. Geophysics 52:810–814

Winterstein DF (1990) Velocity anisotropy terminology for geophysicists. Geophysics 55:1070–1088

Winterstein DF, Paulsson BNP (1990) Velocity anisotropy in shale determined from crosshole seismic and vertical seismic profile data. Geophysics 55:470–479

Winterstein DF, Meadows MA (1991) Shear-wave polarizations and subsurface stress direction at Lost Hills field. Geophysics 56:1331–1348

Wood AB (1941) A text book sound. G Bell and Sons, London

Wood LC, Heiser RC, Treitel S, Riley PL (1978) The debubbling of marine source signatures. Geophysics 43:715–729

Wright J (1986) Reflection coefficients at pore-fluid contacts as a function of offset. Geophysics 51:1858–1860

Wright J (1987) The effect of transverse isotropy on reflection amplitude vs. offset. Geophysics 52:564–567

Wuenschel PC (1976) The vertical array in reflection seismology - some experimental studies. Geophysics 41:219–232

Wuenschel PE (1960) Seismogram synthesis including multiples and transmission coefficients. Geophysics 25:106–129

Wyllie MRJ, Gregory AR, Gardner GHF (1958) An experimental investigation of factors affecting elastic wave velocities in porous media. Geophysics 23:459–493

Wyllie MRJ, Gregory AR, Gardner LW (1956) Elastic wave velocities in heterogeneous, and porous media. Geophysics 21:41–70

Yin H (1993) Acoustic velocity and attenuation in rocks: isotropy, intrinsic anisotropy, and stress induced anisotropy. PhD Thesis, Stanford University

Yomogida K, Benites R (1995) Relation between direct Q and Code Q: a numerical approach. Geophys J Int 123:471–483

Young GB, Braile LW (1976) A computer program for the application of Zoeppritz's amplitude equations and Knott's energy equations. Bull Seism Soc Am 66:1881–1885

Yu G (1985) Offset-amplitude variation and controlled-amplitude processing. Geophysics 50:2697–2708

Zhang J, McMechan GA (1995) Estimation of resolution and covariance for large matrix inversions. Geophys J Int 121:409–426

Zimmerman CJ, Fahmy BA, An effective new method of amplitude versus-offset analysis for marine data. 60th Ann Int Mtg Soc Expl Geophys. Expanded abstracts, pp 1499–1502

Ziolkowski A, Lerwill WE (1979) A simple approach to high resolution seismic profiling for coal. Geophys Prospect 27:360–393

Zoeppritz K (1919) Erdbebenwellen VIIIB. Über Reflexion und Durchgang seismischer Wellen durch Unstetigkeitsflächen. Göttinger Nachr 1:66–84

Subject Index

D

Q

R

T

Printing and Binding: Strauss GmbH, Mörlenbach

MIX
Papier aus verantwortungsvollen Quellen
Paper from responsible sources
FSC® C105338

If you have any concerns about our products,
you can contact us on
ProductSafety@springernature.com

In case Publisher is established outside the EU,
the EU authorized representative is:
Springer Nature Customer Service Center GmbH
Europaplatz 3, 69115 Heidelberg, Germany

Printed by Libri Plureos GmbH
in Hamburg, Germany